To Sheetal

From Vaibhav.

12/2/02

INTERMEDIATE STRUCTURAL ANALYSIS

INTERMEDIATE STRUCTURAL ANALYSIS

C. K. Wang, Ph.D.

Professor of Civil Engineering
University of Wisconsin, Madison

McGRAW-HILL BOOK COMPANY

Auckland Bogotá Guatemala Hamburg Lisbon
London Madrid Mexico New Delhi Panama Paris
San Juan São Paulo Singapore Sydney Tokyo

INTERMEDIATE STRUCTURAL ANALYSIS
INTERNATIONAL EDITION

Copyright © 1983
Exclusive rights by McGraw-Hill Book Co — Singapore for manufacture and export. This book cannot be re-exported from the country to which it is consigned by McGraw-Hill.

6th printing 1989

Copyright © 1983 by McGraw-Hill, Inc.
All rights reserved. No part of this publication may be reproduced or distributed in any form or by any means, or stored in a data base or retrieval system, without the prior written permission of the publisher.

This book was set in Times Roman.
The editors were Julienne V. Brown and Madelaine Eichberg.
The production supervisor was Charles Hess.

Library of Congress Cataloging in Publication Data

Wang, Chu-Kia, date
   Intermediate structural analysis.

   Includes index.
   1. Structures, Theory of.   I. Title.
TA645.W32   624.1'71   81-15615
ISBN 0-07-068135-X   AACR2

When ordering this title use ISBN 0-07-Y66623-7

Printed and bound in Singapore by Fong & Sons Printers Pte Ltd

CONTENTS

Preface   xiii

Chapter 1   Force Method vs. Displacement Method   1

1.1   Statically Determinate Structures vs. Statically Indeterminate Structures   1
1.2   Conditions of Geometry   6
1.3   Force Method: Method of Consistent Deformation   8
1.4   Displacement Method   9
1.5   Methods of Structural Analysis   10
1.6   Fundamental Assumptions   12

Chapter 2   Deformation of Statically Determinate Beams and Rigid Frames   13

2.1   Force Response vs. Deformation Response   13
2.2   Shears and Moments in Beams   15
2.3   Axial Force, Shear, and Moment in Rigid Frames   20
2.4   Common Methods for Determination of Deformation Response   23
2.5   The Unit-Load Method—Derivation of Basic Formula   24
2.6   The Unit-Load Method—Application to Beam Deflections   25
2.7   The Unit-Load Method—Application to Beam Slopes   29
2.8   The Partial-Derivative Method—Castigliano's Theorem   33
2.9   The Partial-Derivative Method—Application to Beam Deflections and Slopes   35
2.10   The Unit-Load Method—Application to Deflections and Slopes of Rigid Frames   39
2.11   Deflections and Slopes of Statically Determinate Rigid Frames Due to Movements of Supports   47

v

2.12	The Moment-Area Method—Derivation of Theorems	49
2.13	The Moment-Area Method—Application to Beam Deflections and Slopes	51
2.14	The Conjugate-Beam Method—Derivation of Theorems	55
2.15	The Conjugate-Beam Method—Application to Beam Deflections and Slopes	60
2.16	The Moment-Area/Conjugate-Beam Method—Application to Deflections and Slopes of Rigid Frames	64
2.17	Exercises	68

Chapter 3 Deflection of Statically Determinate Trusses 71

3.1	Force Response vs. Deformation Response	71
3.2	Methods of Joints and Sections	72
3.3	Methods for Determination of Deflections of Truss Joints	77
3.4	The Unit-Load Method	78
3.5	The Angle-Weights Method	85
3.6	The Joint-Displacement-Equation Method	90
3.7	The Graphical Method—Williot-Mohr Diagram	93
3.8	Exercises	97

Chapter 4 Analysis of Statically Indeterminate Beams and Rigid Frames by the Force Method 100

4.1	Force Method of Analysis	100
4.2	Analysis of Statically Indeterminate Beams by the Force Method	101
4.3	Law of Reciprocal Deflections	107
4.4	Theorem of Least Work	116
4.5	Induced Reactions on Statically Indeterminate Beams Due to Yielding of Supports	121
4.6	Analysis of Statically Indeterminate Rigid Frames by the Force Method	126
4.7	Induced Reactions on Statically Indeterminate Rigid Frames Due to Yielding of Supports	138
4.8	Exercises	142

Chapter 5 Analysis of Statically Indeterminate Trusses by the Force Method 146

5.1	Degree of Indeterminacy	146
5.2	Force Method Using Reactions as Redundants	147
5.3	Force Method Using Axial Forces in Members as Redundants	154
5.4	Force Method Using Both Reactions and Axial Forces in Members as Redundants	165
5.5	Induced Reactions on Statically Indeterminate Trusses Due to Yielding of Supports	169
5.6	Exercises	170

Chapter 6 The Three-Moment Equation — 172
- 6.1 Introduction — 172
- 6.2 Derivation of the Three-Moment Equation — 174
- 6.3 Application of the Three-Moment Equation to Analysis of Continuous Beams Due to Applied Loads — 176
- 6.4 Application of the Three-Moment Equation to Analysis of Continuous Beams Due to Uneven Support Settlements — 183
- 6.5 Exercises — 188

Chapter 7 The Slope-Deflection Method — 190
- 7.1 General Description — 190
- 7.2 Derivation of the Slope-Deflection Equations—Without Rotation of Member Axis — 192
- 7.3 Application to the Analysis of Statically Indeterminate Beams for Applied Loads — 194
- 7.4 Derivation of the Slope-Deflection Equations—With Rotation of Member Axis — 199
- 7.5 Application to the Analysis of Statically Indeterminate Beams for Uneven Support Settlements — 200
- 7.6 Application to the Analysis of Statically Indeterminate Rigid Frames—Without Unknown Joint Translation — 203
- 7.7 Application to the Analysis of Statically Indeterminate Rigid Frames—With Unknown Joint Translation — 210
- 7.8 Application to the Analysis of Statically Indeterminate Rigid Frames Due to Yielding of Supports — 226
- 7.9 Analysis of Gable Frames by the Slope-Deflection Method — 228
- 7.10 Slope-Deflection Equations for Members with Variable Moment of Inertia — 241
- 7.11 Exercises — 256

Chapter 8 The Moment-Distribution Method — 262
- 8.1 Introduction — 262
- 8.2 Basic Concept — 262
- 8.3 Stiffness and Carry-Over Factors — 264
- 8.4 Distribution Factors — 265
- 8.5 Application to the Analysis of Statically Indeterminate Beams for Applied Loads — 267
- 8.6 Check on Moment Distribution — 271
- 8.7 Modified Stiffness Factor and Modified Fixed-End Moment at the Near End When the Far End Is Hinged — 273
- 8.8 Application to the Analysis of Statically Indeterminate Beams for Uneven Support Settlements — 276
- 8.9 Application to the Analysis of Statically Indeterminate Rigid Frames—Without Unknown Joint Translation — 280
- 8.10 Application to the Analysis of Statically Indeterminate Rigid Frames—With Unknown Joint Translation — 289

8.11	Application to the Analysis of Statically Indeterminate Rigid Frames Due to Yielding of Supports	312
8.12	Analysis of Gable Frames by the Moment-Distribution Method	316
8.13	Stiffness and Carry-Over Factors for Members with Variable Moment of Inertia	326
8.14	Fixed-End Moments Due to Member-Axis Rotatation for Members with Variable Moment of Inertia	327
8.15	Moment Distribution Involving Members with Variable Moment of Inertia	329
8.16	Exercises	336

Chapter 9   Matrix Operations   337

9.1	Matrix Notation for Linear Equations	337
9.2	Matrix Multiplication as Means of Elimination of Intermediate Variables in Two Sets of Linear Equations	339
9.3	Rule for Matrix Multiplication	340
9.4	Matrix Inversion as Means of Interchange of Independent and Dependent Variables in a Set of Linear Equations	342
9.5	Method for Matrix Inversion	345
9.6	Solution of Simultaneous Linear Equations by Forward Elimination and Backward Substitution	346
9.7	Solution of Simultaneous Linear Equations by Gauss–Jordan Elimination	349
9.8	Matrix Transposition	351
9.9	Some Useful Properties of Matrices	353
9.10	Exercises	356

Chapter 10   Matrix Displacement Method of Truss Analysis   357

10.1	Degree of Freedom, Number of Independent Unknown Forces, and Degree of Indeterminacy	357
10.2	The Deformation Matrix $[B]$	359
10.3	The Element Stiffness Matrix $[S]$	361
10.4	The Force-Displacement Matrix $[SB]$	361
10.5	The Statics Matrix $[A]$	362
10.6	The Principle of Virtual Work	364
10.7	The Global Stiffness Matrix $[K] = [ASA^T]$	367
10.8	The Local Stiffness Matrix	370
10.9	The Direct Stiffness Method	371
10.10	The Law of Reciprocal Forces and the Law of Reciprocal Displacements	374
10.11	The Joint-Force Matrix $[P]$	375
10.12	Numerical Examples	376
10.13	Effect of Fabrication Errors or Temperature Changes	387
10.14	Effect of Support Settlements	389
10.15	Exercises	394

Chapter 11 Matrix Displacement Method of Beam Analysis — 395

- 11.1 Joints and Elements in Beams vs. Degree of Freedom — 395
- 11.2 Sending Loads on the Elements to the Joints — 396
- 11.3 Degree of Indeterminacy vs. Independent Unknown Forces — 397
- 11.4 The Statics Matrix [A] — 398
- 11.5 The Deformation Matrix [B] — 399
- 11.6 The Principle of Virtual Work — 401
- 11.7 The Element Stiffness Matrix [S] — 402
- 11.8 The Force-Displacement Matrix [SB] — 403
- 11.9 The Global Stiffness Matrix $[K] = [ASA^T]$ — 404
- 11.10 The Local Stiffness Matrix — 405
- 11.11 Numerical Examples — 409
- 11.12 Effect of Support Settlements — 417
- 11.13 Exercises — 420

Chapter 12 Matrix Displacement Method of Rigid-Frame Analysis — 421

- 12.1 Matrix Displacement Method vs. Slope-Deflection Method — 421
- 12.2 Analysis of Rigid Frames without Sidesway — 422
- 12.3 Analysis of Rigid Frames with Sidesway — 429
- 12.4 Analysis of Rigid Frames for Yielding of Supports — 444
- 12.5 Analysis of Gable Frames — 447
- 12.6 Element Stiffness Matrix of Beam Element with Variable Moment of Inertia — 456
- 12.7 Exercises — 457

Chapter 13 Influence Lines and Moving Loads — 458

- 13.1 Definition of Influence Lines — 458
- 13.2 Influence Lines for Statically Determinate Beams — 459
- 13.3 Criterion for Maximum Reaction or Shear in Simple Beams — 467
- 13.4 Criterion for Maximum Bending Moment in Simple Beams — 471
- 13.5 Absolute Maximum Bending Moment in a Simple Beam — 476
- 13.6 Müller–Breslau Influence Theorem for Statically Determinate Beams — 480
- 13.7 Influence Lines for Statically Determinate Trusses — 483
- 13.8 Criterion for Maximum Bending Moment at a Panel Point on the Loaded Chord of a Truss — 484
- 13.9 Criterion for Maximum Bending Moment at a Panel Point on the Unloaded Chord of a Truss — 485
- 13.10 Criterion for Maximum Force in the Web Member of a Truss — 489
- 13.11 Müller–Breslau Influence Theorem for Statically Determinate Trusses — 493
- 13.12 Influence Lines for Statically Indeterminate Beams — 496

13.13	Müller-Breslau Influence Theorem for Statically Indeterminate Beams	498
13.14	Influence Lines for Statically Indeterminate Trusses vs. Müller–Breslau Influence Theorem	500
13.15	Exercises	501

Chapter 14 Approximate Methods of Multistory-Frame Analysis 504

14.1	Vertical- and Lateral-Load Analysis of Multistory Frames	504
14.2	Degree of Indeterminacy vs. Number of Assumptions	505
14.3	Assumptions for Vertical-Load Analysis	507
14.4	Assumptions for Lateral-Load Analysis	508
14.5	Portal Method	510
14.6	Cantilever Method	514
14.7	Alternate Moment and Shear Distribution	519
14.8	Comparison of Methods	524
14.9	Exercises	525

Chapter 15 The Column-Analogy Method 527

15.1	General Introduction	527
15.2	Fixed-End Moments for a Beam Element with Constant Moment of Inertia	529
15.3	Stiffness and Carry-Over Factors for a Beam Element with Constant Moment of Inertia	534
15.4	Fixed-End Moments for a Beam Element with Variable Moment of Inertia	536
15.5	Stiffness and Carry-Over Factors for a Beam Element with Variable Moment of Inertia	542
15.6	Moments in Quadrangular Frames with One Axis of Symmetry	544
15.7	Moments in Closed Frames with One Axis of Symmetry	555
15.8	Moments in Gable Frames with One Axis of Symmetry	560
15.9	Moments in Unsymmetrical Quadrangular Frames	565
15.10	Moments in Unsymmetrical Closed Frames	574
15.11	Exercises	580

Chapter 16 Composite Structures and Rigid Frames with Axial Deformation 584

16.1	General Description	584
16.2	Composite Structure with Truss and Beam Elements—Force Method	585
16.3	Composite Structure with Truss and Beam Elements—Displacement Method	590
16.4	Composite Structure with Truss and Combined Elements—Force Method	594

	16.5	Composite Structure with Truss and Combined Elements—Displacement Method	607
	16.6	Exercises	619

Chapter 17 Secondary Moments in Trusses with Rigid Joints 622

	17.1	General Description	622
	17.2	Methods of Analysis	623
	17.3	Solution as a General Two-Dimensional-Frame Problem	625
	17.4	Iteration Method—From Primary Axial Forces to Secondary Bending Moments	637
	17.5	Iteration Method—From Secondary Bending Moments to Third Axial Forces	644
	17.6	Comparison of Methods	646
	17.7	Exercises	646

Chapter 18 Rigid Frames with Curved Members 649

	18.1	General Description	649
	18.2	Fixed-Condition Forces for a Curved Member	651
	18.3	Analysis of Fixed Arches	653
	18.4	Elastic-Center Method vs. Column-Analogy Method	654
	18.5	Influence Lines for a Symmetrical Fixed Arch	656
	18.6	Influence Lines for an Unsymmetrical Fixed Arch	667
	18.7	Stiffness Matrix of a Curved Member	683
	18.8	Flexibility Matrix of a Curved Member with a Fixed Support	690
	18.9	Flexibility Matrix of a Curved Member with One Hinged Support and One Roller Support	694
	18.10	Effects of Temperature, Shrinkage, Rib Shortening, and Foundation Yielding on Fixed Arches	698
	18.11	Exercises	699

Chapter 19 Displacement Method of Horizontal Grid-Frame Analysis 702

	19.1	Definition of Horizontal Grid Frames	702
	19.2	Methods of Analysis	703
	19.3	Grid Frames with Beam Elements Only	705
	19.4	Grid Frames with Combined Beam and Torsion Elements	710
	19.5	Exercises	718

Chapter 20 Rigid Frames with Semirigid Connections 721

	20.1	General Description	721
	20.2	The Modified Member Flexibilility Matrix	722
	20.3	The Modified Member Stiffness Matrix	724
	20.4	The Modified Fixed-End Moments	725

	20.5	The Displacement Method	728
	20.6	Treatment of Semirigid Connections as Rotational Springs	735
	20.7	Exercises	740

Chapter 21 Effects of Shear Deformations 742

	21.1	General Introduction	742
	21.2	Shape Factors	743
	21.3	Shear Deflections of Statically Determinate Beams	745
	21.4	Relative Significance of Shear Deflections to Bending-Moment Deflections	750
	21.5	Force Method	751
	21.6	Displacement Method	755
	21.7	Exercises	760

Chapter 22 Beams on Elastic Foundation 762

	22.1	General Description	762
	22.2	The Basic Differential Equation	762
	22.3	General Solution of the Differential Equation	763
	22.4	Boundary Conditions of an Unloaded Member	764
	22.5	Flexibility Matrix of a Member on Elastic Foundation	764
	22.6	Stiffness Matrix of a Member on Elastic Foundation	767
	22.7	Fixed-End Shears and Moments Due to Uniform Load	769
	22.8	Fixed-End Shears and Moments due to Straight Haunch Load	770
	22.9	The Local $[SA^T]$ and $[ASA^T]$ Matrices of a Segment in a Beam	772
	22.10	Analysis of Beams on Elastic Foundation by the Displacement Method	773
	22.11	Deflection, Shear, and Moment Ordinates within the Segment	777
	22.12	Exercises	780

Index 783

PREFACE

This book is dedicated to the learning and teaching of structural analysis at the intermediate level. By intermediate level, it is meant that the book covers the material in more detail and variation than an introductory book, but stops short of treatment of such advanced topics as structural stability, limit analysis, second-order analysis, dynamics and stability, and structural optimization. Consequently, this book is suitable for use as a textbook for the second and third courses in structural analysis, when the first course has been in the form of an overall introduction. In cases where at least two courses in structural analysis are to be taken by the same student, however, the book could be used for the first as well as the second course.

About two-thirds of this book are based on this author's *Statically Indeterminate Structures*, published by McGraw-Hill in 1953. This former book has been widely circulated around the globe; the special features of including many numerical examples and line drawings are fervently maintained and upheld in the present book. To repeat one sentence in the preface of the previous book, "The examples in the text may substitute for some, although not all, of the blackboard writing on the part of the teacher so that more time in class may be apportioned to discussions on such items as the relation of analysis to design, the history or bibliography, the comparison of the relative merits of different methods, and the citation of actual structures to which the topic of the day can be applied."

In the years before the advent of the electronic computer, from pocket to desk to room size, traditionally the first course in structural analysis was on statically *determinate* structures, and the subsequent courses, on statically *indeterminate* structures; this was the reason for using *Statically Indeterminate Structures* as the title of the former book. The impact of the computer puts emphasis on the displacement method, which requires no differentiation between determinate and indeterminate structures; moreover the computer is so powerful that complete analysis of a large structural assembly can be readily accomplished. More than ever before, students need thorough understanding of the concepts behind the force method and the displacement method of structural analysis in order that the computer programs may be wisely used with care. Thus, to make the present book a complete treatise on both determinate and indeterminate analysis, sections and chapters are added on the methods of joints and sections for truss analysis, on shears and moments in determinate beams, on influence lines and moving loads, and on approximate methods of multistory-frame analysis.

The bulk of the remaining one-third of the book, which was not in *Statically Indeterminate Structures*, is on the matrix displacement method. The basis for this method is first presented in truss analysis. Then the same numerical examples that were solved by the slope-deflection and moment-distribution methods are again solved by the matrix displacement method in longhand form, thus enabling the students to see what the computer will have to go through. Formerly the secondary moments in trusses with rigid joints were solved only by the iteration method utilizing moment distribution; now the alternative solution as a general two-dimensional-frame problem on the computer is presented as well. Formerly the fixed arch was solved as such only by itself; now it becomes part of an overall problem of including a curved element in a complex rigid frame. Axial deformation in rigid frames is ordinarily ignored in the noncomputer analysis for reason of saving labor; on the contrary, its inclusion greatly simplifies the making of input data in the direct stiffness method of computer analysis. Both approaches are treated in full in the matrix displacement method of two-dimensional-frame analysis.

The last four chapters, on horizontal grid frames, semirigid connections, shear deformations, and beams on elastic foundation are new in this book over the former book *Statically Indeterminate Structures*. Without the computer, the use of the methods, as explained in these four chapters, in real-world applications would be virtually impossible. Now it is a simple matter to put forth the procedural steps in a computer program.

The first six chapters are basic: on statically determinate beams, trusses, and rigid frames; and on the force method of analyzing statically indeterminate beams, trusses, and rigid frames, including the three-moment equation. The choice of sequence may be chapters 1, 2, 3, 4, 5, 6 as presented; or 1, 2, 4, 6, 3, 5; 1, 3, 5, 2, 4, 6; 1, 3, 2, 5, 4, 6 as desired. Chapters 7 and 8 may be taken in reverse order; either may be omitted from the curriculum, or the latter parts of both chapters may be omitted.

Chapter 9 may be omitted if students have already had matrix definitions in mathematics courses. Chapters 10 and 11 cover the basic material in the matrix displacement method of truss and beam analysis. One may then immediately proceed to Chap. 17, wherein the general two-dimensional-frame analysis considering axial-deformation effects is presented along with analysis of trusses with rigid joints. This sequence might constitute the minimum coverage of the matrix displacement method.

Chapters 12 and 16 are recommended for those who prefer to have a more thorough grounding of the matrix displacement method and see how this newer method relates to the conventional consistent-deformation, slope-deflection, and moment-distribution methods.

The remaining chapters, Chap. 13, 14, 15, 18, 19, 20, 21, and 22, are such that one may study them in any order. The material in Chap. 13, on influence lines and moving loads, covers a subject which must be included in the structural analysis education of a civil engineering student.

C. K. Wang

INTERMEDIATE STRUCTURAL ANALYSIS

CHAPTER
ONE
FORCE METHOD VS. DISPLACEMENT METHOD

1.1 Statically Determinate Structures vs. Statically Indeterminate Structures

Most structures fall into one of the following three classifications: beams, rigid frames, or trusses. A *beam* is a structural member subjected to transverse loads only; it is completely analyzed when the shear and moment diagrams are found. A *rigid frame* is a structure composed of members which are connected by rigid joints (welded joints, for instance). A rigid frame is completely analyzed when the variations in axial force, shear, and moment along the lengths of all members are found. A *truss* is a structure in which all members are usually considered to be connected by hinges, thus eliminating shears and moments in the members. A truss is completely analyzed when the axial forces in all members are determined.

Shear and moment diagrams of beams can be drawn when the external reactions are known. In the study of the equilibrium of a coplanar parallel-force system, it has been proved that not more than two unknown forces acting on any one free body can be found by the principles of statics. In the cases of simple, overhanging, or cantilever beams as shown by Fig. 1.1.1a to c, these two unknowns are the reactions R_1 and R_2. In the beam with two internal hinges as shown by Fig. 1.1.1d, there are three pieces joined together at the hinges. The four unknown external reactions and the two interacting hinge reactions can be determined from the six equations of statics, two furnished by each of the three beam segments. Thus, simple, overhanging, and cantilever beams, as well as beams with a number of internal hinges equal to the number of extra reactions beyond two, are statically determinate. If,

1

2 INTERMEDIATE STRUCTURAL ANALYSIS

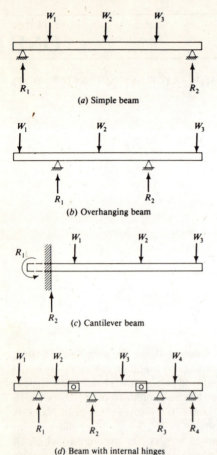

(a) Simple beam

(b) Overhanging beam

(c) Cantilever beam

(d) Beam with internal hinges

Figure 1.1.1 Statically determinate beams.

however, a beam without internal hinges, as is the common case, rests on more than two supports or if in addition one or both end supports are fixed, there are more than two external reactions to be determined. Statics offers only two conditions of equilibrium for a coplanar parallel-force system, and thus only two reactions can thereby be found; any additional reactions are extra, or redundant. Beams with such redundant reactions are called *statically indeterminate beams*. The degree of indeterminacy is given by the number of extra, or redundant, reactions. Thus the beam in Fig. 1.1.2a is statically indeterminate to the second degree because there are four unknown reactions and statics furnishes only two conditions or two equations of equilibrium; the beam in Fig. 1.1.2b is statically indeterminate to the fourth degree; the beam in Fig. 1.1.2c is statically indeterminate to the first degree because it has five

FORCE METHOD VS. DISPLACEMENT METHOD 3

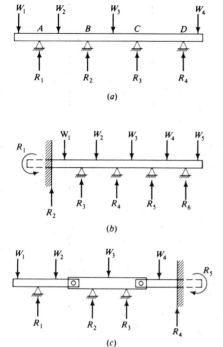

Figure 1.1.2 Statically indeterminate beams.

reactions and two internal hinges. It may be noted that beams are rarely built physically with internal hinges, but such conditions may occur in their behavior when beams are subjected to a set of loads beyond the service condition.

A single-story rigid frame is statically determinate if there are only three external reactions because statics offers only three conditions of equilibrium for a general coplanar-force system. Thus the two rigid frames shown in Fig. 1.1.3 are statically determinate. If, however, a single-story rigid frame has more than three external reactions, it is statically indeterminate, the degree of indeterminacy being equal to the number of redundant reactions. Thus the single-story rigid frame of Fig. 1.1.4a is statically indeterminate to the first degree; of Fig. 1.1.4b, to the third degree; of Fig. 1.1.4c, to the fifth degree. Most common rigid frames are statically indeterminate, as efficiency and strength would require. The degree of indeterminacy of multistory rigid frames goes up by leaps and bounds; they will be treated later in this text. It is sufficient here to learn the definition of statical indeterminacy.

The necessary condition that a truss may be just statically determinate is that it has at least three unknown reactions and the number of members in the truss is $2j - r$, where j is the number of joints and r is the number of reactions. If m is the number of members, the necessary condition for statical

4 INTERMEDIATE STRUCTURAL ANALYSIS

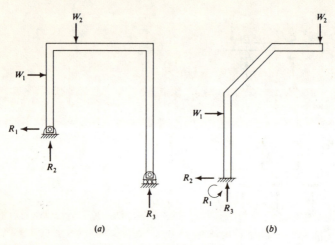

Figure 1.1.3 Statically determinate rigid frames.

determinacy becomes

$$m = 2j - r \tag{1.1.1}$$

The validity of the above equation can be observed by a variation of the equation to the form $m + r = 2j$, in which $m + r$ is the total number of unknowns and $2j$ is the number of available equations of statics by treating each joint as a free body. So long as every joint in a truss is in equilibrium, taking any group of joints or the whole truss as a free body cannot yield any more independent equations of equilibrium. For a truss to be statically determinate and statically stable, however, the m members as dictated by Eq. (1.1.1) must be arranged judiciously, which means that a solution be obtainable for the reactions and the axial forces in the members. Thus the trusses of Fig. 1.1.5a and b are statically determinate and statically stable; while the truss of Fig. 1.1.5c, even though it satisfies Eq. (1.1.1), is statically unstable. When a truss has at least three unknown reactions and the number of members m is in excess of $2j - r$, it is statically indeterminate, and the degree of indeterminacy i becomes

$$i = m - (2j - r) \tag{1.1.2}$$

Thus the truss of Fig. 1.1.6a is statically indeterminate to the second degree; of Fig. 1.1.6b, to the third degree; of Fig. 1.1.6c, also to the third degree. In common situations, trusses are generally composed of a series of triangles with one stacked upon another as shown in Fig. 1.1.7; in this case the first triangle requires three joints and three members, while each successive triangle requires two additional members but only one additional joint. Thus,

$$m - 3 = 2(j - 3)$$

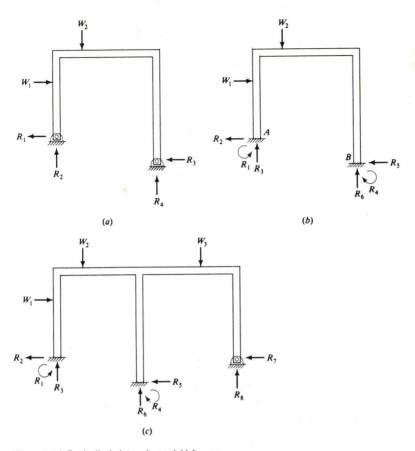

Figure 1.1.4 Statically indeterminate rigid frames.

or

$$m = 2j - 3 \tag{1.1.3}$$

The form shown in Fig. 1.1.7 is obviously statically determinate and statically stable. Consequently the degree of indeterminacy of a truss is equal to the sum of the number of excessive reactions beyond 3 and the number of excessive members beyond $2j - 3$; or

$$i = (r - 3) + [m - (2j - 3)] \tag{1.1.4}$$

Equation (1.1.4), although identical to Eq. (1.1.2), is more convenient to use when both $r - 3$ and $m - (2j - 3)$ are positive. In fact, $r - 3$ is often called the *degree of external indeterminacy*; and $m - (2j - 3)$, which is the number of panels with double diagonals, the *degree of internal indeterminacy*. In this

6 INTERMEDIATE STRUCTURAL ANALYSIS

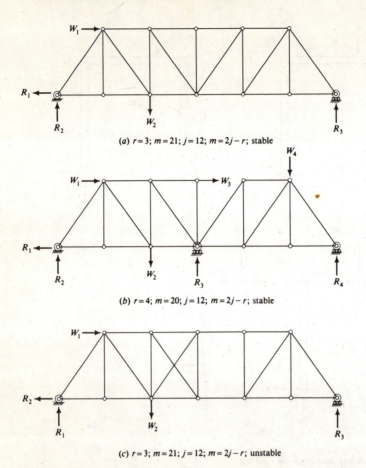

(a) $r=3$; $m=21$; $j=12$; $m=2j-r$; stable

(b) $r=4$; $m=20$; $j=12$; $m=2j-r$; stable

(c) $r=3$; $m=21$; $j=12$; $m=2j-r$; unstable

Figure 1.1.5 Trusses meeting the necessary condition for statical determinacy.

context, one can instantly recognize that the degree of indeterminacy of the truss of Fig. 1.1.6a is $2+0=2$; of Fig. 1.1.6b, $1+2=3$; of Fig. 1.1.6c, $0+3=3$.

1.2 Conditions of Geometry

It can be seen from the preceding discussion that in analyzing statically indeterminate structures it is necessary to have as many extra conditions, in addition to those of statics, as there are redundant reactions (or sum of redundant reactions and redundant members, for trusses). In other words, the number of "nonstatic" conditions must be the same as the degree of indeterminacy. The extra conditions are furnished by the geometry of the deformed structure. For instance, the beam of Fig. 1.1.2a, although ordinarily

FORCE METHOD VS. DISPLACEMENT METHOD 7

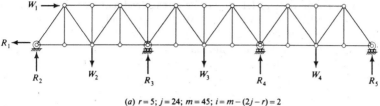

(a) $r=5$; $j=24$; $m=45$; $i=m-(2j-r)=2$

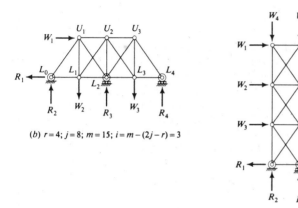

(b) $r=4$; $j=8$; $m=15$; $i=m-(2j-r)=3$

(c) $r=3$; $j=8$; $m=16$; $i=m-(2j-r)=3$

Figure 1.1.6 Statically indeterminate trusses.

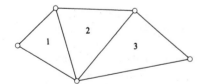

Figure 1.1.7 Stacking of triangles to form a truss.

thought of as a continuous beam, can be considered as an overhanging beam with supports only at A and D, on which forces W_1, W_2, W_3, W_4, R_2, and R_3 are acting. The conditions of geometry to be satisfied by the elastic curve of the statically determinate overhanging beam are that the deflections at B and C must both be zero. These two conditions of geometry are used, at the very beginning, to obtain the unknown forces R_2 and R_3. Alternatively, of course, this same beam can be considered as an overhanging beam with supports only at B and C, on which the forces W_1, W_2, W_3, W_4, R_1, and R_4 are acting. Then R_1 and R_4 are solved first from the conditions of geometry that the deflections at A and D must both be zero.

The rigid frame of Fig. 1.1.4b can be considered as being fixed at A and free at B, on which the forces W_1, W_2, R_4, R_5, and R_6 are acting. The conditions of geometry are that the tangent to the elastic curve at B must remain vertical and that the horizontal and vertical deflections at B must both be zero. Alternatively, this rigid frame can be considered as being hinged at A and supported on rollers at B, on which the forces W_1, W_2, R_1, R_4, and R_5 are acting. Then the conditions of geometry are that the tangent to the elastic curve must remain vertical at both A and B and that the horizontal deflection at B must be zero. One can note that there is a one-to-one correspondence between the unknown redundant reacting force and the condition of geometry.

The truss of Fig. 1.1.6b can be considered as a simple truss with the support at L_2 removed, and with two diagonals L_1U_2 and L_2U_3 also removed; acting on the "equivalent" truss, shown in Fig. 1.2.1b, are the forces W_1, W_2, W_3, R_3, F_{x1}-F_{x1}, and F_{x2}-F_{x2}. The conditions of geometry are that the vertical deflection at L_2 must be zero, the distance between the deflected points L_1 and U_2 must be equal to the elongated (or shortened) length of bar L_1U_2 in which the axial force is equal to F_{x1}, and the distance between the deflected points L_2 and U_3 must be equal to the elongated (or shortened) length of bar L_2U_3 in which the axial force is equal to F_{x2}.

1.3 Force Method: Method of Consistent Deformation

The most basic and general method of analyzing statically indeterminate structures is the method of consistent deformation, usually called the *force method*. The procedure consists in first setting up a basic determinate structure by modifying the given indeterminate structure so that the unknown redundant reactions (or axial forces in the redundant members of a truss) are replaced by unknown redundant forces of equal magnitude. As explained previously, there will be as many conditions of geometry as there are redundant forces (a redundant moment can be considered as a *generalized redundant force*). A system of i simultaneous equations, where i is the degree of indeterminacy, can be established under these conditions of geometry with the redundant forces as unknowns. When these equations are solved, the redundant forces are then known and can be put back to act on the given

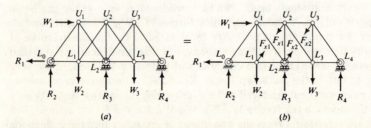

Figure 1.2.1 Equivalent trusses.

indeterminate structure and the remaining reactions solved by the equations of statics. It should be noted that there may be several alternate ways of choosing the basic determinate structure, as explained in Sec. 1.2.

Although the name *method of consistent deformation* has been appropriately used in the past, an alternate name for it might be redundant-force method, or simply force method. The name *force method* has gained popular recognition because of its contrast with the name *displacement method*, which will be defined in the next section. In this textbook, the name force method will be used hereafter.

In establishing the conditions of consistent deformation, sometimes called the conditions of compatibility of strains, the deflections or rotations of the tangents to the elastic curve, in the direction of the redundant forces, of a basic determinate structure are required for the applied loads and for the action of each redundant force. Therefore, before statically indeterminate structures can be dealt with, it is necessary to take up first the various methods of finding the deflections (or rotations of tangents) of statically determinate beams, rigid frames, and trusses. These various methods are treated and illustrated in Chaps. 2 and 3.

It should be pointed out that the deflections of beams, rigid frames, and trusses depend on the sizes of the constituent members in the structure. Therefore, before analyzing a statically indeterminate structure, the sizes of its members must first be assumed, although in most cases only the *relative* sizes are necessary. The procedure in designing a statically indeterminate structure is that of successive approximation; in other words, the structure is first assumed, then analyzed, then the design modified, then reanalyzed, and so on until the last assumed structure needs no further modification. Fortunately, the structure as first modified will in general need no further reanalysis or redesign.

The reader may not be able to grasp fully the entire significance of this section at this time; however, if one reads this section again after working through Chaps. 2 to 5, the concept advanced here should become more obvious, and in fact, quite enlightening.

1.4 Displacement Method

Timoshenko stated that "Navier (1785–1836) was the first to evolve a general method of analyzing statically indeterminate problems."† This approach is indeed the basis of the displacement method that is now widely used in computer applications. The four-bar, pin-connected structure of Fig. 1.4.1a is statically indeterminate to the second degree because there are four unknown bar forces, but only two equations of equilibrium exist at joint E. If the force method is used, the basic determinate structure might contain bars AE and

---

†S. P. Timoshenko, *History of Strength of Materials*, McGraw-Hill Book Company, New York, 1953, p. 75.

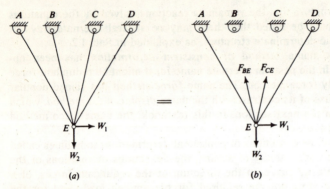

Figure 1.4.1 A four-bar, pin-connected structure.

DE only, whereas the forces in bars BE and CE in Fig. 1.4.1b are first determined from the conditions that the changes in distance between points B and E and points C and E must be just equal to the change in length of bars BE and CE. Thus one may say that the structure of Fig. 1.4.1a is only *statically* indeterminate, but statically and geometrically (deflection-wise), it *is* determinate. Navier maintained that the problem of Fig. 1.4.1a is "determinate" if the horizontal and vertical displacements of pin joint E are taken as the unknowns at the very beginning. Next the elongations of the four bars are expressed in terms of these two unknown displacements. Next the axial forces in the four bars are expressed in terms of the elongations in themselves. Next the two resolution equations of equilibrium for the pin joint E are used to express the applied loads in terms of the axial forces in the bars. In this way, there will always be as many equations of equilibrium as there are unknown joint displacements because of the one-to-one correspondence between each joint load and each joint displacement. In this context, all problems are determinate, and there is no need to differentiate a structure as being statically determinate or indeterminate providing the displacement method is used for its analysis. It is important to remind the reader that when the displacement method is used, sizes of the constituent members in the structure must be part of the given data.

It will be shown in several chapters of this textbook headed by the title *displacement method* that the concept advanced in the preceding paragraph applies not only to trusses but also to continuous beams and rigid frames. Indeed, the entire subject of the finite-element method of structural analysis as it is widely used in electronic computation is based on this fundamental idea.

1.5 Methods of Structural Analysis

Every method of structural analysis can be classified as either a force method or a displacement method.

FORCE METHOD VS. DISPLACEMENT METHOD   11

Force method   For statically determinate structures, the internal forces within the constituent members can be determined by the laws of statics alone at the very beginning, and then the deformed shape of the structure follows. For statically indeterminate structures, the relative sizes of the constituent members are required in the solution for the redundants from the conditions of consistent deformation. The remaining statical unknowns are obtained after the redundants are determined.

Displacement method   In the displacement method, whether the structure is statically determinate or statically indeterminate, the solution procedure is the same; that is, the displacements of the joints in the structure are solved at the very beginning from an equal number of equations of equilibrium. Only then are the internal forces within the constituent members and the external reactions acting on the whole structure determined from the deformed shape of the structure.

The force method of analysis can be derived entirely from the physical conditions of consistent deformation along the lines of action of the redundants, or it can be derived entirely from an elegant theorem which states that the redundants should be such as to keep at a minimum the total strain energy within the structure. This parallel treatment will be emphasized throughout this textbook. In regard to a rigid frame with two fixed supports (e.g., a fixed arch) or a closed cell (e.g., a culvert), there is a special variation of the force method, known as the column-analogy method, which will be dealt with in Chaps. 15 and 18. Another variation of the force method is the use of the three-moment equation in continuous-beam analysis, which is treated in Chap. 6.

Matrix notations have been conveniently used to derive a comprehensive treatment of the displacement method as applied to trusses, continuous beams, and rigid frames. Because of the easiness with which matrix operations can be performed on the electronic computer, the matrix displacement method has become the most modern and efficient method for very large structures, especially in the final-check analysis stage. The matrix displacement methods will be treated in Chaps. 10, 11, 12, 16, and 19. Special shorter versions of the displacement method are the traditional slope-deflection and moment-distribution methods. These two methods are of tremendous importance and are treated in Chaps. 7 and 8.

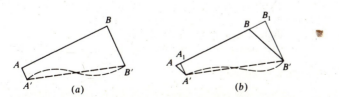

Figure 1.6.1 Effect of end displacements on member length.

In spite of the need for the formal and more exact method of analysis, approximate methods are desirable for purposes of estimation, at least in the search for relative sizes of members. Approximate methods for multistory-frame analysis are contained in Chap. 14.

Some special topics on influence lines and moving loads, secondary stresses in trusses, semirigid connections, shear deformations, and beams on elastic foundation are taken up in Chaps. 13, 17, 20, 21, and 22, respectively.

1.6 Fundamental Assumptions

Throughout this textbook two fundamental assumptions prevail. These two fundamental assumptions may be summarized in the phrase *linear first-order analysis*. The word *linear* refers to the material property that the relationship between stress and strain is linear. The implication is that nowhere in the structure is the stress above the proportional limit. When the response of a structure is sought wherein the material is stressed beyond the proportional limit, the method of analysis falls into the realm of nonlinear analysis, which is certainly dependent on the shape of the stress-strain curve beyond the proportional limit. Nonlinear analysis is beyond the scope of this textbook.

The other fundamental assumption is that the length of a straight segment within the structure is not affected by its curvature during deformation, nor is it affected by displacement of its ends in the transverse direction. As shown by Fig. 1.6.1a, the length of curve $A'B'$ is considered equal to the length of AB, so long as the displacements AA' and BB' are perpendicular to AB. In Fig. 1.6.1b, the length of curve $A'B'$ is considered to be longer than that of AB by the amount $BB_1 - AA_1$, where BB_1 and AA_1 are the components of the displacements AA' and BB' along the direction AB. Theoretically the length of a curve is longer than the chord; the difference is called the "bow effect." Furthermore, the length of chord $A'B'$ in Fig. 1.6.1a is theoretically longer than its projection in the AB direction; this difference is called a "second-order effect." Other notable second-order effects are that equilibrium equations for a structure with large deformations should be written in accordance with the new dimensions of the deformed structure, and that the bending moment in a beam is also affected by the product of the axial force in the beam and the transverse deflection. When none of the second-order effects is considered, it is often said that a first-order analysis has been made. The word *first* means that a first attempt is being made to evaluate the response of a structure due to a static disturbance, such as applied loading, temperature change, or support settlements.

For the present it is sufficient to emphasize that the treatment in this book is confined to linear first-order analysis. Thus a first attempt is being made in the study of structural analysis.

CHAPTER
TWO

DEFORMATION OF STATICALLY DETERMINATE BEAMS AND RIGID FRAMES

2.1 Force Response vs. Deformation Response

As discussed in Chap. 1, reactions to statically determinate beams and rigid frames may be determined by the laws of statics alone. Once the reactions to a beam are known, variation of shear and moment along its entire length may be obtained and plotted as shear and moment diagrams. Likewise, variation of axial force, shear, and moment along the length of each member in a rigid frame may be obtained by treating each member as a free body. The reactions and the internal forces (such as axial force, shear, and moment within a member) constitute the "force response" to a static disturbance, which may be a load system, temperature change, or support settlement.

While axial force is absent in a beam, it does exist in members of a rigid frame. However, in a rigid frame, the change in length due to axial force is usually so small in comparison with the deformation due to bending moment that it is often neglected. Axial deformation in members of a rigid frame will not be considered until Chap. 16. The sliding of one cross section in the transverse direction relative to an adjacent cross section, due to the existence of internal shear force, is again very small in ordinary cases so that the deformation due to shear force will not be considered until Chap. 21. What remains to be considered in the beginning part of this textbook is the shape of the elastic curve due to bending moment within a beam or a member of a rigid frame. The shape of the elastic curve of a beam, or the combined shape of the

14 INTERMEDIATE STRUCTURAL ANALYSIS

elastic curves of the members in a rigid frame, constitutes the "deformation response" of a beam or a rigid frame.

In this chapter, determination of force response of statically determinate beams and rigid frames will be only briefly reviewed since this subject is treated in a first course in mechanics of materials. The main subject of the chapter is the determination of the slope and deflection at any chosen point on the elastic curve of a beam or of a member in a rigid frame.

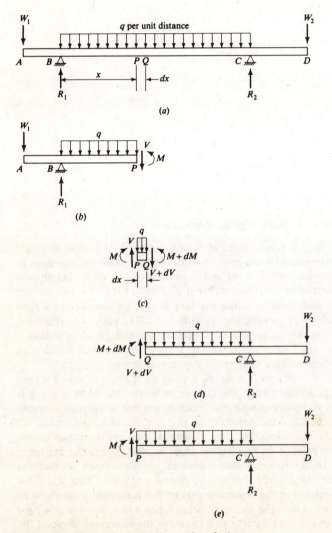

Figure 2.2.1 Shear and moment at a section of a beam.

DEFORMATION OF STATICALLY DETERMINATE BEAMS AND RIGID FRAMES 15

2.2 Shears and Moments in Beams

Consider the statically determinate beam of Fig. 2.2.1a. So far as equilibrium of the entire beam is concerned, there are two unknown reactions R_1 and R_2. It is best to obtain R_1 by using the equilibrium equation that the summation of moments about point C must be zero, then obtain R_2 by using the equation that the summation of moments about point B must be zero, and lastly check by using the equation that the summation of vertical forces must be zero.

Now cut an infinitesimal segment dx bounded by the cross sections at P and Q, as shown in Fig. 2.2.1a. The free-body diagrams of ABP, PQ, and QCD are shown in Fig. 2.2.1b to d. The unknown V is called the *shear force*, or simply the *shear*, at section P; and the unknown M is called the *bending moment*, or simply the *moment*, at section P. From Fig. 2.2.1c, one can note that the adopted sign convention is that (1) shear V is positive if section P tends to slide upward relative to section Q, and (2) moment M is positive if it causes compression in the upper portion of the beam.

The shear V and the moment M at any section of a beam may be obtained by using either the left free body or the right free body. In reference to Fig. 2.2.1b, V is equal to the summation of the known *upward* forces acting on ABP, and M is equal to the summation of the *clockwise* moments of the known forces acting on ABP about P. However, in reference to Fig. 2.2.1e, V is equal to the summation of the known *downward* forces acting on DCP, and M is equal to the summation of the *counterclockwise* moments of the known forces acting on DCP about P. If the preceding rules are strictly followed, the same result for V and M, both in sign and in magnitude, would be obtained regardless of whether the left or the right free body is used.

Example 2.2.1 Determine the shear V and the moment M at section P of the overhanging beam shown in Fig. 2.2.2.

SOLUTION (a) *Determine reactions R_1 and R_2.* Taking moments about point C,

$$10R_1 = 8(12) + 3.6(10)(5) - 10(3)$$
$$R_1 = 24.6 \text{ kN}$$

Taking moments about point B,

$$10R_2 = 3.6(10)(5) + 10(13) - 8(2)$$
$$R_2 = 29.4 \text{ kN}$$

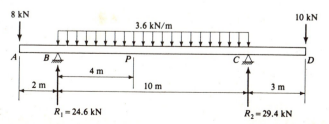

Figure 2.2.2 Beam of Example 2.2.1.

16 INTERMEDIATE STRUCTURAL ANALYSIS

Check by taking the summation of vertical forces,
$$8 + 3.6(10) + 10 \stackrel{?}{=} 24.6 + 29.4$$
$$54 = 54 \quad \text{(OK)}$$

(b) *V and M at P using left free body.*
$$V = \text{sum of upward forces on } ABP$$
$$= +24.6 - 8 - 3.6(4) = +2.2 \text{ kN}$$
$$M = \text{sum of clockwise moments of forces on } ABP \text{ about } P$$
$$= 24.6(4) - 8(6) - 3.6(4)(2) = +21.6 \text{ kN·m}$$

(c) *V and M at P using right free body.*
$$V = \text{sum of downward forces on } DCP$$
$$= +10 + 3.6(6) - 29.4 = +2.2 \text{ kN} \quad \text{(Check)}$$
$$M = \text{sum of counterclockwise moments of forces on } DCP \text{ about } P$$
$$= 29.4(6) - 10(9) - 3.6(6)(3) = +21.6 \text{ kN·m} \quad \text{(Check)}$$

Example 2.2.2. Determine the shear V and the moment M at section P of the cantilever beam shown in Fig. 2.2.3.

SOLUTION (a) *Determine reactions R_1 and R_2.* Taking moments about point A,
$$R_1 = 36(3) + 4(7)(3.5) = 206 \text{ kN·m}$$

Taking the summation of vertical forces,
$$R_2 = 36 + 4(7) = 64 \text{ kN}$$

A check may be made by taking moments about point C, although it does not seem to add much to an independent check. If done, it is
$$206 + 36(4) + 4(7)(3.5) \stackrel{?}{=} 64(7)$$
$$448 = 448 \quad \text{(OK)}$$

(b) *V and M using left free body.*
$$V = \text{sum of upward forces on } ABP$$
$$= +64 - 36 - 4(5) = +8 \text{ kN}$$
$$M = \text{sum of clockwise moments of forces on } ABP \text{ about } P$$
$$= -206 + 64(5) - 36(2) - 4(5)(2.5) = -8 \text{ kN·m}$$

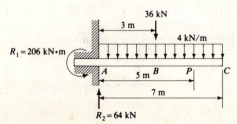

Figure 2.2.3 Beam of Example 2.2.2.

(c) V and M using right free body.

V = sum of downward forces on CP
$= +4(2) = +8$ kN   (Check)
M = sum of counterclockwise moments of forces on CP about P
$= -4(2)(1) = -8$ kN·m   (Check)

Two simple relationships exist between the load intensity q, the shear V, and the moment M at any section P of a beam. These relationships can be derived considering the equilibrium of the infinitesimal segment dx shown in Fig. 2.2.1c. Taking the summation of vertical forces acting on the free body of Fig. 2.2.1c,

$$V = V + dV + q\,dx$$

which upon simplifying becomes

$$\frac{dV}{dx} = -q \qquad (2.2.1)$$

Taking moments about point Q in Fig. 2.2.1c,

$$M + V\,dx = M + dM + q\,dx\left(\frac{dx}{2}\right)$$

which upon simplifying becomes

$$\frac{dM}{dx} = +V \qquad (2.2.2)$$

Equations (2.2.1) and (2.2.2) are two important relationships by means of which shear and moment diagrams as defined subsequently may be obtained.

Shear and moment diagrams are graphical representations of the variation in shear and moment along the entire length of the beam. Since the shear at any section P is the summation of upward forces between the left end of the beam and the point P, the shear diagram can be plotted simply by starting with zero at the left end and stepping or sloping up and down with the forces along the beam until the zero value is again reached at the right end. When a part of the beam is not loaded, the shear is constant; when a part of the beam is uniformly loaded, the shear curve is linear with a downward slope equal to the load intensity q. Equation (2.2.1) verifies this fact.

The increase in moment at any one section over that of a section to left of it is equal to the area of the shear diagram between these two sections. Equation (2.2.2) verifies the previous statement because it may be stated as

$$\int_A^B dM = \int_A^B V\,dx$$

or
$$M_B - M_A = \int_A^B V\,dx \qquad (2.2.3)$$

18 INTERMEDIATE STRUCTURAL ANALYSIS

Thus the moment diagram may be simply plotted by starting with the known value of the moment at the left end and continuing to add the area of the shear diagram until the known value of the moment at the right end is reached. Except for the fixed end of a cantilever, where the moment is always a negative quantity for downward loading, the moment at the ends of a beam is usually equal to zero. Equation (2.2.2) further states that the slope of the moment curve at any point is equal to the shear at that point. One should make use of this concept when attempting to make a freehand sketch of the moment diagram after the moment values at some chosen critical sections have been computed.

Example 2.2.3 Draw the shear and moment diagrams for the beam of Example 2.2.1, repeated here as Fig. 2.2.4a. Assume that the reactions have already been computed.

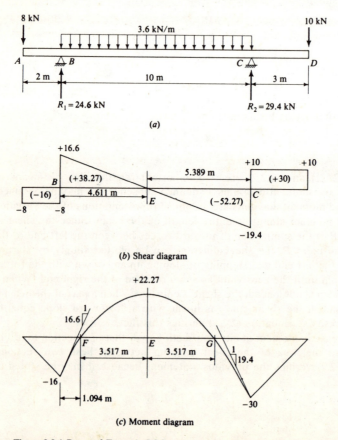

Figure 2.2.4 Beam of Example 2.2.3.

DEFORMATION OF STATICALLY DETERMINATE BEAMS AND RIGID FRAMES 19

SOLUTION (a) *Shear diagram*. The shear diagram is plotted in Fig. 2.2.4b, starting with zero at the left end and arriving again at zero at the right end. Point E, at which shear is equal to zero, is located by the distance BE, which is 16.6 divided by 3.6; or it can be located by the distance CE, which is 19.4 divided by 3.6.

(b) *Moment diagram*. The areas of the parts of the shear diagram are computed to be -16, $+38.27$, -52.27, and $+30$; these values are written in parentheses in Fig. 2.2.4b. The moment diagram of Fig. 2.2.4c is plotted by starting with zero at the left end and successively adding the shear areas to arrive at -16 for B, $-16 + 38.27 = +22.27$ for E, $+22.27 - 52.27 = -30$ for C, and $-30 + 30 = 0$ for D. One may note that the moment curve is linear between A and B, with a downward slope of -8 kN·m/m. Also the slope of the moment curve decreases from $+16.6$ kN·m/m at B to zero at E, and it further decreases from zero at E to -19.4 kN·m/m at C. The moment curve is again linear between C and D, with an upward slope of $+10$ kN·m/m. The distances FE or EG in Fig. 2.2.4c may be computed by noting that the shear area between F and E is equal to $+22.27$; thus

$$\tfrac{1}{2}(3.6)(FE)^2 = 22.27$$

$$FE = 3.517 \text{ m}$$

Example 2.2.4 Draw the shear and moment diagrams for the beam of Example 2.2.2, repeated here as Fig. 2.2.5a. Assume that the reactions have already been computed.

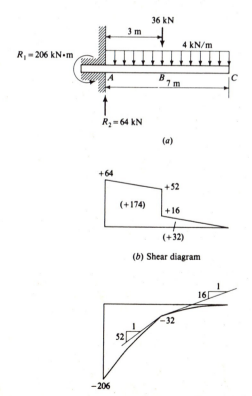

(a)

(b) Shear diagram

(c) Moment diagram   **Figure 2.2.5** Beam of Example 2.2.4.

20 INTERMEDIATE STRUCTURAL ANALYSIS

SOLUTION (a) *Shear diagram.* The shear diagram is plotted in Fig. 2.2.5b, starting with zero at the left end and arriving again at zero at the right end.

(b) *Moment diagram.* The two parts of the shear diagram are computed to be +174 and +32 and written in parentheses in Fig. 2.2.5b. The moment diagram of Fig. 2.2.5c is plotted by starting with −206 kN·m at the left end, since this moment is causing compression in the lower part of the beam and is therefore negative. By successively adding the shear areas, the moment at B is −206 + 174 = −32 and the moment at C is −32 + 32 = 0. One may note that the slope of the moment curve is +64 at A, decreasing to +52 just left of B; it becomes +16 just to right of B, decreasing to zero at C. There is a sudden change of slope right at B because of the presence of the concentrated load there.

2.3 Axial Force, Shear, and Moment in Rigid Frames

Most common rigid frames as built are statically indeterminate; the treatment of statically determinate rigid frames is necessary because it forms a part of the analysis procedure.

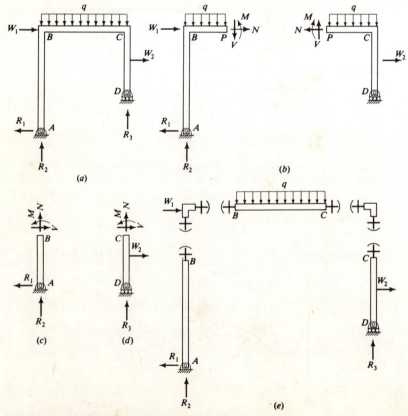

Figure 2.3.1 Axial force, shear, and moment in a rigid frame.

DEFORMATION OF STATICALLY DETERMINATE BEAMS AND RIGID FRAMES 21

Consider the statically determinate rigid frame shown in Fig. 2.3.1a. When this rigid frame is separated into two parts at a typical cross section P as shown in Fig. 2.3.1b, there are three interacting internal forces (one is a moment, but henceforth considered a generalized force): the axial force N, the shear V, and the moment M. N acts in the direction of the longitudinal axis at the centroid and is normal to the cross section; thus either the name *axial force* or the name *normal force* has been used. The axial force N is considered positive if it is tensile, or, in other words, if one part tends to pull on the other. In Fig. 2.3.1b, N is the summation of all known leftward forces acting on ABP, or the summation of all known rightward forces acting on DCP.

The sign convention for the shear V and the moment M, so far as the horizontal member BC is concerned, is the same as for beams. Members AB and DC may be treated as horizontal beams when one views them from the right side, resulting in the positive directions for N, V, and M as shown in Fig. 2.3.1c and d.

One alternate way of determining the variation of axial force, shear, and moment in all members of a rigid frame is to consider each member separately after cutting the entire frame into joints and members as shown by Fig. 2.3.1e. Once this is done, the sign convention for V and M may be obtained by considering each member as being horizontal and by choosing one end as the left end. For a vertical member, the shear diagram would appear reversed when a different end is chosen as the left end, but the moment diagram will always appear on the compression side of the member if the method of obtaining shear and moment diagrams discussed in Sec. 2.2 is used.

Example 2.3.1 Determine the variation of axial force, shear, and moment in all members of the rigid frame shown in Fig. 2.3.2a.

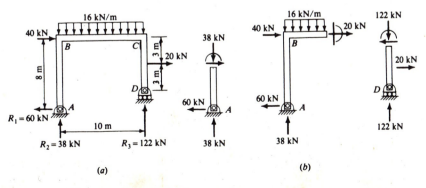

Figure 2.3.2 Rigid frame of Example 2.3.1.

22 INTERMEDIATE STRUCTURAL ANALYSIS

SOLUTION (a) *Determine reactions* R_1, R_2, *and* R_3 (*Fig.* 2.3.2a). Taking the summation of horizontal forces,

$$R_1 = 40 + 20 = 60 \text{ kN}$$

Taking moments about point D,

$$10R_2 = 16(10)(5) - 60(2) - 40(6) - 20(3)$$
$$R_2 = 38 \text{ kN}$$

Taking moments about point A,

$$10R_3 = 16(10)(5) + 40(8) + 20(5)$$
$$R_3 = 122 \text{ kN}$$

Check by taking the summation of vertical forces,

$$16(10) \stackrel{?}{=} 38 + 122$$
$$160 = 160 \quad \text{(OK)}$$

(b) *Axial force.* By inspection of Fig. 2.3.2b, one can see that the axial forces are as follows:

$$N \text{ in } AB = -38 \text{ kN}$$
$$N \text{ in } BC = +20 \text{ kN}$$
$$N \text{ in } CD = -122 \text{ kN}$$

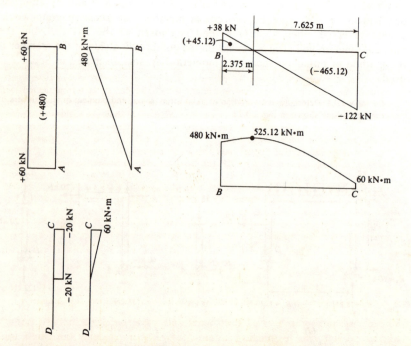

Figure 2.3.3 Shear and moment diagrams for rigid frame of Example 2.3.1.

DEFORMATION OF STATICALLY DETERMINATE BEAMS AND RIGID FRAMES 23

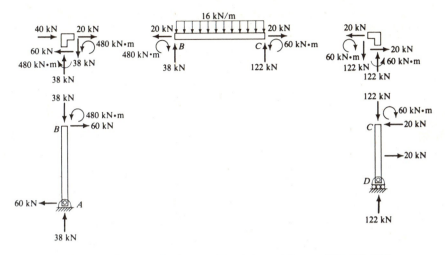

Figure 2.3.4 Free-body diagrams of members and joints for rigid frame of Example 2.3.1.

Figure 2.3.2b is offered here to actually show the mental picture in obtaining the results shown previously by looking directly at Fig. 2.3.2a.

(c) *Shear and moment diagrams.* The shear and moment diagrams for members AB, BC, and CD are shown in Fig. 2.3.3. The shear diagrams are obtained first by looking directly at Fig. 2.3.2a and by using A, B, and D as the left ends of real or imaginary horizontal members AB, BC, and DC, respectively. While the moments at A and D are both known to be zero at the outset, the moment at B of member BC is taken from the moment at B of member BA, which is 480 kN·m causing compression on the outside part of the corner B. Note, in particular, that it is common practice not to show the positive or negative signs on the moment diagrams of rigid frames, because these diagrams will always be plotted on the compression side of the member.

(d) *Alternate method.* Rather than follow the concepts of parts (b) and (c) of the solution, in which the free bodies ABP and DCP are used to obtain the variation of axial force and shear in member BC, it may be desirable for the beginner to draw separately the free-body diagrams of all members and joints as soon as the reactions have been computed. These free-body diagrams are shown in Fig. 2.3.4. From these diagrams, the axial forces in AB, BC, and CD are observed to be −38, +20, and −122 kN. Moreover, the shear and moment diagrams may be drawn following the procedure for beams. These diagrams will be exactly like those shown in Fig. 2.3.3 except that the moments at both ends of each member are known before the moment diagram is plotted by the method of adding successively the shear areas.

2.4 Common Methods for Determination of Deformation Response

Without considering effects of both axial and shear deformation, the deformation response due to bending moment consists of defining the slope and deflection at any point on the elastic curve. In mechanics of materials,

24 INTERMEDIATE STRUCTURAL ANALYSIS

the double-integration method is derived and used to obtain the equations of the elastic curve for various segments between concentrated forces or reactions throughout the entire beam. In structural analysis the complete equation of the elastic curve may not be of interest; rather, the usual requirements are to determine the slope or deflection at only one or several points on the elastic curve. To this end, the methods developed and illustrated in this chapter fall into two categories: the work-energy method or the geometric method. The former includes the unit-load and the partial-derivative methods; and the latter, the moment-area and the conjugate-beam methods.

2.5 The Unit-Load Method—Derivation of Basic Formula

The problem is to find the vertical deflection Δ_C at point C on the simple beam AB of Fig. 2.5.1b subjected to loads W_1, W_2, and W_3. This set of loads causes internal forces in the beam—for instance, a compressive force of F in any fiber such as MN with cross-sectional area dA. This fiber, MN, is

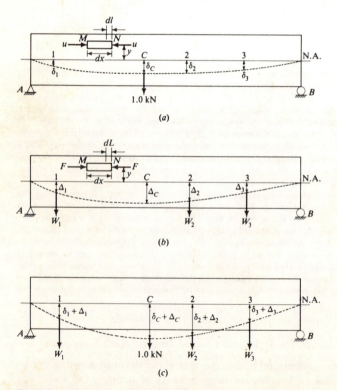

Figure 2.5.1 Basic formula in the unit-load method.

DEFORMATION OF STATICALLY DETERMINATE BEAMS AND RIGID FRAMES

shortened by an amount dL. Of course, the loads also produce deflections all along the beam, such as Δ_1 at W_1, Δ_2 at W_2, and Δ_3 at W_3. The total external work done on the beam, if the loads are gradually applied, is $\frac{1}{2}W_1\Delta_1 + \frac{1}{2}W_2\Delta_2 + \frac{1}{2}W_3\Delta_3$. The total internal energy stored in the beam is equal to $\frac{1}{2}\Sigma F\,dL$. By the law of conservation of work or energy, the total external work done on the beam is equal to the total internal energy stored in the beam, or

$$\tfrac{1}{2}W_1\Delta_1 + \tfrac{1}{2}W_2\Delta_2 + \tfrac{1}{2}W_3\Delta_3 = \tfrac{1}{2}\Sigma F\,dL \tag{2.5.1}$$

Now, if on the same simple beam AB a unit load of 1.0 kN is first gradually applied at C as shown in Fig. 2.5.1a, it will cause deflections of δ_C at point C, δ_1 at point 1, δ_2 at point 2, and δ_3 at point 3. If the loads W_1, W_2, W_3 are added gradually to the beam of Fig. 2.5.1a, on which the unit load at C is already applied, the deflections will be $\delta_C + \Delta_C$ at point C, $\delta_1 + \Delta_1$ at point 1, $\delta_2 + \Delta_2$ at point 2, and $\delta_3 + \Delta_3$ at point 3, as shown by Fig. 2.5.1c. This is so because a linear relationship between force and deflection is implied and the principle of superposition holds.

When the unit load at C is first applied, the relation between external work and internal energy is

$$\tfrac{1}{2}(1.0)(\delta_C) = \tfrac{1}{2}\Sigma u\,dl \tag{2.5.2}$$

where u is the compressive force on any fiber MN with area dA caused by the unit load and dl is the total shortening of this fiber. When the loads W_1, W_2, W_3 are gradually added, the additional external work done on the beam is $\tfrac{1}{2}W_1\Delta_1 + \tfrac{1}{2}W_2\Delta_2 + \tfrac{1}{2}W_3\Delta_3 + (1.0)(\Delta_C)$ because the *constant* 1.0-kN load already on the beam goes through the additional deflection Δ_C and the *gradually* applied loads W_1, W_2, W_3 go through the additional deflections Δ_1, Δ_2, Δ_3. The additional internal energy stored in the beam is $\tfrac{1}{2}\Sigma F\,dL + \Sigma u\,dL$. The total external work done on the beam thus far is therefore $\tfrac{1}{2}(1.0)(\delta_C) + \tfrac{1}{2}W_1\Delta_1 + \tfrac{1}{2}W_2\Delta_2 + \tfrac{1}{2}W_3\Delta_3 + (1.0)(\Delta_C)$, while the total internal energy is $\tfrac{1}{2}\Sigma u\,dl + \tfrac{1}{2}\Sigma F\,dL + \Sigma u\,dL$. Again by the law of conservation of work or energy,

$$\tfrac{1}{2}(1.0)(\delta_C) + \tfrac{1}{2}W_1\Delta_1 + \tfrac{1}{2}W_2\Delta_2 + \tfrac{1}{2}W_3\Delta_3 + (1.0)(\Delta_C) = \tfrac{1}{2}\Sigma u\,dL + \tfrac{1}{2}\Sigma F\,dL + \Sigma u\,dL \tag{2.5.3}$$

Subtracting the sum of Eqs. (2.5.1) and (2.5.2) from Eq. (2.5.3),

$$(1.0)(\Delta_C) = \Sigma u\,dL \tag{2.5.4}$$

Equation (2.5.4), leading to the working formula of Eq. (2.6.3) for beams, is the basic formula in the unit-load method. It can be applied to finding the slope or deflection at any point in a beam, rigid frame, or truss whether dL is caused by applied loads, temperature changes, or fabrication errors.

2.6 The Unit-Load Method—Application to Beam Deflections

The problem is to find the deflection Δ_C at point C on a simple beam AB due to the applied loading of W_1, W_2, and W_3 (see Fig. 2.5.1b). A unit load of 1.0 kN is applied at point C to the unloaded beam AB (see Fig. 2.5.1a). By Eq. (2.5.4), $(1.0)(\Delta_C) = \Sigma u \, dL$. Now let the bending moment at MN in Fig. 2.5.1b be M and the bending moment at MN in Fig. 2.5.1a be m. Let the original length of MN be dx. Then u in Fig. 2.5.1a is

$$u = \frac{my}{I} dA \qquad (2.6.1)$$

and dL in Fig. 2.5.1b is

$$dL = \frac{My}{EI} dx \qquad (2.6.2)$$

Substituting Eqs. (2.6.1) and (2.6.2) into Eq. (2.5.4),

$$(1.0)(\Delta_C) = \Sigma u \, dL = \Sigma \left(\frac{my}{I} dA\right)\left(\frac{my}{EI} dx\right)$$

$$= \int_0^L \int_0^A \frac{Mmy^2 \, dA \, dx}{EI^2}$$

$$= \int_0^L \frac{Mm \, dx}{EI^2} \int_0^A y^2 \, dA = \int_0^L \frac{Mm \, dx}{EI} \qquad (2.6.3)$$

If one considers m as the ratio of the bending moment caused by any load applied at C to the load itself, then m has the dimensional unit for length only and Eq. (2.6.3) may be written as

$$\Delta_C = \int_0^L \frac{Mm \, dx}{EI} \qquad (2.6.4)$$

Equation (2.6.4) is the working formula by which the deflection at any point of a statically determinate beam due to the applied loading can be found.

Example 2.6.1 Find the deflection Δ_B of the cantilever beam shown in Fig. 2.6.1a by the unit-load method.

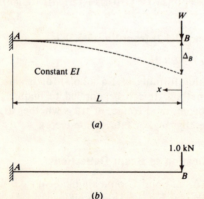

Figure 2.6.1 Cantilever beam of Example 2.6.1.

DEFORMATION OF STATICALLY DETERMINATE BEAMS AND RIGID FRAMES 27

SOLUTION For this problem it is most convenient to choose the origin at B and measure x toward the left. The expressions for M and m are $-Wx$ and $-1.0x$, respectively, valid from $x = 0$ to $x = L$. Using Eq. (2.6.4),

$$\Delta_B = \int_0^L \frac{Mm\,dx}{EI} = \int_0^L \frac{(-Wx)(-1.0x)\,dx}{EI}$$

$$= \frac{W}{EI}\left[\frac{x^3}{3}\right]_0^L = \frac{WL^3}{3EI} \quad \text{downward}$$

Note that the deflection is downward because the unit load has been applied in that direction and a positive answer is obtained.

Example 2.6.2 Find by the unit-load method the deflection Δ_C at midspan of the simple beam AB shown in Fig. 2.6.2a.

SOLUTION For this problem the entire beam may be divided into two segments AC and BC. When the origin for x is chosen at A and B respectively, the expressions for M and m become identical for the two halves of the beam. Thus, using Eq. (2.6.4),

$$\Delta_C = \int_0^L \frac{Mm\,dx}{EI} = 2\int_0^{L/2} \frac{\left(\frac{wL}{2}x - \frac{wx^2}{2}\right)(0.5x)\,dx}{EI}$$

$$= \frac{2}{EI}\left[\frac{wL}{4}\frac{x^3}{3} - \frac{w}{4}\frac{x^4}{4}\right]_0^{L/2} = \frac{2wL^4}{EI}\left(\frac{1}{96} - \frac{1}{256}\right)$$

$$= \frac{5wL^4}{384EI} \quad \text{downward}$$

Example 2.6.3 Find by the unit-load method the deflection Δ_D in centimeters of the simple beam AB shown in Fig. 2.6.3a.

SOLUTION The entire beam is divided into 3 segments: AC, CD, and BD. The origin from which x is measured and the expressions for M and m are shown for each segment in the accompanying table.

Segment	Origin	Limits, m	M, kN·m	m, m
AC	A	0 to 1.8	$80x$	$0.5x$
CD	A	1.8 to 2.7	$80x - 120(x - 1.8)$	$0.5x$
BD	B	0 to 2.7	$40x$	$0.5x$

Using Eq. (2.6.4),

$$\Delta_D = \int_0^{1.8} \frac{(80x)(0.5x)\,dx}{EI} + \int_{1.8}^{2.7} \frac{[80x - 120(x - 1.8)](0.5x)}{EI}$$

$$+ \int_0^{2.7} \frac{(40x)(0.5x)}{EI}$$

$$= \frac{40}{EI}\left[\frac{x^3}{3}\right]_0^{1.8} + \left[\frac{-20x^3}{EI\ 3} + \frac{108x^2}{EI\ 2}\right]_{1.8}^{2.7} + \frac{20}{EI}\left[\frac{x^3}{3}\right]_0^{2.7}$$

$$= \frac{1}{EI}(77.76 + 126.36 + 131.22)$$

$$= \frac{335.34 \text{ kN·m}^3}{EI} = \frac{335.34}{(200)(160)} = 0.01048 \text{ m} = 1.048 \text{ cm} \quad \text{downward}$$

28 INTERMEDIATE STRUCTURAL ANALYSIS

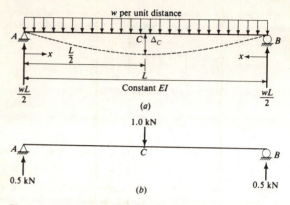

Figure 2.6.2 Simple beam of Example 2.6.2.

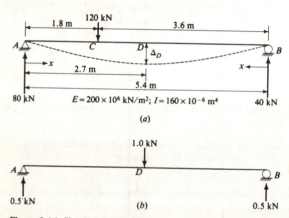

Figure 2.6.3 Simple beam of Example 2.6.3.

Example 2.6.4 Find by the unit-load method the deflection Δ_D of the simple beam AB shown in Fig. 2.6.4a. Note that this beam has a constant wide-flange steel section throughout but is strengthened with two cover plates in the center half of the beam so that the moment of inertia is doubled.

SOLUTION Because there are 3 sections of discontinuity, C, D, and E, in this beam, the entire beam should be divided into 4 segments as follows:

Segment	Origin	Limits	M	m	I
AC	A	0 to $L/4$	$\frac{1}{2}Wx$	$\frac{1}{2}x$	I_c
CD	A	$L/4$ to $L/2$	$\frac{1}{2}Wx$	$\frac{1}{2}x$	$2I_c$
DE	B	$L/4$ to $L/2$	$\frac{1}{2}Wx$	$\frac{1}{2}x$	$2I_c$
EB	B	0 to $L/4$	$\frac{1}{2}Wx$	$\frac{1}{2}x$	I_c

DEFORMATION OF STATICALLY DETERMINATE BEAMS AND RIGID FRAMES 29

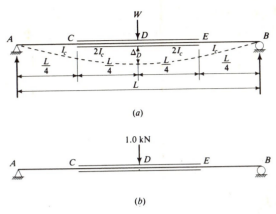

Figure 2.6.4 Simple beam of Example 2.6.4.

Then, using Eq. (2.6.4),

$$\Delta_D = 2\int_0^{L/4} \frac{(\frac{1}{2}Wx)(\frac{1}{2}x)\,dx}{EI_c} + 2\int_{L/4}^{L/2} \frac{(\frac{1}{2}Wx)(\frac{1}{2}x)\,dx}{E(2I_c)}$$

$$= \frac{W}{EI_c}\left[\frac{1}{2}\frac{x^3}{3}\right]_0^{L/4} + \frac{W}{EI_c}\left[\frac{1}{4}\frac{x^3}{3}\right]_{L/4}^{L/2}$$

$$= \frac{W}{EI_c}\left[\frac{1}{384} + \frac{1}{12}\left(\frac{1}{8} - \frac{1}{64}\right)\right] = \frac{3WL^3}{256EI_c} \quad \text{downward}$$

2.7 The Unit-Load Method—Application to Beam Slopes

The slope at any point of a beam is defined to be the tangent function, or the radian measure of the angle between the original beam axis and the tangent to the elastic curve at that point. This angle may be considered as the rotation of the tangent to the elastic curve, or the rotation of the cross section, provided that the tangent is always perpendicular to the cross section. This will not be so when deflection due to shear is included, which will be treated in Chap. 21.

To determine the slope θ_C at any point C of a beam by the unit-load method, it is simply necessary to adapt Eq. (2.6.4) to become

$$\boxed{\theta_C = \int_0^L \frac{Mm\,dx}{EI}} \qquad (2.7.1)$$

where here m is the bending moment due to a unit moment applied at point C. When a positive answer is obtained from using Eq. (2.7.1), the rotation of the tangent is in the same sense as that of the unit moment. The derivation of Eq. (2.7.1) follows the same pattern as does Eq. (2.6.4); note that the work done by a moment couple in going through an angle of rotation is simply the product of that moment couple and the angle of rotation.

Example 2.7.1 Find by the unit-load method the slope θ_B of the cantilever beam shown in Fig. 2.7.1a.

SOLUTION If x is measured toward the left from point B, the expressions for M and m, as seen from Fig. 2.7.1 a and b, are $-Wx$ and -1.0, respectively; thus, using Eq. (2.7.1),

$$\theta_B = \int_0^L \frac{Mm\,dx}{EI} = \int_0^L \frac{(-Wx)(-1.0)\,dx}{EI} = \frac{W}{EI}\left[\frac{x^2}{2}\right]_0^L$$

$$= \frac{WL^2}{2EI} \quad \text{clockwise}$$

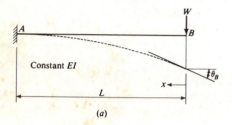

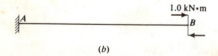

Figure 2.7.1 Cantilever beam of Example 2.7.1.

Example 2.7.2 Find by the unit-load method the slope θ_A or θ_B of the simple beam AB shown in Fig. 2.7.2a.

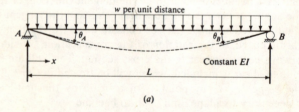

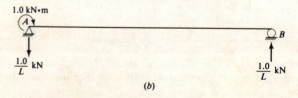

Figure 2.7.2 Simple beam of Example 2.7.2.

DEFORMATION OF STATICALLY DETERMINATE BEAMS AND RIGID FRAMES 31

SOLUTION The expression for m is shorter if x is measured toward the left from point B. Using Eq. (2.7.1),

$$\theta_A = \int_0^L \frac{Mm\, dx}{EI} = \int_0^L \frac{\left(\frac{wL}{2}x - \frac{wx^2}{2}\right)\left(\frac{1.0}{L}x\right)}{EI}$$

$$= \frac{w}{EI}\left[\frac{1}{2}\frac{x^3}{3} - \frac{1}{2L}\frac{x^4}{4}\right]_0^L$$

$$= \frac{wL^3}{EI}\left(\frac{1}{6} - \frac{1}{8}\right) = \frac{wL^3}{24EI} \quad \text{clockwise}$$

$$\theta_B = \frac{wL^3}{24EI} \quad \text{counterclockwise}$$

Example 2.7.3 Find by the unit-load method the slope θ_A in radians of the simple beam AB shown in Fig. 2.7.3a.

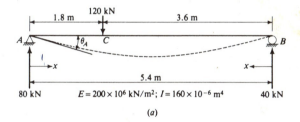

(a)

(b)

Figure 2.7.3 Simple beam of Example 2.7.3.

SOLUTION The entire beam is divided into two segments AC and BC. The origin from which x is measured and the expressions for M and m are shown for each segment in the accompanying table.

Segment	Origin	Limits, m	M, kN·m	m (pure ratio)
AC	A	0 to 1.8	$80x$	$1.0 - \frac{1}{5.4}x$
BC	B	0 to 3.6	$40x$	$\frac{1}{5.4}x$

Using Eq. (2.7.1),

$$\theta_A = \int_0^{1.8} \frac{(80x)[1.0 - (1/5.4)x]\,dx}{EI} + \int_0^{3.6} \frac{(40x)[(1/5.4)x]\,dx}{EI}$$

$$= \frac{80}{EI}\left[\frac{x^2}{2} - \frac{1}{5.4}\frac{x^3}{3}\right]_0^{1.8} + \frac{40}{5.4EI}\left[\frac{x^3}{3}\right]_0^{3.6}$$

$$= \frac{1}{EI}(100.8 + 115.2)$$

$$= \frac{216\text{ kN}\cdot\text{m}^2}{EI} = \frac{216}{(200)(160)}$$

$$= 0.00675 \text{ rad clockwise}$$

Example 2.7.4 Find by the unit-load method the slope θ_A or θ_B of the simple beam AB shown in Fig. 2.7.4a.

SOLUTION To match up Fig. 2.7.4b, on which a unit moment is applied at A, with Fig. 2.7.4a, it is necessary to divide the entire beam into four segments as follows:

Segment	Origin	Limits	M	m	I
AC	A	0 to $L/4$	$\frac{1}{2}Wx$	$1.0 - \frac{1}{L}x$	I_c
CD	A	$L/4$ to $L/2$	$\frac{1}{2}Wx$	$1.0 - \frac{1}{L}x$	$2I_c$
DE	B	$L/4$ to $L/2$	$\frac{1}{2}Wx$	$\frac{1}{L}x$	$2I_c$
EB	B	0 to $L/4$	$\frac{1}{2}Wx$	$\frac{1}{L}x$	I_c

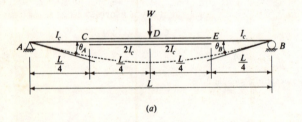

(a)

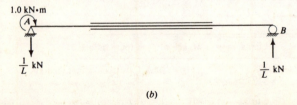

(b)

Figure 2.7.4 Simple beam of Example 2.7.4.

DEFORMATION OF STATICALLY DETERMINATE BEAMS AND RIGID FRAMES 33

Then, using Eq. (2.7.1),

$$\theta_A = \int_0^{L/4} \frac{(\tfrac{1}{2}Wx)[1.0-(1/L)x]\,dx}{EI_c} + \int_{L/4}^{L/2} \frac{(\tfrac{1}{2}Wx)[1.0-(1/L)x]\,dx}{E(2I_c)}$$
$$+ \int_{L/4}^{L/2} \frac{(\tfrac{1}{2}Wx)[(1/L)x]\,dx}{E(2I_c)} + \int_0^{L/4} \frac{(\tfrac{1}{2}Wx)[(1/L)x]\,dx}{EI_c}$$
$$= \frac{5WL^2}{128EI_c} \quad \text{clockwise}$$

$$\theta_B = \frac{5WL^2}{128EI_c} \quad \text{counterclockwise}$$

2.8 The Partial-Derivative Method—Castigliano's Theorem

Castigliano published two theorems in 1879 dealing with linear structures.† Although it is the second theorem which forms the basis of the partial-derivative method of finding the slope and deflection at any point of an elastic curve, both theorems will be treated here for sake of completeness.

Castigliano's first theorem *The partial derivative of the total internal energy in a beam, with respect to the deflection at any point, is equal to the load applied at that point.*

Castigliano's second theorem *The partial derivative of the total internal energy in a beam, with respect to the load applied at any point, is equal to the deflection at that point.*

The validity of the first theorem may be observed from Fig. 2.8.1. In the basic condition described by Fig. 2.8.1a, the loads W_1, W_2, W_3 cause deflections of Δ_1 at point 1, Δ_2 at point 2, and Δ_3 at point 3. Now the incremental loads dW_1, dW_2, dW_3 are in just such a proportion, as shown in

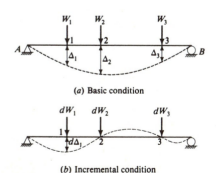

Figure 2.8.1 Castigliano's first theorem.

†S. P. Timoshenko, *History of Strength of Materials*, McGraw-Hill Book Company, New York, 1953, p. 290.

Fig. 2.8.1b, so that they will cause an additional deflection of $d\Delta_1$ only at point 1, and no additional deflections at points 2 and 3. The incremental external work, or the incremental internal energy, dU is

$$dU = W_1\, d\Delta_1 + \tfrac{1}{2} dW_1\, d\Delta_1 \approx W_1\, d\Delta_1 \quad (2.8.1)$$

because the full value of W_1 goes through the additional deflection $d\Delta_1$, and the load dW_1 is gradually added. Applying the definition of partial derivative to Eq. (2.8.1),

$$\frac{\partial U}{\partial \Delta_1} = W_1 \quad (2.8.2)$$

which is Castigliano's first theorem.

The validity of the second theorem may be observed from Fig. 2.8.2. But here a single incremental load dW_1 is added at point 1 only, as shown in Fig. 2.8.2b, causing additional deflections of $d\Delta_1$ at point 1, $d\Delta_2$ at point 2, and $d\Delta_3$ at point 3. The incremental external work, or the incremental internal energy, dU is

$$\begin{aligned} dU &= W_1\, d\Delta_1 + W_2\, d\Delta_2 + W_3\, d\Delta_3 + \tfrac{1}{2} dW_1\, d\Delta_1 \\ &\approx W_1\, d\Delta_1 + W_2\, d\Delta_2 + W_3\, d\Delta_3 \end{aligned} \quad (2.8.3)$$

The total external work, or the total internal energy, $U + dU$ in the beam when both the basic and incremental loads are simultaneously applied, is

$$U + dU = \tfrac{1}{2}(W_1 + dW_1)(\Delta_1 + d\Delta_1) + \tfrac{1}{2}W_2(\Delta_2 + d\Delta_2) + \tfrac{1}{2}W_3(\Delta_3 + d\Delta_3) \quad (2.8.4)$$

For the basic condition alone,

$$U = \tfrac{1}{2}W_1\Delta_1 + \tfrac{1}{2}W_2\Delta_2 + \tfrac{1}{2}W_3\Delta_3 \quad (2.8.5)$$

Subtracting Eq. (2.8.5) from Eq. (2.8.4),

$$\begin{aligned} dU &= \tfrac{1}{2}\Delta_1\, dW_1 + \tfrac{1}{2}W_1\, d\Delta_1 + \tfrac{1}{2}dW_1\, d\Delta_1 + \tfrac{1}{2}W_2\, d\Delta_2 + \tfrac{1}{2}W_3\, d\Delta_3 \\ &\approx \tfrac{1}{2}\Delta_1\, dW_1 + \tfrac{1}{2}W_1\, d\Delta_1 + \tfrac{1}{2}W_2\, d\Delta_2 + \tfrac{1}{2}W_3\, d\Delta_3 \end{aligned} \quad (2.8.6)$$

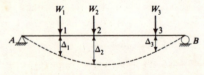

(a) Basic condition

(b) Incremental condition

Figure 2.8.2 Castigliano's second theorem.

DEFORMATION OF STATICALLY DETERMINATE BEAMS AND RIGID FRAMES 35

Substituting Eq. (2.8.3) into Eq. (2.8.6),

$$dU = \tfrac{1}{2}\Delta_1\, dW_1 + \tfrac{1}{2} dU$$
$$dU = \Delta_1\, dW_1 \qquad (2.8.7)$$

It should be noted that Eq. (2.8.7) may be directly obtained by using the notion of complementary energy.† Applying the definition of partial derivative to Eq. (2.8.7),

$$\frac{\partial U}{\partial W_1} = \Delta_1 \qquad (2.8.8)$$

which is Castigliano's second theorem.

2.9 The Partial-Derivative Method—Application to Beam Deflections and Slopes

In order to use Eq. (2.8.8) for determining the deflection at any point, it is necessary to express the internal energy U as a function of the load applied at that point. Thus whether the load at that point is zero or of a known value, that value can be substituted only after the partial differentiation with respect to that load is performed. Further, the expression for U must be obtained in terms of the bending moment. In reference to Fig. 2.5.1b again,

$$U = \Sigma \tfrac{1}{2} F\, dL \qquad (2.9.1)$$

Substituting

$$F = \frac{My}{I}\, dA \qquad \text{and} \qquad dL = \frac{My}{EI}\, dx$$

into Eq. (2.9.1) gives

$$U = \Sigma \tfrac{1}{2}\left(\frac{My}{I}\, dA\right)\left(\frac{My}{EI}\, dx\right) = \int_0^L \frac{M^2\, dx}{2EI^2} \int_0^A y^2\, dA = \int_0^L \frac{M^2\, dx}{2EI} \qquad (2.9.2)$$

Thus the following steps may be used in the partial-derivative method of finding the deflection at point C in a beam:

1. Obtain the expressions of M for various segments of the entire beam in terms of the load W_C applied at point C and all other known loads.
2. Before performing the integration, perform a partial differentiation of Eq. (2.9.2) with respect to the load W_C.
3. Substitute in the integrals the known value of W_C, which may happen to be zero.
4. Perform the integration.

Step 4 may be inserted between steps 1 and 2, but this requires substantially more algebraic manipulation.

†Timoshenko, *History of Strength of Materials*, p. 292.

36 INTERMEDIATE STRUCTURAL ANALYSIS

So far as the determination of the slope at any point C is concerned, the same four steps may be followed, except that the bending moment M is due to the simultaneous action of a moment M_C applied at point C and all other known loads, and the partial differentiation should be made with respect to M_C.

It can be shown that when the first three steps in the partial-derivative method are completed, one arrives at the identical expressions which are written down at the outset in the unit-load method. Taking the partial derivative of Eq. (2.9.2),

$$\Delta_C = \frac{\partial U}{\partial W_C} = \int_0^L \frac{2M(\partial M/\partial W_C)\,dx}{2EI} = \int_0^L \frac{M(\partial M/\partial W_C)\,dx}{EI} \qquad (2.9.3)$$

In Sec. 2.6, it was stated that m may be considered as the ratio of the bending moment caused by any load applied at C to the load itself; in symbols,

$$m = \frac{\partial M}{\partial W_C} \qquad (2.9.4)$$

If Eq. (2.9.4) is substituted into Eq. (2.9.3), the working formula in the unit-load method, Eq. (2.6.4), is obtained. In fact, then, the unit-load method, although independently derived in Sec. 2.5, is again derived here. It is a technique by which the first three steps in the partial-derivative method are bypassed. Nevertheless, Eq. (2.8.8), which is the second theorem of Castigliano, stands in its own right as the most elegant and important theorem in the theory of statically indeterminate structures since it relates to the theorem of least work elaborated in Chap. 4.

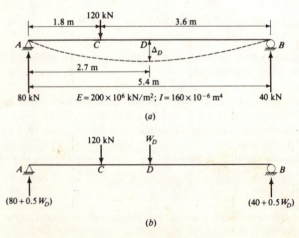

Figure 2.9.1 Simple beam of Example 2.9.1.

DEFORMATION OF STATICALLY DETERMINATE BEAMS AND RIGID FRAMES 37

Example 2.9.1 Solve Example 2.6.3 by the partial-derivative method.

SOLUTION The problem is to find Δ_D in Fig. 2.9.1a. Although there is actually no load acting at point D, a load W_D at D must be added as shown in Fig. 2.9.1b, and later be made equal to zero after the partial differentiation. In reference to Fig. 2.9.1b, the expressions for M are as follows:

Segment	Origin	Limits	M
AC	A	0 to 1.8	$(80 + 0.5W_D)x$
CD	A	1.8 to 2.7	$(80 + 0.5W_D)x - 120(x - 1.8)$
BD	B	0 to 2.7	$(40 + 0.5W_D)x$

Using Eq. (2.9.2),

$$U = \int_0^L \frac{M^2 \, dx}{2EI}$$

$$= \int_0^{1.8} \frac{[(80 + 0.5W_D)x]^2 \, dx}{2EI} + \int_{1.8}^{2.7} \frac{[(80 + 0.5W_D)x - 120(x - 1.8)]^2 \, dx}{2EI}$$

$$+ \int_0^{2.7} \frac{[(40 + 0.5W_D)x]^2 \, dx}{2EI}$$

$$\Delta_D = \frac{\partial U}{\partial W_D}$$

$$= \int_0^{1.8} \frac{2[(80 + 0.5W_D)x](0.5x) \, dx}{2EI}$$

$$+ \int_{1.8}^{2.7} \frac{2[(80 + 0.5W_D)x - 120(x - 1.8)](0.5x) \, dx}{2EI}$$

$$+ \int_0^{2.7} \frac{2[(40 + 0.5W_D)x](0.5x) \, dx}{2EI}$$

Letting $W_D = 0$ in the above expression,

$$\Delta_D = \int_0^{1.8} \frac{(80x)(0.5x) \, dx}{EI} + \int_{1.8}^{2.7} \frac{[80x - 120(x - 1.8)](0.5x) \, dx}{EI}$$

$$+ \int_0^{2.7} \frac{(40x)(0.5x) \, dx}{EI}$$

This expression for Δ_D is exactly the same as that obtained by the unit-load method in Example 2.6.3. Thus

$$\Delta_D = 1.048 \text{ cm downward}$$

Example 2.9.2 Solve Example 2.7.3 by the partial-derivative method.

SOLUTION The problem is to find θ_A in Fig. 2.9.2a. Because the partial differentiation has to be made with respect to a moment applied at A, a clockwise moment M_A is added to act on the beam as shown in Fig. 2.9.2b. In reference to Fig. 2.9.2b, the expressions for M are as follows:

38 INTERMEDIATE STRUCTURAL ANALYSIS

Segment	Origin	Limits	M
AC	A	0 to 1.8	$M_A + \left(80 - \dfrac{M_A}{5.4}\right)x$
BC	B	0 to 3.6	$\left(40 + \dfrac{M_A}{5.4}\right)x$

Using Eq. (2.9.2),

$$U = \int_0^L \frac{M^2\,dx}{2EI}$$

$$= \int_0^{1.8} \frac{[M_A + (80 - M_A/5.4)x]^2\,dx}{2EI} + \int_0^{3.6} \frac{[(40 + M_A/5.4)x]^2\,dx}{2EI}$$

$$\theta_A = \frac{\partial U}{\partial \theta_A}$$

$$= \int_0^{1.8} \frac{2[M_A + (80 - M_A/5.4)x](1 - x/5.4)\,dx}{2EI}$$

$$+ \int_0^{3.6} \frac{2[(40 + M_A/5.4)x](x/5.4)\,dx}{2EI}$$

Letting $M_A = 0$ in the above expression,

$$\theta_A = \int_0^{1.8} \frac{(80x)(1 - x/5.4)\,dx}{EI} + \int_0^{3.6} \frac{(40x)(x/5.4)\,dx}{EI}$$

This expression for θ_A is exactly the same as that obtained by the unit-load method in Example 2.7.3. Thus

$$\theta_A = 0.00675 \text{ rad clockwise}$$

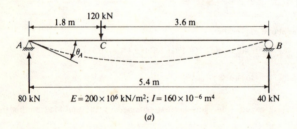

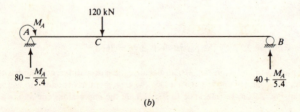

Figure 2.9.2 Simple beam of Example 2.9.2.

2.10 The Unit-Load Method—Application to Deflections and Slopes of Rigid Frames

Unlike an originally straight beam, a rigid frame as presently defined may consist of several originally straight members. In the deformed condition, a point on any one member could have moved to a new location. The component in a certain direction of the total deflection can be determined by applying a unit load in that direction. Consequently, in order to define completely the total deflection, two applications of the unit-load method are needed, usually one in the horizontal direction and the other in the vertical direction. If a unit moment is applied instead of a unit load, the angle of rotation from the original straight member axis to the tangent to the elastic curve can be determined. It should be noted again that the direction of the deflection component, or the rotation of the tangent, is the *same* as or the *opposite* to that of the unit load or the unit moment depending on whether the answer obtained is *positive* or *negative*.

The reader should recall the fundamental first-order assumption, discussed in Sec. 1.6, that the length of any straight segment within the rigid frame is not affected by displacement of its ends in the transverse direction. Moreover, in the present treatment of the deformation response of a rigid frame, the change in length due to axial force is to be neglected, as discussed in Sec. 2.1.

Example 2.10.1 For the rigid frame shown in Fig. 2.10.1, find by the unit-load method the slopes and deflections at points A, B, C, and D.

SOLUTION (a) *Values of M.* In reference to Fig. 2.10.2a, the values of M are as follows:

Segment	Origin	Limits	M	I
AE	A	0 to 4.5	48x	I_c
BE	B	0 to 3	216	I_c
BF	B	0 to 3	216 + 12x	$2I_c$
CF	C	0 to 3	84x	$2I_c$
CD	C	0 to 5	0	I_c

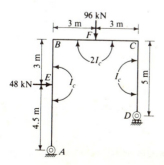

Figure 2.10.1 Rigid frame of Example 2.10.1.

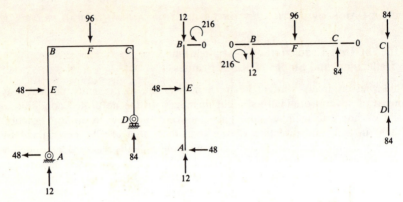

(a) Free-body diagrams from which values of M are obtained

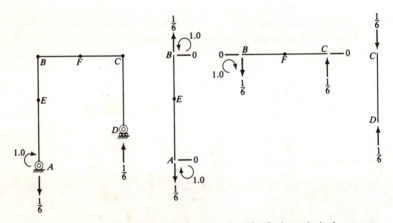

(b) Free-body diagrams from which values of m for θ_A are obtained

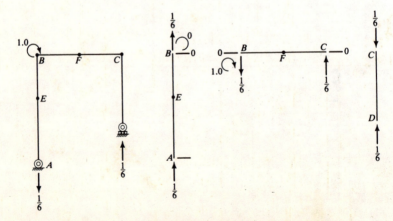

(c) Free-body diagrams from which values of m for θ_B are obtained

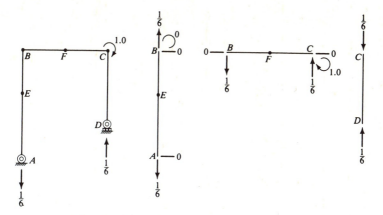

(d) Free-body diagrams from which values of m for θ_c are obtained

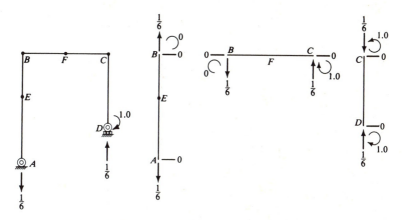

(e) Free-body diagrams from which values of m for θ_D are obtained

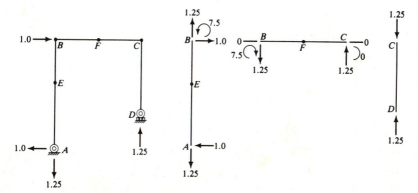

(f) Free-body diagrams from which values of m for Δ_H at B are obtained

42 INTERMEDIATE STRUCTURAL ANALYSIS

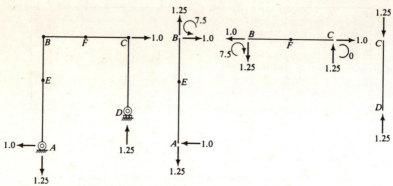

(g) Free-body diagrams from which values of m for Δ_H at C are obtained

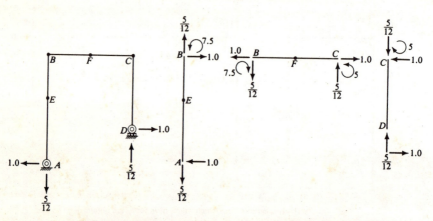

(h) Free-body diagrams from which values of m for Δ_H at D are obtained

Figure 2.10.2 Free-body diagrams for rigid frame of Example 2.10.1.

(b) *Slopes at A, B, C, and D.* Referring to Fig. 2.10.2b to e, and using the same origin for reference as for M, one may see that the values of m for θ_A, θ_B, θ_C, and θ_D are as follows:

Segment	m for θ_A	m for θ_B	m for θ_C	m for θ_D
AE	1.0	0	0	0
BE	1.0	0	0	0
BF	$1.0-\frac{1}{6}x$	$1.0-\frac{1}{6}x$	$-\frac{1}{6}x$	$-\frac{1}{6}x$
CF	$\frac{1}{6}x$	$\frac{1}{6}x$	$-1.0+\frac{1}{6}x$	$-1.0+\frac{1}{6}x$
CD	0	0	0	0

DEFORMATION OF STATICALLY DETERMINATE BEAMS AND RIGID FRAMES 43

One may observe that the expressions of m for θ_C are identical to those for θ_D and thus conclude that the slopes at C and D are equal. This is so because member CD in Fig. 2.10.2a may change direction but must remain straight without bending. Applying Eq. (2.7.1),

$$\theta_A = \int \frac{Mm\,dx}{EI}$$

$$= \int_0^{4.5} \frac{(48x)(+1.0)\,dx}{EI_c} + \int_0^3 \frac{(216)(+1.0)\,dx}{EI_c}$$

$$+ \int_0^3 \frac{(216+12x)(1.0-\tfrac{1}{6}x)\,dx}{E(2I_c)} + \int_0^3 \frac{(84x)(\tfrac{1}{6}x)\,dx}{E(2I_c)}$$

$$= \frac{1}{EI_c}(486 + 648 + 261 + 63) = 1458\frac{\text{kN}\cdot\text{m}^2}{EI_c} \quad \text{clockwise}$$

$$\theta_B = \int \frac{Mm\,dx}{EI}$$

$$= \int_0^3 \frac{(216+12x)(1.0-\tfrac{1}{6}x)\,dx}{E(2I_c)} + \int_0^3 \frac{(84x)(\tfrac{1}{6}x)\,dx}{E(2I_c)}$$

$$= \frac{1}{EI_c}(261 + 63) = 324\frac{\text{kN}\cdot\text{m}^2}{EI_c} \quad \text{clockwise}$$

$$\theta_C = \theta_D = \int \frac{Mm\,dx}{EI}$$

$$= \int_0^3 \frac{(216+12x)(-\tfrac{1}{6}x)\,dx}{E(2I_c)} + \int_0^3 \frac{(84x)(-1.0+\tfrac{1}{6}x)\,dx}{E(2I_c)}$$

$$= \frac{1}{EI_c}(-90 - 126)$$

$$= -\frac{216}{EI_c} \quad \text{or} \quad 216\frac{\text{kN}\cdot\text{m}^2}{EI_c} \quad \text{counterclockwise}$$

(c) *Deflections at A, B, C, and D.* Point A cannot deflect; points B, C, and D can deflect only in the horizontal direction. Referring to Fig. 2.10.2f to h, and using the same origin for reference as for M, one may see that the values of m for Δ_H at B, C, and D are as follows:

Segment	m for Δ_H at B	m for Δ_H at C	m for Δ_H at D
AE	$1.0x$	$1.0x$	$1.0x$
BE	$7.5 - 1.0x$	$7.5 - 1.0x$	$7.5 - 1.0x$
BF	$7.5 - 1.25x$	$7.5 - 1.25x$	$7.5 - \tfrac{5}{12}x$
CF	$1.25x$	$1.25x$	$5 + \tfrac{5}{12}x$
CD	0	0	0

One may observe that the expressions of m for Δ_H at B are identical to those for Δ_H at C and thus conclude that the horizontal deflections of B and C are equal. This is so because any difference in the horizontal deflections at the ends of an originally straight horizontal member would have changed its length. Applying Eq. (2.6.4),

$$\text{or} \quad \begin{aligned} \Delta_H \text{ at } B \\ \Delta_H \text{ at } C \end{aligned} = \int \frac{Mm\,dx}{EI}$$

$$= \int_0^{4.5} \frac{(48x)(1.0x)\,dx}{EI_c} + \int_0^3 \frac{(216)(7.5-1.0x)\,dx}{EI_c}$$

$$+ \int_0^3 \frac{(215+12x)(7.5-1.25x)\,dx}{E(2I_c)}$$

$$+ \int_0^3 \frac{(84x)(1.25x)}{E(2I_c)}$$

$$= \frac{1}{EI_c}(1458 + 3888 + 1957.5 + 472.5)$$

$$= 7776 \frac{\text{kN}\cdot\text{m}^3}{EI_c} \quad \text{to the right}$$

$$\Delta_H \text{ at } D = \int \frac{Mm\,dx}{EI}$$

$$= \int_0^{4.5} \frac{(48x)(1.0x)\,dx}{EI_c} + \int_0^3 \frac{(216)(7.5-1.0x)\,dx}{EI_c}$$

$$+ \int_0^3 \frac{(216+12x)(7.5-\tfrac{5}{12}x)\,dx}{E(2I_c)}$$

$$+ \int_0^3 \frac{(84x)(5+\tfrac{5}{12}x)\,dx}{E(2I_c)}$$

$$= \frac{1}{EI_c}(1458 + 3888 + 2407.5 + 1102.5)$$

$$= 8856 \frac{\text{kN}\cdot\text{m}^3}{EI_c} \quad \text{to the right}$$

Example 2.10.2 For the rigid frame shown in Fig. 2.10.3, find by the unit-load method the slopes and deflections at points B, C, and D.

SOLUTION (a) *Values of M*. In reference to Fig. 2.10.4a, the values of M are as follows:

Segment	Origin	Limits	M	I
AE	A	0 to 4.5	$-504 + 48x$	I_c
BE	B	0 to 3	-288	I_c
FB	F	0 to 3	$-96x$	$2I_c$
CF	C	0 to 3	0	$2I_c$
DC	D	0 to 5	0	I_c

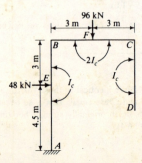

Figure 2.10.3 Rigid frame of Example 2.10.2.

DEFORMATION OF STATICALLY DETERMINATE BEAMS AND RIGID FRAMES 45

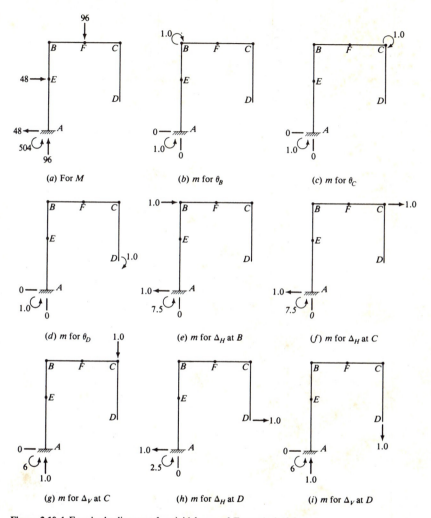

Figure 2.10.4 Free-body diagrams for rigid frame of Example 2.10.2.

(b) *Slopes at B, C, and D.* Referring to Fig. 2.10.4b to d, and using the same origin for reference as for M, one may see that the values of m for θ_B, θ_C, and θ_D are as follows:

Segment	m for θ_B	m for θ_C	m for θ_D
AE	−1.0	−1.0	−1.0
BE	−1.0	−1.0	−1.0
FB	0	−1.0	−1.0
CF	0	−1.0	−1.0
DC	0	0	−1.0

46 INTERMEDIATE STRUCTURAL ANALYSIS

Applying Eq. (2.7.1),

$$\theta_B = \int \frac{Mm\,dx}{EI}$$

$$= \int_0^{4.5} \frac{(-504+48x)(-1.0)\,dx}{EI_c} + \int_0^3 \frac{(-288)(-1.0)\,dx}{EI_c}$$

$$= \frac{1}{EI_c}(1782+864) = 2646\frac{\text{kN}\cdot\text{m}^2}{EI_c} \quad \text{clockwise}$$

$$\theta_C = \int \frac{Mm\,dx}{EI}$$

$$= \int_0^{4.5} \frac{(-504+48x)(-1.0)\,dx}{EI_c} + \int_0^3 \frac{(-288)(-1.0)\,dx}{EI_c}$$

$$+ \int_0^3 \frac{(-96x)(-1.0)\,dx}{E(2I_c)}$$

$$= \frac{1}{EI_c}(1782+864+216) = 2862\frac{\text{kN}\cdot\text{m}^2}{EI_c} \quad \text{clockwise}$$

θ_D = same integrals as those for θ_C

$$= 2862\frac{\text{kN}\cdot\text{m}^2}{EI_c} \quad \text{clockwise}$$

(c) *Deflections at B, C, and D.* Referring to Fig. 2.10.4e to i, and using the same origin for reference as for M, one may see that the values of m for Δ_H at B, Δ_H at C, Δ_V at C, Δ_H at D, and Δ_V at D are as follows:

Segment	m for Δ_H at B	m for Δ_H at C	m for Δ_V at C	m for Δ_H at D	m for Δ_V at D
AE	$-7.5+1.0x$	$-7.5+1.0x$	-6	$-2.5+1.0x$	-6
BE	$-1.0x$	$-1.0x$	-6	$+1.0(5-x)$	-6
FB	0	0	$-1.0(3+x)$	$+5$	$-(3+x)$
CF	0	0	$-1.0x$	$+5$	$-1.0x$
DC	0	0	0	$+1.0x$	0

Applying Eq. (2.6.4),

$$\Delta_H \text{ at } B = \int_0^{4.5} \frac{(-504+48x)(-7.5+1.0x)\,dx}{EI_c} + \int_0^3 \frac{(-288)(-1.0x)\,dx}{EI_c}$$

$$= \frac{1}{EI_c}(9720+1296)$$

$$= 11{,}016\frac{\text{kN}\cdot\text{m}^3}{EI_c} \quad \text{to the right}$$

Δ_H at C = same integrals as those for Δ_H at B

$$= 11{,}016\frac{\text{kN}\cdot\text{m}^3}{EI_c} \quad \text{to the right}$$

$$\Delta_V \text{ at } C = \int_0^{4.5} \frac{(-504+48x)(-6)\,dx}{EI_c} + \int_0^3 \frac{(-288)(-6)\,dx}{EI_c}$$

$$+ \int_0^3 \frac{(-96x)[-1.0(3+x)]\,dx}{E(2I_c)}$$

$$= \frac{1}{EI_c}(10{,}692 + 5184 + 1080)$$

$$= 16{,}956 \frac{\text{kN} \cdot \text{m}^3}{EI_c} \quad \text{downward}$$

$$\Delta_H \text{ at } D = \int_0^{4.5} \frac{(-504 + 48x)(-2.5 + 1.0x)\,dx}{EI_c}$$

$$+ \int_0^3 \frac{(-288)[+1.0(5-x)]\,dx}{EI_c} + \int_0^3 \frac{(-96x)(+5)\,dx}{E(2I_c)}$$

$$= \frac{1}{EI_c}(810 - 3024 - 1080)$$

$$= -3294 \quad \text{or} \quad 3294 \frac{\text{kN} \cdot \text{m}^3}{EI_c} \quad \text{to the left}$$

Δ_V at D = same integrals as those for Δ_V at C

$$= 16{,}956 \frac{\text{kN} \cdot \text{m}^3}{EI_c} \quad \text{downward}$$

2.11 Deflections and Slopes of Statically Determinate Rigid Frames due to Movements of Supports

It can be shown that no internal forces will be set up in statically determinate rigid frames due to movements of supports. If the hinge A of rigid frame $ABCD$ yields 1 cm to the left, as shown in Fig. 2.11.1a, the joints B, C, and D will all move 1 cm to the left and all members will remain straight as originally.

If the hinge A yields 1 cm vertically downward, as shown in Fig. 2.11.1b, the unstressed rigid frame $ABCD$ will take the position $A'B'C'D'$. Since all movements are small in comparison with the size of the frame, the fundamental first-order assumption that the member length does not change when its ends move in the transverse direction still holds. Thus, in Fig. 2.11.1b, $B_1B' = AA' = 1$ cm, and $BB_1 = CC'$. Also, because angles $A'B'C'$ and $B'C'D'$ must remain right angles, the change in the directions of all three members must be the same. Using triangle $B_1C'B'$,

$$\text{Angle } B_1C'B' = \frac{B_1B'}{B_1C'} = \frac{1.0 \text{ cm}}{6 \text{ m}} = \frac{1}{600} \text{ rad}$$

Using triangle $B'A'B_2$,

$$B'B_2 = (A'B_2)(\text{angle } B'AB_2) = (7.5 \text{ m})\left(\frac{1}{600}\right) = 1.25 \text{ cm}$$

which is also equal to BB_1, CC', or DD_1. Then,

$$DD' = DD_1 - D_1D' = 1.25 \text{ cm} - (C'D_1)(\text{angle } D_1C'D')$$

$$= 1.25 - (500)\left(\frac{1}{600}\right) = \frac{5}{12} \text{ cm}$$

If the roller support D yields 1 cm vertically downward, as shown in Fig. 2.11.1c, C_1C' must also be 1 cm, and the angle $C_1B'C' = 1 \text{ cm}/6 \text{ m} =$

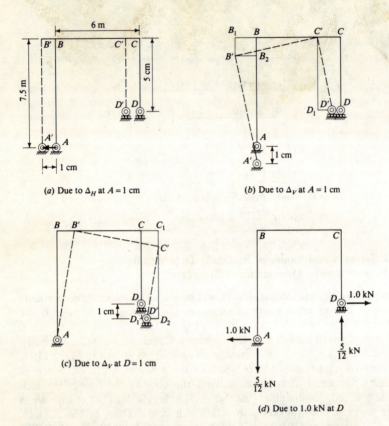

Figure 2.11.1 Effect of support movements on a statically determinate rigid frame.

(1/600) rad. Then

$$BB' = \left(\frac{1}{600} \text{ rad}\right)(750 \text{ cm}) = 1.25 \text{ cm} = CC_1 = D_1D_2$$

Finally,

$$D_1D' = D_1D_2 - D_2D' = 1.25 - \left(\frac{1}{600} \text{ rad}\right)(500 \text{ cm}) = \frac{5}{12} \text{ cm}$$

It is seen that, owing to each case of yielding of the external supports, the roller support must move a certain amount in the horizontal direction in order to keep the frame undeformed. If the support at D is a hinge, or if then the rigid frame becomes statically indeterminate, the members of the rigid frame will have to be bent to accommodate themselves to the new conditions. The problem of finding the internal forces in a statically indeterminate rigid frame due to support movements will be taken up in Chap. 4.

DEFORMATION OF STATICALLY DETERMINATE BEAMS AND RIGID FRAMES

The unit-load method can be used to find the deflection or slope at any point of a statically determinate rigid frame due to movements of supports. For instance, Fig. 2.11.1d is the free-body diagram of the rigid frame $ABCD$ subjected to a horizontal load of 1.0 kN at D. If the hinge A moves 1 cm vertically downward, the joints B, C, and D will also change position. Since the resultant of the four external forces, or reactions acting on the rigid frame, is equal to zero, the external work done by all of them when the body on which they act goes through a rigid-body movement must be equal to zero. The horizontal reaction at A and the vertical reaction at D do no work because these forces are not displaced. Let the movement of D be Δ_H to the right. Then

$$\text{Total external work} = (\tfrac{5}{12}\,\text{kN})(1.0\,\text{cm}) + (1.0\,\text{kN})(\Delta_H \text{ at } D) = 0$$

$$\Delta_H \text{ at } D = -\tfrac{5}{12}\,\text{cm} \quad \text{or} \quad \tfrac{5}{12}\,\text{cm} \quad \text{to the left}$$

This result checks with that of the purely geometrical consideration discussed earlier in this section.

2.12 The Moment-Area Method—Derivation of Theorems

Moment-area theorem 1 *The angle in radians or the change in slope between the tangents at two points on the elastic curve of an originally straight member is equal to the M/EI area between those two points.*

Moment-area theorem 2 *The deflection of a point on the elastic curve from the tangent at another point on the elastic curve, measured in the direction perpendicular to the originally straight member, is equal to the moment of the M/EI area between those two points about the point where the deflection occurs.*

Derivation As discussed in Sec. 1.6, the fundamental assumption is that the length of the elastic curve is equal to that of the original straight axis; or for that matter, all inclined lines in Fig. 2.12.1b are equal in length to their horizontal projections. Let $d\theta$ be the angle between the tangents at points $1'$ and $2'$ on the elastic curve of Fig. 2.12.1b. Since $1'$ and $2'$ are at an infinitesimal distance dx apart, curve $1'\text{-}2'$ is a circular arc with center at point O, as shown in Fig. 2.12.1c. Draw a line through point $2'$ and parallel to line $O\text{-}1'$. It can be seen from Fig. 2.12.1c that

$$d\theta = \frac{\text{elongation of lower extreme fiber}}{c_2}$$

$$= \frac{(Mc_2/EI)\,dx}{c_2} = \frac{M}{EI}\,dx \tag{2.12.1}$$

Integrating Eq. (2.12.1) between the limits A and B,

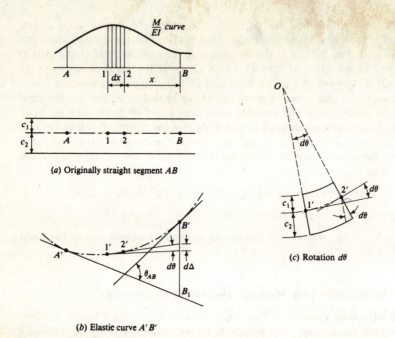

Figure 2.12.1 Derivation of moment-area theorems.

$$\theta_{AB} = \int_A^B d\theta = \int_A^B \frac{M}{EI} dx \qquad (2.12.2)$$

which is moment-area theorem 1.

In Fig. 2.12.1b, prolong the tangents at 1' and 2' until they cut $d\Delta$ on B_1B'. It can be seen that

$$d\Delta = x\, d\theta = x\left(\frac{M}{EI} dx\right) \qquad (2.12.3)$$

Integrating Eq. (2.12.3) between the limits A and B,

$$B_1B' = \int_A^B d\Delta = \int_A^B \frac{Mx\, dx}{EI} \qquad (2.12.4)$$

which is moment-area theorem 2.

Because the moment curve for a beam segment under uniform load is a second-degree parabola, it is useful to derive by means of elementary calculus the properties of such areas as they are shown in Fig. 2.12.2. By means of elementary calculus one may note that the parts A_1 and A_2 are $\frac{1}{3}$ and $\frac{2}{3}$ of the circumscribing rectangle or parallelogram, and the centroids of A_1 and A_2 are at the $\frac{1}{4}$–$\frac{3}{4}$ and $\frac{5}{8}$–$\frac{3}{8}$ division points, respectively.

DEFORMATION OF STATICALLY DETERMINATE BEAMS AND RIGID FRAMES 51

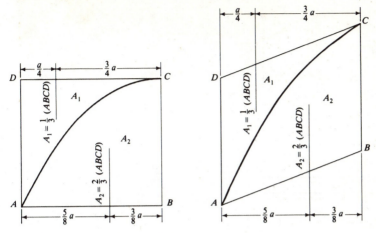

Figure 2.12.2 Properties of a second-degree parabolic area.

2.13 The Moment-Area Method—Application to Beam Deflections and Slopes

In applying the moment-area theorems to find beam deflections and slopes, it is necessary first to draw the moment diagram, or the modified M/I diagram if I is variable, and next to sketch a qualitative picture of the elastic curve consistent with the sign of the bending moment and the support conditions. Only then are the two moment-area theorems to be physically applied to the geometry of the elastic curve. Since the moment diagram is to be plotted on the compression side, the elastic curve is concave on the side where the moment diagram appears.

Example 2.13.1 Find by the moment-area method θ_B and Δ_B of the cantilever beam shown in Fig. 2.13.1a.

SOLUTION Using moment-area theorem 1,

$$\theta_B = \text{change in slope between } A \text{ and } B'$$

$$= \frac{M}{EI} \text{ area between } A \text{ and } B$$

$$= \frac{1}{2}\left(\frac{WL}{EI}\right)(L) = \frac{WL^2}{2EI} \quad \text{clockwise}$$

Using moment-area theorem 2,

$$\Delta_B = \text{deflection of } B' \text{ from tangent at } A$$

$$= \text{moment of } \frac{M}{EI} \text{ area between } A \text{ and } B \text{ about } B$$

$$= \frac{1}{2}\left(\frac{WL}{EI}\right)(L)\left(\frac{2}{3}L\right) = \frac{WL^3}{3EI} \quad \text{downward}$$

52 INTERMEDIATE STRUCTURAL ANALYSIS

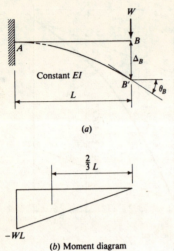

(b) Moment diagram   **Figure 2.13.1** Cantilever beam of Example 2.13.1.

Note that the negative sign of the moment value at A is not used, because the moment-area theorems are only being physically applied to the geometry of the elastic curve as sketched in Fig. 2.13.1a.

Example 2.13.2 Find by the moment-area method θ_A or θ_B, and Δ_C of the simple beam shown in Fig. 2.13.2a.

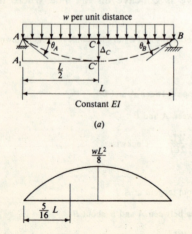

(b) Moment diagram   **Figure 2.13.2** Simple beam of Example 2.13.2.

DEFORMATION OF STATICALLY DETERMINATE BEAMS AND RIGID FRAMES 53

SOLUTION.

θ_A = change in slope between A and C'

$= \dfrac{M}{EI}$ area between A and C

$= \dfrac{2}{3}\left(\dfrac{wL^2}{8EI}\right)\left(\dfrac{L}{2}\right) = \dfrac{wL^3}{24EI}$ clockwise

$\theta_B = \dfrac{wL^3}{24EI}$ counterclockwise

$\Delta_C = A_1A = $ <u>deflection from A from tangent at C'</u>

$= $ moment of $\dfrac{M}{EI}$ <u>area between C and A about A</u>

$= \dfrac{2}{3}\left(\dfrac{wL^2}{8EI}\right)\left(\dfrac{L}{2}\right)\left(\dfrac{5}{16}L\right) = \dfrac{5wL^4}{384EI}$ downward

Example 2.13.3 Find by the moment-area method θ_A, θ_B, and Δ_D of the simple beam shown in Fig. 2.13.3a.

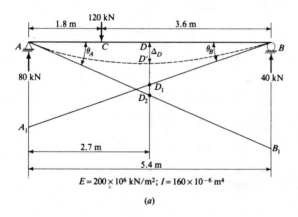

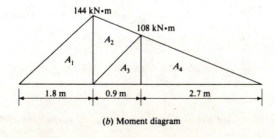

(b) Moment diagram

Figure 2.13.3 Simple beam of Example 2.13.3.

54 INTERMEDIATE STRUCTURAL ANALYSIS

SOLUTION

$$\theta_A = \frac{B_1B}{AB} = \frac{\text{deflection of } B \text{ from tangent at } A}{5.4}$$

$$= \frac{\text{moment of } M/EI \text{ area between } A \text{ and } B \text{ about } B}{5.4}$$

$$= \frac{A_1(4.2) + (A_2 + A_3 + A_4)(2.4)}{5.4EI}$$

$$= \frac{\frac{1}{2}(144)(1.8)(4.2) + \frac{1}{2}(144)(3.6)(2.4)}{5.4EI}$$

$$= \frac{216 \text{ kN} \cdot \text{m}^2}{EI} = \frac{216}{200(160)} = 0.00675 \text{ rad clockwise}$$

$$\theta_B = \frac{A_1A}{5.4} = \frac{\text{deflection of } A \text{ from tangent at } B}{5.4}$$

$$= \frac{\text{moment of } M/EI \text{ area between } B \text{ and } A \text{ about } A}{5.4}$$

$$= \frac{A_1(1.2) + (A_2 + A_3 + A_4)(3.0)}{5.4EI}$$

$$= \frac{\frac{1}{2}(144)(1.8)(1.2) + \frac{1}{2}(144)(3.6)(3.0)}{5.4EI}$$

$$= \frac{172.8 \text{ kN} \cdot \text{m}^2}{EI} = \frac{172.8}{200(160)} = 0.00540 \text{ rad counterclockwise}$$

$$\Delta_D = DD_2 - D_2D'$$

$$= (\theta_A)(AD) - (\text{deflection of } D' \text{ from tangent at } A)$$

$$= \left(\frac{216}{EI}\right)(2.7) - (\text{moment of } M/EI \text{ area between } A \text{ and } D \text{ about } D)$$

$$= \left(\frac{216}{EI}\right)(2.7) - \frac{A_1(1.5) + A_2(0.6) + A_3(0.3)}{EI}$$

$$= \left(\frac{216}{EI}\right)(2.7) - \frac{129.6(1.5) + 64.8(0.6) + 48.6(0.3)}{EI}$$

$$= \frac{335.34 \text{ kN} \cdot \text{m}^3}{200(160)} = 0.01048 \text{ m downward}$$

Alternately,

$$\Delta_D = DD_1 - D_1D'$$

$$= (\theta_B)(BD) - (\text{deflection of } D' \text{ from tangent at } B)$$

$$= \left(\frac{172.8}{EI}\right)(2.7) - (\text{moment of } M/EI \text{ area between } B \text{ and } D \text{ about } D)$$

$$= \left(\frac{172.8}{EI}\right)(2.7) - \frac{A_4(0.9)}{EI} = \left(\frac{172.8}{EI}\right)(2.7) - \frac{145.8(0.9)}{EI}$$

$$= \frac{335.34 \text{ kN} \cdot \text{m}^3}{200(160)} = 0.01048 \text{ m downward} \quad \text{(Check)}$$

Example 2.13.4 Find by the moment-area method θ_A, θ_B, and Δ_D of the simple beam shown in Fig. 2.13.4a.

DEFORMATION OF STATICALLY DETERMINATE BEAMS AND RIGID FRAMES

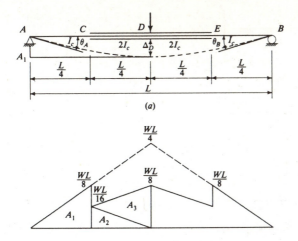

(b) Modified moment diagram

Figure 2.13.4 Simple beam of Example 2.13.4.

SOLUTION For this problem, a modified moment diagram as shown in Fig. 2.13.4b is obtained by dividing the true moment values by the number of I_c's in the respective portions of the beam. In this way the M/EI area is simply equal to the area of the modified moment diagram divided by the constant EI_c.

θ_A = change in slope between A and D'

$$= \frac{1}{EI_c}(A_1 + A_2 + A_3)$$

$$= \frac{1}{EI_c}\left(\frac{1}{2}\frac{WL}{8}\frac{L}{4} + \frac{1}{2}\frac{WL}{16}\frac{L}{4} + \frac{1}{2}\frac{WL}{8}\frac{L}{4}\right) = \frac{5WL^2}{128EI_c} \quad \text{clockwise}$$

$$\theta_B = \frac{5WL^2}{128EI_c} \quad \text{counterclockwise}$$

$\Delta_D = A_1A$ = deflection of A from tangent at D'

$$= \frac{1}{EI_c}(\text{moment of } A_1, A_2, \text{ and } A_3 \text{ about } A)$$

$$= \frac{1}{EI_c}\left(\frac{1}{2}\frac{WL}{8}\frac{L}{4}\frac{L}{6} + \frac{1}{2}\frac{WL}{16}\frac{L}{4}\frac{L}{3} + \frac{1}{2}\frac{WL}{8}\frac{L}{4}\frac{5L}{12}\right) = \frac{3WL^3}{256EI_c} \quad \text{downward}$$

2.14 The Conjugate-Beam Method—Derivation of Theorems

Consider the real cantilever beam of Fig. 2.14.1a, for which the M/EI curve is shown in Fig. 2.14.1b. Imagine now a secondary beam, called a *conjugate beam*, which is an "inverted cantilever" of the "real cantilever," as shown in Fig. 2.14.1c. The M/EI curve on the lower side of the real beam is taken to

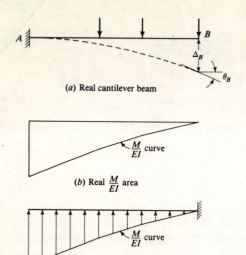

Figure 2.14.1 Conjugate beam of a real cantilever beam.

(a) Real cantilever beam

(b) Real M/EI area

(c) Conjugate beam

act as an upward load on the conjugate beam. Applying moment-area theorem 1,

θ_B clockwise = M/EI area of Fig. 2.14.1b
= positive shear at B of Fig. 2.14.1c (2.14.1)

Applying moment-area theorem 2,

Δ_B downward = moment of M/EI area of Fig. 2.14.1b about B
= positive bending moment at B of Fig. 2.14.1c (2.14.2)

Next consider the slope θ_C and the deflection Δ_C of the real simple beam of Fig. 2.14.2a, for which the M/EI curve is shown in Fig. 2.14.2b. It can be shown that the conjugate beam is also a simple beam, as shown in Fig. 2.14.2c. Using the moment-area method,

θ_C clockwise = $\theta_A - \dfrac{M}{EI}$ area between A and C

$= \dfrac{B_1 B}{L} - \dfrac{M}{EI}$ area between A and C

$= \dfrac{\text{moment of } M/EI \text{ area between } A \text{ and } B \text{ about } B}{L}$

$- \dfrac{M}{EI}$ area between A and C (2.14.3)

DEFORMATION OF STATICALLY DETERMINATE BEAMS AND RIGID FRAMES 57

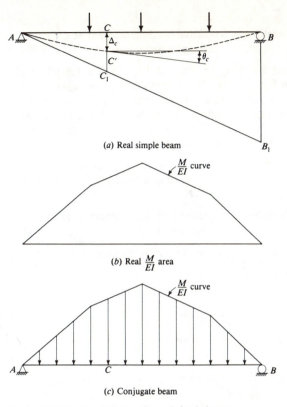

Figure 2.14.2 Conjugate beam of a real simple beam.

$$\Delta_C \text{ downward} = CC_1 - C_1C'$$
$$= \theta_A(AC) - \text{moment of } M/EI \text{ area between } A \text{ and } C \text{ about } C$$
$$= \frac{B_1B}{L}(AC) - \text{moment of } \frac{M}{EI} \text{ area between } A \text{ and } C \text{ about } C$$
$$= \frac{\text{moment of } M/EI \text{ area between } A \text{ and } B \text{ about } B}{L}(AC)$$
$$- \text{moment of } \frac{M}{EI} \text{ area between } A \text{ and } C \text{ about } C \quad (2.14.4)$$

But the last expressions in Eqs. (2.14.3) and (2.14.4) are exactly the positive shear and positive bending moment at point C of the conjugate beam. Thus, again,

$$\theta_C \text{ clockwise in real beam} = \text{positive shear at } C \text{ in conjugate beam} \quad (2.14.5)$$

and

$$\Delta_C \text{ downward in real beam} = \text{positive bending moment at } C \text{ in conjugate beam} \quad (2.14.6)$$

Last consider the real overhanging beam of Fig. 2.14.3a, for which the M/EI curve is shown in Fig. 2.14.3b. The corresponding conjugate beam is shown by Fig. 2.14.3c, in which the real exterior simple support at A is still a simple support, the real interior support at B becomes an unsupported internal hinge, and the real free end at C is replaced by a fixed support. Again it can be shown by the moment-area method that

$$\theta_D \text{ clockwise in real beam} = \text{positive shear at } D \text{ in conjugate beam} \quad (2.14.7)$$

and

$$\Delta_D \text{ downward in real beam} = \text{positive bending moment at } D \text{ in conjugate beam} \quad (2.14.8)$$

regardless whether point D is on segment AB or on segment BC.

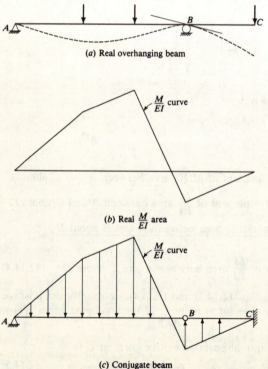

(a) Real overhanging beam

(b) Real $\frac{M}{EI}$ area

(c) Conjugate beam

Figure 2.14.3 Conjugate beam of a real overhanging beam.

DEFORMATION OF STATICALLY DETERMINATE BEAMS AND RIGID FRAMES 59

Thus the two general conjugate-beam theorems can be stated as follows:

Conjugate-beam theorems 1 and 2 *The clockwise slope and the downward deflection at any point in a real beam are equal to the positive shear and positive bending moment at that point in a conjugate beam, respectively; whereas an exterior simple support, an interior support, and a free end in the real beam are to be an exterior simple support, an unsupported internal hinge, and a fixed support in the conjugate beam, respectively and conversely.*

One can see, therefore, that the use of the conjugate beam is only a device for summarizing the procedures in the moment-area method. Actually, of the three types of conjugate beams described in Figs. 2.14.1 to 2.14.3, the middle case is the most convenient to use, because the real and conjugate beams are the same simple beam. Moreover, the boundary points A and B in Fig. 2.14.2 may be replaced by any two points A' and B' on a continuous elastic curve, such as shown in Fig. 2.14.4a. The conjugate beam is the simple beam AB loaded with the M/EI area, but θ_C and Δ_C should both be measured from chord AB, which need not be horizontal. When the slope and position of the chord AB are known, then the true slope and deflection at point C can be obtained.

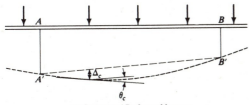

(a) Segment AB of a real beam

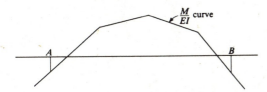

(b) Real $\dfrac{M}{EI}$ area plotted on compression side

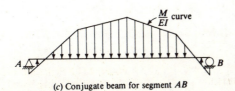

(c) Conjugate beam for segment AB

Figure 2.14.4 Conjugate beam for a real beam segment.

2.15 The Conjugate-Beam Method—Application to Beam Deflections and Slopes

The moment-area and conjugate-beam methods are indeed closely related. The moment-area theorems refer physically to the geometry of the elastic curve, while the conjugate-beam concept makes use of the analogy between slope and shear and between deflection and bending moment. Although each method can stand by itself as entirely general, it is better practice to apply the conjugate-beam theorems between two points of zero (or known) deflections but of unknown slopes, and to apply the moment-area theorems between one point of zero (or known) slope and deflection and another point of unknown slope and deflection.

Example 2.15.1 For the simple beam shown in Fig. 2.15.1a, find by the conjugate-beam method (a) θ_A and θ_B, (b) Δ_D, and (c) the position and amount of maximum deflection.

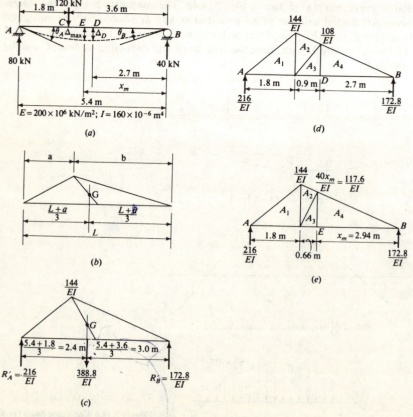

Figure 2.15.1 Simple beam of Example 2.15.1.

DEFORMATION OF STATICALLY DETERMINATE BEAMS AND RIGID FRAMES

SOLUTION (a) θ_A and θ_B. The slopes θ_A and θ_B are simply the reactions to the conjugate beam of Fig. 2.15.1c. The formulas for the location of the centroid of a triangle can be derived and are shown in Fig. 2.15.1b. Referring to Fig. 2.15.1c,

$$\theta_A = R'_A = \frac{216 \text{ kN}\cdot\text{m}^2}{EI} = \frac{216}{200(160)} = 0.00675 \text{ rad clockwise}$$

$$\theta_B = R'_B = \frac{172.8 \text{ kN}\cdot\text{m}^2}{EI} = \frac{172.8}{200(160)} = 0.00540 \text{ rad counterclockwise}$$

(b) Δ_D. Referring to Fig. 2.15.1d and using the left free body,

Δ_D = bending moment at D

$$= \frac{1}{EI}[(216)(2.7) - \tfrac{1}{2}(144)(1.8)(1.5) - \tfrac{1}{2}(144)(0.9)(0.6) - \tfrac{1}{2}(108)(0.9)(0.3)]$$

$$= \frac{335.34 \text{ kN}\cdot\text{m}^3}{EI} = \frac{335.34}{200(160)} = 0.01048 \text{ m downward}$$

Or, using the right free body,

Δ_D = bending moment at D

$$= \frac{1}{EI}[(172.8)(2.7) - \tfrac{1}{2}(108)(2.7)(0.9)]$$

$$= \frac{335.34 \text{ kN}\cdot\text{m}^3}{EI} = \frac{335.34}{200(160)} = 0.01048 \text{ m downward} \quad \text{(Check)}$$

(c) *Position and amount of maximum deflection*. Let the maximum deflection occur at point E of Fig. 2.15.1e. Then the shear at point E of the conjugate beam is zero.

$$V'_E = -\frac{172.8}{EI} + \frac{1}{2}\left(\frac{40x_m}{EI}\right)(x_m) = 0 \qquad x_m = 2.94 \text{ m}$$

$\Delta_{max} = M'_E$

$$= \frac{1}{EI}[(216)(2.46) - \tfrac{1}{2}(144)(1.8)(1.26) - \tfrac{1}{2}(144)(0.66)(0.44) - \tfrac{1}{2}(117.6)(0.66)(0.22)]$$

or $\quad \dfrac{1}{EI}[(172.8)(2.94) - \tfrac{1}{2}(117.6)(2.94)(0.98)]$

$$= \frac{338.62 \text{ kN}\cdot\text{m}^3}{EI} = \frac{338.62}{200(160)} = 0.01058 \text{ m downward}$$

Example 2.15.2 Find by the conjugate-beam method θ_A, θ_B, and Δ_D of the simple beam shown in Fig. 2.15.2a.

SOLUTION Referring to the conjugate beam of Fig. 2.15.2b,

$$\theta_A = R'_A = \frac{5WL^2}{128EI_c} \quad \text{clockwise}$$

$$\theta_B = R'_B = \frac{5WL^2}{128EI_c} \quad \text{counterclockwise}$$

$$\Delta_D = M'_D = R'_A\left(\frac{L}{2}\right) - A_1\left(\frac{L}{3}\right) - A_2\left(\frac{L}{6}\right) - A_3\left(\frac{L}{12}\right)$$

$$= \frac{5WL^2}{128EI_c}\left(\frac{L}{2}\right) - \frac{1}{2}\left(\frac{WL}{8EI_c}\right)\left(\frac{L}{4}\right)\left(\frac{L}{3}\right) - \frac{1}{2}\left(\frac{WL}{16EI_c}\right)\left(\frac{L}{4}\right)\left(\frac{L}{6}\right) - \frac{1}{2}\left(\frac{WL}{8EI_c}\right)\left(\frac{L}{4}\right)\left(\frac{L}{12}\right)$$

$$= \frac{3WL^3}{256EI_c} \quad \text{downward}$$

62 INTERMEDIATE STRUCTURAL ANALYSIS

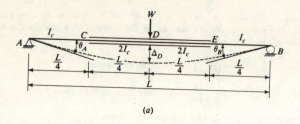

(a)

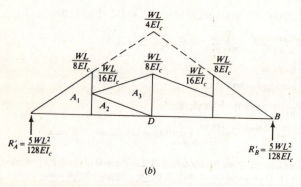

(b)

Figure 2.15.2 Simple beam of Example 2.15.2.

Example 2.15.3 Find Δ_A and Δ_D of the overhanging beam shown in Fig. 2.15.3a by the moment-area/conjugate-beam method.

SOLUTION (a) *Apply conjugate-beam theorem 1 to BC.* As shown by Fig. 2.15.3b, the moment diagram on BC consists of A_1 due to the uniform load, A_2 due to bending moment at B, and A_3 due to bending moment at C. Using conjugate-beam theorem 1 in a physical sense without adhering to a rigorous sign convention but only with reference to Fig. 2.15.3b,

$$\theta_B = \theta_{B1} - \theta_{B2} = \frac{1}{2}\frac{A_1}{EI} - \frac{2}{3}\frac{A_2}{EI} - \frac{1}{3}\frac{A_3}{EI}$$

$$= \frac{1}{2}\left(\frac{393.66}{EI}\right) - \frac{2}{3}\left(\frac{97.2}{EI}\right) - \frac{1}{3}\left(\frac{272.16}{EI}\right)$$

$$= \frac{41.31 \text{ kN·m}^2}{EI} \quad \text{clockwise}$$

$$\theta_C = -\theta_{C1} + \theta_{C2} = -\frac{1}{2}\frac{A_1}{EI} + \frac{1}{3}\frac{A_2}{EI} + \frac{2}{3}\frac{A_3}{EI}$$

$$= -\frac{1}{2}\left(\frac{393.66}{EI}\right) + \frac{1}{3}\left(\frac{97.2}{EI}\right) + \frac{2}{3}\left(\frac{272.16}{EI}\right)$$

$$= \frac{17.01 \text{ kN·m}^2}{EI} \quad \text{clockwise}$$

(b) *Apply moment-area theorem 2 to BA.* Referring to Fig. 2.15.3c,

DEFORMATION OF STATICALLY DETERMINATE BEAMS AND RIGID FRAMES 63

$$\Delta_A = AA_1 - A_1A'$$

$$= \theta_B(AB) - \text{moment of } \frac{M}{EI} \text{ area between } B \text{ and } A \text{ about } A$$

$$= \frac{41.31}{EI}(1.5) - \frac{1}{2}\left(\frac{36}{EI}\right)(1.5)(1.0) = \frac{34.965 \text{ kN} \cdot \text{m}^3}{EI} \quad \text{downward}$$

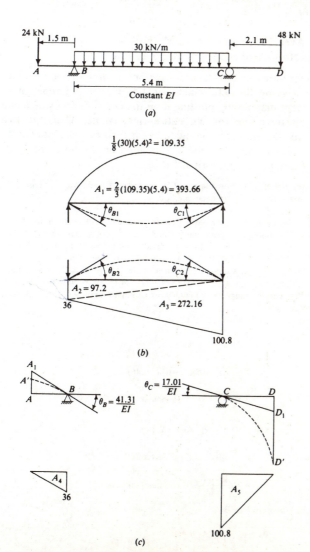

Figure 2.15.3 Overhanging beam of Example 2.15.3.

(c) Apply moment-area theorem 2 to CD. Referring to Fig. 2.15.3c,

$$\Delta_D = DD_1 + D_1D'$$
$$= \theta_C(CD) + \text{moment of } \frac{M}{EI} \text{ area between } C \text{ and } D \text{ about } D$$
$$= \frac{17.01}{EI}(2.1) + \frac{1}{2}\left(\frac{100.8}{EI}\right)(2.1)(1.4) = \frac{183.897 \text{ kN·m}^3}{EI} \quad \text{downward}$$

2.16 The Moment-Area/Conjugate-Beam Method—Application to Deflections and Slopes of Rigid Frames

In applying the moment-area/conjugate-beam method to find deflections and slopes at points on a statically determinate rigid frame, it is important to sketch a correct moment diagram by plotting it on the compression side and a correct qualitative elastic curve for the entire rigid frame. Then the two moment-area theorems and the two conjugate-beam theorems are applied to the geometry of the elastic curve to obtain the slope or the horizontal and vertical deflection of any point on the rigid frame.

Example 2.16.1 For the rigid frame shown in Fig. 2.16.1a, find θ, Δ_H, and Δ_V of points A, B, C, and D by the moment-area/conjugate-beam method.

SOLUTION The free-body diagrams of the entire rigid frame and of each member are shown in Fig. 2.16.1a and b. The moment diagram plotted on the compression side shown in Fig. 2.16.1c is modified to become that of Fig. 2.16.1d by dividing all values along BC by 2 since the moment of inertia of BC is $2I_c$, while that of AB and CD is I_c. The sketch for the elastic curve is shown in Fig. 2.16.1e. In studying this curve, note that B' and C' stay at their original levels and both move the same Δ_H to the right. The tangents at each joint are drawn as shown; the angle between the tangents at joints B' or C' must remain 90°. Note also that member CD remains straight.

Applying conjugate-beam theorem 1 to $B'C'$,

$$\theta_B = R'_B = \frac{1}{EI_c}\left(\frac{5}{6}A_3 + \frac{1}{2}A_4\right)$$
$$= \frac{1}{EI_c}[\tfrac{5}{6}(\tfrac{1}{2})(108)(3) + \tfrac{1}{2}(\tfrac{1}{2})(126)(6)]$$
$$= 324\frac{\text{kN·m}^2}{EI_c} \quad \text{clockwise}$$

$$\theta_C = R'_C = \frac{1}{EI_c}\left(\frac{1}{6}A_3 + \frac{1}{2}A_4\right)$$
$$= \frac{1}{EI_c}[\tfrac{1}{6}(\tfrac{1}{2})(108)(3) + \tfrac{1}{2}(\tfrac{1}{2})(126)(6)]$$
$$= 216\frac{\text{kN·m}^2}{EI_c} \quad \text{counterclockwise}$$

Applying moment-area theorem 1 to $B'A$,

$$\theta_A = \theta_B + \frac{1}{EI_c}(A_1 + A_2)$$
$$= \frac{324}{EI_c} + \frac{1}{EI_c}[\tfrac{1}{2}(216)(4.5) + (216)(3)]$$
$$= 1458\frac{\text{kN·m}^2}{EI_c} \quad \text{clockwise}$$

DEFORMATION OF STATICALLY DETERMINATE BEAMS AND RIGID FRAMES

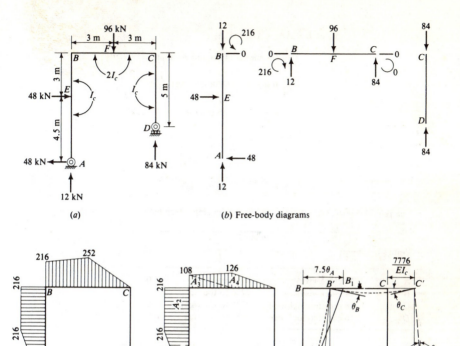

(a)   (b) Free-body diagrams

(c) M diagram plotted on compression side

(d) Modified M diagram plotted on compression side

(e) Elastic curve

Figure 2.16.1 Rigid frame of Example 2.16.1.

Because $C'D'$ is straight,

$$\theta_D = \theta_C = 216 \frac{\text{kN} \cdot \text{m}^2}{EI_c} \quad \text{counterclockwise}$$

Δ_H of $B = BB_1 - B_1B'$

$$= 7.5\theta_A - \frac{1}{EI_c} \quad \text{(moments of } A_1 \text{ and } A_2 \text{ about } B\text{)}$$

$$= 7.5\left(\frac{1458}{EI_c}\right) - \frac{1}{EI_c}[(486)(4.5) + (648)(1.5)]$$

$$= 7776 \frac{\text{kN} \cdot \text{m}^3}{EI_c} \quad \text{to the right}$$

66 INTERMEDIATE STRUCTURAL ANALYSIS

Or
$$\Delta_H \text{ of } B = AA_2 = AA_1 + A_1A_2$$
$$= \frac{1}{EI_c}(\text{moments of } A_1 \text{ and } A_2 \text{ about } A) + 7.5\theta_B$$
$$= \frac{1}{EI_c}[(486)(3) + (648)(6)] + 7.5\left(\frac{324}{EI_c}\right)$$
$$= 7776\frac{\text{kN·m}^3}{EI_c} \qquad \text{to the right} \qquad (\text{Check})$$

$$\Delta_H \text{ of } D = DD_1 + D_1D' = CC' + D_1D'$$
$$= \frac{7776}{EI_c} + 5\theta_C = \frac{7776}{EI_c} + 5\left(\frac{216}{EI_c}\right)$$
$$= 8856\frac{\text{kN·m}^3}{EI_c} \qquad \text{to the right}$$

The required results are summarized in the accompanying table.

Point	$\theta, \frac{\text{kN·m}^2}{EI_c}$	$\Delta_H, \frac{\text{kN·m}^3}{EI_c}$	Δ_V
A	1458 clockwise	0	0
B	324 clockwise	7776 to the right	0
C	216 counterclockwise	7776 to the right	0
D	216 counterclockwise	8856 to the right	0

Example 2.16.2 For the rigid frame shown in Fig. 2.16.2a, find θ, Δ_H, and Δ_V of points A, B, C, and D by the moment-area/conjugate-beam method.

SOLUTION The free-body diagrams of the entire rigid frame and of each member are shown in Fig. 2.16.2a and b. The moment diagram, the modified moment diagram, and the elastic curve are shown in Fig. 2.16.2c to e. The portions $F'C'$ and $C'D'$ of the elastic curve remain straight since they are not acted upon by any moments. It can be seen that Δ_H of B is equal to Δ_H of C, Δ_V of B is zero, and Δ_V of D is equal to Δ_V of C. Note also that for this problem, there is little use of the conjugate-beam method, because the starting point A has zero slope and zero deflection. Applying the moment-area theorems,

$$\theta_B = \theta_A + \frac{1}{EI_c}(A_1 + A_2)$$
$$= 0 + \frac{1}{EI_c}[(288)(7.5) + \tfrac{1}{2}(216)(4.5)]$$
$$= 2646\frac{\text{kN·m}^2}{EI_c} \qquad \text{clockwise}$$

$$\Delta_H \text{ of } B = \frac{1}{EI_c}(\text{moments of } A_1 \text{ and } A_2 \text{ about } B)$$
$$= \frac{1}{EI_c}[(288)(7.5)(3.75) + \tfrac{1}{2}(216)(4.5)(6)]$$
$$= 11{,}016\frac{\text{kN·m}^3}{EI_c} \qquad \text{to the right}$$

$$\theta_C = \theta_B + \frac{1}{EI_c}A_3 = \frac{2646}{EI_c} + \frac{1}{EI_c}[\tfrac{1}{2}(144)(3)]$$
$$= 2862\frac{\text{kN·m}^2}{EI_c} \qquad \text{clockwise}$$

DEFORMATION OF STATICALLY DETERMINATE BEAMS AND RIGID FRAMES

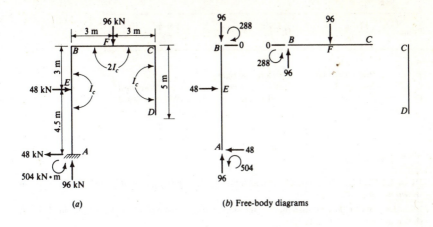

(a)   (b) Free-body diagrams

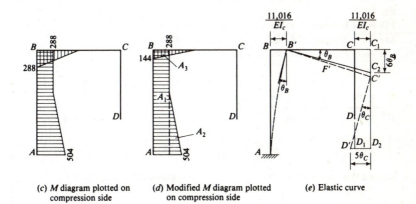

(c) M diagram plotted on compression side

(d) Modified M diagram plotted on compression side

(e) Elastic curve

Figure 2.16.2 Rigid frame of Example 2.16.2.

$$\Delta_V \text{ of } C = C_1C_2 + C_2C' = 6\theta_B + \frac{1}{EI_c}(\text{moment of } A_3 \text{ about } C)$$

$$= 6\left(\frac{2646}{EI_c}\right) + \frac{1}{EI_c}[\tfrac{1}{2}(144)(3)(5)]$$

$$= 16{,}956\frac{\text{kN}\cdot\text{m}^3}{EI_c} \quad \text{downward}$$

$$\theta_D = \theta_C = 2862\frac{\text{kN}\cdot\text{m}^2}{EI_c} \quad \text{clockwise}$$

$$\Delta_H \text{ of } D = D_2D' - D_1D_2 = 5\theta_C - CC_1$$

$$= 5\left(\frac{2862}{EI_c}\right) - \frac{11{,}016}{EI_c} = 3294\frac{\text{kN}\cdot\text{m}^3}{EI_c} \quad \text{to the left}$$

The required results are summarized in the accompanying table.

68 INTERMEDIATE STRUCTURAL ANALYSIS

Point	$\theta, \dfrac{\text{kN} \cdot \text{m}^2}{EI_c}$	$H, \dfrac{\text{kN} \cdot \text{m}^3}{EI_c}$	$V, \dfrac{\text{kN} \cdot \text{m}^3}{EI_c}$
A	0	0	0
B	2646 clockwise	11,016 to the right	0
C	2646 clockwise	11,016 to the right	16,956 downward
D	2862 clockwise	3,294 to the left	16,956 downward

2.17 Exercises

2.1 to 2.4 Find by the unit-load method the slope and deflection at B of the cantilever beam shown in Figs. 2.17.1 to 2.17.4.

2.5 to 2.8 Find by the unit-load method the slope at A and the deflection at C of the simple beam shown in Figs. 2.17.5 to 2.17.8.

2.9 Find by the unit-load method the slope and deflection at A of the overhanging beam shown in Fig. 2.17.9.

2.10 Solve Exercise 2.3 by the partial-derivative method.

2.11 Solve Exercise 2.7 by the partial-derivative method.

2.12 Solve Exercise 2.9 by the partial-derivative method.

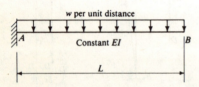

Figure 2.17.1 Exercise 2.1.

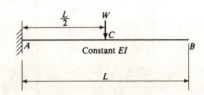

Figure 2.17.2 Exercise 2.2.

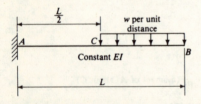

Figure 2.17.3 Exercise 2.3.

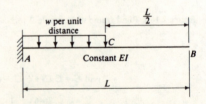

Figure 2.17.4 Exercise 2.4.

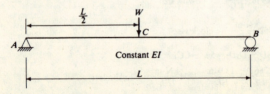

Figure 2.17.5 Exercise 2.5.

DEFORMATION OF STATICALLY DETERMINATE BEAMS AND RIGID FRAMES 69

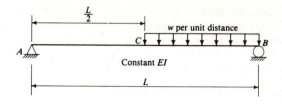

Figure 2.17.6 Exercise 2.6.

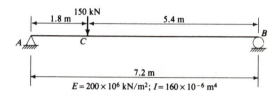

Figure 2.17.7 Exercise 2.7.

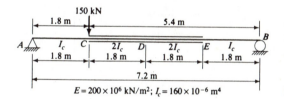

Figure 2.17.8 Exercise 2.8.

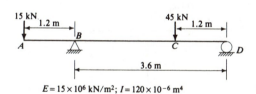

Figure 2.17.9 Exercise 2.9.

2.13 Find by the unit-load method the slopes at A, B, and C and the horizontal deflections at B and D of the rigid frame shown in Fig. 2.17.10.

2.14 Find by the unit-load method the slopes at A, B, and C, and the horizontal and vertical deflections at A and B of the rigid frame shown in Fig. 2.17.11.

2.15 Find by the unit-load method the slopes at A, B, C, D, and E; the horizontal deflections at A and D; and the vertical deflection at A of the rigid frame shown in Fig. 2.17.12.

2.16 Find by the unit-load method the slopes at A, B, and C; the horizontal deflections at A and D; and the vertical deflections at A and B of the rigid frame shown in Fig. 2.17.13.

2.17 Determine Δ_H and Δ_V of each joint of the unloaded rigid frame shown in Exercise 2.13 if hinge A yields 2 cm to the left and 3 cm downward. Check Δ_H of D by the unit-load method.

2.18 Determine Δ_H and Δ_V of each joint of the unloaded rigid frame shown in Exercise 2.15 if the roller support at D yields 2 cm downward.

2.19 to 2.22 Solve Exercises 2.1 to 2.4 by the moment-area method.

2.23 to 2.26 Solve Exercises 2.5 to 2.8 by the moment-area method.

70 INTERMEDIATE STRUCTURAL ANALYSIS

2.27 Solve Exercise 2.9 by the moment-area method.

2.28 to 2.31 Solve Exercises 2.5 to 2.8 by the conjugate-beam method.

2.32 Solve Exercise 2.9 by applying the conjugate-beam method to segment BD and then the moment-area method to segment AB.

2.33 For the rigid frame of Exercise 2.13, find θ, Δ_H, and Δ_V of points A, B, C, and D by the moment-area/conjugate-beam method.

2.34 For the rigid frame of Exercise 2.14, find θ, Δ_H, and Δ_V of points A, B, and C by the moment-area method.

2.35 For the rigid frame of Exercise 2.15, find θ, Δ_H, and Δ_V of points A, B, C, D, and E by the moment-area/conjugate-beam method.

2.36 For the rigid frame of Exercise 2.16, find θ, Δ_H, and Δ_V of points A, B, C, and D by the moment-area method.

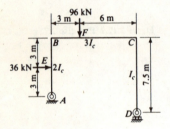

Figure 2.17.10 Exercise 2.13.

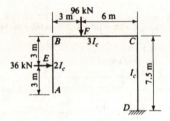

Figure 2.17.11 Exercise 2.14.

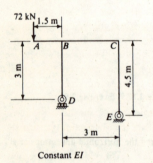

Constant EI

Figure 2.17.12 Exercise 2.15.

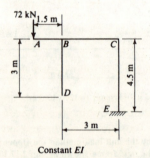

Constant EI

Figure 2.17.13 Exercise 2.16.

CHAPTER
THREE

DEFLECTION OF STATICALLY DETERMINATE TRUSSES

3.1 Force Response vs. Deformation Response

The usual steel truss is a structure composed of individual members so joined together as to form a series of triangles. The joints may be bolted, welded, or fastened together with pins but in the present treatment are assumed to act as smooth hinges. It follows that the members are subjected to axial forces of tension or compression only and are not subjected to bending since the ends are taken as hinged and no loads are applied except at the joints themselves. They therefore remain straight even though the entire truss should assume a different form when loads are applied. The difference between statically determinate and statically indeterminate trusses has been discussed previously in Sec. 1.1. The treatment in this chapter is limited to statically determinate trusses.

When a statically determinate truss as shown in Fig. 3.1.1a is subjected to a set of loads at the joints, there are force and deformation responses to be determined. *Force response* means that reactions, such as R_1, R_2, and R_3 in Fig. 3.1.1a, are to be exerted by the supports on the truss, and axial forces, such as F_1 to F_{13} in Fig. 3.1.1a, are to be induced in the members of the truss. Altogether there are 16 unknowns, and the free-body diagrams of the eight joints will furnish just 16 conditions of equilibrium. *Deformation response* means that the joints will shift to new positions to accommodate the elongations or shortenings of the members, which all remain straight under the assumption of frictionless joints. If the horizontal and vertical deflections of each joint are

72 INTERMEDIATE STRUCTURAL ANALYSIS

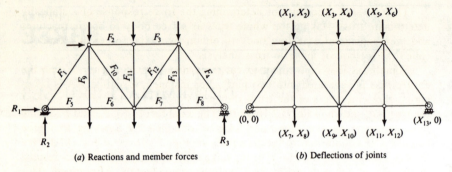

(a) Reactions and member forces (b) Deflections of joints

Figure 3.1.1 Force response and deformation response of a truss.

written in a pair of parentheses as shown in Fig. 3.1.1b, there are altogether 13 unknown deflections X_1 to X_{13}. These 13 deflections must be such as to be consistent with the new lengths of the 13 members of the truss. Thus again, for a statically determinate truss, the number of conditions is equal to the number of unknowns.

The horizontal and vertical deflections of all joints in a truss define completely its deformation response, unlike the descriptions of the deflections and slopes at some chosen points on the elastic curve of a beam or a rigid frame. Because of this distinction, the word *deformation* is used in the title of Chap. 2, whereas the word *deflection* is used in the title of this chapter.

3.2 Methods of Joints and Sections

In common cases the number of unknown reactions on a statically determinate truss is three, although it can be more than three so long as the sum of unknown reactions and unknown axial forces is equal to twice the number of joints. When the number of unknown reactions *is* three, the reactions can be determined before the axial forces in the members by taking the whole truss as a free body and applying the three equations of equilibrium for a coplanar, nonconcurrent force system. Then the axial forces in the members are obtained by successively taking each joint as a free body, beginning with the joint having only two unknown axial forces acting upon it. Usually it is possible to find the next joint with only two unknown axial forces acting on it, and these two unknowns are obtained by applying the two equations of equilibrium to that joint. It is desirable to start from the left end of the truss until about the middle of the span, and then do likewise from the right end. Because the three reactions have been previously obtained by taking the whole truss as a free body, there will be a surplus of three conditions of equilibrium at the joints over the unknown axial forces. These three conditions can be used at the joints near the midspan for checking the correctness of the solution.

DEFLECTION OF STATICALLY DETERMINATE TRUSSES

The two-step procedure of determining the force response of a statically determinate truss—taking the whole truss as a free body to obtain reactions and then each joint as a free body to obtain two axial forces at a time—is called the *method of joints* in truss analysis.

Sometimes it is convenient to cut three members of a truss so that the entire truss is separated into two parts. For instance, a section AA is taken across three members of the truss shown in Fig. 3.2.1a and the two parts of the truss are shown in Fig. 3.2.1b and c. The axial forces F_2, F_6, and F_{10} acting by either part on the other part can be obtained by applying the three equations of equilibrium to the free body of Fig. 3.2.1b or of Fig. 3.2.1c. By always assuming the directions of the unknowns as tensile, the correct results, in magnitude as well as in sign, can be conveniently determined. This technique, called the *method of sections*, is useful when inconvenience is encountered in using the method of joints. The inconvenience arises when it is not possible to find the next joint with only two unknown axial forces, such as in the roof truss of Fig. 3.2.2a, or when none of the two unknown axial forces at a joint is in the horizontal or vertical direction, such as in the bridge truss of Fig. 3.2.2b.

The methods of joints and sections are methods for hand computation. They are useful for a truss of small size, but they are not amenable to digital

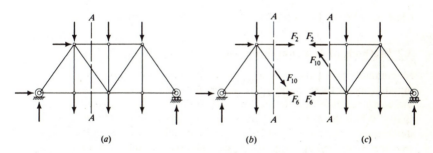

Figure 3.2.1 Method of sections in truss analysis.

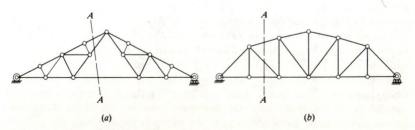

Figure 3.2.2 Cases suitable for method of sections.

74 INTERMEDIATE STRUCTURAL ANALYSIS

computer programming. In digital computation one looks at the overall picture of a general truss by comparing the total number of unknowns against the total number of available conditions and then solves that system of linear equations. This procedure will be examined in Chap. 10.

Example 3.2.1 Solve for the reactions R_1 to R_3, and by the method of joints solve for the member axial forces F_1 to F_{13} of the statically determinate truss shown in Fig. 3.2.3a.

SOLUTION (a) *Reactions R_1 to R_3*. Taking the whole truss as a free body,

$\Sigma F_x = 0$: $\qquad\qquad R_1 = 0$

ΣM about $L_0 = 0$: $\qquad R_3 (4 \text{ panels}) = 180(1) + 270(2) + 360(3)$

$\qquad\qquad\qquad\qquad\qquad R_3 = 450 \text{ kN}$

ΣM about $L_4 = 0$: $\qquad R_2 (4 \text{ panels}) = 180(3) + 270(2) + 360(1)$

$\qquad\qquad\qquad\qquad\qquad R_2 = 360 \text{ kN}$

Check by $\Sigma F_y = 0$: $\qquad 180 + 270 + 360 \stackrel{?}{=} 360 + 450$

$\qquad\qquad\qquad\qquad\qquad 810 = 810 \qquad$ (OK)

(b) *Forces F_1 to F_{13} by method of joints*. In Fig. 3.2.3b, the member forces can be visually obtained and the results written down on the truss diagram as soon as each answer

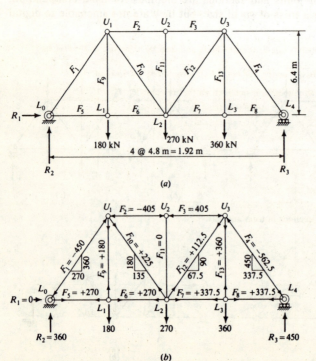

Figure 3.2.3 Truss for Example 3.2.1: Method of joints.

DEFLECTION OF STATICALLY DETERMINATE TRUSSES **75**

is found. The thinking processes are as follows:
1. Using $\Sigma F_y = 0$ at joint L_0, F_1 must push down on it with a vertical component of 360; thus the horizontal component is $\cdot 360(0.75) = 270$, and the total force F_1 is -450, the negative sign indicating compression. Using $\Sigma F_x = 0$ at joint L_0, F_5 must pull on it with 270, or $F_5 = +270$.
2. Using $\Sigma F_x = 0$ at joint L_1, F_6 must pull on it with 270, or $F_6 = +270$. Using $\Sigma F_y = 0$ at joint L_1, F_9 must pull on it with 180, or $F_9 = +180$.
3. Using $\Sigma F_y = 0$ at joint U_1, F_{10} must pull on it with a vertical component of 360 (from F_1) $- 180$ (from F_{10}) $= 180$; thus the horizontal component is $180(0.75) = 135$, and the total force F_{10} is $+225$. Using $\Sigma F_x = 0$ at joint U_1, F_2 must push on it with 270 (from F_1) $+135$ (from F_{10}) $= 405$, or $F_2 = -405$.
4. Using $\Sigma F_x = 0$ at U_2, F_3 must push on it with 405, or $F_3 = -405$. Using $\Sigma F_y = 0$ at joint U_2, $F_{11} = 0$.
5. Now using $\Sigma F_y = 0$ at joint L_4, F_4 must push on it with a vertical component of 450; thus the horizontal component is $450(0.75) = 337.5$, and the total force F_4 is -562.5. Using $\Sigma F_x = 0$ at joint L_4, F_8 must pull on it with 337.5, or $F_8 = +337.5$.
6. Using $\Sigma F_x = 0$ at joint L_3, F_7 must pull on it with 337.5, or $F_7 = +337.5$. Using $\Sigma F_y = 0$ at joint L_3, F_{13} must pull on it with 360, or $F_{13} = +360$.
7. Now F_{12} is the only unknown, but the two conditions at joint U_3 and the two conditions at joint L_2 have not yet been used. Using $\Sigma F_y = 0$ at joint U_3, F_{12} must pull on it with a vertical component of 450 (from F_4) $- 360$ (from F_{13}) $= 90$; thus the horizontal component is $90(0.75) = 67.5$, and the total force F_{12} is $+112.5$. Make first check by $\Sigma F_x = 0$ at joint U_3.
8. Make second and third checks by $\Sigma F_x = 0$ and $\Sigma F_y = 0$ at joint L_2.

Example 3.2.2 By the method of sections, obtain the axial forces F_2, F_6, and F_{10} in Fig. 3.2.3a, using first the left side as the free body and then the right side as the free body.

SOLUTION (*a*) *Left side as the free body.* Referring to Fig. 3.2.4a,

ΣM about $L_2 = 0$ (Note that L_2 is not even on the free body):

$$360(9.6) - 180(4.8) + F_2(6.4) = 0$$

$$F_2 = -405 \text{ kN}$$

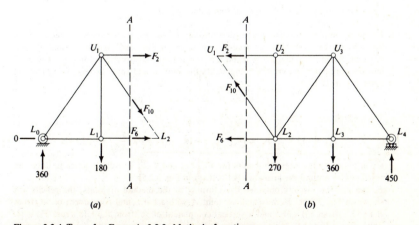

Figure 3.2.4 Truss for Example 3.2.2: Method of sections.

ΣM about $U_1 = 0$:

$$360(4.8) - F_6(6.4) = 0$$
$$F_6 = +270 \text{ kN}$$

$\Sigma F_y = 0$:

$$360 - 180 - 0.8 F_{10} = 0$$
$$F_{10} = +225 \text{ kN}$$

(b) *Right side as the free body.* Referring to Fig. 3.2.4b,

ΣM about $L_2 = 0$:

$$450(9.6) - 360(4.8) + F_2(6.4) = 0$$
$$F_2 = -405 \text{ kN} \quad \text{(Check)}$$

ΣM about $U_1 = 0$ (Note that U_1 is not even on the free body):

$$450(14.4) - 360(9.6) - 270(4.8) - F_6(6.4) = 0$$
$$F_6 = +270 \text{ kN} \quad \text{(Check)}$$

$\Sigma F_y = 0$:

$$450 - 360 - 270 + 0.8 F_{10} = 0$$
$$F_{10} = +225 \text{ kN} \quad \text{(Check)}$$

Example 3.2.3 Solve for the reactions R_1 to R_3, and by the methods of joints and sections as suitable solve for the member axial forces F_1 to F_5 of the statically determinate truss shown in Fig. 3.2.5a.

SOLUTION (a) *Reactions R_1 to R_3.* Taking the whole truss as a free body,

$\Sigma F_x = 0$: $\qquad\qquad\qquad R_1 = 54$ kN to the left as shown

ΣM about $A = 0$: $\qquad\qquad 54(6) = R_3(9.0)$

$\qquad\qquad\qquad\qquad\qquad R_3 = 36$ kN upward as shown

ΣM about $B = 0$ or $\Sigma F_y = 0$: $\qquad R_2 = 36$ kN downward as shown

(b) *Forces F_1 to F_5 by methods of joints and sections.*

1. F_1 can be obtained by cutting a section through F_1 and F_4, even though the section cuts through only two members; the free-body diagram of Fig. 3.2.5b is actually that of joint A alone. Rather than use $\Sigma F_x = 0$ and $\Sigma F_y = 0$ at joint A, resolve the force F_1 at point C into the horizontal and vertical components F_{1x} and F_{1y} and apply the equation M about point $D = 0$. Thus,

$$54(3.6) + F_{1x}(2.4) - 36(4.5) = 0$$

from which

$$F_{1x} = -13.5 \qquad F_{1y} = F_{1x}\left(\frac{60}{4.5}\right) = -18 \qquad F_1 = F_{1x}\left(\frac{7.5}{4.5}\right) = -22.5 \text{ kN}$$

Upon obtaining the negative results for F_{1x}, F_{1y}, and F_1, it is usual practice to draw the small circles around the arrowheads for them in Fig. 3.2.5b.

2. F_4 can now be visually obtained by returning to the method of joints; the results are shown in Fig. 3.2.5c. F_4 must pull on joint A with a horizontal component of 54 (from R_1) + 13.5 (from F_1) = 67.5 and a vertical component of 36 (from R_2) + 18 (from F_1) = 54. Thus, $F_4 = +86.44$ kN, the positive sign indicating tension.

DEFLECTION OF STATICALLY DETERMINATE TRUSSES

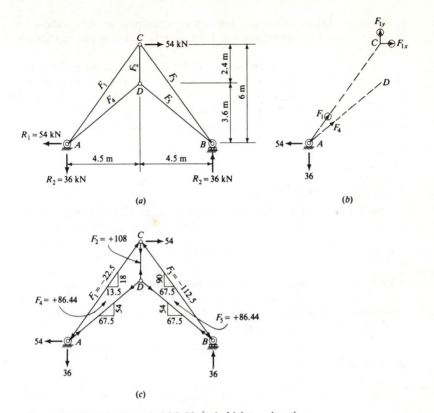

Figure 3.2.5 Truss for Example 3.2.3: Method of joints and sections.

3. F_2 and F_5 are found by using $\Sigma F_x = 0$ and $\Sigma F_y = 0$ at joint D, as shown in Fig. 3.2.5c.
4. F_3 is now the only unknown, and there are four conditions of joint equilibrium left, two at joint B and two at joint C. Use one of the four conditions to find F_3 and the remaining for check.

3.3 Method for Determination of Deflections of Truss Joints

As the lengths of the members in a truss change, the shape of the truss changes, resulting in the deflections of its joints. Physically then, the determination of joint deflections is a problem of geometry. In this category are the angle-weights method, the joint-displacement-equation method, and the graphical (Williot-Mohr) method.

The unit-load method, as derived in the preceding chapter for beams and rigid frames by equating external work to internal energy, can be used to find truss deflections because each member in the truss can be treated as a fiber such as *MN* in Fig. 2.5.1. However, by this method, only one deflection component—that is, either the horizontal component or the vertical component of the total deflection of one joint—can be found in one application of the unit load.

On the other hand, by the angle-weights method, the vertical deflections of all joints in one horizontal chord—that is, either the top chord or the bottom chord—can be found in one operation. By the joint-displacement-equation method or the graphical method, the horizontal and vertical deflections of all joints can be determined at the same time.

The four methods: unit-load, angle-weights, joint-displacement-equation, and the graphical, will be described in the following sections.

3.4 The Unit-Load Method

The basic formula of the unit-load method has been derived and expressed by Eq. (2.5.4), which, when applied to the situation of Fig. 3.4.1, becomes

$$\Delta = \Sigma u_i (\Delta L)_i \tag{3.4.1}$$

in which

Δ = deflection component in the direction of the unit load
u = axial force in a member due to the unit load
ΔL = change in length of a member, due to applied loads, temperature changes, or fabrication errors

When the change in length ΔL is due entirely to applied loads, then by Hooke's law,

$$\Delta L = \frac{FL}{EA} \tag{3.4.2}$$

where F is the axial force in the member, L and A are the length and cross-sectional area, and E is the modulus of elasticity. Substituting Eq. (3.4.2) in Eq. (3.4.1),

$$\Delta = \Sigma \frac{FuL}{EA} \tag{3.4.3}$$

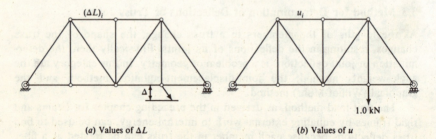

(a) Values of ΔL (b) Values of u

Figure 3.4.1 The unit-load method for truss deflections.

DEFLECTION OF STATICALLY DETERMINATE TRUSSES 79

Equation (3.4.3) is the commonly known formula for truss deflections due to applied loads; while the more basic formula, Eq. (3.4.1), should be used for finding truss deflections due to temperature changes or fabrication errors.

The basic formula, Eq. (3.4.1), can also be derived by the famous principle of virtual work, which can be stated as follows:

Principle of virtual work *The product of a zero force and a nonzero displacement, even if the force and displacement are entirely unrelated phenomena, is still zero.*

This product of a force and an unrelated displacement cannot be "real work," but the product of force and displacement seems like "work," therefore the name *virtual work*.

Consider the two unrelated phenomena in Fig. 3.4.2a and b; one is a real situation in deformation geometry and the other is a fictitious situation in equilibrium. When the composite-equilibrium diagram of Fig. 3.4.2b is separated into the eight free-body diagrams of the joints in Fig. 3.4.2c, the resultant of the forces acting on each joint has to be zero. The product of this zero force and its corresponding nonzero displacement in Fig. 3.4.2a, by the

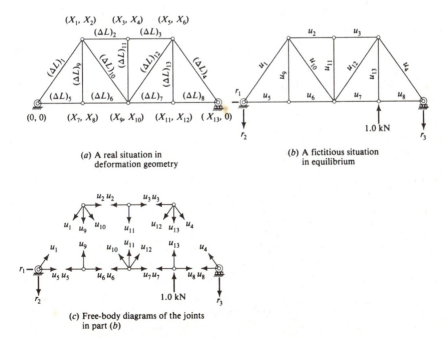

(a) A real situation in deformation geometry

(b) A fictitious situation in equilibrium

(c) Free-body diagrams of the joints in part (b)

Figure 3.4.2 Derivation of the unit-load method by principle of virtual work.

principle of virtual work, must be zero. Thus the total virtual work, or the product of all the forces shown in Fig. 3.4.2c and their corresponding displacements in Fig. 3.4.2a, must be zero; or

$$\sum_{1}^{3} r_i(0) + (1.0)(X_{12}) - \sum_{1}^{13} u_i \, (\Delta L)_i = 0 \qquad (3.4.4)$$

In the above equation, the two u_i forces in each pair are toward each other in Fig. 3.4.2c, but their points of application are separated farther apart by the amount $(\Delta L)_i$ in Fig. 3.4.2a; thus the virtual work is negative. Now one can observe that Eq. (3.4.4) is in fact identical to Eq. (3.4.1).

Actually, the basic formula, Eq. (2.5.4), in the unit-load method for beams and rigid frames can also be perceived by means of the principle of virtual work, only in this case there would be an infinite number of free bodies of elementary pieces in a diagram equivalent to Fig. 3.4.2c, instead of eight free-body diagrams.

Example 3.4.1 By the unit-load method determine the horizontal and vertical deflections of joint L_3 due to the applied loads on the truss shown in Fig. 3.4.3a.

SOLUTION The forces in the members of the truss due to the applied loads, to a 1.0-kN horizontal load applied at joint L_3, and to a 1.0-kN vertical load applied at joint L_3, are found by the methods of joints and sections and are shown in Fig. 3.4.3b to d, respectively. The horizontal and vertical components of the forces in the diagonals are also shown in Fig. 3.4.3b to d so that the equilibrium of each joint can be easily checked by inspection. Note that the F values in Fig. 3.4.3b are in kilonewtons, but each u value in Fig. 3.4.3c and d may be considered as the *ratio* of the axial force in each member to the load applied at the point where deflection is being sought.

The computation involved in the application of Eqs. (3.4.1) to (3.4.3) to obtain the

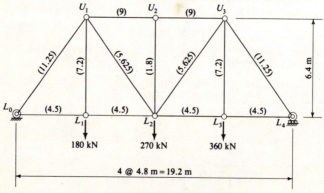

Numbers in parentheses are areas in $10^{-3} \times m^2$
$E = 200 \times 10^6 \, kN/m^2$
(a)

DEFLECTION OF STATICALLY DETERMINATE TRUSSES 81

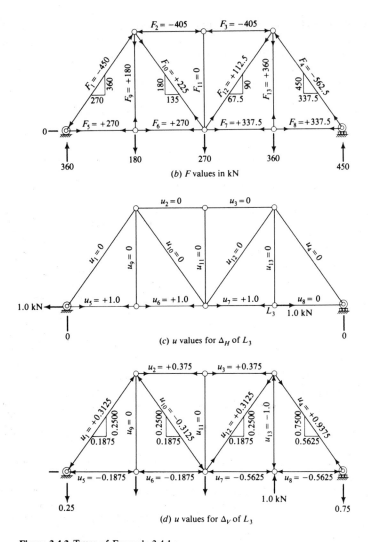

Figure 3.4.3 Truss of Example 3.4.1.

required results is shown in Table 3.4.1. From Table 3.4.1,

$$\Delta_H \text{ of } L_3 = \Sigma(u \text{ for } \Delta_H)\left(\frac{FL}{EA}\right)$$

$$= +4.68 \text{ mm} \quad \text{or} \quad 4.68 \text{ mm to the right}$$

$$\Delta_V \text{ of } L_3 = \Sigma(u \text{ for } \Delta_V)\left(\frac{FL}{EA}\right)$$

$$= -7.60 \text{ mm} \quad \text{or} \quad 7.60 \text{ mm downward}$$

Table 3.4.1 Truss deflections for Example 3.4.1

Member Number	L, m	A, 10^{-3} m^2	F, kN	$\dfrac{FL}{EA}$, mm	u for Δ_H at L_3	u for Δ_V at L_3	(u for Δ_H) $\times FL/EA$	(u for Δ_V) $\times FL/EA$
1	8.0	11.25	−450	−1.60	0.0	+0.3125	0	−0.5000
2	4.8	9.00	−405	−1.08	0.0	+0.3750	0	−0.4050
3	4.8	9.00	−405	−1.08	0.0	+0.3750	0	−0.4050
4	8.0	11.25	−562.5	−2.00	0.0	+0.9375	0	−1.8750
5	4.8	4.50	+270	+1.44	+1.0	−0.1875	+1.44	−0.2700
6	4.8	4.50	+270	+1.44	+1.0	−0.1875	+1.44	−0.2700
7	4.8	4.50	+337.5	+1.80	+1.0	−0.5625	+1.80	−1.0125
8	4.8	4.50	+337.5	+1.80	0.0	−0.5625	0	−1.0125
9	6.4	7.20	+180	+0.80	0.0	0	0	0
10	8.0	5.625	+225	+1.60	0.0	−0.3125	0	−0.5000
11	6.4	1.80	0	0	0.0	0	0	0
12	8.0	5.625	+112.5	+0.80	0.0	+0.3125	0	+0.2500
13	6.4	7.20	+360	+1.60	0.0	−1.0000	0	−1.6000
Total							+4.68	−7.6000

The positive or negative sign of the result of the summation indicates whether the actual direction of the deflection being sought is the same as or opposite to the direction of the applied unit load.

Example 3.4.2 By the unit-load method determine the horizontal and vertical deflections of joint L_3 of the truss shown in Fig. 3.4.3a due to a temperature drop of 50°C in the lower chord only, given the coefficient of expansion or contraction as $\alpha = 11.7 \times 10^{-6}$/°C.

SOLUTION The temperature drop in the lower chord only may be due to the sudden use of the cold storage room under the roof truss during a hot day. In this case, the more fundamental formula, Eq. (3.4.1), is to be used. The elongation (negative for shortening) of members 5, 6, 7, or 8 is equal to

$$(\Delta L)_5 = (\Delta L)_6 = (\Delta L)_7 = (\Delta L)_8 = \alpha t L$$
$$= (11.7 \times 10^{-6})(-50)(4800 \text{ mm})$$
$$= -2.808 \text{ mm}$$

The u values for Δ_H of L_3 and for Δ_V of L_3 are available in Fig. 3.4.3c and d. Thus,

$$\Delta_H \text{ of } L_3 = u_5(\Delta L)_5 + u_6(\Delta L)_6 + u_7(\Delta L)_7 + u_8(\Delta L)_8$$
$$= (+1.0)(-2.808) + (+1.0)(-2.808) + (+1.0)(-2.808) + (0)(-2.808)$$
$$= -8.424 \text{ mm} \quad \text{or} \quad 8.424 \text{ mm to the left}$$

$$\Delta_V \text{ of } L_3 = u_5(\Delta L)_5 + u_6(\Delta L)_6 + u_7(\Delta L)_7 + u_8(\Delta L)_8$$
$$= (-0.1875)(-2.808) + (-0.1875)(-2.808) + (-0.5625)(-2.808) + (-0.5625)(-2.808)$$
$$= +4.212 \text{ mm} \quad \text{or} \quad 4.212 \text{ mm upward}$$

Example 3.4.3 By the unit-load method determine the horizontal and vertical deflections of joint L_3 of the truss shown in Fig. 3.4.3a due to a fabrication error of 6 mm too short in member U_1L_2.

SOLUTION In this problem, ΔL's are zero for all members except member U_1L_2, and

$$\Delta L \text{ of member } U_1L_2 = (\Delta L)_{10} = -6.0 \text{ mm}$$

DEFLECTION OF STATICALLY DETERMINATE TRUSSES 83

Using the u values for Δ_H of L_3 and Δ_V of L_3 in Fig. 3.4.3c and d,

Δ_H of $L_3 = \Sigma u_i (\Delta L)_i = (u_{10}$ for $\Delta_H)(\Delta L)_{10} = (0)(-6.0) = 0$

Δ_V of $L_3 = \Sigma u_i (\Delta L)_i = (u_{10}$ for $\Delta_V)(\Delta L)_{10} = (-0.3125)(-6.0)$

$\phantom{\Delta_V \text{ of } L_3} = +1.875$ mm   or   1.875 mm upward

Example 3.4.4 By the unit-load method determine the increase in distance between the joints L_1 and U_2 due to the applied loads on the truss of Fig. 3.4.3a.

SOLUTION In order to determine the increase in distance between the joints L_1 and U_2 when the shape of the truss changes owing to the applied loads of Fig. 3.4.3a, it is possible to first find the deflection component of joint L_1 along the extended direction of U_2L_1 and then the deflection component of joint U_2 along the extended direction of L_1U_2. The sum of the two deflection components is the increase in distance. However, this combined result can be more easily obtained by applying a pair of unit loads at L_1 and U_2 as shown in Fig. 3.4.4. Since the u forces in Fig. 3.4.4 are the combined forces due to both unit loads, the use of these u values in Eqs. (3.4.1) or (3.4.3) will give the increase in distance between joints L_1 and U_2. Thus, using the ΔL values in Table 3.4.1,

$\dfrac{\text{Increase in distance}}{\text{between } L_1 \text{ and } U_2} = \Sigma u_i (\Delta L)_i$

$= (+0.6)(\Delta L)_2 + (+0.6)(\Delta L)_6 + (+0.8)(\Delta L)_9 + (-1.0)(\Delta L)_{10}$
$\quad + (+0.8)(\Delta L)_{11}$

$= (+0.6)(-1.08) + (+0.6)(+1.44) + (+0.8)(+0.80)$
$\quad + (-1.0)(+1.60) + (+0.8)(0)$

$= -0.744$ mm

The negative sign of the result shows that there is actually a decrease in distance of 0.744 mm between the joints L_1 and U_2.

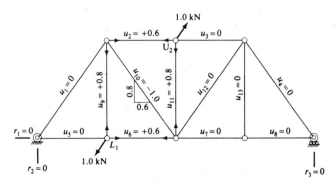

Figure 3.4.4 Values of u for increase in distance between joints L_1 and U_2.

Example 3.4.5 By the unit-load method determine the horizontal deflection of the roller support B due to the load of 54 kN applied at joint C of the truss shown in Fig. 3.4.5a.

SOLUTION The force response of the truss to the applied load is taken from the results of Example 3.2.3 (Fig. 3.2.5) and shown again as Fig. 3.4.5b. To obtain the horizontal deflection

84 INTERMEDIATE STRUCTURAL ANALYSIS

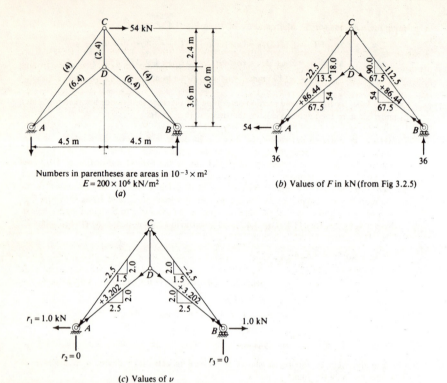

Figure 3.4.5 Truss of Example 3.4.5.

of the roller support B, a horizontal unit load is applied at B, and its force response is shown in Fig. 3.4.5c. The required computation involved in the use of Eqs. (3.4.1) to (3.4.3) is shown in Table 3.4.2; Thus,

$$\Delta_H \text{ of } B = \Sigma u_i \left(\frac{FL}{EA}\right)_i = +7.818 \text{ mm} \quad \text{or} \quad 7.818 \text{ mm to the right}$$

Table 3.4.2 Horizontal deflection of B in Example 3.4.5

Member	L, m	A, 10^{-3} m^2	F, kN	ΔL, mm	u	$u \Delta L$
AC	7.5	4.0	− 22.5	−0.211	−2.5	+0.528
CD	2.4	2.4	+108.0	+0.540	+4.0	+2.160
BC	7.5	4.0	−112.5	−1.055	−2.5	+2.638
AD	5.763	6.4	+ 86.44	+0.389	+3.202	+1.246
BD	5.763	6.4	+ 86.44	+0.389	+3.202	+1.246
Total						+7.818

3.5 The Angle-Weights Method

Before taking up the angle-weights method of finding truss deflections, it is necessary first to derive a set of cyclic formulas to express the changes in the sizes of the three interior angles of a triangle due to small changes in the lengths of its three sides. If the three sides L_A, L_B, L_C opposite to angles A, B, C of triangle ABC, as shown in Fig. 3.5.1, have their lengths increased by small amounts of $\epsilon_A L_A$, $\epsilon_B L_B$, $\epsilon_C L_C$, the increases ΔA, ΔB, ΔC of the angles A, B, C are:

$$\Delta A = (\epsilon_A - \epsilon_B) \cot C + (\epsilon_A - \epsilon_C) \cot B \quad (3.5.1a)$$

$$\Delta B = (\epsilon_B - \epsilon_C) \cot A + (\epsilon_B - \epsilon_A) \cot C \quad (3.5.1b)$$

$$\Delta C = (\epsilon_C - \epsilon_A) \cot B + (\epsilon_C - \epsilon_B) \cot A \quad (3.5.1c)$$

Equation (3.5.1) is cyclic in nature; it can be stated in words as follows:

The increase in the angle is equal to the sum of two products, each obtained from multiplying the difference of the strain of the opposite side minus the strain of the adjacent side by the cotangent of the included angle.

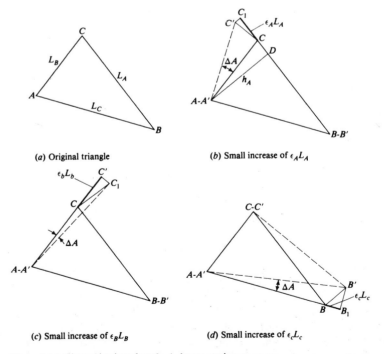

(a) Original triangle

(b) Small increase of $\epsilon_A L_A$

(c) Small increase of $\epsilon_B L_B$

(d) Small increase of $\epsilon_c L_c$

Figure 3.5.1 Change in size of angle A due to strains ϵ_A, ϵ_B, ϵ_C.

Proof The original triangle ABC is shown in Fig. 3.5.1a. The effects on the size of angle A due to the elongations in L_A, L_B, and L_C are shown separately in Fig. 3.5.1b to d. In each case, the deformed triangle $A'B'C'$ is obtained by following the first-order assumption that the length of a member is not changed by the transverse displacements of its ends. Thus, in Fig. 3.5.1b, CC_1 is an extension of BC and is equal to $\epsilon_A L_A$, $C_1 C'$ is perpendicular to BC_1, and CC' is perpendicular to AC. The angle at C' inside the small triangle CC_1C' is equal to angle C of the original triangle because their sides are mutually perpendicular. Then,

$$CC' = \frac{CC_1}{\sin C'} = \frac{\epsilon_A L_A}{\sin C}$$

The increase in size of angle A is

$$\Delta A = \frac{CC'}{AC} = \frac{CC'}{L_B} = \frac{\epsilon_A L_A}{L_B \sin C} = \frac{\epsilon_A L_A}{h_A} = \frac{\epsilon_A (CD + BD)}{h_A}$$

$$= \frac{\epsilon_A (h_A \cot C + h_A \cot B)}{h_A} = \epsilon_A (\cot C + \cot B) \tag{3.5.2}$$

In reference to Fig. 3.5.1c, CC_1 is an extension of AC and is equal to $\epsilon_B L_B$, $C_1 C'$ is perpendicular to AC_1, and CC' is perpendicular to BC. Using the small triangle CC_1C',

$$C_1 C' = CC_1 \cot C' = CC_1 \cot C = \epsilon_B L_B \cot C$$

and

$$\Delta A = -\frac{C_1 C'}{AC_1} = -\frac{C_1 C'}{AC} = -\frac{\epsilon_B L_B \cot C}{L_B} = -\epsilon_B \cot C \tag{3.5.3}$$

In reference to Fig. 3.5.1d, BB_1 is an extension of AB and is equal to $\epsilon_C L_C$, $B_1 B'$ is perpendicular to AB_1, and BB' is perpendicular to CB. Using the small triangle BB_1B',

$$B_1 B' = BB_1 \cot B' = BB_1 \cot B = \epsilon_C L_C \cot B$$

and

$$\Delta A = -\frac{B_1 B'}{AB_1} = -\frac{B_1 B'}{AB} = -\frac{\epsilon_C L_C \cot B}{L_C} = -\epsilon_C \cot B \tag{3.5.4}$$

The right side of Eq. (3.5.1a) is the sum of the right sides of Eqs. (3.5.2) to (3.5.4). Equations (3.5.1b and c) can be similarly proved.

Now if it is required to find Δ_V of B, Δ_V of C, and Δ_V of D at joints B, C, and D of the truss shown in Fig. 3.5.2a due to known values of strains in the members, the geometric procedure described subsequently can be used. Note that joints B, C, and D move horizontally as well, but only the vertical deflections are plotted directly under joints B, C, and D in Fig. 3.5.2a. For joint B, determine the decrease in the sizes of angles ABF, FBG, and GBC; the total decrease ΔB in angle ABC is equal to the sum of these three parts.

DEFLECTION OF STATICALLY DETERMINATE TRUSSES 87

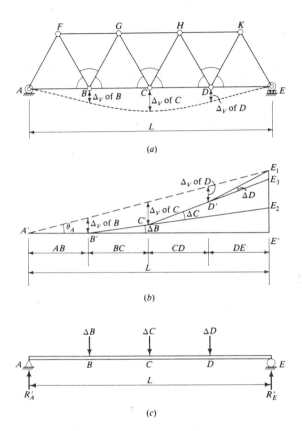

Figure 3.5.2 The angle-weights method.

Similarly, ΔC and ΔD represent the decreases in the otherwise 180° angles BCD and CDE, respectively. Graphically the shape of the lower chord should become that of the broken line $A'B'C'D'E'$ in Fig. 3.5.2b.

Actually the closing line $A'E'$ should stay horizontal; therefore Δ_V of B, Δ_V of C, and Δ_V of D are the vertical intercepts between this closing line $A'E'$ and the broken line $A'B'C'D'E'$. From Fig. 3.5.2b,

$$\theta_A = \frac{E_1 E'}{L} = \frac{E_1 E_2 + E_2 E_3 + E_3 E'}{L}$$

$$= \frac{(\Delta B)(BE) + (\Delta C)(CE) + (\Delta D)(DE)}{L} \qquad (3.5.5)$$

Note that the length of every inclined line in Fig. 3.5.2b is taken to be equal to its horizontal projection. From the geometry of Fig. 3.5.2b,

88 INTERMEDIATE STRUCTURAL ANALYSIS

$$\Delta_V \text{ of } B = \theta_A(AB) \qquad (3.5.6a)$$

$$\Delta_V \text{ of } C = \theta_A(AC) - (\Delta B)(BC) \qquad (3.5.6b)$$

$$\Delta_V \text{ of } D = \theta_A(AD) - (\Delta B)(BD) - (\Delta C)(CD) \qquad (3.5.6c)$$

If an imaginary simple beam AE or a conjugate simple beam AE, as shown in Fig. 3.5.2c, is loaded by the decreases ΔB, ΔC, ΔD in the original 180° angles at joints B, C, D,

$$R'_A = \frac{(\Delta B)(BE) + (\Delta C)(CE) + (\Delta D)(DE)}{L}$$

which is identical to the expression for θ_A in Eq. (3.5.5). The bending moments M'_B, M'_C, and M'_D of this simple beam will take identical expressions to Eqs. (3.5.6a to c); thus,

$$\Delta_V \text{ of } B = M'_B \text{ in conjugate beam} \qquad (3.5.7a)$$

$$\Delta_V \text{ of } C = M'_C \text{ in conjugate beam} \qquad (3.5.7b)$$

$$\Delta_V \text{ of } D = M'_D \text{ in conjugate beam} \qquad (3.5.7c)$$

The decreases in the otherwise 180° angles at B, C, and D of the truss are hereby applied as concentrated loads on the conjugate beam; therefore the name *angle-weights method*. This method is convenient to use when the vertical deflections of all joints on an originally horizontal chord of a truss are to be determined.

Example 3.5.1 By the angle-weights method determine the vertical deflections of all lower-chord joints due to the applied loads on the truss shown in Fig. 3.5.3a.

Table 3.5.1 Angle changes in Example 3.5.1

Joint	Angle changes in 10^{-3}	Summation of angle changes
L_1	$L_0L_1U_1 = (-0.200 - 0.300)(\frac{3}{4}) + (-0.200 - 0.125)(\frac{4}{3}) = -0.80833$	-0.78333×10^{-3}
	$U_1L_1L_2 = (+0.200 - 0.125)(\frac{4}{3}) + (+0.200 - 0.300)(\frac{3}{4}) = +0.02500$	
L_2	$L_1L_2U_1 = (+0.125 - 0.300)(0) + (+0.125 - 0.200)(\frac{4}{3}) = -0.10000$	-0.46250×10^{-3}
	$U_1L_2U_2 = (-0.225 - 0.200)(\frac{3}{4}) + (-0.225 - 0.000)(0) = -0.31875$	
	$U_2L_2U_3 = (-0.225 - 0.000)(0) + (-0.225 - 0.100)(\frac{3}{4}) = -0.24375$	
	$U_3L_2L_3 = (+0.250 - 0.100)(\frac{4}{3}) + (+0.250 - 0.375)(0) = +0.20000$	
L_3	$L_2L_3U_3 = (+0.100 - 0.375)(\frac{3}{4}) + (+0.100 - 0.250)(\frac{4}{3}) = -0.40625$	-1.54167×10^{-3}
	$U_3L_3L_4 = (-0.250 - 0.250)(\frac{4}{3}) + (-0.250 - 0.375)(\frac{3}{4}) = -1.13542$	

DEFLECTION OF STATICALLY DETERMINATE TRUSSES 89

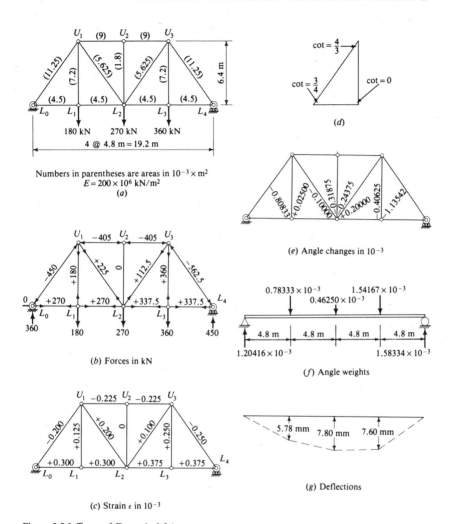

Figure 3.5.3 Truss of Example 3.5.1.

SOLUTION The reactions and member forces have been solved in Example 3.2.1; these are shown again in Fig. 3.5.3b. The strain of each member is equal to F/EA, as shown in Fig. 3.5.3c. By the use of Eq. (3.5.1), the changes in the angles around joints L_1, L_2, and L_3 are computed in Table 3.5.1 (refer to Fig. 3.5.3d for the cotangent function of angles). The answers are entered in Fig. 3.5.3e. The total decreases in the otherwise 180° angles are applied as downward angle weights on the conjugate beam shown in Fig. 3.5.3f. The moment diagram for this conjugate beam, plotted on the tension side this time to show downward deflection, is the deflected shape of the lower chord so far as vertical deflections are concerned.

3.6 The Joint-Displacement-Equation Method

The joint-displacement-equation method is an algebraic method by which the horizontal and vertical deflections of all joints in a truss due to the changes in length of its members can be computed. The method is based on one equation, called the *joint-displacement equation*, which is derived below.

Let AB in Fig. 3.6.1a be the member of an originally undeformed truss. Owing to the changes in length of the members of the truss, the joint A is displaced by X_1 units to the right and X_2 units up, and the joint B by X_3 units to the right and X_4 units up. Let L be the original length of AB and ΔL be the increase in its length. Let α be the counterclockwise angle measured at A from the horizontal direction to the right to the direction from A to B. Note that the angle α may be anywhere between 0 and 360°.

If H and V are the horizontal and vertical projections of the original length L, then $H + X_3 - X_1$ and $V + X_4 - X_2$ would be the horizontal and vertical projections of the new length $L + \Delta L$. Following this the relationship

$$L + \Delta L = \sqrt{(H + X_3 - X_1)^2 + (V + X_4 - X_2)^2} \qquad (3.6.1)$$

must hold true. The first-order assumption violates this relationship, contending that so long as the joint displacements X_1 to X_4 are small, transverse displacements of the ends of a straight segment in a structure do not change its length. According to this assumption, the change in length of a member in a truss can be expressed by the displacements of its ends through a linear equation.

Returning to Fig. 3.6.1, if point A' of the new member $A'B'$ is superimposed on A, the horizontal and vertical components of BB' would be $X_3 - X_1$ and $X_4 - X_2$, respectively, as shown in Fig. 3.6.1b. Resolving BB' into its longitudinal component BB_1 along the extension of AB and its transverse component BB_2 perpendicular to AB, and using the first-order assumption,

$$\Delta L = BB_1 = (X_3 - X_1) \cos \alpha + (X_4 - X_2) \sin \alpha \qquad (3.6.2)$$

Equation (3.6.2) is the all-important joint-displacement equation. In this equation, $\cos \alpha$ and $\sin \alpha$ are dimensionless, while ΔL, $X_3 - X_1$, and $X_4 - X_2$ are quantities of the same small order. The equation is most useful when three of the X_1 to X_4 values are known, and the fourth one is to be found.

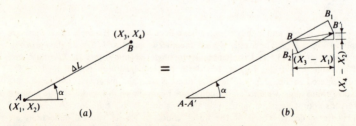

Figure 3.6.1 Derivation of the joint-displacement equation.

DEFLECTION OF STATICALLY DETERMINATE TRUSSES 91

The use of the joint-displacement equation to find the horizontal and vertical deflections of all joints in a truss will be described in the following example.

Example 3.6.1 By the joint-displacement-equation method determine the horizontal and vertical deflections of all joints of the truss subjected to the member elongations shown in Fig. 3.6.2a.

SOLUTION The location of joint L_0 does not change with the deformation of the truss, while the direction of every member may change. In the first step of applying the joint-displacement-equation method, however, any joint may be assumed to be fixed in location and any member through this joint to be fixed in direction. When the joint displacements of all joints are determined relative to this fixed joint and the fixed direction, the whole deformed truss may be translated and rotated so as to fulfill certain physical requirements; for instance, the translation will bring the joint L'_0 back to the hinge, and the rotation will bring the joint L'_4 to the horizontal line through the hinge.

It is generally more expedient to choose a point near the midspan as the fixed point, or the reference point. For this problem, choose joint U_1 as the reference point, and choose member U_1L_2 through it as the member fixed in direction, or as the reference member. The joint displacements relative to the reference point and reference member are shown in Fig. 3.6.2b: most of these are obtained by inspection, using the known elongations in the horizontal and vertical members; others marked as b_1, b_2, b_3, and b_4 are solved from the joint-displacement equation, using the elongations in members L_0U_1, L_2U_3, U_3L_3, and U_3L_4 as the conditions, respectively.

The horizontal and vertical displacements of the reference point U_1 are $(0,0)$. As a reference member, U_1L_2 is assumed to be fixed in direction; thus joint L_2 must extend out in the direction from U_1 to L_2 an amount of 1.60, which is the known elongation of U_1L_2. This means that the vertical and horizontal displacements of joint L_2 are $(+0.96, -1.28)$. Going from U_1 toward the right, the horizontal displacement of U_2 has to be -1.08 (negative means to the left) to accommodate the known shortening of 1.08 in member U_1U_2; and that of U_3 has to be -2.16 because U_2U_3 is known to shorten another 1.08. Going from U_1 downward, the vertical displacement of L_1 has to be -0.80 (negative means downward) to accommodate the known elongation of 0.80 in member U_1L_1. Going from L_2 toward the left, the horizontal displacement of L_1 has to be -0.48 to accommodate the known elongation of 1.44 in member L_1L_2; and that of L_0 has to be -1.92 because L_0L_1 lengthens by another 1.44. Going from L_2 toward the right, the horizontal displacement of L_3 has to be $+2.76$ to accommodate the known elongation of 1.80 in member L_2L_3; and that of L_4 has to be $+4.56$ because L_3L_4 lengthens by another 1.80. Going from L_2 upward, the vertical displacement of U_2 has to be -1.28, which is the same as that of L_2 because the length of U_2L_2 does not change.

At this time, the unknown vertical displacement of L_0 is marked as b_1; of U_3, as b_2; of L_3, as b_3; and of L_4, as b_4. Applying the joint-displacement equation from U_1 toward L_0,

$$-1.60 = (-1.92 - 0)(-0.6) + (b_1 - 0)(-0.8)$$
$$b_1 = +3.44$$

from L_2 toward U_3,

$$+0.80 = (-2.16 - 0.96)(+0.6) + (b_2 + 1.28)(+0.8)$$
$$b_2 = +2.06$$

and from U_3 toward L_4,

$$-2.00 = (+4.56 + 2.16)(+0.6) + (b_4 - 2.06)(-0.8)$$
$$b_4 = +9.60$$

92 INTERMEDIATE STRUCTURAL ANALYSIS

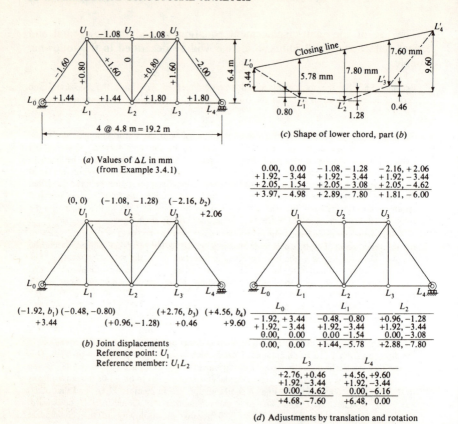

Figure 3.6.2 Truss of Example 3.6.1.

Going from U_3 downward, the vertical displacement of joint L_3 has to be 0.46 upward to accommodate the known elongation of 1.60 in member U_3L_3; or

$$b_3 = +0.46$$

The above values of b_1 to b_4 are entered into Fig. 3.6.2b as soon as each is obtained.

The joint displacements as noted in Fig. 3.6.2b are consistent with the known changes in lengths of all members, except joint L_0 should not have displaced at all and the vertical displacement of L_4 has to be zero. These two physical requirements can be met by first giving the entire truss a translation, that is, 1.92 mm to the right and 3.44 mm downward. In Fig. 3.6.2d, the first line represents the results obtained in Fig. 3.6.2b; the second line shows the displacements due to translation, these being the same for every joint. Now the vertical displacement of L_4 is still $9.60 - 3.44 = 6.16$ mm upward, which is too high, so the entire truss is to be rotated about point L_0 through a small clockwise angle of 6.16 mm/19.2 m in order to bring L_4 down to zero.

When a radial arm like OP in Fig. 3.6.3 rotates around the point O through a small angle $d\theta$, the displacement PP' can be taken perpendicular to OP and equal to $R\,d\theta$. The

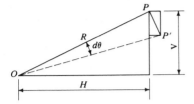

Figure 3.6.3 Displacement PP' due to a small rotation.

horizontal component of PP' is, by similar triangles,

$$\Delta_H \text{ (to the right)} = PP'\left(\frac{V}{R}\right) = R\,d\theta\left(\frac{V}{R}\right) = V\,d\theta \quad (3.6.3a)$$

$$\Delta_V \text{ (downward)} = PP'\left(\frac{H}{R}\right) = R\,d\theta\left(\frac{H}{R}\right) = H\,d\theta \quad (3.6.3b)$$

Equations (3.6.3a and b) are used to establish the third line in Fig. 3.6.2d. The positive and negative signs at each joint are first written down by observation. For instance, because the entire truss rotates about L_0 in a clockwise direction, U_3 moves downward to the right in a direction perpendicular to an imaginary line drawn through L_0 and U_3. The vertical displacements of L_1, L_2, and L_3 are $\frac{1}{4}$, $\frac{2}{4}$, and $\frac{3}{4}$ of 6.16 mm, all in the downward direction. The vertical displacements of U_1, U_2, and U_3 must be the same as those of L_1, L_2, and L_3, respectively, since the rotation is the rigid-body motion of the original undeformed truss. Likewise, the horizontal displacements of U_1, U_2, and U_3 must be the same, the direction is to the right, and the amount is

$$V\,d\theta = (6.4 \text{ m})\left(\frac{6.16 \text{ mm}}{19.2 \text{ m}}\right) = 2.05 \text{ mm}$$

The final vertical and horizontal displacements of all joints are, then, the sum of three parts: first due to member elongations on the basis of a reference point and a reference member through it, second due to translation, and third due to rotation. These final results are shown in the fourth line of Fig. 3.6.2d.

If only the vertical deflections of all lower-chord joints are required, then, at the completion of the results shown in Fig. 3.6.2b, a sketch of the shape of the lower chord as shown by Fig. 3.6.2c is all that is needed. On this figure, a horizontal reference line is first drawn, and the vertical deflections of $+3.44$, -0.80, -1.28, $+0.46$, and $+9.60$ taken from Fig. 3.6.2b are plotted above or below this horizontal line to arrive at L_0', L_1', L_2', L_3', and L_4'. Then the closing line $L_0'L_4'$ is drawn and the vertical intercepts between this closing line and the various points L_0', L_1', L_2', L_3' and L_4' are the required vertical deflections, which are 0, 5.78 mm, 7.80 mm, 7.60 mm, and 0, respectively.

3.7 The Graphical Method—Williot-Mohr Diagram

The angle-weights method and the joint-displacement-equation method are algebraic methods using only geometric relations. Thus it is easy to infer that the horizontal and vertical deflections of all joints can be found by a graphical solution. Theoretically it is possible to draw the shape of the deformed truss by using the new lengths of members as the sides of the component triangles. But the changed lengths are only a little longer or shorter than the original lengths, a fact which makes the deformed truss almost coincide with the

94 INTERMEDIATE STRUCTURAL ANALYSIS

original truss. This difficulty can be avoided by using a method suggested by Williot in 1877.† In this method two different scales are used in plotting the original length L and the changes in length, ΔL.

Consider the truss shown in Fig. 3.7.1a, of which the original shape is $ABCDE$. Assume that the changes in lengths of the seven members are $+3$, -4, -5, $+6$, $+7$, -8, and $+9$ small units, as written on the members in Fig. 3.7.1a. The new locations of the joints are A', B', C', D', and E', when joint C' is placed over C and member $C'B'$ is placed over CB. These new positions

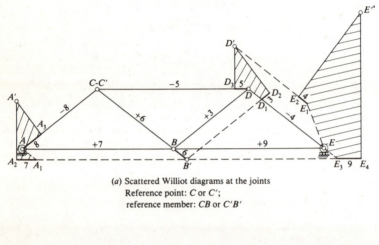

(a) Scattered Williot diagrams at the joints
 Reference point: C or C';
 reference member: CB or $C'B'$

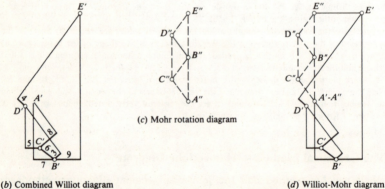

(b) Combined Williot diagram (c) Mohr rotation diagram (d) Williot-Mohr diagram

Figure 3.7.1 Williot-Mohr diagram.

† S. P. Timoshenko, *History of Strength of Materials*, McGraw-Hill Book Company, New York, 1953, p. 314.

can be determined graphically by drawing the new triangles $C'B'A'$, $C'B'D'$, and $D'B'E'$ in succession as follows:

1. *For triangle $C'B'A'$.* Place C' over C. Extend CB to B', making BB' equal to 6 units. Draw $B'A_1$ parallel and equal to BA, and extend $B'A_1$ by 7 units to A_2. Contract 8 units from A toward C to arrive at A_3. The perpendicular lines erected at A_2 and A_3 intersect at A'.
2. *For triangle $C'B'D'$.* Draw $B'D_1$ parallel and equal to BD, and extend $B'D_1$ by 3 units to D_2. Contract 5 units from D toward C to arrive at D_3. The perpendicular lines erected at D_2 and D_3 intersect at D'.
3. *For triangle $D'B'E'$.* Draw $D'E_1$ parallel and equal to DE, and contract 4 units from E_1 toward D' to arrive at E_2. Draw $B'E_3$ parallel and equal to BE, and extend $B'E_3$ by 9 units to E_4. The perpendicular lines erected at E_2 and E_4 intersect at E'.

On completion of the graphical work shown in Fig. 3.7.1a, one can see that the shaded areas around joints A, D, and E are not dependent on the scale used to plot the original truss $ABCDE$, and that these areas may be made to a different scale. The reason is that perpendiculars instead of circular arcs are drawn when the point of intersection of the two new sides of a triangle is being sought. This is consistent with the first-order assumption that transverse displacement does not change the length of a member.

The scattered displacement diagrams represented by the shaded areas in Fig. 3.7.1a can be combined into one diagram as shown in Fig. 3.7.1b. C' is first spotted as the reference point. Since $C'B'$ is chosen to be the reference member, B' in Fig. 3.7.1a should relatively move 6 units downward to the right in the direction of CB away from the reference point. This is done in Fig. 3.7.1b to obtain B'. Since A' in Fig. 3.7.1a should move 7 units to the left relative to B' and 8 units upward to the right toward C', in Fig. 3.7.1b the 7 units and 8 units are plotted to left of B' and upward to right from C', with their perpendiculars to intersect at A'. Next, since D' in Fig. 3.7.1a should move 3 units upward to the right from B' and 5 units to the left relative to C', in Fig. 3.7.1b the 3 units and 5 units are plotted upward to the right from B' and 5 units to the left of C', with their perpendiculars to intersect at D'. Last, since E' in Fig. 3.7.1a should move 9 units to the right relative to B' and 4 units upward to the left toward D', in Fig. 3.7.1b the 9 units and 4 units are plotted to the right of B' and upward to the left from D', with their perpendiculars to intersect at E'. The deflection of each joint can now be measured in Fig. 3.7.1b from the reference point toward the single-prime point, on the conditions that C' does not move and $C'B'$ does not rotate.

The point E' is relatively higher than the point A' in Fig. 3.7.1b, while the physical requirement is that they should remain at the same elevation. Mohr in 1887 suggested a graphical method to obtain the deflections of all joints, both in magnitude and direction, due to a rigid-body rotation of the undeformed truss.† First, the magnitude of the deflection at a joint is the product of its distance from the pivot and the small angle of rotation. Second, the direction of the deflection at a joint is always perpendicular to the imaginary straight line joining the pivot and the joint in question. One can conclude then that the magnitude and direction of these deflections can be measured from a geometrically similar truss which is perpendicular to the

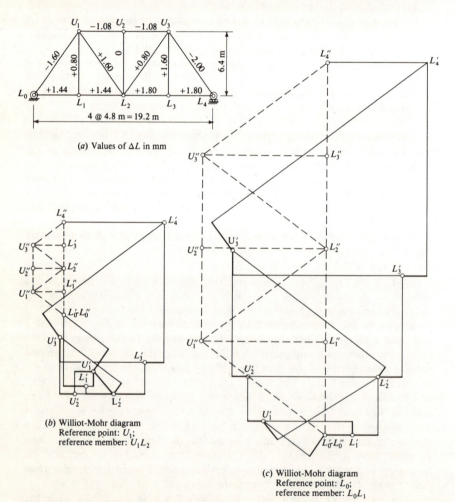

(a) Values of ΔL in mm

(b) Williot-Mohr diagram
Reference point: U_1;
reference member: U_1L_2

(c) Williot-Mohr diagram
Reference point: L_0;
reference member: L_0L_1

Figure 3.7.2 Solution for Example 3.7.1.

†J. I. Parcel and R. R. B. Moorman, *Analysis of Statically Indeterminate Structures*, John Wiley & Sons, Inc., New York, 1955, p. 107.

DEFLECTION OF STATICALLY DETERMINATE TRUSSES   97

original shape of the truss. Thus the deflection of each joint is measured from the double-prime point to the pivot point A'' on the Mohr rotation diagram of Fig. 3.7.1c.

When the Williot and Mohr diagrams are combined as shown in Fig. 3.7.1d, the final deflection of each joint can be simply measured from the double-prime point to the single-prime point. This final deflection is the sum of three points: (1) due to rotation, from the double-prime point to A'' (or A'); (2) due to translation, from A' (or A'') to the reference point; and (3) due to changes in lengths of members, from the reference point to the single-prime point. One can see now that the three parts of the deflection in the graphical method are exactly identical to the three parts in the joint-displacement-equation method.

Example 3.7.1 By the graphical method determine the horizontal and vertical deflections of all joints of the truss subjected to the member elongations shown in Fig. 3.7.2a.

SOLUTION Two graphical solutions are shown in Fig. 3.7.2b and c. The solution shown in Fig. 3.7.2b takes much less space than that of Fig. 3.7.2c when the same scale is used, or in a limited available space a larger scale can be used if a good choice of reference member is made. The best choice is to use a member with zero or little rotation as the reference member.

3.8 Exercises

3.1 By the unit-load method determine the horizontal and vertical deflections of all lower-chord joints due to the applied loads on the truss shown in Fig. 3.8.1.

3.2 By the unit-load method determine the increase in distance between the joints U_2 and L_3 due to the applied loads on the truss shown in Fig. 3.8.1.

3.3 By the unit-load method determine the horizontal and vertical deflections of joint L_4 of the truss shown in Fig. 3.8.2 due to a temperature drop of 40°C in the lower chord only. Coefficient of expansion or contraction is $11.7 \times 10^{-6}/°C$.

3.4 By the unit-load method determine the horizontal and vertical deflections of joint L_4 of the truss shown in Fig. 3.8.2 if member U_2U_3 were fabricated 12 mm too long.

3.5 By the unit-load method determine the horizontal deflection of the roller support B due to the load of 81 kN applied at C of the truss shown in Fig. 3.8.3.

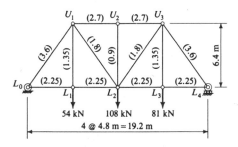

Numbers in parentheses are areas in $10^{-3} \times m^2$
$E = 200 \times 10^6$ kN/m^2

Figure 3.8.1 Exercises 3.1 and 3.2.

98 INTERMEDIATE STRUCTURAL ANALYSIS

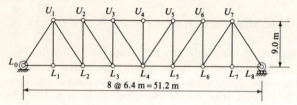

Figure 3.8.2 Exercises 3.3 and 3.4.

3.6 By the unit-load method determine the horizontal and vertical deflections of joint U_3 due to the load of 45 kN applied at U_3 of the truss shown in Fig. 3.8.4.

3.7 By the angle-weights method determine the vertical deflections of all lower-chord joints due to the applied loads on the truss of Exercise 3.1.

3.8 By the angle-weights method determine the vertical deflections of all lower-chord joints of the truss described in Exercise 3.3.

3.9 By the angle-weights method determine the vertical deflections of all lower chord joints of the truss described in Exercise 3.4.

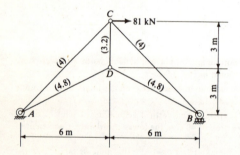

Numbers in parentheses are areas in $10^{-3} \times m^2$
$E = 200 \times 10^6$ kN/m^2

Figure 3.8.3 Exercise 3.5.

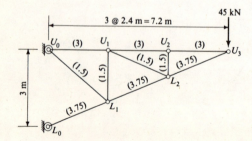

Numbers in parentheses are areas in $10^{-3} \times m^2$
$E = 200 \times 10^6$ kN/m^2

Figure 3.8.4 Exercise 3.6.

DEFLECTION OF STATICALLY DETERMINATE TRUSSES 99

3.10 By the joint-displacement-equation method determine the horizontal and vertical deflections of all joints due to the applied loads on the truss of Exercise 3.1, using U_1 as the reference point and U_1L_2 as the reference member.

3.11 Solve Exercise 3.10, using L_2 as the reference point and L_2U_2 as the reference member.

3.12 Solve Exercise 3.10, using L_0 as the reference point and L_0L_1 as the reference member.

3.13 By the graphical method determine the horizontal and vertical deflections of all joints due to the applied loads on the truss of Exercise 3.1, using U_1 as the reference point and U_1L_2 as the reference member.

3.14 Solve Exercise 3.13, using L_2 as the reference point and L_2U_2 as the reference member.

3.15 Solve Exercise 3.13, using L_0 as the reference point and L_0L_1 as the reference member.

CHAPTER
FOUR
ANALYSIS OF STATICALLY INDETERMINATE BEAMS AND RIGID FRAMES BY THE FORCE METHOD

4.1 Force Method of Analysis

It has been demonstrated in Chaps. 2 and 3 that the force response of statically determinate beams, rigid frames, and trusses can be determined solely by the laws of statics, and in the solution procedure the properties of members, such as moments of inertia for beams and rigid frames or areas for trusses, are not required. The deformation response can be obtained after the force response, but in the solution procedure, member properties are required.

When a structure—whether it be a beam, a rigid frame, or a truss—is statically indeterminate, the force response cannot be determined by the laws of statics alone. In these situations, some unknown reactions, or member forces, equal in number to the degree of indeterminacy, can be regarded as unknown forces acting on a basic determinate structure, and their magnitudes can be obtained at the very beginning from the conditions of consistent deformation. In establishing these conditions of geometry, member properties are then required. Thus the force response of statically indeterminate structures does depend on the member properties. While the name *method of consistent deformation* is more descriptive, the name *force method* has gained popularity because of its contrast with the displacement method, which can be so effectively programmed on a digital computer.

For beams and one-story rigid frames very often the excess reactions are chosen as the redundants, although bending moments at some convenient points may be chosen as the redundants as will be described in Chap. 6. The

ANALYSIS OF STATICALLY INDETERMINATE BEAMS 101

condition of geometry corresponding to each reacting force (or reacting moment) is that the deflection (or the slope in case of a reacting moment) at its point of application must be zero. Thus in the force method of analysis, there are always as many conditions of geometry as there are unknown redundant reacting forces and moments. These redundant reacting forces and moments can both be called generalized forces, hence the name force method.

4.2 Analysis of Statically Indeterminate Beams by the Force Method

For the coplanar parallel-force system acting on a beam, there are two independent conditions of statics, with one more for each internal hinge present in the beam. The number of excess reactions over that of independent equations of equilibrium is the degree of indeterminacy, or

$$NI = NR - 2 - NIH \qquad (4.2.1)$$

in which NI is the degree of indeterminacy, NR is the total number of reactions, and NIH is the number of internal hinges in the beam.

In the force method of analysis of statically indeterminate beams, the first step is to choose the redundant reactions, remove the physical restraints associated with the redundant reactions, and obtain a basic determinate beam subjected to the combined action of the applied loads and the unknown redundant reacting forces. If a simple support is removed and the reaction is replaced by an unknown reacting force, the condition of geometry is that the deflection there must be zero. If a fixed support is changed into a simple support and the original moment reaction is replaced by an unknown reacting moment, the condition of geometry is that the slope there must be zero. If a fixed support is completely removed and the original restraint is replaced by an unknown reacting force and an unknown reacting moment, the two conditions of geometry are that both the deflection and the slope there must be zero.

There are usually alternate ways of choosing the redundants, resulting in different types of basic determinate beams on which the applied loads and the redundant forces act simultaneously. In the following examples, alternate solutions are shown.

Example 4.2.1 Analyze by the force method and draw shear and moment diagrams for the beam of Fig. 4.2.1a.

SOLUTION From Eq. (4.2.1), the degree of indeterminacy is equal to

$$NI = NR - 2 - NIH = 3 - 2 - 0 = 1$$

Choose R_3 as the redundant, remove the support at B, and obtain the basic beam, which is a cantilever beam fixed at A only as shown in Fig. 4.2.1b and c. The condition of geometry is that the deflection at B, under the combined action of the applied load (Fig. 4.2.1b) and the redundant (Fig. 4.2.1c), must be zero; or

$$\Delta_{B1} = \Delta_{B2}$$

102 INTERMEDIATE STRUCTURAL ANALYSIS

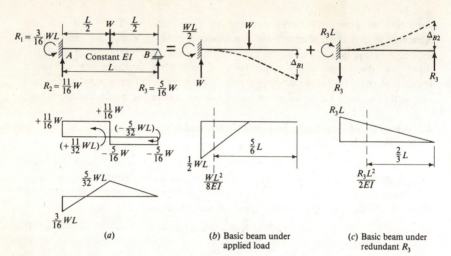

(a) (b) Basic beam under applied load (c) Basic beam under redundant R_3

Figure 4.2.1 Beam of Example 4.2.1.

By the moment-area method (the unit-load method could be used instead),

$$\Delta_{B1} = \frac{WL^2}{8EI}\left(\frac{5}{6}L\right) = \frac{5WL^3}{48EI} \qquad \Delta_{B2} = \frac{R_3L^2}{2EI}\left(\frac{2}{3}L\right) = \frac{R_3L^3}{3EI}$$

Equating Δ_{B1} to Δ_{B2} and solving for R_3,

$$R_3 = \frac{5}{16}W$$

With R_3 now known, return to Fig. 4.2.1a, and use the two equations of statics to find R_1 and R_2:

$$R_1 = \frac{3}{16}WL \qquad R_2 = \frac{11}{16}W$$

The shear and moment diagrams for the given beam can then be drawn as shown in Fig. 4.2.1a.

ALTERNATE SOLUTION Choose R_1 as the redundant, change the fixed support at A into a simple support, and obtain the basic beam, which is a simple beam supported at A and B as shown in Fig. 4.2.2b and c. The condition of geometry is that the slope at A under the combined action of the applied load (Fig. 4.2.2b) and the redundant (Fig. 4.2.2c) must be zero; or

$$\theta_{A1} = \theta_{A2}$$

By the conjugate-beam method (the unit-load method could be used instead),

θ_{A1} = reaction at A of conjugate beam in Fig. 4.2.2b

$$= \frac{1}{2}\frac{WL^2}{8EI} = \frac{WL^2}{16EI}$$

θ_{A2} = reaction at A of conjugate beam in Fig. 4.2.2c

$$= \frac{2R_1L}{3 2EI} = \frac{R_1L}{3EI}$$

ANALYSIS OF STATICALLY INDETERMINATE BEAMS 103

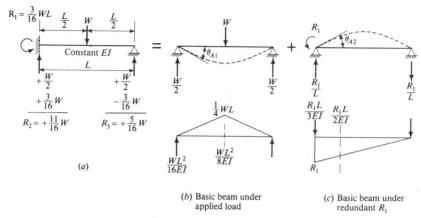

(b) Basic beam under applied load

(c) Basic beam under redundant R_1

Figure 4.2.2 Alternate solution for Example 4.2.1.

Equating θ_{A1} to θ_{A2} and solving for R_1,

$$R_1 = \frac{3}{16}WL$$

With R_1 now known, return to Fig. 4.2.2a, and use the two equations of statics to find R_2 and R_3,

$$R_2 = \frac{11}{16}W \qquad R_3 = \frac{5}{16}W$$

Note that these two reactions, R_2 and R_3, can also be obtained conveniently by superposition. In Fig. 4.2.2a, the reactions due to W alone are shown as $+W/2$ (positive means upward) at A and at B, and those to balance the counterclockwise end moment of $\frac{3}{16}WL$ are $+\frac{3}{16}W$ at A and $-\frac{3}{16}W$ at B to form a clockwise reacting couple of $\frac{3}{16}WL$. The combined values for the reactions are thus $R_2 = +\frac{11}{16}W$ and $R_3 = +\frac{5}{16}W$.

Example 4.2.2 Analyze by the force method and draw shear and moment diagrams for the beam of Fig. 4.2.3a.

SOLUTION From Eq. (4.2.1), the degree of indeterminacy is equal to

$$NI = NR - 2 - NIH = 4 - 2 - 0 = 2$$

By symmetry, R_3 and R_4 must each be equal to $\frac{1}{2}wL$ upward, and the only unknown is the fixed-end moment R_1 or R_2 at each end. R_1 must be counterclockwise to hold the tangent horizontal at A, and R_2 must be clockwise to hold the tangent horizontal at B. The simple beam AB is chosen as the basic beam, which is subjected to the combined action of the applied load (Fig. 4.2.3b) and the equal but opposite end moments R_1 and R_2 (Fig. 4.2.3c). The condition of geometry is that the slope must be zero at both A and B; or

$$\theta_{A1} = \theta_{A2}$$

By the conjugate-beam method (the unit-load method could be used instead),

θ_{A1} = reaction at A of conjugate beam in Fig. 4.2.3b

$$= \frac{1}{2}\frac{wL^3}{12EI} = \frac{wL^3}{24EI}$$

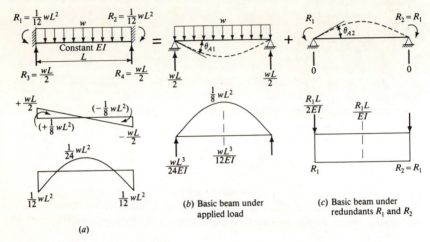

Figure 4.2.3 Beam of Example 4.2.2.

θ_{A2} = reaction at A of conjugate beam in Fig. 4.2.3c

$$= \frac{1}{2}\frac{R_1 L}{EI} = \frac{R_1 L}{2EI}$$

Equating θ_{A1} to θ_{A2} and solving for R_1,

$$R_1 = \frac{1}{12}wL^2 \qquad R_2 = \frac{1}{12}wL^2$$

With R_1 and R_2 now known, return to Fig. 4.2.3a and complete the shear and moment diagrams as shown there.

ALTERNATE SOLUTION A cantilever beam fixed at A and free at B may be chosen as the basic beam, which is subjected to the combined action of the applied load, the redundant R_2, and the redundant R_4, as shown in Fig. 4.2.4b to d. Two conditions of geometry exist; the one corresponding to R_2 is that the slope must be zero at B, and the other corresponding to R_4 is that the deflection must be zero at B. Thus,

$$\theta_{B1} + \theta_{B2} = \theta_{B3}$$
$$\Delta_{B1} + \Delta_{B2} = \Delta_{B3}$$

By the moment-area method (the unit-load method could be used instead),

$$\theta_{B1} = \frac{wL^3}{6EI} \qquad \Delta_{B1} = \frac{wL^3}{6EI}\left(\frac{3}{4}L\right) = \frac{wL^4}{8EI}$$

$$\theta_{B2} = \frac{R_2 L}{EI} \qquad \Delta_{B2} = \frac{R_2 L}{EI}\left(\frac{L}{2}\right) = \frac{R_2 L^2}{2EI}$$

$$\theta_{B3} = \frac{R_4 L^2}{2EI} \qquad \Delta_{B3} = \frac{R_4 L^2}{2EI}\left(\frac{2}{3}L\right) = \frac{R_4 L^3}{3EI}$$

The conditions of geometry become

$$\frac{wL^3}{6EI} + \frac{R_2 L}{EI} = \frac{R_4 L^2}{2EI}$$

$$\frac{wL^4}{8EI} + \frac{R_2 L^2}{2EI} = \frac{R_4 L^3}{3EI}$$

ANALYSIS OF STATICALLY INDETERMINATE BEAMS 105

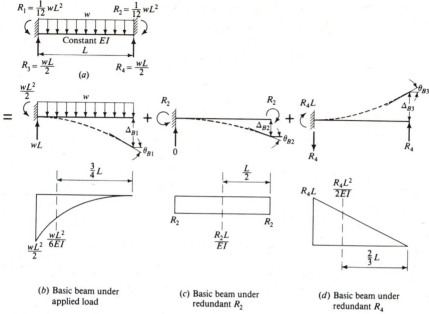

(b) Basic beam under applied load

(c) Basic beam under redundant R_2

(d) Basic beam under redundant R_4

Figure 4.2.4 Alternate solution for Example 4.2.2.

Solving,

$$R_2 = \frac{1}{12}wL^2 \qquad R_4 = \frac{1}{2}wL$$

With R_2 and R_4 now known, return to Fig. 4.2.4a to obtain R_1 and R_3 from the two equations of statics.

Example 4.2.3 Analyze the beam of Fig. 4.2.5a by the force method.

SOLUTION From Eq. (4.2.1), the degree of indeterminacy is equal to

$$NI = NR - 2 - NIH = 4 - 2 - 0 = 2$$

The simple beam AB is chosen as the basic beam, which is subjected to the applied load, the reacting moment R_1 at A, and the reacting moment R_2 at B, as shown in Fig. 4.2.5b to d. The conditions of geometry are

$$\theta_{A1} = \theta_{A2} + \theta_{A3}$$
$$\theta_{B1} = \theta_{B2} + \theta_{B3}$$

By the conjugate-beam method (the unit-load method could be used instead),

$$\theta_{A1} = \frac{Wab}{2EI}\left(\frac{L+b}{3L}\right) \qquad \theta_{B1} = \frac{Wab}{2EI}\left(\frac{L+a}{3L}\right)$$

$$\theta_{A2} = \frac{R_1 L}{2EI}\left(\frac{2}{3}\right) \qquad \theta_{B2} = \frac{R_1 L}{2EI}\left(\frac{1}{3}\right)$$

$$\theta_{A3} = \frac{R_2 L}{2EI}\left(\frac{1}{3}\right) \qquad \theta_{B3} = \frac{R_2 L}{2EI}\left(\frac{2}{3}\right)$$

106 INTERMEDIATE STRUCTURAL ANALYSIS

Substituting the above values in the two conditions of geometry,

$$\frac{R_1 L}{3EI} + \frac{R_2 L}{6EI} = \frac{Wab(L+b)}{6LEI}$$

$$\frac{R_1 L}{6EI} + \frac{R_2 L}{3EI} = \frac{Wab(L+a)}{6LEI}$$

Solving for R_1 and R_2,

$$R_1 = \frac{Wab^2}{L^2} \qquad R_2 = \frac{Wba^2}{L^2}$$

With R_1 and R_2 now known, return to Fig. 4.2.5a to obtain R_3 and R_4 by the two equations of statics; thus,

$$R_3 = \frac{Wb}{L} + \frac{R_1 - R_2}{L} = \frac{Wb}{L} + \frac{Wab}{L^3}(b-a) = \frac{Wb^2(3a+b)}{L^3}$$

$$R_4 = \frac{Wa}{L} - \frac{R_1 - R_2}{L} = \frac{Wa}{L} - \frac{Wab}{L^3}(b-a) = \frac{Wa^2(3b+a)}{L^3}$$

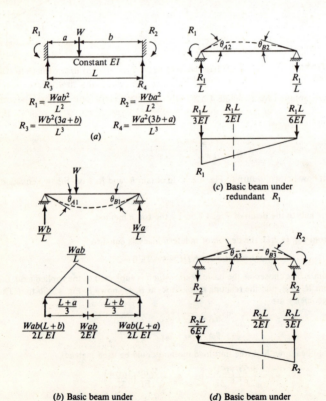

Figure 4.2.5 Beam of Example 4.2.3.

ANALYSIS OF STATICALLY INDETERMINATE BEAMS **107**

ALTERNATE SOLUTION A cantilever beam fixed at A and free at B may be chosen as the basic beam, which is subjected to the combined action of the applied load, the redundant R_2, and the redundant R_4. The two conditions of geometry are that both the slope and the deflection at B should be zero. The details of the solution are not shown.

4.3 Law of Reciprocal Deflections

The law of reciprocal deflections, although it was first discovered by Maxwell in 1864 in his work on trusses,† applies in fact to beams and rigid frames as well. Later a general reciprocal virtual-work theorem, of which the law of reciprocal deflections may be regarded as a special case, was proved by Betti in 1872.‡ It is stated as follows:

> **General reciprocal virtual-work theorem** *The virtual work done by a P-force system in going through the deformation of a Q-force system is equal to the virtual work done by the Q-force system in going through the deformation of the P-force system.*

Proof Because this theorem applies to any type of structure, consider as an example the rigid frame $ABCD$ shown in Fig. 4.3.1. Let the P-force system include three forces P_1, P_2, P_3 and the reactions at the supports, and the Q-force system include two forces Q_1, Q_2 and the reactions at the supports. Let Δ_{PP1}, Δ_{PP2}, Δ_{PP3} be the deflections in the P directions due to the P-force system, and Δ_{QP1}, Δ_{QP2} be the deflections in the Q directions, also due to the

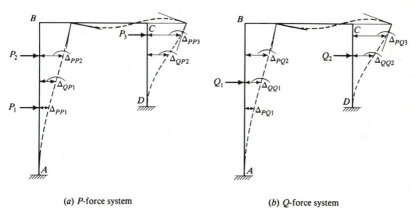

(*a*) *P*-force system (*b*) *Q*-force system

Figure 4.3.1 Reciprocal virtual-work theorem.

---

†S. P. Timoshenko, *History of Strength of Materials*, McGraw-Hill Book Company, New York, 1953, p. 207.
‡Ibid, p. 320.

P-force system, as shown in Fig. 4.3.1a. Let Δ_{QQ1}, Δ_{QQ2} be the deflections in the Q directions due to the Q-force system, and Δ_{PQ1}, Δ_{PQ2}, Δ_{PQ3} be the deflections in the P directions, also due to the Q-force system, as shown in Fig. 4.3.1b. The real work done by applying the P forces first and then adding the Q forces is equal to, using linear relationship between forces and deflections,

$$W = \sum_{i=1}^{i=3} \frac{1}{2} P_i \Delta_{PPi} + \sum_{i=1}^{i=2} \frac{1}{2} Q_i \Delta_{QQi} + \sum_{i=1}^{i=3} P_i \Delta_{PQi} \qquad (4.3.1)$$

Note that the P forces are already on the structure when the Q forces are gradually added; therefore the real work done by each P force is equal to the product of its full value and the additional deflection at its point of application due to the Q forces. Likewise, if the Q forces are applied first and the P forces are added, the real work done on the structure is

$$W = \sum_{i=1}^{i=2} \frac{1}{2} Q_i \Delta_{QQi} + \sum_{i=1}^{i=3} \frac{1}{2} P_i \Delta_{PPi} + \sum_{i=1}^{i=2} Q_i \Delta_{QPi} \qquad (4.3.2)$$

Since the total internal energy within the structure must be the same when both P and Q forces are on it regardless of the sequence of application, the total external work done on it must also be the same. Equating (4.3.1) to (4.3.2) and canceling,

$$\sum_{i=1}^{i=3} P_i \Delta_{PQi} = \sum_{i=1}^{i=2} Q_i \Delta_{QPi} \qquad (4.3.3)$$

The forces and the deflections after each summation sign in Eq. (4.3.3), when looked upon without reference to Eqs. (4.3.1) and (4.3.2), are derived from separated phenomena shown by Fig. 4.3.1a and Fig. 4.3.1b, therefore the name *virtual work* is used.

In the special case wherein there is only one unit force in the P system, and also only one unit force in the Q system, as shown for the beam in Fig. 4.3.2, the application of the general reciprocal virtual-work theorem gives

$$\delta_{PQ} = \delta_{QP} \quad \text{for} \quad \begin{matrix} P = 1.0 \\ Q = 1.0 \end{matrix} \qquad (4.3.4)$$

Equation (4.3.4) is the *law of reciprocal deflections*, stated as follows:

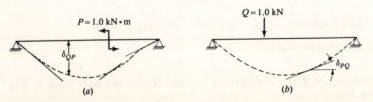

Figure 4.3.2 Law of reciprocal deflections: Case 1.

ANALYSIS OF STATICALLY INDETERMINATE BEAMS 109

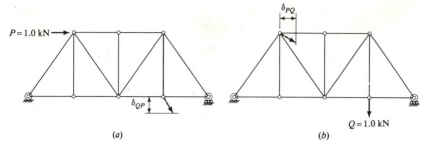

Figure 4.3.3 Law of reciprocal deflections: Case 2.

The deflection at Q due to a unit load at P is equal to the deflection at P due to a unit load at Q.

Equation (4.3.4) may also be verified by actually trying to obtain δ_{QP} and δ_{PQ} independently of each other by the unit-load method. For the case of Fig. 4.3.2,

$$\delta_{QP} = \int \frac{Mm\,dx}{EI} = \int \frac{m_P m_Q\,dx}{EI}$$

and

$$\delta_{PQ} = \int \frac{Mm\,dx}{EI} = \int \frac{m_Q m_P\,dx}{EI}$$

For the case of Fig. 4.3.3,

$$\delta_{QP} = \Sigma \frac{FuL}{EA} = \Sigma \frac{u_P u_Q L}{EA}$$

$$\delta_{PQ} = \Sigma \frac{FuL}{EA} = \Sigma \frac{u_Q u_P L}{EA}$$

The law of reciprocal deflections also applies between a unit P moment and a unit Q force, as shown by Fig. 4.3.4. In this case, δ_{QP} is the *downward* deflection at Q due to a unit *counterclockwise* moment applied at P, and δ_{PQ} is the *counterclockwise* slope at P due to a unit *downward* load applied at Q. Note the words *downward* and *counterclockwise*, each italicized twice in the previous sentence.

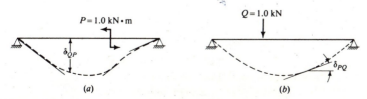

Figure 4.3.4 Law of reciprocal deflections: Case 3.

When δ_{PQ} and δ_{QP} are each independently obtained by the geometric method—that is, the moment-area and conjugate-beam method for beams and rigid frames, or the angle-weights, joint-displacement equation, and graphical method for trusses—their identity serves to verify the correctness of the solution. However, no such independent check can be provided by the unit-load method, because δ_{PQ} and δ_{QP} involve identical integrals or summations.

Since the force method of analysis requires the computation of deflections and slopes, the use of the law of reciprocal deflections can help to reduce or simplify the amount of work involved, especially when the structure is subjected to concentrated forces only, or to a moving unit load as in influence-line computations described in Chap. 13.

Example 4.3.1 Analyze by the force method and draw shear and moment diagrams for the beam of Fig. 4.3.5a.

SOLUTION The beam as given has three supports, while two are enough for static equilibrium; thus it is statically indeterminate to the first degree. The support at B is removed, and acting on the basic beam AC are the applied load and the redundant reaction R_2. The condition of geometry is that the deflection at B of the basic beam, due to the combined action of the 120-kN load and the redundant R_2, is zero; or, as shown in Fig. 4.3.5b and c,

$$\Delta_{B1} = \Delta_{B2}$$

Instead of finding Δ_{B1} and Δ_{B2} each independently, the law of reciprocal deflections can be used to advantage by first solving the problem shown in Fig. 4.3.5d. Applying the conjugate-beam method to Fig. 4.3.5d,

$$\delta_{DB} = \text{bending moment at } D \text{ of conjugate beam}$$

$$= \frac{5.6}{EI}(3) - \frac{1}{2}\left(\frac{1.2}{EI}\right)(3)(1) = \frac{15}{EI}$$

$$\delta_{BB} = \text{bending moment at } B \text{ of conjugate beam}$$

$$= \frac{6.4}{EI}(4) - \frac{1}{2}\left(\frac{2.4}{EI}\right)(4)\left(\frac{4}{3}\right) = \frac{19.2}{EI}$$

By the law of reciprocal deflections,

$$\Delta_{B1} = 120\delta_{BD} = 120\delta_{DB} = 120\left(\frac{15}{EI}\right) = \frac{1800}{EI}$$

Also,

$$\Delta_{B2} = R_2 \delta_{BB} = R_2 \left(\frac{19.2}{EI}\right)$$

Equating Δ_{B1} to Δ_{B2} and solving for R_2,

$$R_2 = 93.75 \text{ kN}$$

Returning to the given beam shown again in Fig. 4.3.5e,

$R_1 = 84 \text{ (due to load)} - 0.4 R_2 = 84 - 0.4(93.75) = 46.50 \text{ kN}$

$R_3 = 36 \text{ (due to load)} - 0.6 R_2 = 36 - 0.6(93.75)$

$= -20.25 \text{ kN}$ or $20.25 \text{ kN downward}$

The shear and moment diagrams are then drawn as shown in Fig. 4.3.5e.

ANALYSIS OF STATICALLY INDETERMINATE BEAMS 111

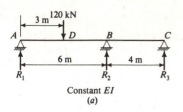

Constant EI
(a)

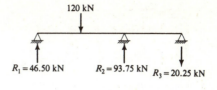

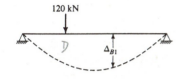

(b) Basic beam under applied load

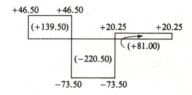

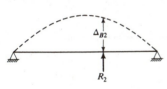

(c) Basic beam under redundant R_2

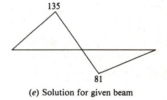

(e) Solution for given beam

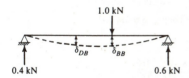

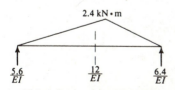

(d) Basic beam under unit load at B

Figure 4.3.5 Beam of Example 4.3.1.

ALTERNATE SOLUTION   If the reaction at C is chosen as the redundant, the basic beam is the overhanging beam ABC supported at A and B. The condition of geometry is that the deflection at the free end C of the basic beam due to the combined action of the applied load and the redundant R_3 (which is assumed downward to begin with, because Δ_{C1} in Fig. 4.3.6b

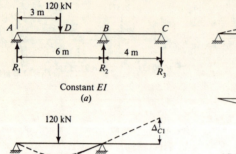
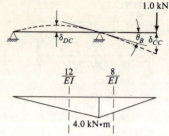
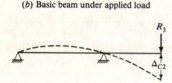
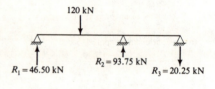

(b) Basic beam under applied load

(d) Basic beam under unit load at C

(c) Basic beam under redundant R_3

(e) Solution for given beam

Figure 4.3.6 Alternate solution for Example 4.3.1.

is upward) is zero; or, as shown in Fig. 4.3.6b and c,

$$\Delta_{C1} = \Delta_{C2}$$

Again, the problem shown in Fig. 4.3.6d is solved first to obtain δ_{DC} and δ_{CC}.

δ_{DC} = bending moment at D of conjugate beam AB

$$= \frac{4}{EI}(3) - \frac{1}{2}\left(\frac{2.0}{EI}\right)(3)(1) = \frac{9}{EI}$$

$$\delta_{CC} = 4\theta_B + \frac{8}{EI}\left(\frac{8}{3}\right) = 4\left(\frac{8}{EI}\right) + \frac{8}{EI}\left(\frac{8}{3}\right) = \frac{160}{3EI}$$

By the law of reciprocal deflections,

$$\Delta_{C1} = 120\delta_{CD} = 120\delta_{DC} = 120\left(\frac{9}{EI}\right) = \frac{1080}{EI}$$

Also,

$$\Delta_{C2} = R_3\delta_{CC} = R_3\left(\frac{160}{3EI}\right)$$

Equating Δ_{C1} to Δ_{C2} and solving for R_3,

$$R_3 = 20.25 \text{ kN downward as assumed}$$

Returning to the given beam shown again in Fig. 4.3.6e,

$$R_1 = +60 \text{ (due to load)} - \frac{4R_3}{6} = +60 - \frac{4(20.25)}{6} = 46.50 \text{ kN upward}$$

$$R_2 = +60 \text{ (due to load)} + \frac{10R_3}{6} = +60 + \frac{10(20.25)}{6} = 93.75 \text{ kN upward}$$

ANALYSIS OF STATICALLY INDETERMINATE BEAMS

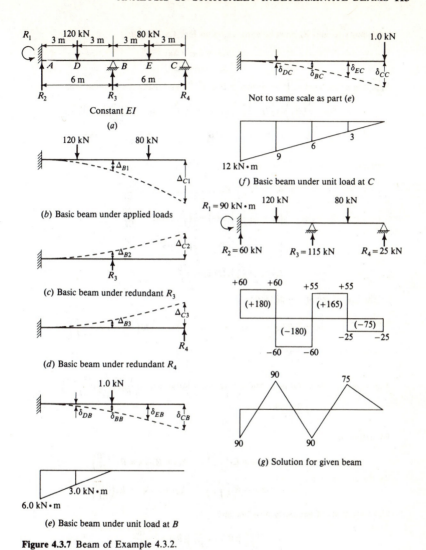

Figure 4.3.7 Beam of Example 4.3.2.

Example 4.3.2 Analyze by the force method and draw shear and moment diagrams for the beam of Fig. 4.3.7a.

SOLUTION The given beam has four reactions R_1 to R_4; it is statically indeterminate to the second degree. By removing the supports at B and C the cantilever beam fixed at A and free at C is chosen as the basic beam. The conditions of geometry are that the deflections at both B and C of the basic beam under the combined action of the applied loads and the redundants R_3 and R_4 must be zero; or, from Fig. 4.3.7b to d,

$$\Delta_{B1} = \Delta_{B2} + \Delta_{B3}$$
$$\Delta_{C1} = \Delta_{C2} + \Delta_{C3}$$

Applying a unit load at B as shown in Fig. 4.3.7e and using the moment-area method,

$$\delta_{DB} = \frac{1}{2}\left(\frac{6}{EI}\right)(3)(2) + \frac{1}{2}\left(\frac{3}{EI}\right)(3)(1) = \frac{22.5}{EI} \qquad \delta_{BB} = \frac{1}{2}\left(\frac{6}{EI}\right)(6)(4) = \frac{72}{EI}$$

$$\delta_{EB} = \frac{1}{2}\left(\frac{6}{EI}\right)(6)(7) = \frac{126}{EI} \qquad \delta_{CB} = \frac{1}{2}\left(\frac{6}{EI}\right)(6)(10) = \frac{180}{EI}$$

Applying a unit load at C as shown in Fig. 4.3.7f and using the moment-area method,

$$\delta_{DC} = \frac{1}{2}\left(\frac{12}{EI}\right)(3)(2) + \frac{1}{2}\left(\frac{9}{EI}\right)(3)(1) = \frac{49.5}{EI}$$

$$\delta_{BC} = \frac{1}{2}\left(\frac{12}{EI}\right)(6)(4) + \frac{1}{2}\left(\frac{6}{EI}\right)(6)(2) = \frac{180}{EI}$$

$$\delta_{EC} = \frac{1}{2}\left(\frac{12}{EI}\right)(9)(6) + \frac{1}{2}\left(\frac{3}{EI}\right)(9)(3) = \frac{364.5}{EI}$$

$$\delta_{CC} = \frac{1}{2}\left(\frac{12}{EI}\right)(12)(8) = \frac{576}{EI}$$

By the law of reciprocal deflections,

$$\delta_{B1} = 120\delta_{BD} + 80\delta_{BE} = 120\delta_{DB} + 80\delta_{EB} = 120\left(\frac{22.5}{EI}\right) + 80\left(\frac{126}{EI}\right)$$

$$= \frac{12,780}{EI}$$

$$\delta_{C1} = 120\delta_{CD} + 80\delta_{CE} = 120\delta_{DC} + 80\delta_{EC} = 120\left(\frac{49.5}{EI}\right) + 80\left(\frac{364.5}{EI}\right)$$

$$= \frac{35,100}{EI}$$

By definition,

$$\Delta_{B2} = R_3\delta_{BB} = R_3\left(\frac{72}{EI}\right) \qquad \Delta_{C2} = R_3\delta_{CB} = R_3\left(\frac{180}{EI}\right)$$

$$\Delta_{B3} = R_4\delta_{BC} = R_4\left(\frac{180}{EI}\right) \qquad \Delta_{C3} = R_4\delta_{CC} = R_4\left(\frac{576}{EI}\right)$$

The conditions of geometry now become

$$\left(\frac{72}{EI}\right)(R_3) + \left(\frac{180}{EI}\right)(R_4) = \frac{12,780}{EI}$$

$$\left(\frac{180}{EI}\right)(R_3) + \left(\frac{576}{EI}\right)(R_4) = \frac{35,100}{EI}$$

Solving the above two equations for R_3 and R_4,

$$R_3 = 115 \text{ kN upward} \qquad R_4 = 25 \text{ kN upward}$$

Returning to the given beam with the known values of R_3 and R_4, and solving for R_1 and R_2 by the two equations of statics,

$$R_1 = 90 \text{ kN·m counterclockwise} \qquad R_2 = 60 \text{ kN upward}$$

The shear and moment diagrams are shown in Fig. 4.3.7g.

ANALYSIS OF STATICALLY INDETERMINATE BEAMS 115

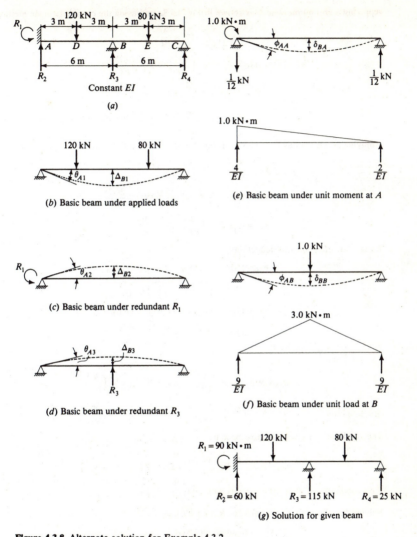

Figure 4.3.8 Alternate solution for Example 4.3.2.

ALTERNATE SOLUTION Suppose that the simple beam AC is chosen as the basic beam, as shown in Fig. 4.3.8. The redundants are R_1 and R_3. The conditions of geometry are that the slope at A and the deflection at B of the basic beam under the combined action of the applied loads and the redundants R_1 and R_3 should both be zero; or, from Fig. 4.3.8b to d,

$$\theta_{A1} = \theta_{A2} + \theta_{A3}$$
$$\Delta_{B1} = \Delta_{B2} + \Delta_{B3}$$

116 INTERMEDIATE STRUCTURAL ANALYSIS

Applying a unit moment at A as shown in Fig. 4.3.8e and using the conjugate-beam method,

$$\phi_{AA} = \frac{2}{3}\frac{1}{2}\left(\frac{1.0}{EI}\right)(12) = \frac{4}{EI}$$

$$\delta_{DA} = \frac{2}{EI}(9) - \frac{1}{2}\left(\frac{0.75}{EI}\right)(9)(3) = \frac{7.875}{EI}$$

$$\delta_{BA} = \frac{2}{EI}(6) - \frac{1}{2}\left(\frac{0.5}{EI}\right)(6)(2) = \frac{9}{EI}$$

$$\delta_{EA} = \frac{2}{EI}(3) - \frac{1}{2}\left(\frac{0.25}{EI}\right)(3)(1) = \frac{5.625}{EI}$$

Applying a unit load at B as shown in Fig. 4.3.8f and using the conjugate-beam method,

$$\phi_{AB} = \frac{9}{EI}$$

$$\delta_{DB} = \delta_{EB} = \frac{9}{EI}(3) - \frac{1}{2}\left(\frac{15}{EI}\right)(3)(1) = \frac{24.75}{EI}$$

$$\delta_{BB} = \frac{9}{EI}(6) - \frac{1}{2}\left(\frac{3.0}{EI}\right)(6)(2) = \frac{36}{EI}$$

By the law of reciprocal deflections,

$$\theta_{A1} = 120\phi_{AD} + 80\phi_{AE} = 120\delta_{DA} + 80\delta_{EA}$$

$$= 120\left(\frac{7.875}{EI}\right) + 80\left(\frac{5.625}{EI}\right) = \frac{1395}{EI}$$

$$\Delta_{B1} = 120\delta_{BD} + 80\delta_{BE} = 120\delta_{DB} + 80\delta_{EB}$$

$$= 120\left(\frac{24.75}{EI}\right) + 80\left(\frac{24.75}{EI}\right) = \frac{4950}{EI}$$

By definition,

$$\theta_{A2} = R_1\phi_{AA} = R_1\left(\frac{4}{EI}\right) \quad \Delta_{B2} = R_1\delta_{BA} = R_1\left(\frac{9}{EI}\right)$$

$$\theta_{A3} = R_3\phi_{AB} = R_3\left(\frac{9}{EI}\right) \quad \Delta_{B3} = R_3\delta_{BB} = R_3\left(\frac{36}{EI}\right)$$

The conditions of geometry now become

$$R_1\left(\frac{4}{EI}\right) + R_3\left(\frac{9}{EI}\right) = \frac{1395}{EI}$$

$$R_1\left(\frac{9}{EI}\right) + R_3\left(\frac{36}{EI}\right) = \frac{4950}{EI}$$

Solving the above two equations for R_1 and R_3,

$R_1 = 90$ kN·m counterclockwise $\quad R_3 = 115$ kN upward

Returning to the given beam with the known values of R_1 and R_3 and solving for R_2 and R_4 by the two equations of statics,

$R_2 = 60$ kN upward $\quad R_4 = 25$ kN upward

4.4 Theorem of Least Work

The *theorem of least work* applies to any statically indeterminate structure, whether it be a beam, a rigid frame, or a truss. It may be stated as follows:

ANALYSIS OF STATICALLY INDETERMINATE BEAMS   117

For any statically indeterminate structure, the redundants should be such as to make the total internal energy within the structure a minimum.

The validity of the theorem of least work comes directly from Castigliano's second theorem, which was presented in Sec. 2.8. The previous derivation makes use of a beam as a specific instance; actually, both Castigliano's theorems apply to all structures.

Continuing the use of simple instances to explain a general theorem, consider the beam of Fig. 4.4.1a, which is statically indeterminate to the first degree. Obviously the simple determinate beam of Fig. 4.4.1b, subjected to the combined action of the applied loads and the redundant reacting force R_2, is *equivalent* to the given beam. The condition of geometry with which R_2 can be determined is that the deflection at B of the equivalent beam should be zero. This deflection, by Castigliano's second theorem or the partial-derivative method, is $\Delta_B = \partial U/\partial R_2$; so the condition for determining R_2 is $\partial U/\partial R_2 = 0$, or R_2 is such as to make the total internal energy a minimum. In other words, when nature has its free choice, it will always tend to conserve energy. The same idea can be extended to the beam of Fig. 4.4.2a, of which the equivalent beam is shown in Fig. 4.4.2b. The conditions of geometry with which R_1 and R_3 can be determined are $\theta_A = \partial U/\partial R_1 = 0$ and $\Delta_B = \partial U/\partial R_3 = 0$, or the redundants R_1 and R_3 are such as to make the total internal energy a minimum.

The application of the theorem of least work to the analysis of statically indeterminate beams will be illustrated by two examples. It will be found that

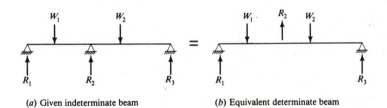

(a) Given indeterminate beam   (b) Equivalent determinate beam

Figure 4.4.1 Theorem of least work, one redundant.

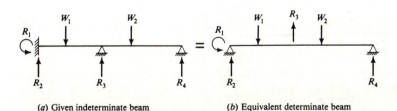

(a) Given indeterminate beam   (b) Equivalent determinate beam

Figure 4.4.2 Theorem of least work, two redundants.

118 INTERMEDIATE STRUCTURAL ANALYSIS

the solution procedure is substantially identical to that in the method of consistent deformation, particularly if the partial-derivative method or the unit-load method, instead of the moment-area/conjugate-beam method, is used to compute the deflections and slopes of the basic beam. Thus in reality, the method of least work is a variation of the force method, wherein the redundant reacting forces or moments are determined at the very beginning.

Example 4.4.1 Solve Example 4.2.1 by the method of least work.

SOLUTION The given beam is shown again in Fig. 4.4.3a. The cantilever beam shown in Fig. 4.4.3b is chosen as the equivalent determinate beam, which is subjected to both W and R_3 as the loads on it. The internal energy within the beam is, from Eq. (2.9.2),

$$U = \int \frac{M^2\,dx}{2EI} = \int_0^{L/2} \frac{(R_3 x)^2\,dx}{2EI} + \int_0^{L/2} \frac{[R_3(L/2+x) - Wx]^2\,dx}{2EI}$$

The first integral covers the portion BC using B as origin and the second integral covers the

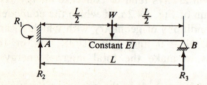

(a) Given indeterminate beam

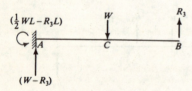

(b) Cantilever beam as equivalent determinate beam

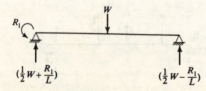

(c) Simple beam as equivalent determinate beam

Figure 4.4.3 Beam of Example 4.4.1.

ANALYSIS OF STATICALLY INDETERMINATE BEAMS **119**

portion CA using C as origin. Applying the theorem of least work,

$$\frac{\partial U}{\partial R_3} = \int_0^{L/2} \frac{(R_3x)(x)\,dx}{EI} + \int_0^{L/2} \frac{[R_3(L/2+x) - Wx](L/2+x)\,dx}{EI}$$

$$= \left[\frac{R_3 x^3}{3EI}\right]_0^{L/2} + \left[\frac{R_3(L/2+x)^3}{3EI} - \frac{WLx^2}{4EI} - \frac{Wx^3}{3EI}\right]_0^{L/2}$$

$$= \frac{R_3 L^3}{3EI} - \frac{5WL^3}{48EI} = 0$$

Solving the above equation, which is identical to the consistent-deformation equation in Example 4.2.1,

$$R_3 = \frac{5}{16}W \quad \text{upward}$$

Returning to the original beam with R_3 now known and using the two equations of statics,

$R_1 = \frac{1}{2}WL - R_3 L = \frac{1}{2}WL - \frac{5}{16}WL = \frac{3}{16}WL \quad \text{counterclockwise}$

$R_2 = W - R_3 = W - \frac{5}{16}W = \frac{11}{16}W \quad \text{upward}$

ALTERNATE SOLUTION The simple beam shown in Fig. 4.4.3c is chosen as the equivalent determinate beam, which is subjected to both W and R_1 as the loads on it. The internal energy within the beam is

$$U = \int_0^{L/2} \frac{[-R_1 + (\frac{1}{2}W + R_1/L)x]^2\,dx}{2EI} + \int_0^{L/2} \frac{[(\frac{1}{2}W - R_1/L)x]^2\,dx}{2EI}$$

The first integral covers the portion AC using A as origin and the second integral covers the portion BC using B as origin. Applying the theorem of least work,

$$\frac{\partial U}{\partial R_1} = \int_0^{L/2} \frac{[-R_1 + (\frac{1}{2}W + R_1/L)x](-1 + x/L)\,dx}{EI} + \int_0^{L/2} \frac{[(\frac{1}{2}W - R_1/L)x](-x/L)\,dx}{EI}$$

$$= \frac{R_1 L}{3EI} - \frac{WL^3}{16EI} = 0$$

Solving the above equation, which is identical to the consistent-deformation equation in Example 4.2.1,

$$R_1 = \frac{3}{16}WL \quad \text{counterclockwise}$$

Returning to the original beam with R_1 now known and using the two equations of statics,

$$R_2 = \frac{1}{2}W + \frac{R_1}{L} = \frac{1}{2}W + \frac{3}{16}W = \frac{11}{16}W \quad \text{upward}$$

$$R_3 = \frac{1}{2}W - \frac{R_1}{L} = \frac{1}{2}W - \frac{3}{16}W = \frac{5}{16}W \quad \text{upward}$$

Example 4.4.2 Solve Example 4.3.2 by the method of least work.

SOLUTION The given beam is shown again in Fig. 4.4.4a. The cantilever beam shown in Fig. 4.4.4b is chosen as the equivalent determinate beam which is subjected to the two known forces and the redundant reacting forces R_3 and R_4. The internal energy within the beam is, from Eq. (2.9.2),

120 INTERMEDIATE STRUCTURAL ANALYSIS

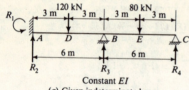

(a) Given indeterminate beam

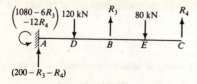

(b) Cantilever beam as equivalent determinate beam

Figure 4.4.4 Beam of Example 4.4.2.

$$U = \int \frac{M^2 \, dx}{2EI}$$

$$= \int_0^3 \frac{(R_4 x)^2 \, dx}{2EI} + \int_0^3 \frac{[R_4(3+x) - 80x]^2 \, dx}{2EI}$$

$$+ \int_0^3 \frac{[R_4(6+x) - 80(3+x) + R_3 x]^2 \, dx}{2EI}$$

$$+ \int_0^3 \frac{[R_4(9+x) - 80(6+x) + R_3(3+x) - 120x]^2 \, dx}{2EI}$$

The four integrals above cover the segments CE, EB, BD, and DA, with C, E, B, and D as the origin of reference, respectively. Applying the theorem of least work,

$$\frac{\partial U}{\partial R_3} = \int_0^3 \frac{(R_4 x)(0) \, dx}{EI} + \int_0^3 \frac{[R_4(3+x) - 80x](0) \, dx}{EI}$$

$$+ \int_0^3 \frac{[R_4(6+x) - 80(3+x) + R_3 x](x) \, dx}{EI}$$

$$+ \int_0^3 \frac{[R_4(9+x) - 80(6+x) + R_3(3+x) - 120x](3+x) \, dx}{EI}$$

$$= \left(\frac{72}{EI}\right)(R_3) + \left(\frac{180}{EI}\right)(R_4) - \frac{12{,}780}{EI} = 0$$

$$\frac{\partial U}{\partial R_4} = \int_0^3 \frac{(R_4 x)(x) \, dx}{EI} + \int_0^3 \frac{[R_4(3+x) - 80x](3+x) \, dx}{EI}$$

$$+ \int_0^3 \frac{[R_4(6+x) - 80(3+x) + R_3 x](6+x) \, dx}{EI}$$

$$+ \int_0^3 \frac{[R_4(9+x) - 80(6+x) + R_3(3+x) - 120x](9+x) \, dx}{EI}$$

$$= \left(\frac{180}{EI}\right)(R_3) + \left(\frac{576}{EI}\right)(R_4) - \frac{35{,}100}{EI} = 0$$

Note that the coefficient of R_4 in $\partial W/\partial R_3 = 0$ is $180/EI$, and the coefficient of R_3 in $\partial W/\partial R_4 = 0$ is also $180/EI$. This is a natural consequence of the law of reciprocal deflections, and it serves well here as a check on the numerical work. Note also that these two equations are identical to those obtained in Example 4.3.2. Solving these two equations,

$$R_3 = 115 \text{ kN} \qquad R_4 = 25 \text{ kN}$$

Returning to the original beam with the above known values of R_3 and R_4 and using the two equations of statics,

$$R_1 = 90 \text{ kN·m} \qquad R_2 = 60 \text{ kN}$$

ALTERNATE SOLUTION An alternate solution may be performed by using the simple beam AC as the equivalent determinate beam, which is subjected to the two known forces and the redundants R_1 and R_3. Then R_1 and R_3 are first obtained from the conditions $\partial U/\partial R_1 = 0$ and $\partial U/\partial R_3 = 0$. The details are not shown.

4.5 Induced Reactions on Statically Indeterminate Beams due to Yielding of Supports

Often in structural design of bridges or buildings, it is necessary to provide for probable unequal vertical settlements of the supports, and in the case of fixed supports, probable rotations of the footings due to geological causes. For statically determinate beams, or indeed for all other types of statically determinate structures as well, yielding of supports, whether it be vertical settlement or rotational slip, will not produce any change in the shape of the structure at all, but the entire structure may be displaced as a rigid body. Three cases showing the effects of yielding of supports of statically determinate beams are shown in Fig. 4.5.1.

For statically indeterminate beams or statically indeterminate structures in general, there will be change in the shape of the structure due to yielding of supports, and, of course, there is dislocation of the structure to adjust itself to the support settlement or rotation. Associated with the change in the shape of the structure, there are induced reactions which form a balanced force system by themselves since there are no other external forces acting on the structure.

If more than one vertical settlement or rotational slip occurs, it is more convenient to determine the induced reactions due to one yielding at a time and then combine the effects due to separate causes. The procedure for determining such induced reactions is first to choose a basic determinate beam by removing the excess restraints *including* the yielding supports and replacing them by unknown redundant reactions. The condition of consistent deformation at the yielding support will be that the deflection or slope there should be equal to the predicted amount of yielding, while the conditions at all other redundants, if any,

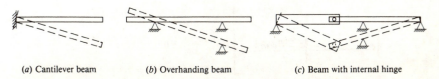

(a) Cantilever beam   (b) Overhanding beam   (c) Beam with internal hinge

Figure 4.5.1 Yielding of supports of statically determinate beams.

requiring zero deflection or slope remain the same, as treated in the earlier part of this chapter. In fact, the reason for choosing the word *consistent* over the word *zero* in the term *conditions of consistent deformation* is exactly explained here. After the redundant reactions are solved, they can be placed to act on the equivalent basic beam and the remaining two reactions obtained by using the two equations of statics.

Example 4.5.1 By the force method determine all induced reactions due to a vertical settlement of 4.5 mm at the support B of the beam shown in Fig. 4.5.2a.

SOLUTION As shown in Fig. 4.5.2b and c, the cantilever beam AC is chosen as the basic determinate beam, on which only the redundant reacting forces R_3 and R_4 are acting. The values of δ_{BB}, $\delta_{BC} = \delta_{CB}$, and δ_{CC} are taken from Example 4.3.2; they are

$$\delta_{BB} = \frac{72 \text{ kN} \cdot \text{m}^3}{EI} = \frac{72(1000)}{(200 \times 10^6)(160 \times 10^{-6})} = 2.25 \text{ mm}$$

$$\delta_{BC} = \delta_{CB} = \frac{180 \text{ kN} \cdot \text{m}^3}{EI} = \frac{180(1000)}{(200 \times 10^6)(160 \times 10^{-6})} = 5.625 \text{ mm}$$

$$\delta_{CC} = \frac{576 \text{ kN} \cdot \text{m}^3}{EI} = \frac{576(1000)}{(200 \times 10^6)(160 \times 10^{-6})} = 18 \text{ mm}$$

The conditions of consistent deformation are, from Fig. 4.5.2b and c,

$$R_3 \delta_{BB} - R_4 \delta_{BC} = 4.5 \text{ mm}$$

$$R_3 \delta_{CB} - R_4 \delta_{CC} = 0$$

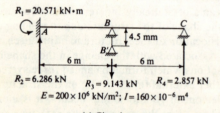

(a) Given beam

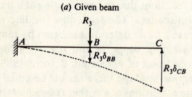

(b) Basic beam under redundant R_3

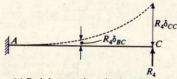

(c) Basic beam under redundant R_4

Figure 4.5.2 Beam of Example 4.5.1.

ANALYSIS OF STATICALLY INDETERMINATE BEAMS 123

Substituting the values of δ_{BB}, $\delta_{BC} = \delta_{CB}$, and δ_{CC},

$$2.25R_3 - 5.625R_4 = 4.5$$
$$5.625R_3 - 18R_4 = 0$$

Solving,

$$R_3 = 9.143 \text{ kN downward as assumed}$$
$$R_4 = 2.857 \text{ kN upward as assumed}$$

The two remaining induced reactions, R_1 and R_2, may be obtained by using the two equations of statics; thus,

$$R_1 = 6R_3 - 12R_4 = 6(9.143) - 12(2.857) = 20.574 \text{ kN·m counterclockwise}$$
$$R_2 = R_3 - R_4 = 9.143 - 2.857 = 6.286 \text{ kN upward}$$

ALTERNATE SOLUTION As shown in Fig. 4.5.3b and c, the simple beam AC may be chosen as the basic determinate beam, on which only the redundant reacting forces R_1 and R_3 are acting. The values of ϕ_{AA}, δ_{BA}, ϕ_{AB}, and δ_{BB} are taken from Example 4.3.2; they are

$$\phi_{AA} = \frac{4 \text{ kN·m}^2}{EI} = \frac{4}{(200 \times 10^6)(160 \times 10^{-6})} = 0.125 \times 10^{-3} \text{ rad}$$

$$\delta_{BA} = \frac{9 \text{ kN·m}^3}{EI} = \frac{9}{(200 \times 10^6)(160 \times 10^{-6})} = 0.28125 \times 10^{-3} \text{ m}$$

$$\phi_{AB} = \frac{9 \text{ kN·m}^2}{EI} = \frac{9}{(200 \times 10^6)(160 \times 10^{-6})} = 0.28125 \times 10^{-3} \text{ rad}$$

$$\delta_{BB} = \frac{36 \text{ kN·m}^3}{EI} = \frac{36}{(200 \times 10^6)(160 \times 10^{-6})} = 1.125 \times 10^{-3} \text{ m}$$

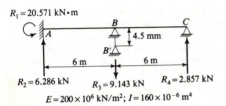

(a) Given beam

(b) Basic beam under redundant R_2

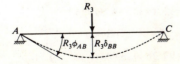

(c) Basic beam under redundant R_3

Figure 4.5.3 Alternate solution for Example 4.5.1.

Note that $\delta_{BA} = \phi_{AB} = 0.28125 \times 10^{-3}$; the meaning is that a 1.0-kN·m moment *counterclockwise* applied at A causes an *upward* deflection of $\delta_{BA} = 0.28125 \times 10^{-3}$ m at B, and a 1.0-kN *downward* load applied at B causes a *clockwise* rotation of $\phi_{AB} = 0.28125 \times 10^{-3}$ rad at A. The conditions of consistent deformation are, from Fig. 4.5.3b and c,

$$R_3 \delta_{BB} - R_1 \delta_{BA} = 0.0045 \text{ m}$$
$$R_3 \phi_{AB} - R_1 \phi_{AA} = 0$$

Substituting the values of δ_{BB}, $\delta_{BA} = \phi_{AB}$, and ϕ_{AA},

$$1.125 R_3 - 0.28125 R_1 = 4.5$$
$$0.28125 R_3 - 0.125 R_1 = 0$$

Solving,

$$R_1 = 20.571 \text{ kN counterclockwise as assumed}$$
$$R_3 = 9.143 \text{ kN downward as assumed}$$

The two remaining induced reactions, R_2 and R_4, may be obtained by using the two equations of statics; thus,

$$R_2 = \frac{1}{2}R_3 + \frac{R_1}{12} = \frac{1}{2}(9.143) + \frac{20.571}{12} = 6.286 \text{ kN upward}$$

$$R_4 = \frac{1}{2}R_3 - \frac{R_1}{12} = \frac{1}{2}(9.143) - \frac{20.571}{12} = 2.857 \text{ kN upward}$$

Example 4.5.2 In Example 4.5.1, the induced reactions due to a 4.5-mm vertical settlement at support B of the beam have been determined by the force method, as shown in Fig. 4.5.4a. Draw shear and moment diagrams, and treating each span separately, determine the slopes at A and B for span AB and the slopes at B and C for span BC.

SOLUTION The emphasis in this example is to show that the reaction to the conjugate beam is the angle between the *chord*, which is the straight line joining the two points on the elastic curve corresponding to the ends of the conjugate beam, and the tangent to the elastic curve. For this problem the moment diagram on span AB is separated into two parts as shown in Fig. 4.5.4d, of which the M/EI areas are 1.607×10^{-3} rad and 1.929×10^{-3} rad, respectively. For convenience, the elastic curve related to the chord, due to each M/EI area, is also shown in Fig. 4.5.4d. Using conjugate-beam theorem 1,

$$\phi_{AB} = 1.286 \times 10^{-3} - 0.536 \times 10^{-3} = 0.750 \times 10^{-3} \text{ counterclockwise}$$
$$\phi_{BA} = 1.071 \times 10^{-3} - 0.643 \times 10^{-3} = 0.428 \times 10^{-3} \text{ counterclockwise}$$

From the geometry shown in Fig. 4.5.4d,

$$\theta_A = \frac{BB'}{AB} - \phi_{AB} = \frac{4.5}{6000} - 0.750 \times 10^{-3} = 0 \quad \text{(Check)}$$

$$\theta_B = \frac{4.5}{6000} - \phi_{BA} = 0.750 \times 10^{-3} - 0.428 \times 10^{-3} = 0.322 \times 10^{-3} \text{ clockwise}$$

Referring to span BC shown in Fig. 4.5.4e,

$$\theta_B = \phi_{BC} - \frac{4.5}{6000} = 1.071 \times 10^{-3} - 0.750 \times 10^{-3}$$
$$= 0.321 \times 10^{-3} \text{ clockwise} \quad \text{(Check)}$$

$$\theta_C = \frac{4.5}{6000} + \phi_{CB} = 0.750 \times 10^{-3} + 0.536 \times 10^{-3}$$
$$= 1.286 \times 10^{-3} \text{ counterclockwise}$$

ANALYSIS OF STATICALLY INDETERMINATE BEAMS **125**

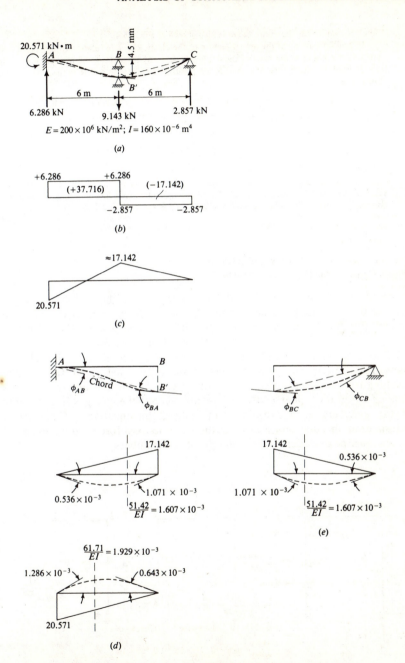

Figure 4.5.4 Equilibrium and continuity checks for beam in Example 4.5.1.

The elastic curve for the entire beam is also shown in Fig. 4.5.4a.
Instead of looking at the actual picture of the elastic curve in relation to the chord and obtaining the slopes at the ends of the conjugate beam by inspection, the formula $\theta = R + \phi$ may be used, in which case θ, ϕ, and R are positive if clockwise and R is the slope of the chord. Then,

$$\theta_A = R_{AB} + \phi_{AB} = (+0.750 \times 10^{-3}) + (+0.536 \times 10^{-3} - 1.286 \times 10^{-3}) = 0$$

$$\theta_B = R_{AB} + \phi_{BA} = (+0.750 \times 10^{-3}) + (-1.607 \times 10^{-3} + 0.643 \times 10^{-3})$$
$$= +0.322 \times 10^{-3}$$

$$\theta_B = R_{BC} + \phi_{BC} = (-0.750 \times 10^{-3}) + (+1.071 \times 10^{-3}) = +0.321 \times 10^{-3}$$

$$\theta_C = R_{BC} + \phi_{CB} = (-0.750 \times 10^{-3}) + (-0.536 \times 10^{-3}) = -1.286 \times 10^{-3}$$

The "physical-inspection" approach is more instructive at this time, but the "sign-convention" approach will be more adaptable to an automatic procedure that may be used in a computer program. This dual approach will be used extensively in Chap. 11.

4.6 Analysis of Statically Indeterminate Rigid Frames by the Force Method

As discussed earlier in Sec. 2.1., the axial deformation due to axial forces in the members of a rigid frame is usually so small in comparison with the deformation due to bending moment that it is to be neglected. Consideration of axial deformation will be postponed until Chap. 16. For the present, the conditions of consistent deformation used in the force method of analyzing statically indeterminate rigid frames relate to the deformed rigid frame resulting from bending moment only. Although it has been found that there is little effect on the values of the redundants even if the changes in the length of the members were considered in making the equations of consistent deformation, this does not mean that there are no axial forces, and these axial forces must be considered in the *design* of the members.

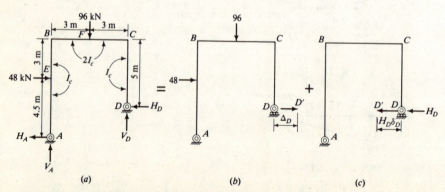

Figure 4.6.1 Compatibility condition for redundant H_D in Example 4.6.1.

ANALYSIS OF STATICALLY INDETERMINATE BEAMS 127

Example 4.6.1 Analyze by the force method the rigid frame with two hinged supports shown in Fig. 4.6.1a.

SOLUTION For one-story rigid frames the degree of indeterminacy is equal to the number of unknown reactions minus 3. In this problem there are four reactions to be determined; therefore the given rigid frame is statically indeterminate to the first degree. The logical basic structure is a rigid frame with one hinged support and one support on horizontal rollers. When H_D is chosen as the redundant, the condition of consistent deformation is that the horizontal deflection of the roller support D of the basic rigid frame, when it is subjected to the combined action of the applied loads and the redundant H_D, must be zero; or as shown in Fig. 4.6.1,

$$\Delta_D = H_D \delta_D$$

where δ_D is the deflection to the left if H_D is equal to 1.0 kN. The condition of consistent deformation has often been called the "compatibility condition," since the horizontal deflection at the roller support D of the basic structure must be *compatible* with the zero horizontal deflection requirement at D of the given indeterminate structure. Note the title *compatibility condition* for Fig. 4.6.1.

Either the unit-load method or the moment-area/conjugate-beam method can be used for determining Δ_D and δ_D in Fig. 4.6.1; the solution by the unit-load method is shown here. By referring to the free-body diagrams in Fig. 4.6.2, expressions for M and m in various parts of the rigid frame may be observed. Thus,

$$EI_c \Delta_D = \int_0^{4.5} (48x)(1.0x)\,dx + \int_0^3 (216)(4.5+x)\,dx$$
$$\text{part } AE, \text{ origin at } A \qquad \text{part } EB, \text{ origin at } E$$

$$+ \tfrac{1}{2}\int_0^3 (216+12x)(7.5-\tfrac{5}{12}x)\,dx + \tfrac{1}{2}\int_0^3 (84x)(5+\tfrac{5}{12}x)\,dx$$
$$\text{part } BF, \text{ origin at } B \qquad \text{part } CF, \text{ origin at } C$$

$$+ \int_0^5 (0)(-1.0x)\,dx$$
$$\text{part } DC, \text{ origin at } D$$

$$= 1458 + 3888 + 2407.5 + 1102.5 + 0 = 8856 \text{ kN} \cdot \text{m}^3$$

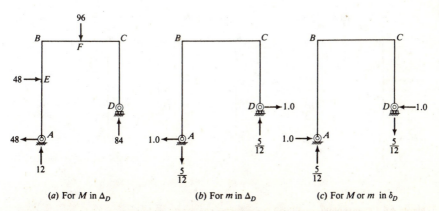

(a) For M in Δ_D (b) For m in Δ_D (c) For M or m in δ_D

Figure 4.6.2 Free-body diagrams for M and m in Δ_D and δ_D.

128 INTERMEDIATE STRUCTURAL ANALYSIS

$$EI_c\delta_D = \int_0^{7.5}(-1.0x)^2\,dx + \tfrac{1}{2}\int_0^6(-7.5+\tfrac{5}{12}x)^2\,dx$$
$$\text{part } AB\text{, origin at } A \qquad \text{part } BC\text{, origin at } C$$
$$+\int_0^5(+1.0x)^2\,dx$$
$$\text{part } DC\text{, origin at } D$$
$$= 140.625 + 18.75 + 41.667 = 301.042 \text{ kN·m}^3$$

From the compatibility condition,

$$H_D = \frac{EI_c\Delta_D}{EI_c\delta_D} = \frac{8856}{301.042} = 29.42 \text{ kN to left as assumed}$$

Returning to the given rigid frame of Fig. 4.6.1a and using the three equations of statics,

$\Sigma F_x = 0$: $\quad H_A = 48 - H_D = 48 - 29.42 = 18.58$ kN to left

$\Sigma M_D = 0$: $\quad V_A = \dfrac{96(3) - 48(2) - 18.58(2.5)}{6} = 24.26$ kN upward

$\Sigma M_A = 0$: $\quad V_D = \dfrac{96(3) + 48(4.5) - 29.42(2.5)}{6} = 71.74$ kN upward

Check by $\Sigma F_y = 0$:

$$V_A + V_D \stackrel{?}{=} 96$$
$$24.26 + 71.74 = 96 \quad \text{(OK)}$$

ALTERNATE SOLUTION   When V_A is used as the redundant, the basic determinate structure is hinged at D and supported on vertical rollers (rollers which can roll in a vertical plane) at A. As shown in Fig. 4.6.3, the compatibility condition is that the downward deflection Δ_A in Fig. 4.6.3b is equal to the upward deflection $V_A\delta_A$ in Fig. 4.6.3c wherein δ_A is the upward deflection for V_A equal to 1.0 kN; or

$$\Delta_A = V_A\delta_A$$

The use of the moment-area/conjugate-beam method to determine Δ_D and δ_D would be an interesting exercise, in which the primary unknown has to be the slope at D so that the horizontal deflection at A is zero. The unit-load method using the M and m expressions

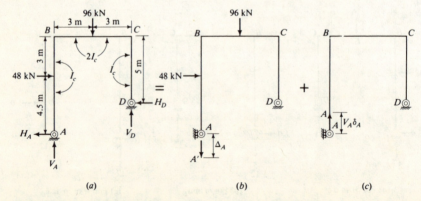

Figure 4.6.3 Compatibility condition for redundant V_A in Example 4.6.1.

ANALYSIS OF STATICALLY INDETERMINATE BEAMS 129

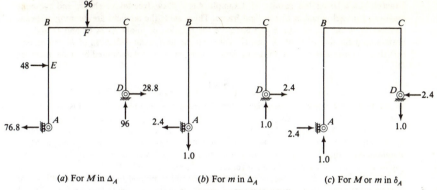

(a) For M in Δ_A (b) For m in Δ_A (c) For M or m in δ_A

Figure 4.6.4 Free-body diagrams for M and m in Δ_A and δ_A.

obtained from the free-body diagrams in Fig. 4.6.4 may be more straightforward in this case; thus,

$$EI_c\Delta_A = \int_0^{4.5} (76.8x)(2.4x)\,dx + \int_{4.5}^{7.5} [(76.8x - 48)(x - 4.5)](2.4x)\,dx$$
$$\text{part } AE, \text{ origin at } A \qquad \text{part } EB, \text{ origin at } A$$

$$+ \tfrac{1}{2}\int_0^3 [76.8(7.5) - 48(3)][2.4(7.5) - 1.0x]\,dx$$
$$\text{part } BF, \text{ origin at } B$$

$$+ \tfrac{1}{2}\int_0^3 [28.8(5) + 96x][2.4(5) + 1.0x]\,dx + \int_0^5 (-28.8x)(-2.4x)\,dx$$
$$\text{part } CF, \text{ origin at } C \qquad\qquad \text{part } DC, \text{ origin at } D$$

$$= 5598.72 + 16{,}951.68 + 10{,}692 + 5940 + 2880$$
$$= 42{,}062.4 \text{ kN·m}^3$$

$$EI_c\delta_A = \int_0^{7.5} (-2.4x)^2\,dx + \tfrac{1}{2}\int_0^6 [-2.4(7.5) + 1.0x]^2\,dx$$
$$\text{part } AB, \text{ origin at } A \qquad \text{part } BC, \text{ origin at } B$$

$$+ \int_0^5 (2.4x)^2\,dx$$
$$\text{part } DC, \text{ origin at } D$$

$$= 810 + 684 + 240 = 1734 \text{ kN·m}^3$$

From the compatibility condition,

$$V_A = \frac{EI_c\Delta_A}{EI_c\delta_A} = \frac{42{,}062.4}{1734} = 24.26 \text{ kN upward as assumed}$$

Returning to the given rigid frame of Fig. 4.6.3a and using the three equations of statics,

$\Sigma F_y = 0$: $V_D = 96 - 24.26 = 71.74$ kN upward

$\Sigma M_D = 0$: $H_A = \dfrac{96(3) - 48(2) - 24.26(6)}{2.5} = 18.58$ kN to left

$\Sigma M_A = 0$: $H_D = \dfrac{96(3) + 48(4.5) - 71.74(6)}{2.5} = 29.42$ kN to left

Check by $\Sigma F_x = 0$: $H_A + H_D \stackrel{?}{=} 48$

                   $18.58 + 29.42 = 48$ (OK)

130 INTERMEDIATE STRUCTURAL ANALYSIS

Example 4.6.2 Using the results of Example 4.6.1, draw free-body, shear, and moment diagrams for each member of the rigid frame. Then, using the moment-area/conjugate-beam method, draw the elastic curve and compute the rotation of joints A, B, C, and D and the horizontal deflection of joints B and C. Note that the horizontal deflection of B determined by using member AB is equal to the horizontal deflection of C determined by using member DC.

SOLUTION Start with the free body AB in Fig. 4.6.5a; the moment to act at B is found to be $48(3) - 18.58(7.5) = 4.65$ clockwise. Then, by using the free body CD in Fig. 4.6.5b, the moment to act at C is $29.42(5) = 147.10$ counterclockwise. After placing the counterclockwise moment of 4.65 and the clockwise moment of 147.10 to act at B and C of the free body BC in Fig. 4.6.5c, the end reactions are computed by superposition of the simple-beam reactions with those due to end moments; the values of 24.26 at B and 71.74 at C thus obtained check exactly with the axial forces in columns AB and CD.

To obtain θ_B and θ_C by applying the conjugate-beam method to member BC, it is more convenient to separate the moment diagram into three parts, one due to load on the span and the other two due to end moments, as shown in Fig. 4.6.5d; thus

$$\theta_B = \frac{108}{EI_c} - \frac{4.7}{EI_c} - \frac{73.6}{EI_c} = \frac{29.7}{EI_c} \text{ clockwise}$$

$$\theta_C = \frac{147.1}{EI_c} + \frac{2.3}{EI_c} - \frac{108}{EI_c} = \frac{41.4}{EI_c} \text{ clockwise}$$

Applying the moment-area theorems to member AB, as shown in Fig. 4.6.5e,

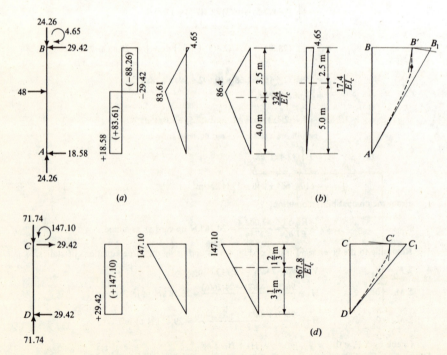

(a)

(b)

(c)

(d)

ANALYSIS OF STATICALLY INDETERMINATE BEAMS 131

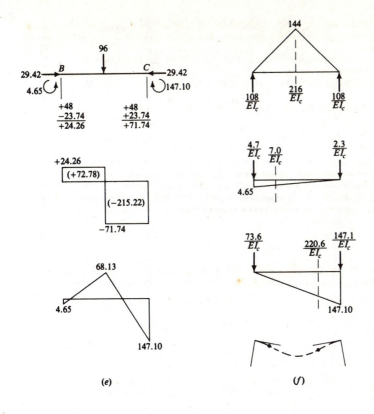

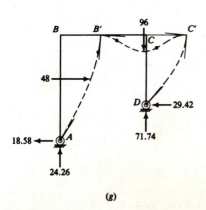

Figure 4.6.5 Equilibrium and compatibility checks for rigid frame in Example 4.6.1.

132 INTERMEDIATE STRUCTURAL ANALYSIS

$$\theta_A = \theta_B + \frac{324}{EI_c} - \frac{17.4}{EI_c} = \frac{29.7 + 324 - 17.4}{EI_c} = \frac{336.3}{EI_c} \text{ clockwise}$$

$$\Delta_H \text{ of } B = 7.5\theta_A - \frac{324}{EI_c}(3.5) + \frac{17.4}{EI_c}(2.5) = \frac{1432}{EI_c} \text{ to the right}$$

Applying the moment-area theorems to member CD, as shown in Fig. 4.6.5f,

$$\theta_D = \theta_C + \frac{367.8}{EI_c} = \frac{41.4 + 367.8}{EI_c} = \frac{408.9}{EI_c} \text{ clockwise}$$

$$\Delta_H \text{ of } C = 5\theta_D - \frac{367.8}{EI_c}(1\tfrac{2}{3}) = \frac{1432}{EI_c} \text{ to the right} \quad \text{(Check)}$$

The fact that Δ_H of C is found to be equal to Δ_H of B constitutes the compatibility check. The number of compatibility checks is always equal to the degree of indeterminacy.

The final elastic curve with all the forces acting on it is shown in Fig. 4.6.5g. The line of action of the 96-kN force is displaced; however, its original location must be used in the equilibrium equations, as always in first-order analysis.

Example 4.6.3 Analyze by the force method the rigid frame with two fixed supports shown in Fig. 4.6.6a.

SOLUTION The given rigid frame in Fig. 4.6.6a is statically indeterminate to the third degree. The rigid frame fixed at A and free at D is chosen as the basic determinate structure,

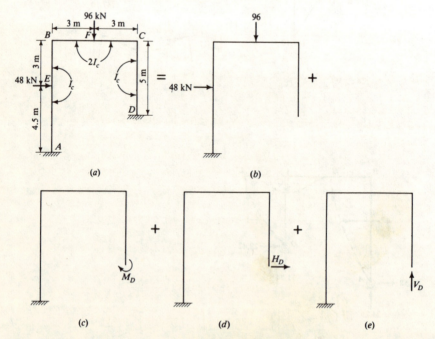

Figure 4.6.6 Compatibility conditions for rigid frame in Example 4.6.2.

ANALYSIS OF STATICALLY INDETERMINATE BEAMS 133

on which M_D, H_D, and V_D act as redundants in addition to the applied loads, as shown in Fig. 4.6.6b to e. When the number of redundants is one or two, it is instructive to show physically the conditions of consistent deformation, as has been done up to this point. For this problem it may be more convenient to set up the positive directions of the unknown redundants as *clockwise* for M_D, to the *right* for H_D, and *upward* for V_D. Also, the three compatibility conditions can be stated as follows:

1. Sum of *clockwise* rotations at D in Fig. 4.6.6b to $e = 0$
2. Sum of horizontal deflections to the *right* at D in Fig. 4.6.6b to $e = 0$
3. Sum of vertical deflections *upward* at D in Fig. 4.6.6b to $e = 0$

If M, m_θ, m_H, and m_V are defined as the bending moments to be obtained from the free-body diagrams shown in Fig. 4.6.7 and the unit-load method is applied, the compatibility conditions become

$$\int \frac{Mm_\theta\,dx}{EI} + M_D \int \frac{m_\theta^2\,dx}{EI} + H_D \int \frac{m_H m_\theta\,dx}{EI} + V_D \int \frac{m_V m_\theta\,dx}{EI} = 0$$

$$\int \frac{Mm_H\,dx}{EI} + M_D \int \frac{m_\theta m_H\,dx}{EI} + H_D \int \frac{m_H^2\,dx}{EI} + V_D \int \frac{m_V m_H\,dx}{EI} = 0$$

$$\int \frac{Mm_V\,dx}{EI} + M_D \int \frac{m_\theta m_V\,dx}{EI} + H_D \int \frac{m_H m_V\,dx}{EI} + V_D \int \frac{m_V^2\,dx}{EI} = 0$$

If any bending moment which causes compression on the outside of the rigid frame *ABCD* in Fig. 4.6.7 is designated as being positive, the values of the integrals are

$$\int \frac{Mm_\theta\,dx}{EI} = \frac{1}{EI_c}\left[\frac{1}{2}\int_0^3 (-96x)(-1.0)\,dx + \int_0^3 (-288)(-1.0)\,dx\right.$$
$$\text{part } FB \qquad\qquad \text{part } BE$$
$$\left. + \int_0^{4.5} (-288 - 48x)(-1.0)\,dx\right]$$
$$\text{part } EA$$

$$= \frac{1}{EI_c}(216 + 864 + 1782) = \frac{2862}{EI_c}$$

$$\int \frac{Mm_H\,dx}{EI} = \frac{1}{EI_c}\left[\frac{1}{2}\int_0^3 (-96x)(+5)\,dx + \int_0^3 (-288)(5-x)\,dx\right.$$
$$\text{part } FB \qquad\qquad \text{part } BE$$
$$\left. + \int_0^{4.5} (-288 - 48x)(2-x)\,dx\right]$$
$$\text{part } EA$$

$$= \frac{1}{EI_c}(-1080 - 3024 + 810) = -\frac{3294}{EI_c}$$

$$\int \frac{Mm_V\,dx}{EI} = \frac{1}{EI_c}\left[\frac{1}{2}\int_0^3 (-96x)(3+x)\,dx + \int_0^3 (-288)(+6)\,dx\right.$$
$$\text{part } FB \qquad\qquad \text{part } BE$$
$$\left. + \int_0^{4.5} (-288 - 48x)(+6)\,dx\right]$$
$$\text{part } EA$$

$$= \frac{1}{EI_c}(-1080 - 5184 - 10{,}692) = -\frac{16{,}956}{EI_c}$$

$$\int \frac{m_\theta^2\,dx}{EI} = \frac{1}{EI_c}\left[\int_0^5 (-1.0)^2\,dx + \frac{1}{2}\int_0^6 (-1.0)^2\,dx + \int_0^{7.5} (-1.0)^2\,dx\right]$$
$$\text{part } DC \qquad \text{part } CB \qquad \text{part } BA$$

$$= \frac{1}{EI_c}(5 + 3 + 7.5) = \frac{15.5}{EI_c}$$

134 INTERMEDIATE STRUCTURAL ANALYSIS

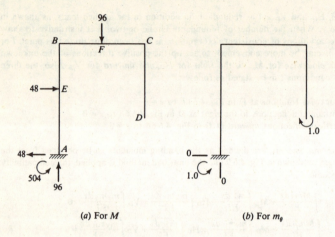

(a) For M (b) For m_θ

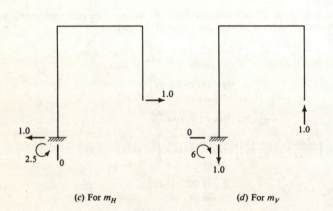

(c) For m_H (d) For m_V

Figure 4.6.7 Free-body diagrams for M, m_θ, m_H, and m_V in Example 4.6.2.

$$\int \frac{m_H^2\, dx}{EI} = \frac{1}{EI_c} \left[\int_0^5 \underbrace{(+1.0x)^2\, dx}_{\text{part } DC} + \frac{1}{2}\int_0^6 \underbrace{(+5)^2\, dx}_{\text{part } CB} + \int_0^{7.5} \underbrace{(5-x)^2\, dx}_{\text{part } BA} \right]$$

$$= \frac{1}{EI_c}(41.667 + 75 + 46.875) = \frac{163.542}{EI_c}$$

$$\int \frac{m_V^2\, dx}{EI} = \frac{1}{EI_c} \left[\frac{1}{2}\int_0^6 \underbrace{(+1.0x)^2\, dx}_{\text{part } CB} + \int_0^{7.5} \underbrace{(+6)^2\, dx}_{\text{part } BA} \right]$$

$$= \frac{1}{EI_c}(36 + 270) = \frac{306}{EI_c}$$

ANALYSIS OF STATICALLY INDETERMINATE BEAMS 135

$$\int \frac{m_\theta m_H \, dx}{EI} = \frac{1}{EI_c} \left[\int_0^5 (-1.0)(+1.0x) \, dx + \frac{1}{2} \int_0^6 (-1.0)(+5) \, dx \right.$$
$$\text{part } DC \qquad \text{part } CB$$
$$\left. + \int_0^{7.5} (-1.0)(5-x) \, dx \right]$$
$$\text{part } BA$$

$$= \frac{1}{EI_c}(-12.5 - 15 - 9.375) = -\frac{36.875}{EI_c}$$

$$\int \frac{m_\theta m_V \, dx}{EI} = \frac{1}{EI_c} \left[\frac{1}{2} \int_0^6 (-1.0)(+1.0x) \, dx + \int_0^{7.5} (-1.0)(+6) \, dx \right]$$
$$\text{part } CB \qquad \text{part } BA$$

$$= \frac{1}{EI_c}(-9 - 45) = -\frac{54}{EI_c}$$

$$\int \frac{m_H m_V \, dx}{EI} = \frac{1}{EI_c} \left[\frac{1}{2} \int_0^6 (+5)(+1.0x) \, dx + \int_0^{7.5} (5-x)(+6) \, dx \right]$$
$$\text{part } CB \qquad \text{part } BA$$

$$= \frac{1}{EI_c}(45 + 56.25) = \frac{101.25}{EI_c}$$

Substituting the values of the integrals into the compatibility conditions and canceling EI_c in every denominator,

$$+2862 + 15.5M_D - 36.875H_D - 54V_D = 0$$
$$-3294 - 36.875M_D + 163.542H_D + 101.25V_D = 0$$
$$-16,956 - 54M_D + 101.25H_D + 306V_D = 0$$

Solving these three equations for M_D, H_D, and V_D,

$M_D = -65.26$ or   65.26 kN·m counterclockwise
$H_D = -27.353$ or   27.353 kN to the left
$V_D = +52.945$ or   52.945 kN upward

Returning to the given rigid frame (see Fig. 4.6.8g) with the above known values of M_D, H_D, and V_D and solving the remaining unknown reactions by the three equations of statics,

$\Sigma F_x = 0$: $H_A = 48 - 27.353 = 20.647$ kN to the left
$\Sigma F_y = 0$: $V_A = 96 - 52.945 = 43.055$ kN upward
$\Sigma M_A = 0$: $M_A = 96(3) + 48(4.5) - 27.353(2.5) - 52.945(6)$
                $= 52.68$ kN·m counterclockwise

Check by $\Sigma M_D = 0$:

$$96(3) + 52.68 + 65.26 \stackrel{?}{=} 48(2) + 20.647(2.5) + 43.055(6)$$
$$405.94 \approx 405.95 \quad (OK)$$

ALTERNATE SOLUTION An alternate solution may be made by choosing a different basic structure, such as using M_A, H_A, and V_A as the redundants or using M_A, M_D, and H_D as the redundants. In general, other than for more practice in using the force method, it is much more direct and useful to check the correctness of the solution by making the equilibrium and compatibility checks shown in the next example.

Example 4.6.4 Using the results of Example 4.6.3, draw free-body, shear, and moment diagrams for each member of the rigid frame. Then, using the moment-area/conjugate-beam method,

136 INTERMEDIATE STRUCTURAL ANALYSIS

draw the elastic curve and compute the rotation of joints B and C and the horizontal deflection of joints B and C. Note the *three* compatibility checks: (1) rotation at B determined from member AB is equal to rotation at B determined from member BC, (2) rotation at C determined from member DC is equal to rotation at C determined from member BC, and (3) horizontal deflection of B determined from member AB is equal to horizontal deflection of C determined from member DC.

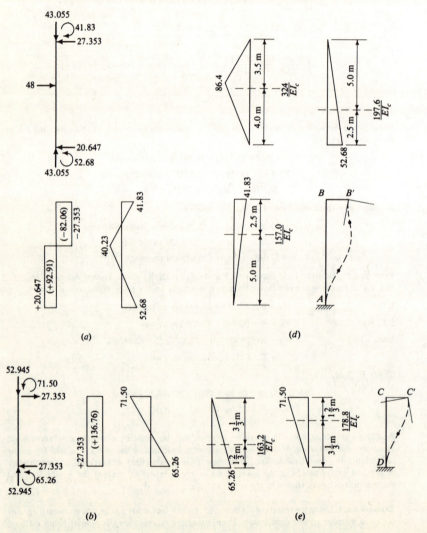

Figure 4.6.8 Equilibrium and compatibility checks for rigid frame in Example 4.6.2.

ANALYSIS OF STATICALLY INDETERMINATE BEAMS 137

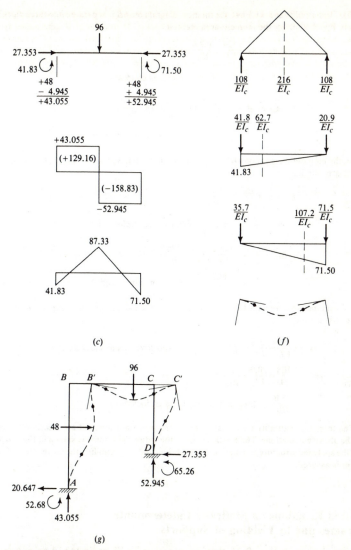

Figure 4.6.8 Continued.

SOLUTION (a) *Equilibrium checks.* First, the moment at B in Fig. 4.6.8a is found to be 41.83 clockwise. Second, the moment at C in Fig. 4.6.8b is found to be 71.50 counterclockwise. Third, by placing the moments of 41.83 counterclockwise and 71.50 clockwise at B and C of Fig. 4.6.8c, the end reactions are found by superposition to be 43.055 and 52.945, which check with the axial forces in AB and CD. The shear and moment diagrams are then drawn for all three members.

(b) *Compatibility checks.* First, the moment diagram on AB is separated into three parts as shown in Fig. 4.6.8d. By the moment-area method,

$$\theta_B = \theta_A - \frac{324}{EI_c} + \frac{197.6}{EI_c} + \frac{157.0}{EI_c}$$

$$= +\frac{30.6}{EI_c} \quad \text{(positive means clockwise)}$$

$$\Delta_H \text{ of } B = -\frac{324}{EI_c}(3.5) + \frac{197.6}{EI_c}(5) + \frac{157.0}{EI_c}(2.5)$$

$$= +\frac{246.5}{EI_c} \quad \text{(positive means to right)}$$

Second, the moment diagram on DC is separated into two parts as shown in Fig. 4.6.8e. By the moment-area method,

$$\theta_C = \theta_D + \frac{163.2}{EI_c} - \frac{178.8}{EI_c}$$

$$= -\frac{15.6}{EI_c} \quad \text{(negative means counterclockwise)}$$

$$\Delta_H \text{ of } C = +\frac{163.2}{EI_c}\left(3\frac{1}{3}\right) - \frac{178.8}{EI_c}\left(1\frac{2}{3}\right)$$

$$= +\frac{246.0}{EI_c} \quad \text{(positive means to right)}$$

Third, the moment diagram on BC is separated into three parts as shown in Fig. 4.6.8f. By the conjugate-beam method,

$$\theta_B = +\frac{108}{EI_c} - \frac{41.8}{EI_c} - \frac{35.7}{EI_c} = +\frac{30.5}{EI_c} \quad \text{(positive means clockwise)}$$

$$\theta_C = -\frac{108}{EI_c} + \frac{20.9}{EI_c} + \frac{71.5}{EI_c}$$

$$= -\frac{15.6}{EI_c} \quad \text{(negative means counterclockwise)}$$

The three compatibility checks as stated in the problem statement are clearly observed from the above calculations. The elastic curve of the entire rigid frame is shown in Fig. 4.6.8g. When the analyzed structure satisfies both equilibrium and compatibility, the correctness of the solution is assured.

4.7 Induced Reactions on Statically Indeterminate Rigid Frames due to Yielding of Supports

It has been shown in Sec. 2.11 that no reactions will be induced to act on a statically determinate rigid frame due to movement or yielding of supports; however, the entire rigid frame may be displaced as a whole in a so-called rigid-body movement. The yielding of supports of statically indeterminate rigid frames, however, either in the form of a linear displacement or a rotational slip (for fixed supports only) always induces reactions, and therefore there may be axial forces, shears, and moments in the members of the rigid frame. The procedure for finding the induced reactions is first to derive a

ANALYSIS OF STATICALLY INDETERMINATE BEAMS 139

basic determinate structure by removing the excess supports *including* those which are subjected to yielding and then to treat the redundant reactions as unknown forces acting on the basic structure. The conditions of consistent deformation (or compatibility conditions) should then include those at the yielding supports, where the deflection or rotation should be equal to the predicted amount of yielding, and those at the unyielding redundant supports, where the deflection or rotation should still be zero. After the redundants are solved, they can be placed back to act on the basic structure as loads and the remaining reactions solved by the equations of statics. The variation in axial force, shears, and moments in the members can then be found as usual.

Example 4.7.1 By the force method determine all reactions induced to act on the rigid frame of Fig. 4.7.1a by a rotational slip of 0.002 rad clockwise of joint D and a vertical settlement of 15 mm at joint D. Use $E = 200 \times 10^6$ kN/m^2 and $I = 400 \times 10^{-6}$ m^4.

SOLUTION The rigid frame fixed at A and free at D is chosen as the basic determinate structure on which the redundant reactions M_D, H_D, and V_D are acting, as shown in Fig. 4.7.1b to d. By referring to the compatibility conditions in Example 4.6.3, and adapting to the requirements of this problem,

$$+\frac{15.5}{EI_c}M_D - \frac{36.875}{EI_c}H_D - \frac{54}{EI_c}V_D = +0.002$$

$$-\frac{36.875}{EI_c}M_D + \frac{163.542}{EI_c}H_D + \frac{101.25}{EI_c}V_D = 0$$

$$-\frac{54}{EI_c}M_D + \frac{101.25}{EI_c}H_D + \frac{306}{EI_c}V_D = -0.015$$

Substituting the value of EI_c,

$$+15.5M_D - 36.875H_D - 54V_D = +160$$
$$-36.875M_D + 163.542H_D + 101.25V_D = 0$$
$$-54M_D + 101.25H_D + 306V_D = -1200$$

Solving the above three equations,

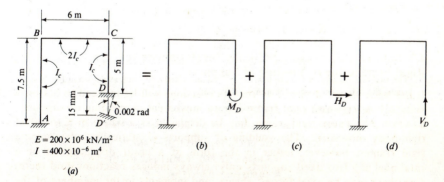

Figure 4.7.1 Compatibility conditions for rigid frame in Example 4.7.1.

$M_D = +1.965$ or   1.965 kN·m clockwise

$H_D = +3.3405$ or   3.3405 kN to the right

$V_D = -4.6802$ or   4.6802 kN downward

Returning to the given rigid frame with the known values of M_D, H_D, and V_D and solving by the three equations of statics,

$$H_A = 3.3405 \text{ kN to the left}$$
$$V_A = 4.6802 \text{ kN upward}$$
$$M_A = 3.3405(2.5) + 4.6802(6) + 1.965$$
$$= 38.397 \text{ kN·m counterclockwise}$$

Example 4.7.2 Using the results of Example 4.7.1, draw free-body, shear, and moment diagrams for each member of the rigid frame. Then, using the moment-area method, draw the elastic curve for the entire rigid frame. Starting from the fixed support at A, go through member AB to joint B, member BC to joint C, and member CD to joint D. Show that the rotation at D is 0.002 rad clockwise, horizontal deflection at D is zero, and vertical deflection at D is 15 mm downward.

SOLUTION First using member AB, the moment at B is found to be 13.343 clockwise, as shown in Fig. 4.7.2a. Then using member DC, the moment at C is found to be 14.738 clockwise, as shown in Fig. 4.7.2b. Placing the counterclockwise moments of 13.343 and 14.738 at B and C of member BC as shown in Fig. 4.7.2c, the end reactions are found to be 4.6802 upward at B and 4.6802 downward at C, both of which check with the axial forces in members AB and CD.

The moment diagram on AB is separated into two parts as shown in Fig. 4.7.2d; the M/EI areas of the parts are 0.625×10^{-3} rad and 1.800×10^{-3} rad. Using the two moment-area theorems,

$$\theta_B = (+0.625 + 1.800) \times 10^{-3}$$
$$= +2.425 \times 10^{-3} \quad \text{(positive means clockwise)}$$
$$\Delta_H \text{ of } B = 0.625(2.5) + 1.800(5.0)$$
$$= +10.56 \text{ mm} \quad \text{(positive means to right)}$$

The two M/EI areas on BC are 0.250×10^{-3} rad and 0.276×10^{-3} rad, as shown in Fig. 4.7.2f. Again using the two moment-area theorems,

$$\theta_C = \theta_B + 0.250 \times 10^{-3} - 0.276 \times 10^{-3} = (+2.425 + 0.250 - 0.276) \times 10^{-3}$$
$$= +2.399 \times 10^{-3} \text{ rad} \quad \text{(positive means clockwise)}$$
$$\Delta_V \text{ of } C = 6000\theta_B + 0.250(4.0) - 0.276(2.0)$$
$$= 6000(2.425 \times 10^{-3}) + 0.250(4.0) - 0.276(2.0)$$
$$= +14.999 \text{ mm} \quad \text{(positive means downward; check with 15 mm)}$$

The two M/EI areas on CD are 0.061×10^{-3} rad and 0.460×10^{-3} rad, as shown in Fig. 4.7.2e. Using the two moment-area theorems a third time,

$$\theta_D = \theta_C + 0.061 \times 10^{-3} = (+2.399 + 0.061 - 0.461) \times 10^{-3}$$
$$= +1.999 \times 10^{-3} \text{ rad} \quad \text{(positive means clockwise; check with 0.002 rad)}$$
$$\Delta_H \text{ of } D = \Delta_H \text{ of } C - 5\theta_C - 0.061(1\tfrac{2}{3}) + 0.460(3\tfrac{1}{3})$$
$$= +10.56 - 5(2.399) - 0.061(1\tfrac{2}{3}) + 0.460(1\tfrac{1}{3})$$
$$= -0.003 \text{ mm} \approx 0 \quad \text{(Check)}$$

The elastic curve of the entire rigid frame is shown in Fig. 4.7.2g.

ANALYSIS OF STATICALLY INDETERMINATE BEAMS 141

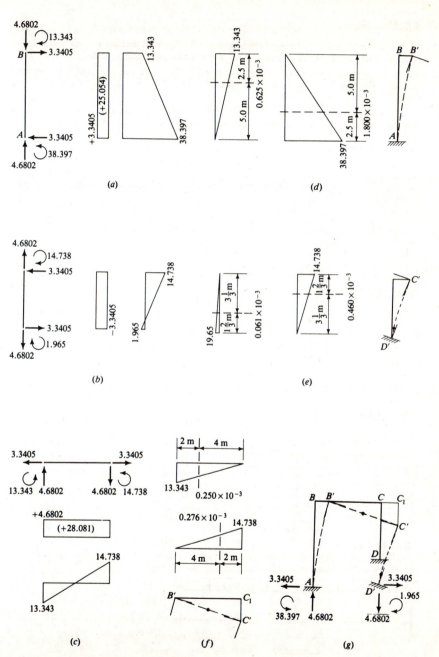

Figure 4.7.2 Equilibrium and compatibility checks for rigid frame in Example 4.7.1.

4.8 Exercises

4.1 By the force method analyze the beam shown in Fig. 4.8.1 and draw shear and moment diagrams. First use R_3 as the redundant, then check solution by using R_1 as the redundant.

4.2 By the force method analyze the beam shown in Fig. 4.8.2 and draw shear and moment diagrams. First use R_1 and R_2 ($=R_1$) together as the redundants, and then check the solution by using R_1 and R_3 as the redundants.

4.3 By the force method analyze the beam shown in Fig. 4.8.3 and draw shear and moment diagrams. First use R_2 as the redundant, and then check the solution by using R_3 as the redundant. Note that the shear and moment diagrams for the right span should be identical to those of Example 4.2.1.

4.4 By the force method analyze the beam shown in Fig. 4.8.4 and draw shear and moment diagrams. First use R_2 as the redundant, and then check the solution by using R_1 as the redundant. Note that the shear and moment diagrams for the left span should be identical to those of Exercise 4.1.

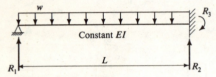

Figure 4.8.1 Exercise 4.1.

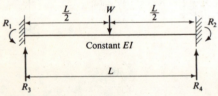

Figure 4.8.2 Exercise 4.2.

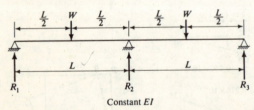

Figure 4.8.3 Exercise 4.3.

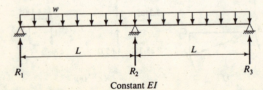

Figure 4.8.4 Exercise 4.4.

ANALYSIS OF STATICALLY INDETERMINATE BEAMS 143

4.5 By the force method analyze the beam shown in Fig. 4.8.5 and draw shear and moment diagrams. First use R_2 as the redundant, and then check the solution by using R_1 as the redundant. In each case, make use of the law of reciprocal deflections.

4.6 By the force method analyze the beam shown in Fig. 4.8.6 and draw shear and moment diagrams. First use R_1 and R_2 as the redundants, and then check the solution by using R_2 and R_4 as the redundants. In each case, make use of the law of reciprocal deflections.

4.7 and 4.8 By the force method analyze the beam shown in Figs. 4.8.7 and 4.8.8 and draw shear and moment diagrams. Use R_2 and R_3 as the redundants and make use of the law of reciprocal deflections.

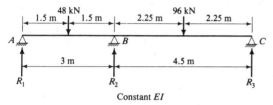

Figure 4.8.5 Exercise 4.5.

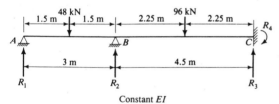

Figure 4.8.6 Exercise 4.6.

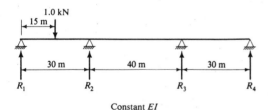

Figure 4.8.7 Exercise 4.7.

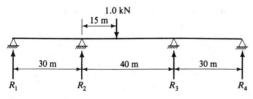

Figure 4.8.8 Exercise 4.8.

4.9 Solve Exercise 4.1 by using the theorem of least work.

4.10 Solve Exercise 4.8 by using the theorem of least work.

4.11 By the force method determine all induced reactions due to a vertical settlement of 7.5 mm at the support B of the beam in Exercise 4.5. Determine the slopes at A, B, and C of the final elastic curve. Use $E = 200 \times 10^6$ kN/m^2 and $I = 160 \times 10^{-6}$ m^4.

4.12 By the force method determine all induced reactions due to a vertical settlement of 7.5 mm at the support B of the beam in Exercise 4.6. Determine the slopes at A, B, and C of the final elastic curve. Use $E = 200 \times 10^6$ kN/m^2 and $I = 160 \times 10^{-6}$ m^4.

4.13 and **4.14** By the force method analyze the rigid frame with two hinged supports shown in Figs. 4.8.9 and 4.8.10. Draw free-body, shear, and moment diagrams and make the equilibrium checks. Draw the elastic curve and make the compatibility check.

4.15 and **4.16** By the force method analyze the rigid frame with two fixed supports shown in Figs. 4.8.11 and 4.8.12. Draw free-body, shear, and moment diagrams and make the equilibrium checks. Draw the elastic curve and make the three compatibility checks.

4.17 Determine all reactions induced to act on the rigid frame of Exercise 4.14 by a 15-mm vertical settlement at the hinged support E. Draw free-body, shear, and moment diagrams and make the equilibrium checks. Draw the elastic curve and make the compatibility check. Use $E = 200 \times 10^6$ kN/m^2 and $I = 500 \times 10^{-6}$ m^4.

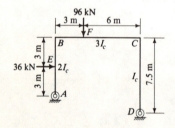

Figure 4.8.9 Exercise 4.13.

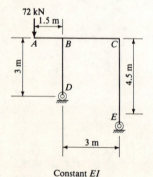

Constant EI

Figure 4.8.10 Exercise 4.14.

4.18 Determine all reactions induced to act on the rigid frame of Exercise 4.15 by a rotational slip of 0.002 rad clockwise of joint A and a vertical settlement of 15 mm also at joint A. Draw free-body, shear, and moment diagrams and make the equilibrium checks. Draw the elastic curve and make the three compatibility checks. Use $E = 200 \times 10^6$ kN/m^2 and $I_c = 500 \times 10^{-6}$ m^4.

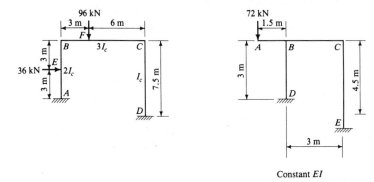

Figure 4.8.11 Exercise 4.15. **Figure 4.8.12** Exercise 4.16.

CHAPTER
FIVE
ANALYSIS OF STATICALLY INDETERMINATE TRUSSES BY THE FORCE METHOD

5.1 Degree of Indeterminacy

A general discussion on the difference between statically determinate and indeterminate trusses was given in Chap. 1. Then methods of determining the force and deformation response of statically determinate trusses were treated in Chap. 3. Before taking up the analysis of statically indeterminate trusses, it is desirable to review and extend some of the ideas regarding the degree of indeterminacy of a truss.

What makes a truss statically indeterminate is the fact that the total number of unknown reactions and axial forces in members required to define the force response exceeds the total number of independent equations of statics. The former can be easily counted, and the latter is equal to twice the number of joints. Therefore the most general expression for the degree of indeterminacy of a truss is

$$i = (r + m) - 2j \tag{5.1.1}$$

in which r is the number of reactions, m is the number of members, and j is the number of joints.

In the force method of analysis, the approach is first to choose a number of redundants equal to the degree of indeterminacy. Caution must be exercised to ensure that when the restraints corresponding to the redundants are removed, there remains a truss which is both statically determinate and stable. Whether the number of reactions is exactly three or more than three, the redundants may be chosen entirely from the unknown member forces. However, when the number of reactions is more than three, the redundants

ANALYSIS OF TRUSSES BY THE FORCE METHOD 147

may include the excess reactions over three, which is the minimum number required for statical stability.

A statically indeterminate truss which has more than three reactions but exactly $(2j-3)$ members is externally indeterminate; one which has just three reactions but more than $(2j-3)$ members is internally indeterminate; and one which has more than three reactions and also more than $(2j-3)$ members is both externally and internally indeterminate. In choosing the redundants, it is usually more convenient to include the excess reactions over three and the rest from the member forces. But, as has been said before, it is possible to use only member forces as redundants even if the truss is both externally and internally indeterminate. This latter approach might be adopted, in fact, when an overall general computer program is to be written.

When a reaction is chosen as a redundant, the support in the direction of the reaction is removed and its action is replaced by an unknown reacting force. The compatibility condition is that the deflection in the direction of the reaction must be zero. When a member force is chosen as a redundant, that member may be either entirely removed or simply cut at some point along the member. In the former case, the action of the member on the basic truss is replaced by a pair of forces pulling on the joints, and the compatibility condition is that the increase in distance between the joints is equal to the elongation of the member itself. In the latter case, both parts of the cut member remain with the basic truss; thus the action of the member is replaced by a pair of pulling forces applied at the cut ends, and the compatibility condition becomes that the total overlap produced at the cut ends must be zero. When the number of redundant member forces is two or more, it is generally more convenient to cut them rather than remove them to arrive at the compatibility conditions.

5.2 Force Method Using Reactions as Redundants

When a truss has more than three reactions and exactly $(2j-3)$ members, it is externally indeterminate. The three examples in this section are in this category. They are solved by using reactions as redundants. Of course, they can be solved alternately by using member forces as redundants.

Example 5.2.1 Analyze the scissors truss shown in Fig. 5.2.1a by using the horizontal reaction at the hinged support B as the redundant.

SOLUTION Using Eq. (5.1.1), $i = (r+m) - 2j = (4+5) - 2(4) = 1$. Also $m = 5$ is exactly equal to $2j - 3 = 5$; therefore the truss is externally indeterminate. After the restraint offered by the redundant H_B is removed, the basic truss has a roller support at B instead of a hinged support. The compatibility condition is that the horizontal deflection at B of the basic truss, when subjected to the combined action of the applied load and the redundant H_B, must be zero; or

$$\Delta_B \text{ in Fig. } 5.2.1b = H_B \delta_B \text{ in Fig. } 5.2.1c$$

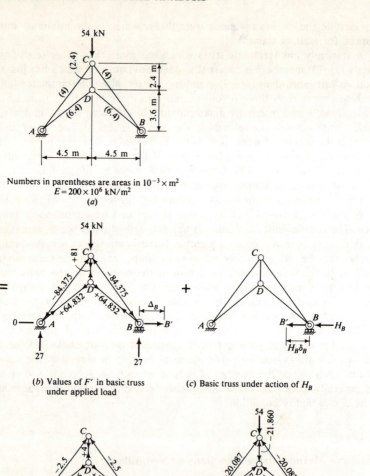

(a) Numbers in parentheses are areas in $10^{-3} \times m^2$
$E = 200 \times 10^6 \text{ kN/m}^2$

(b) Values of F' in basic truss under applied load

(c) Basic truss under action of H_B

(d) Values of u in basic truss

(e) Values of F in given truss

Figure 5.2.1 Truss of Example 5.2.1.

where δ_B is the horizontal deflection to left if $H_B = 1.0$ kN to left. It is the same as the horizontal deflection to right if $H_B = 1.0$ kN to right.

Using the u forces as defined in Fig. 5.2.1d and applying the unit-load method,

$$\Delta_B = \Sigma \frac{F'uL}{EA} = +7.4440 \text{ mm}$$

ANALYSIS OF TRUSSES BY THE FORCE METHOD

and
$$\delta_B = \Sigma \frac{u^2 L}{EA} = +0.28948 \times 10^{-3} \text{ mm}$$

Using the compatibility condition,
$$H_B = \frac{\Delta_B}{\delta_B} = \frac{+7.4440}{+0.28948 \times 10^{-3}}$$
$$= +25.715 \text{ kN} \quad \text{(positive means to left)} \tag{5.2.1}$$

Applying superposition,
$$F_i \text{ in given truss} = F'_i - H_B u_i \quad i = 1 \text{ to } 5 \tag{5.2.2}$$

The numerical work involved in Eqs. (5.2.1) and (5.2.2) is shown in Table 5.2.1. Although not quite independent, a check can be made by seeing that

$$\text{Outward deflection of } B \text{ in Fig. 5.2.1}a = \Sigma \frac{(F \text{ in Fig. 5.2.1}e)uL}{EA} = 0$$

This is done in the last column of Table 5.2.1.

Example 5.2.2 Analyze the truss with cross diagonals and two hinged supports shown in Fig. 5.2.2a by using the horizontal reaction at A as the redundant.

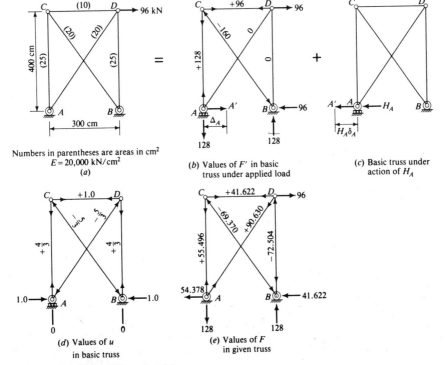

Figure 5.2.2 Truss of Example 5.2.2.

Table 5.2.1 Numerical computation for Example 5.2.1

Member	L, m	$A, 10^{-3}$ m^2	F', kN	u	$\dfrac{F'uL}{EA}$, mm	$\dfrac{u^2L}{EA}, 10^{-3}$ mm	$F = F' - H_Bu$, kN	$\dfrac{FuL}{EA}$, mm
AC	7.5	4.0	−84.375	−2.5	+1.9775	+0.05859	−20.087	+0.4708
BC	7.5	4.0	−84.375	−2.5	+1.9775	+0.05859	−20.087	+0.4708
AD	5.7628	6.4	+64.832	+3.2016	+0.9345	+0.04615	−17.497	−0.2522
BD	5.7628	6.4	+64.832	+3.2016	+0.9345	+0.04615	−17.497	−0.2522
CD	2.4	2.4	+81	+4	+1.6200	+0.08000	−21.860	−0.4372
Σ					+7.4440	+0.28948		0.0000

ANALYSIS OF TRUSSES BY THE FORCE METHOD   151

SOLUTION Using Eq. (5.1.1), $i = (r + m) - 2j = (4 + 5) - 2(4) = 1$. Also, $m = 5$ is exactly equal to $2j - 3 = 5$; therefore the truss is externally indeterminate. After the restraint offered by the redundant H_A is removed, the basic truss has a roller support at A instead of a hinged support. The compatibility condition is that the horizontal deflection at A of the basic truss, when subjected to the combined action of the applied load and the redundant H_A, must be zero; or

$$\Delta_A \text{ in Fig. } 5.2.2b = H_A \delta_A \text{ in Fig. } 5.2.2c$$

where δ_A is the horizontal deflection to left if $H_A = 1.0$ kN to left. It is the same as the horizontal deflection to right if $H_A = 1.0$ kN to right.

Using the u forces as defined in Fig. 5.2.2d and applying the unit-load method,

$$\Delta_A = \frac{F'uL}{EA} = +0.61386 \text{ cm}$$

and

$$\delta_B = \frac{u^2 L}{EA} = +11.2888 \times 10^{-3} \text{ cm}$$

Using the compatibility condition,

$$H_A = \frac{\Delta_A}{\delta_A} = \frac{+0.61386}{+11.2888 \times 10^{-3}}$$

$$= +54.378 \text{ kN} \qquad \text{(positive means to left)} \qquad (5.2.3)$$

Applying superposition,

$$F_i \text{ in given truss} = F'_i - H_A u_i \qquad i = 1 \text{ to } 5 \qquad (5.2.4)$$

The numerical work involved in Eqs. (5.2.3) and (5.2.4) is shown in Table 5.2.2. Although not quite independent, a check can be made by seeing that

$$\text{Inward deflection of } A \text{ in Fig. } 5.2.2a = \Sigma \frac{(F \text{ in Fig. } 5.2.2e) uL}{EA} = 0$$

This is done in the last column of Table 5.2.2.

Example 5.2.3 Analyze the truss shown in Fig. 5.2.3a by using the reactions at joints L_2 and L_6 as the redundants.

SOLUTION Because the truss is made up of a series of triangles stacked with one upon another, it is not necessary to count in order to reach the conclusion that m is exactly equal to $2j - 3$. With five unknown reactions in total, the given truss is externally indeterminate to the second degree. With V_2 and V_6 as the redundants, the basic truss has a hinged support at L_0 and a roller support at L_8, as shown in Fig. 5.2.3b. The compatibility conditions are

$$\delta_{22} V_2 + \delta_{26} V_6 = \delta_{23}$$
$$\delta_{62} V_2 + \delta_{66} V_6 = \delta_{63}$$

in which δ_{23} and δ_{63} are the downward deflections at L_2 and L_6 due to a unit downward load at L_3, δ_{22} and δ_{62} are the upward deflections at L_2 and L_6 due to a unit upward load at L_2, and δ_{26} and δ_{66} are the upward deflections at L_2 and L_6 due to a unit upward load at L_6.

It can be shown that all the δ quantities in the compatibility equations can be derived, with the help of the law of reciprocal deflections, from the shape of the lower chord due to a unit downward load applied at L_2 of the basic truss. This work is done, with the results shown in Fig. 5.2.3c and d. First the member forces due to a unit downward load applied at L_2 are determined by the method of joints; then the member elongations are determined; and finally all the joint displacements are found by using $U_4 L_4$ as the reference member. The

Table 5.2.2 Numerical computation for Example 5.2.2

Member	L, cm	A, cm^2	F', kN	u	$\dfrac{F'uL}{EA}$, cm	$\dfrac{u^2L}{EA}, 10^{-3}$ cm	$F = F' - H_A u$	$\dfrac{FuL}{EA}$, cm
AC	400	25	+128	+4/3	+0.13653	+1.4222	+55.496	+0.05920
BD	400	25	0	+4/3	0	+1.4222	−72.504	−0.07734
AD	500	20	0	−5/3	0	+3.4722	+90.630	−0.18881
BC	500	20	−160	−5/3	+0.33333	+3.4722	−69.370	+0.14452
CD	300	10	+96	+1.0	+0.14400	+1.5000	+41.622	+0.06243
Σ					+0.61386	+11.2888		0.00000

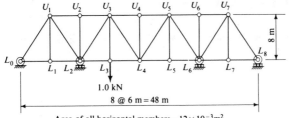

Area of all horizontal members = $12 \times 10^{-3} m^2$
Area of all vertical members = $3 \times 10^{-3} m^2$
Area of all diagonal members = $6 \times 10^{-3} m^2$
$E = 200 \times 10^6$ kN/m²

(a) Given truss

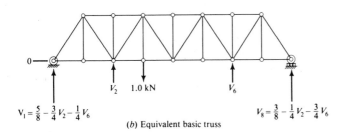

$V_1 = \frac{5}{8} - \frac{3}{4} V_2 - \frac{1}{4} V_6$ $V_8 = \frac{3}{8} - \frac{1}{4} V_2 - \frac{3}{4} V_6$

(b) Equivalent basic truss

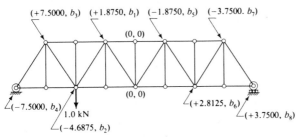

(c) Joint displacements in m × 10^{-6} of basic truss due to 1.0 kN applied at L_2 using $U_4 L_4$ as reference member

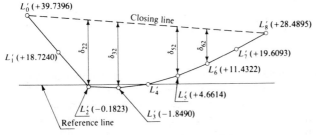

(d) Joint displacements in m × 10^{-6} above or below reference line through L_4' of basic truss due to 1.0 kN applied at L_2

Figure 5.2.3 Truss of Example 5.2.3.

quantities b_1 to b_8 in Fig. 5.2.3c are computed by successively using the joint-displacement equation:

$$b_1 = -1.8490 \times 10^{-6} \text{ m} \qquad b_5 = +4.6614 \times 10^{-6} \text{ m}$$
$$b_2 = -0.1823 \times 10^{-6} \text{ m} \qquad b_6 = +11.4322 \times 10^{-6} \text{ m}$$
$$b_3 = +18.7240 \times 10^{-6} \text{ m} \qquad b_7 = +19.6093 \times 10^{-6} \text{ m}$$
$$b_4 = +39.7396 \times 10^{-6} \text{ m} \qquad b_8 = +28.4895 \times 10^{-6} \text{ m}$$

When the shape of the lower chord is plotted in Fig. 5.2.3d using the eight b values shown, the deflections δ_{22}, δ_{32}, δ_{52}, and δ_{62} are the distances of L'_2, L'_3, L'_5 and L'_6 below the closing line L'_0-L'_8; thus

$$\delta_{22} = 37.1094 \times 10^{-6} \text{ m}$$
$$\delta_{32} = 37.3698 \times 10^{-6} \text{ m}$$
$$\delta_{52} = 28.0469 \times 10^{-6} \text{ m}$$
$$\delta_{62} = 19.8698 \times 10^{-6} \text{ m}$$

Using the law of reciprocal deflections,

$$\delta_{26} = \delta_{62} \qquad \delta_{23} = \delta_{32} \qquad \delta_{25} = \delta_{52}$$

and, by symmetry,

$$\delta_{22} = \delta_{66} \qquad \delta_{63} = \delta_{25}$$

Now the compatibility equations become, numerically,

$$+37.1094 \times 10^{-6} V_2 + 19.8698 \times 10^{-6} V_6 = +37.3698 \times 10^{-6}$$
$$+19.8698 \times 10^{-6} V_2 + 37.1094 \times 10^{-6} V_6 = +28.0469 \times 10^{-6}$$

Solving the above equations,

$V_2 = +0.8444$ kN   (positive means upward)

$V_6 = +0.3036$ kN   (positive means upward)

Returning with the known values of V_2 and V_6 to the truss of Fig. 5.2.3b,

$V_0 = \tfrac{5}{8} - \tfrac{3}{4} V_2 - \tfrac{1}{4} V_6 = 0.6250 - 0.75(0.8444) - 0.25(0.3036)$

$\quad = -0.0842$ kN   (negative means downward)

$V_8 = \tfrac{3}{8} - \tfrac{1}{4} V_2 - \tfrac{3}{4} V_6 = 0.3750 - 0.25(0.8444) - 0.75(0.3036)$

$\quad = -0.0638$ kN   (negative means downward)

With all reactions now known, the member forces may be obtained by the method of joints as usual.

5.3 Force Method Using Axial Forces in Members as Redundants

When a truss is statically indeterminate, whether the number of reactions is equal exactly to three or is more than three, the redundants can always be chosen entirely from the member forces. Of course, when the number of reactions is equal exactly to three, the redundants cannot have reactions in themselves in any case.

When a member force has been chosen as a redundant, one approach is to remove it entirely from the basic truss, in which case its action is replaced by a pair of unknown forces pulling on the joints. The compatibility condition is

ANALYSIS OF TRUSSES BY THE FORCE METHOD 155

that the relative movement of the joints away from each other is equal to the elongation of the member. The alternate approach is only to cut the redundant member and let both parts of it remain in the basic truss. The unknown is a pair of forces pulling at the cut ends, and the compatibility condition is that the total overlap at the cut due to the combined action of the applied loads and all the redundant forces must be zero. The first approach clearly illustrates the physical meaning of the "self-lengthening" effect, and the second approach makes the procedure more automatic for multiple redundant trusses. Both approaches will be used in Example 5.3.1, but only the cut-member method will be used in Examples 5.3.2 and 5.3.3.

Example 5.3.1 Analyze the scissors truss of Example 5.2.1 by using the force in member CD as the redundant.

SOLUTION (a) *Member CD removed.* As shown in Fig. 5.3.1, the compatibility condition is that the increase in distance between joints C and D in Fig. 5.3.1b minus the decrease in distance in Fig. 5.3.1c is equal to the self-lengthening of member CD; or

$$\Delta \text{ (Fig. 5.3.1}b) - F_{CD}\delta \text{ (Fig. 5.3.1}c) = \frac{F_{CD}L_{CD}}{EA_{CD}}$$

By the unit-load method,

$$\Delta \text{ (Fig. 5.3.1}b) = \sum_1^4 \frac{F'uL}{EA} = -0.39550 \text{ mm}$$

if the u values are as defined in Fig. 5.3.1d. δ_B in Fig. 5.3.1c is the relative movement toward each other if $F_{CD} = 1.0$ kN pulling on the joints. It is the same as the relative movement apart if $F_{CD} = 1.0$ kN pushing on the joints. Therefore,

$$\delta = \sum_1^4 \frac{u^2L}{EA} = +13.0926 \times 10^{-3} \text{ mm}$$

Then using the compatibility condition,

$$F_{CD} = \frac{\Delta}{\delta + (L_{CD}/EA_{CD})} = \frac{-0.39550}{+13.0926 \times 10^{-3} + [2.4/(200 \times 10^6)(2.4)]}$$

$$= \frac{-0.39550}{+18.0926 \times 10^{-6}}$$

$$= -21.860 \text{ kN} \qquad \text{(negative means compression)} \qquad (5.3.1)$$

Applying superposition,

$$F_i \text{ in given truss} = F'_i - F_{CD}u_i \qquad i = 1 \text{ to } 4 \qquad (5.3.2)$$

The numerical work involved in Eqs. (5.3.1) and (5.3.2) is shown in Table 5.3.1. The last column in this table shows the check that

Relative distance apart between joints C and D when CD is removed from Fig. 5.3.1e

$$= \frac{F_{CD}L_{CD}}{EA_{CD}} = -0.10930 \text{ mm}$$

(b) *Member CD cut.* As shown in Fig. 5.3.2, the compatibility condition is that the sum of the overlap at the cut ends within CD in Fig. 5.3.2b and that in Fig. 5.3.2c must be zero. By the unit-load method,

$$\sum_1^5 \frac{F'uL}{EA} + \sum_1^5 \frac{(F_{CD}u)(u)(L)}{EA} = 0$$

156 INTERMEDIATE STRUCTURAL ANALYSIS

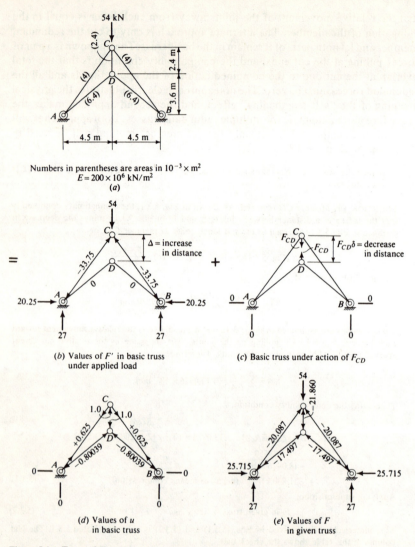

Figure 5.3.1 Truss of Example 5.3.1; solution by removing member CD.

from which

$$F_{CD} = -\frac{\sum_1^5 \dfrac{F'uL}{EA}}{\sum_1^5 \dfrac{u^2 L}{EA}} = -\frac{+0.39550}{+18.0926 \times 10^{-3}}$$

$$= -21.860 \text{ kN} \quad \text{(negative means compression)} \tag{5.3.3}$$

Table 5.3.1 Numerical computation for Example 5.3.1 (member CD removed)

Member	L, m	$A, 10^{-3}$ m^2	F', kN	u	$\dfrac{F'uL}{EA}$, mm	$\dfrac{u^2 L}{EA}, 10^{-3}$ mm	$F = F' - F_{CD}u$, kN	$\dfrac{FuL}{EA}$, mm
AC	7.5	4.0	−33.75	+0.62500	−0.19775	+3.6621	−20.087	−0.11770
BC	7.5	4.0	−33.75	+0.62500	−0.19775	+3.6621	−20.087	−0.11770
AD	5.7628	6.4	0	−0.80039	0	+2.8842	−17.497	+0.06305
BD	5.7628	6.4	0	−0.80039	0	+2.8842	−17.497	+0.06305
Σ					−0.39550	+13.0926		−0.10930

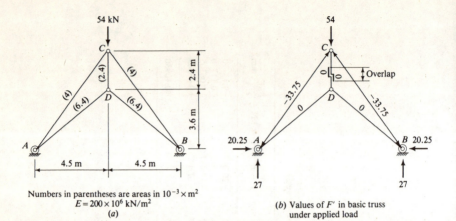
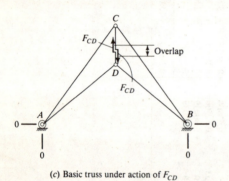
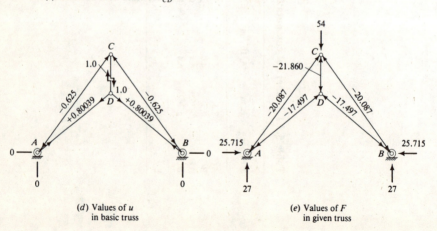

Figure 5.3.2 Truss of Example 5.3.1; solution by cutting member CD.

ANALYSIS OF TRUSSES BY THE FORCE METHOD 159

if the u values are as defined in Fig. 5.3.2d. Applying superposition,

$$F_i = F'_i + F_{CD} u_i \quad i = 1 \text{ to } 5 \quad (5.3.4)$$

The numerical work involved in Eqs. (5.3.3) and (5.3.4) is shown in Table 5.3.2. The last column in this table shows the check that if member CD is cut, the overlap there is zero so long as the forces as shown in Fig. 5.3.2e are maintained in the members.

Once the physical explanation is made for the self-lengthening effect in the method wherein the redundant member is removed, one may note that the cut member approach is in a way more elegantly conceptual although the negative sign following the first equals sign in Eq. (5.3.3) may be disturbing. However, there is no negative sign in Eq. (5.3.4) in contrast to its presence in Eq. (5.3.2). The reason is the fact that the u values in Eqs. (5.3.1) and (5.3.2) are due to a pair of forces *pushing* on the joints, while the u values in Eqs. (5.3.3) and (5.3.4) are due to a pair of forces *pulling* at the cut ends.

Example 5.3.2 Analyze the truss with cross diagonals and two hinged supports in Example 5.2.2 by using the force in member BC as the redundant.

SOLUTION As shown in Fig. 5.3.3, the compatibility condition is that the overlap at the cut ends in member BC in Fig. 5.3.3b plus the overlap there in Fig. 5.3.3c must be zero; or

$$\sum_1^5 \frac{F'uL}{EA} + \sum_1^5 \frac{(F_{BC}u)(u)(L)}{EA} = 0$$

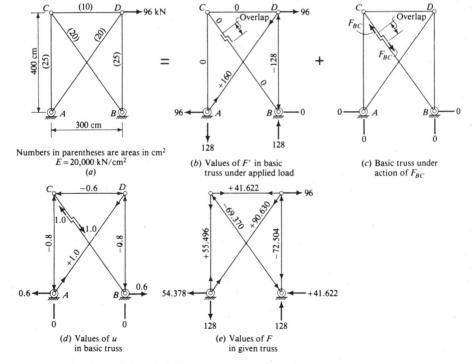

Figure 5.3.3 Truss of Example 5.3.2.

Table 5.3.2 Numerical computation for Example 5.3.1 (member CD cut)

Member	L, m	A, 10^{-3} m^2	F' kN	u	$\dfrac{F'uL}{EA}$, mm	$\dfrac{u^2 L}{EA}$, 10^{-3} mm	$F = F' + F_{CD}u$, kN	$\dfrac{FuL}{EA}$, mm
AC	7.5	4.0	−33.75	−0.62500	+0.19775	+3.6621	−20.087	−0.11770
BC	7.5	4.0	−33.75	−0.62500	+0.19775	+3.6621	−20.087	−0.11770
AD	5.7628	6.4	0	+0.80039	0	+2.8842	−17.497	+0.06305
BD	5.7628	6.4	0	+0.80039	0	+2.8842	−17.497	+0.06305
CD	2.4	2.4	0	+1.0	0	+5.0000	−21.860	−0.10930
Σ					+0.39550	+18.0926		0.00000

ANALYSIS OF TRUSSES BY THE FORCE METHOD **161**

from which

$$F_{BC} = -\frac{\sum_1^5 \dfrac{F'uL}{EA}}{\sum_1^5 \dfrac{u^2L}{EA}} = -\frac{+281.92 \times 10^{-3}}{+4.064 \times 10^{-3}}$$

$$= -69.370 \text{ kN} \quad \text{(negative means compression)} \quad (5.3.5)$$

if the u values are as defined in Fig. 5.3.3d. Applying superposition,

$$F_i = F'_i + F_{BC} u_i \qquad i = 1 \text{ to } 5 \quad (5.3.6)$$

The numerical work involved in Eqs. (5.3.5) and (5.3.6) is shown in Table 5.3.3. The last column in this table shows the check that if member BC is cut, the overlap there is zero so long as the forces as shown in Fig. 5.3.3e are maintained in the members.

Example 5.3.3 Analyze the truss shown in Fig. 5.3.4a by using the forces in members BC and DE as the redundants.

SOLUTION Using Eq. (5.1.1), $i = (r + m) - 2j = (4 + 10) - 2(6) = 2$. There are multiple choices for the two redundants. Using two parallel diagonals such as BC and DE as the redundants is probably the most convenient. Other choices include (a) using horizontal reaction at support F and member force AD, (b) using member forces AB and CD, or (c) using member forces AB and DE. Unacceptable choices are (a) using horizontal reaction at support F and member force DE, because there is no member force at E to balance a horizontal reaction; or (b) using member forces AB and BC because joint B will not be able to take a horizontal force.

If member forces BC and DE are used as the redundants, the two compatibility conditions are

$$\sum_1^{10} \frac{Fu_1 L}{EA} = 0 \quad \text{and} \quad \sum_1^{10} \frac{Fu_2 L}{EA} = 0 \quad (5.3.7)$$

in which F is the final force in each member of the given indeterminate truss, and u_1 and u_2 are as defined in Fig. 5.3.4e and f. Substituting

$$F = F' + u_1 F_{BC} + u_2 F_{DE} \quad (5.3.8)$$

into the compatibility equations gives

$$\sum_1^{10} \frac{F' u_1 L}{EA} + F_{BC} \sum_1^{10} \frac{u_1^2 L}{EA} + F_{DE} \sum_1^{10} \frac{u_1 u_2 L}{EA} = 0 \quad (5.3.9a)$$

and

$$\sum_1^{10} \frac{F' u_2 L}{EA} + F_{BC} \sum_1^{10} \frac{u_1 u_2 L}{EA} + F_{DE} \sum_1^{10} \frac{u_2^2 L}{EA} = 0 \quad (5.3.9b)$$

Using the numerical values for the summations obtained in Table 5.3.4, Eqs. (5.3.9a and b) become

$$+36.76 + 6.564 F_{BC} + 0.270 F_{DE} = 0$$

$$-83.72 + 0.270 F_{BC} + 6.294 F_{DE} = 0$$

Solving the two preceding equations for F_{BC} and F_{DE},

$$F_{BC} = -6.158 \text{ kN} \quad \text{(negative means compression)}$$
$$F_{DE} = +13.566 \text{ kN} \quad \text{(positive means tension)}$$

The F values computed by using Eq. (5.3.8) and shown in the last column of Table 5.3.4 are entered on the truss diagram of Fig. 5.3.4g, where the equilibrium equations $\Sigma F_x = 0$ and

Table 5.3.3 Numerical computation for Example 5.3.2

Member	L, cm	A, cm^2	F', kN	u	$\dfrac{F'uL}{EA}$, 10^{-3} cm	$\dfrac{u^2 L}{EA}$, 10^{-3} cm	$F = F' + F_{BC}u$, kN	$\dfrac{FuL}{EA}$, 10^{-3} cm
AC	400	25	0	−0.8	0	+0.512	+55.496	− 35.517
BD	400	25	−128	−0.8	+81.92	+0.512	−72.504	+ 46.402
AD	500	20	+160	+1.0	+200	+1.250	+90.630	+113.287
BC	500	20	0	+1.0	0	+1.250	−69.370	− 86.712
CD	300	10	0	−0.6	0	+0.540	+41.622	− 37.460
Σ					+281.92	+4.064		+ 0.000

ANALYSIS OF TRUSSES BY THE FORCE METHOD 163

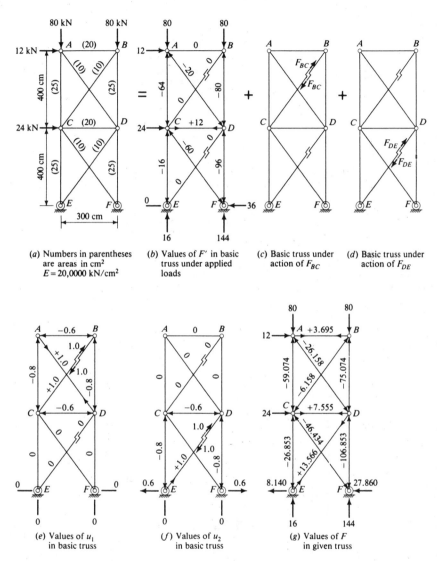

Figure 5.3.4 Truss of Example 5.3.3.

$\Sigma F_y = 0$ may be checked out for each joint. Although not shown in Table 5.3.4, the final F values may be used to check whether the original compatibility equations (5.3.7) are satisfied. However, these checks are not absolutely independent. For a good independent check, two different sets of u_1 and u_2 values determined from a different basic truss should be used with the final F values in the two compatibility equations (5.3.7).

Table 5.3.4 Numerical computation for Example 5.3.3

Member	L, cm	A, cm²	F', kN	u_1	u_2	$\dfrac{F'u_1 L}{EA}$, 10^{-3} cm	$\dfrac{F'u_2 L}{EA}$, 10^{-3} cm	$\dfrac{u_1^2 L}{EA}$, 10^{-3} cm	$\dfrac{u_1 u_2 L}{EA}$, 10^{-3} cm	$\dfrac{u_2^2 L}{EA}$, 10^{-3} cm	F, kN
AB	300	20	0	−0.6	0	0	0	+0.270	0	0	+ 3.695
CD	300	20	+12	−0.6	−0.6	− 5.40	− 5.40	+0.270	+0.270	+0.270	+ 7.555
AC	400	25	−64	−0.8	0	+40.96	0	+0.512	0	0	− 59.074
BD	400	25	+80	−0.8	0	+51.20	0	+0.512	0	0	− 75.074
CE	400	25	−16	0	−0.8	0	+ 10.24	0	0	+0.512	− 26.853
DF	400	25	−96	0	−0.8	0	+ 61.44	0	0	+0.512	−106.853
AD	500	10	−20	+1.0	0	−50	0	+2.500	0	0	− 26.158
BC	500	10	0	+1.0	0	0	0	+2.500	0	0	+ 6.158
CF	500	10	−60	0	+1.0	0	−150	0	0	+2.500	− 46.434
DE	500	10	0	0	+1.0	0	0	0	0	+2.500	+ 13.566
Σ						+36.76	− 83.72	+6.564	+0.270	+6.294	

5.4 Force Method Using Both Reactions and Axial Forces in Members as Redundants

When a truss has more than three reactions and at the same time has more than $(2j-3)$ members, it is both externally and internally indeterminate. The redundants may be chosen entirely from the member forces while keeping all reactions in the basic truss. A procedure more convenient and agreeable to analysts is to use $(r-3)$ redundant reactions and $(m-2j+3)$ redundant axial forces. In this way the basic truss will always have a hinged support and a roller support and contain $(2j-3)$ members.

Example 5.4.1 Analyze the truss shown in Fig. 5.4.1a by using the reaction at C and the axial force in member BF as the redundants.

SOLUTION Using Eq. (5.1.1), $i = (r+m) - 2j = (4+10) - 2(6) = 2$. The basic truss is a simple truss with a hinged support at A and a roller support at D; it has a stack of four triangles with uncut sides from left to right. Although member BF is cut, the parts on both sides of the cut stay with the basic truss. The equivalent basic truss shown in Fig. 5.4.1b is subjected to the two applied loads, the redundant reacting force R_C, and the pair of pulling forces F_{BF}.

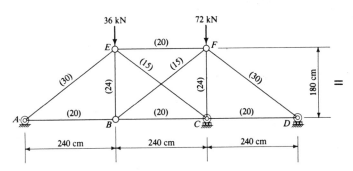

(a) Numbers in parentheses are areas in cm²
$E = 20{,}000$ kN/cm²

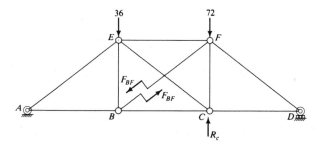

(b) Equivalent basic truss

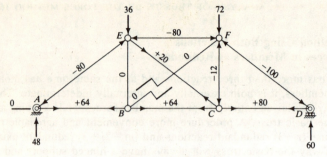

(c) Values of F' in basic truss under applied loads

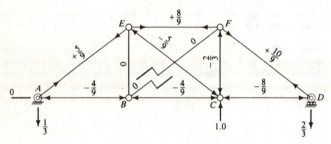

(d) Values of u_1 in basic truss

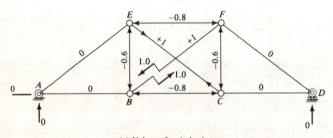

(e) Values of u_2 in basic truss

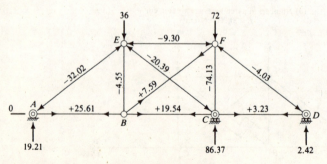

(f) Values of F in given truss

Figure 5.4.1 Truss of Example 5.4.1.

Table 5.4.1. Numerical computation for Example 5.4.1

Member	L, cm	A, cm²	F' kN	u_1	u_2	$\dfrac{F'u_1L}{EA}$, 10^{-3} cm	$\dfrac{F'u_2L}{EA}$, 10^{-3} cm	$\dfrac{u_1^2L}{EA}$, 10^{-3} cm	$\dfrac{u_1u_2L}{EA}$, 10^{-3} cm	$\dfrac{u_2^2L}{EA}$, 10^{-3} cm	F, kN
AB	240	20	+64	−4/9	0	−17.07	0	+0.1185	0	0	+25.61
BC	240	20	+64	−4/9	−0.8	−17.07	−30.72	+0.1185	+0.2133	+0.3840	+19.54
CD	240	20	+80	−8/9	0	−42.67	0	+0.4741	0	0	+3.23
EF	240	20	−80	+8/9	−0.8	−42.67	+38.40	+0.4741	−0.4267	+0.3840	−9.30
BE	180	24	0	0	−0.6	0	0	0	0	+0.1350	−4.55
CF	180	24	−12	−2/3	−0.6	+3.00	+2.70	+0.1667	+0.1500	+0.1350	−74.13
AE	300	30	−80	+5/9	0	−22.22	0	+0.1543	0	0	−32.02
BF	300	15	0	0	+1.0	0	0	0	0	+1.0000	+7.59
CE	300	15	+20	−5/9	+1.0	−11.11	+20.00	+0.3086	−0.5555	+1.0000	−20.39
DF	300	30	−100	+10/9	0	−55.55	0	+0.6173	0	0	−4.03
Σ						−205.36	+30.38	+2.4321	−0.6189	+3.0380	

acting on the cut ends of member BF. The two compatibility conditions are (1) the upward deflection at C is equal to zero, and (2) the overlap at the cut is equal to zero; or

$$\sum_1^{10} \frac{Fu_1L}{EA} = 0 \quad \text{and} \quad \sum_1^{10} \frac{Fu_2L}{EA} = 0 \tag{5.4.1}$$

in which F is the final force in each member of the given indeterminate truss, and u_1 and u_2 are as defined in Fig. 5.4.1d and e. Substituting

$$F = F' + u_1R_C + u_2F_{BF} \tag{5.4.2}$$

into the compatibility equations gives

$$\sum_1^{10} \frac{F'u_1L}{EA} + R_C \sum_1^{10} \frac{u_1^2L}{EA} + F_{BF} \sum_1^{10} \frac{u_1u_2L}{EA} = 0 \tag{5.4.3a}$$

and

$$\sum_1^{10} \frac{F'u_2L}{EA} + R_C \sum_1^{10} \frac{u_1u_2L}{EA} + F_{BF} \sum_1^{10} \frac{u_2^2L}{EA} = 0 \tag{5.4.3b}$$

Using the numerical values for the summations obtained in Table 5.4.1, Eqs. (5.4.3a and b) become

$$-205.36 + 2.4321R_C - 0.6189F_{BF} = 0$$
$$+30.38 - 0.6189R_C + 3.0380F_{BF} = 0$$

Solving the above two equations for R_C and F_{BF},

$$R_C = +86.37 \text{ kN} \quad \text{(positive means upward)}$$
$$F_{BF} = +7.59 \text{ kN} \quad \text{(positive means tension)}$$

The F values computed by using Eq. (5.4.2) and shown in the last column of Table 5.4.1 are entered on the truss diagram of Fig. 5.4.1f, where the equilibrium equations $\Sigma F_x = 0$ and $\Sigma F_y = 0$ may be checked out for each joint. Compatibility checks using the same redundants in Eqs. (5.4.1) are not quite independent because mistakes in values of F', u_1, and u_2 may cancel each other. For independent compatibility checks two different redundants ought to be taken, and then the final F values and the two new sets of u_1 and u_2 values can be substituted into Eqs. (5.4.1).

Another method for completely checking the correctness of the solution is to take the results of the analysis shown in Fig. 5.4.1f and then determine the horizontal and vertical deflections of each joint by a geometric method such as the joint-displacement-equation method. In this process, one will find that the vertical deflection of joint C will turn out to be zero and the decrease in distance, say between joints C and E, is equal to the shortening of member CE because all joint displacements *can* be obtained without knowing that the support C and member CE do exist. The member elongations and joint displacements are shown in Fig. 5.4.2.

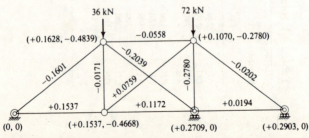

Figure 5.4.2 Member elongations and joint displacements in millimeters for truss in Example 5.4.1.

ANALYSIS OF TRUSSES BY THE FORCE METHOD 169

5.5 Induced Reactions on Statically Indeterminate Trusses due to Yielding of Supports

As discussed previously, the yielding of the supports of a statically indeterminate structure induces external reactions and thus internal forces in the structure. The procedure for obtaining the reactions by the force method, or the method of consistent deformation, is first to choose a basic determinate structure by including the reaction at the yielding support as a redundant and then apply the condition of consistent deformation that the deflection in the direction of that redundant is equal to the amount of yielding. The usual compatibility conditions will apply to the remaining redundants.

Example 5.5.1 Determine the reactions V_0, V_2, V_6, and V_8 as caused by a 15-mm settlement of the support at L_2 in the truss of Example 5.2.3.

SOLUTION It has been computed in Example 5.2.3 that, for the simple truss supported at L_0 and L_8 only,

1. A downward load of 1.0 kN applied at L_2 causes a downward deflection of 37.1094×10^{-3} mm at L_2 and a downward deflection of 19.8698×10^{-3} mm at L_6, and, by symmetry,
2. A downward load of 1.0 kN applied at L_6 causes a downward deflection of 19.8698×10^{-3} mm at L_2 and a downward deflection of 37.1094×10^{-3} mm at L_6.

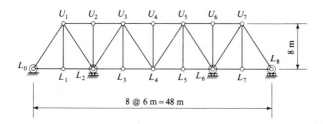

Area of all horizontal members = 12×10^{-3} m^2
Area of all vertical members = 3×10^{-3} m^2
Area of all diagonal members = 6×10^{-3} m^2
$E = 200 \times 10^6$ kN/m^2
(a)

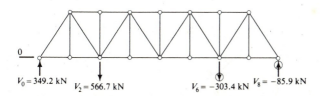

$V_0 = 349.2$ kN $V_2 = 566.7$ kN $V_6 = -303.4$ kN $V_8 = -85.9$ kN

(b) Induced reactions due to 15-mm settlement at L_2

Figure 5.5.1 Truss of Example 5.5.1.

In the present problem, let the unknown reactions be V_2 kN downward at L_2 and V_6 kN downward at L_6, as shown in Fig. 5.5.1b. Applying the condition of consistent deformation to the vertical deflections at L_2 and L_6,

$$37.1094 \times 10^{-3} V_2 + 19.8698 \times 10^{-3} V_6 = 15.0$$

and

$$19.8698 \times 10^{-3} V_2 + 37.1094 \times 10^{-3} V_6 = 0$$

Solving the above two equations for V_2 and V_6,

$V_2 = +566.7$ kN   (positive means downward)

$V_6 = -303.4$ kN   (negative means upward)

Using the two equations of statics for the entire truss shown in Fig. 5.5.1b,

$V_0 = 349.2$ kN upward

$V_8 = 85.9$ kN downward

The axial forces in the members resulting from the four reactions can then be determined by the methods of joints or sections.

5.6 Exercises

5.1 By the force method analyze the truss shown in Fig. 5.6.1 by using (a) the horizontal reaction at A as the redundant, and (b) the force in member CD as the redundant.

5.2 By the force method analyze the truss of Example 5.2.3 by using the vertical reactions at L_0 and L_8 as the redundants.

5.3 By the force method determine the reactions on the truss shown in Fig. 5.6.2 by using V_2 and V_4 as the redundants.

5.4 By the force method analyze the truss shown in Fig. 5.6.3 by using (a) the force in member U_1L_2 as the redundant, and (b) the force in member U_1U_2 as the redundant.

5.5 By the force method analyze the truss shown in Fig. 5.6.4 by using the forces in members L_1U_2 and U_2L_3 as the redundants. Check the solution by using two different members as the redundants.

5.6 By the force method analyze the truss of Example 5.3.3 by using the forces in members AD and CF as the redundants.

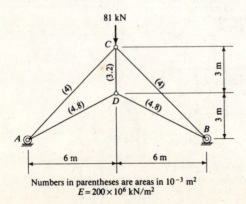

Numbers in parentheses are areas in 10^{-3} m^2
$E = 200 \times 10^6$ kN/m^2

Figure 5.6.1 Exercise 5.1.

ANALYSIS OF TRUSSES BY THE FORCE METHOD 171

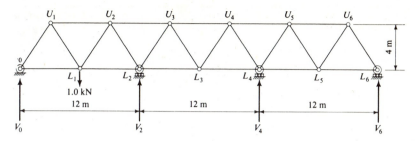

Area of all horizontal members = 3.6×10^{-3} m^2
Area of all diagonal members = 2.4×10^{-3} m^2
$E = 200 \times 10^6$ kN/m^2

Figure 5.6.2 Exercise 5.3.

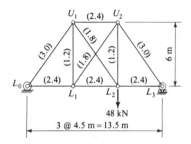

Numbers in parentheses are areas in 10^{-3} m^2
$E = 200 \times 10^6$ kN/m^2

Figure 5.6.3 Exercise 5.4.

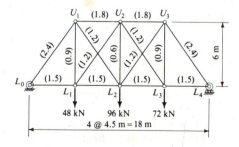

Numbers in parentheses are areas in 10^{-3} m^2
$E = 200 \times 10^6$ kN/m^2

Figure 5.6.4 Exercise 5.5.

5.7 By the force method analyze the truss of Example 5.4.1 by using the reaction at D and the force in member EC as the redundants.

5.8 By the force method determine all reactions induced to act on the truss of Exercise 5.3 by a vertical settlement of 15 mm at the second support.

CHAPTER
SIX
THE THREE-MOMENT EQUATION

6.1 Introduction

Statically indeterminate beams have been defined as those beams for which the equations of statics are not sufficient in determining the force and deformation response when loads are applied or when supports settle unevenly. In Chap. 4, the degree of indeterminacy of a statically indeterminate beam was shown to be

$$NI = NR - 2 - NIH \tag{4.2.1}$$

in which NI is the degree of indeterminacy, NR is the total number of reactions, and NIH is the number of internal hinges in the beam. Beams are not usually built with internal hinges, although in limit analysis,† wherein the responses of a beam are investigated after the yielding-moment capacities are reached at some sections of the beam, the beam behaves as if hinges were developed at these critical sections. Beams with internal hinges will not be treated in this chapter.

The highest possible degree of indeterminacy for a beam of one span (or of two supports) is 2 when both ends of the span are fixed. Beams with more than one span are called *continuous beams*, because they "continue" over the intermediate supports. The beams shown in Fig. 6.1.1 are continuous beams of four spans. The degrees of indeterminacy of the three beams in that figure are 3, 4, and 5, respectively.

---

†For example, see C.-K. Wang, *Matrix Methods of Structural Analysis*, 2d ed., American Publishing Company, Madison, Wis., 1970, Chap. 11.

THE THREE-MOMENT EQUATION **173**

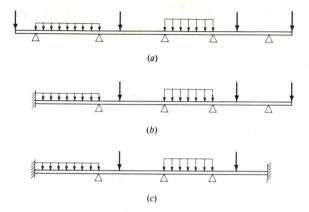

Figure 6.1.1 Continuous beams.

Continuous beams are common in bridge and building construction. The span lengths may be unequal, and the moments of inertia of the cross section may be different from one span to another. The analysis of such beams by the force method using reactions as redundants would require the computation of a large number of deflections or slopes in a basic beam with several variations in moments of inertia of its cross section. A more convenient approach is to use the statically unknown bending moments at the supports as the redundants. For the unknown moment at a fixed support, the compatibility condition is that the slope there must be zero. For the unknown moment at an intermediate support, the compatibility condition is that the slope of the elastic curve at the right end of the span to the left of the support must be equal to the slope of the elastic curve at the left end of the span to the right of the support. In this way, each span can be considered by itself as a simple beam with constant moment of inertia, acted upon by the loads on it and the moments at both ends if there are any. More efficiently, the compatibility condition corresponding to the unknown bending moment at any intermediate support can be expressed in terms of the loads on the two adjacent spans and the bending moments at *three* successive supports, including the one before and the one after the support being considered. Since this compatibility condition involves three bending moments at supports, it is called the *three-moment equation.* Clapeyron in 1857 described the analysis of continuous beams by use of the three-moment equation†; Mohr in 1860 further adapted the method to include the effects of uneven support settlements.‡ With the advent of computer methods, one will find that the solution of the three-moment equations, when written for a large number of spans, amounts to the

---

†S. P. Timoshenko, *History of Strength of Materials*, McGraw-Hill Book Company, New York, 1953, p. 145.
‡Ibid, p. 146.

174 INTERMEDIATE STRUCTURAL ANALYSIS

inversion of a matrix with only tridiagonal elements, a common subroutine in computer programming.

The purpose of this chapter is to derive the three-moment equation and illustrate its application to the analysis of statically indeterminate beams due to applied loads or due to uneven support settlements.

6.2 Derivation of the Three-Moment Equation

The three-moment equation expresses the relation between the bending moments at three successive supports of a continuous beam, subjected to loads applied on the two adjacent spans, with or without uneven settlements of supports. This relation can be derived on the basis of the *continuity* of the elastic curve over the middle support; that is, the slope of the elastic curve at the right end of the left span must be equal to the slope of the elastic curve at the left end of the right span.

Let AB and BC in Fig. 6.2.1a be the two adjacent spans in an originally horizontal beam. Owing to uneven settlements, supports A and C are at

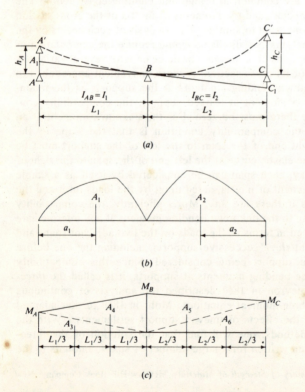

Figure 6.2.1 Moment diagrams on two adjacent spans of a continuous beam.

THE THREE-MOMENT EQUATION 175

higher elevations than support B by the amounts h_A and h_C, respectively; thus the elastic curve passes through points A', B, and C'. Let M_A, M_B, and M_C be the bending moments at A, B, and C, these moments being positive if they cause compression in the upper parts of the beam.

Now consider Fig. 6.2.2, where the moment diagram on span AB is broken into two parts: Fig. 6.2.2b represents the moment diagram due to loads applied on AB when it is considered as a simple beam, and Fig. 6.2.2c represents the moment diagram resulting from the moments M_A and M_B at the supports. By superposition, the entire moment diagram is shown in Fig. 6.2.2a.

Returning to Fig. 6.2.1, note that the moment diagrams on spans AB and BC are each broken into two parts: the parts A_1 and A_2 due to loads on the respective spans and the parts A_3, A_4 and A_5, A_6 due to end moments M_A, M_B on span AB and M_B, M_C on span BC. The simple-beam moment diagrams due to loads applied on the spans are known in advance, and the objective of the analysis is to find the bending moments M_A, M_B, and M_C at the supports.

A relation between M_A, M_B, and M_C may be derived from the compatibility condition that the beam is continuous at B, or the tangent at B to the elastic curve BA' is on the same straight line as the tangent at B to the elastic curve BC', as shown in Fig. 6.2.1a. In other words the joint B can be considered a rigid joint (monolithic in reinforced concrete construction, welded in steel construction); thus the two tangents at B to the elastic curves on both sides of B must remain at 180° to each other. Since the tangent A_1BC_1 in Fig. 6.2.1a must be a straight line,

$$\frac{AA_1}{L_1} = \frac{CC_1}{L_2} \tag{6.2.1}$$

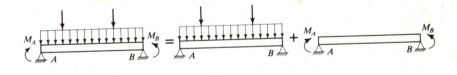

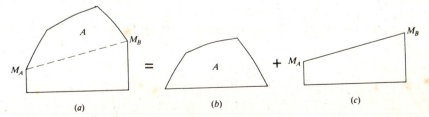

(a) (b) (c)

Figure 6.2.2 Superposition of moment diagrams on a typical span.

in which

$$AA_1 = h_A - A_1A' = h_A - (\text{deflection of } A' \text{ from the tangent at } B)$$

$$= h_A - \frac{1}{EI_1}\left(A_1a_1 + \frac{1}{3}A_3L_1 + \frac{2}{3}A_4L_1\right)$$

$$= h_A - \frac{1}{EI_1}\left(A_1a_1 + \frac{1}{6}M_AL_1^2 + \frac{1}{3}M_BL_1^2\right) \quad (6.2.2)$$

and

$$CC_1 = C_1C' - h_C = (\text{deflection of } C' \text{ from the tangent at } B) - h_C$$

$$= \frac{1}{EI_2}\left(A_2a_2 + \frac{2}{3}A_5L_2 + \frac{1}{3}A_6L_2\right) - h_C$$

$$= \frac{1}{EI_2}\left(A_2a_2 + \frac{1}{3}M_BL_2^2 + \frac{1}{6}M_CL_2^2\right) - h_C \quad (6.2.3)$$

Substituting Eqs. (6.2.2) and (6.2.3) into Eq. (6.2.1),

$$\frac{h_A}{L_1} - \frac{1}{L_1EI_1}\left(A_1a_1 + \frac{1}{6}M_AL_1^2 + \frac{1}{3}M_BL_1^2\right) = \frac{1}{L_2EI_2}\left(A_2a_2 + \frac{1}{3}M_BL_2^2 + \frac{1}{6}M_CL_2^2\right) - \frac{h_C}{L_2}$$

Multiplying every term in the above equation by $6E$ and reducing,

$$M_A\left(\frac{L_1}{I_1}\right) + 2M_B\left(\frac{L_1}{I_1} + \frac{L_2}{I_2}\right) + M_C\left(\frac{L_2}{I_2}\right) = -\frac{6A_1a_1}{L_1I_1} - \frac{6A_2a_2}{L_2I_2} + \frac{6Eh_A}{L_1} + \frac{6Eh_C}{L_2} \quad (6.2.4)$$

Equation (6.2.4) is the three-moment equation.

6.3 Application of the Three-Moment Equation to Analysis of Continuous Beams due to Applied Loads

The three-moment equation can be used to analyze statically indeterminate beams. For example, assume that the problem is to analyze the continuous beam of Fig. 6.3.1, which is subjected to the applied loads as shown. The beam is statically indeterminate to the third degree, but the redundancy will be removed when the bending moments at all supports become known. The moments at supports A and E can easily be found by inspection, following the laws of statics. To determine the moments at supports B, C, and D, three three-moment equations can be written on the basis of continuity at supports B, C, and D. In other words, the bending moments M_B, M_C, and M_D are

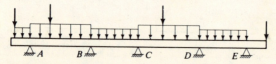

Figure 6.3.1 A typical continuous beam with known moments at extreme supports.

THE THREE-MOMENT EQUATION **177**

chosen as the redundants, and the conditions of consistent deformation are those of continuity, which can be represented by the three-moment equation. Thus there are always as many conditions of continuity as there are unknown bending moments at supports. Once the bending moments at all supports are known, each span can be treated separately as being subjected to the loads applied on the span and the end moments. Reactions at the ends of each span can be found by the laws of statics, and shear and moment diagrams can be drawn accordingly.

If the continuous beam has a fixed end, as shown in Fig. 6.3.2, the bending moment at the fixed support is one of the unknown redundants. The compatibility condition corresponding to the unknown fixed-end moment is that the slope of the tangent at A is zero. This condition can be met by adding an imaginary span $A_0 A$ of any length L_0 simply supported at A_0 and having an infinitely large moment of inertia for its cross section. In this way a three-moment equation using the fixed support A as the middle support can be written. Since the imaginary span $A_0 A$ has infinitely large moment of inertia, the moment diagram on it, whatever it may be, can yield no M/EI area, hence no elastic curve. So long as $A_0 A$ remains undeformable, the common tangent at A is a horizontal straight line.

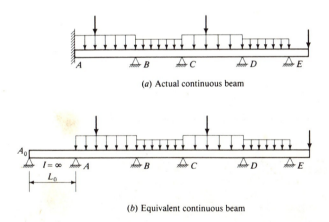

(a) Actual continuous beam

(b) Equivalent continuous beam

Figure 6.3.2 A typical continuous beam with unknown moment at fixed end.

Example 6.3.1 Analyze the continuous beam shown in Fig. 6.3.3a by using the three-moment equation. Draw shear and moment diagrams. Sketch the elastic curve.

SOLUTION The moment diagrams on AB, BC, and CD, obtained by considering each span as a simple beam subjected to the applied loads, are shown in Fig. 6.3.3b. Note that, for span BC, separate moment diagrams are drawn for the uniform load and for the concentrated load.

By inspection, $M_A = 0$ and $M_D = -36$ kN·m (negative because it causes compression in the lower part of the beam at D).

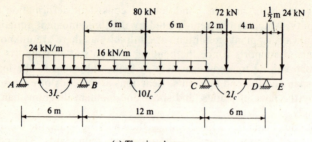

(a) The given beam

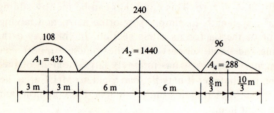

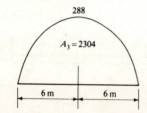

(b) Moment diagrams on simple spans due to applied loads

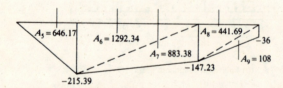

(c) Moment diagrams on simple spans due to end moments

Figure 6.3.3 Continuous beam of Example 6.3.1.

Applying the three-moment equation to

Spans AB and BC:

$$M_A\left(\frac{6}{3I_c}\right) + 2M_B\left(\frac{6}{3I_c} + \frac{12}{10I_c}\right) + M_C\left(\frac{12}{10I_c}\right) = -\frac{6(432)(3)}{6(3I_c)} - \frac{6(1440)(6)}{12(10I_c)} - \frac{6(2304)(6)}{12(10I_c)}$$

THE THREE-MOMENT EQUATION 179

Spans BC and CD:

$$M_B\left(\frac{12}{10I_c}\right) + 2M_C\left(\frac{12}{10I_c} + \frac{6}{2I_c}\right) + M_D\left(\frac{6}{2I_c}\right) = -\frac{6(1440)(6)}{12(10I_c)} - \frac{6(2304)(6)}{12(10I_c)} - \frac{6(288)(10/3)}{6(2I_c)}$$

Simplifying,

$$6.4M_B + 1.2M_C = -1555.2$$
$$1.2M_B + 8.4M_C = -1495.2$$

Solving,

$$M_B = -215.39 \text{ kN·m}$$
$$M_C = -147.23 \text{ kN·m}$$

Before proceeding any further, it is advantageous at this point to check the correctness of the values determined above for M_B and M_C. This can be done by checking θ_B in spans AB and BC and θ_C in spans BC and CD. Note that the total moment diagram on each span is the sum of those in Fig. 6.3.3b and c.

Applying the conjugate-beam method to

Span AB:

$$\theta_A = \frac{1}{E(3I_c)}\left(\frac{1}{2}A_1 - \frac{1}{3}A_5\right)$$

$$= +\frac{0.20}{EI_c} \quad \text{(positive means clockwise)}$$

$$\theta_B = \frac{1}{E(3I_c)}\left(-\frac{1}{2}A_1 + \frac{2}{3}A_5\right)$$

$$= +\frac{71.59}{EI_c} \quad \text{(positive means clockwise)}$$

Span BC:

$$\theta_B = \frac{1}{E(10I_c)}\left[\frac{1}{2}(A_2 + A_3) - \frac{2}{3}A_6 - \frac{1}{3}A_7\right]$$

$$= +\frac{71.60}{EI_c} \quad \text{(positive means clockwise)}$$

$$\theta_C = \frac{1}{E(10I_c)}\left[-\frac{1}{2}(A_2 + A_3) + \frac{1}{3}A_6 + \frac{2}{3}A_7\right]$$

$$= -\frac{85.23}{EI_c} \quad \text{(negative means counterclockwise)}$$

Span CD:

$$\theta_C = \frac{1}{E(2I_c)}\left(+\frac{10/3}{6}A_4 - \frac{2}{3}A_8 - \frac{1}{3}A_9\right)$$

$$= -\frac{85.23}{EI_c} \quad \text{(negative means counterclockwise)}$$

$$\theta_D = \frac{1}{E(2I_c)}\left(-\frac{8/3}{6}A_4 + \frac{1}{3}A_8 + \frac{2}{3}A_9\right)$$

$$= +\frac{45.62}{EI_c} \quad \text{(positive means clockwise)}$$

It is seen from the above computations that the same values of θ_B are obtained by using the moment diagrams on spans AB and BC, respectively. Likewise, the same values of θ_C are obtained by using the moment diagrams on spans BC and CD. This indicates that the three-moment equations have been established and solved correctly, but there is no proof that the areas A_1 through A_4 have been computed correctly. Values of θ_A and θ_D are also computed so that they may be shown on the elastic curve.

The reactions are determined as shown in Fig. 6.3.4a. The total reaction at the end of

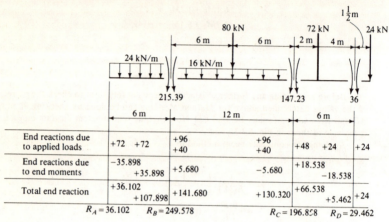

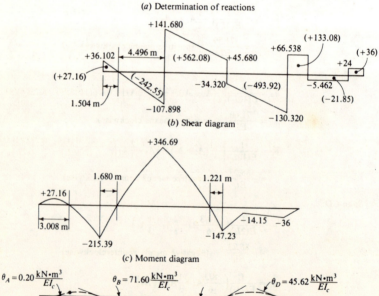

Figure 6.3.4 Solution for continuous beam in Example 6.3.1.

each span is equal to the sum of the reaction due to loads applied on the span and that due to the moments at the ends of the span. For instance, the sum of the end moment acting on span BC is $215.39 - 147.23 = 68.16$ kN·m *counterclockwise*, which requires a clockwise reaction couple, or an upward reaction of $68.16/12 = 5.680$ kN at B and a downward reaction of 5.680 kN at C. The total reaction to the continuous beam at support B is equal to the sum of the end reactions at B to spans BA and BC, or $R_B = 107.898 + 141.680 = 249.578$ kN. After all the reactions are determined, the shear diagram is drawn as shown in Fig. 6.3.4b. The point of zero shear on span AB is at $36.102/24 = 1.504$ m from support A. The area of each branch of the shear diagram is computed and indicated on the shear diagram. The moment diagram is plotted as shown in Fig. 6.3.4c, the relation that the change in moment between any two points is equal to the area of the shear diagram between those two points being used successively between convenient points. By so doing, M_B and M_C are checked back to be -215.39 and -147.23, thus indicating that the reactions found previously are correct. Also note that the moment diagram is linear where the shear is constant and the slope of the moment curve at any point is equal to the shear at that point. The *qualitative* elastic curve is shown in Fig. 6.3.4d.

Example 6.3.2 Analyze the continuous beam shown in Fig. 6.3.5a by using the three-moment equation. Draw shear and moment diagrams. Sketch the elastic curve.

SOLUTION The moment diagrams on AB, BC, and CD, obtained by considering each span as a simple beam subjected to the applied loads, are shown in Fig. 6.3.5b. Since support A is fixed, an imaginary span A_0A, of length L_0 and with $I = \infty$, is added. By inspection, $M_{A_0} = 0$ and $M_D = -36$ kN·m.

Applying the three-moment equation to

Spans A_0A and AB:

$$M_{A_0}\left(\frac{L_0}{\infty}\right) + 2M_A\left(\frac{L_0}{\infty} + \frac{6}{3I_c}\right) + M_B\left(\frac{6}{3I_c}\right) = -\frac{6(432)(3)}{6(3I_c)}$$

Spans AB and BC:

$$M_A\left(\frac{6}{3I_c}\right) + 2M_B\left(\frac{6}{3I_c} + \frac{12}{10I_c}\right) + M_C\left(\frac{12}{10I_c}\right) = -\frac{6(432)(3)}{6(3I_c)} - \frac{6(1440)(6)}{12(10I_c)} - \frac{6(2304)(6)}{12(10I_c)}$$

Spans BC and CD:

$$M_B\left(\frac{12}{10I_c}\right) + 2M_C\left(\frac{12}{10I_c} + \frac{6}{2I_c}\right) + M_D\left(\frac{6}{2I_c}\right) = -\frac{6(1440)(6)}{12(10I_c)} - \frac{6(2304)(6)}{12(10I_c)} - \frac{6(288)(10/3)}{6(2I_c)}$$

Simplifying,

$$4M_A + 2M_B = -432$$
$$2M_A + 6.4M_B + 1.2M_C = -1555.2$$
$$1.2M_B + 8.4M_C = -1495.2$$

Solving,

$$M_A = -0.36 \text{ kN·m}$$
$$M_B = -215.28 \text{ kN·m}$$
$$M_C = -147.24 \text{ kN·m}$$

It can be seen that the moment required to hold the tangent at A in the horizontal position is very small, because θ_A is only $+0.20$ kN·m^2/EI_c in Fig. 6.3.4d, when there is a simple support at A as in the previous example.

To check the values determined above for M_A, M_B, and M_C, the conjugate-beam method is again used to find θ_A, θ_B, and θ_C. The moment diagram on each span is the sum of those in Fig. 6.3.5b and c.

182 INTERMEDIATE STRUCTURAL ANALYSIS

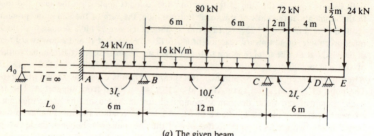

(a) The given beam

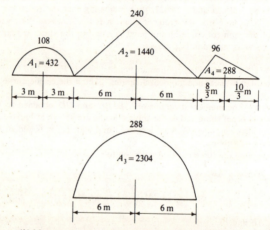

(b) Moment diagrams on simple spans due to applied loads

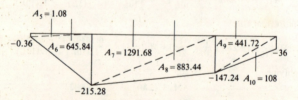

(c) Moment diagrams on simple spans due to end moments

Figure 6.3.5 Continuous beam of Example 6.3.2.

Applying the conjugate-beam method to

Span AB:

$$\theta_A = \frac{1}{E(3I_c)}\left(\frac{1}{2}A_1 - \frac{2}{3}A_5 - \frac{1}{3}A_6\right) = 0$$

$$\theta_B = \frac{1}{E(3I_c)}\left(-\frac{1}{2}A_1 + \frac{1}{3}A_5 + \frac{2}{3}A_6\right)$$

$$= +\frac{71.64}{EI_c} \quad \text{(positive means clockwise)}$$

Span BC:

$$\theta_B = \frac{1}{E(10I_c)}\left[\frac{1}{2}(A_2 + A_3) - \frac{2}{3}A_7 - \frac{1}{3}A_8\right]$$

$$= +\frac{71.64}{EI_c} \quad \text{(positive means clockwise)}$$

$$\theta_C = \frac{1}{E(10I_c)}\left[-\frac{1}{2}(A_2 + A_3) + \frac{1}{3}A_7 + \frac{2}{3}A_8\right]$$

$$= -\frac{85.25}{EI_c} \quad \text{(negative means counterclockwise)}$$

Span CD:

$$\theta_C = \frac{1}{E(2I_c)}\left(\frac{10/3}{6}A_4 - \frac{2}{3}A_9 - \frac{1}{3}A_{10}\right)$$

$$= -\frac{85.24}{EI_c} \quad \text{(negative means counterclockwise)}$$

$$\theta_D = \frac{1}{E(2I_c)}\left(-\frac{8/3}{6}A_4 + \frac{1}{3}A_9 + \frac{2}{3}A_{10}\right)$$

$$= +\frac{45.62}{EI_c} \quad \text{(positive means clockwise)}$$

The reactions, shear and moment diagrams, and the sketch for the elastic curve can then be found in the same way as in the previous example.

6.4 Application of the Three-Moment Equation to Analysis of Continuous Beams due to Uneven Support Settlements

The three-moment equation, Eq. (6.2.4), may be used to analyze continuous beams due to the simultaneous action of applied loads and uneven support settlements. In practice, it is often more convenient to consider the effect of uneven support settlements separately so that the designer may understand how much bending is caused by the applied loads and how much by the predicted amount of uneven support settlements. Obviously the beam must be designed to provide the strength capable of withstanding the bending due to all possible critical combinations of load patterns with support settlements.

Example 6.4.1 Using the three-moment equation, analyze the continuous beam shown in Fig. 6.4.1a for a 15-mm settlement of support B. Draw shear and moment diagrams. Sketch the elastic curve.

SOLUTION When there are only uneven settlements of supports but no applied loads, the three-moment equation becomes

$$M_A\left(\frac{L_1}{I_1}\right) + 2M_B\left(\frac{L_1}{I_1} + \frac{L_2}{I_2}\right) + M_C\left(\frac{L_2}{I_2}\right) = +\frac{6Eh_A}{L_1} + \frac{6Eh_C}{L_2}$$

In using the above equation, one must note that h_A and h_C are the amounts by which supports A and C are *higher* than the support B.

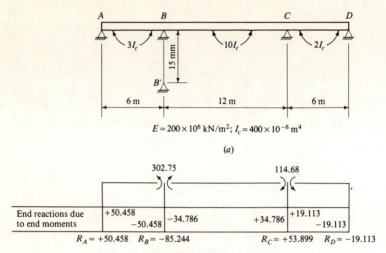

(a)

(b) Determination of reactions

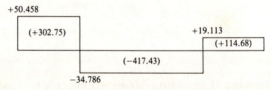

(c) Shear diagram

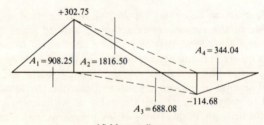

(d) Moment diagram

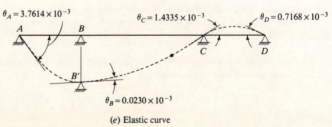

(e) Elastic curve

Figure 6.4.1 Continuous beam of Example 6.4.1.

THE THREE-MOMENT EQUATION

Applying the three-moment equation to

Spans AB and BC:

$$M_A\left(\frac{6}{3I_c}\right) + 2M_B\left(\frac{6}{3I_c} + \frac{12}{10I_c}\right) + M_C\left(\frac{12}{10I_c}\right) = +\frac{6E(+0.015)}{6} + \frac{6E(+0.015)}{12}$$

$$2M_A + 6.4M_B + 1.2M_C = +0.0225EI_c$$

Spans BC and CD:

$$M_B\left(\frac{12}{10I_c}\right) + 2M_C\left(\frac{12}{10I_c} + \frac{6}{2I_c}\right) + M_D\left(\frac{6}{2I_c}\right) = +\frac{6E(-0.015)}{12}$$

$$1.2M_B + 8.4M_C + 3M_D = -0.0075EI_c$$

Since $M_A = M_D = 0$ and $EI_c = 80{,}000$ kN·m, the three-moment equations become

$$6.4M_B + 1.2M_C = +1800$$
$$1.2M_B + 8.4M_C = -600$$

Solving,

$$M_B = +302.75 \text{ kN·m}$$
$$M_C = -114.68 \text{ kN·m}$$

The reactions, together with shear and moment diagrams, are shown in Fig. 6.4.1b to d. The sketch of the elastic curve is shown in Fig. 6.4.1e.

To check the correctness of M_B and M_C, the slopes at B and C of the elastic curve should each be computed by using the moment diagram on the span to the left and then to the right and seeing that the same result is obtained. For completeness the slopes at A and D are also computed so that the elastic curve in Fig. 6.4.1e may be better plotted.

Applying the conjugate-beam method to

Span AB:

$$\theta_A = R_{AB} + \frac{1}{3EI_c}\left(\frac{1}{3}A_1\right) = +\frac{0.015}{6} + \frac{1}{3(80{,}000)}\left[\frac{1}{3}(908.25)\right]$$

$$= +3.7614 \times 10^{-3}$$

$$\theta_B = R_{AB} + \frac{1}{3EI_c}\left(-\frac{2}{3}A_1\right) = +\frac{0.015}{6} + \frac{1}{3(80{,}000)}\left[-\frac{2}{3}(908.25)\right]$$

$$= -0.0229 \times 10^{-3}$$

Span BC:

$$\theta_B = R_{BC} + \frac{1}{10EI_c}\left(\frac{2}{3}A_2 - \frac{1}{3}A_3\right)$$

$$= -\frac{0.015}{12} + \frac{1}{10(80{,}000)}\left[\frac{2}{3}(1816.50) - \frac{1}{3}(688.08)\right]$$

$$= -0.0230 \times 10^{-3}$$

$$\theta_C = R_{BC} + \frac{1}{10EI_c}\left(-\frac{1}{3}A_2 + \frac{2}{3}A_3\right)$$

$$= -\frac{0.015}{12} + \frac{1}{10(80{,}000)}\left[-\frac{1}{3}(1816.50) + \frac{2}{3}(688.08)\right]$$

$$= -1.4335 \times 10^{-3}$$

186 INTERMEDIATE STRUCTURAL ANALYSIS

Span CD:

$$\theta_C = R_{CD} + \frac{1}{2EI_c}\left(-\frac{2}{3}A_4\right) = 0 + \frac{1}{2(80{,}000)}\left[-\frac{2}{3}(344.04)\right]$$
$$= -1.4335 \times 10^{-3}$$

$$\theta_D = R_{CD} + \frac{1}{2EI_c}\left(\frac{1}{3}A_4\right) = 0 + \frac{1}{2(80{,}000)}\left[\frac{1}{3}(344.04)\right]$$
$$= +0.7168 \times 10^{-3}$$

Example 6.4.2 Using the three-moment equation, analyze the continuous beam shown in Fig. 6.4.2 for a 15-mm settlement of support B. Draw shear and moment diagrams. Sketch the elastic curve.

SOLUTION Applying the three-moment equation to

Spans A_0A and AB:

$$M_{A_0}\left(\frac{L_0}{\infty}\right) + 2M_A\left(\frac{L_0}{\infty} + \frac{6}{3I_c}\right) + M_B\left(\frac{6}{3I_c}\right) = +\frac{6E(-0.015)}{6}$$
$$4M_A + 2M_B = -0.0150 EI_c$$

Spans AB and BC:

$$M_A\left(\frac{6}{3I_c}\right) + 2M_B\left(\frac{6}{3I_c} + \frac{12}{10I_c}\right) + M_C\left(\frac{12}{10I_c}\right) = +\frac{6E(+0.015)}{6} + \frac{6E(+0.015)}{12}$$
$$2M_A + 6.4M_B + 1.2M_C = +0.0225 EI_c$$

Spans BC and CD:

$$M_B\left(\frac{12}{10I_c}\right) + 2M_C\left(\frac{12}{10I_c} + \frac{6}{2I_c}\right) + M_D\left(\frac{6}{2I_c}\right) = +\frac{6E(-0.015)}{12}$$
$$1.2M_B + 8.4M_C + 3M_D = -0.0075 EI_c$$

Setting $M_D = 0$ and $EI_c = 80{,}000$ kN·m in the three-moment equations,

$$4M_A + 2M_B = -1200$$
$$2M_A + 6.4M_B + 1.2M_C = +1800$$
$$1.2M_B + 8.4M_C = -600$$

Solving,

$$M_A = -537.70 \text{ kN·m}$$
$$M_B = +475.41 \text{ kN·m}$$
$$M_C = -139.34 \text{ kN·m}$$

The reactions, shear and moment diagrams, and the elastic curve are shown in Fig. 6.4.2.

To check the correctness of M_A, M_B, and M_C, the three compatibility conditions can be verified by seeing that (1) $\theta_A = 0$, (2) θ_B computed from span BA = θ_B computed from span BC, and (3) θ_C computed from span CB = θ_C computed from span CD. Applying the conjugate-beam method to

Span AB:

$$\theta_A = R_{AB} + \frac{1}{3EI_c}\left(-\frac{2}{3}A_1 + \frac{1}{3}A_2\right)$$
$$= +\frac{0.015}{6} + \frac{1}{3(80{,}000)}\left[-\frac{2}{3}(1613.10) + \frac{1}{3}(1426.23)\right]$$
$$= 0$$

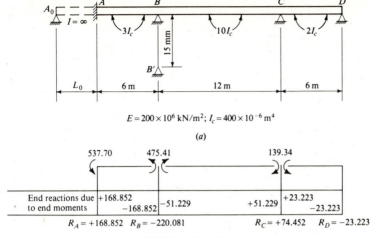

$E = 200 \times 10^6 \text{ kN/m}^2; I_c = 400 \times 10^{-6} \text{ m}^4$

(a)

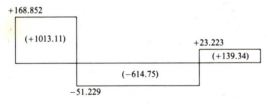

(b) Determination of reactions

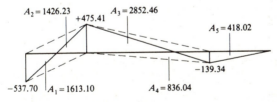

(c) Shear diagram

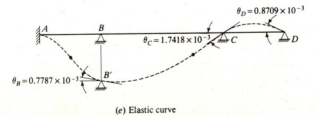

(d) Moment diagram

(e) Elastic curve

Figure 6.4.2 Continuous beam of Example 6.4.2.

$$\theta_B = R_{AB} + \frac{1}{3EI_c}\left(\frac{1}{3}A_1 - \frac{2}{3}A_2\right)$$

$$= +\frac{0.015}{6} + \frac{1}{3(80,000)}\left[\frac{1}{3}(1613.10) - \frac{2}{3}(1426.23)\right]$$

$$= +0.7787 \times 10^{-3}$$

Span BC:

$$\theta_B = R_{BC} + \frac{1}{10EI_c}\left(\frac{2}{3}A_3 - \frac{1}{3}A_4\right)$$

$$= -\frac{0.015}{12} + \frac{1}{10(80,000)}\left[\frac{2}{3}(2852.46) - \frac{1}{3}(836.04)\right]$$

$$= +0.7787 \times 10^{-3}$$

$$\theta_C = R_{BC} + \frac{1}{10EI_c}\left(-\frac{1}{3}A_3 + \frac{2}{3}A_4\right)$$

$$= -\frac{0.015}{12} + \frac{1}{10(80,000)}\left[-\frac{1}{3}(2852.46) + \frac{2}{3}(836.04)\right]$$

$$= -1.7418 \times 10^{-3}$$

Span CD:

$$\theta_C = R_{CD} + \frac{1}{2EI_c}\left(-\frac{2}{3}A_5\right)$$

$$= 0 + \frac{1}{2(80,000)}\left[-\frac{2}{3}(418.02)\right]$$

$$= -1.7418 \times 10^{-3}$$

$$\theta_D = R_{CD} + \frac{1}{2EI_c}\left(\frac{1}{3}A_5\right)$$

$$= 0 + \frac{1}{2(80,000)}\left[\frac{1}{3}(418.02)\right]$$

$$= +0.8709 \times 10^{-3}$$

6.5 Exercises

6.1 Analyze the continuous beam shown in Fig. 6.5.1 by using the three-moment equation. Draw shear and moment diagrams. Sketch the elastic curve. Also check the continuity of slopes at the two intermediate supports.

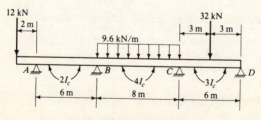

$E = 20 \times 10^6$ kN/m²; $I_C = 40 \times 10^{-6}$ m⁴

Figure 6.5.1 Exercise 6.1.

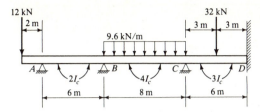

$E = 20 \times 10^6 \text{ kN/m}^2$; $I_C = 40 \times 10^{-6}$ m^4 **Figure 6.5.2** Exercise 6.2.

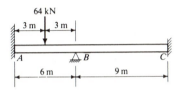

Constant EI **Figure 6.5.3** Exercise 6.5.

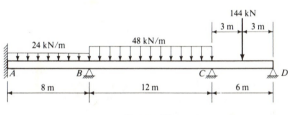

Constant EI

Figure 6.5.4 Exercise 6.6.

6.2 Analyze the continuous beam shown in Fig. 6.5.2 by using the three-moment equation. Draw shear and moment diagrams. Sketch the elastic curve. Also check the three compatibility conditions for the elastic curve.

6.3 Using the three-moment equation, analyze the continuous beam of Exercise 6.1 for a 4.5-mm settlement of support B without the applied loads. Draw shear and moment diagrams. Sketch the elastic curve. Compute the slopes of the elastic curve at all supports.

6.4 Using the three-moment equation, analyze the continuous beam of Exercise 6.2 for a 4.5-mm settlement of support B without the applied loads. Draw shear and moment diagrams. Sketch the elastic curve. Compute the slopes of the elastic curve at all supports.

6.5 and 6.6 Analyze the continuous beams shown in Figs. 6.5.3 and 6.5.4 by using the three-moment equation. Draw shear and moment diagrams. Sketch the elastic curve. Check the continuity of slopes at all supports.

CHAPTER
SEVEN
THE SLOPE-DEFLECTION METHOD

1% due to Axial loads (generally)

7.1 General Description

The slope-deflection method is a general method by which all beams and rigid frames, whether statically determinate or indeterminate, can be analyzed under the assumption that all deformation is due to the effect of bending moment only. Certainly the method has more appeal for the analysis of statically indeterminate beams and rigid frames; however, because of the method's generality, a computer program can be formulated, thus making it useful in the solution of both determinate and indeterminate problems.

The most striking feature of the slope-deflection method is that the rotational and translational displacements of the rigid joints are taken as the *primary* unknowns, and their values are determined prior to those of the bending moments at the member ends. In this way the degree of indeterminacy is no longer relevant because the so-called compatibility conditions in the force method are always satisfied to begin with when the joint displacements (rotations and translations) are taken as the unknowns. It will be shown that, by means of the slope-deflection equations to be derived in Secs. 7.2 and 7.4, all member-end moments and shears can be expressed in terms of the unknown joint displacements. It will also be shown that for each unknown joint rotation or translation, there is a corresponding condition of joint-moment or joint-force equilibrium. Consequently, there are always as many conditions of equilibrium as there are unknown joint displacements. Once the joint displacements are found, the slope-deflection equations are called upon again to give the member-end moments.

Required in the analysis of beams and rigid frames by the slope-deflection method are the relative values of the moment of inertia of the members in the

THE SLOPE-DEFLECTION METHOD **191**

case of analysis for applied loads, and their absolute values in the case of analysis for yielding of supports. For determinate structures, this requirement places an unnecessary burden on making preliminary assumptions for the member sizes, but for indeterminate structures this same requirement holds in the force method of analysis described in Chaps. 4 and 6.

As a simple example to show the merit of the slope-deflection method, consider the rigid frame in Fig. 7.1.1a, loaded as shown. Although the fact is not pertinent to the slope-deflection method, one may note that this rigid frame is statically indeterminate to the sixth degree. The force method could be used, but the amount of work involved would make that method too laborious. Because this rigid frame is kept from horizontal movement by the fixed support at A and from vertical movement by the fixed bases at D and E, and since axial deformation of the members is to be neglected, all five joints must remain in their original locations. (The case in which some joints may change positions when the rigid frame is deformed will be taken up later.) Clockwise joint rotations θ_B and θ_C are considered to be *positive*, as shown in Fig. 7.1.1a. The free-body diagrams of all members are shown in Fig. 7.1.1b. Note that, at any one end of each member, there may be three forces: direct

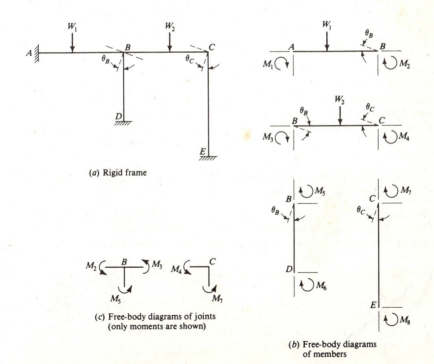

(a) Rigid frame

(c) Free-body diagrams of joints (only moments are shown)

(b) Free-body diagrams of members

Figure 7.1.1 Typical rigid frame without joint translation.

pull or thrust, end shear, and end moment (a moment is considered as a generalized force). Let the eight end moments acting at the ends of the four members be denoted by M_1 to M_8. *Clockwise* moments acting at the member ends are considered to be *positive*, as shown in Fig. 7.1.1b.

It is possible, by means of the slope-deflection equations to be derived in the next section, to express the two end moments acting on each member in terms of the two end rotations and the loads on the member. Thus the moments M_1 to M_8 in Fig. 7.1.1b can be expressed in terms of the two unknown joint rotations θ_B and θ_C. The free-body diagrams of joints B and C are shown in Fig. 7.1.1c. Of course, the action of the member on the joint consists of a force in the direction of the member axis, a force transverse to the member axis, and a moment, each being the opposite of the action of the joint on the member. In Fig. 7.1.1c, only the moments are shown. These moments are shown in their *positive* direction, which is *counterclockwise*. For equilibrium, summation of all moments acting on each joint must be zero; thus

$$M_2 + M_3 + M_5 = 0$$
$$M_4 + M_7 = 0$$

The above two equations are necessary and sufficient to determine the values of θ_B and θ_C. All end moments can then be found by substituting the known joint rotations back into the slope-deflection equations. Finally, by using simple statics, the axial force, shears, and moments in each member can be determined. To summarize, one can see that the continuity requirements are satisfied at the outset by taking the joint rotations as the unknowns, and the conditions of statics, requiring that the sum of moments acting on each joint with unknown rotation be zero, are used to solve for the unknown joint rotations. When the results of an analysis satisfy both statics and compatibility, their correctness is assured.

7.2 Derivation of the Slope-Deflection Equations—Without Rotation of Member Axis

For convenience in treatment, the slope-deflection equations for the simple case of Fig. 7.2.1a will be derived in this section, and those for the general case of Fig. 7.2.1b in Sec. 7.4. In this simple case, a straight line joining the

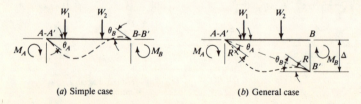

(a) Simple case (b) General case

Figure 7.2.1 Slope-deflection equations.

ends of the elastic curve $A'B'$ either coincides with the original member axis AB or is still parallel to it; thus there is no rotation of the member axis. In the general case, a straight line joining the ends of the elastic curve $A'B'$ makes a clockwise angle R equal to Δ/L from the original member axis AB. The name *slope-deflection equations*, however, comes from the general case, because they express the end moments M_A and M_B in terms of the end slopes θ_A and θ_B and the relative transverse deflection Δ of the ends; while Δ is equal to zero in the simple case.

For the simple case of no member-axis rotation, the problem is now to derive expressions for M_A and M_B in terms of θ_A and θ_B and the loads acting on AB. This condition of Fig. 7.2.2a can be separated into two conditions shown by Fig. 7.2.2b and c: Fig. 7.2.2b is called the *fixed condition*, in which the moments M_{0A} and M_{0B} are capable of maintaining zero slopes at A and B with loads acting on AB; and Fig. 7.2.2c is called the *joint-force condition*, in which the moments M'_A and M'_B are capable of maintaining the slopes θ_A and θ_B without loads acting on AB. Thus,

$$M_A = M_{0A} + M'_A \qquad (7.2.1a)$$

$$M_B = M_{0B} + M'_B \qquad (7.2.1b)$$

The fixed-end moments M_{0A} and M_{0B} can be separately determined by the force method described in Chap. 4; their expressions for a prismatic member subjected to a uniform load or a single concentrated load are shown in Fig. 7.2.3.

The end moments M'_A and M'_B required to maintain the slopes θ_A and θ_B can be obtained by referring to Fig. 7.2.4, to which the conjugate-beam

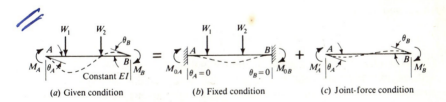

(a) Given condition (b) Fixed condition (c) Joint-force condition

Figure 7.2.2 Basis for slope-deflection equations: Simple case.

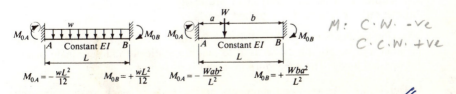

Figure 7.2.3 Fixed-end moments for uniform load and for single concentrated load.

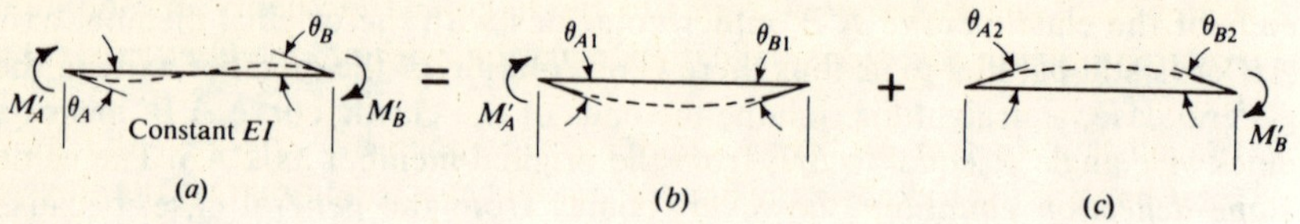

Figure 7.2.4 Statics and deformation of an unloaded member in bending, without member-axis rotation.

method is applied. Thus,

$$\theta_A = +\theta_{A1} - \theta_{A2} = +\frac{M'_A L}{3EI} - \frac{M'_B L}{6EI} \quad (7.2.2a)$$

$$\theta_B = -\theta_{B1} + \theta_{B2} = -\frac{M'_A L}{6EI} + \frac{M'_B L}{3EI} \quad (7.2.2b)$$

Solving Eqs. (7.2.2a and b) for M'_A and M'_B,

$$M'_A = +\frac{4EI}{L}\theta_A + \frac{2EI}{L}\theta_B \quad (7.2.3a)$$

$$M'_B = +\frac{2EI}{L}\theta_A + \frac{4EI}{L}\theta_B \quad (7.2.3b)$$

Substituting Eqs. (7.2.3a and b) into Eqs. (7.2.1a and b),

$$M_A = M_{0A} + \frac{2EI}{L}(2\theta_A + \theta_B) \quad (7.2.4a)$$

$$M_B = M_{0B} + \frac{2EI}{L}(2\theta_B + \theta_A) \quad (7.2.4b)$$

Or, in general,

$$\left(M_{\text{near end}} = M_{0\,(\text{near end})} + \frac{2EI}{L}(2\theta_{\text{near end}} + \theta_{\text{far end}})\right) \quad (7.2.5)$$

Equation (7.2.5) is the slope-deflection equation for a member in bending without member-axis rotation; it says that the moment at any end of a member in bending is equal to the fixed-end moment due to loads acting on the member plus $2EI/L$ times the sum of twice the slope at the near end and the slope at the far end.

7.3 Application to the Analysis of Statically Indeterminate Beams for Applied Loads

The slope-deflection equation, Eq. (7.2.5), for a member in bending without member-axis rotation may be used to analyze statically indeterminate beams for applied loads. The procedure involved is as follows:

1. Determine the fixed-end moments at the ends of each span, using the formulas for uniform and concentrated loads shown in Fig. 7.2.3.

THE SLOPE-DEFLECTION METHOD 195

2. Express all end moments in terms of the fixed-end moments and the joint rotations, using the slope-deflection equation without member-axis rotation.
3. Establish a system of linear simultaneous equations in which the joint rotations are the unknowns, using the equilibrium condition that the sum of the counterclockwise moments acting on each joint should be zero.
4. Solve for the unknown joint rotations.
5. Substitute the known values of the joint rotations back into the slope-deflection equations to obtain the end moments.
6. Determine all reactions, draw shear and moment diagrams, and sketch the elastic curve.

Example 7.3.1 Analyze the continuous beam of Example 6.3.1 by the slope-deflection method. Draw shear and moment diagrams. Sketch the elastic curve.

SOLUTION (a) *Fixed-end moments.* The given beam is shown in Fig. 7.3.1a. If the slopes at A, B, C, and D are held to be zero, the given beam may be separated into three fixed-end beams, shown in Fig. 7.3.1b, and a cantilever beam, which is not shown in Fig. 7.3.1b. The cantilever portion DE is not regarded as a bona fide member, so no slope-deflection equations will be written for it. The fixed-end moments M_{01} to M_{06} are, following the sign convention that a clockwise moment acting on the member end is positive,

$$M_{01} = -\frac{24(6)^2}{12} = -72 \text{ kN·m} \qquad M_{02} = +72 \text{ kN·m}$$

$$M_{03} = -\frac{16(12)^2}{12} - \frac{80(6)(6)^2}{12^2} = -312 \text{ kN·m} \qquad M_{04} = +312 \text{ kN·m}$$

$$M_{05} = -\frac{72(2)(4)^2}{6^2} = -64 \text{ kN·m} \qquad M_{06} = +\frac{72(4)(2)^2}{6^2} = +32 \text{ kN·m}$$

(b) *Slope-deflection equations.*

$$M_1 = M_{01} + \frac{2E(3I_c)}{6}(2\theta_A + \theta_B) = -72 + 2EI_c\theta_A + EI_c\theta_B$$

$$M_2 = M_{02} + \frac{2E(3I_c)}{6}(2\theta_B + \theta_A) = +72 + 2EI_c\theta_B + EI_c\theta_A$$

$$M_3 = M_{03} + \frac{2E(10I_c)}{12}(2\theta_B + \theta_C) = -312 + 3.333EI_c\theta_B + 1.667EI_c\theta_C$$

$$M_4 = M_{04} + \frac{2E(10I_c)}{12}(2\theta_C + \theta_B) = +312 + 3.333EI_c\theta_C + 1.667EI_c\theta_B$$

$$M_5 = M_{05} + \frac{2E(2I_c)}{6}(2\theta_C + \theta_D) = -64 + 1.333EI_c\theta_C + 0.667EI_c\theta_D$$

$$M_6 = M_{06} + \frac{2E(2I_c)}{6}(2\theta_D + \theta_C) = +32 + 1.333EI_c\theta_D + 0.667EI_c\theta_C$$

(c) *Simultaneous equations in θ_A, θ_B, θ_C, and θ_D.* The free-body diagrams of the members AB, BC, and CD, as well as those of the joints A, B, C, and D, are shown in Fig. 7.3.1c. Note that the end moments M_1 to M_6 are not yet known (even M_1 and M_6 should obviously be equal to zero and $+36$, but these are the joint-moment conditions at joints A and D), so they must be shown in their positive directions. For rotational equilibrium of the joints,

$$M_1 = 0 \qquad M_2 + M_3 = 0 \qquad M_4 + M_5 = 0 \qquad M_6 - 36 = 0$$

196 INTERMEDIATE STRUCTURAL ANALYSIS

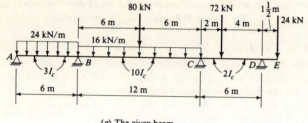

(a) The given beam

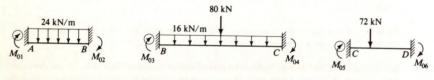

(b) Three members in the fixed condition

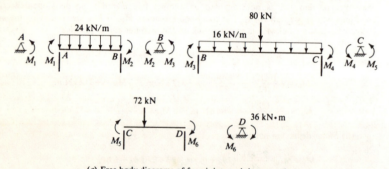

(c) Free-body diagrams of four joints and three members

Figure 7.3.1 Beam of Example 7.3.1.

Substituting the slope-deflection equations for M_1 to M_6 in the above equations and rearranging,

$$+2.000 EI_c \theta_A + 1.000 EI_c \theta_B = + 72.0$$
$$+1.000 EI_c \theta_A + 5.333 EI_c \theta_B + 1.667 EI_c \theta_C = +240.0$$
$$+1.667 EI_c \theta_B + 4.666 EI_c \theta_C + 0.667 EI_c \theta_D = -248.0$$
$$+0.667 EI_c \theta_C + 1.333 EI_c \theta_D = + 4.0$$

Note that if a diagonal is drawn as shown downward to the right on the left side of the above four equations, not only are the coefficients on this diagonal predominant in their own equations, but the other coefficients are symmetrical with respect to this diagonal. This can be proved to be always true by the nature of the slope-deflection equations and the joint-moment conditions. In order to observe this phenomenon, it is important to arrange the unknowns in the order of θ_A, θ_B, θ_C, and θ_D along the horizontal direction, and the joint-moment conditions in the order of joints A, B, C, and D in the vertical direction. In

Chap. 11, one will find a rigorous proof for this symmetry and the physical interpretation of it.

(d) *Solution of the simultaneous equations.* The simultaneous equations in θ_A, θ_B, θ_C, and θ_D may be solved by the forward elimination and backward substitution procedure. The results are

$$EI_c\theta_D = +45.62 \qquad EI_c\theta_C = -85.23$$
$$EI_c\theta_B = +71.60 \qquad EI_c\theta_A = + 0.20$$

Note that the results are exactly the same as those obtained in Example 6.3.1 and shown in Fig. 6.3.4d.

(e) *Back Substitution.*

$$M_1 = -72 + 2(+0.20) + 71.60 = 0$$
$$M_2 = +72 + 2(+71.60) + 0.20 = +215.4$$
$$M_3 = -312 + 3.333(+71.60) + 1.667(-85.23) = -215.4$$
$$M_4 = +312 + 3.333(-85.23) + 1.667(+71.60) = +147.3$$
$$M_5 = -64 + 1.333(-85.23) + 0.667(+45.62) = -147.2$$
$$M_6 = +32 + 1.333(+45.62) + 0.667(-85.23) = +36.0$$

Note that the results for M_1 to M_6 do satisfy the four joint-moment conditions (1) $M_1 = 0$, (2) $M_2 + M_3 = 0$, (3) $M_4 + M_5 = 0$, and (4) $M_6 - 36 = 0$.

(f) *Reactions, shear and moment diagrams, and elastic curve.* These have been done in Example 6.3.1 and shown in Fig. 6.3.4. Note, however, when placing the moments computed in part (e) above on the free-body diagrams of Fig. 6.3.4a, a positive moment acts clockwise at the member end and a negative moment acts counterclockwise at the member end. This sign convention is often called the _slope-deflection sign convention_, different from the _designer's sign convention_, which states that any bending moment causing compression on the top part of the cross section is positive. At the left end of the member, both sign conventions call a clockwise moment positive; yet at the right end, the slope-deflection sign convention calls a clockwise moment positive, while the designer's sign convention would call it negative because it causes compression on the lower part of the beam there.

Example 7.3.2 Analyze the continuous beam of Example 6.3.2 by the slope-deflection method. Draw shear and moment diagrams. Sketch the elastic curve.

SOLUTION The given beam is shown in Fig. 7.3.2a; the only difference between this beam and the one of the previous example is that the support at A is now fixed instead of its being a simple support. Therefore θ_A is zero for this beam. Letting $\theta_A = 0$ in the slope-deflection equations shown in part (b) of Example 7.3.1,

$$M_1 = -72 + EI_c\theta_B$$
$$M_2 = +72 + 2EI_c\theta_B$$
$$M_3 = -312 + 3.333EI_c\theta_B + 1.667EI_c\theta_C$$
$$M_4 = +312 + 3.333EI_c\theta_C + 1.667EI_c\theta_B$$
$$M_5 = -64 + 1.333EI_c\theta_C + 0.667EI_c\theta_D$$
$$M_6 = +32 + 1.333EI_c\theta_D + 0.667EI_c\theta_C$$

In fact, the three simultaneous equations in θ_B, θ_C, and θ_D for this problem are identical to the second, third, and fourth equations of Example 7.3.1 except that the terms involving

198 INTERMEDIATE STRUCTURAL ANALYSIS

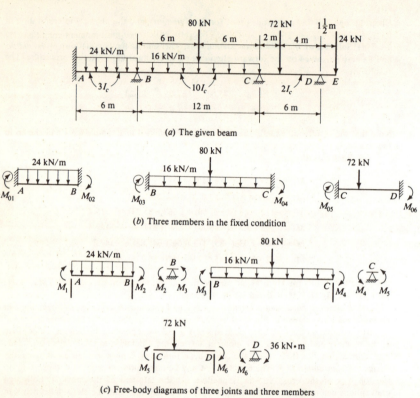

(a) The given beam

(b) Three members in the fixed condition

(c) Free-body diagrams of three joints and three members

Figure 7.3.2 Beam of Example 7.3.2.

θ_A are deleted; thus

$$5.333 EI_c\theta_B + 1.667 EI_c\theta_C = +240.0$$
$$1.667 EI_c\theta_B + 4.666 EI_c\theta_C + 0.667 EI_c\theta_D = -248.0$$
$$ 0.667 EI_c\theta_C + 1.333 EI_c\theta_D = + 4.0$$

Solving the above three equations,

$$EI_c\theta_B = +71.64 \qquad EI_c\theta_C = -85.25 \qquad EI_c\theta_D = +45.63$$

Substituting the values of θ_B, θ_C, and θ_D found above back into the slope-deflection equations,

$$M_1 = -72 + 71.64 = -0.36$$
$$M_2 = +72 + 2(+71.64) = +215.3$$
$$M_3 = -312 + 3.333(+71.64) + 1.667(-85.25) = -215.3$$
$$M_4 = +312 + 3.333(-85.25) + 1.667(+71.64) = +147.3$$
$$M_5 = -64 + 1.333(-85.25) + 0.667(+45.63) = -147.2$$
$$M_6 = +32 + 1.333(+45.63) + 0.667(-85.25) = +36.0$$

THE SLOPE-DEFLECTION METHOD 199

Note again that the above results for M_1 to M_6 satisfy the three joint-moment conditions (1) $M_2 + M_3 = 0$, (2) $M_4 + M_5 = 0$, and (3) $M_6 - 36.0 = 0$, as required by the free-body diagrams of joints B, C, and D in Fig. 7.3.2c. The reactions, shear and moment diagrams, and the elastic curve can be obtained as usual.

One must note at this point that even though the end moments as finally obtained satisfy equilibrium at the joints, there is no guarantee of the correctness of the analysis, because the fixed-end moments themselves may have been incorrectly computed and the $2EI/L$ values may have been incorrectly incorporated into the slope-deflection equations. For absolute assurance, it will be necessary to recompute the joint rotations from the final moment diagram and see that these joint rotations satisfy the continuity requirements at all joints.

7.4 Derivation of the Slope-Deflection Equations— With Rotation of Member Axis

The slope-deflection equations derived in Sec. 7.2 are only for the simple case of Fig. 7.2.1a where there is no rotation of the member axis. In the general case, one end of the member, say the right end, may be deflected by an amount Δ more than the left end so that the member axis has rotated through a clockwise angle R equal to Δ/L, as shown by Fig. 7.2.1b. The derivation of the slope-deflection equations for the general case follows the same pattern as that for the simple case. Again the given condition of Fig. 7.4.1a is separated into the fixed condition of Fig. 7.4.1b and the joint-force condition of Fig. 7.4.1c; thus

$$M_A = M_{0A} + M'_A \qquad (7.4.1a)$$

$$M_B = M_{0B} + M'_B \qquad (7.4.1b)$$

However, as shown in Fig. 7.4.1c, it is the end rotations ϕ_A and ϕ_B from the rotated member axis $A'B'$ to the elastic-curve tangents that are caused by the end moments M'_A and M'_B, not the end slopes θ_A and θ_B from the original member axis to the elastic-curve tangents. This fact has been emphasized in Chap. 2 in the application of the conjugate-beam theorem 1 and also in

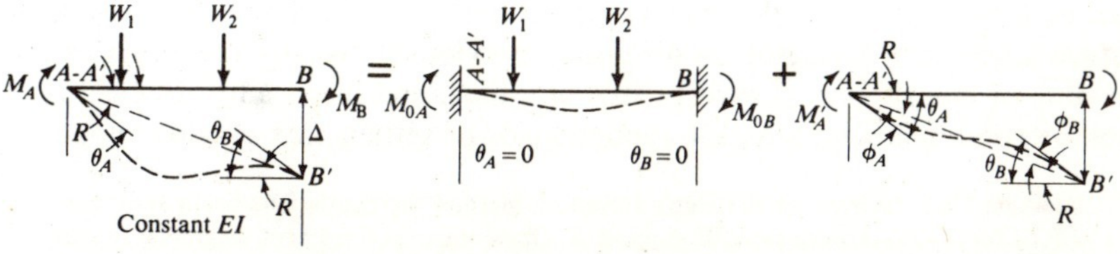

(a) Given condition   (b) Fixed condition   (c) Joint-force condition

Figure 7.4.1 Basis for slope-deflection equations: General case.

Example 4.5.2. Consequently, Eqs. (7.2.2a and b) become

$$\phi_A = \theta_A - R = +\frac{M'_A L}{3EI} - \frac{M'_B L}{6EI} \qquad (7.4.2a)$$

$$\phi_B = \theta_B - R = -\frac{M'_A L}{6EI} + \frac{M'_B L}{3EI} \qquad (7.4.2b)$$

and Eqs. (7.2.3a and b) become

$$M'_A = \frac{4EI}{L}(\theta_A - R) + \frac{2EI}{L}(\theta_B - R) \qquad (7.4.3a)$$

$$M'_B = \frac{2EI}{L}(\theta_A - R) + \frac{4EI}{L}(\theta_B - R) \qquad (7.4.3b)$$

Substituting Eqs. (7.4.3a and b) into Eqs. (7.4.1a and b),

$$M_A = M_{0A} + \frac{2EI}{L}(2\theta_A + \theta_B - 3R) \qquad (7.4.4a)$$

$$M_B = M_{0B} + \frac{2EI}{L}(2\theta_B + \theta_A - 3R) \qquad (7.4.4b)$$

Or, in general,

$$\boxed{M_{\text{near end}} = M_{0\ (\text{near end})} + \frac{2EI}{L}(2\theta_{\text{near end}} + \theta_{\text{far end}} - 3R)} \qquad (7.4.5)$$

Equation (7.4.5) is the slope-deflection equation for a member in bending with member-axis rotation. Of course, when there is no rotation of the member axis, R is equal to zero and Eq. (7.4.5) for the general case reverts to Eq. (7.2.5) of the simple case.

7.5 Application to the Analysis of Statically Indeterminate Beams for Uneven Support Settlements

The general slope-deflection equation, Eq. (7.4.5), can be used to analyze statically indeterminate beams due to the combined action of applied loads and uneven support settlements. In such cases, there are fixed-end moments due to loads acting on the members, as well as some known values of R before analysis begins. But, as discussed previously, usually the effect of the yielding of one support is investigated at a time, and the results obtained may then be combined with those of applied loads or settlements at other supports.

Example 7.5.1 Analyze by the slope-deflection method the continuous beam shown in Fig. 7.5.1a for a 15-mm settlement of support B. Draw shear and moment diagrams. Sketch the elastic curve. (Note that this problem is identical to Example 6.4.1.)

SOLUTION One may note at the outset that for this problem, the end moments M_1 to M_6 as shown in Fig. 7.5.1b will have to depend on the actual values of E and I_c, but the joint

THE SLOPE-DEFLECTION METHOD **201**

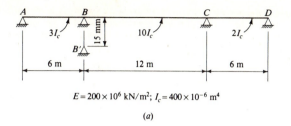

$E = 200 \times 10^6$ kN/m²; $I_c = 400 \times 10^{-6}$ m⁴

(a)

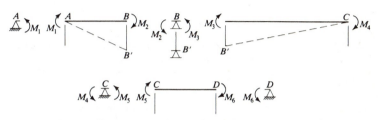

(b) Free-body diagrams of four joints and three members

Figure 7.5.1 Beam of Example 7.5.1.

rotations θ_A, θ_B, θ_C, and θ_D depend only on the relative magnitudes of the moment of inertia of the members and not on the actual values of E and I_c at all. For convenience, however, the actual values of EI/L of all spans will be used in the slope-deflection equations; thus

EI/L of $AB = 200(1200)/6 = 40{,}000$ kN·m

EI/L of $BC = 200(4000)/12 = 66{,}667$ kN·m

EI/L of $CD = 200(800)/6 = 26{,}667$ kN·m

The known R values are (note that clockwise rotation of the member axis is positive),

$$R_{AB} = +\frac{0.015}{6} = +0.0025 \quad R_{BC} = -\frac{0.015}{12} = -0.00125 \quad R_{CD} = 0$$

The slope-deflection equations are

$M_1 = 2(40{,}000)(2\theta_A + \theta_B - 0.0075) = 160{,}000\theta_A + 80{,}000\theta_B - 600$

$M_2 = 2(40{,}000)(2\theta_B + \theta_A - 0.0075) = 160{,}000\theta_B + 80{,}000\theta_A - 600$

$M_3 = 2(66{,}667)(2\theta_B + \theta_C + 0.00375) = 266{,}667\theta_B + 133{,}333\theta_C + 500$

$M_4 = 2(66{,}667)(2\theta_C + \theta_B + 0.00375) = 266{,}667\theta_C + 133{,}333\theta_B + 500$

$M_5 = 2(26{,}667)(2\theta_C + \theta_D) = 106{,}666\theta_C + 53{,}333\theta_D$

$M_6 = 2(26{,}667)(2\theta_D + \theta_C) = 106{,}667\theta_D + 53{,}333\theta_C$

Substituting the above slope-deflection equations into the joint conditions (1) $M_1 = 0$, (2) $M_2 + M_3 = 0$, (3) $M_4 + M_5 = 0$, and (4) $M_6 = 0$,

$160{,}000\theta_A + 80{,}000\theta_B \qquad\qquad\qquad\qquad = +600$

$80{,}000\theta_A + 426{,}667\theta_B + 133{,}333\theta_C \qquad\qquad = +100$

$\qquad\qquad + 133{,}333\theta_B + 373{,}333\theta_C + 53{,}333\theta_D = -500$

$\qquad\qquad\qquad\qquad + 53{,}333\theta_C + 106{,}666\theta_D = \quad 0$

On the left side of the preceding four equations note the symmetry in the coefficients of the joint rotations with respect to the principal diagonal downward to the right.

The solution of the simultaneous equations follows the usual pattern of forward elimination and backward substitution; thus

$$\theta_D = +0.7167 \times 10^{-3} \quad \theta_C = -1.4335 \times 10^{-3}$$
$$\theta_B = -0.0229 \times 10^{-3} \quad \theta_A = +3.7615 \times 10^{-3}$$

Substituting the above θ values back into the slope-deflection equations,

$$M_1 = +0.01 \quad M_2 = -302.74 \quad M_3 = +302.76$$
$$M_4 = +114.68 \quad M_5 = -114.68 \quad M_6 = 0.00$$

The above results for M_1 to M_6 clearly satisfy the four joint conditions. For an absolute independent check, however, it is necessary to compute the slopes from the moment diagram and see that the two compatibility conditions at joints B and C are satisfied.

The reactions, shear and moment diagrams, and the elastic curve have been worked out in Example 6.4.1 and shown in Fig. 6.4.1.

Example 7.5.2 Analyze by the slope-deflection method the continuous beam shown in Fig. 7.5.2a for a 15-mm settlement of support B. Draw shear and moment diagrams. Sketch the elastic curve. (Note that this problem is identical to Example 6.4.2.)

SOLUTION The only difference between the beam in this problem and the one in the previous example is that the support at A is now fixed instead of its being a simple support.

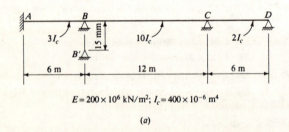

(b) Free-body diagrams of three joints and three members

Figure 7.5.2 Beam of Example 7.5.2.

Letting $\theta_A = 0$ in the slope-deflection equations of the previous example,

$$M_1 = 80,000\theta_B - 600$$
$$M_2 = 160,000\theta_B - 600$$
$$M_3 = 266,667\theta_B + 133,333\theta_C + 500$$
$$M_4 = 266,667\theta_C + 133,333\theta_B + 500$$
$$M_5 = 106,667\theta_C + 53,333\theta_D$$
$$M_6 = 106,667\theta_D + 53,333\theta_C$$

There are only three joint conditions in this problem instead of the four in the previous example because joint A is now fixed and it can furnish any nonzero moment M_1. The joint conditions (1) $M_2 + M_3 = 0$, (2) $M_4 + M_5 = 0$, and (3) $M_6 = 0$ as shown in Fig. 7.5.2b now become

$$426,667\theta_B + 133,333\theta_C \qquad\qquad = +100$$
$$133,333\theta_B + 373,333\theta_C + 53,333\theta_D = -500$$
$$\qquad\qquad + 53,333\theta_C + 106,666\theta_D = 0$$

Solving the preceding three equations,

$$\theta_D = +0.8709 \times 10^{-3} \qquad \theta_C = -1.7418 \times 10^{-3} \qquad \theta_B = +0.7787 \times 10^{-3}$$

Substituting the above θ values back into the slope-deflection equations,

$$M_1 = -537.70 \qquad M_2 = -475.41 \qquad M_3 = +475.41$$
$$M_4 = +139.34 \qquad M_5 = -139.34 \qquad M_6 = 0.00$$

The above results for M_1 to M_6 clearly satisfy the three joint conditions. For an absolute independent check, however, it is necessary to compute the slopes from the moment diagram and see that the three compatibility conditions at joints A, B, and C are satisfied.

7.6 Application to the Analysis of Statically Indeterminate Rigid Frames—Without Unknown Joint Translation

The slope-deflection method is extremely suitable for the analysis of statically indeterminate rigid frames. It is, of course, applicable to statically determinate rigid frames as well, but almost all rigid frames actually built in practice are statically indeterminate. Unlike the 180° rigid joints at the supports of continuous beams, more than two member ends may enter the same rigid joint, in which case the equilibrium condition corresponding to the unknown rotation of that joint will involve more than two end moments. For example, the equilibrium condition for the rigid joint shown in Fig. 7.6.1 is

$$M_2 + M_3 + M_{16} + M_{17} = 0 \qquad\qquad (7.6.1)$$

As discussed previously in Chaps. 2 and 4, common rigid-frame analysis is based on the assumption that the axial deformation, being many times smaller compared with bending deflection, may be ignored. Under this

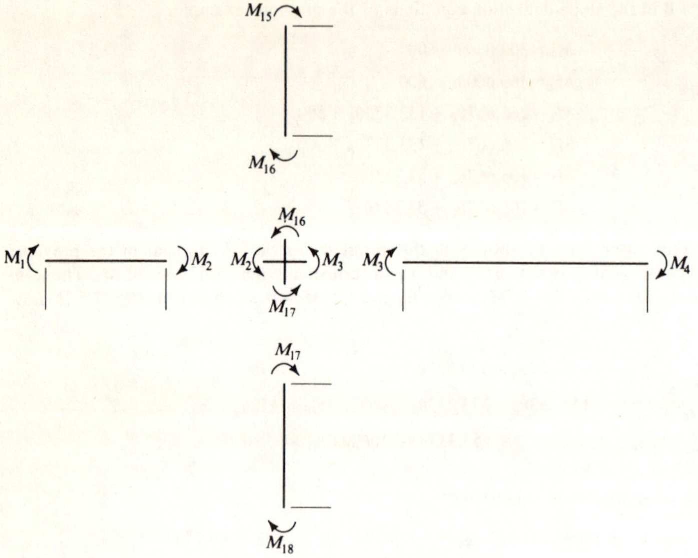

Figure 7.6.1 Joint-moment condition in slope-deflection method.

assumption the geometry of many rigid frames is such that none of the joints can undergo unknown change of position under applied loads or yielding of supports. Thus the unknown displacements involve only joint rotations, and all end moments can be expressed in terms of these unknowns through the slope-deflection equations. Then the joint conditions such as the one shown by Eq. (7.6.1) can be expressed in terms of the unknown rotations. Since there are always as many joint conditions as there are unknown joint rotations, these unknown joint rotations can be solved. By substituting the values of the joint rotations back into the slope-deflection equations, end moments are obtained. With all end moments known, the axial force, shears, and moments in all members can be found by applying the laws of statics to the individual members. Finally, the free-body, shear and moment diagrams, as well as the elastic curve, may be drawn.

Example 7.6.1 Analyze the rigid frame shown in Fig. 7.6.2a by the slope-deflection method. Determine the axial force, shears, and moments in all members. Sketch the elastic curve.

SOLUTION The load applied at A of the given rigid frame may be transferred to B with the action of a moment of 54 kN·m acting counterclockwise on joint B, as shown in Fig. 7.6.2b. Analysis may then be performed on the equivalent frame of Fig. 7.6.2b. Without axial deformation, joint B must remain 5 m to left of the fixed support C and 5 m above the fixed support D; thus joint B cannot move under any applied loads, and R is zero for all

THE SLOPE-DEFLECTION METHOD **205**

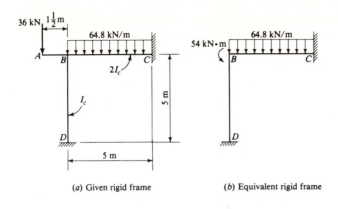

(a) Given rigid frame   (b) Equivalent rigid frame

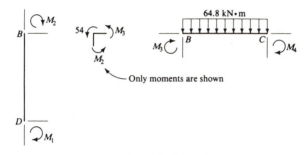

(c) Free-body diagrams

Figure 7.6.2 Rigid frame of Example 7.6.1.

members. Writing the slope-deflection equations and noting that $\theta_C = \theta_D = 0$,

$$M_1 = 0 + \frac{2EI_c}{5}(2\theta_D + \theta_B) = 0.40EI_c\theta_B$$

$$M_2 = 0 + \frac{2EI_c}{5}(2\theta_B + \theta_D) = 0.80EI_c\theta_B$$

$$M_3 = -\frac{64.8(5)^2}{12} + \frac{2E(2I_c)}{5}(2\theta_B + \theta_C) = -135 + 1.60EI_c\theta_B$$

$$M_4 = +\frac{64.8(5)^2}{12} + \frac{2E(2I_c)}{5}(2\theta_C + \theta_B) = +135 + 0.80EI_c\theta_B$$

The joint condition may be observed from the free-body diagram of joint B in Fig. 7.6.2c, or

$$M_2 + M_3 + 54 = 0$$
$$0.80EI_c\theta_B - 135 + 1.60EI_c\theta_B + 54 = 0$$
$$EI_c\theta_B = +33.75$$

Substituting the value of $EI_c\theta_B$ found above back into the slope-deflection equations,

$$M_1 = 0.40(+33.75) = +13.5$$
$$M_2 = 0.80(+33.75) = +27.0$$
$$M_3 = -135 + 1.60(+33.75) = -81.0$$
$$M_4 = +135 + 0.80(+33.75) = +162.0$$

The end moments obtained above are used to work out the complete solution as shown in Fig. 7.6.3. The moment diagram of Fig. 7.6.3b is plotted on the compression side; no positive or negative signs are indicated on it. A qualitative sketch of the elastic curve is shown in Fig. 7.6.3c to conform with the moment diagram and with the known fact that the rotation of joint B is clockwise.

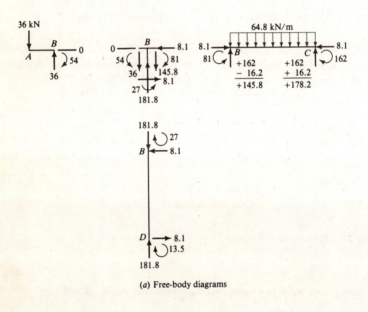

(a) Free-body diagrams

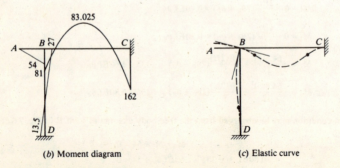

(b) Moment diagram   (c) Elastic curve

Figure 7.6.3 Solution of rigid frame in Example 7.6.1.

THE SLOPE-DEFLECTION METHOD 207

Example 7.6.2 Analyze the rigid frame shown in Fig. 7.6.4a by the slope-deflection method. Determine the axial force, shears, and moments in all members. Sketch the elastic curve.

SOLUTION Joints A, B, C, and D cannot have linear displacements; so $R = 0$ for all members. The unknown joint rotations are θ_A, θ_B, and θ_C. Using the end-moment number designation in Fig. 7.6.4b,

$$M_1 = -\frac{96(10)}{8} + \frac{2E(2I_c)}{10}(2\theta_A + \theta_B) = -120 + 0.80EI_c\theta_A + 0.40EI_c\theta_B$$

$$M_2 = +\frac{96(10)}{8} + \frac{2E(2I_c)}{10}(2\theta_B + \theta_A) = +120 + 0.80EI_c\theta_B + 0.40EI_c\theta_A$$

$$M_3 = -\frac{120(4)(6)^2}{10^2} + \frac{2E(2I_c)}{10}(2\theta_B + \theta_C) = -172.8 + 0.80EI_c\theta_B + 0.40EI_c\theta_C$$

$$M_4 = +\frac{120(6)(4)^2}{10^2} + \frac{2E(2I_c)}{10}(2\theta_C + \theta_B) = +115.2 + 0.80EI_c\theta_C + 0.40EI_c\theta_B$$

$$M_5 = 0 + \frac{2E(1.5I_c)}{6}(2\theta_C) = +1.00EI_c\theta_C$$

$$M_6 = 0 + \frac{2E(1.5I_c)}{6}(\theta_C) = +0.50EI_c\theta_C$$

Using the joint conditions (1) $M_1 = 0$, (2) $M_2 + M_3 = 0$, and (3) $M_4 + M_5 = 0$, the following simultaneous equations in θ_A, θ_B, and θ_C are obtained:

$$+0.80EI_c\theta_A + 0.40EI_c\theta_B \qquad\qquad = +120.0$$
$$+0.40EI_c\theta_A + 1.60EI_c\theta_B + 0.40EI_c\theta_C = + 52.8$$
$$\qquad\qquad +0.40EI_c\theta_B + 1.80EI_c\theta_C = -115.2$$

Note again the symmetry of the coefficients with respect to the main diagonal of the preceding three equations. Solving these equations,

$$EI_c\theta_C = -67.12 \qquad EI_c\theta_B = +14.03 \qquad EI_c\theta_A = +142.98$$

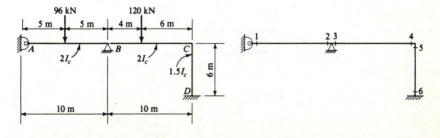

(a) The given rigid frame   (b) End-moment number designation

Figure 7.6.4 Rigid frame of Example 7.6.2.

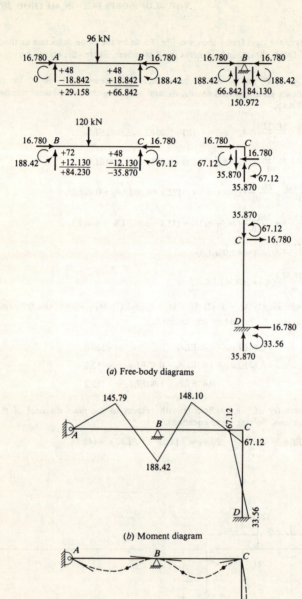

(a) Free-body diagrams

(b) Moment diagram

(c) Elastic curve

Figure 7.6.5 Solution of rigid frame in

THE SLOPE-DEFLECTION METHOD 209

Substituting the above values of joint rotations into the slope-deflection equations,

$$M_1 = -120 + 0.80(+142.98) + 0.40(+14.03) = 0.00$$
$$M_2 = +120 + 0.80(+14.03) + 0.40(+142.98) = +188.42$$
$$M_3 = -172.8 + 0.80(+14.03) + 0.40(-67.12) = -188.42$$
$$M_4 = +115.2 + 0.80(-67.12) + 0.40(+14.03) = +67.12$$
$$M_5 = +1.00(-67.12) = -67.12$$
$$M_6 = +0.50(-67.12) = -33.56$$

The free-body diagrams, the moment diagram, and the elastic curve are shown in Fig. 7.6.5.

Example 7.6.3 Analyze the rigid frame shown in Fig. 7.6.6a by the slope-deflection method. Determine the axial force, shears, and moments in all members. Sketch the elastic curve.

SOLUTION Joints D, E, and F are fixed. Joints A, B, and C cannot move in the vertical direction, but each may shift the same distance in the horizontal direction. In the present problem, however, on account of the symmetry both in the properties of the rigid frame itself and in the applied loads, joints A, B, and C will not have any horizontal displacement. Thus R is equal to zero for all members. In addition, θ_B has to be zero by reason of symmetry, and θ_C has to be equal in magnitude but opposite in sign to θ_A. Thus there is really only one unknown joint rotation, which is θ_A. Using the end-moment number designation in Fig. 7.6.6b,

$$M_1 = -256 + \frac{2E(4I_c)}{8}(2\theta_A + \theta_B) = -256 + 2EI_c\theta_A$$

$$M_2 = +256 + \frac{2E(4I_c)}{8}(2\theta_B + \theta_A) = +256 + EI_c\theta_A$$

$$M_3 = -256 + \frac{2E(4I_c)}{8}(2\theta_B + \theta_C) = -256 - EI_c\theta_A$$

$$M_4 = +256 + \frac{2E(4I_c)}{8}(2\theta_C + \theta_B) = +256 - 2EI_c\theta_A$$

$$M_5 = 0 + \frac{2EI_c}{6}(2\theta_A + \theta_D) = +0.6667 EI_c\theta_A$$

$$M_6 = 0 + \frac{2EI_c}{6}(2\theta_D + \theta_A) = +0.3333 EI_c\theta_A$$

$$M_7 = 0 + \frac{2EI_c}{6}(2\theta_B + \theta_E) = 0$$

$$M_8 = 0 + \frac{2EI_c}{6}(2\theta_E + \theta_B) = 0$$

$$M_9 = 0 + \frac{2EI_c}{6}(2\theta_C + \theta_F) = -0.6667 EI_c\theta_A$$

$$M_{10} = 0 + \frac{2EI_c}{6}(2\theta_F + \theta_C) = -0.3333 EI_c\theta_A$$

From the joint condition $M_1 + M_5 = 0$,

$$-256 + 2EI_c\theta_A + 0.6667 EI_c\theta_A = 0$$
$$EI_c\theta_A = +96$$

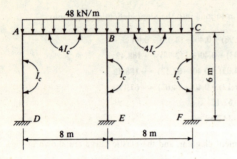

(a) The given rigid frame

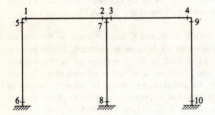

(b) End-moment number designation

Figure 7.6.6 Rigid frame of Example 7.6.3.

Substituting the value of θ_A found above back into the slope-deflection equations,

$M_1 = -64 \quad M_2 = +352 \quad M_3 = -352 \quad M_4 = +64 \quad M_5 = +64$

$M_6 = +32 \quad M_7 = 0 \quad M_8 = 0 \quad M_9 = -64 \quad M_{10} = -32$

The free-body diagrams, the moment diagram, and the elastic curve are shown in Fig. 7.6.7.

7.7 Application to the Analysis of Statically Indeterminate Rigid Frames—With Unknown Joint Translation

It has been stated repeatedly that for common rigid frames the axial deformation of the member is usually so small when compared with deformation due to bending moment that it is altogether ignored. Coupled with the first-order assumption that transverse displacements of the ends of any member do not affect its length, the problem that remains with rigid-frame analysis by the slope-deflection method is to determine, by observation or otherwise, whether the joints can have translation (linear displacement) in some direction. If so, such translations are unknown quantities in addition to unknown joint rotations in the making of the slope-deflection equations. Furthermore, for each unknown translation, there ought to be a corresponding equation of equilibrium so that there can always be as many equations of equilibrium as there are unknown joint rotations or translations.

THE SLOPE-DEFLECTION METHOD 211

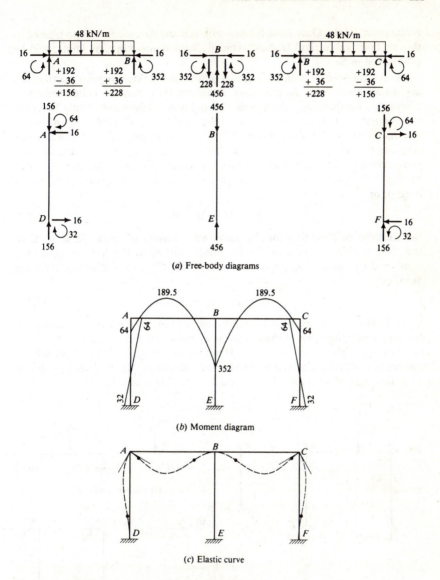

(a) Free-body diagrams

(b) Moment diagram

(c) Elastic curve

Figure 7.6.7 Solution of rigid frame in Example 7.6.3.

When all members in a rigid frame are either horizontal or vertical, it is called a rigid frame with rectangular joints. When nonhorizontal or non-vertical members enter a rigid joint, the angle between the original member axes there will be different from 90°, resulting in a rigid frame with nonrectangular joints. Only rigid frames with rectangular joints will be treated in this section; gable rigid frames, typically with nonrectangular joints, will be considered in Sec. 7.10.

For rigid frames with rectangular joints, the unknown joint translations are usually horizontal in direction; therefore they may be called unknown *sideways*. Furthermore, the number of unknown sideways should be equal to the number of stories in the rectangular rigid frame. Consider, for instance, the one-story rigid frame of Fig. 7.7.1a. The only possible unknown translation is that of the sideway to the right of joints A, B, or C. Just like the equilibrium condition for the unknown *clockwise* rotation of joint B, as shown by Fig. 7.7.1b, that the sum of *counterclockwise* moments acting on joint B is zero, or

$$M_2 + M_3 + M_7 = 0 \tag{7.7.1}$$

the equilibrium condition for the unknown sideway of joints A, B, or C to the *right* is that the sum of the horizontal forces to the *left* acting on the combined free-body diagram of joints A, B, and C, as shown by Fig. 7.7.1c, is zero, or

$$-W_1 - H_5 - H_7 - H_9 = 0 \tag{7.7.2}$$

wherein H_5, H_7, and H_9 may be expressed in terms of the end moments from the free-body diagrams of the columns, as shown in Fig. 7.7.1d.

For the typical two-story rigid frame shown in Fig. 7.7.2a, there are six unknown joint rotations and two unknown sideways which are designated as Δ_1 to the *right* of joints A, B, or C and Δ_2 to the *right* of joints D, E, or F. The

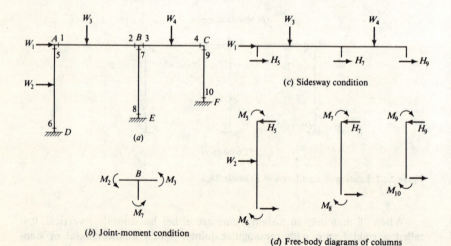

Figure 7.7.1 A typical one-story rigid frame.

THE SLOPE-DEFLECTION METHOD **213**

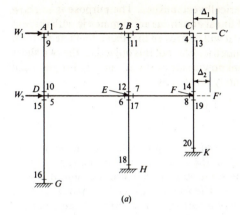

(a)

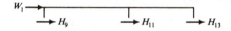

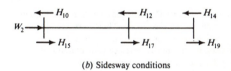

(b) Sidesway conditions

Figure 7.7.2 A typical two-story rigid frame.

two equilibrium conditions corresponding to the two unknown sidesways are obtained from equating the horizontal forces to the *left* acting on the combined free body of joints A, B, and C and then of joints D, E, and F, to zero. Thus,

$$-W_1 - H_9 - H_{11} - H_{13} = 0 \qquad (7.7.3a)$$

and $\qquad -W_2 + H_{10} + H_{12} + H_{14} - H_{15} - H_{17} - H_{19} = 0 \qquad (7.7.3b)$

Note that the subscripts used in the preceding two equations are the numbers assigned to each of the 20 member ends.

The conditions expressed by Eqs. (7.7.2) or (7.7.3) can be called the horizontal shear conditions, or simply the *shear conditions*, because if there are no horizontal forces acting between the member ends on the columns themselves, the H forces are the shears in the columns. In the following examples, the shear conditions are established by equating to zero the sum of the forces toward the left acting on a horizontal line of joints. This procedure will be "literally" obeyed until the simultaneous equations in the unknown joint rotations and translations are numerically established. The purpose is to show that the coefficients in the system of linear equations are symmetric with respect to the principal diagonal. In Chap. 12 it will be proved that these coefficients form a square matrix which is always symmetric. Without this objective, the two shear conditions corresponding to the unknown sidesways Δ_1 and Δ_2 for the rigid frame of Fig. 7.7.2 could be alternately written as

$$W_1 = H_{10} + H_{12} + H_{14} \tag{7.7.4a}$$

and
$$W_1 + W_2 = H_{16} + H_{18} + H_{20} \tag{7.7.4b}$$

As shown by Fig. 7.7.3, Eqs. (7.7.4a and b) simply state that the sum of all horizontal forces to the right acting on the rigid frame from the top down to the bases of a set of columns in the same story is equal to the sum of shears acting to the left at the column bases. The use of Eqs. (7.7.4a and b) instead of Eqs. (7.7.3a and b) will in no way affect the correctness of the solution; however, the symmetry of the coefficients with respect to the principal diagonal may not be observed in the simultaneous equations.

Example 7.7.1 Analyze the rigid frame shown in Fig. 7.7.4a by the slope-deflection method. Draw shear and moment diagrams. Sketch the elastic curve. (Note that this problem is identical to Example 4.6.1.)

SOLUTION Joints A, B, C, and D can all rotate. Joints B and C may shift an equal amount to the right; this unknown sidesway is Δ, as shown by Fig. 7.7.4b. Thus there are altogether five unknown displacements (note that rotations and translations are both called *displacements* to conform with the terminology used in the matrix displacement method of Chap. 12). Writing the slope-deflection equations for the six member-end moments,

$$M_1 = -\frac{48(4.5)(3)^2}{7.5^2} + \frac{2EI_c}{7.5}\left(2\theta_A + \theta_B - \frac{3\Delta}{7.5}\right)$$

$$= -34.56 + 0.53333EI_c\theta_A + 0.26667EI_c\theta_B - 0.10667EI_c\Delta$$

$$M_2 = +\frac{48(3)(4.5)^2}{7.5^2} + \frac{2EI_c}{7.5}\left(2\theta_B + \theta_A - \frac{3\Delta}{7.5}\right)$$

$$= +51.84 + 0.53333EI_c\theta_B + 0.26667EI_c\theta_A - 0.10667EI_c\Delta$$

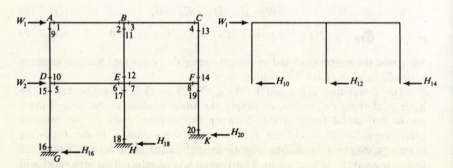

Figure 7.7.3 Alternate free-body diagrams for sidesway.

THE SLOPE-DEFLECTION METHOD

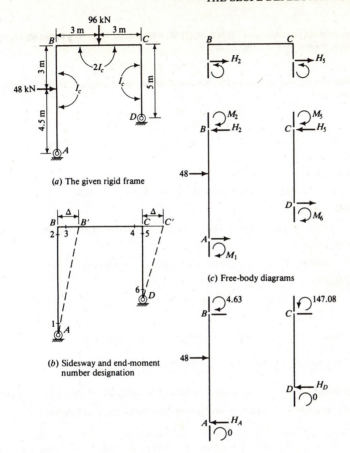

(a) The given rigid frame

(b) Sidesway and end-moment number designation

(c) Free-body diagrams

(d) Check for shear condition

Figure 7.7.4 Rigid frame of Example 7.7.1.

$$M_3 = -\frac{96(6)}{8} + \frac{2E(2I_c)}{6}(2\theta_B + \theta_C)$$

$$= -72 + 1.33333 EI_c\theta_B + 0.66667 EI_c\theta_C$$

$$M_4 = +\frac{96(6)}{8} + \frac{2E(2I_c)}{6}(2\theta_C + \theta_B)$$

$$= +72 + 1.33333 EI_c\theta_C + 0.66667 EI_c\theta_B$$

$$M_5 = 0 + \frac{2EI_c}{5}\left(2\theta_C + \theta_D - \frac{3\Delta}{5}\right)$$

$$= 0.80000 EI_c\theta_C + 0.40000 EI_c\theta_D - 0.24000 EI_c\Delta$$

$$M_6 = 0 + \frac{2EI_c}{5}\left(2\theta_D + \theta_C - \frac{3\Delta}{5}\right)$$

$$= 0.80000 EI_c\theta_D + 0.40000 EI_c\theta_C - 0.24000 EI_c\Delta$$

The joint-moment conditions are (1) $M_1 = 0$, (2) $M_2 + M_3 = 0$, (3) $M_4 + M_5 = 0$, and (4) $M_6 = 0$. The shear condition is, from Fig. 7.7.4c,

$$-H_2 - H_5 = 0$$

Expressions for H_2 and H_5 may be obtained from the free-body diagrams of the columns in Fig. 7.7.4c; thus

$$H_2 = +\frac{48(4.5)}{7.5} + \frac{M_1 + M_2}{7.5} \qquad H_5 = \frac{M_5 + M_6}{5}$$

The five simultaneous equations in which the unknowns are θ_A, θ_B, θ_C, θ_D, and Δ become

(1) from $M_1 = 0$,

$$+ 0.53333 EI_c \theta_A + 0.26667 EI_c \theta_B - 0.10667 EI_c \Delta = +34.56$$

(2) from $M_2 + M_3 = 0$,

$$+ 0.26667 EI_c \theta_A + 1.86666 EI_c \theta_B + 0.66667 EI_c \theta_C - 0.10667 EI_c \Delta = -20.16$$

(3) from $M_4 + M_5 = 0$,

$$+ 0.66667 EI_c \theta_B + 2.13333 EI_c \theta_C + 0.40000 EI_c \theta_D - 0.24000 EI_c \Delta = -72$$

(4) from $M_6 = 0$,

$$+ 0.40000 EI_c \theta_C + 0.80000 EI_c \theta_D - 0.24000 EI_c \Delta = 0$$

(5) from $-H_2 - H_5 = 0$,

$$-\left[\frac{48(4.5)}{7.5} + \frac{17.28 + 0.80000 EI_c \theta_A + 0.80000 EI_c \theta_B - 0.21333 EI_c \Delta}{7.5}\right]$$
$$-\left(\frac{1.20000 EI_c \theta_C + 1.20000 EI_c \theta_D - 0.48000 EI_c \Delta}{5}\right) = 0$$
$$-0.10667 EI_c \theta_A - 0.10667 EI_c \theta_B - 0.24000 EI_c \theta_C - 0.24000 EI_c \theta_D + 0.12444 EI_c \Delta = +31.104$$

The coefficients in the five simultaneous equations may be tabulated as follows:

$EI_c \theta_A$	$EI_c \theta_B$	$EI_c \theta_C$	$EI_c \theta_D$	$EI_c \Delta$	=
+0.53333	+0.26667			−0.10667	+34.56
+0.26667	+1.86666	+0.66667		−0.10667	+20.16
	+0.66667	+2.13333	+0.40000	−0.24000	−72.00
		+0.40000	+0.80000	−0.24000	0
−0.10667	−0.10667	−0.24000	−0.24000	+0.12444	+31.104

In Chap. 12, it will be proved that the 5×5 matrix to the left of the equals sign in the accompanying table has to be symmetric about the principal diagonal, and the column of constants below the equals sign contains the equivalent moments acting on joints A, B, C, and D and the equivalent horizontal force acting on the combined free body of joints B and C, wherein the equivalent system is defined as one which will cause the same joint rotations and sidesway as the original applied loads.

The method for solving a system of linear equations will be treated in Chap. 9. With

THE SLOPE-DEFLECTION METHOD 217

many kinds of electronic calculators easily accessible, the actual solution of the problem is no longer a difficult task either in concept or in execution. Solving for the joint displacements,

$$EI_c\Delta = +1432.7 \quad EI_c\theta_D = +409.12 \quad EI_c\theta_C = +41.40$$
$$EI_c\theta_B = +29.82 \quad EI_c\theta_A = +336.42$$

Substituting the above values into the slope-deflection equations,

$$M_1 = -0.01 \quad M_2 = +4.63 \quad M_3 = -4.64$$
$$M_4 = +147.08 \quad M_5 = -147.08 \quad M_6 = +0.01$$

Note that a quadrangular frame with two hinged supports is quite flexible in sidesway so that the large rotations at A and D and the large sidesway require a large number of significant figures in the slope-deflection equations in order to arrive at good values for the end moments. The end moments obviously satisfy the four joint-moment conditions. The shear condition may be checked by seeing if $H_A + H_D$ in Fig. 7.7.4d is equal to the horizontal load of 48 kN; or

$$H_A + H_D \stackrel{?}{=} 48$$

$$\left[\frac{4.8(3)}{7.5} - \frac{4.63}{7.5}\right] + \left(\frac{147.08}{5}\right) \stackrel{?}{=} 48$$

$$18.583 + 29.416 \approx 48 \quad \text{(OK)}$$

The solution is correct so long as the values of EI/L and the fixed-end moments are correctly entered into the slope-deflection equations. For an absolute independent check, the joint rotations and translations must be computed again from the final moment diagram and the compatibility condition (one for this singly indeterminate structure) checked as in Example 4.6.2. The free-body, shear and moment diagrams of the members, and the elastic curve are available in Example 4.6.1 or Fig. 4.6.5.

Example 7.7.2 Analyze the rigid frame shown in Fig. 7.7.5a by the slope-deflection method. Draw shear and moment diagrams. Sketch the elastic curve. (Note that this problem is identical to Example 4.6.2.)

SOLUTION The only difference between the rigid frame in Fig. 7.7.5a and that of the previous example is that the supports at A and D are now fixed instead of being hinged. Consequently θ_A and θ_D are both known to be zero to begin with, and the joint conditions $M_1 = 0$ and $M_6 = 0$ no longer apply. The three simultaneous equations in which θ_B, θ_C, and Δ are the unknowns will simply be identical to the second, third, and fifth equations in the previous example, without the first and fourth columns before the equals sign. Thus

$EI_c\theta_B$	$EI_c\theta_C$	$EI_c\Delta$	=
+1.86666	+0.66667	−0.10667	+20.16
+0.66667	+2.13333	−0.24000	−72.00
−0.10667	−0.24000	+0.12444	+31.104

Solving the three simultaneous equations in the table,

$$EI_c\Delta = +245.96 \quad EI_c\theta_C = -15.586 \quad EI_c\theta_B = +30.422$$

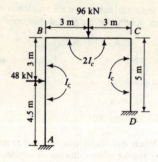
(a) The given rigid frame

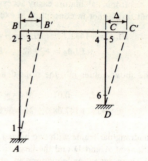
(b) Sidesway and end-moment number designation

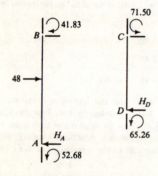
(c) Check for shear condition

Figure 7.7.5 Rigid frame of Example 7.7.2.

Substituting the above values into the slope-deflection equations, as shown in the previous example, except that $\theta_A = \theta_D = 0$,

$$M_1 = -34.56 + 0.26667(+30.422) - 0.10667(+245.96) = -52.68$$
$$M_2 = +51.84 + 0.53333(+30.422) - 0.10667(+245.96) = +41.83$$
$$M_3 = -72 + 1.33333(+30.422) + 0.66667(-15.586) = -41.83$$
$$M_4 = +72 + 1.33333(-15.586) + 0.66667(+30.422) = +71.50$$
$$M_5 = 0 + 0.80000(-15.586) - 0.24000(+245.96) = -71.50$$
$$M_6 = 0 + 0.40000(-15.586) - 0.24000(+245.96) = -65.26$$

Note that the two joint-moment conditions, (1) $M_2 + M_3 = 0$ and (2) $M_4 + M_5 = 0$, are obviously satisfied. The shear condition may be checked by seeing if $H_A + H_D$ in Fig. 7.7.5c is equal to the horizontal load of 48 kN; or

$$H_A + H_D \stackrel{?}{=} 48$$
$$\left[\frac{4.8(3)}{7.5} + \frac{52.68 - 41.83}{7.5}\right] + \left(\frac{71.50 + 65.26}{5}\right) \stackrel{?}{=} 48$$
$$20.647 + 27.352 \approx 48 \quad \text{(OK)}$$

The free-body, shear and moment diagrams of the members, and the elastic curve are available in Example 4.6.2 or Fig. 4.6.8.

Example 7.7.3 Analyze the rigid frame shown in Fig. 7.7.6a by the slope-deflection method. Draw shear and moment diagrams. Sketch the elastic curve.

SOLUTION Using the sidesway and end-moment number designation of Fig. 7.7.6b, the 10 slope-deflection equations are

$$M_1 = -\frac{96(8)}{8} + \frac{2E(3I_c)}{8}(2\theta_A + \theta_B) = -96 + 1.50EI_c\theta_A + 0.75EI_c\theta_B$$

$$M_2 = +\frac{96(8)}{8} + \frac{2E(3I_c)}{8}(2\theta_B + \theta_A) = +96 + 1.50EI_c\theta_B + 0.75EI_c\theta_A$$

$$M_3 = 0 + \frac{2E(3I_c)}{8}(2\theta_B + \theta_C) = +1.50EI_c\theta_B + 0.75EI_c\theta_C$$

$$M_4 = 0 + \frac{2E(3I_c)}{8}(2\theta_C + \theta_B) = +1.50EI_c\theta_C + 0.75EI_c\theta_B$$

$$M_5 = 0 + \frac{2E(2I_c)}{10}\left(2\theta_A + \theta_D - \frac{3\Delta}{10}\right) = +0.80EI_c\theta_A - 0.12EI_c\Delta$$

$$M_6 = 0 + \frac{2E(2I_c)}{10}\left(2\theta_D + \theta_A - \frac{3\Delta}{10}\right) = +0.40EI_c\theta_A - 0.12EI_c\Delta$$

$$M_7 = 0 + \frac{2E(2I_c)}{8}\left(2\theta_B + \theta_E - \frac{3\Delta}{8}\right) = +1.00EI_c\theta_B - 0.1875EI_c\Delta$$

$$M_8 = 0 + \frac{2E(2I_c)}{8}\left(2\theta_E + \theta_B - \frac{3\Delta}{8}\right) = +0.50EI_c\theta_B - 0.1875EI_c\Delta$$

$$M_9 = 0 + \frac{2E(2I_c)}{4}\left(2\theta_C + \theta_F - \frac{3\Delta}{4}\right) = +0.200EI_c\theta_C - 0.75EI_c\Delta$$

$$M_{10} = 0 + \frac{2E(2I_c)}{4}\left(2\theta_F + \theta_C - \frac{3\Delta}{4}\right) = +1.00EI_c\theta_C - 0.75EI_c\Delta$$

The joint-moment conditions corresponding to the unknown joint rotations θ_A, θ_B, and θ_C are (1) $M_1 + M_5 = 0$, (2) $M_2 + M_3 + M_7 = 0$, and (3) $M_4 + M_9 = 0$. The shear condition corresponding to the unknown sidesway Δ can be obtained from taking the joints A, B, and C together as a free body and using the direction toward the left as positive; thus from Fig. 7.7.6c,

$$-H_5 - H_7 - H_9 = 0$$

$$-\frac{M_5 + M_6}{10} - \frac{M_7 + M_8}{8} - \frac{M_9 + M_{10}}{4} = 0$$

The three joint-moment conditions and the shear condition when expressed in terms of the unknown joint displacements, are as shown in the accompanying table.

$EI_c\theta_A$	$EI_c\theta_B$	$EI_c\theta_C$	$EI_c\Delta$	=
+2.30	+0.75		−0.12	+96
+0.75	+4.00	+0.75	−0.1875	−96
	+0.75	+3.50	−0.75	0
−0.12	−0.1875	−0.75	+0.445875	0

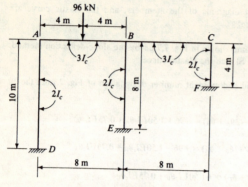

(a) The given rigid frame

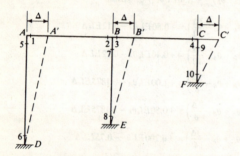

(b) Sidesway and end-momnet number designation

(c) Shear condition

(d) Check for shear condition

Figure 7.7.6 Rigid frame of Example 7.7.3.

Solving the four simultaneous equations in the table,
$$EI_c\Delta = +19.529 \qquad EI_c\theta_C = +11.79 \qquad EI_c\theta_B = -35.48 \qquad EI_c\theta_A = +54.33$$
Substituting the above values of the joint displacements back into the slope-deflection equations,

$$M_1 = -41.12 \qquad M_2 = +83.53 \qquad M_3 = -44.38 \qquad M_4 = -8.92$$
$$M_5 = +41.12 \qquad M_6 = +19.39 \qquad M_7 = -39.14 \qquad M_8 = -21.40$$
$$M_9 = +8.93 \qquad M_{10} = -2.86$$

The end moments obtained above obviously satisfy the joint-moment conditions (1) $M_1 + M_5 = 0$, (2) $M_2 + M_3 + M_7 = 0$, and (3) $M_4 + M_9 = 0$, with round-off error of ± 0.01. The shear condition may be checked by considering the entire rigid frame as a free body and using the equilibrium condition $\Sigma F_x = 0$; or, referring to Fig. 7.7.6d,

$$+H_D - H_E + H_F \stackrel{?}{=} 0$$
$$+\frac{41.12 + 19.39}{10} - \frac{39.14 + 21.40}{8} + \frac{8.93 + 2.86}{4} \stackrel{?}{=} 0$$
$$+6.051 - 7.568 + 1.518 \approx 0 \qquad \text{(OK)}$$

As has been said repeatedly, the solution thus far has to be correct provided the fixed-end moments and the EI/L values are entered correctly in the 10 slope-deflection equations.

The free-body diagrams, the moment diagram, and the elastic curve are shown in Fig. 7.7.7. In order to make a good sketch of the elastic curve, the joint rotation and the joint translation must be in the correct proportion. To achieve this, the angles can be plotted by their tangent functions; thus if θ_A in Fig. 7.7.7c is 54.33 kN·m^2/EI_c, the distance B_1B' in that figure should be equal to $8\theta_A$ or 434.6 kN·m^3/EI_c. Note again that the moment diagram is plotted on the compression side, to which the curvature of the elastic curve must conform.

Example 7.7.4 Analyze the rigid frame shown in Fig. 7.7.8a by the slope-deflection method. Draw shear and moment diagrams. Sketch the elastic curve.

SOLUTION Using the sidesway and end-moment number designation of Fig. 7.7.8b, the 12 slope-deflection equations for the six members are as follows (for convenience, the symbol EI_c has been dropped in front of the θ's and Δ's):

$$M_1 = 0 + \frac{2(3)}{4.8}(2\theta_A + \theta_B) = 2.50\theta_A + 1.25\theta_B$$

$$M_2 = 0 + \frac{2(3)}{4.8}(2\theta_B + \theta_A) = 2.50\theta_B + 1.25\theta_A$$

$$M_3 = 0 + \frac{2(3)}{4.8}(2\theta_C + \theta_D) = 2.50\theta_C + 1.25\theta_D$$

$$M_4 = 0 + \frac{2(3)}{4.8}(2\theta_D + \theta_C) = 2.50\theta_D + 1.25\theta_C$$

$$M_5 = 0 + \frac{2(2)}{6.4}\left[2\theta_A + \theta_C - \frac{3(\Delta_1 - \Delta_2)}{6.4}\right] = 1.25\theta_A + 0.625\theta_C - 0.29297\Delta_1 + 0.29297\Delta_2$$

$$M_6 = 0 + \frac{2(2)}{6.4}\left[2\theta_C + \theta_A - \frac{3(\Delta_1 - \Delta_2)}{6.4}\right] = 1.25\theta_C + 0.625\theta_A - 0.29297\Delta_1 + 0.29297\Delta_2$$

$$M_7 = 0 + \frac{2(2)}{6.4}\left[2\theta_B + \theta_D - \frac{3(\Delta_1 - \Delta_2)}{6.4}\right] = 1.25\theta_B + 0.625\theta_D - 0.29297\Delta_1 + 0.29297\Delta_2$$

$$M_8 = 0 + \frac{2(2)}{6.4}\left[2\theta_D + \theta_B - \frac{3(\Delta_1 - \Delta_2)}{6.4}\right] = 1.25\theta_D + 0.625\theta_B - 0.29297\Delta_1 + 0.29297\Delta_2$$

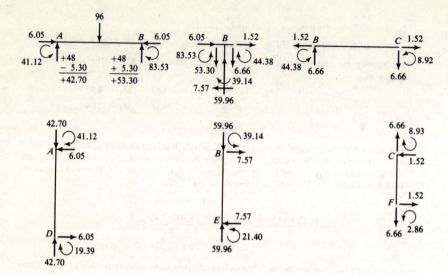

(a) Free-body diagrams

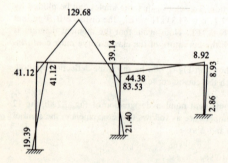

(b) Moment diagram

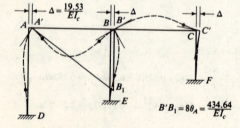

(c) Elastic curve

Figure 7.7.7 Solution of rigid frame in Example 7.7.3.

THE SLOPE-DEFLECTION METHOD **223**

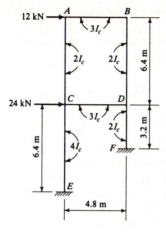

(a) The given rigid frame

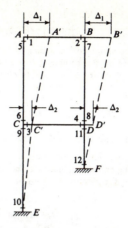

(b) Sidesway and end-moment number designation

$$H_6 = \frac{M_5 + M_6}{6.4} \qquad H_8 = \frac{M_7 + M_8}{6.4}$$

$12 \rightarrow A \qquad\qquad B$

$H_5 = \frac{M_5 + M_6}{6.4} \qquad H_7 = \frac{M_7 + M_8}{6.4}$

$-12 - H_5 - H_7 = 0$

$24 \rightarrow C \qquad\qquad D$

$H_9 = \frac{M_9 + M_{10}}{6.4} \qquad H_{11} = \frac{M_{11} + M_{12}}{3.2}$

$-24 + H_6 + H_8 - H_9 - H_{11} = 0$

(c) Shear conditions

Figure 7.7.8 Rigid frame of Example 7.7.4.

$$M_9 = 0 + \frac{2(4)}{6.4}\left(2\theta_C + \theta_E - \frac{3\Delta_2}{6.4}\right) = 2.50\theta_C - 0.58594\Delta_2$$

$$M_{10} = 0 + \frac{2(4)}{6.4}\left(2\theta_E + \theta_C - \frac{3\Delta_2}{6.4}\right) = 1.25\theta_C - 0.58594\Delta_2$$

$$M_{11} = 0 + \frac{2(2)}{3.2}\left(2\theta_D + \theta_F - \frac{3\Delta_2}{3.2}\right) = 2.50\theta_D - 1.17188\Delta_2$$

$$M_{12} = 0 + \frac{2(2)}{3.2}\left(2\theta_F + \theta_D - \frac{3\Delta_2}{3.2}\right) = 1.25\theta_D - 1.17188\Delta_2$$

The joint-moment conditions are, for $M_1 + M_5 = 0$,

$$+3.75\theta_A + 1.25\theta_B + 0.625\theta_C - 0.29297\Delta_1 + 0.29297\Delta_2 = 0$$

for $M_2 + M_7 = 0$,

$$+1.25\theta_A + 3.75\theta_B + 0.625\theta_D - 0.29297\Delta_1 + 0.29297\Delta_2 = 0$$

for $M_3 + M_6 + M_9 = 0$,

$$+0.625\theta_A + 6.25\theta_C + 1.25\theta_D - 0.29297\Delta_1 - 0.29297\Delta_2 = 0$$

and for $M_4 + M_8 + M_{11} = 0$,

$$+0.625\theta_B + 1.25\theta_C + 0.625\theta_D - 0.29297\Delta_1 - 0.87891\Delta_2 = 0$$

The shear condition corresponding to the unknown sidesway Δ_1 is that the summation of the forces acting on the combined free body of joints A and B to the *left* (opposite to the positive direction for Δ_1) is equal to zero; thus, from Fig. 7.7.8c,

$$-12 - H_5 - H_7 = 0$$

$$-12 - \frac{M_5 + M_6}{6.4} - \frac{M_7 + M_8}{6.4} = 0$$

$$-0.29297\theta_A - 0.29297\theta_B - 0.29297\theta_C - 0.29297\theta_D + 0.183106\Delta_1 - 0.183106\Delta_2 = +12$$

The shear condition corresponding to the unknown sidesway Δ_2 is that the summation of the forces acting on the combined free-body diagram of joints C and D to the *left* (opposite to the positive direction for Δ_2) is equal to zero; thus, from Fig. 7.7.8c,

$$-24 + H_6 + H_8 - H_9 - H_{11} = 0$$

$$-24 + \frac{M_5 + M_6}{6.4} + \frac{M_7 + M_8}{6.4} - \frac{M_9 + M_{10}}{6.4} - \frac{M_{11} + M_{12}}{3.2} = 0$$

$$+0.29297\theta_A + 0.29297\theta_B - 0.29297\theta_C - 0.87891\theta_D - 0.183106\Delta_1 + 1.098638\Delta_2 = +24$$

Note again the symmetry of the coefficients of the θ's and Δ's in the six simultaneous equations with respect to the principal diagonal.

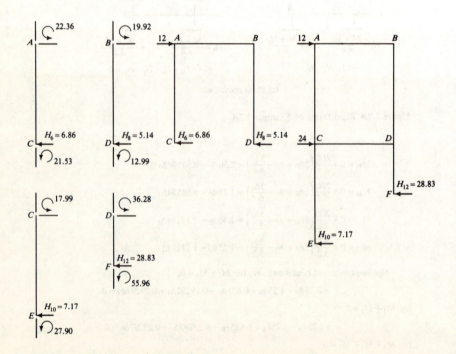

Figure 7.7.9 Check for shear conditions in Example 7.7.4.

THE SLOPE-DEFLECTION METHOD 225

Solving the six simultaneous equations,

$$\theta_A = +6.6138\frac{kN \cdot m^2}{EI_c} \qquad \theta_B = +4.6592\frac{dN \cdot m^2}{EI_c}$$

$$\theta_C = +7.9347\frac{kN \cdot m^2}{EI_c} \qquad \theta_D = +15.7440\frac{kN \cdot m^2}{EI_c}$$

$$\Delta_1 = +186.010\frac{kN \cdot m^3}{EI_c} \qquad \Delta_2 = +64.550\frac{kN \cdot m^3}{EI_c}$$

Substituting the above values of the joint rotations and translations back into the slope-deflection equations,

$$M_1 = +22.36 \qquad M_2 = +19.92 \qquad M_3 = +39.52 \qquad M_4 = +49.28$$
$$M_5 = -22.36 \qquad M_6 = -21.53 \qquad M_7 = -19.92 \qquad M_8 = -12.99$$
$$M_9 = -17.99 \qquad M_{10} = -27.90 \qquad M_{11} = -36.28 \qquad M_{12} = -55.96$$

The end moments obtained above obviously satisfy the joint conditions of (1) $M_1 + M_5 = 0$, (2) $M_2 + M_7 = 0$, (3) $M_3 + M_6 + M_9 = 0$, and (4) $M_4 + M_8 + M_{11} = 0$. The check for shear conditions may be made by using the alternate free-body diagrams for sidesway as described in Fig. 7.7.3. Thus the base shears at points 6 and 8 in Fig. 7.7.9 should be equal to the lateral load acting on AB, and the base shears at points 10 and 12 in the same figure should be equal to the sum of the lateral loads acting on AB and CD.

The free-body diagram, the moment diagram, and the elastic curve of the entire rigid frame are shown in Fig. 7.7.10.

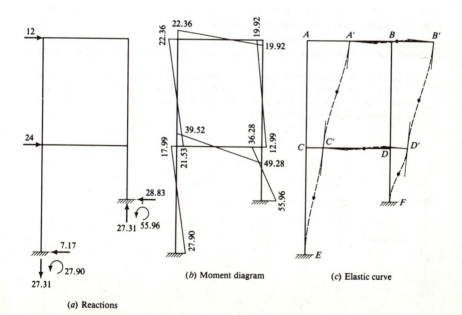

(a) Reactions (b) Moment diagram (c) Elastic curve

Figure 7.7.10 Solution of rigid frame in Example 7.7.4.

7.8 Application to the Analysis of Statically Indeterminate Rigid Frames due to Yielding of Supports

The induced reactions and the axial forces, shears, and moments in the members of a statically indeterminate rigid frame due to yielding of supports can be found by the slope-deflection method. Take, for instance, the rigid frame shown in Fig. 7.8.1. It is required to determine all end moments due to a rotational slip of α rad of joint D, and a horizontal movement Δ_1 to the right and a vertical settlement Δ_2 of joint E. Before writing the slope-deflection equations it is necessary to locate the possible displaced positions of the joints A, B, C, D, E, and F. The positions of D', E', and F' are given data, as shown in Fig. 7.8.1. Joints A, B, and C may shift an unknown amount of sideway Δ_3 to the right, but joint B must also move down an amount Δ_2 so that the length of member BE may not change. From Fig. 7.8.1, the R values for the five members are

$$R_{AB} = +\frac{\Delta_2}{L_{AB}} \qquad R_{BC} = -\frac{\Delta_2}{L_{BC}}$$

$$R_{AD} = +\frac{\Delta_3}{L_{AD}} \qquad R_{BE} = +\frac{\Delta_3 - \Delta_1}{L_{BE}} \qquad R_{CF} = +\frac{\Delta_3}{L_{CF}}$$

The unknown joint displacements are θ_A, θ_B, θ_C, and Δ_3, since it is known in advance that $\theta_D = +\alpha$, $\theta_E = 0$, and $\theta_F = 0$. The conditions of equilibrium corresponding to the four unknowns are the three joint-moment conditions and the shear condition. Generally the effect of one yielding is to be investigated at one time; the preceding discussion and the subsequent example, both containing more than one yielding, serve only to explain the procedure.

Example 7.8.1 By the slope-deflection method determine all reactions induced to act on the rigid frame of Fig. 7.8.2a by a rotational slip of 0.002 rad clockwise of joint D and a vertical settlement of 15 mm at joint D. Use $E = 200 \times 10^6 \text{ kN/m}^2$ and $I_c = 400 \times 10^{-6} \text{ m}^4$. (Note that this problem is identical to Example 4.7.1.)

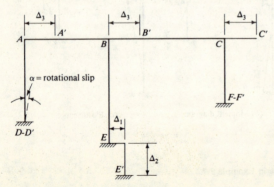

Figure 7.8.1 Rigid frame with yielding of supports.

THE SLOPE-DEFLECTION METHOD 227

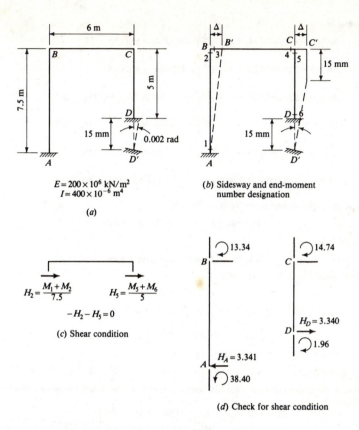

Figure 7.8.2 Rigid frame of Example 7.8.1.

SOLUTION Since there are no loads applied on the members, the fixed-end moments are all equal to zero. The possible displaced positions of the joints are shown in Fig. 7.8.2b. Joints B and C may shift a common sidesway Δ to the right, but joint C must also move 15 mm downward, the same amount as joint D. The values of EI/L of members AB, BC, and CD are, respectively,

$$EI/L \text{ of } AB = \frac{200(400)}{7.5} = 10{,}667 \text{ kN·m}$$

$$EI/L \text{ of } BC = \frac{200(800)}{6} = 26{,}667 \text{ kN·m}$$

$$EI/L \text{ of } CD = \frac{200(400)}{5} = 16{,}000 \text{ kN·m}$$

Using the end-moment number designations shown in Fig. 7.8.2b, the slope-deflection

228 INTERMEDIATE STRUCTURAL ANALYSIS

equations are

$$M_1 = 0 + 2(10{,}667)\left(2\theta_A + \theta_B - \frac{3\Delta}{7.5}\right) = 21{,}333\theta_B - 8533.3\Delta$$

$$M_2 = 0 + 2(10{,}667)\left(2\theta_B + \theta_A - \frac{3\Delta}{7.5}\right) = 42{,}667\theta_B - 8533.3\Delta$$

$$M_3 = 0 + 2(26{,}667)\left[2\theta_B + \theta_C - \frac{3(0.015)}{6}\right] = 106{,}667\theta_B + 53{,}333\theta_C - 400$$

$$M_4 = 0 + 2(26{,}667)\left[2\theta_C + \theta_B - \frac{3(0.015)}{6}\right] = 106{,}667\theta_C + 53{,}333\theta_B - 400$$

$$M_5 = 0 + 2(16{,}000)\left(2\theta_C + 0.002 - \frac{3\Delta}{5}\right) = 64{,}000\theta_C + 64 - 19{,}200\Delta$$

$$M_6 = 0 + 2(16{,}000)\left[2(+0.002) + \theta_C - \frac{3\Delta}{5}\right] = 128 + 32{,}000\theta_C - 19{,}200\Delta$$

The joint-moment conditions are, for $M_2 + M_3 = 0$,

$$+149{,}333\theta_B + 53{,}333\theta_C - 8533.3\Delta = +400$$

and for $M_4 + M_5 = 0$,

$$+53{,}333\theta_B + 170{,}667\theta_C - 19{,}200\Delta = +336$$

The shear condition is, from Fig. 7.8.2c,

$$-H_2 - H_5 = 0$$

$$-\frac{M_1 + M_2}{7.5} - \frac{M_5 + M_6}{5} = 0$$

$$-8533.3\theta_B - 19{,}200\theta_C + 9955.5\Delta = +38.4$$

Solving the three simultaneous equations for θ_B, θ_C, and Δ,

$$\theta_B = +2.4253 \times 10^{-3} \text{ rad} \qquad \theta_C = +2.3992 \times 10^{-3} \text{ rad} \qquad \Delta = +10.563 \times 10^{-3} \text{ m}$$

Substituting the above values of joint rotations and translation back into the slope-deflection equations,

$$M_1 = -38.40 \qquad M_2 = +13.34 \qquad M_3 = -13.34$$
$$M_4 = -14.74 \qquad M_5 = +14.74 \qquad M_6 = +1.96$$

The end moments obtained above obviously satisfy the joint-moment conditions of (1) $M_2 + M_3 = 0$ and (2) $M_4 + M_5 = 0$. The check for shear condition may be made by computing H_A and H_D in Fig. 7.8.2d and seeing that H_A is equal to H_D when the entire rigid frame is taken as a free body.

The complete free-body, shear and moment diagrams, the elastic curve, and the compatibility check have been worked out in Example 4.7.1 and shown in Fig. 4.7.2.

7.9 Analysis of Gable Frames by the Slope-Deflection Method

In the slope-deflection method, the primary unknowns are the joint displacements, which may include rotations and translations. For each unknown joint displacement, there is always a corresponding equilibrium condition. For any rigid frame to be analyzed, it is a simple matter to count the number of rigid joints which are *free* to rotate, and this number may be called the *degree of freedom in rotation*. The equilibrium condition corresponding to each unknown *clockwise* joint rotation is simply that the sum of the *counterclockwise* moments acting by the adjoining members on the joint is equal to zero.

Next it is necessary to determine whether each rigid joint can have unknown translation. For instance, the typical one-span gable frame shown in Fig. 7.9.1a has five rigid joints A, B, C, D, and E. The horizontal displacement u to the right and the vertical displacement v upward of each joint are shown in the parentheses in that figure. In order to satisfy the assumption that axial deformation is to be ignored, joints B and D cannot have vertical displacement because if they did, the lengths of columns AB and DE would change. The unknown joint translations are therefore four in number; namely, u_B, u_C, v_C, and u_D. However, u_B, u_C, v_C, and u_D must be such that they do not change the lengths of the rafters BC and CD; thus only two of the four unknown joint translations are free independent unknowns. The number of free independent unknown joint translations in a rigid frame is called the *degree of freedom in sidesway*. The word *sidesway* derives from the fact that in common rectangular rigid frames, the unknown translations are usually the deflections of the horizontal members in the horizontal direction.

In building up the slope-deflection equations, the independent joint translations, equal in number to the degree of freedom in sidesway, must be chosen from the unknown joint translations. For the gable frame of Fig. 7.9.1, one can choose any two of the four translations u_B, u_C, v_C, and u_D and use

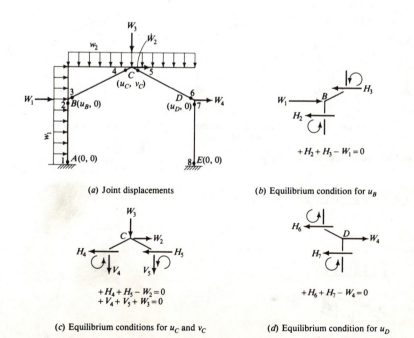

Figure 7.9.1 Sidesway conditions in gable frames.

230 INTERMEDIATE STRUCTURAL ANALYSIS

them in the slope-deflection equations. In order to obtain the symmetry about the main diagonal for the coefficients of the joint displacements in the simultaneous equations, the equilibrium equations as shown in Fig. 7.9.1b to d must be strictly followed. If u_B, u_C, or u_D is chosen as an independent unknown, the corresponding equilibrium equation is that the sum of the forces acting to the *left* (opposite to the positive direction of u_B, u_C or u_D) on joint B, C, or D must be zero. If v_C is chosen, the corresponding equilibrium equation is that the sum of the forces acting *downward* (opposite to the positive direction of v_C) on joint C must be zero. However, if symmetry is not desired, one can make use of any two of the four equilibrium equations shown in Fig. 7.9.1, not necessarily those corresponding to the independent joint translations chosen to be used in the slope-deflection equations. The proof for the symmetry as described here will be made in Chap. 12.

Example 7.9.1 Analyze the gable frame shown in Fig. 7.9.2a by the slope-deflection method. Draw shear and moment diagrams. Sketch the elastic curve.

SOLUTION (a) *Fixed-end moments.* The magnitude of the fixed-end moment at either end of a horizontal span with length L, when subjected to a uniform vertical load of w per unit horizontal distance, is equal to $wL^2/12$, as shown in Fig. 7.9.3a. For the inclined member in

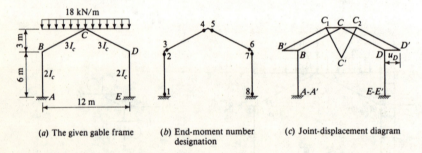

(a) The given gable frame (b) End-moment number designation (c) Joint-displacement diagram

Figure 7.9.2 Gable frame of Example 7.9.1.

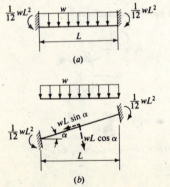

Figure 7.9.3 Fixed-end moments for an inclined member.

THE SLOPE-DEFLECTION METHOD 231

Fig. 7.9.3b, the total load of wL is resolved into two components, one along the member and the other perpendicular to the member. The transverse load per unit distance in the inclined direction is $(wL \cos \alpha)/(L \sec \alpha) = w \cos^2 \alpha$, and the magnitude of the fixed-end moment (FEM) at either end is

$$\text{FEM} = \frac{1}{12}(w \cos^2 \alpha)(L \sec \alpha)^2 = \frac{1}{12}wL^2$$

which is identical to that in Fig. 7.9.3a. For the present problem, using the end-moment number designation of Fig. 7.9.1b, the fixed-end moments are

$$M_{03} = -\frac{18(6)^2}{12} = -54 \text{ kN·m} \quad M_{04} = +54 \text{ kN·m}$$

$$M_{05} = -54 \text{ kN·m} \quad M_{06} = +54 \text{ kN·m}$$

(b) *Joint-displacement diagram.* As far as joint rotations are concerned, the unknowns are θ_B, θ_C, and θ_D. Because of symmetry, θ_C has to be zero and θ_D has to be equal in magnitude but opposite in sign to θ_B. For joint translations, the possible changes in position of the joints are shown in the joint-displacement diagram of Fig. 7.9.2c. Here again, because of symmetry, let $BB' = DD' = u_D$. The parallelograms BCC_1B' and DCC_2D' are drawn in Fig. 7.9.2c so that $B'C_1$ and $D'C_2$ are equal to BC and DC, respectively. Perpendiculars (not arcs) are erected at C_1 and C_2 to C_1B' and C_2D' until they intersect at C'. Thus, following the first-order assumption, the lengths of $B'C'$ and $C'D'$ are equal to those of C_1B' and C_2D', respectively. The rotations of the member axes are

$$R_{12} = -\frac{u_D}{6} \quad R_{34} = +\frac{C_1C'}{B'C_1} = +\frac{\sqrt{5}u_D}{3\sqrt{5}} = +\frac{u_D}{3}$$

$$R_{56} = -\frac{C_2C'}{C_2D'} = -\frac{\sqrt{5}u_D}{3\sqrt{5}} = -\frac{u_D}{3} \quad R_{78} = +\frac{u_D}{6}$$

(c) *Slope-deflection equations.* The slope-deflection equations for M_1 to M_8 are

$$M_1 = 0 + \frac{2E(2I_c)}{6}\left(2\theta_A + \theta_B + \frac{3u_D}{6}\right) = \frac{2}{3}EI_c\theta_B + \frac{1}{3}EI_c u_D$$

$$M_2 = 0 + \frac{2E(2I_c)}{6}\left(2\theta_B + \theta_A + \frac{3u_D}{6}\right) = \frac{4}{3}EI_c\theta_B + \frac{1}{3}EI_c u_D$$

$$M_3 = -54 + \frac{2E(3I_c)}{3\sqrt{5}}\left(2\theta_B + \theta_C - \frac{3u_D}{3}\right) = -54 + 1.7888 EI_c\theta_B - 0.8944 EI_c u_D$$

$$M_4 = +54 + \frac{2E(3I_c)}{3\sqrt{5}}\left(2\theta_C + \theta_B - \frac{3u_D}{3}\right) = +54 + 0.8944 EI_c\theta_B - 0.8944 EI_c u_D$$

$$M_5 = -54 + \frac{2E(3I_c)}{3\sqrt{5}}\left(2\theta_C + \theta_D + \frac{3u_D}{3}\right) = -54 - 0.8944 EI_c\theta_B + 0.8944 EI_c u_D$$

$$M_6 = +54 + \frac{2E(3I_c)}{3\sqrt{5}}\left(2\theta_D + \theta_C + \frac{3u_D}{3}\right) = +54 - 1.7888 EI_c\theta_B + 0.8944 EI_c u_D$$

$$M_7 = 0 + \frac{2E(2I_c)}{6}\left(2\theta_D + \theta_E - \frac{3u_D}{6}\right) = -\frac{4}{3}EI_c\theta_B - \frac{1}{3}EI_c u_D$$

$$M_8 = 0 + \frac{2E(2I_c)}{6}\left(2\theta_E + \theta_D - \frac{3u_D}{6}\right) = -\frac{2}{3}EI_c\theta_B - \frac{1}{3}EI_c u_D$$

(d) *Simultaneous equations in θ_B and u_D.* The equilibrium conditions corresponding to θ_B and u_D are (1) sum of *counterclockwise* moments acting on joint B is equal to zero, and (2) sum of horizontal forces to the *left* acting on joint D is equal to zero. Thus, from $M_2 + M_3 = 0$ in Fig. 7.9.4a,

$$+3.1221 EI_c\theta_B - 0.5611 EI_c u_D = +54$$

232 INTERMEDIATE STRUCTURAL ANALYSIS

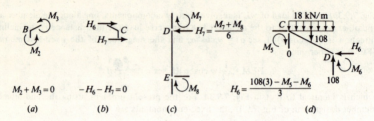

$M_2 + M_3 = 0$	$-H_6 - H_7 = 0$		
(a)	(b)	(c)	(d)

Figure 7.9.4 Equilibrium conditions for gable frame in Example 7.9.1.

and from $-H_6 - H_7 = 0$ in Fig. 7.9.4b to d,

$$-\frac{324 - M_5 - M_6}{3} - \frac{M_7 + M_8}{6} = 0$$

which becomes

$$-0.5611 EI_c \theta_B + 0.70740 EI_c u_D = +108$$

Note the symmetry of the coefficients of θ_B and u_D in the two simultaneous equations about the main diagonal. Solving,

$$EI_c \theta_B = +52.17 \qquad EI_c u_D = +194.04$$

(e) *Back substitution.* Substituting the values of $EI_c \theta_B$ and $EI_c u_D$ obtained above back into the slope-deflection equations,

$$M_1 = +99.46 \qquad M_2 = +134.24 \qquad M_3 = -134.23 \qquad M_4 = -72.89$$
$$M_5 = +72.89 \qquad M_6 = +134.23 \qquad M_7 = -134.24 \qquad M_8 = -99.46$$

(f) *Free-body, shear, and moment diagrams.* The free-body, shear, and moment diagrams of the four members are shown in Fig. 7.9.5. Note that the free-body diagram of member BC (also of member CD) is shown in two ways, first with horizontal and vertical forces and then with longitudinal and transverse forces. The shear and moment diagrams are plotted from using the transverse forces. The fact that all shear and moment diagrams do close shows that equilibrium of the entire gable frame is satisfied.

(g) *Sketch of the elastic curve.* The compatibility checks are obtained by computing the joint rotations and translations from the moment diagram, as shown in Fig. 7.9.6. Separating the moment diagram on AB into two parts and applying the two moment-area theorems,

$$\theta_B = -\frac{149.19}{EI_c} + \frac{201.36}{EI_c} = +\frac{52.17}{EI_c} \qquad \text{(positive means clockwise)}$$

$$BB' = +\frac{149.14}{EI_c}(4) - \frac{201.36}{EI_c}(2) = +\frac{194.04}{EI_c} \qquad \text{(positive means to left)}$$

Separating the moment diagram on BC into three parts and applying the two moment-area theorems,

$$\theta_C = \theta_B - \frac{120.75}{EI_c} + \frac{150.08}{EI_c} - \frac{81.49}{EI_c}$$

$$= +\frac{202.25}{EI_c} - \frac{202.24}{EI_c} \approx 0 \qquad \text{(Check)}$$

$$C_1 C' = 6.7082 \theta_B - \frac{120.75}{EI_c}\left(\frac{1}{2}\right)(6.7082) + \frac{150.08}{EI_c}\left(\frac{2}{3}\right)(6.7082) - \frac{81.49}{EI_c}\left(\frac{1}{3}\right)(6.7082)$$

$$= \frac{433.92}{EI_c}$$

THE SLOPE-DEFLECTION METHOD 233

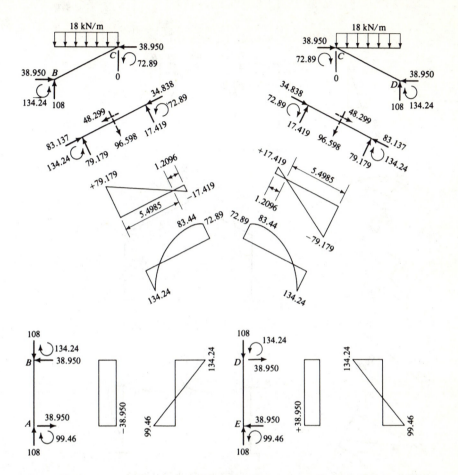

Figure 7.9.5 Free-body, shear, and moment diagrams of members in gable frame of Example 7.9.1.

The compatibility checks are obtained because θ_C is found to be equal to zero, and triangle CC_1C' in Fig. 7.9.6d is such that C' is directly under C, wherein $CC' = 388.08$ kN·m^3/EI_c.

Example 7.9.2 Analyze the gable frame shown in Fig. 7.9.7a by the slope-deflection method. Draw shear and moment diagrams. Sketch the elastic curve.

SOLUTION (a) *Fixed-end moments.* Member BC is inclined, but it is subjected to a horizontal uniform load. The fixed-end moments are equal to $\frac{1}{12}$ times the intensity of horizontal loading times the square of the vertical projection of the member. Using the end-moment number designation shown in Fig. 7.9.7b, the fixed-end moments are

$$M_{01} = -\tfrac{1}{12}(9)(6)^2 = -27.0 \qquad M_{02} = +27.0$$
$$M_{03} = -\tfrac{1}{12}(9)(3)^2 = -6.75 \qquad M_{04} = +6.75$$

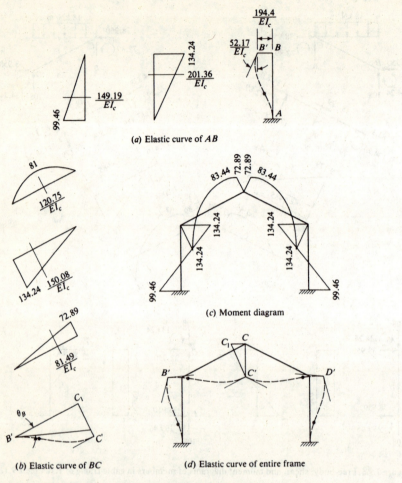

Figure 7.9.6 Elastic curve of gable frame in Example 7.9.1.

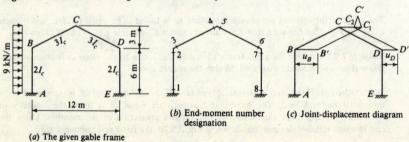

(a) The given gable frame

(b) End-moment number designation

(c) Joint-displacement diagram

Figure 7.9.7 Gable frame of Example 7.9.2.

THE SLOPE-DEFLECTION METHOD 235

(b) *Joint-displacement diagram.* Because of the unsymmetrical loading, the horizontal deflection u_B may not be equal to the horizontal deflection u_D. In Fig. 7.9.7c, let $BB' = u_B$ and $DD' = u_D$. Parallelograms $BB'C_1C$ and $DD'C_2C$ are drawn next. Then perpendiculars to C_1B' and C_2D' are erected at C_1 and C_2, respectively, until they intersect at C'. One can see now that the length of $B'C'$ is equal to $B'C_1$ and then to BC; and the length of $C'D'$ is equal to C_2D' and then to CD. Choosing u_B and u_D as the free independent unknown translations, the rotations of the member axes are

$$R_{12} = +\frac{u_B}{6} \qquad R_{34} = -\frac{C_1C'}{C_1B'} = -\frac{(u_B - u_D)(\sqrt{5}/2)}{3\sqrt{5}} = -\frac{u_B - u_D}{6}$$

$$R_{56} = +\frac{C_2C'}{C_2D'} = +\frac{(u_B - u_D)(\sqrt{5}/2)}{3\sqrt{5}} = +\frac{u_B - u_D}{6} \qquad R_{78} = +\frac{u_D}{6}$$

(c) *Slope-deflection equations.* The slope-deflection equations for M_1 to M_8 are

$$M_1 = -27.0 + \frac{2E(2I_c)}{6}\left(2\theta_A + \theta_B - \frac{3u_B}{6}\right) = -27.0 + \frac{2}{3}EI_c\theta_B - \frac{1}{3}EI_c u_B$$

$$M_2 = +27.0 + \frac{2E(2I_c)}{6}\left(2\theta_B + \theta_A - \frac{3u_B}{6}\right) = +27.0 + \frac{4}{3}EI_c\theta_B - \frac{1}{3}EI_c u_B$$

$$M_3 = -6.75 + \frac{2E(3I_c)}{3\sqrt{5}}\left[2\theta_B + \theta_C + \frac{3(u_B - u_D)}{6}\right]$$

$$= -6.75 + 1.7888 EI_c\theta_B + 0.8944 EI_c\theta_C + 0.44721 EI_c u_B - 0.44721 EI_c u_D$$

$$M_4 = +6.75 + \frac{2E(3I_c)}{3\sqrt{5}}\left[2\theta_C + \theta_B + \frac{3(u_B - u_D)}{6}\right]$$

$$= +6.75 + 1.7888 EI_c\theta_C + 0.8944 EI_c\theta_B + 0.44721 EI_c u_B - 0.44721 EI_c u_D$$

$$M_5 = 0 + \frac{2E(3I_c)}{3\sqrt{5}}\left[2\theta_C + \theta_D - \frac{3(u_B - u_D)}{6}\right]$$

$$= +1.7888 EI_c\theta_C + 0.8944 EI_c\theta_D - 0.44721 EI_c u_B + 0.44721 EI_c u_D$$

$$M_6 = 0 + \frac{2E(3I_c)}{3\sqrt{5}}\left[2\theta_D + \theta_C - \frac{3(u_B - u_D)}{6}\right]$$

$$= +1.7888 EI_c\theta_D + 0.8944 EI_c\theta_C - 0.44721 EI_c u_B + 0.44721 EI_c u_D$$

$$M_7 = 0 + \frac{2E(2I_c)}{6}\left(2\theta_D + \theta_E - \frac{3u_D}{6}\right) = +\frac{4}{3}EI_c\theta_D - \frac{1}{3}EI_c u_D$$

$$M_8 = 0 + \frac{2E(2I_c)}{6}\left(2\theta_E + \theta_D - \frac{3u_D}{6}\right) = +\frac{2}{3}EI_c\theta_D - \frac{1}{3}EI_c u_D$$

(d) *Simultaneous equations in* θ_B, θ_C, θ_D, u_B, *and* u_D. The equilibrium conditions corresponding to θ_B, θ_C, and θ_D are, from Fig. 7.9.8a to c, (1) $M_2 + M_3 = 0$, (2) $M_4 + M_5 = 0$, and (3) $M_6 + M_7 = 0$. The equilibrium condition corresponding to u_B is that the sum of horizontal forces to the left (opposite to the positive direction for u_B) acting on joint B is equal to zero; and that corresponding to u_D is that the sum of horizontal forces to the left (opposite to the positive direction for u_D) acting on joint D is equal to zero. Thus from Fig. 7.9.8d,

$$-H_2 + H_3 = 0 \qquad -H_6 - H_7 = 0$$

The expressions for H_2 and H_7 may be obtained from the free-body diagrams of AB and DE in Fig. 7.9.8, or

$$H_2 = 27 + \frac{M_1 + M_2}{6} \qquad H_7 = \frac{M_7 + M_8}{6}$$

236 INTERMEDIATE STRUCTURAL ANALYSIS

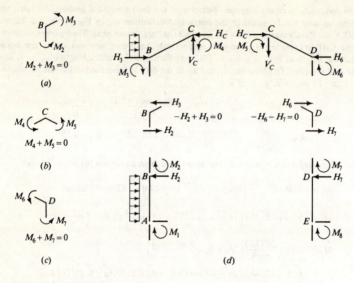

Figure 7.9.8 Equilibrium conditions for gable frame in Example 7.9.2.

The expressions for H_3 and H_6, however, can be obtained only from considering the equilibrium of the rafters BC and CD. Because u_C and v_C are not chosen as the free independent unknown joint translations, they are dependent quantities. Consequently there should be both horizontal and vertical equilibrium of forces acting on joint C. Since there are no externally applied concentrated forces acting on C in the present problem, the notations H_C and V_C have been adopted for the forces acting on CB and CD, as shown in Fig. 7.9.8d. H_C and V_C can be solved by writing the moment equilibrium equations for the rafters BC and CD. Using ΣM about B equal to zero for free body BC,

$$+3H_C + 6V_C = M_3 + M_4 + 27(1.5)$$

Using ΣM about D equal to zero for free body DC,

$$-3H_C + 6V_C = M_5 + M_6$$

Solving the preceding two equations for H_C,

$$H_C = \frac{(M_3 + M_4) - (M_5 + M_6)}{6} + 6.75$$

Using $\Sigma F_x = 0$ for free body BC,

$$H_3 = H_C - 27 = \frac{(M_3 + M_4) - (M_5 + M_6)}{6} - 20.25$$

Using $\Sigma F_x = 0$ for free body CD,

$$H_6 = H_C = \frac{(M_3 + M_4) - (M_5 + M_6)}{6} + 6.75$$

The five simultaneous equations in θ_B, θ_C, θ_D, u_B and u_D can now be established using

THE SLOPE-DEFLECTION METHOD 237

the three joint-moment equilibrium conditions and the two sidesway equilibrium conditions; these are as shown in the accompanying table.

$EI_c\theta_B$	$EI_c\theta_C$	$EI_c\theta_D$	$EI_c u_B$	$EI_c u_D$	=
+3.1221	+0.8944		+0.11388	−0.44721	−20.25
+0.8944	+3.5776	+0.8944			− 6.75
	+0.8944	+3.1221	−0.44721	+0.11388	0
+0.11387		−0.44720	+0.40925	−0.29815	+47.25
−0.44720		+0.11387	−0.29814	+0.40925	+ 6.75

Note the symmetry of the coefficients of the joint displacements in the five tabulated equations with respect to the principal diagonal downward to the right. Solving,

$$EI_c\theta_B = +33.56 \quad EI_c\theta_C = -23.56 \quad EI_c\theta_D = +53.12$$
$$EI_c u_B = +409.5 \quad EI_c u_D = +336.7$$

(e) *Back substitution.* Substituting the values of the joint displacements obtained above back into the slope-deflection equations,

$$M_1 = -141.13 \quad M_2 = -64.75 \quad M_3 = +64.76 \quad M_4 = +27.18$$
$$M_5 = -27.19 \quad M_6 = +41.40 \quad M_7 = -41.41 \quad M_8 = -76.82$$

There are some obvious round-off errors in the end-moment values shown.

(f) *Free-body, shear and moment diagrams.* The free-body, shear and moment diagrams of the four members are shown in Fig. 7.9.9. The horizontal forces acting on columns AB and DE are first solved; H_A is 61.31 kN and H_E is 19.70 kN. Considering the entire rigid frame as a free body, the sum of H_A and H_D must be equal to the total horizontal wind load of 81 kN. The fact that $H_A + H_D = 61.31 + 19.70 = 81.01$ kN shows a very small round-off error. The force V_C may be computed from the free body BC or from the free body CD. In each case the value obtained is 12.22 kN, which shows a good equilibrium check. The two equilibrium checks described above constitute the alternate checks on the two sidesway equilibrium conditions. Note also that two free-body diagrams are shown for each of the two inclined members, one with horizontal and vertical forces and the other with transverse and longitudinal forces. The shear and moment diagrams are based on the transverse forces.

(g) *Sketch of the elastic curve.* The three compatibility checks, equal in number to the degree of indeterminacy for this gable frame, are obtained when the joint rotations and translations are computed from the moment diagram, as shown in Fig. 7.9.10. Separating the moment diagram on AB into three parts and applying the two moment-area theorems,

$$\theta_B = -\frac{81}{EI_c} - \frac{97.14}{EI_c} + \frac{211.70}{EI_c} = +\frac{33.56}{EI_c} \quad \text{(positive means clockwise)}$$

$$BB' = -\frac{81}{EI_c}(3) - \frac{97.14}{EI_c}(2) + \frac{211.70}{EI_c}(4)$$

$$= +\frac{409.5}{EI_c} \quad \text{(positive means to right)}$$

238 INTERMEDIATE STRUCTURAL ANALYSIS

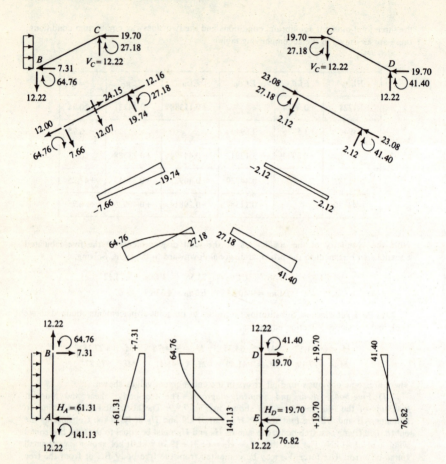

Figure 7.9.9 Free-body, shear and moment diagrams of members in gable frame of Example 7.9.2.

Separating the moment diagram on BC into three parts and applying the two moment-area theorems,

$$\theta_C = \theta_B - \frac{15.09}{EI_c} - \frac{72.40}{EI_c} + \frac{30.39}{EI_c}$$

$$= -\frac{23.54}{EI_c} \quad \text{(negative means counterclockwise)}$$

$$C_1C' = C_3C'' - C_1C_3$$

$$= \frac{15.09}{EI_c}\left(\frac{1}{2}\right)(6.7082) + \frac{72.40}{EI_c}\left(\frac{2}{3}\right)(6.7082) - \frac{30.39}{EI_c}\left(\frac{1}{3}\right)(6.7082) - 6.7082\theta_B$$

$$= \frac{81.3}{EI_c}$$

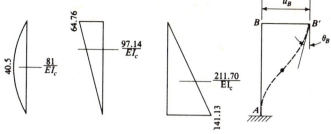

(a) Elastic curve of AB

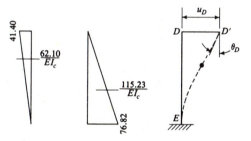

(c) Elastic curve of DE

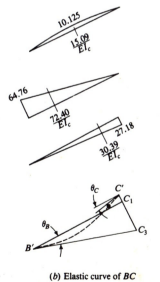

(b) Elastic curve of BC

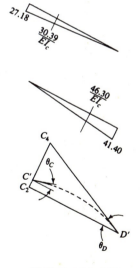

(d) Elastic curve of CD

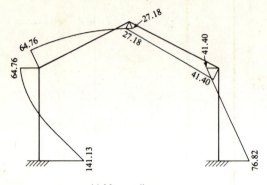

(e) Moment diagram

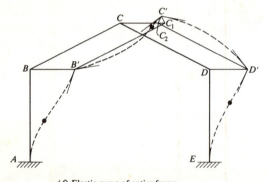

(f) Elastic curve of entire frame

Figure 7.9.10 Elastic curve of gable frame in Example 7.9.2.

From Fig. 7.9.10f,

$$u_C = CC_1 - C_1C' \frac{1}{\sqrt{5}} = \frac{40.95}{EI_c} - \frac{81.3}{EI_c}\left(\frac{1}{\sqrt{5}}\right)$$

$$= +\frac{373.1}{EI_c} \quad \text{(positive means to right)}$$

$$v_C = C_1C' \frac{2}{\sqrt{5}} = \frac{81.3}{EI_c}\left(\frac{2}{\sqrt{5}}\right) = +\frac{72.7}{EI_c} \quad \text{(positive means upward)}$$

Separating the moment diagram on DE into two parts and applying the two moment-area theorems,

$$\theta_D = -\frac{62.10}{EI_c} + \frac{115.23}{EI_c} = +\frac{53.13}{EI_c} \quad \text{(positive means clockwise)}$$

$$DD' = -\frac{62.10}{EI_c}(2) + \frac{115.23}{EI_c}(4) = +\frac{336.7}{EI_c} \quad \text{(positive means to right)}$$

Separating the moment diagram on DC into two parts and applying the two moment-area theorems,

$$\theta_C = \theta_D - \frac{30.39}{EI_c} - \frac{46.30}{EI_c}$$

$$= -\frac{23.56}{EI_c} \quad \text{(negative means counterclockwise)}$$

$$C_2C' = C_2C_4 - C_4C'$$

$$= 6.7082\theta_D - \frac{30.39}{EI_c}\left(\frac{1}{3}\right)(6.7082) - \frac{46.30}{EI_c}\left(\frac{2}{3}\right)(6.7082) = \frac{81.2}{EI_c}$$

From Fig. 7.9.10f,

$$u_C = CC_2 + C_2C'\frac{1}{\sqrt{5}}$$

$$= \frac{336.7}{EI_c} + \frac{81.2}{EI_c}\left(\frac{1}{\sqrt{5}}\right) = +\frac{373.0}{EI_c} \quad \text{(positive means to right)}$$

$$v_C = C_2C'\frac{2}{\sqrt{5}} = \frac{81.2}{EI_c}\left(\frac{2}{\sqrt{5}}\right) = +\frac{72.6}{EI_c} \quad \text{(positive means upward)}$$

Note that θ_C, u_C, and v_C are computed first from the path ABC and then from the path EDC. The fact that the same three results are obtained constitutes the three compatibility checks.

7.10 Slope-Deflection Equations for Members with Variable Moment of Inertia

The slope-deflection equations, Eqs. (7.4.4a and b),

$$M_A = M_{0A} + \frac{2EI}{L}(2\theta_A + \theta_B - 3R)$$

$$M_B = M_{0B} + \frac{2EI}{L}(2\theta_B + \theta_A - 3R)$$

have been derived on the basis that the member AB has a constant cross section with moment of inertia equal to I. Now if the member AB has variable cross sections, the moment of inertia will be variable; in other words, I has different values at different sections along the beam. Oftentimes the moment of inertia at any section of distance x from the left end is expressed by $n_x I_c$, where n_x is a ratio larger than unity and I_c is the smallest moment of inertia that exists along the span. If the moment of inertia changes abruptly at two or three points along the span, such as in a wide-flange steel beam with cover plates, n_x is a constant for each finite length. However, n_x may be changing continuously from point to point such as in a built-up, large steel girder with a straight top flange but a curved bottom flange (Fig. 7.10.1a), or in a reinforced concrete beam with straight or parabolic haunches (Fig. 7.10.1b).

To provide a more direct linkage between the slope-deflection method and the matrix displacement method described in Chap. 12, the slope-deflection equations for a member ij (instead of member AB) with variable moment of

242 INTERMEDIATE STRUCTURAL ANALYSIS

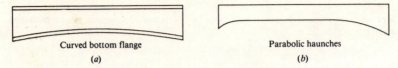

(a) Curved bottom flange

(b) Parabolic haunches

Figure 7.10.1 Typical beam spans with variable moment of inertia.

inertia will be written in the form

$$M_i = M_{0i} + s_{ii}\frac{EI_c}{L}(\theta_A - R) + s_{ij}\frac{EI_c}{L}(\theta_B - R) \quad (7.10.1a)$$

$$M_j = M_{0j} + s_{ji}\frac{EI_c}{L}(\theta_A - R) + s_{jj}\frac{EI_c}{L}(\theta_B - R) \quad (7.10.1b)$$

Note that a distinction is made between the member end i and the joint A into which member end i enters; also between the member end j and the joint B. The fixed-end moments M_{0i} and M_{0j} are themselves affected by n_x, which defines how the moment of inertia varies along the span. The four quantities s_{ii}, s_{ij}, s_{ji}, and s_{jj} are called the *stiffness coefficients*, because s_{ii} and s_{ij} measure the end moments needed at end i to maintain unit values of end rotations ϕ_i and ϕ_j, respectively, while s_{ji} and s_{jj} measure the end moments needed at end j to maintain unit values of end rotations ϕ_i and ϕ_j, also respectively. Thus the first subscript refers to the location of the end moment, and the second subscript refers to the location of the end rotation. Note in Fig. 7.10.2a that

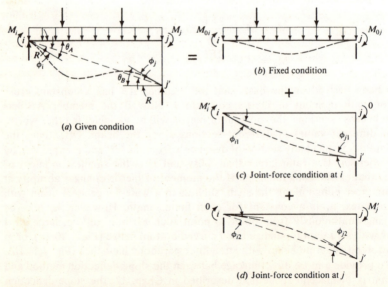

Figure 7.10.2 Basis for slope-deflection equations.

the end rotations ϕ_i and ϕ_j are each equal to the end slopes θ_A and θ_B, minus the member-axis rotation R.

In Chap. 15 it will be shown that the stiffness coefficients and the fixed-end moments for a member with variable moment of inertia can be determined directly by the column-analogy method. For the present, the stiffness coefficients are to be obtained indirectly from the flexibility coefficients soon to be described, and the fixed-end moments are also to be obtained indirectly from the end rotations of the elastic curve if the member ij is simply supported at both ends.

Following the same procedure used in Sec. 7.4, the given condition for member ij in Fig. 7.10.2a can be separated into the fixed condition of Fig. 7.10.2b, and the joint-force condition of Fig. 7.10.2c and d. For equivalency in end moments,

$$M_i = M_{0i} + M'_i \qquad (7.10.2a)$$

$$M_j = M_{0j} + M'_j \qquad (7.10.2b)$$

and for equivalency in end rotations,

$$\phi_i = \phi_{i1} + \phi_{i2} \qquad (7.10.3a)$$

$$\phi_j = \phi_{j1} + \phi_{j2} \qquad (7.10.3b)$$

Note that ϕ_{j1} in Fig. 7.10.2c and ϕ_{i2} in Fig. 7.10.2d should have negative values.

Define the four flexibility coefficients f_{ii}, f_{ij}, f_{ji}, and f_{jj} such that when a unit clockwise moment is applied at the end i of a simple span ij, the clockwise end rotations at i and j are $f_{ii}(L/EI_c)$ and $f_{ji}(L/EI_c)$, and when a unit clockwise moment is applied at the end j of the same simple span ij, the clockwise end rotations at i and j are $f_{ij}(L/EI_c)$ and $f_{jj}(L/EI_c)$, as shown in Fig. 7.10.3. In this case, the first subscript refers to the location of the end rotation, and the second subscript refers to the location of the end moment. If member ij has constant moment of inertia, $f_{ii} = f_{jj} = +\frac{1}{3}$ and $f_{ij} = f_{ji} = -\frac{1}{6}$. For members with variable moment of inertia, the flexibility coefficients can be determined either by the conjugate-beam method or the unit-load method. The conjugate beam should be loaded by the $M/(n_x EI_c)$ diagram as shown in Fig. 7.10.3. When the unit-load method is used, expressions for $f_{ij}(L/EI_c)$ and $f_{ji}(L/EI_c)$ are observed to be identical, each equal to $(m_i m_j\, dx)/EI$, in which m_i and m_j are bending moments due to unit moment applied at i and j, respectively. Thus, f_{ij} and f_{ji} are always identical as the natural consequence of the law of reciprocal deflections (see Sec. 4.3).

Following the definition of the flexibility coefficients, Eqs. (7.10.3a and b) become:

$$\phi_i = f_{ii}\frac{L}{EI_c}M'_i + f_{ij}\frac{L}{EI_c}M'_j \qquad (7.10.4a)$$

$$\phi_j = f_{ji}\frac{L}{EI_c}M'_i + f_{jj}\frac{L}{EI_c}M'_j \qquad (7.10.4b)$$

244 INTERMEDIATE STRUCTURAL ANALYSIS

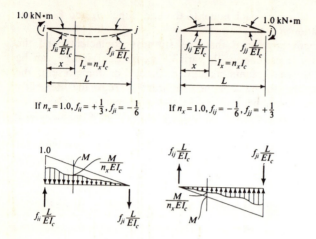

Figure 7.10.3 Flexibility coefficients of a member with variable moment of inertia.

Solving Eqs. (7.10.4a and b) simultaneously for M'_i and M'_j,

$$M'_i = s_{ii}\frac{EI_c}{L}\phi_i + s_{ij}\frac{EI_c}{L}\phi_j \qquad (7.10.5a)$$

$$M'_j = s_{ji}\frac{EI_c}{L}\phi_i + s_{jj}\frac{EI_c}{L}\phi_j \qquad (7.10.5b)$$

in which

$$s_{ii} = +\frac{f_{jj}}{f_{ii}f_{jj} - f_{ij}f_{ji}} \qquad s_{ij} = -\frac{f_{ij}}{f_{ii}f_{jj} - f_{ij}f_{ji}}$$

$$s_{ji} = -\frac{f_{ji}}{f_{ii}f_{jj} - f_{ij}f_{ji}} \qquad s_{jj} = +\frac{f_{ii}}{f_{ii}f_{jj} - f_{ij}f_{ji}} \qquad (7.10.6)$$

The general slope-deflection equations for a member with variable moment of inertia, Eqs. (7.10.1a and b), are obtained by substituting Eqs. (7.10.5a and b) into Eqs. (7.10.2a and b) and letting $\phi_i = \theta_A - R$ and $\phi_j = \theta_B - R$.

Thus the four stiffness coefficients s_{ii}, s_{ij}, s_{ji}, and s_{jj} can be obtained from the four flexibility coefficients f_{ii}, f_{ij}, f_{ji}, and f_{jj}. Since $f_{ij} = f_{ji}$, $s_{ij} = s_{ji}$. For a member with constant moment of inertia, $f_{ii} = f_{jj} = +\frac{1}{3}$, $f_{ij} = f_{ji} = -\frac{1}{6}$, $s_{ii} = s_{jj} = +4$, and $s_{ij} = s_{ji} = +2$, in which case Eqs. (7.10.1a and b) become Eqs. (7.4.4a and b).

The fixed-end moments for a member with variable moment of inertia can be determined by the force method by using a simple span as the basic determinate structure. The two compatibility conditions are, from observing

THE SLOPE-DEFLECTION METHOD 245

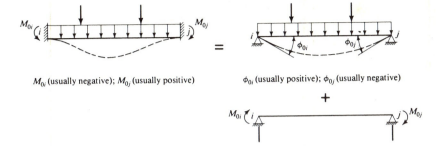

M_{0i} (usually negative); M_{0j} (usually positive) ϕ_{0i} (usually positive); ϕ_{0j} (usually negative)

Figure 7.10.4 Fixed-end moments as redundants on a simple span.

Fig. 7.10.4,

$$\phi_{0i} + f_{ii}\frac{L}{EI_c}M_{0i} + f_{ij}\frac{L}{EI_c}M_{0j} = 0 \qquad (7.10.7a)$$

$$\phi_{0j} + f_{ji}\frac{L}{EI_c}M_{0i} + f_{jj}\frac{L}{EI_c}M_{0j} = 0 \qquad (7.10.7b)$$

Solving Eqs. (7.10.7a and b) for M_{0i} and M_{0j},

$$M_{0i} = -s_{ii}\frac{EI_c}{L}\phi_{0i} - s_{ij}\frac{EI_c}{L}\phi_{0j} \qquad (7.10.8a)$$

$$M_{0j} = -s_{ji}\frac{EI_c}{L}\phi_{0i} - s_{jj}\frac{EI_c}{L}\phi_{0j} \qquad (7.10.8b)$$

The preceeding two equations can be visualized physically by noting that the fixed-end moments are there to maintain the opposites of the end rotations in case the member is simply supported at both ends. In common cases of downward loading on a member, ϕ_{0i} is positive and ϕ_{0j} is negative.

Summarizing, for a member with variable moment of inertia, the flexibility coefficients should first be computed by the conjugate-beam method or the unit-load method or both, then the stiffness coefficients are computed from Eq. (7.10.6), and finally the fixed-end moments are computed from Eqs. (7.10.8a and b), wherein ϕ_{0i} and ϕ_{0j} are to be obtained again by either the conjugate-beam method or the unit-load method or both.

Example 7.10.1 Analyze the continuous beam shown in Fig. 7.10.5a by the slope-deflection method. Draw shear and moment diagrams. Sketch the elastic curve.

SOLUTION (a) *Flexibility coefficients*. The variation in the moment of inertia along span 1-2 is identical to that along span 3-4, except in the opposite-handed way. Thus the flexibility coefficients need be determined for span 1-2 only. For span 1-2, the flexibility coefficients

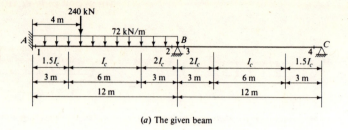

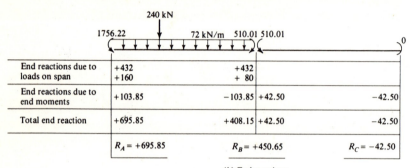

End reactions due to loads on span	+432 +160		+432 + 80			
End reactions due to end moments	+103.85		−103.85	+42.50		−42.50
Total end reaction	+695.85		+408.15	+42.50		−42.50

$R_A = +695.85$ $R_B = +450.65$ $R_C = -42.50$

(b) End reactions

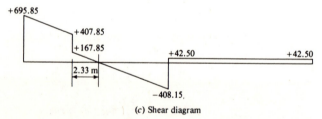

(c) Shear diagram

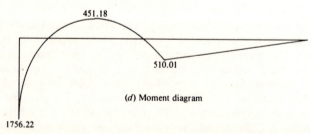

(d) Moment diagram

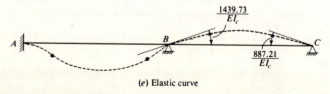

(e) Elastic curve

Figure 7.10.5 Continuous beam of Example 7.10.1.

THE SLOPE-DEFLECTION METHOD **247**

can be obtained by either the conjugate-beam method or the unit-load method. Referring to the two conjugate beams in Fig. 7.10.6 and the computations shown in Table 7.10.1a and b,

$$R_{11} = +\frac{3.19792}{EI_c} = f_{11}\frac{L}{EI_c} \qquad f_{11} = +\frac{3.19792}{L} = +0.266493$$

$$R_{21} = -\frac{1.73958}{EI_c} = f_{21}\frac{L}{EI_c} \qquad f_{21} = -\frac{1.73958}{L} = -0.144965$$

$$R_{12} = -\frac{1.73958}{EI_c} = f_{12}\frac{L}{EI_c} \qquad f_{12} = -\frac{1.73958}{L} = -0.144965$$

$$R_{22} = +\frac{2.82292}{EI_c} = f_{22}\frac{L}{EI_c} \qquad f_{22} = +\frac{2.82292}{L} = +0.235243$$

Using the expressions for m_1 and m_2 observed from the free-body diagrams in Fig. 7.10.6 and applying the unit-load method,

$$f_{11}\frac{L}{EI_c} = \int_0^L \frac{m_1^2\,dx}{EI}$$

$$= \frac{1}{EI_c}\left[\frac{1}{2.0}\int_0^3\left(\frac{1}{12}x\right)^2 dx + \frac{1}{1.0}\int_3^9\left(\frac{1}{12}x\right)^2 dx + \frac{1}{1.5}\int_9^{12}\left(\frac{1}{12}x\right)^2 dx\right]$$

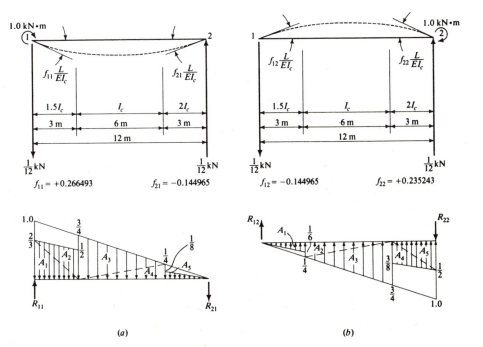

Figure 7.10.6 Flexibility coefficients for span 1 = 2 in beam of Example 7.10.1.

248 INTERMEDIATE STRUCTURAL ANALYSIS

Table 7.10.1a Reactions on conjugate beam in Fig. 7.10.6a

Part	$A\left(\dfrac{1}{EI_c}\right)$	b = distance from 2	a = distance from 1	$R_{11} = +\dfrac{Ab}{L}\left(\dfrac{1}{EI_c}\right)$	$R_{21} = -\dfrac{Aa}{L}\left(\dfrac{1}{EI_c}\right)$
A_1	$\frac{1}{2}(\frac{2}{3})(3) = 1.00000$	11.0	1.0	+0.91667	−0.08333
A_2	$\frac{1}{2}(\frac{1}{2})(3) = 0.75000$	10.0	2.0	+0.62500	−0.12500
A_3	$\frac{1}{2}(\frac{3}{4})(6) = 2.25000$	7.0	5.0	+1.31250	−0.93750
A_4	$\frac{1}{2}(\frac{1}{4})(6) = 0.75000$	5.0	7.0	+0.31250	−0.43750
A_5	$\frac{1}{2}(\frac{1}{8})(3) = 0.18750$	2.0	10.0	+0.03125	−0.15625
Total	− − 4.93750			+3.19792	−1.73958

Table 7.10.1b Reactions on conjugate beam in Fig. 7.10.6b

Part	$A\left(\dfrac{1}{EI_c}\right)$	b = distance from 2	a = distance from 1	$R_{12} = -\dfrac{Ab}{L}\left(\dfrac{1}{EI_c}\right)$	$R_{22} = +\dfrac{Aa}{L}\left(\dfrac{1}{EI_c}\right)$
A_1	$\frac{1}{2}(\frac{1}{6})(3) = 0.25000$	10.0	2.0	−0.20833	+0.04167
A_2	$\frac{1}{2}(\frac{1}{4})(6) = 0.75000$	7.0	5.0	−0.43750	+0.31250
A_3	$\frac{1}{2}(\frac{3}{4})(6) = 2.25000$	5.0	7.0	−0.93750	+1.31250
A_4	$\frac{1}{2}(\frac{3}{8})(3) = 0.56250$	2.0	10.0	−0.09375	+0.46875
A_5	$\frac{1}{2}(\frac{1}{2})(3) = 0.75000$	1.0	11.0	−0.06250	+0.68750
Total	− − 4.56250			−1.73958	+2.82292

$$f_{12}\frac{L}{EI_c} = f_{21}\frac{L}{EI_c} = \int_0^L \frac{m_1 m_2\, dx}{EI}$$

$$= \frac{1}{EI_c}\left[\frac{1}{1.5}\int_0^3 \left(1-\frac{x}{12}\right)\left(-\frac{1}{12}x\right) dx + \frac{1}{1.0}\int_3^9 \left(1-\frac{x}{12}\right)\left(-\frac{1}{12}x\right) dx \right.$$

$$\left. + \int_9^{12}\left(1-\frac{x}{12}\right)\left(-\frac{1}{12}x\right) dx\right]$$

$$f_{22}\frac{L}{EI_c} = \int_0^L \frac{m_2^2\, dx}{EI}$$

$$= \frac{1}{EI_c}\left[\frac{1}{1.5}\int_0^3 \left(-\frac{1}{12}x\right)^2 dx + \frac{1}{1.0}\int_3^9 \left(-\frac{1}{12}x\right)^2 dx + \frac{1}{2.0}\int_9^{12}\left(-\frac{1}{12}x\right)^2 dx\right]$$

The three preceding equations give

$$f_{11} = +0.266493 \qquad f_{12} = f_{21} = -0.144965 \qquad f_{22} = +0.235243$$

which are identical to the results of the conjugate-beam method.

THE SLOPE-DEFLECTION METHOD 249

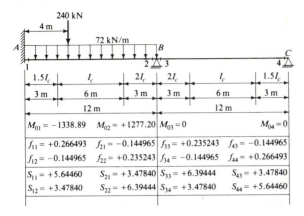

Figure 7.10.7 Summary of FEM, flexibility coefficients, and stiffness coefficients for beam in Example 7.10.1.

The summary of the flexibility coefficients for both spans 1-2 and 3-4 is shown in Fig. 7.10.7.

(b) *Stiffness coefficients.* The stiffness coefficients for span 1-2 are computed by using Eq. (7.10.6); thus,

$$f_{11}f_{12} - f_{12}f_{21} = (+0.266493)(+0.235243) - (-0.144965)^2 = 0.04167576$$

$$s_{11} = +\frac{f_{22}}{f_{11}f_{22} - f_{12}f_{21}} = +\frac{+0.235243}{0.04167576} = +5.64460$$

$$s_{12} = -\frac{f_{12}}{f_{11}f_{22} - f_{12}f_{21}} = -\frac{-0.144965}{0.04167576} = +3.47840$$

$$s_{21} = -\frac{f_{21}}{f_{11}f_{22} - f_{12}f_{21}} = -\frac{-0.144965}{0.04167576} = +3.47840$$

$$s_{22} = +\frac{f_{11}}{f_{11}f_{12} - f_{12}f_{21}} = +\frac{+0.266493}{0.04167576} = +6.39444$$

The summary of the stiffness coefficients for both spans 1-2 and 3-4 is shown in Fig. 7.10.7.

(c) *Fixed-end moments.* The fixed-end moments for span 1-2 can be computed by using Eqs. (7.10.8a and b). The end rotations ϕ_{01} and ϕ_{02} due to the uniform load can be computed first by the conjugate-beam method, as shown in Fig. 7.10.8a and Table 7.10.2a, and then checked by the unit-load method. Note that in Fig. 7.10.8a only the natural moment diagram is shown, the moment areas A_1 to A_6 are divided by $n_x EI_c$ to arrive at the M/EI areas in Table 7.10.2a. Also the natural areas A_1 and A_6 are each identical to that of a right parabolic area, which is the moment diagram of a simple beam equal in span to the horizontal projection of the end points of the inclined parabola. Also note that the bending moment at any point in a simple beam which divides the span length L into a and b is equal to $\frac{1}{2}wab$, wherein w is the intensity of the uniform load. From Table 7.10.2a,

$$\phi_{01}(\text{uniform load}) = +\frac{4600.125}{EI_c} \qquad \phi_{02}(\text{uniform load}) = -\frac{4417.875}{EI_c}$$

Similarly from Fig. 7.10.8b and Table 7.10.2b,

$$\phi_{01}(\text{conc. load}) = +\frac{1903.333}{EI_c} \qquad \phi_{02}(\text{conc. load}) = -\frac{1516.666}{EI_c}$$

250 INTERMEDIATE STRUCTURAL ANALYSIS

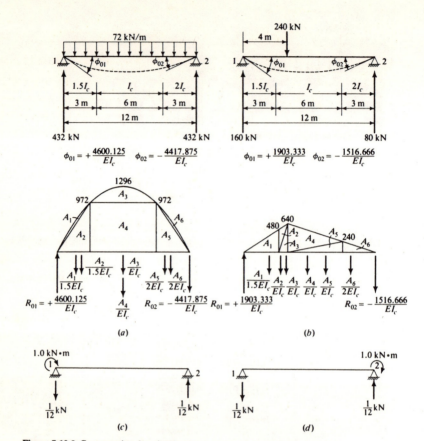

Figure 7.10.8 Computation for simple-span end rotations in Example 7.10.1.

By the unit-load method, in reference to the free-body diagrams in Fig. 7.10.8,

$$\phi_{01}(\text{uniform load}) = \frac{1}{EI_c}\left[\frac{1}{2.0}\int_0^3 (432x - 36x^2)(\tfrac{1}{12}x)\,dx + \frac{1}{1.0}\int_3^9 (432x - 36x^2)(\tfrac{1}{12}x)\,dx\right.$$

$$\left. + \frac{1}{1.5}\int_9^{12} (432x - 36x^2)(\tfrac{1}{12}x)\,dx\right]$$

$$= +\frac{4600.125}{EI_c}$$

$$\phi_{02}(\text{uniform load}) = \frac{1}{EI_c}\left[\frac{1}{1.5}\int_0^3 (432x - 36x^2)(-\tfrac{1}{12}x)\,dx + \frac{1}{1.0}\int_3^9 (432x - 36x^2)(-\tfrac{1}{12}x)\,dx\right.$$

$$\left. + \frac{1}{2.0}\int_9^{12} (432x - 36x^2)(-\tfrac{1}{12}x)\,dx\right]$$

$$= -\frac{4417.875}{EI_c}$$

THE SLOPE-DEFLECTION METHOD

Table 7.10.2a Reactions on conjugate beam in Fig. 7.10.8a

Part	$\left(\dfrac{M}{EI}\text{area}\right)\left(\dfrac{1}{EI_c}\right)$	b = distance from 2	a = distance from 1	$+\left(\dfrac{M}{EI}\text{area}\right)\left(\dfrac{b}{L}\right)$	$-\left(\dfrac{M}{EI}\text{area}\right)\left(\dfrac{a}{L}\right)$
1	$\dfrac{2(81)(3)}{3\ 1.5} = 108$	10.5	1.5	+ 94.5	− 13.5
2	$\dfrac{1(972)(3)}{2\ 1.5} = 972$	10.0	2.0	+ 810.0	− 162.0
3	$\dfrac{2(324)(6)}{3\ 1.0} = 1296$	6.0	6.0	+ 648.0	− 648.0
4	$\dfrac{(972)(6)}{1.0} = 5832$	6.0	6.0	+2916.0	−2916.0
5	$\dfrac{1(972)(3)}{2\ 2.0} = 729$	2.0	10.0	+ 121.5	− 607.5
6	$\dfrac{2(81)(3)}{3\ 2.0} = 81$	1.5	10.5	+ 10.125	− 70.875
Total	9018			$R_{01} = +\dfrac{4600.125}{EI_c}$	$R_{02} = -\dfrac{4417.875}{EI_c}$

Table 7.10.2b Reactions on conjugate beam in Fig. 7.10.8b

Part	$\left(\dfrac{M}{EI}\text{area}\right)\left(\dfrac{1}{EI_c}\right)$	b = distance from 2	a = distance from 1	$+\left(\dfrac{M}{EI}\text{area}\right)\left(\dfrac{b}{L}\right)$	$-\left(\dfrac{M}{EI}\text{area}\right)\left(\dfrac{a}{L}\right)$
1	$\dfrac{1(480)(3)}{2\ 1.5} = 480$	10.0	2.0	+400	− 80
2	$\dfrac{1(480)(1)}{2\ 1.0} = 240$	$8\frac{2}{3}$	$3\frac{1}{3}$	+173.333	− 66.666
3	$\dfrac{1(640)(1)}{2\ 1.0} = 320$	$8\frac{1}{3}$	$3\frac{2}{3}$	+222.222	− 97.777
4	$\dfrac{1(640)(5)}{2\ 1.0} = 1600$	$6\frac{1}{3}$	$5\frac{2}{3}$	+844.444	−755.555
5	$\dfrac{1(240)(5)}{2\ 1.0} = 600$	$4\frac{2}{3}$	$7\frac{1}{3}$	+233.333	−366.666
6	$\dfrac{1(240)(3)}{2\ 2.0} = 180$	2.0	12.0	+ 30	−150
Total	3420			$R_{01} = +\dfrac{1903.333}{EI_c}$	$R_{02} = -\dfrac{1516.666}{EI_c}$

$$\phi_{01} \text{ (conc. load)} = \frac{1}{EI_c} \left[\frac{1}{2.0} \int_0^3 (80x)(\tfrac{1}{12}x) \, dx + \frac{1}{1.0} \int_3^8 (80x)(\tfrac{1}{12}x) \, dx \right.$$
$$+ \frac{1}{1.5} \int_0^3 (160x)(1 - \tfrac{1}{12}x) \, dx$$
$$\left. + \frac{1}{1.5} \int_3^4 (160x)(1 - \tfrac{1}{12}x) \, dx \right]$$
$$= + \frac{1903.333}{EI_c}$$

$$\phi_{02} \text{ (conc. load)} = \frac{1}{EI_c} \left[\frac{1}{1.5} \int_0^3 (160x)(-\tfrac{1}{12}x) \, dx + \frac{1}{1.0} \int_3^4 (160x)(-\tfrac{1}{12}x) \, dx \right.$$
$$+ \frac{1}{2.0} \int_0^3 (80x)(-1 + \tfrac{1}{12}x) \, dx$$
$$\left. + \frac{1}{1.0} \int_3^8 (80x)(-1 + \tfrac{1}{12}x) \, dx \right]$$
$$= -\frac{1516.666}{EI_c}$$

Using Eqs. (7.10.8a and b),

$$M_{01} \text{ (uniform load)} = -s_{11} \frac{EI_c}{L} \phi_{01} - s_{12} \frac{EI_c}{L} \phi_{02}$$
$$= -\frac{5.64460}{12}(+4600.125) - \frac{3.47840}{12}(-4417.875)$$
$$= -883.23$$

$$M_{02} \text{ (uniform load)} = -s_{21} \frac{EI_c}{L} \phi_{01} - s_{22} \frac{EI_c}{L} \phi_{02}$$
$$= -\frac{3.47840}{12}(+4600.125) - \frac{6.39444}{12}(4417.875)$$
$$= +1020.73$$

$$M_{01} \text{ (conc. load)} = -\frac{5.64460}{12}(+1903.333) - \frac{3.47840}{12}(-1516.666)$$
$$= -455.66$$

$$M_{02} \text{ (conc. load)} = -\frac{3.47840}{12}(+1903.333) - \frac{6.39444}{12}(-1516.666)$$
$$= +256.47$$

The total fixed-end moments for span 1-2 are

$$M_{01} = -883.23 - 455.66 = -1338.89 \text{ kN} \cdot \text{m}$$
$$M_{02} = +1020.73 + 256.47 = +1277.20 \text{ kN} \cdot \text{m}$$

(d) *Solution by the slope-deflection method.* For the present problem, the member-axis rotation R is zero for both spans. Using the end-moment number designation and the summary of fixed-end moments and stiffness coefficients shown in Fig. 7.10.7, the slope-deflection equations, Eqs. (7.10.1), for M_1 to M_4 are

$$M_1 = M_{01} + s_{11} \frac{EI_c}{L}(\theta_A - R) + s_{12} \frac{EI_c}{L}(\theta_B - R)$$
$$= -1338.89 + 5.64460 \frac{EI_c}{12} \theta_A + 3.47840 \frac{EI_c}{12} \theta_B$$
$$= -1338.89 + 0.289867 EI_c \theta_B$$

THE SLOPE-DEFLECTION METHOD 253

$$M_2 = M_{02} + s_{21}\frac{EI_c}{L}(\theta_A - R) + s_{22}\frac{EI_c}{L}(\theta_B - R)$$

$$= +1277.20 + 3.47480\frac{EI_c}{L}\theta_A + 6.39444\frac{EI_c}{12}\theta_B$$

$$= +1277.20 + 0.532870 EI_c\theta_B$$

$$M_3 = M_{03} + s_{33}\frac{EI_c}{L}(\theta_B - R) + s_{34}\frac{EI_c}{L}(\theta_C - R)$$

$$= 0 + 6.39444\frac{EI_c}{12}\theta_B + 3.47840\frac{EI_c}{12}\theta_C$$

$$= +0.532870 EI_c\theta_B + 0.289867 EI_c\theta_C$$

$$M_4 = M_{04} + s_{43}\frac{EI_c}{L}(\theta_B - R) + s_{44}\frac{EI_c}{L}(\theta_C - R)$$

$$= 0 + 3.47840\frac{EI_c}{12}\theta_B + 5.64460\frac{EI_c}{12}\theta_C$$

$$= +0.289867 EI_c\theta_C + 0.470383 EI_c\theta_C$$

The joint-moment conditions corresponding to the two unknown joint rotations are (1) $M_2 + M_3 = 0$, and (2) $M_4 = 0$. The two simultaneous equations become

$$1.065740 EI_c\theta_B + 0.289867 EI_c\theta_C = -1277.20$$

$$0.289867 EI_c\theta_B + 0.470383 EI_c\theta_C = 0$$

Solving,

$$EI_c\theta_B = -1439.72 \qquad EI_c\theta_C = +887.21$$

Substituting the preceding values of joint rotations back into the slope-deflection equations,

$$M_1 = -1756.22 \qquad M_2 = +510.02 \qquad M_3 = -510.01 \qquad M_4 = 0.00$$

Using these values of the end moments, the end reactions are computed as shown in Fig. 7.10.5b, and the shear and moment diagrams are obtained as shown in Fig. 7.10.5c and d.

(e) *Compatibility checks.* To check the correctness of the solution, the slopes of the elastic curve at both ends of each span should be computed by using the simple-span end rotations due to the applied loads and the flexibility coefficients; thus,

$$\theta_1 = \phi_1 + R_{12}$$

$$= \phi_{01}\text{ (uniform load)} + \phi_{01}\text{ (conc. load)} + f_{11}\frac{L}{EI_c}M_1 + f_{12}\frac{L}{EI_c}M_2 + R_{12}$$

$$= +\frac{4600.125}{EI_c} + \frac{1903.333}{EI_c} + (+0.266493)\left(\frac{12}{EI_c}\right)(-1756.22) + (-0.144965)\left(\frac{12}{EI_c}\right)(+510.02) + 0$$

$$= \frac{1}{EI_c}(4600.125 + 1903.333 - 5616.25 - 887.22) = -\frac{0.012}{EI_c} \approx 0$$

$$\theta_2 = \phi_2 + R_{12}$$

$$= \phi_{02}\text{ (uniform load)} + \phi_{02}\text{ (conc. load)} + f_{21}\frac{L}{EI_c}M_1 + f_{22}\frac{L}{EI_c}M_2 + R_{12}$$

$$= -\frac{4417.875}{EI_c} - \frac{1516.666}{EI_c} + (-0.144965)\left(\frac{12}{EI_c}\right)(-1756.22) + (+0.235245)\left(\frac{12}{EI_c}\right)(+510.02) + 0$$

$$= \frac{1}{EI_c}(-4417.875 - 1516.666 + 3055.08 + 1439.72) = -\frac{1439.72}{EI_c}$$

$$\theta_3 = \phi_3 + R_{34} = \phi_{03} + f_{33}\frac{L}{EI_c}M_3 + f_{34}\frac{L}{EI_c}M_4 + R_{34}$$

$$= 0 + (+0.235243)\left(\frac{12}{EI_c}\right)(-510.01) + 0 + 0 = -\frac{1439.72}{EI_c}$$

$$\theta_4 = \phi_4 + R_{34} = \phi_{04} + f_{43}\frac{L}{EI_c}M_3 + f_{44}\frac{L}{EI_c}M_4 + R_{34}$$

$$= 0 + (-0.144965)\left(\frac{12}{EI_c}\right)(-510.01) + 0 + 0 = +\frac{887.21}{EI_c}$$

The fact that $\theta_1 = \theta_A = 0$ and $\theta_2 = \theta_3 = \theta_B = -(1439.72/EI_c)$ constitutes the two compatibility checks for this beam, which is statically indeterminate to the second degree. The final elastic curve is shown in Fig. 7.10.5e.

(f) *Discussion.* One must note that the equilibrium and compatibility checks ensure only the correctness of the solution procedure but not the correctness of the simple-span end rotations due to applied loads and the flexibility coefficients. In Chap. 15, the column-analogy method, by which the fixed-end moments and stiffness coefficients are determined directly, will be described, thus providing another means of checking these quantities to be used in the slope-deflection equations.

Example 7.10.2 By the slope-deflection method analyze the continuous beam shown in Fig. 7.10.9a for a vertical settlement of 15 mm at support B. Draw shear and moment diagrams. Sketch the elastic curve.

SOLUTION For either span 1-2 or span 3-4, the value of EI_c/L is as follows:

$$\frac{EI_c}{L} = \frac{(200\times10^6)(3600\times10^{-6})}{12} = 60{,}000 \text{ kN·m}$$

Because the continuous beam in this problem is identical to that in the previous example, the stiffness coefficients shown in Fig. 7.10.7 are used again to write out the following slope-deflection equations:

$$M_1 = M_{01} + s_{11}\frac{EI_c}{L}(\theta_A - R_{12}) + s_{12}\frac{EI_c}{L}(\theta_B - R_{12})$$

$$= 0 + 5.64460(60{,}000)\left(0 - \frac{0.015}{12}\right) + 3.47840(60{,}000)\left(\theta_B - \frac{0.015}{12}\right)$$

$$= -684.22 + 208{,}704\theta_B$$

$$M_2 = M_{02} + s_{21}\frac{EI_c}{L}(\theta_A - R_{12}) + s_{22}\frac{EI_c}{L}(\theta_B - R_{12})$$

$$= 0 + 3.47840(60{,}000)\left(0 - \frac{0.015}{12}\right) + 6.39444(60{,}000)\left(\theta_B - \frac{0.015}{12}\right)$$

$$= -740.46 + 383{,}666\theta_B$$

$$M_3 = M_{03} + s_{33}\frac{EI_c}{L}(\theta_B - R_{34}) + s_{34}\frac{EI_c}{L}(\theta_C - R_{34})$$

$$= 0 + 6.39444(60{,}000)\left(\theta_B + \frac{0.015}{12}\right) + 3.47840(60{,}000)\left(\theta_C + \frac{0.015}{12}\right)$$

$$= +740.46 + 383{,}666\theta_B + 208{,}704\theta_C$$

$$M_4 = M_{04} + s_{43}\frac{EI_c}{L}(\theta_B - R_{34}) + s_{44}\frac{EI_c}{L}(\theta_C - R_{34})$$

$$= 0 + 3.47840(60{,}000)\left(\theta_B + \frac{0.015}{12}\right) + 5.64460(60{,}000)\left(\theta_C + \frac{0.015}{12}\right)$$

$$= +684.22 + 208{,}704\theta_B + 338{,}676\theta_C$$

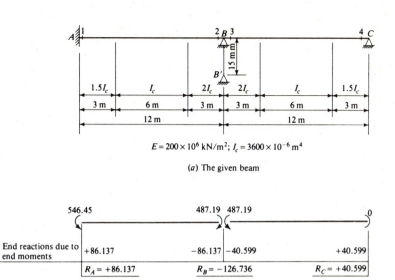

(a) The given beam

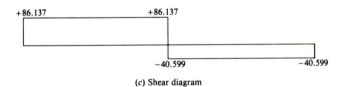

(b) End reactions

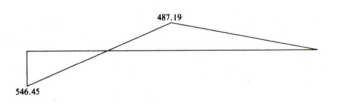

(c) Shear diagram

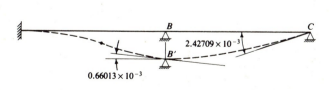

(d) Moment diagram

(e) Elastic curve

Figure 7.10.9 Continuous beam of Example 7.10.2.

The joint-moment conditions $M_2 + M_3 = 0$ and $M_4 = 0$ yield the following two simultaneous equations:
$$+767{,}332\theta_B + 208{,}704\theta_C = 0$$
$$+208{,}704\theta_B + 338{,}676\theta_C = -684.22$$

Solving,
$$\theta_B = +0.66013 \times 10^{-3} \text{ rad} \qquad \theta_C = -2.42709 \times 10^{-3} \text{ rad}$$

Substituting the above values of joint rotations back into the slope-deflection equations,
$$M_1 = -546.45 \qquad M_2 = -487.19 \qquad M_3 = +487.19 \qquad M_4 = 0.00$$

Using the above values of the end moments, the end reactions are computed as shown in Fig. 7.10.9b, and the shear and moment diagrams are obtained as shown in Fig. 7.10.9c and d.

For compatibility checks, the slopes of the elastic curve at points 1, 2, 3, and 4 are computed using the flexibility coefficients shown in Fig. 7.10.7; thus,

$$\theta_1 = \phi_{01} + f_{11}\frac{L}{EI_c}M_1 + f_{12}\frac{L}{EI_c}M_2 + R_{12}$$
$$= 0 + (+0.266493)\left(\frac{1}{60{,}000}\right)(-546.45) + (-0.144965)\left(\frac{1}{60{,}000}\right)(-487.19) + \left(+\frac{0.015}{12}\right)$$
$$= 0 - 2.42709 \times 10^{-3} + 1.17707 \times 10^{-3} + 1.25000 \times 10^{-3}$$
$$= -0.00002 \times 10^{-3} \approx 0$$

$$\theta_2 = \phi_{02} + f_{21}\frac{L}{EI_c}M_1 + f_{12}\frac{L}{EI_c}M_2 + R_{12}$$
$$= 0 + (-0.144965)\left(\frac{1}{60{,}000}\right)(-546.45) + (+0.235243)\left(\frac{1}{60{,}000}\right)(-487.19) + \left(+\frac{0.015}{12}\right)$$
$$= 0 + 1.32027 \times 10^{-3} - 1.91013 \times 10^{-3} + 1.25000 \times 10^{-3}$$
$$= +0.66014 \times 10^{-3}$$

$$\theta_3 = \phi_{03} + f_{33}\frac{L}{EI_c}M_3 + f_{34}\frac{L}{EI_c}M_4 + R_{34}$$
$$= 0 + (+0.235243)\left(\frac{1}{60{,}000}\right)(+487.19) + 0 + \left(-\frac{0.015}{12}\right)$$
$$= 0 + 1.91013 \times 10^{-3} + 0 - 1.25000 = +0.66013 \times 10^{-3}$$

$$\theta_4 = \phi_{04} + f_{43}\frac{L}{EI_c}M_3 + f_{44}\frac{L}{EI_c}M_4 + R_{34}$$
$$= 0 + (-0.144965)\left(\frac{1}{60{,}000}\right)(+487.19) + 0 + \left(-\frac{0.015}{12}\right)$$
$$= 0 - 1.17707 \times 10^{-3} + 0 - 1.25000 = -2.42707 \times 10^{-3}$$

The fact that $\theta_1 = \theta_A = 0$ and $\theta_2 = \theta_3 = \theta_B = +0.66013 \times 10^{-3}$ constitutes the two compatibility checks. However, as discussed in the previous example, the solution is correct only if the flexibility coefficients have been correctly computed in the first place from the way in which the moment of inertia varies along the span. The final elastic curve is shown in Fig. 7.10.9e.

7.11 Exercises

7.1 Solve Exercise 6.1 by the slope-deflection method.
7.2 Solve Exercise 6.2 by the slope-deflection method.

THE SLOPE-DEFLECTION METHOD 257

7.3 Solve Exercise 6.3 by the slope-deflection method.
7.4 Solve Exercise 6.4 by the slope-deflection method.
7.5 Solve Exercise 6.5 by the slope-deflection method.
7.6 Solve Exercise 6.6 by the slope-deflection method.
7.7 to 7.13 Analyze the rigid frames shown in Figs. 7.11.1 to 7.11.7 by the slope-deflection method. Draw shear and moment diagrams. Sketch the elastic curves.
7.14 Solve Exercise 4.13 by the slope-deflection method.
7.15 Solve Exercise 4.14 by the slope-deflection method.
7.16 Solve Exercise 4.15 by the slope-deflection method.
7.17 Solve Exercise 4.16 by the slope-deflection method.

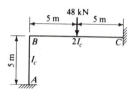

Figure 7.11.1 Exercise 7.7.

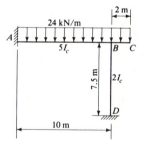

Figure 7.11.2 Exercise 7.8.

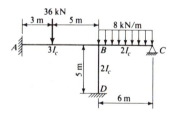

Figure 7.11.3 Exercise 7.9.

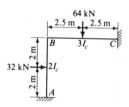

Figure 7.11.4 Exercise 7.10.

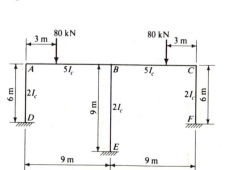

Figure 7.11.5 Exercise 7.11.

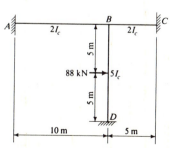

Figure 7.11.6 Exercise 7.12.

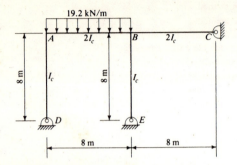

Figure 7.11.7 Exercise 7.13.

Figure 7.11.8 Exercise 7.18.

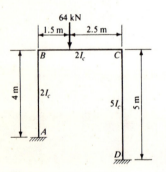

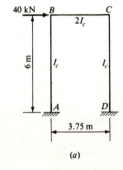

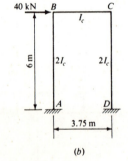

Figure 7.11.9 Exercise 7.19.

Figure 7.11.10 Exercise 7.20.

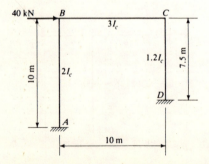

Figure 7.11.11 Exercise 7.21.

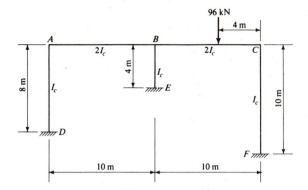

Figure 7.11.12 Exercise 7.22.

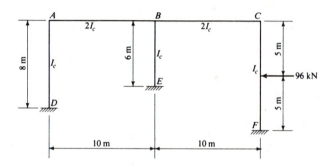

Figure 7.11.13 Exercise 7.23.

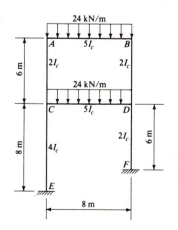

Figure 7.11.14 Exercise 7.24.

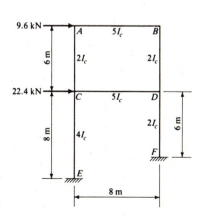

Figure 7.11.15 Exercise 7.25.

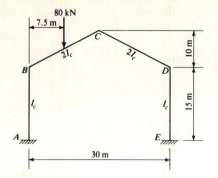

Figure 7.11.16 Exercise 7.28.

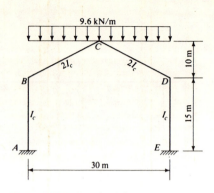

Figure 7.11.17 Exercise 7.29.

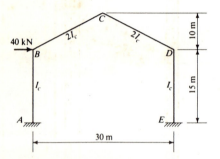

Figure 7.11.18 Exercise 7.30.

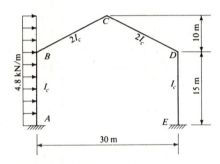

Figure 7.11.19 Exercise 7.31.

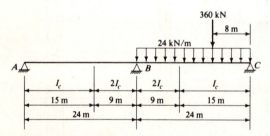

Figure 7.11.20 Exercise 7.32.

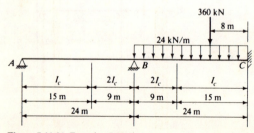

Figure 7.11.21 Exercise 7.33.

7.18 to **7.25** Analyze the rigid frames shown in Fig. 7.11.8 to 7.11.15 by the slope-deflection method. Draw shear and moment diagrams. Sketch the elastic curves.

7.26 Solve Exercise 4.17 by the slope-deflection method.

7.27 Solve Exercise 4.18 by the slope-deflection method.

7.28 to **7.31** Analyze the gable frames shown in Figs. 7.11.16 to 7.11.19 by the slope-deflection method. Draw shear and moment diagrams. Sketch the elastic curves.

7.32 and **7.33** Analyze the continuous beams shown in Figs. 7.11.20 and 7.11.21 by the slope-deflection method. Draw shear and moment diagrams. Sketch the elastic curves.

7.34 and **7.35** Analyze the continuous beams in Exercises 7.32 and 7.33 for a vertical settlement of 15 mm at support B. Use $E = 200 \times 10^6 \text{ kN/m}^2$ and $I_c = 3600 \times 10^{-6} \text{ m}^4$.

CHAPTER
EIGHT

THE MOMENT-DISTRIBUTION METHOD

8.1 Introduction

The moment-distribution method was originally presented by Prof. Hardy Cross† in the 1930s and is considered one of the most important contributions ever made to the structural analysis of continuous beams and rigid frames. Essentially it is a method to solve the simultaneous equations in the slope-deflection method by successive approximations, accurate to as many significant figures as desired. As will be described in Chap. 12, the slope-deflection method, once cast in matrix notation, can be conveniently programmed on an electronic computer, by which a large number of simultaneous equations can be solved quickly and inexpensively. Nevertheless, the moment-distribution method can itself be programmed on an electronic computer and competes well in many situations with other methods. In preliminary analysis and design of a small structure or of parts of a large structure, the moment-distribution method remains unsurpassed for its simplicity and for the physical feeling that it conveys to the analyst about the force and deformation response of the structure.

8.2 Basic Concept

When a continuous beam is subjected to applied loads or support settlements, there is no *unknown* member-axis rotation in its deformation response,

---

†Hardy Cross, "Analysis of Continuous Frames by Distributing Fixed-End Moments," *Proc. ASCE*, May 1930; see also *Tran. ASCE*, vol. 96, pap. 1793, 1932.

THE MOMENT-DISTRIBUTION METHOD 263

whereas the joints in a rigid frame may or may not have the freedom of undergoing unknown amounts of translation. Although the moment-distribution method can be used, by a multiple-stage indirect process, to analyze rigid frames with unknown joint translations as well, it is best to describe its basic concept on the premise that the structure has no unknown member-axis rotation.

The deformation response of a continuous beam or a rigid frame without unknown joint translation is completely defined by the unknown joint rotations, such as θ_B, θ_C, and θ_D in Fig. 8.2.1a and c. Physically, it is conceivable that locking moments can be applied to joints B, C, and D so as to maintain zero slopes at B, C, and D, as shown by the fixed conditions in Fig. 8.2.1b and d. In fact, the magnitudes and directions of these locking moments are known in advance in terms of the applied loads or the support settlements. When the locking moment at one of the joints is released, that joint will rotate. This rotation induces changes not only in the moments at the member ends entering the released joint but also in the locking moments at the immediately adjacent joints on both sides of the released joint. If each joint is successively released and locked back and then this process is repeated, a time

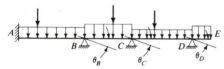

(a) Given condition, applied loads

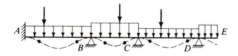

(b) Fixed condition, applied loads

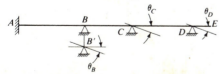

(c) Given condition, support settlement

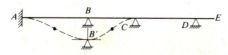

(d) Fixed condition, support settlement

Figure 8.2.1 Fixed conditions in moment-distribution method.

will be reached at which every joint has attained its full needed value in the final deformation response. Then the locking moments would have been dissipated, or *distributed*, throughout the structure by means of successive amounts of joint rotations; hence the name *moment distribution*.

8.3 Stiffness and Carry-Over Factors

In order to fully develop the procedures in the moment-distribution method, it is useful to consider the following problem:

If a clockwise moment M_A is applied at the hinged end of a straight member of constant moment of inertia, which is hinged at one end and fixed at the other, as shown in Fig. 8.3.1a, find the rotation θ_A at the hinged support and the moment M_B at the fixed support.

Separating the bending-moment diagram for Fig. 8.3.1a into those shown in Fig. 8.3.1b and c, applying the conjugate-beam theorem to find θ_{B1} and θ_{B2}, and then equating θ_B to zero,

$$\theta_B = -\theta_{B1} + \theta_{B2} = -\frac{M_A L}{6EI} + \frac{M_B L}{3EI} = 0$$

from which

$$\boxed{M_B = +\tfrac{1}{2} M_A} \tag{8.3.1}$$

Again applying the conjugate-beam theorem to find θ_{A1} and θ_{A2},

$$\theta_A = \theta_{A1} - \theta_{A2} = +\frac{M_A L}{3EI} - \frac{M_B L}{6EI}$$

Substituting Eq. (8.3.1) into the preceding equation and solving for M_A,

$$\boxed{M_A = \frac{4EI}{L} \theta_A} \tag{8.3.2}$$

Thus, for a span AB which is hinged at A and fixed at B, a clockwise rotation θ_A can be effected by applying a clockwise moment $M_A = (4EI/L)\theta_A$ at A, which in turn induces a clockwise moment $M_B = \tfrac{1}{2} M_A$ at B. The

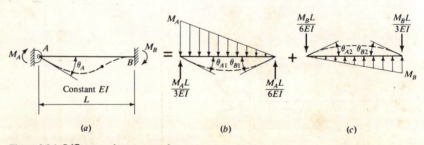

Figure 8.3.1 Stiffness and carry-over factors.

expression $4EI/L$ is called the *stiffness factor*, which is defined as the moment at the near end to cause a unit rotation at the near end when the far end is fixed. The number $+\frac{1}{2}$ is the *carry-over factor*, which is defined as the ratio of the moment at the fixed far end to the moment at the rotating near end. Note that clockwise rotations of the joints and clockwise moments acting at the member ends are considered positive in the moment-distribution method just as in the slope-deflection method.

8.4 Distribution Factors

Consider the continuous beam ABC shown in Fig. 8.4.1a, for which the fixed-end moments are shown in Fig. 8.4.1b. If the joints B and C are to be locked against rotation, the locking moments as shown in Fig. 8.4.1c must be applied, wherein the locking moment of 80 kN·m at B is the algebraic sum of the two fixed-end moments at B on spans BA and BC. Note at this time that the fixed-end moments are shown in their proper places in Table 8.4.1.

When joint B is released, it will rotate in the counterclockwise direction, opposite to the direction of the locking moment. This counterclockwise rotation will induce counterclockwise moments at B to act on BA and BC in amounts proportional to the stiffness factors of the respective spans, with a sum of 80 kN·m. The stiffness factor of span BA is $4E(5I_c)/10 = 2EI_c$, and that of span

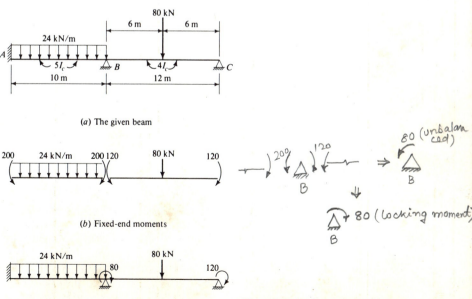

Figure 8.4.1 Basic concept in moment-distribution method.

Table 8.4.1 Moment distribution for continuous beam in Fig. 8.4.1

Joint		A	B		C
Member		AB	BA	BC	CB
$4EI/L$		$2EI_c$	$2EI_c$	$1.3333EI_c$	$1.3333EI_c$
Distribution factor		....	0.6000	0.4000	1.0000
Cycle 1	FEM	−200	+200	−120	+120
	BAL	0	− 48	− 32	−120
Cycle 2	CO	− 24	0	− 60	− 16
	BAL	0	+ 36	+ 24	+ 16
Cycle 3	CO	+ 18	0	+ 8	+ 12
	BAL	0	− 4.8	− 3.2	− 12
Cycle 4	CO	− 2.4	0	− 6	− 1.6
	BAL	0	+ 3.6	+ 2.4	+ 1.6
Cycle 5†	CO	+ 1.8	0	+ 0.8	+ 1.2
	BAL	0	− 0.48	− 0.32	− 1.2
Total (at end of 5 cycles)		−206.6	+186.32	−186.32	0

†The process can be continued to any desired degree of accuracy.

BC is $4E(4I_c)/12 = 1.3333EI_c$, as shown in Table 8.4.1. Thus, $2EI_c/(2EI_c + 1.3333EI_c) = (0.6000)(80 \text{ kN·m})$, or 48 kN·m counterclockwise will act on BA; and $1.3333EI_c/(2EI_c + 1.3333EI_c) = (0.4000)(80 \text{ kN·m})$, or 32 kN·m counterclockwise will act on BC. The numbers 0.6000 and 0.4000 are called the *distribution factors* (DF), which may be defined as the fractions by which the total unbalance at the joint is to be distributed to the member ends meeting at the joint. The counterclockwise rotation of joint B, which induces 48 kN·m counterclockwise at BA and 32 kN·m counterclockwise at BC, also induces the carry-over moments of $\frac{1}{2}(48) = 24$ kN·m counterclockwise at A and $\frac{1}{2}(32) = 16$ kN·m counterclockwise at C. Since support A is naturally fixed, it can readily assume the carry-over moment. However, the locking moment at C will be changed from 120 kN·m clockwise to $120 − 16 = 104$ kN·m clockwise. If, rather than changing the value of 120 kN·m in the original lock, an auxiliary lock is added to exert the 16 kN·m counterclockwise, then the next release of joint C can be made only to the original lock. The distribution factor to CB at joint C is 100 percent, or the balancing moment is $−120$ kN·m at C. Thus the first cycle of moment distribution is completed (see Table 8.4.1). To summarize, all member ends are first locked up against rotation by fixed-end moments (FEM) of $−200$,

+200, −120, +120. Then joints B and C are released in succession, and the balancing moments (BAL) are 0, −48, −32, −120. The carry-over moments (CO) of −24, 0, −60, −16 are considered to be a new set of fixed-end moments required to lock all joints in the positions reached at the end of the first cycle.

In the second cycle of moment distribution, the joints B and C are each released for the second time. The unbalance at joint B is −60, the balancing moments of +36 and +24 are numerically 0.6000 and 0.4000 times the unbalance. The balancing moment to the unbalance of −16 at joint C is +16. The carry-over moments of +18, 0, +8, +12 are considered to be the new locking moments at the beginning of the third cycle. This same process can be repeated for as many cycles as desired to bring the balancing moments to very small magnitudes. Thus any degree of accuracy can be obtained, and the required number of cycles decreases as the required accuracy decreases. The final or total balanced moments are obtained by adding all numbers in the respective columns.

8.5 Application to the Analysis of Statically Indeterminate Beams for Applied Loads

The analysis of statically indeterminate beams for applied loads by the moment-distribution method can be performed by making up a moment-distribution table as shown by Table 8.4.1. Some additional remarks are as follows:

1. The analyst should decide at the beginning the degree of accuracy desired and use the same number of decimal places in all entries in the table.
2. The analyst may use judgment to stop the moment distribution at the end of any cycle, but no later than when the last digit in every column has changed by less than 1 in the cycle just completed.
3. The sign of the balancing moments at each joint is opposite to that of the unbalance, which is the algebraic sum of the FEMs in the first cycle, or of the carry-overs in all other cycles.
4. The analyst should make sure that the numerical sum of the balancing moments is exactly equal to the numerical value of the unbalance at every balancing.
5. In carry-overs, where an odd number is to be divided by 2, the usual practice is to use the nearest even number.
6. The procedure for the check on moment distribution will be described in Sect. 8.6.

Example 8.5.1 Analyze the continuous beam of Example 7.3.1, shown again in Fig. 8.5.1, by the moment-distribution method.

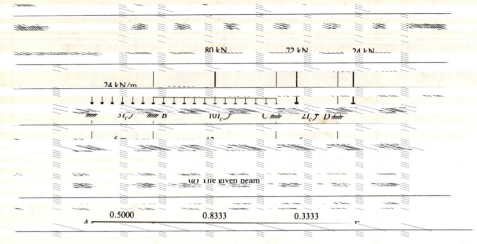

(b) Values of EI/L in terms of EI_c

(c) Values of FEM in kN·m

Figure 8.5.1 Continuous beam of Example 8.5.1.

SOLUTION (a) *FEM and values of EI/L.* The notations M_{FAB} and M_{FBA} are used to mean the fixed-end moments at A and B of span AB, respectively.

$$M_{FAB} = -\frac{24(6)^2}{12} = -72 \text{ kN·m} \qquad M_{FBA} = +72 \text{ kN·m}$$

$$M_{FBC} = -\frac{16(12)^2}{12} - \frac{80(12)}{8} \qquad M_{FCB} = +312 \text{ kN·m}$$

$$M_{FCD} = -\frac{72(6)}{6} \qquad M_{FDC} = +\frac{72 \cdot 6}{6}$$

The overhang DE is not a bona fide member; it is treated as a cantilever beam fixed at D in the fixed condition. The fixed-end moment at D acts counterclockwise on DE; therefore it is negative.

$$M_{FDE} = -24(1.5) = -36 \text{ kN·m}$$

Since the sidesway is prevented, the condition may be dropped. The values of EI/L for spans AB, BC, and CD are:

$$\frac{EI}{L}(AB) = \frac{E(3I_c)}{6} = 0.5000 EI_c$$

$$\frac{EI}{L}(BC) = \frac{E(10I_c)}{12} = 0.8333 EI_c$$

$$\frac{EI}{L}(CD) = \frac{E(2I_c)}{6} = 0.3333 EI_c$$

The values of EI/L and FEM are shown in Fig. 8.5.1b and c.

Table 8.5.2 Moment distribution for continuous beam in Example 8.5.2

Joint		A	B		C		D	
Member		AB	BA	BC	CB	CD	DC	Overhang
EI/L (in terms of EI_c)		0.5000	0.5000	0.8333	0.8333	0.3333	0.3333	
Cycle	DF	....	0.3750	0.6250	0.7143	0.2857	1.0000	
1	FEM	−72.00	+72.00	−312.00	+312.00	−64.00	+32.00	−36.00
	BAL	0.00	+90.00	+150.00	−177.15	−70.85	+4.00	
2	CO	+45.00	0.00	−88.58	+75.00	+2.00	−35.42	
	BAL	0.00	+33.22	+55.36	−55.00	−22.00	+35.42	
3	CO	+16.61	0.00	−27.50	+27.68	+17.71	−11.00	
	BAL	0.00	+10.31	+17.19	−32.42	−12.97	+11.00	
4	CO	+5.16	0.00	−16.21	+8.60	+5.50	−6.48	
	BAL	0.00	+6.08	+10.13	−10.07	−4.03	+6.48	
5	CO	+3.04	0.00	−5.04	+5.06	+3.24	2.02	
	BAL	0.00	+1.89	+3.15	−5.93	−2.37	+2.02	
6	CO	+0.94	0.00	−2.96	+1.58	+1.01	−1.18	
	BAL	0.00	+1.11	+1.85	−1.85	−0.74	+1.18	
7	CO	+0.56	0.00	−0.92	+0.92	+0.59	−0.37	
	BAL	0.00	+0.34	+0.58	−1.08	−0.43	+0.37	
8	CO	+0.17	0.00	−0.54	+0.29	+0.18	−0.22	
	BAL	0.00	+0.20	+0.34	−0.34	−0.13	+0.22	
9	CO	+0.10	0.00	−0.17	+0.17	+0.11	−0.06	
	BAL	0.00	+0.06	+0.11	−0.20	−0.08	+0.06	
10	CO	+0.03	0.00	−0.10	+0.06	+0.03	−0.04	
	BAL	0.00	+0.04	+0.06	−0.06	−0.03	+0.04	
11	CO	+0.02	0.00	−0.03	+0.03	+0.02	−0.02	
	BAL	0.00	+0.01	+0.02	−0.04	−0.01	+0.02	
Total balanced M		−0.37	+215.26	−215.26	+147.25	−147.25	+36.00	−36.00
Check	Change	+71.63	+143.26	+96.74	−164.75	−83.25	+4.00	
	$-\frac{1}{2}$(change)	−71.63	−35.82	+82.38	48.37	−2.00	+41.62	
	Sum	0.00	+107.44	+179.12	−213.12	−85.25	+45.62	
	$EI_c\theta$	0.00	+71.63	+71.65	−85.25	−85.25	+45.62	

THE MOMENT-DISTRIBUTION METHOD

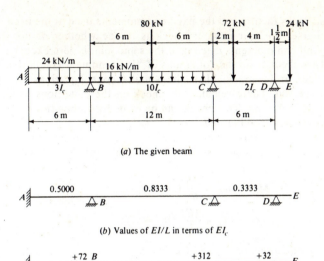

(a) The given beam

(b) Values of EI/L in terms of EI_c

(c) Value of FEM in kN·m

Figure 8.5.2 Continuous beam of Example 8.5.2.

the signs in any one line before writing the numerical values. This helps the analyst to concentrate on one thing at a time. Note also that the balancing moments placed at the member ends meeting in one joint in any one cycle must add exactly to the total unbalance at that joint, to the last digit. This will ensure that the sum of the final balanced moments meeting at the same joint is exactly zero. The check on moment distribution shown in Table 8.5.1 will be explained in Sec. 8.6.

Example 8.5.2 Analyze the continuous beam of Example 7.3.2, shown again in Fig. 8.5.2, by the moment-distribution method.

SOLUTION The given beam in this example is identical to that of the previous example except that the left end is now fixed instead of being a simple support. The values of EI/L and FEM shown in Fig. 8.5.2b and c are the same as those in Fig. 8.5.1b and c.

Table 8.5.2 shows the moment distribution, which stops at the end of the eleventh cycle because this cycle changes the value of every end moment by 0.01 or less, except at A of AB, but the carry-over to there from the eleventh cycle would have been equal to zero. Note that four dots are shown for the distribution factor in column AB because the fixed joint A never needs to be released and the name *distribution factor* does not apply there. Thus the balancing moments under column AB are all equal to zero. The check on moment distribution shown in Table 8.5.2 will be explained in Sec. 8.6.

8.6 Check on Moment Distribution

The moment-distribution table begins with the relative stiffness factors EI/L and fixed-end moments and ends with the final end moments. A check can be made to ensure that the correct end moments have been obtained on the basis of the relative stiffness factors and the fixed-end moments used at the head of the table. Of course, the first obvious check is to see whether or not the algebraic sum of all end moments meeting at the same joint is zero except at a fixed support. This check is on the conditions of "joint-moment equilibrium," as they have been used to establish the simultaneous equations in the slope-deflection method. There remains the check on the conditions of continuity of slopes at each joint. This check can be made by finding the slope of the elastic curve at each member end.

Because the member-axis rotations are zero for all spans, the slope-deflection equations for a typical span AB may be written as, from Eqs. (7.2.4a and b),

$$M_A = M_{0A} + \frac{2EI}{L}(2\theta_A + \theta_B)$$

$$M_B = M_{0B} + \frac{2EI}{L}(2\theta_B + \theta_A)$$

Solving the above equations for θ_A and θ_B,

$$\theta_A = \frac{(M_A - M_{0A}) - \frac{1}{2}(M_B - M_{0B})}{3EI/L} \tag{8.6.1a}$$

$$\theta_B = \frac{(M_B - M_{0B}) - \frac{1}{2}(M_A - M_{0A})}{3EI/L} \tag{8.6.1b}$$

Equations (8.6.1a and b) may be rewritten in a single equation as

$$\theta_{\text{near end}} = \frac{(\text{Change}_{\text{near end}}) + (-\frac{1}{2})(\text{Change}_{\text{far end}})}{3EI/L} \tag{8.6.2}$$

which, if stated in words, means that the slope at either end of a span is equal to the sum of the change in moment (final end moment minus fixed-end moment) at that end and $-\frac{1}{2}$ times the change in moment at the far end, divided by $3EI/L$ of that span. By means of Eq. (8.6.2) the slopes at the ends of each span can be computed. The check on the conditions of continuity will be that the slopes at the ends of all members meeting at the same joint should be equal to each other and that the slope at a fixed support should be equal to zero.

The first line in the check on moment distribution, as shown in Tables 8.5.1 and 8.5.2, contains the change in moment at that end; the second line contains $-\frac{1}{2}$ times the change in moment at the far end; the third line contains the sum of the two values above it; and the fourth line contains the quotient obtained by dividing the number on the third line by $3EI/L$.

The number of continuity checks which can be observed is equal to the degree of indeterminacy of the structure. In Table 8.5.1, the two continuity checks are (1) $\theta_{BA} = +71.59/EI_c$ and $\theta_{BC} = +71.60/EI_c$, and (2) $\theta_{CB} = -85.24/EI_c$ and $\theta_{CD} = -85.21/EI_c$. In Table 8.5.2, the three continuity checks are (1) $\theta_A = 0$, (2) $\theta_{BA} = +71.63/EI_c$ and $\theta_{BC} = +71.65/EI_c$, and (3) $\theta_{CB} = -85.25/EI_c$ and $\theta_{CD} = -85.25/EI_c$.

THE MOMENT-DISTRIBUTION METHOD 273

It is to be noted that the aforementioned check has nothing to do with the correctness of the values of EI/L and fixed-end moments used at the head of the moment-distribution table. If these quantities are wrong, the final end moments will also be wrong even though they have met the test of the check.

8.7 Modified Stiffness Factor and Modified Fixed-End Moment at the Near End When the Far End Is Hinged

An examination of the moment distribution shown in Tables 8.5.1 and 8.5.2 shows that when there is a simple or hinged support at the extreme end, the alternate locking and release are recorded by a pair of equal and opposite numbers in each cycle, except in the first cycle when there is an overhang beyond the exterior support. This seems to be a wasteful portion in the distribution process. Actually the exterior simple or hinged support can be treated as such at the outset by permitting the joint there to rotate at all times.

Consider the two single-span beams shown in Fig. 8.7.1. The reacting moment at B can be solved by the moment-distribution method. If there is an

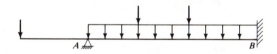

FEM	+30	−80		+100
BAL		+50		0
CO		0		+25
BAL		0		0
Total	+30	−30		+125

(a)

FEM	−80		+100
BAL	+80		0
CO	0		+40
BAL	0		0
Total	0		+140

(b)

Figure 8.7.1 Modified fixed-end moment.

274 INTERMEDIATE STRUCTURAL ANALYSIS

overhang beyond the simple support A as shown in Fig. 8.7.1a, the clockwise moment acting on the overhang is entered as, say, $+30$; the fixed-end moments on span AB are, say, -80 and $+100$. After a two-cycle moment distribution, the reacting moment at B is found to be 125 acting clockwise at B. If there is no overhang, as shown in Fig. 8.7.1b, the reacting moment at B is 140 acting clockwise at B. The value $+125$ or $+140$, as shown in Fig. 8.7.1a and b, will be called the *modified fixed-end moment* at the near end when the far end is a simple or hinged support.

In Fig. 8.7.2a are shown four members, AB, CB, DB, and EB, meeting at a rigid joint B. There is a simple or hinged support at A, but the supports at C, D, and E are fixed. A releasing moment acting on joint B will cause it to rotate, inducing moments of $4EI/L$ times θ_B at the near end of members BC, BD, and BE and moments of $2EI/L$ times θ_B at the fixed ends C, D, and E. If the support at A is treated as a simple support, the moment induced to act at B of member BA will be the M_B shown in Fig. 8.7.2b, and the moment at A will always be zero. Applying the conjugate-beam method to span AB of Fig. 8.7.2b,

$$\theta_B = \frac{M_B L}{3EI}$$

from which

$$M_B = \frac{3EI}{L} \theta_B = \frac{3}{4} \frac{4EI}{L} \theta_B \qquad (8.7.1)$$

Thus the moment required to cause a unit rotation at the near end when the far end is a simple or hinged support is equal to $3EI/L$. The quantity $3EI/L$ is called the *modified stiffness factor*, and it is $\frac{3}{4}$ times the "regular stiffness factor." If the value of EI/L has been used as the relative stiffness factor in computing the distribution factors in the moment-distribution method, then for any member end whose far end is hinged, the modified relative stiffness

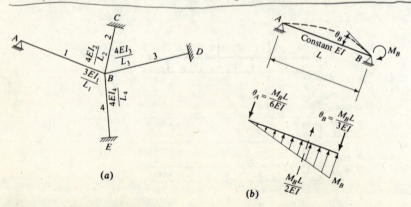

(a)　　　　(b)

Figure 8.7.2 Modified stiffness factor.

Table 8.7.1 Moment distribution for continuous beam in Example 8.7.1†

Joint		A	B		C		D	
Member		AB	BA	BC	CB	CD	DC	Overhang
EI/L (in terms of EI_c)		0.5000	0.5000	0.8333	0.8333	0.3333	0.3333	
$0.75 EI/L$			0.3750			0.2500		
Modified DF		....	0.3104	0.6896	0.7692	0.2308	....	
FEM		−72.00	+72.00	−312.00	+312.00	−64.00	+32.00	−36.00
BAL		+72.00	0.00			0.00	+4.00	
CO		0.00	+36.00			+2.00	0.00	
BAL		0.00	0.00			0.00	0.00	
Cycle 1	Mod. FEM	0.00	+108.00	−312.00	+312.00	−62.00	+36.00	
	BAL		+63.32	+140.68	−192.30	−57.70		
2	CO		0.00	−96.15	+70.34	0.00		
	BAL		+29.84	+66.31	−54.10	−16.24		
3	CO		0.00	−27.05	+33.16	0.00		
	BAL		+8.40	+18.65	−25.51	−7.65		
4	CO		0.00	−12.76	+9.32	0.00		
	BAL		+3.96	+8.80	−7.17	−2.15		
5	CO		0.00	−3.58	+4.40	0.00		
	BAL		+1.11	+2.47	−3.38	−1.02		
6	CO		0.00	−1.69	+1.24	0.00		
	BAL		+0.52	+1.17	−0.95	−0.29		
7	CO		0.00	−0.48	+0.58	0.00		
	BAL		+0.15	+0.33	−0.45	−0.13		
8	CO		0.00	−0.22	+0.16	0.00		
	BAL		+0.07	+0.15	−0.12	−0.04		
9	CO		0.00	−0.06	+0.08	0.00		
	BAL		+0.02	+0.04	−0.06	−0.02		
10	CO		0.00	−0.03	−0.02	0.00		
	BAL		+0.01	+0.02	+0.02	0.00		
Total balanced M		0.00	+215.40	−215.40	+147.24	−147.24	+36.00	−36.00

†For check, see Table 8.5.1.

factor should be equal to $0.75EI/L$. Returning to the situation described by Fig. 8.7.2a, if the support at A is treated as a simple or hinged support at all times, the distribution factors used to balance the joint B should be in the ratio of $0.75(EI/L$ of $BA)$, EI/L of BC, EI/L of BD, and EI/L of BE. These distribution factors, where one of the far member ends is treated as a simple or hinged support, are called the *modified distribution factors*.

The simple supports at A and D of Example 8.5.1, or the simple support at D of Example 8.5.2, may be treated as such before the first cycle of moment distribution. However, the modified fixed-end moment at the exterior side of the first interior support, should be determined and used, and the modified distribution factors at the first interior support should be computed by reducing the stiffness factor at the interior end of the exterior span to $\frac{3}{4}$ of its regular value. This alternate moment distribution by the "modified stiffness method" is illustrated in the following two examples.

Example 8.7.1 Rework the moment distribution shown in Table 8.5.1 by treating the exterior simple supports at A and D as such at the outset.

SOLUTION In this particular case, 10 cycles are needed in this supposedly more simplified modified stiffness method shown in Table 8.7.1, as opposed to 7 cycles in the regular stiffness method of Table 8.5.1, although there is still more blank space in Table 8.7.1 than in Table 8.5.1. Note that in Table 8.7.1 the values of $0.75EI/L$ are entered only under BA and CD. The modified distribution factors are $0.3750/(0.3750 + 0.8333) = 0.3104$ to BA and $0.8333/(0.3750 + 0.8333) = 0.6896$ to BC; they are $0.8333/(0.8333 + 0.2500) = 0.7692$ to BC and $0.2500/(0.8333 + 0.2500) = 0.2308$ to CD. Four dots are shown for the modified distribution factor under columns AB and DC, because joints A and D are treated as simple supports to begin with and no alternate locking and release will ever be applied to them. However, the modified fixed-end moments of $+108$ at B of BA and of -62 at C of CD must be determined in advance of the first cycle of moment distribution by carrying out the two "auxiliary" moment distributions, one for each end, as shown in the beginning part of Table 8.7.1. Note particularly the vertical arrows drawn in columns AB and DC. By drawing these arrows immediately after the modified FEM horizontal line is filled out, the analyst is reminded not to carry any more moment to the columns AB and DC. The check on the moment distribution of Table 8.7.1 should be identical to that of Table 8.5.1.

Example 8.7.2 Rework the moment distribution shown in Table 8.5.2 by treating the exterior simple support at D as such at the outset.

SOLUTION In this case, 10 cycles of moment distribution are needed in the modified stiffness method shown in Table 8.7.2, as opposed to 11 cycles in the regular stiffness method of Table 8.5.2. The "auxiliary" moment distribution is done only for the right end of the beam to obtain the modified fixed-end moment of -62 at C of CD. The remainder of Table 8.7.2 should require no further explanation.

8.8 Application to the Analysis of Statically Indeterminate Beams for Uneven Support Settlements

The moment-distribution method can be used to analyze statically indeterminate beams for uneven support settlements. The physical concept involved is that fixed-end moments as shown in Fig. 8.8.1 are applied to the member

THE MOMENT-DISTRIBUTION METHOD

Table 8.7.2 Moment distribution for continuous beam in Example 8.7.2†

Joint		A	B		C		D	
Member		AB	BA	BC	CB	CD	DC	Overhang
EI/L (in terms of EI_c)		0.5000	0.5000	0.8333	0.8333	0.3333	0.3333	
$0.75 EI/L$						0.2500		
Modified DF		....	0.3750	0.6250	0.7692	0.2308	....	
FEM		−72.00	+72.00	−312.00	+312.00	− 64.00	+32.00	−36.00
BAL						0.00	+ 4.00	
CO						+ 2.00	0.00	
BAL						0.00	0.00	
Cycle 1	Mod. FEM	−72.00	+ 72.00	−312.00	+312.00	− 62.00	+36.00	−36.00
	BAL	0.00	+ 90.00	+150.00	−192.30	− 57.70	↑	
2	CO	+45.00	0.00	− 96.15	+ 75.00	0.00		
	BAL	0.00	+ 36.06	+ 60.09	− 57.69	− 17.31		
3	CO	+18.03	0.00	− 28.84	+ 30.04	0.00		
	BAL	0.00	+ 10.82	+ 18.02	− 23.11	− 6.93		
4	CO	+ 5.41	0.00	− 11.56	+ 9.01	0.00		
	BAL	0.00	+ 4.34	+ 7.22	− 6.93	− 2.08		
5	CO	+ 2.17	0.00	− 3.46	+ 3.61	0.00		
	BAL	0.00	+ 1.30	+ 2.16	− 2.78	− 0.83		
6	CO	+ 0.65	0.00	− 1.39	+ 1.08	0.00		
	BAL	0.00	+ 0.52	+ 0.87	− 0.83	− 0.25		
7	CO	+ 0.26	0.00	− 0.42	+ 0.44	0.00		
	BAL	0.00	+ 0.16	+ 0.26	− 0.34	− 0.10		
8	CO	+ 0.08	0.00	− 0.17	+ 0.13	0.00		
	BAL	0.00	+ 0.06	+ 0.11	− 0.10	− 0.03		
9	CO	+ 0.03	0.00	− 0.05	+ 0.06	0.00		
	BAL	0.00	+ 0.02	+ 0.03	− 0.05	− 0.01		
10	CO	+ 0.01	0.00	− 0.02	− 0.02	0.00		
	BAL	0.00	+ 0.01	+ 0.01	+ 0.02	0.00	↓	
Total balanced M		− 0.36	+215.29	−215.29	+147.24	−147.24	+36.00	−36.00

†For check, see Table 8.5.2.

ends to hold the end slopes to zero but meanwhile to allow the supports to settle unevenly. Then the joints are released (or balanced); the carry-over moments become the next unbalanced moments; the joints are again balanced; and so on. In other words, the fixed-end moments due to member-axis rotation $R = \Delta/L$ are treated in exactly the same manner as those due to applied loads.

Formulas for M_{0A} and M_{0B} due to a clockwise member-axis rotation $R = \Delta/L$ can be derived by the moment-area method, as shown in Fig. 8.8.1. By moment-area theorem 1, the change in slope between A and B is equal to the M/EI area between A and B; thus the absolute values of M_{0A} and M_{0B}

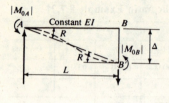

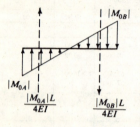

Figure 8.8.1 Fixed-end moments due to member-axis rotation.

must be equal, as may be observed from the moment diagram. By moment-area theorem 2, the distance BB' is equal to the moment of the M/EI area between A and B about B; this moment is that of a couple equal to the product of the triangular area in Fig. 8.8.1 and the distance between the centroids of the two triangles; thus

$$\Delta = \frac{|M_{0A} \text{ or } M_{0B}|L}{4EI} \frac{2}{3} L$$

from which

$$M_{0A} = M_{0B} = -\frac{6EI\Delta}{L^2} = -\frac{6EIR}{L} \tag{8.8.1}$$

The negative sign in the above equation is added in order to adhere to the sign convention used in the slope-deflection and moment-distribution methods.

Example 8.8.1 Analyze the continuous beam of Example 7.5.1, shown again in Fig. 8.8.2, by the moment-distribution method.

SOLUTION (a) *FEM and values of EI/L.*

$$M_{FAB} = M_{FBA} = -\frac{6EIR}{L} = -\frac{6(200)(1200)}{6}\left(+\frac{0.015}{6}\right) = -600 \text{ kN·m}$$

$$M_{FBC} = M_{FCB} = -\frac{6EIR}{L} = -\frac{6(200)(4000)}{12}\left(-\frac{0.015}{12}\right) = +500 \text{ kN·m}$$

$$M_{FCD} = M_{FDC} = 0$$

$$\frac{EI}{L}(AB) = \frac{200(1200)}{6} = 40{,}000 \text{ kN·m} \qquad \frac{EI}{L}(BC) = \frac{200(4000)}{12} = 66{,}667 \text{ kN·m}$$

$$\frac{EI}{L}(CD) = \frac{200(800)}{6} = 26{,}667 \text{ kN·m}$$

THE MOMENT-DISTRIBUTION METHOD 279

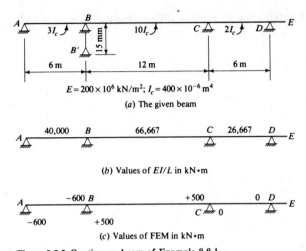

Figure 8.8.2 Continuous beam of Example 8.8.1.

Note that the values of EI/L can all be expressed in terms of EI_c or in absolute dimensional units such as kilonewton-meters. The latter choice is preferred in this problem because the slopes obtained in the check procedure in the moment-distribution table would be exact in radians. The values of EI/L and FEM are shown in Fig. 8.8.2b and c.

(b) *Moment distribution.* The moment distribution can be performed using either the regular stiffness method or the modified stiffness method, as shown in Table 8.8.1a and b.

Example 8.8.2 Analyze the continuous beam of Example 7.5.2, shown again in Fig. 8.8.3, by the moment-distribution method.

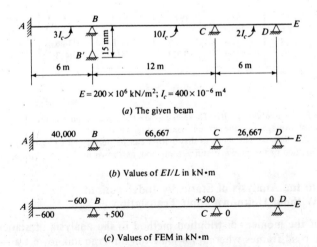

Figure 8.8.3 Continuous beam of Example 8.8.2.

Table 8.8.1a Moment distribution by regular stiffness method, Example 8.8.1

Joint		A	B		C		D
Member		AB	BA	BC	CB	CD	DC
EI/L, kN·m		40,000	40,000	66,667	66,667	26,667	26,667
Cycle	DF	1.0000	0.3750	0.6250	0.7143	0.2857	1.0000
1	FEM	−600.0	−600.0	+500.0	+500.0	0.0	0.0
	BAL	+600.0	+ 37.5	+ 62.5	−357.2	−142.8	0.0
2	CO	+ 18.8	+300.0	−178.6	+ 31.2	0.0	−71.4
	BAL	− 18.8	− 45.5	− 75.9	− 22.3	− 8.9	+71.4
3	CO	− 22.8	− 9.4	− 11.2	− 38.0	+ 35.7	− 4.4
	BAL	+ 22.8	+ 7.7	+ 12.9	+ 1.6	+ 0.7	+ 4.4
4	CO	+ 3.8	+ 11.4	+ 0.8	+ 6.4	+ 2.2	+ 0.4
	BAL	− 3.8	− 4.6	− 7.6	− 6.1	− 2.5	− 0.4
5	CO	− 2.3	− 1.9	− 3.0	− 3.8	− 0.2	− 1.2
	BAL	+ 2.3	+ 1.8	+ 3.1	+ 2.8	+ 1.2	+ 1.2
6	CO	+ 0.9	+ 1.2	+ 1.4	+ 1.6	+ 0.6	+ 0.6
	BAL	− 0.9	− 1.0	− 1.6	− 1.6	− 0.6	− 0.6
7	CO	− 0.5	− 0.5	− 0.8	− 0.8	− 0.3	− 0.3
	BAL	+ 0.5	+ 0.5	+ 0.8	+ 0.8	+ 0.3	+ 0.3
Total balanced M		0.0	−302.8	+302.8	+114.6	−114.6	0.0
Check	Change	+600.0	+297.2	−197.2	−385.4	−114.6	0.0
	½(change)	−148.6	−300.0	+192.7	+ 98.6	0.0	+57.3
	Sum	+451.4	− 2.8	− 4.5	−286.8	−114.6	+57.3
	θ, 10^{-3} rad	+3.762	−0.023	−0.023	−1.434	−1.433	+0.716

SOLUTION (a) *FEM and values of EI/L.* These values are the same as those in the previous example; they are shown again on the present beam in Fig. 8.8.3b and c.

(b) *Moment distribution.* The moment distribution can be performed using either the regular stiffness method or the modified stiffness method, as shown in Table 8.8.2a and b.

8.9 Application to the Analysis of Statically Indeterminate Rigid Frames—Without Unknown Joint Translation

The application of the moment-distribution method to the analysis of statically indeterminate rigid frames wherein no joint can undergo unknown translation is very similar to that for continuous beams as described in Secs. 8.5

THE MOMENT-DISTRIBUTION METHOD

Table 8.8.1b Moment distribution by modified stiffness method, Example 8.8.1†

Joint		A	B		C		D
Member		AB	BA	BC	CB	CD	DC
EI/L, kN·m		40,000	40,000	66,667	66,667	26,667	26,667
$0.75 EI/L$, kN·m			30,000			20,000	
Modified DF		....	0.3103	0.6897	0.7692	0.2308	....
FEM		−600.0	−600.0	+500.0	+500.0	0.0	0.0
BAL		+600.0	0.0			0.0	0.0
CO		0.0	+300.0			0.0	0.0
BAL		0.0	0.0			0.0	0.0
Cycle 1	Mod. FEM	0.0	−300.0	+500.0	+500.0	0.0	0.0
	BAL	↑	− 62.1	−137.9	−384.6	−115.4	↑
2	CO		0.0	−192.3	− 69.0	0.0	
	BAL		+ 59.7	+132.6	+ 53.1	+ 15.9	
3	CO		0.0	+ 26.6	+ 66.3	0.0	
	BAL		− 8.2	− 18.4	− 51.0	− 15.3	
4	CO		0.0	− 25.5	− 9.2	0.0	
	BAL		+ 7.9	+ 17.6	+ 7.1	+ 2.1	
5	CO		0.0	+ 3.6	+ 8.8	0.0	
	BAL		− 1.1	− 2.5	− 6.8	− 2.0	
6	CO		0.0	− 3.4	− 1.2	0.0	
	BAL		+ 1.0	+ 2.4	+ 0.9	+ 0.3	
7	CO		0.0	+ 0.4	+ 1.2	0.0	
	BAL		− 0.1	− 0.3	− 0.9	− 0.3	
8	CO		0.0	− 0.4	− 0.2	0.0	
	BAL	↓	+ 0.1	+ 0.3	+ 0.2	0.0	↓
Total balanced M		0.0	−302.8	+302.8	+114.7	−114.7	0.0

†For check, see Table 8.8.1a.

and 8.8, except that in the case of rigid frames there are frequently more than two members meeting in one joint. In such cases, the unbalance at the beginning of each cycle is the algebraic sum of the FEMs (for the first cycle) or of the carry-over moments clustered around the joint. This unbalance is then distributed to the several member ends according to the distribution factors. Many analysts like to write the column of numbers at each member

Table 8.8.2a Moment distribution by regular stiffness method, Example 8.8.2

Joint		A	B		C		D
Member		AB	BA	BC	CB	CD	DC
EI/L, kN·m		40,000	40,000	66,667	66,667	26,667	26,667
Cycle	DF	....	0.3750	0.6250	0.7143	0.2857	1.0000
1	FEM	−600.0	−600.0	+500.0	+500.0	0.0	0.0
	BAL	0.0	+ 37.5	+ 62.5	−357.2	−142.8	0.0
2	CO	+ 18.8	0.0	−178.6	+ 31.2	0.0	−71.4
	BAL	0.0	+ 67.0	+111.6	− 22.3	− 8.9	+71.4
3	CO	+ 33.5	0.0	− 11.2	+ 55.8	+ 35.7	− 4.4
	BAL	0.0	+ 4.2	+ 7.0	− 65.4	− 26.1	+ 4.4
4	CO	+ 2.1	0.0	− 32.7	+ 3.5	+ 2.2	− 13.0
	BAL	0.0	+ 12.3	+ 20.4	− 4.1	− 1.6	+13.0
5	CO	+ 6.2	0.0	− 2.0	+ 10.2	+ 6.5	− 0.8
	BAL	0.0	+ 0.8	+ 1.2	− 11.9	− 4.8	+ 0.8
6	CO	+ 0.4	0.0	− 6.0	+ 0.6	+ 0.4	− 2.4
	BAL	0.0	+ 2.2	+ 3.8	− 0.7	− 0.3	+ 2.4
7	CO	+ 1.1	0.0	− 0.4	+ 1.9	+ 1.2	− 0.2
	BAL	0.0	+ 0.2	+ 0.2	− 2.2	− 0.9	+ 0.2
8	CO	+ 0.1	0.0	− 1.1	+ 0.1	+ 0.1	− 0.4
	BAL	0.0	+ 0.4	+ 0.7	− 0.1	− 0.1	+ 0.4
9	CO	+ 0.2	0.0	0.0	+ 0.4	+ 0.2	0.0
	BAL	0.0	0.0	0.0	− 0.4	− 0.2	0.0
Total balanced M		−537.6	−475.4	+475.4	+139.4	−139.4	0.0
Check	Change	+ 62.4	+124.6	− 24.6	−360.6	−139.4	0.0
	$-\frac{1}{2}$(change)	− 62.3	− 31.2	+180.3	+ 12.3	0.0	+69.7
	Sum	≈0	+ 93.4	+155.7	−348.3	−139.4	+69.7
	θ, 10^{-3} rad	≈0	+0.778	+0.778	−1.742	−1.742	+0.871

end in the moment-distribution process directly above or below the member in a scale drawing of the rigid frame, thus having a better mental picture of the locking and releasing sequence. The tabular form as suggested and used in this chapter does not have this advantage, although it is more compact and a computer program can be written to generate this kind of tabulation as its output.

Table 8.8.2b Moment distribution by modified stiffness method, Example 8.8.2†

Joint		A	B		C		D
Member		AB	BA	BC	CB	CD	DC
EI/L, kN·m		40,000	40,000	66,667	66,667	26,667	26,667
$0.75 EI/L$, kN·m						20,000	
Modified DF		....	0.3750	0.6250	0.7692	0.2308	....
FEM		−600.0	−600.0	+500.0	+500.0	0.0	0.0
BAL						0.0	0.0
CO						0.0	0.0
BAL						0.0	0.0
Cycle 1	Mod. FEM	−600.0	−600.0	+500.0	+500.0	0.0	0.0
	BAL	0.0	+ 37.5	+ 62.5	−384.6	−115.4	↑
2	CO	+ 18.0	0.0	−192.3	+ 31.2	0.0	
	BAL	0.0	+ 72.1	+120.2	− 24.0	− 7.2	
3	CO	+ 36.0	0.0	− 12.0	+ 60.1	0.0	
	BAL	0.0	+ 4.5	+ 7.5	− 46.2	− 13.9	
4	CO	+ 2.2	0.0	− 23.1	+ 3.8	0.0	
	BAL	0.0	+ 8.7	+ 14.4	− 2.9	− 0.9	
5	CO	+ 4.4	0.0	− 1.4	+ 7.2	0.0	
	BAL	0.0	+ 0.5	+ 0.9	− 5.5	− 1.7	
6	CO	+ 0.2	0.0	− 2.8	+ 0.4	0.0	
	BAL	0.0	+ 1.0	+ 1.8	− 0.3	− 0.1	
7	CO	+ 0.5	0.0	− 0.2	+ 0.9	0.0	
	BAL	0.0	+ 0.1	+ 0.1	− 0.7	− 0.2	
8	CO	0.0	0.0	− 0.4	0.0	0.0	
	BAL	0.0	+ 0.2	+ 0.2	0.0	0.0	
9	CO	+ 0.1	0.0	0.0	+ 0.1	0.0	
	BAL	0.0	0.0	0.0	− 0.1	0.0	↓
Total balanced M		−537.8	−475.4	+475.4	+139.4	−139.4	0.0

†For check, see Table 8.8.2a.

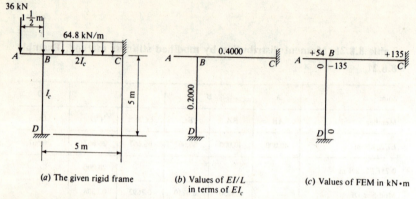

(a) The given rigid frame (b) Values of EI/L in terms of EI_c (c) Values of FEM in kN·m

Figure 8.9.1 Rigid frame of Example 8.9.1.

Example 8.9.1 Analyze the rigid frame of Example 7.6.1, shown again in Fig. 8.9.1, by the moment-distribution method.

SOLUTION The fixed-end moment acting at B on BA is clockwise and therefore it is positive.

$$M_{FBA} = +36(1.5) = +54 \text{ kN·m}$$

$$M_{FBC} = -\frac{64.8(5)^2}{12} = -135 \text{ kN·m} \qquad M_{FCB} = +135 \text{ kN·m}$$

$$M_{FBD} = M_{FDB} = 0$$

$$\frac{EI}{L}(BC) = \frac{E(2I_c)}{5} = 0.4000 EI_c \qquad \frac{EI}{L}(BD) = \frac{E(I_c)}{5} = 0.2000 EI_c$$

The values of EI/L and FEM are shown in Fig. 8.9.1b and c. Note that the FEMs are written below the member at the left end and above the member at the right end; a vertical member is treated as a horizontal member looking from the right.

The moment distribution is shown in Table 8.9.1. The joints are arranged in alphabetical order from left to right. The members radiating from each joint are also arranged in order of the far-end alphabet. Since there is no exterior simple or hinged support in this rigid frame, there is no opportunity to apply the modified stiffness method. Note also that the carry-overs are not necessarily between two adjacent columns. The moment distribution comes to a complete stop at the end of the second cycle; this always happens when there is only one unknown joint rotation for the entire rigid frame, since there will be no need to solve a linear equation of one unknown by successive approximations.

Example 8.9.2 Analyze the rigid frame of Example 7.6.2, shown again in Fig. 8.9.2, by the moment-distribution method.

SOLUTION The fixed-end moments are:

$$M_{FAB} = -\frac{96(10)}{8} = -120 \text{ kN·m} \qquad M_{FBA} = +120 \text{ kN·m}$$

$$M_{FBC} = -\frac{120(4)(6)^2}{10^2} = -172.8 \text{ kN·m} \qquad M_{FCB} = +\frac{120(6)(4)^2}{10^2} = +115.2 \text{ kN·m}$$

$$M_{FCD} = M_{FDC} = 0$$

Table 8.9.1 Moment distribution for rigid frame in Example 8.9.1

Joint			B		C	D
Member		Overhang	BC	BD	CB	DB
EI/L (in terms of EI_c)			0.4000	0.2000	0.4000	0.2000
Cycle	DF		0.6667	0.3333	....	....
1	FEM	+54.00	−135.00	0.00	+135.00	0.00
	BAL		+ 54.00	+27.00	0.00	0.00
2	CO		0.00	0.00	+ 27.00	+13.50
	BAL		0.00	0.00	0.00	0.00
Total balanced M		+54.00	− 81.00	+27.00	+162.00	+13.50
Check	Change		+ 54.00	+27.00	+ 27.00	+13.50
	$-\frac{1}{2}$(change)		− 13.50	− 6.75	− 27.00	−13.50
	Sum		+ 40.50	+20.25	0.00	0.00
	$EI_c\theta$		+ 33.75	+33.75	0.00	0.00

The values of EI/L are:

$$\frac{EI}{L}(AB) = \frac{E(2I_c)}{10} = 0.2000 EI_c \qquad \frac{EI}{L}(BC) = \frac{E(2I_c)}{10} = 0.2000 EI_c$$

$$\frac{EI}{L}(CD) = \frac{E(1.5 I_c)}{6} = 0.2500 EI_c$$

The values of EI/L and FEM are shown in Fig. 8.9.2b and c.

Because joint A is hinged, either the regular stiffness factor or the modified stiffness factor may be used for end B of member BA; both versions of the moment distribution are shown in Table 8.9.2a and b. In this problem, the modified stiffness method has a distinct advantage over the regular stiffness method, as it should in most cases. One may see more readily now that once the moment-distribution table is filled out to where the FEM values are, the analyst does not need to know the shape of the structure anymore. At this time the rest of the tabulation, including the check on moment distribution, can be entrusted to the electronic computer to produce the entire moment-distribution table.

Example 8.9.3 Analyze the rigid frame of Example 7.6.3, shown again in Fig. 8.9.3, by the moment-distribution method.

SOLUTION When this rigid frame was analyzed by the slope-deflection method in Example 7.6.3, it was recognized at the outset that (1) θ_B must be equal to zero and there is no bending of column BE, and (2) θ_A and θ_C must be equal in magnitude but opposite in sign. The solution by moment distribution can be equally simplified by considering only the two-member rigid frame of Fig. 8.9.3b and c, because what happens to AB and AD in Fig. 8.9.3b and c applies as well to BC and CF in Fig. 8.9.3a, and the axial force in BE of the original rigid frame would simply be equal to twice the vertical reaction at B of Fig. 8.9.3b or c. The moment distribution for the rigid frame in Fig. 8.9.3b and c is shown in Table 8.9.3.

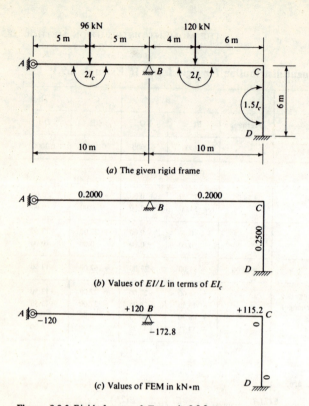

Figure 8.9.2 Rigid frame of Example 8.9.2.

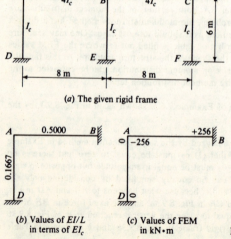

Figure 8.9.3 Rigid frame of Example 8.9.3.

Table 8.9.2a Moment distribution by regular stiffness method, Example 8.9.2

Joint		A	B		C		D
Member		AB	BA	BC	CB	CD	DC
EI/L (in terms of EI_c)		0.2000	0.2000	0.2000	0.2000	0.2500	0.2500
Cycle	DF	1.0000	0.5000	0.5000	0.4444	0.5556	....
1	FEM	−120.00	+120.00	−172.80	+115.20	0.00	0.00
	BAL	+120.00	+ 26.40	+ 26.40	− 51.20	−64.00	0.00
2	CO	+ 13.20	+ 60.00	− 25.60	+ 13.20	0.00	−32.00
	BAL	− 13.20	− 17.20	− 17.20	− 5.87	− 7.33	0.00
3	CO	− 8.60	− 6.60	− 2.94	− 8.60	0.00	− 3.66
	BAL	+ 8.60	+ 4.77	+ 4.77	+ 3.82	+ 4.78	0.00
4	CO	+ 2.38	+ 4.30	+ 1.91	+ 2.38	0.00	+ 2.39
	BAL	− 2.38	− 3.10	− 3.11	− 1.06	− 1.32	0.00
5	CO	− 1.55	− 1.19	− 0.53	− 1.56	0.00	− 0.66
	BAL	+ 1.55	+ 0.86	+ 0.86	+ 0.69	+ 0.87	0.00
6	CO	+ 0.43	+ 0.78	+ 0.34	+ 0.43	0.00	+ 0.44
	BAL	− 0.43	− 0.56	− 0.56	− 0.19	− 0.24	0.00
7	CO	− 0.28	− 0.22	− 0.10	− 0.28	0.00	− 0.12
	BAL	+ 0.28	+ 0.16	+ 0.16	+ 0.12	+ 0.16	0.00
8	CO	+ 0.08	+ 0.14	+ 0.06	+ 0.08	0.00	+ 0.08
	BAL	− 0.08	− 0.10	− 0.10	− 0.04	− 0.04	0.00
9	CO	− 0.05	− 0.04	− 0.02	− 0.05	0.00	− 0.02
	BAL	+ 0.05	+ 0.03	+ 0.03	+ 0.02	+ 0.03	0.00
10	CO	+ 0.02	+ 0.02	+ 0.01	+ 0.02	0.00	+ 0.02
	BAL	− 0.02	− 0.02	− 0.01	− 0.01	− 0.01	0.00
Total balanced M		0.00	+188.43	−188.43	+ 67.10	− 67.10	−33.53
Check	Change	+120.00	+ 68.43	− 15.63	− 48.10	− 67.10	−33.53
	−½(change)	− 34.22	− 60.00	+ 24.05	+ 7.82	+16.76	+33.55
	Sum	+ 85.78	+ 8.43	+ 8.42	− 40.28	−50.34	≈ 0
	$EI_c\theta$	+142.97	+ 14.05	+ 14.03	− 67.13	−67.12	≈ 0

Table 8.9.2b Moment distribution by modified stiffness method, Example 8.9.2†

Joint		A	B		C		D
Member		AB	BA	BC	CB	CD	DC
EI/L (in terms of EI_c)		0.2000	0.2000	0.2000	0.2000	0.2500	0.2500
0.75EI/L			0.1500				
Modified DF		....	0.4286	0.5714	0.4444	0.5556	....
FEM BAL		−120.00 +120.00	+120.00 0.00	−172.80	+115.20	0.00	0.00
CO BAL		0.00 0.00	+60.00 0.00				
Cycle 1	Mod. FEM BAL	0.00	+180.00 − 3.08	−172.80 − 4.12	+115.20 − 51.20	0.00 −64.00	0.00 0.00
2	CO BAL		0.00 + 10.97	− 25.60 + 14.63	− 2.06 + 0.92	0.00 + 1.14	−32.00 0.00
3	CO BAL		0.00 − 0.20	+ 0.46 − 0.26	+ 7.32 − 3.25	0.00 − 4.07	+ 0.57 0.00
4	CO BAL		0.00 + 0.69	− 1.62 + 0.93	− 0.13 + 0.06	0.00 + 0.07	− 2.04 0.00
5	CO BAL		0.00 − 0.01	+ 0.03 − 0.02	+ 0.46 − 0.20	0.00 − 0.26	+ 0.04 0.00
6	CO BAL		0.00 + 0.04	− 0.10 + 0.06	− 0.01 0.00	0.00 + 0.01	− 0.13 0.00
7	CO BAL		0.00 0.00	0.00 0.00	+ 0.03 − 0.01	0.00 − 0.02	0.00 0.00
8	CO BAL		0.00 0.00	0.00 0.00	0.00 0.00	0.00 0.00	+ 0.01 0.00
Total balanced M		0.00	+188.41	−188.41	+ 67.13	−67.13	−33.55

†For check, see Table 8.9.2a.

Table 8.9.3 Moment distribution for rigid frame in Example 8.9.3

Joint		A		B	D
Member		AB	AD	BA	DA
EI/L (in terms of EI_c)		0.5000	0.1667	0.5000	0.1667
Cycle	DF	0.7500	0.2500	....	....
1	FEM	−256	0	+256	0
	BAL	+192	+64	0	0
2	CO	0	0	+ 96	+32
	BAL	0	0	0	0
Total balanced M		− 64	+64	+352	+32
Check	Change	+192	+64	+ 96	+32
	$-\frac{1}{2}$(change)	− 48	−16	− 96	−32
	Sum	+144	+48	0	0
	$EI_c\theta$	+ 96	+96	0	0

8.10 Application to the Analysis of Statically Indeterminate Rigid Frames—With Unknown Joint Translation

The application of the moment-distribution method to the analysis of statically indeterminate rigid frames wherein some joints may undergo unknown translation, or of rectangular rigid frames with unknown sideways, will be described first by means of a simple case for which the degree of freedom in sidesway is equal to 1. The three major steps involved in the analysis of the rigid frame shown in Fig. 8.10.1a are as follows:

1. Prevent the sidesway of member BC by adding a support at C, as shown in Fig. 8.10.1b. This rigid frame without unknown joint translation can be analyzed by moment distribution. The sum of H_{A1} and H_{D1} in Fig. 8.10.1b should add up to less than W_1 because there is an additional horizontal reaction at C for the horizontal equilibrium of the entire rigid frame.
2. Lock the joints B and C against rotation but allow them to displace an arbitrary amount Δ' to the right, thus inducing a set of fixed-end moments on columns AB and CD, as shown in Fig. 8.10.1c. This is in fact the way by which the rigid frame with the additional support at C is to be solved by moment distribution if the support at C yields an amount Δ' to the right. After the fixed-end moments due to the arbitrary sidesway Δ' are distributed, the induced reactions H_{A2} and H_{D2} in Fig. 8.10.1c can be determined from the set of balanced moments at the column ends.

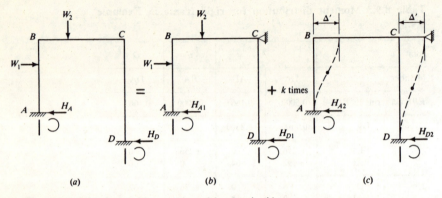

Figure 8.10.1 Rigid frame with one degree of freedom in sidesway.

3. The true sidesway Δ for the rigid frame in Fig. 8.10.1a is equal to $k\Delta'$, wherein the value of k should be such as to satisfy the shear condition

$$(H_{A1} + H_{D1}) + k(H_{A2} + H_{D2}) = W_1 \qquad (8.10.1)$$

In arranging the required computations, there should be two moment-distribution tables, one each for the first two steps described above. After the value of k has been determined from the shear condition, such as Eq. (8.10.1), a third table may be made up compiling the fixed-end moments and the balanced moments in the first table with k times the fixed-end moments and the balanced moments in the second table. The slopes at the member ends can still be computed by using Eq. (8.6.2), provided the change in moment is equal to the combined balanced moment minus the combined fixed-end moment in the third table. Equation (8.6.2) has been derived on the basis that there is no member-axis rotation. However, this equation is still valid if the member axis does rotate. This fact can be proved by solving the general slope-deflection equations

$$M_A = M_{0A} + \frac{2EI}{L}(2\theta_A + \theta_B - 3R)$$

$$M_B = M_{0B} + \frac{2EI}{L}(2\theta_B + \theta_A - 3R)$$

for θ_A and θ_B. The expressions for θ_A and θ_B are

$$\theta_A = \frac{[M_A - (M_{0A} - 6EIR/L)] + (-\tfrac{1}{2})[M_B - (M_{0B} - 6EIR/L)]}{3EI/L} \qquad (8.10.2a)$$

$$\theta_B = \frac{[M_B - (M_{0B} - 6EIR/L)] + (-\tfrac{1}{2})[M_A - (M_{0A} - 6EIR/L)]}{3EI/L} \qquad (8.10.2b)$$

Equations (8.10.2a and b) can be written in a single equation, in the same form as Eq. (8.6.2), as

$$\theta_{\text{near end}} = \frac{(\text{Change}_{\text{near end}}) + (-\frac{1}{2})(\text{Change}_{\text{far end}})}{3EI/L}$$

except that this time the change is equal to the final balanced moment minus the combined fixed-end moment due to the applied loads on the member and due to member-axis rotation.

If the degree of freedom in sideway is more than 1, then there should be one moment distribution for each arbitrary amount of sideway, such as $\Delta'_1, \Delta'_2, \Delta'_3$, etc. The correct sideways are $k_1\Delta'_1, k_2\Delta'_2, k_3\Delta'_3$, etc., and the values of k_1, k_2, k_3, etc. will be determined by solving the shear conditions as simultaneous equations. Thus there will still be simultaneous equations to be solved, but the number of simultaneous equations is equal to the degree of freedom in sideway.

Morris has proposed a method of alternate moment and shear distribution in which the locking and releasing of the joints alternate with the prevention and permission of sideways.† So in one operation the unbalanced moments are distributed according to the moment-distribution factors, and in the next operation the shear conditions are satisfied by introducing a set of fixed-end moments generated by sideways. Gradually the structure will arrive at its final compatible state through alternate amounts of joint rotations and joint translations. One application of this method will be described in Sec. 14.7.

Example 8.10.1 Analyze the rigid frame of Example 7.7.1, shown again in Fig. 8.10.2, by the moment-distribution method.

SOLUTION (a) *Values of EI/L.* The values of EI/L (Fig. 8.10.2b) for the three members are:

$$\frac{EI}{L}(AB) = \frac{E(I_c)}{7.5} = 0.1333EI_c \qquad \frac{EI}{L}(BC) = \frac{E(2I_c)}{6} = 0.3333EI_c$$

$$\frac{EI}{L}(CD) = \frac{E(I_c)}{5} = 0.2000EI_c$$

(b) *Moment distribution for the applied loads with sideway prevented.* For this rectangular rigid frame, the only unknown joint translation is the sideway of member BC, which is designated as Δ to the right. When this sideway is prevented, the fixed-end moments due to the applied loads are (Fig. 8.10.2c):

$$M_{FAB} = -\frac{48(4.5)(3)^2}{7.5^2} = -34.56 \text{ kN·m} \qquad M_{FBA} = +\frac{48(3)(4.5)^2}{7.5^2} = +51.84 \text{ kN·m}$$

$$M_{FBC} = -\frac{96(6)}{8} = -72 \text{ kN·m} \qquad M_{FCB} = +72 \text{ kN·m}$$

$$M_{FCD} = M_{FDC} = 0$$

The moment distribution for the above fixed-end moments by the modified stiffness method is shown in Table 8.10.1a. The balanced moments acting on columns AB and CD are shown in Fig. 8.10.3a, and the values of H_{A1} and H_{D1} are computed by considering the moment

---

†H. Sutherland and H. L. Bowman, *Structural Theory*, John Wiley & Sons, Inc., New York, 1950, p. 261; also *Trans. ASCE*, vol. 96, 1932, p. 66.

292 INTERMEDIATE STRUCTURAL ANALYSIS

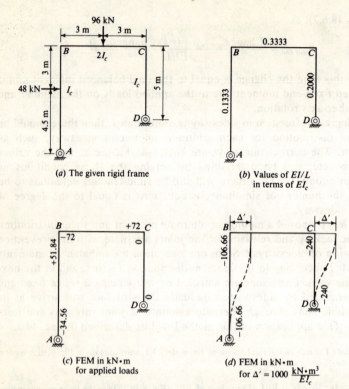

Figure 8.10.2 Rigid frame of Example 8.10.1.

equilibrium of the columns; thus

$$H_{A1} = \frac{48(3)}{7.5} - \frac{76.48}{7.5} = 19.200 - 10.177 = 9.003 \text{ kN}$$

$$H_{D1} = \frac{26.14}{5} = 5.228 \text{ kN}$$

(c) *Moment distribution for arbitrary* $\Delta' = 1000 \text{ kN·m}^3/EI_c$. The fixed-end moments due to Δ' can be computed by using Eq. (8.8.1); the values for $\Delta' = 1000 \text{ kN·m}^3/EI_c$ are shown in Fig. 8.10.2d.

$$M_{FAB} = M_{FBA} = -\frac{6E(I_c)\Delta'}{7.5^2} = -\frac{6(1000)}{7.5^2} = -106.66 \text{ kN·m}$$

$$M_{FCD} = M_{FDC} = -\frac{6E(I_c)\Delta'}{5^2} = -\frac{6(1000)}{5^2} = -240 \text{ kN·m}$$

The moment distribution for the above fixed-end moments by the modified stiffness method is shown in Table 8.10.1b. Note that the number of significant figures used in the FEM values of Table 8.10.1a and b are comparable. The balanced moments acting on columns AB and CD are shown in Fig. 8.10.3b, and the values of H_{A2} and H_{D2} are computed from the free-body diagrams of the columns in Fig. 8.10.3b; thus

$$H_{A2} = \frac{50.15}{7.5} = 6.687 \text{ kN} \qquad H_{D2} = \frac{84.39}{5} = 16.878 \text{ kN}$$

Table 8.10.1a Moment distribution for applied loads on rigid frame of Example 8.10.1

Joint		A	B		C		D
Member		AB	BA	BC	CB	CD	DC
EI/L (in terms of EI_c)		0.1333	0.1333	0.3333	0.3333	0.2000	0.2000
$0.75 EI/L$			0.1000			0.1500	
Modified DF		....	0.2308	0.7692	0.6897	0.3103	....
FEM		−34.56	+51.84	−72.00	+72.00	0.00	0.00
BAL		+34.56	0.00			0.00	0.00
CO		0.00	+17.28			0.00	0.00
BAL		0.00	0.00			0.00	0.00
Cycle 1	Mod. FEM	0.00	+69.12	−72.00	+72.00	0.00	0.00
	BAL		+ 0.66	+ 2.22	−49.66	−22.34	
2	CO		0.00	−24.83	+ 1.11	0.00	
	BAL		+ 5.73	+19.10	− 0.77	− 0.34	
3	CO		0.00	− 0.38	+ 9.55	0.00	
	BAL		+ 0.09	+ 0.29	− 6.59	− 2.96	
4	CO		0.00	− 3.30	+ 0.14	0.00	
	BAL		+ 0.76	+ 2.54	− 0.10	− 0.04	
5	CO		0.00	− 0.05	+ 1.27	0.00	
	BAL		+ 0.01	+ 0.04	− 0.88	− 0.39	
6	CO		0.00	− 0.44	+ 0.02	0.00	
	BAL		+ 0.10	+ 0.34	− 0.01	− 0.01	
7	CO		0.00	0.00	+ 0.17	0.00	
	BAL		0.00	0.00	− 0.12	− 0.05	
8	CO		0.00	− 0.06	0.00	0.00	
	BAL		+ 0.01	+ 0.05	0.00	0.00	
9	CO		0.00	0.00	+ 0.02	0.00	
	BAL		0.00	0.00	− 0.01	− 0.01	
Total balanced M		0.00	+76.48	−76.48	+26.14	−26.14	0.00
Check	Change	+34.56	+24.64	− 4.48	−45.86	−26.14	0.00
	$-\frac{1}{2}$(change)	−12.32	−17.28	+22.93	+ 2.24	0.00	+13.07
	Sum	+22.24	+ 7.36	+18.45	−43.62	−26.14	+13.07
	$EI_c\theta$	+55.60	+18.40	+18.45	−43.62	−43.57	+21.78

Table 8.10.1b Moment distribution for sidesway of 1000 kN·m^3/EI_c in rigid frame of Example 8.10.1

Joint		A	B		C		D
Member		AB	BA	BC	CB	CD	DC
EI/L (in terms of EI_c)		0.1333	0.1333	0.3333	0.3333	0.2000	0.2000
$0.75 EI/L$			0.1000			0.1500	
Modified DF		....	0.2308	0.7692	0.6897	0.3103	....
FEM		−106.66	−106.66	0.00	0.00	−240.00	−240.00
BAL		+106.66	0.00			0.00	+240.00
CO		0.00	+53.33			+120.00	0.00
BAL		0.00	0.00			0.00	0.00
Cycle 1	Mod. FEM	0.00	−53.33	0.00	0.00	−120.00	0.00
	BAL	↑	+12.31	+41.02	+82.76	+37.24	↑
2	CO		0.00	+41.38	+20.51	0.00	
	BAL		−9.55	−31.83	−14.15	−6.36	
3	CO		0.00	−7.08	−15.92	0.00	
	BAL		+1.63	+5.45	+10.98	+4.94	
4	CO		0.00	+5.49	+2.92	0.00	
	BAL		−1.27	−4.22	−1.88	−0.84	
5	CO		0.00	−0.94	−2.11	0.00	
	BAL		+0.22	+0.72	+1.46	+0.65	
6	CO		0.00	+0.73	+0.36	0.00	
	BAL		−0.17	−0.56	−0.25	−0.11	
7	CO		0.00	−0.12	−0.28	0.00	
	BAL		+0.03	+0.09	+0.19	+0.09	
8	CO		0.00	+0.10	+0.04	0.00	
	BAL		−0.02	−0.08	−0.03	+0.01	
9	CO		0.00	−0.02	−0.04	0.00	
	BAL	↓	0.00	+0.02	+0.03	0.01	↓
Total balanced M		0.00	−50.15	+50.15	+84.39	−84.39	0.00
Check	Change	+106.66	+56.51	+50.15	+84.39	+155.61	+240.00
	−½(change)	−28.26	−53.33	−42.20	−25.08	−120.00	−77.80
	Sum	+78.40	+3.18	+7.95	+59.31	+35.61	+162.20
	$EI_c\theta$	+196.00	+7.95	+7.95	+59.31	+59.35	+270.33

THE MOMENT-DISTRIBUTION METHOD 295

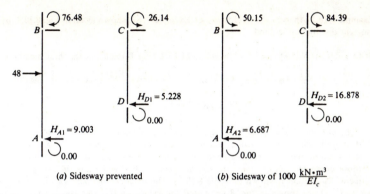

(a) Sidesway prevented (b) Sidesway of 1000 $\frac{kN \cdot m^3}{EI_c}$

Figure 8.10.3 Shear condition for rigid frame in Example 8.10.1.

(d) *Correct amount of sidesway* Δ. The shear condition from which the correct amount of sidesway $\Delta = k\Delta'$ can be determined is, from observing Fig. 8.10.3,

$$(H_{A1} + H_{D1}) + k(H_{A2} + H_{D2}) = 48$$
$$(9.003 + 5.228) + k(6.687 + 16.878) = 48$$
$$k = 1.433$$

$$\Delta = k\Delta' = 1.433(1000) = 1433 \frac{kN \cdot m^3}{EI_c}$$

Table 8.10.1c shows the combination of the fixed-end moments and balanced moments in Table 8.10.1a with 1.433 times the corresponding values in Table 8.10.1b. The correct member-end slopes are computed by using Eq. (8.6.2). The end slopes are $\theta_A = +336.5/EI_c$, $\theta_B = +29.8/EI_c$, $\theta_C = (+41.5$ or $+41.4)/EI_c$, and $\theta_D = +409.2/EI_c$.

Table 8.10.1c Combination of moments due to applied loads and sidesway for rigid frame in Example 8.10.1

Joint	A	B		C		D
Member	AB	BA	BC	CB	CD	DC
FEM (1)	− 34.56	+ 51.84	−72.00	+ 72.00	0.00	0.00
FEM (2)	−152.84	−152.84	0.00	0.00	−343.92	−343.92
Total FEM	−187.40	−101.00	−72.00	+ 72.00	−343.92	−343.92
BAL M (1)	0.00	+ 76.48	−76.48	+ 26.14	− 26.14	0.00
BAL M (2)	0.00	− 71.86	+71.86	+120.93	−120.93	0.00
Total BAL M	0.00	+ 4.62	− 4.62	+147.07	−147.07	0.00
Change	+187.40	+105.62	+67.38	+ 75.07	+196.85	+343.92
$-\frac{1}{2}$(change)	− 52.81	− 93.70	−37.54	− 33.69	−171.96	− 98.42
Sum	+134.59	+ 11.92	+29.84	+ 41.68	+ 24.89	+245.50
$EI_c\theta$	+336.5	+ 29.8	+29.8	+ 41.4	+ 41.5	+409.2

Example 8.10.2 Analyze the rigid frame of Example 7.7.2, shown again in Fig. 8.10.4, by the moment-distribution method.

SOLUTION (a) *Values of EI/L*. The values of EI/L (Fig. 8.10.4b) for the three members are:

$$\frac{EI}{L}(AB) = \frac{E(I_c)}{7.5} = 0.1333 EI_c \qquad \frac{EI}{L}(BC) = \frac{E(2I_c)}{6} = 0.3333 EI_c$$

$$\frac{EI}{L}(CD) = \frac{E(I_c)}{5} = 0.2000 EI_c$$

(b) *Moment distribution for the applied loads with sidesway prevented*. This rigid frame is identical to that in the previous example except that the supports at A and D are now fixed instead of being hinged. The fixed-end moments due to the applied loads are the same as those computed before; they are shown again on this rigid frame in Fig. 8.10.4c. The moment distribution is shown in Table 8.10.2a. The balanced moments acting on columns AB and CD are shown in Fig. 8.10.5a, and the values of H_{A1} and H_{D1} are computed by considering the moment equilibrium of the columns; thus

$$H_{A1} = \frac{48(3)}{7.5} - \frac{65.56 - 27.70}{7.5} = 19.200 - 5.048 = 14.152 \text{ kN}$$

$$H_{D1} = \frac{33.43 + 16.72}{5} = 10.030 \text{ kN}$$

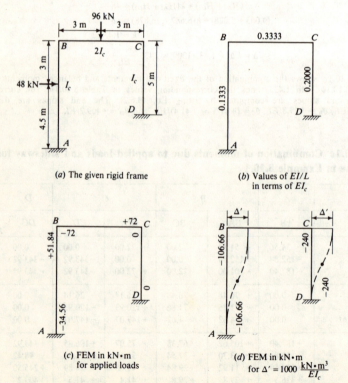

(a) The given rigid frame

(b) Values of EI/L in terms of EI_c

(c) FEM in kN·m for applied loads

(d) FEM in kN·m for $\Delta' = 1000 \dfrac{\text{kN·m}^3}{EI_c}$

Figure 8.10.4 Rigid frame of Example 8.10.2.

THE MOMENT-DISTRIBUTION METHOD 297

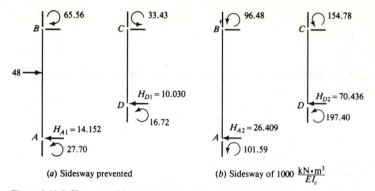

(a) Sidesway prevented (b) Sidesway of 1000 $\frac{kN \cdot m^3}{EI_c}$

Figure 8.10.5 Shear condition for rigid frame in Example 8.10.2.

(c) *Moment distribution for arbitrary* $\Delta' = 1000 \, kN \cdot m^3/EI_c$. The fixed-end moments due to $\Delta' = 1000 \, kN \cdot m^3/EI_c$ are $-106.66 \, kN \cdot m$ on both ends of column AB, and $-240 \, kN \cdot m$ on both ends of column CD. These values are quoted from the previous example and are shown again in Fig. 8.10.4d. From the results of the moment distribution shown in Fig. 8.10.2b, the balanced moments acting on columns AB and CD are shown in Fig. 8.10.5b. The values of H_{A2} and H_{D2} are computed from the free-body diagrams of the columns in Fig. 8.10.5b; thus

$$H_{A2} = \frac{96.48 + 101.59}{7.5} = 26.409 \, kN \qquad H_{D2} = \frac{154.78 + 197.40}{5} = 70.436 \, kN$$

(d) *Correct amount of sidesway* Δ. The shear condition from which the correct amount of sidesway $\Delta = k\Delta'$ can be determined is, from observing Fig. 8.10.5,

$$(H_{A1} + H_{D1}) + k(H_{A2} + H_{D2}) = 48$$
$$(14.152 + 10.030) + k(26.409 + 70.436) = 48$$
$$k = 0.246$$
$$\Delta = k\Delta' = 0.246(1000) = 246 \, \frac{kN \cdot m^3}{EI_c}$$

Table 8.10.2c shows the combination of the fixed-end moments and balanced moments in Table 8.10.2a with 0.246 times the corresponding values in Table 8.10.2b. The correct member-end slopes are computed by using Eq. (8.6.2). The end slopes are $\theta_A = 0$, $\theta_B = (+30.40$ or $+30.43)/EI_c$, $\theta_C = (-15.59$ or $-15.58)/EI_c$, and $\theta_D = 0$.

Example 8.10.3 Analyze the rigid frame of Example 7.7.3, shown again in Fig. 8.10.6, by the moment-distribution method.

SOLUTION (a) *Values of EI/L*. The values of EI/L for the five members are (Fig. 8.10.6b):

$$\frac{EI}{L}(AB) = \frac{EI}{L}(BC) = \frac{E(3I_c)}{8} = 0.3750 EI_c \qquad \frac{EI}{L}(AD) = \frac{E(2I_c)}{10} = 0.2000 EI_c$$

$$\frac{EI}{L}(BE) = \frac{E(2I_c)}{8} = 0.2500 EI_c \qquad \frac{EI}{L}(CF) = \frac{E(2I_c)}{4} = 0.5000 EI_c$$

(b) *Moment distribution for the applied load with sidesway prevented.* For the single concentrated load, there are fixed-end moments only on span AB (Fig. 8.9.6c):

$$M_{FAB} = -\frac{96(8)}{8} = -96 \, kN \cdot m \qquad M_{FBA} = +96 \, kN \cdot m$$

Table 8.10.2a Moment distribution for applied loads on rigid frame of Example 8.10.2

Joint		A	B		C		D
Member		AB	BA	BC	CB	CD	DC
EI/L (in terms of EI_c)		0.1333	0.1333	0.3333	0.3333	0.2000	0.2000
Cycle	DF	....	0.2857	0.7143	0.6250	0.3750	....
1	FEM	−34.56	+51.84	−72.00	+72.00	0.00	0.00
	BAL	0.00	+ 5.76	+14.40	−45.00	−27.00	0.00
2	CO	+ 2.88	0.00	−22.50	+ 7.20	0.00	−13.50
	BAL	0.00	+ 6.43	+16.07	− 4.50	− 2.70	0.00
3	CO	+ 3.22	0.00	− 2.25	+ 8.04	0.00	− 1.35
	BAL	0.00	+ 0.64	+ 1.61	− 5.02	− 3.02	0.00
4	CO	+ 0.32	0.00	− 2.51	+ 0.80	0.00	− 1.51
	BAL	0.00	+ 0.72	+ 1.79	− 0.50	− 0.30	0.00
5	CO	+ 0.36	0.00	− 0.25	+ 0.90	0.00	− 0.15
	BAL	0.00	+ 0.07	+ 0.18	− 0.56	− 0.34	0.00
6	CO	+ 0.04	0.00	− 0.28	+ 0.09	0.00	− 0.17
	BAL	0.00	+ 0.08	+ 0.20	− 0.06	− 0.03	0.00
7	CO	+ 0.04	0.00	− 0.03	+ 0.10	0.00	− 0.02
	BAL	0.00	+ 0.01	+ 0.02	− 0.06	− 0.04	0.00
8	CO	0.00	0.00	− 0.03	+ 0.01	0.00	− 0.02
	BAL	0.00	+ 0.01	+ 0.02	− 0.01	0.00	0.00
Total balanced M		−27.70	+65.56	−65.56	+33.43	−33.43	−16.72
Check	Change	+ 6.86	+13.72	+ 6.44	−38.57	−33.43	−16.72
	$-\frac{1}{2}$(change)	− 6.86	− 3.43	+19.28	− 3.22	+ 8.36	+16.72
	Sum	0.00	+10.29	+25.72	−41.79	−25.07	0.00
	$EI_c\theta$	0.00	+25.72	+25.72	−41.79	−41.78	0.00

Table 8.10.2b Moment distribution for sideway of $1000 \text{ kN} \cdot \text{m}^3/EI_c$ in rigid frame of Example 8.10.2

Joint		A	B		C		D
Member		AB	BA	BC	CB	CD	DC
EI/L (in terms of EI_c)		0.1333	0.1333	0.3333	0.3333	0.2000	0.2000
Cycle	DF	....	0.2857	0.7143	0.6250	0.3750	....
1	FEM	−106.66	−106.66	0.00	0.00	−240.00	−240.00
	BAL	0.00	+30.47	+76.19	+150.00	+90.00	0.00
2	CO	+15.24	0.00	+75.00	+38.10	0.00	+45.00
	BAL	0.00	−21.43	−53.47	−23.81	−14.29	0.00
3	CO	−10.72	0.00	−11.90	−26.78	0.00	−7.14
	BAL	0.00	+3.40	+8.50	+16.74	+10.04	0.00
4	CO	+1.70	0.00	+8.37	+4.25	0.00	+5.02
	BAL	0.00	−2.39	−5.98	−2.66	−1.59	0.00
5	CO	−1.20	0.00	−1.33	−2.99	0.00	−0.80
	BAL	0.00	+0.38	+0.95	+1.87	+1.12	0.00
6	CO	+0.19	0.00	+0.94	+0.48	0.00	+0.56
	BAL	0.00	−0.27	−0.67	−0.30	−0.18	0.00
7	CO	−0.14	0.00	−0.15	−0.34	0.00	−0.09
	BAL	0.00	+0.04	+0.11	+0.21	+0.13	0.00
8	CO	+0.02	0.00	+0.10	+0.06	0.00	+0.06
	BAL	0.00	−0.03	−0.07	−0.04	−0.02	0.00
9	CO	−0.02	0.00	−0.02	−0.04	0.00	−0.01
	BAL	0.00	+0.01	+0.01	+0.03	+0.01	0.00
Total balanced M		−101.59	−96.48	+96.48	+154.78	−154.78	−197.40
Check	Change	+5.07	+10.18	+96.48	+154.78	+85.22	+42.60
	$-\frac{1}{2}$(change)	−5.09	−2.54	−77.39	−48.24	−21.30	−42.61
	Sum	≈0	+7.64	+19.09	+106.54	+63.92	≈0
	$EI_c\theta$	≈0	+19.10	+19.09	+106.54	+106.53	≈0

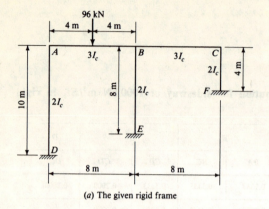
(a) The given rigid frame

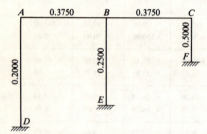
(b) Values of EI/L in terms of EI_c

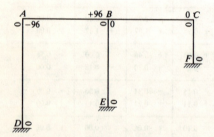
(c) FEM in kN·m for applied loads

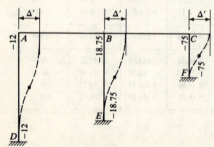
(d) FEM in kN·m for $\Delta' = 100\ \dfrac{\text{kN}\cdot\text{m}^3}{EI_c}$

Figure 8.10.6 Rigid frame of Example 8.10.3.

Table 8.10.2c Combination of moments due to applied loads and sidesway for rigid frame in Example 8.10.2

Joint	A	B		C		D
Member	AB	BA	BC	CB	CD	DC
FEM (1)	−34.56	+51.84	−72.00	+72.00	0.00	0.00
FEM (2)	−26.23	−26.23	0.00	0.00	−59.02	−59.02
Total FEM	−60.79	+25.61	−72.00	+72.00	−59.02	−59.02
BAL M (1)	−27.70	+65.56	−65.56	+33.43	−33.43	−16.72
BAL M (2)	−24.98	−23.73	+23.73	+38.06	−38.06	−48.55
Total BAL M	−52.68	+41.83	−41.83	+71.49	−71.49	−65.27
Change	+ 8.11	+16.22	+30.17	0.51	−12.47	− 6.25
−½(change)	− 8.11	− 4.06	+ 0.26	−15.08	+ 3.12	+ 6.24
Sum	0.00	+12.16	+30.43	−15.59	− 9.35	≈ 0
$EI_c\theta$	0.00	+30.40	+30.43	−15.59	−15.58	≈ 0

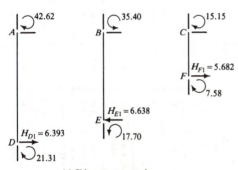

(a) Sidesway prevented

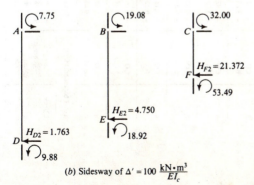

(b) Sidesway of $\Delta' = 100 \ \dfrac{\text{kN} \cdot \text{m}^3}{EI_c}$

Figure 8.10.7 Shear condition for rigid frame in Example 8.10.3.

Table 8.10.3a Moment distribution for the applied load on rigid frame of Example 8.10.3

Joint		A		B			C		D	E	F
Member		AB	AD	BA	BC	BE	CB	CF	DA	EB	FC
EI/L (in terms of EI_c)		0.3750	0.2000	0.3750	0.3750	0.2500	0.3750	0.5000	0.2000	0.2500	0.5000
Cycle	DF	0.6522	0.3478	0.3750	0.3750	0.2500	0.4286	0.5714			
1	FEM	−96.00	0.00	+96.00	0.00	0.00	0.00	0.00	0.00	0.00	0.00
	BAL	+62.61	+33.39	−36.00	−36.00	−24.00	0.00	0.00	0.00	0.00	0.00
2	CO	−18.00	0.00	+31.30	0.00	0.00	−18.00	0.00	+16.70	−12.00	0.00
	BAL	+11.74	+6.26	−11.74	−11.74	−7.82	+7.72	+10.28	0.00	0.00	0.00
3	CO	−5.87	0.00	+5.87	+3.86	0.00	−5.87	0.00	+3.13	−3.91	+5.14
	BAL	+3.83	+2.04	−3.65	−3.65	−2.43	+2.52	+3.35	0.00	0.00	0.00
4	CO	−1.82	0.00	+1.92	+1.26	0.00	−1.82	0.00	+1.02	−1.22	+1.68
	BAL	+1.19	+0.63	−1.19	−1.19	−0.80	+0.78	+1.04	0.00	0.00	0.00
5	CO	−0.60	0.00	+0.60	+0.39	0.00	−0.60	0.00	+0.32	−0.40	+0.52
	BAL	+0.39	+0.21	−0.37	−0.37	−0.25	+0.26	+0.34	0.00	0.00	0.00
6	CO	−0.18	0.00	+0.20	+0.13	0.00	−0.18	0.00	+0.10	−0.12	+0.17
	BAL	+0.12	+0.06	−0.13	−0.12	−0.08	+0.08	+0.10	0.00	0.00	0.00
7	CO	−0.06	0.00	+0.06	+0.04	0.00	−0.06	0.00	+0.03	−0.04	+0.05
	BAL	+0.04	+0.02	−0.04	−0.04	−0.02	+0.03	+0.03	0.00	0.00	0.00
8	CO	−0.02	0.00	+0.02	+0.02	0.00	−0.02	0.00	+0.01	−0.01	+0.02
	BAL	+0.01	+0.01	−0.02	−0.02	0.00	+0.01	+0.01	0.00	0.00	0.00
Total balanced M		−42.62	+42.62	+82.83	−47.43	−35.40	−15.15	+15.15	+21.31	−17.70	+7.58
Check	Change	+53.38	+42.62	−13.17	−47.43	−35.40	−15.15	+15.15	+21.31	−17.70	+7.58
	−½(change)	+6.58	−10.66	−26.69	+7.58	+8.85	+23.72	−3.79	−21.31	+17.70	−7.58
	Sum	+59.96	+31.96	−39.86	−39.85	−26.55	+8.57	+11.36	0.00	0.00	0.00
	$EI_c\theta$	+53.30	+53.27	−35.43	−35.42	−35.40	+7.56	+7.57	0.00	0.00	0.00

Table 8.10.3b Moment distribution for sidesway of $100 \text{ kN} \cdot \text{m}^3/EI_c$ in rigid frame of Example 8.10.3

Joint	A		B			C		D	E	F
Member	AB	AD	BA	BC	BE	CB	CF	DA	EB	FC
EI/L (in terms of EI_c)	0.3750	0.2000	0.3750	0.3750	0.2500	0.3750	0.5000	0.2000	0.2500	0.5000
Cycle DF	0.6522	0.3478	0.3750	0.3750	0.2500	0.4286	0.5714	...	...	...
1 FEM	0.00	−12.00	0.00	0.00	−18.75	0.00	−75.00	−12.00	−18.75	−75.00
BAL	+7.83	+4.17	+7.03	+7.03	+4.69	+22.14	+42.86	0.00	0.00	0.00
2 CO	+3.52	0.00	+3.92	+16.07	0.00	+3.52	0.00	+2.08	+2.34	+21.43
BAL	−2.30	−1.22	−7.50	−7.49	−5.00	−1.51	−2.01	0.00	0.00	0.00
3 CO	−3.75	0.00	−1.15	−0.76	0.00	−3.74	0.00	−0.61	−2.50	−1.00
BAL	+2.45	+1.30	+0.72	+0.71	+0.48	+1.60	+2.14	0.00	0.00	0.00
4 CO	+0.36	0.00	+1.22	+0.80	0.00	+0.36	0.00	+0.65	+0.24	+1.07
BAL	−0.23	−0.13	−0.76	−0.76	−0.50	−0.15	−0.21	0.00	0.00	0.00
5 CO	−0.38	0.00	−0.12	−0.08	0.00	−0.38	0.00	−0.06	−0.25	−0.10
BAL	+0.25	+0.13	+0.08	+0.07	+0.05	+0.16	+0.22	0.00	0.00	0.00
6 CO	+0.04	0.00	+0.12	+0.08	0.00	+0.04	0.00	+0.06	+0.02	+0.11
BAL	−0.03	−0.01	−0.08	−0.07	−0.05	−0.02	−0.02	0.00	0.00	0.00
7 CO	−0.04	0.00	−0.02	−0.01	0.00	−0.04	0.00	0.00	−0.02	−0.01
BAL	+0.03	+0.01	+0.01	+0.01	+0.01	+0.02	+0.02	0.00	0.00	0.00
8 CO	0.00	0.00	+0.02	+0.01	0.00	0.00	0.00	0.00	0.00	0.00
BAL	0.00	0.00	−0.01	−0.01	−0.01	0.00	0.00	0.00	0.00	0.00
Total balanced M	+7.75	−7.75	+3.48	+15.60	−19.08	+32.00	−32.00	−9.88	−18.92	−53.49
Check Change	+7.75	+4.25	+3.48	+15.60	−0.33	+32.00	+43.00	+2.12	−0.17	+21.51
$-\tfrac{1}{2}$(change)	−1.74	−1.06	−3.88	−16.00	+0.08	−7.80	−10.76	−2.12	+0.16	−21.50
Sum	+6.01	+3.19	−0.40	−0.40	−0.25	+24.20	+32.24	0.00	≈0	≈0
$EI_c\theta$	+5.34	+5.32	−0.35	−0.35	−0.33	+21.51	+21.49	0.00	≈0	≈0

Table 8.10.3c Combination of moments due to applied load and sidesway for rigid frame in Example 8.10.3

Joint	A		B			C		D	E	F
Member	AB	AD	BA	BC	BE	CB	CF	DA	EB	FC
FEM (1)	−96.00	0.00	+96.00	0.00	0.00	0.00	0.00	0.00	0.00	0.00
FEM (2)	0.00	− 2.34	0.00	0.00	− 3.66	0.00	−14.62	− 2.34	− 3.66	−14.62
Total FEM	−96.00	− 2.34	+96.00	0.00	− 3.66	0.00	−14.62	− 2.34	− 3.66	−14.62
BAL M (1)	−42.62	+42.62	+82.83	−47.43	−35.40	−15.15	+15.15	+21.31	−17.70	+ 7.58
BAL M (2)	+ 1.51	− 1.51	+ 0.68	+ 3.04	− 3.72	+ 6.24	− 6.24	− 1.93	− 3.69	−10.43
Total BAL M	−41.11	+41.11	+83.51	−44.39	−39.12	− 8.91	+ 8.91	+19.38	−21.39	− 2.85
Change	+54.89	+43.45	−12.49	−44.39	−35.46	− 8.91	+23.53	+21.72	−17.73	+11.77
−½(change)	+ 6.24	−10.86	−27.44	+ 4.46	+ 8.86	+22.20	− 5.88	−21.72	+17.73	−11.76
Sum	+61.13	+32.59	−39.93	−39.93	−26.60	+13.29	+17.65	0.00	0.00	+ 0.01
$EI_c\theta$	+54.34	+54.32	−35.49	−35.49	−35.47	+11.81	+11.77	0.00	0.00	≈ 0

THE MOMENT-DISTRIBUTION METHOD 305

The moment distribution is shown in Table 8.10.3a. The balanced moments acting on the three columns are shown in Fig. 8.10.7a; the three horizontal reactions at the three column bases are:

$$H_{D1} = \frac{42.62 + 21.31}{10} = 6.393 \text{ kN to right}$$

$$H_{E1} = \frac{35.40 + 17.70}{8} = 6.638 \text{ kN to left}$$

$$H_{F1} = \frac{15.15 + 7.58}{4} = 5.682 \text{ kN to right}$$

(c) *Moment distribution for arbitrary* $\Delta' = 100 \text{ kN·m}^3/EI_c$. The fixed-end moments for the three columns due to a horizontal displacement of ABC toward the right by an arbitrary amount Δ' are:

$$M_{FAD} = M_{FDA} = -\frac{6E(2I_c)\Delta'}{10^2} = -0.12 EI_c \Delta'$$

$$M_{FBE} = M_{FEB} = -\frac{6E(2I_c)\Delta'}{8^2} = -0.1875 EI_c \Delta'$$

$$M_{FCF} = M_{FFC} = -\frac{6E(2I_c)\Delta'}{4^2} = -0.75 EI_c \Delta'$$

If Δ' is set at $100 \text{ kN·m}^3/EI_c$, the fixed-end moments of -12, -18.75, and -75 are comparable to the values of -96 and $+96$, which are the fixed-end moments due to the applied load. The moment distribution of the fixed-end moments due to sidesway is shown in Table 8.10.3b. The horizontal reactions at the three column bases are computed from the free-body diagrams in Fig. 8.10.7b; thus

$$H_{D2} = \frac{7.75 + 9.88}{10} = 1.763 \text{ kN to left}$$

$$H_{E2} = \frac{19.08 + 18.92}{8} = 4.750 \text{ kN to left}$$

$$H_{F2} = \frac{32.00 + 53.49}{4} = 21.372 \text{ kN to left}$$

(d) *Correct amount of sidesway* Δ. In considering the horizontal force equilibrium of the entire rigid frame as given in Fig. 8.10.6a, the sum of the horizontal reactions at the three column bases must be zero since there is no horizontal load acting on the rigid frame at all. Thus, from observing Fig. 8.10.7,

$$(H_{D1} - H_{E1} + H_{F1}) - k(H_{D2} + H_{E2} + H_{F2}) = 0$$
$$(+6.393 - 6.638 + 5.682) - k(1.763 + 4.750 + 21.372) = 0$$
$$k = 0.1950$$

$$\Delta = k\Delta' = 0.1950(100) = 19.50 \frac{\text{kN·m}^3}{EI_c}$$

The combination of the fixed-end moments, as well as of the balanced moments, due to the applied load with those due to the correct amount of sidesway is shown in Table 8.10.3c. The member-end slopes are computed by using Eq. (8.6.2); their values are: $\theta_A = (+54.34$ or $+54.32)/EI_c$, $\theta_B = (-35.49, -35.49,$ or $-35.47)/EI_c$, $\theta_C = (+11.81$ or $+11.77)/EI_c$, $\theta_D = 0$, $\theta_E = 0$, and $\theta_F = 0$.

Example 8.10.4 Analyze the rigid frame of Example 7.7.4, shown again in Fig. 8.10.8, by the moment-distribution method.

306 INTERMEDIATE STRUCTURAL ANALYSIS

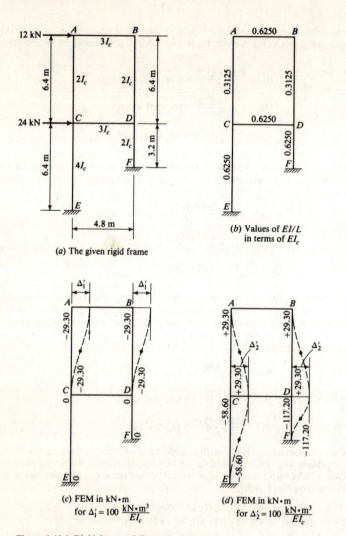

(a) The given rigid frame

(b) Values of EI/L in terms of EI_c

(c) FEM in kN·m for $\Delta_1' = 100 \dfrac{\text{kN} \cdot \text{m}^3}{EI_c}$

(d) FEM in kN·m for $\Delta_2' = 100 \dfrac{\text{kN} \cdot \text{m}^3}{EI_c}$

Figure 8.10.8 Rigid frame of Example 8.10.4.

SOLUTION (a) *Values of EI/L*. The values of EI/L for six members are (Fig. 8.10.8b):

$$\frac{EI}{L}(AB) = \frac{EI}{L}(CD) = \frac{E(3I_c)}{4.8} = 0.6250 EI_c \qquad \frac{EI}{L}(AC) = \frac{EI}{L}(BD) = \frac{E(2I_c)}{6.4} = 0.3125 EI_c$$

$$\frac{EI}{L}(CE) = \frac{E(4I_c)}{6.4} = 0.6250 EI_c \qquad \frac{EI}{L}(DF) = \frac{E(2I_c)}{3.2} = 0.6250 EI_c$$

Note that even though the shape of the structure has no symmetry about a vertical center line, it so happens that the relative stiffness factors are symmetrical about the vertical center

THE MOMENT-DISTRIBUTION METHOD 307

line, because the short column DF is half as long as the long column CE and its moment of inertia is also half that of CE.

(b) *Analysis procedure.* The degree of freedom in sidesway of the given rigid frame is 2; there are unknown sidesways of Δ_1 of AB to the right and of Δ_2 of CD to the right. The two shear conditions are (1) the sum of the horizontal shears to the left at the bases of the two upper columns must be equal to the horizontal load on AB, and (2) the sum of the horizontal shears to the left at the bases of the two lower columns must be equal to the sum of the two horizontal loads on AB and CD. When two supports are added at B and D to prevent the horizontal deflections there, the horizontal loads of 12 kN and 24 kN acting at A and C will be directly resisted by these two supports. Consequently, the only internal forces introduced into the system are axial compressive forces of 12 kN and 24 kN in members AC and BD; there is no bending anywhere. Thus the analysis procedure by the moment-distribution method involves first a moment distribution for the fixed-end moments due to an arbitrary sidesway Δ_1' of beam AB only, and then for an arbitrary sidesway Δ_2' of beam CD only. The correct amount of sidesway of AB is $\Delta_1 = k_1 \Delta_1'$ and that of CD is $\Delta_2 = k_2 \Delta_2'$. The unknown values of k_1 and k_2 can be found from the two shear conditions.

(c) *Moment distribution for sidesway* Δ_1' *of beam AB.* The fixed-end moments due to sidesway Δ_1' of AB to the right are (Fig. 8.10.8c):

$$M_{FAC} = M_{FCA} = M_{FBD} = M_{FDB} = -\frac{6E(2I_c)\Delta_1'}{6.4^2} = -0.29297 EI_c \Delta_1'$$

For the common accuracy of three or four significant figures, using fixed-end moments of -29.30 kN·m for $EI_c \Delta_1' = 100$ kN·m^3 seems to be reasonable. From the balanced moments in the moment distribution shown in Table 8.10.4a, the free-body diagrams of the four columns are shown in Fig. 8.10.9a. Note that the moment-distribution table itself gives identical

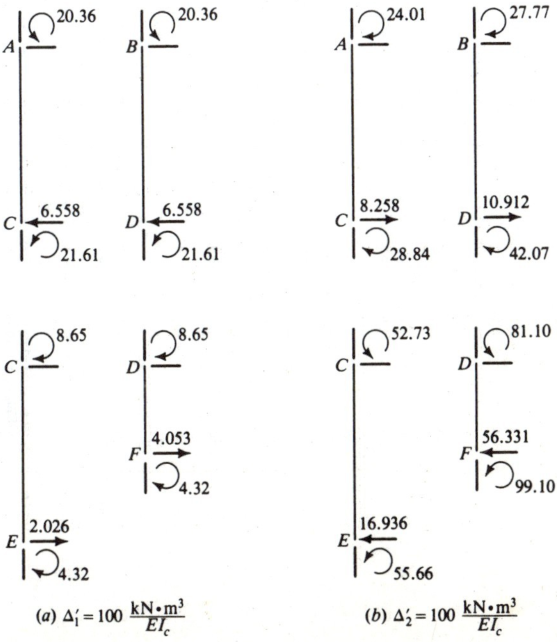

(a) $\Delta_1' = 100\ \dfrac{\text{kN}\cdot\text{m}^3}{EI_c}$ (b) $\Delta_2' = 100\ \dfrac{\text{kN}\cdot\text{m}^3}{EI_c}$

Figure 8.10.9 Shear conditions for rigid frame in Example 8.10.4.

Table 8.10.4a Moment distribution for sideway of beam AB in rigid frame of Example 8.10.4

Joint		A			B			C			D		E	F
Member		AB	AC	BA	BD	CA	CD	CE	DB	DC	DF	EC	FD
EI/L (in terms of EI_c)		0.6250	0.3125	0.6250	0.3125	0.3125	0.6250	0.6250	0.3125	0.6250	0.6250	0.6250	0.6250
Cycle	DF	0.6667	0.3333	0.6667	0.3333	0.2000	0.4000	0.4000	0.2000	0.4000	0.4000		
1	FEM	0.00	−29.30	0.00	−29.30	−29.30	0.00	0.00	−29.30	0.00	0.00	0.00	0.00
	BAL	+19.53	+9.77	+19.53	+9.77	+5.86	+11.72	+11.72	+5.86	+11.72	+11.72	0.00	0.00
2	CO	+9.76	+2.93	+9.76	+2.93	+4.88	+5.86	0.00	+4.88	+5.86	0.00	+5.86	+5.86
	BAL	−8.46	−4.23	−8.46	−4.23	−2.15	−4.30	−4.29	−2.15	−4.30	−4.29	0.00	0.00
3	CO	−4.23	−1.08	−4.23	−1.08	−2.12	−2.15	0.00	−2.12	−2.15	0.00	−2.14	−2.14
	BAL	+3.54	+1.77	+3.54	+1.77	+0.85	+1.71	+1.71	+0.85	+1.71	+1.71	0.00	0.00
4	CO	+1.77	+0.42	+1.77	+0.42	+0.88	+0.86	0.00	+0.88	+0.86	0.00	+0.86	+0.86
	BAL	−1.46	−0.73	−1.46	−0.73	−0.35	−0.69	−0.70	−0.35	−0.69	−0.70	0.00	0.00
5	CO	−0.73	−0.18	−0.73	−0.18	−0.36	−0.34	0.00	−0.36	−0.34	0.00	−0.35	−0.35
	BAL	+0.61	+0.30	+0.61	+0.30	+0.14	+0.28	+0.28	+0.14	+0.28	+0.28	0.00	0.00
6	CO	+0.30	+0.07	+0.30	+0.07	+0.15	+0.14	0.00	+0.15	+0.14	0.00	+0.14	+0.14
	BAL	−0.25	−0.12	−0.25	−0.12	−0.06	−0.12	−0.11	−0.06	−0.12	−0.11	0.00	0.00
7	CO	−0.12	−0.03	−0.12	−0.03	−0.06	−0.06	0.00	−0.06	−0.06	0.00	−0.06	−0.06
	BAL	+0.10	+0.05	+0.10	+0.05	+0.02	+0.05	+0.05	+0.02	+0.05	+0.05	0.00	0.00
8	CO	+0.05	+0.01	+0.05	+0.01	+0.02	+0.02	0.00	+0.02a	+0.02	0.00	+0.02	+0.02
	BAL	−0.04	−0.02	−0.04	−0.02	0.00	−0.02	−0.02	0.00	−0.02	−0.02	0.00	0.00
9	CO	−0.02	0.00	−0.02	0.00	−0.01	−0.01	0.00	−0.01	−0.01	0.00	−0.01	−0.01
	BAL	+0.01	+0.01	+0.01	+0.01	0.00	+0.01	+0.01	0.00	+0.01	+0.01	0.00	0.00
Total balanced M		+20.36	−20.36	+20.36	−20.36	−21.61	+12.96	+8.65	−21.61	+12.96	+8.65	+4.32	+4.32
Check	Change	+20.36	+8.94	+20.36	+8.94	+7.69	+12.96	+8.65	+7.69	+12.96	+8.65	+4.32	+4.32
	−½(change)	−10.18	−3.64	−10.18	−3.84	−4.47	−6.48	−2.16	−4.47	−6.48	−2.16	−4.32	−4.32
	Sum	+10.18	+5.10	+10.18	+5.10	+3.22	+6.48	+6.49	+3.22	+6.48	+6.49	0.00	0.00
	$EI_c\theta$	+5.43	+15.44	+5.43	+5.44	+3.43	+3.45	+3.46	+3.43	+3.45	+3.46	0.00	0.00

308

Table 8.10.4b Moment distribution for sidesway of beam CD in rigid frame of Example 8.10.4

Joint	A		B			C			D			E		F
Member	AB	AC	BA	BD	CA	CD	CE	DB	DC	DF	EC		FD
EI/L (in terms of EI_c)	0.6250	0.3125	0.6250	0.3125	0.3125	0.6250	0.6250	0.3125	0.6250	0.6250	0.6250		0.6250
Cycle DF	0.6667	0.3333	0.6667	0.3333	0.2000	0.4000	0.4000	0.2000	0.4000	0.4000	....		....
1 FEM	0.00	+29.30	0.00	+29.30	+29.30	0.00	−58.60	+29.30	0.00	−117.20	−58.60		−117.20
BAL	−19.53	−9.77	−19.53	−9.77	+5.86	+11.72	+11.72	+17.58	+35.16	+35.16	0.00		0.00
2 CO	−9.76	+2.93	−9.76	+8.79	+4.88	+17.58	0.00	−4.88	+5.86	0.00	+5.86		+17.58
BAL	+4.55	+2.28	+0.65	+0.32	−2.54	−5.08	−5.08	−0.20	−0.39	−0.39	0.00		0.00
3 CO	+0.32	−1.27	+2.28	−0.10	+1.14	−0.20	0.00	+0.16	−2.54	0.00	−2.54		−0.20
BAL	+0.63	+0.32	−1.45	−0.73	−0.19	−0.38	−0.37	+0.48	+0.95	+0.95	0.00		0.00
4 CO	−0.72	−0.10	+0.32	+0.24	+0.16	+0.48	0.00	−0.35	−0.19	0.00	−0.18		+0.48
BAL	+0.55	+0.27	−0.37	−0.19	−0.13	−0.25	−0.26	+0.11	+0.22	+0.22	0.00		0.00
5 CO	−0.18	−0.06	+0.28	+0.06	+0.14	+0.11	0.00	−0.10	−0.12	0.00	−0.13		+0.11
BAL	+0.18	+0.06	−0.23	−0.11	−0.05	−0.10	−0.10	+0.04	+0.09	+0.09	0.00		0.00
6 CO	−0.12	−0.02	+0.09	+0.02	+0.03	+0.04	0.00	−0.06	−0.05	0.00	−0.05		+0.04
BAL	+0.09	+0.05	−0.07	−0.04	−0.01	−0.03	−0.03	+0.02	+0.05	+0.04	0.00		0.00
7 CO	−0.04	0.00	+0.04	+0.01	+0.02	+0.02	0.00	−0.02	−0.02	0.00	−0.02		+0.02
BAL	+0.03	+0.01	−0.03	−0.02	−0.01	−0.02	−0.01	+0.01	+0.01	+0.02	0.00		0.00
8 CO	−0.02	0.00	+0.02	0.00	0.00	0.00	0.00	−0.01	−0.01	0.00	0.00		+0.01
BAL	+0.01	+0.01	−0.01	−0.01	0.00	0.00	0.00	0.00	+0.01	+0.01	0.00		0.00
Total balanced M	−24.01	+24.01	−27.77	+27.77	+28.84	+23.89	−52.73	+42.07	+39.03	−81.10	−55.66		−99.16
Check Change	−24.01	−5.29	−27.77	−1.53	−0.46	+23.89	+5.87	+12.77	+39.03	+36.10	+2.94		+18.04
−½(change)	+13.88	+0.23	+12.00	−6.38	+2.64	−19.52	−1.47	+0.76	−11.94	−9.02	+2.94		+18.05
Sum	−10.13	−5.06	−15.77	−7.91	+2.18	+4.37	+4.40	+13.53	+27.09	+27.08	0		≈ 0
$EI_c\theta$	−5.40	−5.40	−8.41	−8.44	+2.32	+2.33	+2.35	+14.43	+14.45	+14.44	0.00		≈ 0

309

Table 8.10.4c Combination of moments due to sidesway of *AB* and of *CD* in rigid frame of Example 8.10.4

Joint	A		B			C			D			E	F
Member	AB	AC	BA	BD	CA	CD	CE	DB	DC	DF	EC	FD
FEM (1)	0.00	−54.45	0.00	−54.45	−54.45	0.00	0.00	−54.45	0.00	0.00	0.00	0.00
FEM (2)	0.00	+18.91	0.00	+18.91	+18.91	0.00	−37.83	+18.91	0.00	−75.65	−37.83	−75.65
Total FEM	0.00	−35.54	0.00	−35.54	−35.54	0.00	−37.83	−35.54	0.00	−75.65	−37.83	−75.65
BAL M (1)	+37.84	−37.84	+37.84	−37.84	−40.16	+24.08	+16.08	−40.16	+24.08	+16.08	+ 8.03	+ 8.03
BAL M (2)	−15.50	+15.50	−17.92	+17.92	+18.62	+15.42	−34.04	+27.16	+25.19	−52.35	−35.93	−64.01
Total BAL M	+22.34	−22.34	+19.92	−19.92	−21.54	+39.50	−17.96	−13.00	+49.27	−36.27	−27.90	−55.98
Change	+22.34	+13.20	+19.92	+15.62	+14.00	+39.50	+19.87	+22.54	+49.27	+39.38	+ 9.93	+19.67
−½(change)	− 9.96	− 7.00	−11.17	−11.27	− 6.60	−24.64	− 4.96	− 7.81	−19.75	− 9.84	− 9.94	−19.69
Sum	+12.38	+ 6.20	+ 8.75	+ 4.35	+ 7.40	+14.86	+14.91	+14.73	+29.52	+29.54	− 0.01	− 0.02
$EI_c\theta$	+ 6.60	+ 6.61	+ 4.66	+ 4.64	+ 7.89	+ 7.92	+ 7.95	+15.71	+15.74	+15.75	≈ 0	≈ 0

moments at the ends of left or right columns, although the free-body diagram of the lower right column is different from that of the lower left column.

(d) *Moment distribution for sideway* Δ_2' *of beam CD*. The fixed-end moments due to sideway $\Delta_2' = 100 \text{ kN·m}^3/EI_c$ of CD to the right are (Fig. 8.10.8d):

$$M_{FAC} = M_{FCA} = M_{FBD} = M_{FDB} = +\frac{6E(2I_c)\Delta_2'}{6.4^2} = +0.29297EI_c\Delta_2' = +29.30 \text{ kN·m}$$

$$M_{FCE} = M_{FEC} = -\frac{6E(4I_c)\Delta_2'}{6.4^2} = -0.58594EI_c\Delta_2' = -58.60 \text{ kN·m}$$

$$M_{FDF} = M_{FFD} = -\frac{6E(2I_c)\Delta_2'}{3.2^2} = -1.17188EI_c\Delta_2' = -117.20 \text{ kN·m}$$

The round-off values of -58.60 and -117.20 are used because they should be 2 and 4 times the 29.30 used for columns AC and BD. From the balanced moments in the moment distribution shown in Table 8.10.4b, the free-body diagrams of the four columns are shown in Fig. 8.10.9b. Note that the short column DF has the largest shear force in it.

(e) *Correct amounts of sideways* Δ_1 *and* Δ_2. The two shear conditions are, from observing Fig. 8.10.9,

$$+k_1(6.558 + 6.558) - k_2(8.258 + 10.912) = +12.0$$

$$-k_1(2.026 + 4.053) + k_2(16.936 + 56.331) = +24.0$$

or

$$+13.116k_1 - 19.170k_2 = +12.0$$

$$-6.079k_1 + 73.267k_2 = +24.0$$

Solving the above two equations for k_1 and k_2,

$$k_1 = +1.8584 \qquad \Delta_1 = k_1\Delta_1' = 1.8584(100) = 185.84 \frac{\text{kN·m}^3}{EI_c}$$

$$k_2 = +0.6455 \qquad \Delta_2 = k_2\Delta_2' = 0.6455(100) = 64.55 \frac{\text{kN·m}^3}{EI_c}$$

Table 8.10.4c shows the combination of the two sets of fixed-end moments and balanced moments. Each combined value is equal to the sum of 1.8584 times the value in Table 8.10.4a and 0.6455 times the value in Table 8.10.4b. The member-end slopes are computed from using Eq. (8.6.2); they are: $\theta_A = (+6.60 \text{ or } +6.61)/EI_c$, $\theta_B = (+4.66 \text{ or } +4.64)/EI_c$, $\theta_C = (+7.89, +7.92, \text{ or } +7.95)/EI_c$, $\theta_D = (+15.71, +15.74, \text{ or } +15.75)/EI_c$, $\theta_E = 0$, and $\theta_F = 0$.

(f) *Discussion in regard to reciprocal virtual-work theorem.* Based on the results of the two moment distributions, one due to $\Delta_1' = 100 \text{ kN·m}^3/EI_c$ and the other due to $\Delta_2' = 100 \text{ kN·m}^3/EI_c$, the free-body diagrams of the columns are shown in Fig. 8.10.9. Had the free-body diagrams of the entire rigid frame been worked out using the balanced moments in Tables 8.10.4a and b, they would be as shown in Fig. 8.10.10. One can note that the horizontal force at C to cause a $100 \text{ kN·m}^3/EI_c$ horizontal deflection to the right at A is 19.195 kN to the left, and that the horizontal force at A to cause a $100 \text{ kN·m}^3/EI_c$ horizontal deflection to the right at C is 19.170 kN to the left. The fact that 19.195 kN is not exactly equal to 19.170 kN is due to round-off errors, because they should be theoretically equal according to the reciprocal virtual-work theorem proved in Sec. 4.3. Applying this theorem, represented by Eq. (4.3.3), to the P and Q systems in Fig. 8.10.10,

$$P * \Delta Q = Q * \Delta P$$

$$(-19.195)(+100) = (-19.170)(+100)$$

There is only one force in the P system (19.195 kN to the left) that has a virtual displacement ($100 \text{ kN·m}^3/EI_c$ to the right) in the Q system; also there is only one force in the Q system (19.170 kN to the left) that has a virtual displacement ($100 \text{ kN·m}^3/EI_c$ to the right) in the P system. This phenomenon is often called the *Law of reciprocal forces*.

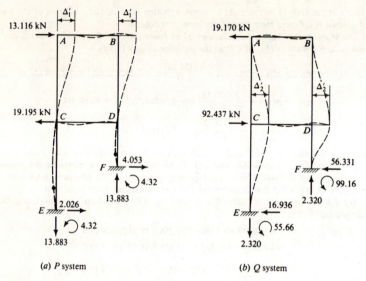

(a) P system (b) Q system

Figure 8.10.10 Law of reciprocal forces.

In reference to Fig. 8.10.10, the two shear conditions may be stated as follows: (1) external force at $A = +12 = +13.116k_1 - 19.170k_2$, and (2) external force at $C = +24 = -19.195k_1 + 92.437k_2$. Then the coefficients of k_1 and k_2 in the preceding two simultaneous equations are symmetrical with respect to the main diagonal downward to the right as a natural consequence of the law of reciprocal forces.

8.11 Application to the Analysis of Statically Indeterminate Rigid Frames due to Yielding of Supports

The moment-distribution method can readily be used to analyze statically indeterminate rigid frames due to yielding of supports. If the yielding is a rotational slip at a certain support, the fixed-end moments are the moments required to maintain this rotational slip. If the yielding is a translational settlement, the fixed-end moments are the moments required to hold the tangents at the joints fixed in direction while allowing the settlement to take place. Thus the end moments required to maintain a rotation $d\theta$ rad in the clockwise direction at end A of member AB are (Fig. 8.11.1a)

$$M_{FAB} = +\frac{4EI}{L}d\theta \qquad M_{FBA} = +\frac{2EI}{L}d\theta \qquad (8.11.1)$$

The fixed-end moments required to hold the tangents at A and B of member AB fixed when B settles an amount Δ downward relative to A are (Fig. 8.11.1b)

$$M_{FAB} = -\frac{6EI\Delta}{L^2} \qquad M_{FBA} = -\frac{6EI\Delta}{L^2} \qquad (8.11.2)$$

THE MOMENT-DISTRIBUTION METHOD 313

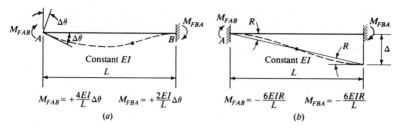

Figure 8.11.1 Fixed-end moments due to rotational slip and translational settlement.

Equations (8.11.1) and (8.11.2) can easily be verified by applying the general slope-deflection equations, Eqs. (7.4.4a and b):

$$M_A = M_{0A} + \frac{2EI}{L}(2\theta_A + \theta_B - 3R) \qquad M_B = M_{0B} + \frac{2EI}{L}(2\theta_B + \theta_A - 3R)$$

Furthermore, Eq. (8.11.1) can be obtained from the definitions of stiffness and carry-over factors in Sec. 8.3; and Eq. (8.11.2) is the same as Eq. (8.8.1), which was derived by applying the moment-area theorems.

Example 8.11.1 Analyze the rigid frame of Example 7.8.1, shown again in Fig. 8.11.2, by the moment-distribution method.

SOLUTION (a) *Values of EI/L.* The analysis is for a 15-mm vertical settlement and a 0.002-rad-clockwise rotational slip, both of support D. The values of EI/L in absolute units are computed for the three members and shown in Fig. 8.11.2b; they are:

$$\frac{EI}{L}(AB) = \frac{200(400)}{7.5} = 10{,}667 \text{ kN·m}$$

$$\frac{EI}{L}(BC) = \frac{200(800)}{6} = 26{,}667 \text{ kN·m}$$

$$\frac{EI}{L}(CD) = \frac{200(400)}{5} = 16{,}000 \text{ kN·m}$$

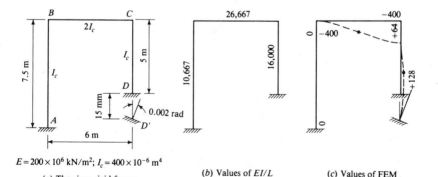

$E = 200 \times 10^6$ kN/m²; $I_c = 400 \times 10^{-6}$ m⁴

(a) The given rigid frame
(b) Values of EI/L in kN·m
(c) Values of FEM in kN·m

Figure 8.11.2 Rigid frame of Example 8.11.1.

314 INTERMEDIATE STRUCTURAL ANALYSIS

(b) *Moment distribution for the yielding of supports with sidesway prevented.* When the support at D settles 15 mm vertically downward and simultaneously rotates 0.002 rad clockwise, the fixed-end moments required to maintain the shape of the elastic curve at $\theta_B = 0$, $\theta_C = 0$, and sidesway $= 0$, are (Fig. 8.11.2c):

$$M_{FAB} = M_{FBA} = 0$$

$$M_{FBC} = M_{FCB} = -\frac{6(200)(800)(0.015)}{6^2} = -400 \text{ kN} \cdot \text{m}$$

$$M_{FDC} = +\frac{4(200)(400)(0.002)}{5} = +128 \text{ kN} \cdot \text{m}$$

$$M_{FCD} = +\frac{2(200)(400)(0.002)}{5} = +64 \text{ kN} \cdot \text{m}$$

The moment distribution for the above fixed-end moments is shown in Table 8.11.1a. The balanced moments acting on columns AB and CD and the horizontal reactions at A and D are shown in Fig. 8.11.3a, wherein

$$H_{A1} = \frac{95.0 + 47.6}{7.5} = 19.013 \text{ kN to right}$$

$$H_{D1} = \frac{145.5 + 168.7}{5} = 62.840 \text{ kN to right}$$

(c) *Moment distribution for an arbitrary sidesway* $\Delta' = 1000 \text{ kN} \cdot \text{m}^3/EI_c$. The fixed-end moments are

$$M_{FAB} = M_{FBA} = -\frac{6EI\Delta'}{7.5^2} = -\frac{6(1000)}{7.5^2} = -106.66 \text{ kN} \cdot \text{m}$$

$$M_{FCD} = M_{FDC} = -\frac{6EI\Delta'}{5^2} = -\frac{6(1000)}{5^2} = -240 \text{ kN} \cdot \text{m}$$

The moment distribution for the above fixed-end moments has been shown in Table 8.10.2b; parts of it are repeated here as Table 8.11.1b. The balanced moments acting on columns AB and CD and the horizontal reactions at A and D are shown in Fig. 8.11.3b, wherein

$$H_{A2} = 26.409 \text{ kN to left} \qquad H_{D2} = 70.436 \text{ kN to left}$$

(d) *Correct amount of sidesway* Δ. The shear condition is that the sum of the horizontal

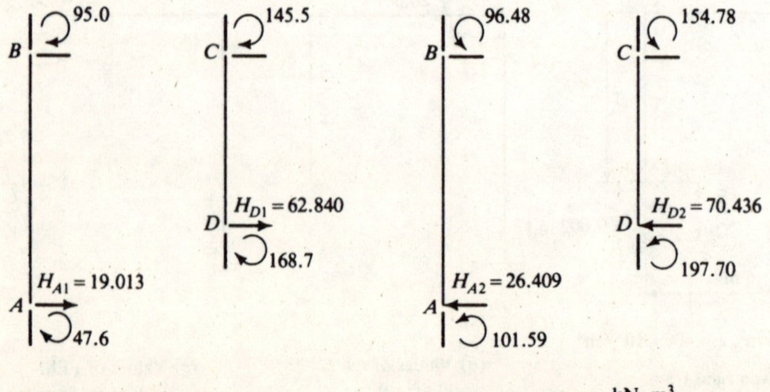

(a) Sidesway prevented

(b) Sidesway of $1000 \dfrac{\text{kN} \cdot \text{m}^3}{EI_c} = 12.5$ mm

Figure 8.11.3 Shear condition for rigid frame in Example 8.11.1.

THE MOMENT-DISTRIBUTION METHOD

Table 8.11.1a Moment distribution for FEM due to yielding of supports in rigid frame of Example 8.11.1

Joint		A	B		C		D
Member		AB	BA	BC	CB	CD	DC
EI/L, kN·m		10,667	10,667	26,667	26,667	16,000	16,000
Cycle	DF	....	0.2857	0.7143	0.6250	0.3750	....
1	FEM	0.0	0.0	−400.0	−400.0	+ 64.0	+128.0
	BAL	0.0	+114.3	+285.7	+210.0	+126.0	0.0
2	CO	+57.2	0.0	+105.0	+142.8	0.0	+ 63.0
	BAL	0.0	− 30.0	− 75.0	− 89.2	− 53.6	0.0
3	CO	−15.0	0.0	− 44.6	− 37.5	0.0	− 26.8
	BAL	0.0	+ 12.8	+ 31.8	+ 23.4	+ 14.1	0.0
4	CO	+ 6.4	0.0	+ 11.7	+ 15.9	0.0	+ 7.0
	BAL	0.0	− 3.3	− 8.4	− 9.9	− 6.0	0.0
5	CO	− 1.6	0.0	− 5.0	− 4.2	0.0	− 3.0
	BAL	0.0	+ 1.4	+ 3.6	+ 2.6	+ 1.6	0.0
6	CO	+ 0.7	0.0	+ 1.3	+ 1.8	0.0	+ 0.8
	BAL	0.0	− 0.4	− 0.9	− 1.1	− 0.7	0.0
7	CO	− 0.2	0.0	− 0.6	− 0.4	0.0	− 0.4
	BAL	0.0	+ 0.2	+ 0.4	+ 0.2	+ 0.2	0.0
8	CO	+ 0.1	0.0	+ 0.1	+ 0.2	0.0	+ 0.1
	BAL	0.0	0.0	− 0.1	− 0.1	− 0.1	0.0
Total balanced M		+47.6	+ 95.0	− 95.0	−145.5	+145.5	+168.7
Check	Change	+47.6	+ 95.0	+305.0	+254.5	+ 81.5	+ 40.7
	−½(change)	−47.5	− 23.8	−127.2	−152.5	− 20.4	− 40.8
	Sum	+ 0.1	+ 71.2	+177.8	+102.0	+ 61.1	− 0.1
	θ, 10^{-3} rad	≈0	+2.225	+2.223	+1.275	+1.273	≈0

reactions at A and D must be equal to zero; thus, from observing Fig. 8.11.3,

$$(H_{A1} + H_{D1}) = k(H_{A2} + H_{D2})$$
$$(19.013 + 62.840) = k(26.409 + 70.436)$$
$$k = 0.8452$$
$$\Delta = k\Delta' = \frac{0.8452(1000)}{EI_c} = \frac{0.8452(1000)}{200(400)} = 0.010565 \text{ m}$$

Table 8.11.1c shows the combination of the fixed-end moments and balanced moments in Table 8.11.1a with 0.8452 times the corresponding values in Table 8.11.1b. The correct member-end slopes are computed by using Eq. (8.6.2). The end slopes are $\theta_A = 0$, $\theta_B = (+2.428 \text{ or } +2.424) \times 10^{-3}$ rad, $\theta_C = (+2.401 \text{ or } +2.398) \times 10^{-3}$ rad, and $\theta_D = 0$.

Table 8.11.1b FEM and balanced moments for sidesway in rigid frame of Example 8.11.1 (from Table 8.10.2b)

Joint	A	B		C		D
Member	AB	BA	BC	CB	CD	DC
FEM	−106.66	−106.66	0.00	0.00	−240.00	−240.00
BAL M	−101.59	−96.48	+96.48	+154.78	−154.78	−197.40

Table 8.11.1c Combination of moments due to yielding of supports and sidesway for rigid frame of Example 8.11.1

Joint	A	B		C		D
Member	AB	BA	BC	CB	CD	DC
FEM (1)	0.0	0.0	−400.0	−400.0	+ 64.0	+128.0
FEM (2)	−90.1	− 90.1	0.0	0.0	−202.8	−202.8
Total FEM	−90.1	− 90.1	−400.0	−400.0	−138.8	− 74.8
BAL M (1)	+47.6	+ 95.0	− 95.0	−145.5	+145.5	+168.7
BAL M (2)	−85.9	− 81.5	+ 81.5	+130.8	−130.8	−166.8
Total BAL M	−38.3	+ 13.5	− 13.5	− 14.7	+ 14.7	+ 1.9
Change	+51.8	+103.6	+386.5	+385.3	+153.5	+ 76.9
−½(change)	−51.8	− 25.9	−192.6	−193.2	− 38.4	− 76.8
Sum	0.0	+ 77.7	+193.9	+192.1	+115.1	+ 0.1
θ, 10^{-3} rad	0.0	+2.428	+2.424	+2.401	+2.398	≈ 0

8.12 Analysis of Gable Frames by the Moment-Distribution Method

When a single-span gable frame with two fixed supports such as the one shown in Fig. 8.12.1a is subjected to an unsymmetrical loading, there are five unknown joint displacements, of which three are the joint rotations at B, C, and D, and the remaining two are to be chosen from the horizontal deflections of joints B, C, and D and the vertical deflection of joint C. In the slope-deflection method of solution, there are five simultaneous equations to be solved; this task is not an attractive one, especially considering the time needed to set up the system of equations. The moment-distribution method offers an alternative method of solution, but one still needs to solve two simultaneous equations, as described in Fig. 8.12.1. As stated in the introduction to this chapter, the chief advantage of the moment-distribution method lies in the fact that the analyst will have a physical feeling of the structural response and may become more familiar with the structure during the analysis so as to be able to improve the parameters used in the design.

THE MOMENT-DISTRIBUTION METHOD 317

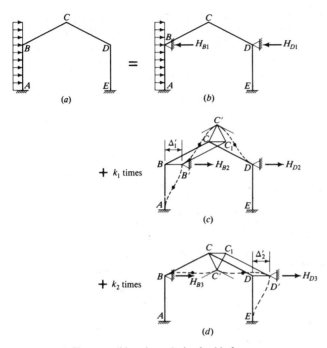

Figure 8.12.1 Shear conditions in analysis of gable frames.

The basic approach used in the analysis of the gable frame with two degrees of freedom in sidesway involves the addition of two supports to prevent the two independent sidesways selected from the four unknown joint translations. In Fig. 8.12.1b, the sidesways of joints B and D are prevented by the respective supports. The moment distribution for the fixed-end moments due to the applied load on the rigid frame of Fig. 8.12.1b is no different from that for a four-span continuous beam. From the results of this moment distribution, the required reactions H_{B1} and H_{D1} can be determined. Then the supports at B and D are permitted to yield arbitrary amounts Δ_1' and Δ_2', one at a time, as shown in Fig. 8.12.1c and d. Fixed-end moments are applied to hold the joints against rotation while the yielding of Δ_1' or Δ_2' occurs. After these fixed-end moments are distributed, the induced reactions H_{B2} and H_{D2} in the case of Δ_1' and H_{B3} and H_{D3} in the case of Δ_2' can be determined from statics using the balanced moments. Calling the correct sidesways $\Delta_1 = k_1 \Delta_1'$ and $\Delta_2 = k_2 \Delta_2'$, the two "shear" conditions from which k_1 and k_2 can be solved are:

$$H_{B1} = k_1 H_{B2} + k_2 H_{B3} \qquad (8.12.1a)$$
$$H_{D1} = k_1 H_{D2} + k_2 H_{D3} \qquad (8.12.1b)$$

When both the gable frame itself and the applied loads are symmetrical with respect to a vertical line through the apex, only half of the entire frame needs to be involved in the moment distribution. The two-member rigid frame shown in Fig. 8.12.2a behaves like the left half of the entire rigid frame; it has a fixed support at A and a roller support at C with zero rotation of joint C. In Fig. 8.12.2b, the sidesway at B is prevented by the additional support; and in Fig. 8.12.2c this support is permitted to yield an arbitrary amount Δ', with a consequent vertical displacement CC' of the roller support so as to keep the member length of BC. The shear condition by which the correct sidesway $\Delta = k\Delta'$ can be obtained is

$$H_{B1} = kH_{B2} \tag{8.12.2}$$

The moment distribution for the two-member rigid frame in Fig. 8.12.2b and c is no different from that of a two-span continuous beam with both ends fixed, because even in Fig. 8.12.2c there is still no further unknown joint translation beyond those in the fixed condition.

Example 8.12.1 Analyze the gable frame of Example 7.9.1, shown again in Fig. 8.12.3, by the moment-distribution method.

SOLUTION (a) *Values of EI/L.* The values of EI/L for the two members in the equivalent half frame shown in Fig. 8.12.3b are:

$$\frac{EI}{L}(AB) = \frac{E(2I_c)}{6} = 0.3333EI_c \qquad \frac{EI}{L}(BC) = \frac{E(3I_c)}{6.7082} = 0.4472EI_c$$

(b) *Moment distribution for the applied load with sidesway prevented.* When sidesway is prevented by a lateral support at B, as shown in Fig. 8.12.3c, all three joints on the half frame cannot have translation. The fixed-end moments are

$$M_{FBC} = -\frac{18(6)^2}{12} = -54 \text{ kN} \cdot \text{m} \qquad M_{FCB} = +54 \text{ kN} \cdot \text{m}$$

From the moment distribution shown in Table 8.12.1a, the balanced moments are shown on the free-body diagrams of column AB and rafter BC in Fig. 8.12.4a. Note that in considering BC as the free body, the vertical force acting at C must be zero because of the symmetry. From the free-body diagram of joint B,

$$H_{B1} = 117.71 \text{ kN to right}$$

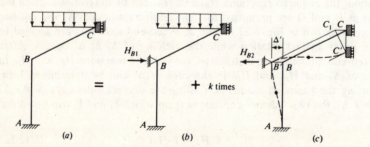

Figure 8.12.2 Shear condition for symmetrical gable frame symmetrically loaded.

THE MOMENT-DISTRIBUTION METHOD 319

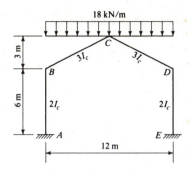

(a) The given rigid frame

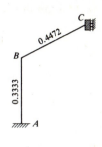

(b) Values of EI/L in terms of EI_c

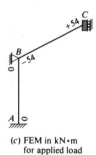

(c) FEM in kN·m for applied load

(d) FEM in kN·m for $\Delta' = 1000 \dfrac{\text{kN} \cdot \text{m}^3}{EI_c}$

Figure 8.12.3 Gable frame of Example 8.12.1.

(c) *Moment distribution for arbitrary sidesway* $\Delta' = 1000$ kN·m³/EI_c of joint B to the left. Since the horizontal force required at B to prevent sidesway is 117.71 kN to the right, the actual sidesway, when this restraining force is removed, must be to the left. In Fig. 8.12.3d, the yielding of the support at D is assumed to be an arbitrary amount $\Delta' = 1000$ kN·m³/EI_c to the left. To maintain all member lengths unchanged, joint C has to move to C' as shown. The fixed-end moments required to keep all end slopes to zero are

$$M_{FAB} = M_{FBA} = +\frac{6E(2I_c)\Delta'}{6^2} = +\frac{6(2)(1000)}{6^2} = +333.3 \text{ kN·m}$$

$$M_{FBC} = M_{FCB} = -\frac{6E(3I_c)(C_1C')}{6.7082^2} = -\frac{18(1000\sqrt{5})}{6.7082^2} = -894.4 \text{ kN·m}$$

The moment distribution for the above fixed-end moments is shown in Table 8.12.1b. From the free-body diagrams shown in Fig. 8.12.4b, the induced horizontal reaction at B is found to be

$$H_{B2} = 606.5 \text{ kN to left}$$

Table 8.12.1a Moment distribution for applied load on gable frame in Example 8.12.1

Joint		A	B		C
Member		AB	BA	BC	CB
EI/L (in terms of EI_c)		0.3333	0.3333	0.4472	0.4472
Cycle	DF	....	0.4270	0.5730	....
1	FEM	0.00	0.00	−54.00	+54.00
	BAL	0.00	+23.06	+30.94	0.00
2	CO	+11.53	0.00	0.00	+15.47
	BAL	0.00	0.00	0.00	0.00
Total balanced M		+11.53	+23.06	−23.06	+69.47
Check	Change	+11.53	+23.06	+30.94	+15.47
	$-\frac{1}{2}$(change)	−11.53	− 5.76	− 7.74	−15.47
	Sum	0.00	+17.30	+23.20	0.00
	$EI_c\theta$	0.00	+17.30	+17.29	0.00

Table 8.12.1b Moment distribution for sidesway in gable frame of Example 8.12.1

Joint		A	B		C
Member		AB	BA	BC	CB
EI/L (in terms of EI_c)		0.3333	0.3333	0.4472	0.4472
Cycle	DF	....	0.4270	0.5730	....
1	FEM	+333.3	+333.3	−894.4	−894.4
	BAL	0.0	+239.6	+321.5	0.0
2	CO	+119.8	0.0	0.0	+160.8
	BAL	0.0	0.0	0.0	0.0
Total balanced M		+453.1	+572.9	−572.9	−733.6
Check	Change	+119.8	+239.6	+321.5	+160.8
	$-\frac{1}{2}$(change)	−119.8	− 59.9	− 80.4	−160.8
	Sum	0.0	+179.7	+241.1	0.0
	$EI_c\theta$	0.0	+179.7	+179.7	0.0

Table 8.12.1c Combination of moments due to applied load and sidesway in gable frame of Example 8.12.1

Joint	A	B		C
Member	AB	BA	BC	CB
BAL M (1)	+11.53	+ 23.06	− 23.06	+ 69.47
BAL M (2)	+87.93	+111.18	−111.18	−142.37
Total BAL M	+99.46	+134.24	−134.24	− 72.90
$EI_c\theta$ (1)	0.00	+ 17.30	+ 17.29	0.00
$EI_c\theta$ (2)	0.00	+ 34.87	+ 34.87	0.00
$EI_c\theta$	0.00	+ 52.17	+ 52.16	0.00

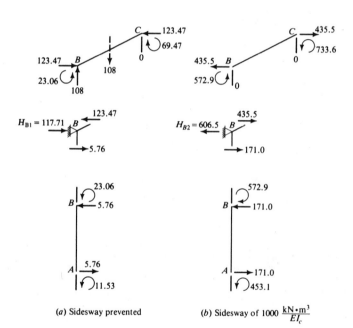

(a) Sidesway prevented (b) Sidesway of 1000 $\dfrac{kN \cdot m^3}{EI_c}$

Figure 8.12.4 Shear condition for gable frame in Example 8.12.1.

(d) *Correct amount of sidesway* Δ. Using the shear condition described in Eq. (8.12.2),

$$H_{B1} = kH_{B2}$$
$$117.71 = k(606.5)$$
$$k = 0.1941$$

$$\Delta = k\Delta' = 0.1941(1000) = 194.1 \, \frac{kN \cdot m^3}{EI_c}$$

The combined balanced moments and end slopes shown in Table 8.12.1c are the sums of the corresponding values in Table 8.12.1a and 0.1941 times those in Table 8.12.1b. These balanced moments and end slopes agree very closely with those in the slope-deflection solution of Example 7.9.1.

Example 8.12.2 Analyze the gable frame of Example 7.9.2, shown again in Fig. 8.12.5, by the moment-distribution method.

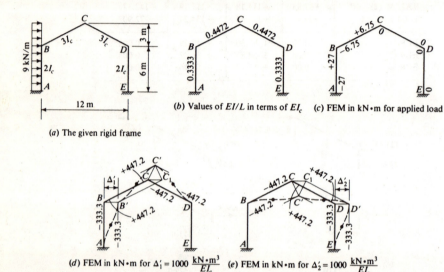

Figure 8.12.5 Gable frame of Example 8.12.2.

SOLUTION The concept in the solution procedure for a gable frame subjected to an unsymmetrical loading has been explained in this section. The complete solution of the given rigid frame shown in Fig. 8.12.5a is described by Table 8.12.2a to 'd and Figs. 8.12.5 and 8.12.6. From the balanced moments in Table 8.12.2a to c, the horizontal reactions at the supports B

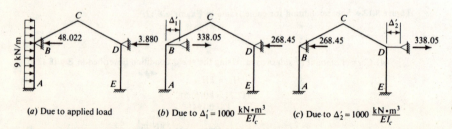

Figure 8.12.6 Shear conditions for gable frame in Example 8.12.2.

Table 8.12.2a Moment distribution for applied load on gable frame in Example 8.12.2

Joint	A	B		C		D		E
Member	AB	BA	BC	CB	CD	DC	DE	ED
EI/L (in terms of EI_c)	0.3333	0.3333	0.4472	0.4472	0.4472	0.4472	0.3333	0.3333
Cycle DF	....	0.4270	0.5730	0.5000	0.5000	0.5730	0.4270	....
1 FEM	−27.00	+27.00	− 6.75	+6.75	0.00	0.00	0.00	0.00
BAL	0.00	− 8.65	−11.60	−3.38	−3.37	0.00	0.00	0.00
2 CO	− 4.32	0.00	− 1.69	−5.80	0.00	−1.69	0.00	0.00
BAL	0.00	+ 0.72	+ 0.97	+2.90	+2.90	+0.97	+0.72	0.00
3 CO	+ 0.36	0.00	+ 1.45	+0.48	+0.48	+1.45	0.00	+0.36
BAL	0.00	− 0.62	− 0.83	−0.48	−0.48	−0.83	−0.62	0.00
4 CO	− 0.31	0.00	− 0.24	−0.42	−0.42	−0.24	0.00	−0.31
BAL	0.00	+ 0.10	+ 0.14	+0.42	+0.42	+0.14	+0.10	0.00
5 CO	+ 0.05	0.00	+ 0.21	+0.07	+0.07	+0.21	0.00	+0.05
BAL	0.00	− 0.09	− 0.12	−0.07	−0.07	−0.12	−0.09	0.00
6 CO	− 0.04	0.00	− 0.04	−0.06	−0.06	−0.04	0.00	−0.04
BAL	0.00	+ 0.02	+ 0.02	+0.06	+0.06	+0.02	+0.02	0.00
7 CO	+ 0.01	0.00	+ 0.03	+0.01	+0.01	+0.03	0.00	+0.01
BAL	0.00	− 0.01	− 0.02	−0.01	−0.01	−0.02	−0.01	0.00
Total balanced M	−31.25	+18.47	−18.47	+0.47	−0.47	−0.12	+0.12	+0.07
Check Change	− 4.25	− 8.53	−11.72	−6.28	−0.47	−0.12	+0.12	+0.07
−½(change)	+ 4.26	+ 2.12	+ 3.14	+5.86	+0.06	+0.24	−0.04	−0.06
Sum	+ 0.01	− 6.41	− 8.58	−0.42	−0.41	+0.12	+0.08	+0.01
$EI_c\theta$	≈ 0	− 6.41	− 6.40	−0.31	−0.31	+0.09	+0.08	≈ 0

323

Table 8.12.2b Moment distribution for sidesway of 1000 kN·m³/EI_c at top of left column in gable frame of Example 8.12.2

Joint	A	B		C		D		E
Member	AB	BA	BC	CB	CD	DC	DE	ED
EI/L (in terms of EI_c)	0.3333	0.3333	0.4472	0.4472	0.4472	0.4472	0.3333	0.3333
Cycle DF	....	0.4270	0.5730	0.5000	0.5000	0.5730	0.4270	....
1 FEM	−333.3	−333.3	+447.2	+447.2	−447.2	−447.2	0.0	0.0
BAL	0.0	− 48.6	− 65.3	0.0	0.0	+256.2	+191.0	0.0
2 CO	− 24.3	0.0	0.0	− 32.6	+128.1	0.0	0.0	+ 95.5
BAL	0.0	0.0	0.0	− 47.8	− 47.7	0.0	0.0	0.0
3 CO	0.0	0.0	− 23.9	0.0	0.0	− 23.9	0.0	0.0
BAL	0.0	+ 10.2	+ 13.7	0.0	0.0	+ 13.7	+ 10.2	0.0
4 CO	+ 5.1	0.0	0.0	+ 6.8	+ 6.8	0.0	0.0	+ 5.1
BAL	0.0	0.0	0.0	− 6.8	− 6.8	0.0	0.0	0.0
5 CO	0.0	0.0	− 3.4	0.0	0.0	− 3.4	0.0	0.0
BAL	0.0	+ 1.4	+ 2.4	0.0	0.0	+ 2.0	+ 1.4	0.0
6 CO	+ 0.7	0.0	0.0	+ 1.0	+ 1.0	0.0	0.0	+ 0.7
BAL	0.0	0.0	0.0	− 1.0	− 1.0	0.0	0.0	0.0
7 CO	0.0	0.0	− 0.5	0.0	0.0	− 0.5	0.0	0.0
BAL	0.0	+ 0.2	+ 0.3	0.0	0.0	+ 0.3	+ 0.2	0.0
8 CO	+ 0.1	0.0	0.0	+ 0.1	+ 0.1	0.0	0.0	+ 0.1
BAL	0.0	0.0	0.0	− 0.1	− 0.1	0.0	0.0	0.0
Total balanced M	−351.7	−370.1	+370.1	+366.8	−366.8	−202.8	+202.8	+101.4
Check Change	− 18.4	− 36.8	− 77.1	− 80.4	+ 80.4	+244.4	+202.8	+101.4
−½(change)	+ 18.4	+ 9.2	+ 40.2	+ 38.6	−122.2	− 40.2	− 50.7	−101.4
Sum	0.0	− 27.6	− 36.9	− 41.8	− 41.8	+204.2	+152.1	0.0
$EI_c\theta$	0.0	− 27.6	− 27.5	− 31.2	− 31.2	+152.2	+152.1	0.0

324

Table 8.12.2c Moment distribution for sidesway of 1000 kN·m³/EI_c at top of right column in gable frame of Example 8.12.2 (refer to Table 8.12.2b)

Joint	A	B		C		D		E
Member	AB	BA	BC	CB	CD	DC	DE	ED
FEM	0.0	0.0	−447.2	−447.2	+447.2	+447.2	−333.3	−333.3
Total balanced M	+101.4	+202.8	−202.8	−366.8	+366.8	+370.1	−370.1	−351.7
$EI_c\theta$	0.0	+152.1	+152.1	−31.2	−31.2	−27.5	−27.6	0.0

Table 8.12.2d Combination of moments due to applied load and sidesway in gable frame of Example 8.12.2

Joint	A	B		C		D		E
Member	AB	BA	BC	CB	CD	DC	DE	ED
BAL M (1)	− 31.25	+ 18.47	− 18.47	+ 0.47	− 0.47	− 0.12	+ 0.12	+ 0.07
BAL M (2)	−143.93	−151.46	+151.46	+150.11	−150.11	− 83.00	+ 83.00	+ 41.50
BAL M (3)	+ 34.12	+ 68.24	− 68.24	−123.42	+123.42	+124.53	−124.53	−118.34
Total BAL M	−141.06	− 64.75	+ 64.75	+ 27.16	− 27.16	+ 41.41	− 41.41	− 76.77
$EI_c\theta$ (1)	0.00	− 6.41	− 6.40	− 0.31	− 0.31	+ 0.09	+ 0.08	0.00
$EI_c\theta$ (2)	0.00	− 11.30	− 11.25	− 12.77	− 12.77	+ 62.29	+ 62.25	0.00
$EI_c\theta$ (3)	0.00	+ 51.18	+ 51.18	− 10.50	− 10.50	− 9.25	− 9.29	0.00
Total $EI_c\theta$	0.00	+ 33.47	+ 33.53	− 23.58	− 23.58	− 53.13	− 53.04	0.00

and C are summarized in Fig. 8.12.6a to c. The two shear conditions become

$$+338.05k_1 - 268.45k_2 = 48.022$$
$$-268.45k_1 + 338.05k_2 = 3.880$$

Note that the coefficients of k_1 and k_2 on the left sides of the above two equations are symmetrical with respect to the main diagonal downward to the right, again as a natural consequence of the reciprocal virtual-work theorem, or, in this case, the law of reciprocal forces. Solving the two shear equations for k_1 and k_2,

$$k_1 = +0.40925 \qquad k_2 = +0.33647$$

Thus the correct amounts of sidesway are:

$$\Delta_1 = k_1\Delta_1' = (+0.40925)(1000) = +409.25\,\frac{\text{kN}\cdot\text{m}^3}{EI_c}$$

$$\Delta_2 = k_2\Delta_2' = (+0.33647)(1000) = +336.47\,\frac{\text{kN}\cdot\text{m}^3}{EI_c}$$

The combined values of the balanced moments and end slopes shown in Table 8.12.2d are each equal to the sum of three parts; namely, (1) those in Table 8.12.2a, (2) 0.40925 times those in Table 8.12.2b, and (3) 0.33647 times those in Table 8.12.2c. These final balanced moments and end slopes agree very closely with those in the slope-deflection solution of Example 7.9.2.

8.13 Stiffness and Carry-Over Factors for Members with Variable Moment of Inertia

In Sec. 7.10, the flexibility and stiffness coefficients of a member with variable moment of inertia were defined. For a member ij with both ends hinged, if clockwise moments M_i and M_j are applied to the ith and jth ends respectively, the clockwise rotations ϕ_i and ϕ_j are [Eqs. (7.10.4a and b)]

$$\phi_i = f_{ii}\frac{L}{EI_c}M_i + f_{ij}\frac{L}{EI_c}M_j \qquad (8.13.1a)$$

$$\phi_j = f_{ji}\frac{L}{EI_c}M_i + f_{jj}\frac{L}{EI_c}M_j \qquad (8.13.1b)$$

in which f_{ii}, $f_{ij} = f_{ji}$, f_{jj} are defined as the flexibility coefficients, and for a member with constant moment of inertia, $f_{ii} = f_{jj} = +\frac{1}{3}$, $f_{ij} = f_{ji} = -\frac{1}{6}$. Conversely, for a member ij with both ends hinged, the clockwise moments M_i and M_j required to maintain clockwise rotations ϕ_i and ϕ_j at the ith and jth ends respectively are [Eqs. (7.10.5a and b)]

$$M_i = s_{ii}\frac{EI_c}{L}\phi_i + s_{ij}\frac{EI_c}{L}\phi_j \qquad (8.13.2a)$$

$$M_j = s_{ji}\frac{EI_c}{L}\phi_i + s_{jj}\frac{EI_c}{L}\phi_j \qquad (8.13.2b)$$

in which s_{ii}, $s_{ij} = s_{ji}$, s_{jj} are defined as the stiffness coefficients, and for a member with constant moment of inertia, $s_{ii} = s_{jj} = +4$, $s_{ij} = s_{ji} = +2$.

Since either of the two preceding sets of equations can be obtained from the other by solving simultaneously the two equations in the same set,

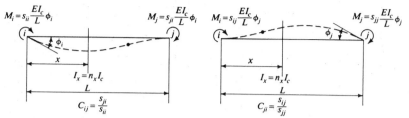

Figure 8.13.1 Stiffness and carry-over factors for members with variable moment of inertia.

relationships exist between the flexibility and stiffness coefficients such that [Eq. (7.10.6)]

$$s_{ii} = +\frac{f_{jj}}{f_{ii}f_{jj} - f_{ij}f_{ji}} \qquad s_{ij} = -\frac{f_{ij}}{f_{ii}f_{jj} - f_{ij}f_{ji}}$$

$$s_{ji} = -\frac{f_{ji}}{f_{ii}f_{jj} - f_{ij}f_{ji}} \qquad s_{jj} = +\frac{f_{ii}}{f_{ii}f_{jj} - f_{ij}f_{ji}}$$

(8.13.3)

In Sec. 8.3, the stiffness factor was defined as the moment at the near end to cause a unit rotation at the near end when the far end is fixed, while the carry-over factor is the ratio of the moment at the fixed far end to the moment at the rotating near end. In Fig. 8.13.1 are shown two elastic curves for the member with variable moment of inertia: in Fig. 8.13.1a the member has a ϕ_i, but ϕ_j is zero; and in Fig. 8.13.1b the member has a ϕ_j, but ϕ_i is zero. The moments at the ends of the member in Fig. 8.13.1a can be readily obtained by inserting a zero value for ϕ_j in Eqs. (8.13.2a and b); and in Fig. 8.13.1b, a zero value for ϕ_i. Thus, following the definition of stiffness and carry-over factors, from Fig. 8.13.1a,

$$\text{Stiffness factor at the } i\text{th end} = s_{ii}\frac{EI_c}{L} \qquad (8.13.4a)$$

$$\text{Carry-over factor from } i\text{th end to } j\text{th end} = C_{ij} = \frac{s_{ji}}{s_{ii}} \qquad (8.13.4b)$$

and, from Fig. 8.13.1b,

$$\text{Stiffness factor at the } j\text{th end} = s_{jj}\frac{EI_c}{L} \qquad (8.13.5a)$$

$$\text{Carry-over factor from } j\text{th end to } i\text{th end} = C_{ji} = \frac{s_{ij}}{s_{jj}} \qquad (8.13.5b)$$

8.14 Fixed-End Moments due to Member-Axis Rotation for Members with Variable Moment of Inertia

The fixed-end moments M_{0i} and M_{0j} due to a clockwise member-axis rotation R for a member with variable moment of inertia, as shown by Fig. 8.14.1a, may be conveniently obtained by applying the slope-deflection equations

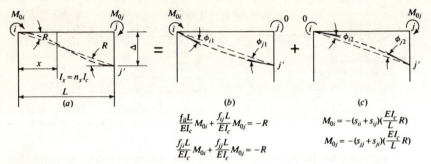

$$\frac{f_{ii}L}{EI_c} M_{0i} + \frac{f_{ij}L}{EI_c} M_{0j} = -R$$

$$\frac{f_{ji}L}{EI_c} M_{0i} + \frac{f_{jj}L}{EI_c} M_{0j} = -R$$

$$M_{0i} = -(s_{ii} + s_{ij})(\frac{EI_c}{L} R)$$

$$M_{0j} = -(s_{jj} + s_{ji})(\frac{EI_c}{L} R)$$

Figure 8.14.1 Fixed-end moments due to member-axis rotation for members with variable moment of inertia.

[Eqs. (7.10.1a and b)]

$$M_i = M_{0i} + s_{ii} \frac{EI_c}{L} (\theta_A - R) + s_{ij} \frac{EI_c}{L} (\theta_B - R)$$

$$M_j = M_{0j} + s_{ji} \frac{EI_c}{L} (\theta_A - R) + s_{jj} \frac{EI_c}{L} (\theta_B - R)$$

The M_i and M_j on the left sides of the preceding two equations become the M_{0i} and M_{0j} in Fig. 8.14.1a; the M_{0i} and M_{0j} on the right sides should be zero because there is no applied load on the member; and the θ_A and θ_B on the right sides are zero. Thus,

$$M_{0i} \text{ (Fig. 8.14.1a)} = -(s_{ii} + s_{ij}) \frac{EI_c}{L} R$$

$$= -s_{ii}(1 + C_{ij}) \frac{EI_c}{L} R \quad (8.14.1a)$$

$$M_{0j} \text{ (Fig. 8.14.1a)} = -(s_{jj} + s_{ji}) \frac{EI_c}{L} R$$

$$= -s_{jj}(1 + C_{ji}) \frac{EI_c}{L} R \quad (8.14.1b)$$

These two equations can also be obtained by basic considerations of the definitions of the flexibility coefficients. Resolve the condition of Fig. 8.14.1a into two parts as shown by Fig. 8.14.1b and c; then the clockwise rotations ϕ_i and ϕ_j from the member axis ij' to the tangents to the elastic curve at i and j are

$$\phi_i = \phi_{i1} + \phi_{i2} = f_{ii} \frac{L}{EI_c} M_{0i} + f_{ij} \frac{L}{EI_c} M_{0j} \quad (8.14.2a)$$

$$\phi_j = \phi_{j1} + \phi_{j2} = f_{ji} \frac{L}{EI_c} M_{0i} + f_{jj} \frac{L}{EI_c} M_{0j} \quad (8.14.2b)$$

But ϕ_i and ϕ_j, the clockwise rotations at i and j measured from the member axis ij' to the tangents to the elastic curve at i and j, are each equal to $-R$ in

Fig. 8.14.1a. Making this substitution in Eqs. (8.14.2a and b) and solving for M_{0i} and M_{0j},

$$M_{0i} = \frac{-f_{jj} + f_{ij}}{f_{ii}f_{jj} - f_{ij}f_{ji}} \frac{EI_c}{L} R = -(s_{ii} + s_{ij}) \frac{EI_c}{L} R$$

$$M_{0j} = \frac{-f_{ii} + f_{ji}}{f_{ii}f_{jj} - f_{ij}f_{ji}} \frac{EI_c}{L} R = -(s_{jj} + s_{ji}) \frac{EI_c}{L} R$$

which are the same as Eqs. (8.14.1a and b).

8.15 Moment Distribution Involving Members with Variable Moment of Inertia

When a moment distribution involves a member with variable moment of inertia, the stiffness factor at each end of that member is no longer equal to $4EI/L$ but must be computed from Eq (8.13.4a) or Eq. (8.13.5a); also, the carry-over factors from i to j and from j to i are no longer equal to $+\frac{1}{2}$ but must be computed from Eqs. (8.13.4b) and (8.13.5b). Furthermore, the fixed-end moments due to member-axis rotation are no longer equal to $-6EIR/L$ but must be computed from Eqs. (8.14.1a and b). Other than the necessary cautions required in applying the aforementioned formulas, the procedures in moment distribution are the same as those previously discussed.

So far as the check procedure in moment distribution is concerned, the member-end slopes θ_A and θ_B can be solved from the general slope-deflection equations [Eqs. (7.10.1a and b)]

$$M_i = M_{0i} + s_{ii} \frac{EI_c}{L} (\theta_A - R) + s_{ij} \frac{EI_c}{L} (\theta_B - R)$$

$$M_j = M_{0j} + s_{ji} \frac{EI_c}{L} (\theta_A - R) + s_{jj} \frac{EI_c}{L} (\theta_B - R)$$

Solving the above two equations for θ_A and θ_B,

$$\theta_A = \frac{\left\{M_i - \left[M_{0i} - (s_{ii} + s_{ij}) \frac{EI_c}{L} R\right]\right\} + (-C_{ji})\left\{M_j - \left[M_{0j} - (s_{jj} + s_{ji}) \frac{EI_c}{L} R\right]\right\}}{s_{ii}(1 - C_{ij}C_{ji})(EI_c/L)}$$

(8.15.1a)

$$\theta_B = \frac{\left\{M_j - \left[M_{0j} - (s_{jj} + s_{ji}) \frac{EI_c}{L} R\right]\right\} + (-C_{ij})\left\{M_i - \left[M_{0i} - (s_{ii} + s_{ij}) \frac{EI_c}{L} R\right]\right\}}{s_{jj}(1 - C_{ij}C_{ji})(EI_c/L)}$$

(8.15.1b)

Equations (8.15.1a and b) can be expressed in a single formula as

$$\theta_{near} = \frac{(\text{Change}_{near\ end}) + (-C_{far\ to\ near})(\text{Change}_{far\ end})}{s_{near\ end}(1 - C_{ij}C_{ji})(EI_c/L)}$$

(8.15.2)

330 INTERMEDIATE STRUCTURAL ANALYSIS

in which the "change" is equal to the final end moment minus the combined fixed-end moment due to applied load on the member and due to member-axis rotation.

So far as the modified stiffness factors are concerned, since the modified stiffness factor has been defined as the moment at the near end to cause a unit rotation at the near end when the far end is hinged, the following derivations can be made:

Case 1 When $\theta_i \neq 0$ and $M_j = 0$,

$$M_i = s'_{ii} \frac{EI_c}{L} \theta_i \quad (8.15.3)$$

in which s'_{ii} is the modified stiffness factor at the ith end. But by definition of the flexibility coefficient f_{ii},

$$\theta_i = f_{ii} \frac{M_i L}{EI_c} \quad (8.15.4)$$

Substituting Eq. (8.15.4) into Eq. (8.15.3),

$$s'_{ii} = \frac{1}{f_{ii}} = \frac{s_{ii}s_{jj} - s_{ij}s_{ji}}{s_{jj}} = s_{ii}(1 - C_{ij}C_{ji}) \quad (8.15.5)$$

Case 2 When $\theta_j \neq 0$ and $M_i = 0$,

$$M_j = s'_{jj} \frac{EI_c}{L} \theta_j \quad (8.15.6)$$

in which s'_{jj} is the modified stiffness factor at the jth end. But by definition of the flexibility coefficient f_{jj},

$$\theta_j = f_{jj} \frac{M_j L}{EI_c} \quad (8.15.7)$$

Substituting Eq. (8.15.7) into Eq. (8.15.6),

$$s'_{jj} = \frac{1}{f_{jj}} = \frac{s_{ii}s_{jj} - s_{ij}s_{ji}}{s_{ii}} = s_{jj}(1 - C_{ij}C_{ji}) \quad (8.15.8)$$

Equations (8.15.5) and (8.15.8) may be summarized in one statement; the modified stiffness factor is the product of the regular stiffness factor and a reduction factor equal to 1 minus the product of the carry-over factors.

Example 8.15.1 Analyze the continuous beam of Example 7.10.1, shown again in Fig. 8.15.1, by the moment-distribution method.

SOLUTION The fixed-end moments due to the applied loads on span AB, and the stiffness coefficients for both spans AB and BC are taken from Example 7.10.1 and repeated in Fig. 8.15.1b and c. The stiffness and carry-over factors for spans AB and BC are as follows:

THE MOMENT-DISTRIBUTION METHOD 331

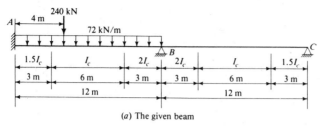

(a) The given beam

| $s_{AA} = +5.64460$ | $s_{BA} = +3.47840$ | $s_{BB} = +6.39444$ | $s_{CB} = +3.47840$ |
| $s_{AB} = +3.47840$ | $s_{BB} = +6.39444$ | $s_{BA} = +3.47840$ | $s_{CC} = +5.64460$ |

(b) Stiffness coefficients

-1338.89 $+1277.20$ 0
  0

(c) Fixed-end moments

Figure 8.15.1 Continuous beam of Example 8.15.1.

Span AB:

$$S_A = s_{AA}\frac{EI}{L} = 5.64460\frac{EI_c}{12} = 0.4704 EI_c$$

$$S_B = s_{BB}\frac{EI}{L} = 6.39444\frac{EI_c}{12} = 0.5329 EI_c$$

$$C_{AB} = \frac{s_{BA}}{s_{AA}} = \frac{3.47840}{5.64460} = 0.6162$$

$$C_{BA} = \frac{s_{AB}}{s_{BB}} = \frac{3.47840}{6.39444} = 0.5440$$

Span BC:

$$S_B = s_{BB}\frac{EI}{L} = 6.39444\frac{EI_c}{12} = 0.5329 EI_c$$

$$S_C = s_{CC}\frac{EI}{L} = 5.64460\frac{EI_c}{12} = 0.4704 EI_c$$

$$C_{BC} = \frac{s_{CB}}{s_{BB}} = \frac{3.47840}{6.39444} = 0.5440$$

$$C_{CB} = \frac{s_{BC}}{s_{CC}} = \frac{3.47840}{5.64460} = 0.6162$$

The moment distribution by the regular stiffness method and the check procedure are shown in Table 8.15.1a. Note particularly the carry-overs in this table; for instance, the -348 in column AB is 0.5440 times the -639 in column BA, and the $+214$ in column BC is 0.6162 times the $+347$ in column CB. In the check procedure, Eq. (8.15.2) is used to compute the slopes at the four member ends.

When the support at C is treated as a simple support, the modified moment distribution stops at the end of the second cycle, as shown in Table 8.15.1b. In this problem, because there is no load on span BC, the modified fixed-end moment at B of span BC is still equal to zero.

Table 8.15.1a Moment distribution by regular stiffness method, Example 8.15.1

Joint		A	B		C
Member		AB	BA	BC	CB
S (in terms of EI_c)		0.4704	0.5329	0.5329	0.4704
DF		....	0.5000	0.5000	1.0000
Cycle	COF	0.6162	0.5440	0.5440	0.6162
1	FEM	−1339	+1277	0	0
	BAL	0	− 639	− 638	0
2	CO	− 348	0	0	−347
	BAL	0	0	0	+347
3	CO	0	0	+ 214	0
	BAL	0	− 107	− 107	0
4	CO	− 58	0	0	− 58
	BAL	0	0	0	+ 58
5	CO	0	0	+ 36	0
	BAL	0	− 18	− 18	0
6	CO	− 10	0	0	− 10
	BAL	0	0	0	+ 10
7	CO	0	0	+ 6	0
	BAL	0	− 3	− 3	0
8	CO	− 1	0	0	− 1
	BAL	0	0	0	+ 1
Total balanced M		−1756	+ 510	− 510	0
Check	Change	− 417	− 767	− 510	0
	−(COF)(change)	+ 417	+ 257	0	+277
	Sum	0	− 510	− 510	+277
	$S(1-CC)$	0.3127	0.3543	0.3543	0.3127
	$EI_c\theta$	0	−1439	−1439	+886

THE MOMENT-DISTRIBUTION METHOD 333

Table 8.15.1b Moment distribution by modified stiffness method, Example 8.15.1†

Joint		A	B		C
Member		AB	BA	BC	CB
S (in terms of EI_c)		0.4704	0.5329	0.5329	0.4704
$S' = S(1 - CC)$				0.3543	
Modified DF		....	0.6007	0.3993	....
COF		0.6162	0.5440	0.5440	0.6162
FEM BAL		−1339	+1277	0 0	0 0
CO BAL				0 0	0 0
Cycle 1	Mod. FEM BAL	−1339 0	+1277 − 767	0 −510	0 ↑
2	CO BAL	− 417 0	0 0	0 0	↓
Total balanced M		−1756	+ 510	−510	0

†For check, See Table 8.15.1a

Example 8.15.2 Analyze the continuous beam of Example 7.10.2, shown again in Fig. 8.15.2, by the moment-distribution method.

SOLUTION The stiffness coefficients for both spans AB and BC are taken from Example 7.10.1 and repeated in Fig. 8.15.2b. The fixed-end moments due to the known member-axis rotations of spans AB and BC are computed by using Eqs. (8.14.1a and b); thus

$$M_{FAB} = -(s_{AA} + s_{AB})\left(\frac{EI}{L} R\right)$$

$$= -(5.64460 + 3.47840)\left[\frac{200(3600)}{12}\right]\left(+\frac{0.015}{12}\right)$$

$$= -684.22 \text{ kN·m}$$

$$M_{FBA} = -(s_{BB} + s_{BA})\left(\frac{EI}{L} R\right)$$

$$= -(6.39444 + 3.47840)\left[\frac{200(3600)}{12}\right]\left(+\frac{0.015}{12}\right)$$

$$= -740.46 \text{ kN·m}$$

Table 8.15.2a Moment distribution by regular stiffness method, Example 8.15.2

Joint		A	B		C
Member		AB	BA	BC	CB
S, kN·m		338,690	383,690	383,690	338,690
DF		....	0.5000	0.5000	1.0000
Cycle	COF	0.6162	0.5440	0.5440	0.6162
1	FEM	−684.2	−740.5	+740.5	+684.2
	BAL	0.0	0.0	0.0	−684.2
2	CO	0.0	0.0	−421.6	0.0
	BAL	0.0	+210.8	+210.8	0.0
3	CO	+114.7	0.0	0.0	+114.7
	BAL	0.0	0.0	0.0	−114.7
4	CO	0.0	0.0	−70.7	0.0
	BAL	0.0	+35.4	+35.3	0.0
5	CO	+19.2	0.0	0.0	+19.2
	BAL	0.0	0.0	0.0	−19.2
6	CO	0.0	0.0	−11.8	0.0
	BAL	0.0	+5.9	+5.9	0.0
7	CO	+3.2	0.0	0.0	+3.2
	BAL	0.0	0.0	0.0	−3.2
8	CO	0.0	0.0	−2.0	0.0
	BAL	0.0	+1.0	+1.0	0.0
9	CO	+0.5	0.0	0.0	+0.5
	BAL	0.0	0.0	0.0	−0.5
10	CO	0.0	0.0	−0.3	0.0
	BAL	0.0	+0.2	+0.1	0.0
11	CO	+0.1	0.0	0.0	0.0
	BAL	0.0	0.0	0.0	0.0
Total balanced M		−546.5	−487.2	+487.2	0.0
Check	Change	+137.7	+253.3	−253.3	−684.2
	−(COF)(change)	−137.8	−84.8	+421.6	+137.8
	Sum	−0.1	+168.5	+168.3	−546.4
	$S(1-CC)$	225,160	255,080	255,080	225,160
	$\theta, 10^{-3}$ rad	≈0	+0.661	+0.660	−2.427

Table 8.15.2b Moment distribution by modified stiffness method, Example 8.15.2†

Joint		A	B		C
Member		AB	BA	BC	CB
S, kN·m		338,690	383,690	383,690	338,690
$S' = S(1 - CC)$				255,080	
Modified DF		....	0.6007	0.3993	....
COF		0.6162	0.5440	0.5440	0.6162
FEM BAL		−684.2	−740.5	+740.5 0.0	+684.2 −684.2
CO BAL				−421.6 0.0	0.0 0.0
Cycle 1	Mod. FEM BAL	−684.2 0.0	−740.5 +253.3	+318.9 +168.3	0.0
2	CO BAL	+137.8 0.0	0.0 0.0	0.0 0.0	
Total balanced M		−546.4	−487.2	+487.2	0.0

†For check, see Table 8.15.2a.

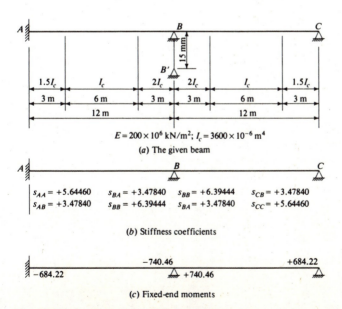

(a) The given beam

$s_{AA} = +5.64460 \quad s_{BA} = +3.47840 \quad s_{BB} = +6.39444 \quad s_{CB} = +3.47840$
$s_{AB} = +3.47840 \quad s_{BB} = +6.39444 \quad s_{BA} = +3.47840 \quad s_{CC} = +5.64460$

(b) Stiffness coefficients

−684.22   −740.46   +684.22
          +740.46

(c) Fixed-end moments

Figure 8.15.2 Continuous beam of Example 8.15.2.

$$M_{FBC} = -(s_{BB} + s_{BC})\left(\frac{EI}{L}R\right)$$

$$= -(6.39444 + 3.47840)\left[\frac{200(3600)}{12}\right]\left(-\frac{0.015}{12}\right)$$

$$= +740.46 \text{ kN} \cdot \text{m}$$

$$M_{FCB} = -(s_{CC} + s_{CB})\left(\frac{EI}{L}R\right)$$

$$= -(5.64460 + 3.47840)\left[\frac{200(3600)}{12}\right]\left(-\frac{0.015}{12}\right)$$

$$= +684.22 \text{ kN} \cdot \text{m}$$

The preceding fixed-end moments are shown in Fig. 8.15.2c.

The moment distribution by both the regular stiffness and modified stiffness methods are shown in Table 8.15.2a and b. The check procedure, however, is shown only in Table 8.15.2a.

8.16 Exercises

8.1 to **8.35** Solve the 35 exercises in Chapter 7 by the moment-distribution method. In those exercises where there is at least one exterior simple or hinged support, the moment distribution may be performed by either the regular stiffness method or the modified stiffness method or both methods.

CHAPTER
NINE
MATRIX OPERATIONS

9.1 Matrix Notation for Linear Equations

A set of linear equations expresses an array of dependent variables in terms of an array of independent variables by stating that each dependent variable is equal to the sum of the products of some coefficient and each independent variable. In the following equations

$$\begin{aligned} x_1 &= a_{11}y_1 + a_{12}y_2 + \cdots + a_{1m}y_m \\ x_2 &= a_{21}y_1 + a_{22}y_2 + \cdots + a_{2m}y_m \\ &\vdots \\ x_n &= a_{n1}y_1 + a_{n2}y_2 + \cdots + a_{nm}y_m \end{aligned} \qquad (9.1.1)$$

the m independent variables are $y_1, y_2, \ldots, y_m$ and the n dependent variables are $x_1, x_2, \ldots, x_n$. The coefficients are in the form a_{ij}, which is the contribution of the jth independent variable into the ith dependent variable. Thus if the values of the independent variables $y_1, y_2, \ldots, y_m$ are assigned, the values of the dependent variables $x_1, x_2, \ldots, x_n$ can be computed by using the given set of linear equations, Eqs. (9.1.1).

Equations (9.1.1), shown in the ordinary algebraic form, may be written in matrix notation as

$$\begin{Bmatrix} x_1 \\ x_2 \\ \vdots \\ x_n \end{Bmatrix} = \begin{bmatrix} a_{11} & a_{12} & \cdots & a_{1m} \\ a_{21} & a_{22} & \cdots & a_{2m} \\ \vdots & & & \vdots \\ a_{n1} & a_{n2} & \cdots & a_{nm} \end{bmatrix} \begin{Bmatrix} y_1 \\ y_2 \\ \vdots \\ y_m \end{Bmatrix} \qquad (9.1.2)$$

338 INTERMEDIATE STRUCTURAL ANALYSIS

One can see that the left side of Eq. (9.1.2) is identical to the left side of Eqs. (9.1.1) except that a pair of braces has been added. The coefficients in Eqs. (9.1.1), however, have been arranged in Eq. (9.1.2) as a rectangular block of numbers enclosed in a pair of brackets. This rectangular block of members is called a *matrix*. The remaining part of Eq. (9.1.2) is simply the enclosure of a vertical arrangement of the independent variables in a pair of braces again. In fact the vertical arrangement of the x's (or of the y's) in Eq. (9.1.2) is itself a matrix; except for this special rectangular block there is only one column of numbers. A matrix with only a single column is called a *column matrix*. It is common to enclose column matrices in braces and rectangular matrices in brackets.

Equation (9.1.2) can be written in an abbreviated form as

$$\{X\}_{n\times 1} = [A]_{n\times m}\{Y\}_{m\times 1} \qquad (9.1.3a)$$

although it will still be necessary to define the matrices $\{X\}$, $[A]$, and $\{Y\}$ as

$$\{X\}_{n\times 1} = \begin{Bmatrix} x_1 \\ x_2 \\ \cdot \\ x_n \end{Bmatrix} \qquad [A]_{n\times m} = \begin{bmatrix} a_{11} & a_{12} & \cdots & a_{1m} \\ a_{21} & a_{22} & \cdots & a_{2m} \\ \cdots\cdots\cdots\cdots\cdots \\ a_{n1} & a_{n2} & \cdots & a_{nm} \end{bmatrix} \qquad \{Y\}_{m\times 1} = \begin{Bmatrix} y_1 \\ y_2 \\ \cdot \\ y_m \end{Bmatrix}$$

$$(9.1.3b)$$

In this textbook the matrix notation for a set of linear equations such as Eqs. (9.1.1) will be exhibited as

$$\{X\}_{n\times 1} = [A]_{n\times m}\{Y\}_{m\times 1} \qquad (9.1.4a)$$

in which

$$[A]_{n\times m} = \begin{array}{|c|c|c|c|c|} \hline {}_x\backslash{}^y & 1 & 2 & \cdots & m \\ \hline 1 & a_{11} & a_{12} & \cdots & a_{1m} \\ \hline 2 & a_{21} & a_{22} & \cdots & a_{2m} \\ \hline \cdot & \cdots & \cdots & \cdots & \cdots \\ \hline n & a_{n1} & a_{n2} & \cdots & a_{nm} \\ \hline \end{array} \qquad (9.1.4b)$$

Note the addition in Eq. (9.1.4b) of the extreme left column showing the row numbers of the dependent variable, and of the extreme upper row showing the column numbers of the independent variable. Thus the reader may look at this rectangular diagram and immediately interpret its meaning to be that of the ordinary algebraic form, Eqs. (9.1.1).

Example 9.1.1 Write into the matrix form the set of linear equations shown below in the ordinary algebraic form

$$x_1 = 12y_1 - 2y_2 + 7y_3$$
$$x_2 = 4y_1 + 15y_2 - 3y_3$$
$$x_3 = 9y_1 + 5y_2 + 10y_3$$
$$x_4 = 6y_1 - 8y_2 + 11y_3$$

SOLUTION

$$\{X\}_{4\times 1} = [A]_{4\times 3}\{Y\}_{3\times 1}$$

in which

$$[A]_{4\times 3} = $$

y \ x	1	2	3
1	+12	−2	+7
2	+4	+15	−3
3	+9	+5	+10
4	+6	−8	+11

9.2 Matrix Multiplication as Means of Elimination of Intermediate Variables in Two Sets of Linear Equations

Consider two sets of linear equations: the first set expresses l values of x as linear functions of m values of y, and the second set expresses the m values of y as linear functions of n values of z. In the ordinary algebraic form, these two sets of linear equations are:

$$x_1 = a_{11}y_1 + a_{12}y_2 + \cdots + a_{1m}y_m$$
$$x_2 = a_{21}y_1 + a_{22}y_2 + \cdots + a_{2m}y_m$$
$$\vdots$$
$$x_l = a_{l1}y_1 + a_{l2}y_2 + \cdots + a_{lm}y_m$$

(9.2.1)

and

$$y_1 = b_{11}z_1 + b_{12}z_2 + \cdots + b_{1n}z_n$$
$$y_2 = b_{21}z_1 + b_{22}z_2 + \cdots + b_{2n}z_n$$
$$\vdots$$
$$y_m = b_{m1}z_1 + b_{m2}z_2 + \cdots + b_{mn}z_n$$

(9.2.2)

Now suppose that it is desired to express the l values of x directly as linear functions of n values of z. This can be achieved by substituting into Eqs. (9.2.1) the expressions for the y's in Eqs. (9.2.2). The results will be in the ordinary algebraic form

$$x_1 = c_{11}z_1 + c_{12}z_2 + \cdots + c_{1n}z_n$$
$$x_2 = c_{21}z_1 + c_{22}z_2 + \cdots + c_{2n}z_n$$
$$\vdots$$
$$x_l = c_{l1}z_1 + c_{l2}z_2 + \cdots + c_{ln}z_n$$
(9.2.3)

Equations (9.2.1), (9.2.2), and (9.2.3) can be written in matrix notation as

$$\{X\}_{l\times1} = [A]_{l\times m}\{Y\}_{m\times1} \quad (9.2.4)$$

$$\{Y\}_{m\times1} = [B]_{m\times n}\{Z\}_{n\times1} \quad (9.2.5)$$

$$\{X\}_{l\times1} = [C]_{l\times n}\{Z\}_{n\times1} \quad (9.2.6)$$

By substituting Eq. (9.2.5) into Eq. (9.2.4), the following matrix equation is obtained.

$$\{X\}_{l\times1} = [A]_{l\times m}[B]_{m\times n}\{Z\}_{n\times1} \quad (9.2.7)$$

When Eq. (9.2.7) is compared with Eq. (9.2.6), one sees that

$$[C]_{l\times n} = [A]_{l\times m}[B]_{m\times n} \quad (9.2.8)$$

Equation (9.2.8) can be interpreted to mean that the matric [C] is the product of a *premultiplier* matrix [A] and a *postmultiplier* matrix [B]. Thus matrix multiplication is hereby shown to be the means of eliminating the intermediate variables y_1 to y_m so that the variables x_1 to x_l are expressed directly in terms of the variables z_1 to z_n. From this explanation one can conclude that in any matrix multiplication the number of columns in the premultiplier must be equal to the number of rows in the postmultiplier.

9.3 Rule for Matrix Multiplication

The problem is now to develop a systematic way of computing the elements in the [C] matrix as defined in the preceding section. In the ordinary algebraic form, let the first system be

$$x_1 = +12y_1 - 2y_2 + 7y_3$$
$$x_2 = + 4y_1 + 15y_2 - 3y_3$$
$$x_3 = + 9y_1 + 5y_2 + 10y_3$$
$$x_4 = + 6y_1 - 8y_2 + 11y_3$$
(9.3.1)

and the second system be

$$y_1 = + 6z_1 - 11z_2$$
$$y_2 = +10z_1 + 7z_2$$
$$y_3 = + 9z_1 + 5z_2$$
(9.3.2)

Substituting Eqs. (9.3.2) in Eqs. (9.3.1),

$$x_1 = +12(6z_1 - 11z_2) - 2(10z_1 + 7z_2) + 7(9z_1 + 5z_2) = +115z_1 - 111z_2$$
$$x_2 = + 4(6z_1 - 11z_2) + 15(10z_1 + 7z_2) - 3(9z_1 + 5z_2) = +147z_1 + 46z_2$$
$$x_3 = + 9(6z_1 - 11z_2) + 5(10z_1 + 7z_2) + 10(9z_1 + 5z_2) = +194z_1 - 14z_2$$
$$x_4 = + 6(6z_1 - 11z_2) - 8(10z_1 + 7z_2) + 11(9z_1 + 5z_2) = + 55z_1 - 67z_2$$
(9.3.3)

The arithmetic involved in Eqs. (9.3.3) may be performed in a diagrammatic arrangement of matrix multiplication as shown in Fig. 9.3.1. From this figure it may be observed that the element c_{ij} is equal to the sum of three products, each obtained by multiplying the element on the ith row of the $[A]$ matrix starting from the extreme left and the corresponding element on the jth column of the $[B]$ matrix starting from the top. For instance,

$$c_{32} = (+9)(-11) + (+5)(+7) + (+10)(+5) = -14$$

The pairing is done from the outer side toward the inner side, as shown by the two dashed lines with arrowheads in Fig. 9.3.1, thus the name *inner product rule*. The inner product rule for matrix multiplication in which $[A]_{l \times m}$ is the premultiplier, $[B]_{m \times n}$ is the postmultiplier, and the product $[C]_{l \times n}$ is equal to

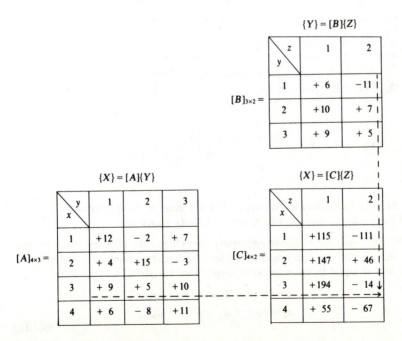

Figure 9.3.1 Matrix multiplication

$[A]_{l \times m}[B]_{m \times n}$, may be stated as

$$c_{ij} = \sum_{k=1}^{k=m} a_{ik} * b_{kj} \quad (9.3.4)$$

The justification for the inner product rule can be observed simply by comparing the ways in which, say, c_{32} is obtained in Eqs. (9.3.3) and in Fig. 9.3.1.

Example 9.3.1 Given the $[A]$, $[B]$, and $[C]$ matrices as shown in Fig. 9.3.1, and given $z_1 = -5$, $z_2 = +3$, find $\{X\}_{4 \times 1}$: (a) by $\{X\}_{4 \times 1} = [C]_{4 \times 2}\{Z\}_{2 \times 1}$, and (b) by $\{Y\}_{3 \times 1} = [B]_{3 \times 2}\{Z\}_{2 \times 1}$ and then $\{X\}_{4 \times 1} = [A]_{4 \times 3}\{Y\}_{3 \times 1}$.

SOLUTION

(a) $\{X\}_{4 \times 1} = [C]_{4 \times 2}\{Z\}_{2 \times 1} = \begin{bmatrix} +115 & -111 \\ +147 & +46 \\ +144 & -14 \\ +55 & -67 \end{bmatrix} \begin{Bmatrix} -5 \\ +3 \end{Bmatrix}$

$= \begin{Bmatrix} (+115)(-5) + (-111)(+3) = -908 \\ (+147)(-5) + (+46)(+3) = -597 \\ (+144)(-5) + (-14)(+3) = -1012 \\ (+55)(-5) + (-67)(+3) = -476 \end{Bmatrix}$

Note that the above matrix multiplication could have been shown on a diagram such as in Fig. 9.3.1; but for such a simple case and to save space, the usual mathematical representation is used. In fact, the arithmetic shown to the left of the equals sign within the last column matrix shown above will not be similarly shown in the remainder of this textbook.

(b) $\{Y\}_{3 \times 1} = [B]_{3 \times 2}\{Z\}_{2 \times 1} = \begin{bmatrix} +6 & -11 \\ +10 & +7 \\ +9 & +5 \end{bmatrix} \begin{Bmatrix} -5 \\ +3 \end{Bmatrix} = \begin{Bmatrix} -63 \\ -29 \\ -30 \end{Bmatrix}$

$\{X\}_{4 \times 1} = [A]_{4 \times 3}\{Y\}_{3 \times 1} = \begin{bmatrix} +12 & -2 & +7 \\ +4 & +15 & -3 \\ +9 & +5 & +10 \\ +6 & -8 & +11 \end{bmatrix} \begin{Bmatrix} -63 \\ -29 \\ -30 \end{Bmatrix} = \begin{Bmatrix} -908 \\ -597 \\ -1012 \\ -476 \end{Bmatrix}$

9.4 Matrix Inversion as Means of Interchange of Independent and Dependent Variables in a Set of Linear Equations

In the following set of linear equations, the x's are the independent variables and the y's are the dependent variables; further, the number of independent variables is equal to the number of dependent variables:

$$\begin{aligned} x_1 &= a_{11}y_1 + a_{12}y_2 + \cdots + a_{1n}y_n \\ x_2 &= a_{21}y_1 + a_{22}y_2 + \cdots + a_{2n}y_n \\ &\vdots \\ x_n &= a_{n1}y_1 + a_{n2}y_2 + \cdots + a_{nn}y_n \end{aligned} \quad (9.4.1a)$$

In the matrix form Eqs. (9.4.1a) become

$$\{X\}_{n\times1} = [A]_{n\times n}\{Y\}_{n\times1}$$

$$[A]_{m\times n} = \begin{array}{|c|c|c|c|c|} \hline {}_{x}\diagdown{}^{y} & 1 & 2 & \cdots & n \\ \hline 1 & a_{11} & a_{21} & \cdots & a_{1n} \\ \hline 2 & a_{21} & a_{22} & \cdots & a_{2n} \\ \hline \cdot & \cdots & \cdots & \cdots & \cdots \\ \hline n & a_{n1} & a_{n2} & \cdots & a_{nn} \\ \hline \end{array}$$

Most times it is possible (occasionally impossible as will be discussed later in this section) to reverse the role of the independent and dependent variables in Eqs. (9.4.1a) such that

$$\begin{aligned} y_1 &= b_{11}x_1 + b_{12}x_2 + \cdots + b_{1n}x_n \\ y_2 &= b_{21}x_1 + b_{22}x_2 + \cdots + b_{2n}x_n \\ &\vdots \\ y_n &= b_{n1}x_1 + b_{n2}x_2 + \cdots + b_{nn}x_n \end{aligned} \qquad (9.4.2a)$$

In the matrix form Eqs. (9.4.2a) become

$$\{Y\}_{n\times1} = [B]_{n\times n}\{X\}_{n\times1}$$

$$[B]_{n\times n} = \begin{array}{|c|c|c|c|c|} \hline {}_{y}\diagdown{}^{x} & 1 & 2 & \cdots & n \\ \hline 1 & b_{11} & b_{12} & \cdots & b_{1n} \\ \hline 2 & b_{21} & b_{22} & \cdots & b_{2n} \\ \hline \cdot & \cdots & \cdots & \cdots & \cdots \\ \hline n & b_{n1} & b_{n2} & \cdots & b_{nn} \\ \hline \end{array} \qquad (9.4.2b)$$

When an ordinary symbolic algebraic equation like $x = ay$ is to be solved for y, one can write either $y = x/a$ or $y = a^{-1}x$; but for the matrix equation $\{X\}_{n\times1} = [A]_{n\times n}\{Y\}_{n\times1}$, only the form

$$\{Y\}_{n\times1} = [A^{-1}]_{n\times n}\{X\}_{n\times1} \qquad (9.4.2c)$$

is permissible. Thus the $[B]$ matrix defined in Eq. (9.4.2b) is the same as the $[A^{-1}]$ matrix in Eq. (9.4.2c), or

$$[B]_{n\times n} = [A^{-1}]_{n\times n} \qquad (9.4.3)$$

The $[A^{-1}]$ matrix is called the *inverse* of the $[A]$ matrix, and the procedure of finding the contents of the $[A^{-1}]$ matrix from those of the $[A]$ matrix is called *matrix inversion*.

If Eq. (9.4.2c) is substituted into Eq. (9.4.1b),

$$\{X\}_{n\times 1} = [A]_{n\times n}\{Y\}_{n\times 1} = [A]_{n\times n}[A^{-1}]_{n\times n}\{X\} = [AA^{-1}]_{n\times n}\{X\}_{n\times 1}$$
(9.4.4a)

which must mean, in the ordinary algebraic form,

$$\begin{aligned} x_1 &= 1.0x_1 + 0.0x_2 + \cdots + 0.0x_n \\ x_2 &= 0.0x_1 + 1.0x_2 + \cdots + 0.0x_n \\ &\vdots \\ x_n &= 0.0x_1 + 0.0x_2 + \cdots + 1.0x_n \end{aligned}$$
(9.4.4b)

so long as each x is equal only to itself and is independent of the other values of x. Thus the $[AA^{-1}]_{n\times n}$ matrix is in the form of

$$\{X\}_{n\times 1} = [AA^{-1}]_{n\times n}\{X\}_{n\times 1}$$

$[AA^{-1}]_{n\times n} = $

x \ x	1	2	...	n
1	+1.0	0.0	...	0.0
2	0.0	+1.0	...	0.0
.	...	...	...	...
n	0.0	0.0	...	+1.0

$= [I]_{n\times n}$ (9.4.4c)

In Eq. (9.4.4c) the symbol $[I]$ is used for the *unit matrix* (sometimes also called the *identity matrix*), which is a square matrix containing $+1.0$ as each element on the main diagonal and zero at all positions off the main diagonal. Thus the product of a square matrix and its own inverse, if there is one, is always equal to a unit matrix.

Consider now the following set of linear equations:

$$\begin{aligned} x_1 &= 4y_1 - 3y_2 \\ x_2 &= 8y_1 - 6y_2 \end{aligned}$$
(9.4.5)

One can see that x_2 is always equal to twice x_1 regardless of the assigned values of the y's. In such a case, it would be impossible to find any set of values of y if the assigned values of x_1 and x_2 were not in the ratio of $1/2$. Even if the ratio of x_1 and x_2 is equal to $1/2$, the two equations in Eqs. (9.4.5) become identical, and one cannot solve for two unknown y's from a single equation. Consequently, Eqs. (9.4.5) are irreversible and the matrix

$$[A]_{2\times 2} = \begin{bmatrix} +4 & -3 \\ +8 & -6 \end{bmatrix}$$

has no inverse. For those readers who have knowledge of linear or matrix algebra, of course it is well known that a square matrix which has no inverse is called a *singular matrix* and that the determinant of a singular matrix is zero.

9.5 Method for Matrix Inversion

There are several ways of finding the inverse of a nonsingular square matrix; only one will be presented in this section. Let the given matrix be

$$[A]_{4\times 4} = \begin{array}{|c|c|c|c|c|} \hline {}_x\!\diagdown\!{}^y & 1 & 2 & 3 & 4 \\ \hline 1 & a_{11} & a_{12} & a_{13} & a_{14} \\ \hline 2 & a_{21} & a_{22} & a_{23} & a_{24} \\ \hline 3 & a_{31} & a_{32} & a_{33} & a_{34} \\ \hline 4 & a_{41} & a_{42} & a_{43} & a_{44} \\ \hline \end{array} \tag{9.5.1}$$

and its inverse be

$$[A^{-1}]_{4\times 4} = [B]_{4\times 4} = \begin{array}{|c|c|c|c|c|} \hline {}_y\!\diagdown\!{}^x & 1 & 2 & 3 & 4 \\ \hline 1 & b_{11} & b_{12} & b_{13} & b_{14} \\ \hline 2 & b_{21} & b_{22} & b_{23} & b_{24} \\ \hline 3 & b_{31} & b_{32} & b_{33} & b_{34} \\ \hline 4 & b_{41} & b_{42} & b_{43} & b_{44} \\ \hline \end{array} \tag{9.5.2}$$

Then in the ordinary algebraic form, Eq. (9.5.1) becomes

$$\begin{aligned} a_{11}y_1 + a_{12}y_2 + a_{13}y_3 + a_{14}y_4 &= x_1 \\ a_{21}y_1 + a_{22}y_2 + a_{23}y_3 + a_{24}y_4 &= x_2 \\ a_{31}y_1 + a_{32}y_2 + a_{33}y_3 + a_{34}y_4 &= x_3 \\ a_{41}y_1 + a_{42}y_2 + a_{43}y_3 + a_{44}y_4 &= x_4 \end{aligned} \tag{9.5.3}$$

and Eq. (9.5.2) becomes

$$\begin{aligned} y_1 &= b_{11}x_1 + b_{12}x_2 + b_{13}x_3 + b_{14}x_4 \\ y_2 &= b_{21}x_1 + b_{22}x_2 + b_{23}x_3 + b_{24}x_4 \\ y_3 &= b_{31}x_1 + b_{32}x_2 + b_{33}x_3 + b_{34}x_4 \\ y_4 &= b_{41}x_1 + b_{42}x_2 + b_{43}x_3 + b_{44}x_4 \end{aligned} \tag{9.5.4}$$

If, in Eqs. (9.5.3), only x_1 is equal to $+1$ and $x_2 = x_3 = x_4 = 0$, then the solution of the four simultaneous equations should yield

$$y_1 = b_{11}$$
$$y_2 = b_{21}$$
$$y_3 = b_{31}$$
$$y_4 = b_{41}$$
(9.5.5)

because Eqs. (9.5.5) are what one will get if x_1 is made equal to $+1$ and $x_2 = x_3 = x_4 = 0$ in Eqs. (9.5.4). Consequently, the jth column in the inverse matrix can be obtained by solving simultaneously the set of linear equations like Eqs. (9.5.3) after making x_j equal to $+1$ and all the other x's equal to zero. To obtain all columns in the inverse matrix, it is necessary only to solve simultaneously Eqs. (9.5.3) n times; each time the coefficients of the y's to the left side of the equals sign remain unchanged, but the column vectors to the right of the equals sign should take the form of a unit matrix.

Two common methods of solving simultaneous linear equations will be presented in Secs. 9.6 and 9.7. When no solution of the simultaneous equations described previously for the purpose of matrix inversion can be found, then one can conclude that the given matrix is a singular matrix.

9.6 Solution of Simultaneous Linear Equations by Forward Elimination and Backward Substitution

Consider the following set of four simultaneous linear equations in four unknowns:

$$
\begin{aligned}
+15x_1 - 3x_2 - 8x_3 + x_4 &= -50 \\
+ 2x_1 + 10x_2 - 6x_3 + 5x_4 &= +54 \\
+ 4x_1 - 5x_2 + 12x_3 - 3x_4 &= -89 \\
+ 3x_1 - 8x_2 + x_3 - 9x_4 &= -61
\end{aligned}
$$
(9.6.1)

These four equations can be reduced to three equations without x_1's by eliminating x_1's between the topmost equation and each of the three remaining equations. The topmost equation, used three times in the elimination process, is called the *pivotal equation*, and the coefficient of x_1 in this equation is called the *pivot*. Subtracting $+2/15$, $+4/15$, and $+3/15$ times the pivotal equation from the second, third, and fourth equations in Eqs. (9.6.1), respectively,

$$
\begin{aligned}
+10.4x_2 - 4.9333x_3 + 4.8666x_4 &= +60.666 \\
- 4.2x_2 + 14.1333x_3 - 3.2666x_4 &= -75.666 \\
- 7.4x_2 + 2.6000x_3 - 9.2000x_4 &= -51.000
\end{aligned}
$$
(9.6.2)

Equations (9.6.2) can be reduced to two equations without x_2's by eliminating x_2's between the topmost equation and each of the two remaining equations. The pivot is $+10.4$. Subtracting $-4.2/10.4$ and $-7.4/10.4$ times the pivotal equation from the second and third equations in Eqs. (9.6.2), respectively,

$$+12.1410x_3 - 1.3012x_4 = -51.666$$
$$- 0.9103x_3 - 5.7372x_4 = -7.8333$$
(9.6.3)

Equations (9.6.3) can be reduced to one equation without x_3's by eliminating x_3's between the top equation and the bottom equation. The pivot is $+12.1410$. Subtracting $-0.9103/12.1410$ times the pivotal equation from the second equation in Eqs. (9.6.3),

$$-5.8347x_4 = -11.6692 \qquad (9.6.4)$$

Solving Eq. (9.6.4) for x_4,

$$x_4 = +2.0000 \qquad (9.6.5)$$

Substituting the value of x_4 in Eq. (9.6.5) into the topmost (or pivotal) equation in Eqs. (9.6.3) and solving for x_3,

$$x_3 = -4.0000 \qquad (9.6.6)$$

Substituting the values of x_4 and x_3 in Eqs. (9.6.5) and (9.6.6) into the topmost (or pivotal) equation in Eqs. (9.6.2) and solving for x_2,

$$x_2 = +3.0000 \qquad (9.6.7)$$

Substituting the values of x_4, x_3, and x_2 in Eqs. (9.6.5), (9.6.6), and (9.6.7) into the topmost (or pivotal) equation in Eqs. (9.6.1) and solving for x_1,

$$x_1 = -5.0000 \qquad (9.6.8)$$

In the foregoing procedure, Eqs. (9.6.1) to (9.6.5) constitute the forward elimination and Eqs. (9.6.6) to (9.6.8) constitute the backward substitution. The results of $x_1 = -5$, $x_2 = +3$, $x_3 = -4$, and $x_4 = +2$ happen to be the same values by which the original set of equations was established in setting up the numerical example. Often round-off errors may occur in the process. In fact, the accuracy can be improved, or the round-off errors reduced, if just before each forward elimination, the coefficient of any x with the largest absolute value is chosen as the pivot. Then the horizontal row containing the pivot is used as the pivotal equation in order to eliminate the vertical column containing the pivot. The pivotal equation is not only used in the forward elimination to combine with each of the remaining equations but also in the backward substitution to obtain the value of the variable in the pivotal column.

In Eqs. (9.6.1), the coefficient $+15$ at the upper left corner has the largest absolute value of the 16 eligible candidates for the pivot, and it has been used to obtain Eqs. (9.6.2). But in Eqs. (9.6.2), the coefficient $+14.1333$ has the largest absolute value of the nine candidates for the pivot. Had it been used as the pivot instead of the coefficient $+10.4$, the second equation in Eqs. (9.6.2) would be used as the pivotal equation to eliminate x_3's in the first and third equations in Eqs. (9.6.2). The results of this elimination would be

$$-8.9340x_2 + 3.1679x_4 = +34.254$$
$$-6.0387x_2 - 8.5991x_4 = -37.080 \qquad (9.6.9)$$

In the 4×4 original square matrix containing 16 positions, the locations of the four pivots used in the four forward eliminations may appear anywhere so long as there is only one pivot in each row or column; some possible combinations are shown in Fig. 9.6.1. In Fig. 9.6.1a, the pivots proceed along the main diagonal in sequence; in Fig. 9.6.1b, the pivots stay on the main diagonal but are not in sequence; and in Fig. 9.6.1c, the pivots scatter around the entire panel.

In Table 9.6.1, a unit matrix is installed beyond the column of constants that have been used in Eqs. (9.6.1) to (9.6.8). The forward eliminations are performed using pivots on the main diagonal in sequence; and the backward substitutions yield the results of the solutions for the simultaneous equations not only for the column vector of $-50, +54, -89$, and -61, but also for the unit matrix as described in the method for matrix inversion. From Table 9.6.1, it can be seen that

$$\begin{bmatrix} +15 & -3 & -8 & +1 \\ +2 & +10 & -6 & +5 \\ +4 & -5 & +12 & -3 \\ +3 & -8 & +1 & -9 \end{bmatrix}^{-1} \begin{bmatrix} +0.042440 & +0.051941 & +0.052936 & +0.015927 \\ -0.048140 & +0.164962 & +0.044430 & +0.071487 \\ -0.020541 & +0.019637 & +0.080988 & -0.018368 \\ +0.054656 & -0.127139 & -0.012849 & -0.171388 \end{bmatrix}$$

$$(9.6.10)$$

As a check on the correctness of the results shown in Eq. (9.6.10), one may multiply the original matrix by its inverse and see if a unit matrix is indeed obtained. In this particular case,

$$[AA^{-1}] = \begin{bmatrix} +1.000004 & -0.000006 & -0.000003 & 0. \\ +0.000006 & +0.999985 & -0.000001 & -0.000008 \\ 0. & +0.000015 & +0.999997 & +0.000021 \\ -0.000005 & +0.000015 & -0.000003 & +1.000009 \end{bmatrix} \qquad (9.6.11)$$

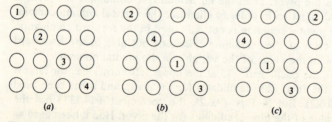

Figure 9.6.1 Possible pivot locations in 4×4 panel.

Table 9.6.1 Solution of simultaneous linear equations by forward elimination and backward substitution

	x_1	x_2	x_3	x_4	=				
Eqs. (9.6.1)	+15	−3	−8	+1	−50	+1	0	0	0
	+2	+10	−6	+5	+54	0	+1	0	0
	+4	−5	+12	−3	−89	0	0	+1	0
	+3	−8	+1	−9	−61	0	0	0	+1
Eqs. (9.6.2)		+10.4	−4.9333	+4.8666	+60.666	−0.1333	+1	0	0
		−4.2	+14.1333	−3.2666	−75.666	−0.26666	0	+1	0
		−7.4	+2.6	−9.2	−51	−0.2	0	0	+1
Eqs. (9.6.3)			+12.141	−1.3012	−51.166	−0.32051	+0.40385	+1	0
			−0.9102	−5.7372	−7.8333	−0.29487	+0.71154	0	+1
Eq. (9.6.4)				−5.8347	−11.6692	−0.31890	+0.74182	+0.74969	+1
Eq. (9.6.5)				$x_4 =$	+2.0000	+0.054656	−0.127139	−0.012849	−0.171388
Eq. (9.6.6)			$x_3 =$		−4.0000	−0.020541	+0.019637	+0.080988	−0.018368
Eq. (9.6.7)		$x_2 =$			+3.0000	−0.048140	+0.164962	+0.044430	+0.071487
Eq. (9.6.8)	$x_1 =$				−5.0000	+0.042440	+0.051941	+0.052936	+0.015927

The round-off errors in Eq. (9.6.11) are rather insignificant since the pivot values +15, +10.4, and +12.141 used in the forward eliminations are quite large in relation to the values of other eligible elements.

9.7 Solution of Simultaneous Linear Equations by Gauss-Jordan Elimination

In the preceding section, a systematic procedure has been described for the solution of a set of simultaneous linear equations by forward elimination and backward substitution. In this method two passes are required, one forward and one backward. Also, before each forward elimination, a choice can be made from a rectangular panel of eligible elements for the element with the largest absolute value to be used as the pivot. The mathematicians Gauss and Jordan have come up with a method in which only one pass is required, that of forward elimination without backward substitution. In fact, the first elimination in the two methods is identical. However, after this first elimination and each of the successive eliminations, instead of dropping out the pivotal equation, it is retained at all times. Furthermore, in the elimination process, the previous pivotal equations are treated in the same way as all the other equations.

350 INTERMEDIATE STRUCTURAL ANALYSIS

If the pivots are taken along the main diagonal in sequence, the Gauss-Jordan eliminations for a typical 4×4 matrix and one column of constants may be shown as follows:

$$(a) \begin{bmatrix} a_{11} & a_{12} & a_{13} & a_{14} & a_{15} \\ a_{21} & a_{22} & a_{23} & a_{24} & a_{25} \\ a_{31} & a_{32} & a_{33} & a_{34} & a_{35} \\ a_{41} & a_{42} & a_{43} & a_{44} & a_{45} \end{bmatrix} \xrightarrow{\text{1st elimination}} \begin{bmatrix} +1 & b_{12} & b_{13} & b_{14} & b_{15} \\ 0 & b_{22} & b_{23} & b_{24} & b_{25} \\ 0 & b_{32} & b_{33} & b_{34} & b_{35} \\ 0 & b_{42} & b_{43} & b_{44} & b_{45} \end{bmatrix} (b)$$

$$\xrightarrow{\text{2d elimination}} \begin{bmatrix} +1 & 0 & c_{13} & c_{14} & c_{15} \\ 0 & +1 & c_{23} & c_{24} & c_{25} \\ 0 & 0 & c_{33} & c_{34} & c_{35} \\ 0 & 0 & c_{43} & c_{44} & c_{45} \end{bmatrix} (c)$$

$$\xrightarrow{\text{3d elimination}} \begin{bmatrix} +1 & 0 & 0 & d_{14} & d_{15} \\ 0 & +1 & 0 & d_{24} & d_{25} \\ 0 & 0 & +1 & d_{34} & d_{35} \\ 0 & 0 & 0 & d_{44} & d_{45} \end{bmatrix} (d)$$

$$\xrightarrow{\text{4th elimination}} \begin{bmatrix} +1 & 0 & 0 & 0 & e_{15} \\ 0 & +1 & 0 & 0 & e_{25} \\ 0 & 0 & +1 & 0 & e_{35} \\ 0 & 0 & 0 & +1 & e_{45} \end{bmatrix} (e)$$

The explanations for the elimination processes are as follows:

1. Part (a) shows the given matrix of coefficients and the column of constants.
2. The first row of part (b) is equal to the first row of part (a) divided by the pivot a_{11}. The second to fourth rows of part (b) are obtained by subtracting a_{21}, a_{31}, and a_{44} times the first row in part (b) from the corresponding rows in part (a).
3. The second row of part (c) is equal to the second row of part (b) divided by the pivot b_{22}. The first, third, and fourth rows of part (c) are obtained by subtracting b_{12}, b_{32}, and b_{42} times the second row in part (c) from the corresponding rows in part (b).
4. The third row of part (d) is equal to the third row of part (c) divided by the pivot c_{33}. The first, second, and fourth rows of part (d) are obtained by subtracting c_{13}, c_{23}, and c_{43} times the third row in part (d) from the corresponding rows in part (c).
5. The fourth row of part (e) is equal to the fourth row of part (d) divided by the pivot d_{44}. The first three rows in part (e) are obtained by subtracting

d_{14}, d_{24}, and d_{34} times the fourth row in part (e) from the corresponding rows in part (d).

Thus at the end of the Gauss-Jordan eliminations, the given matrix becomes a unit matrix and the given column of constants is the solution of the set of simultaneous linear equations. As in the method of forward elimination and backward substitution, the pivots in the Gauss-Jordan eliminations need not stay on the main diagonal in sequence, but each pivot may be chosen from the eligible elements on the main diagonal on the basis of the largest absolute value, or from the eligible elements remaining on the square panel, such as shown in Fig. 9.6.1c. In such cases, however, the results of the solution will not appear in sequence in the rows of the original column of constants. For instance, if the pivots are located as shown in Fig. 9.6.1c, the results will be as follows:

$$\begin{bmatrix} 0. & 0. & 0. & +1.0 & x_4 \\ +1.0 & 0. & 0. & 0. & x_1 \\ 0. & +1.0 & 0. & 0. & x_2 \\ 0. & 0. & +1.0 & 0. & x_3 \end{bmatrix}$$

Table 9.7.1 shows the solution of the set of simultaneous linear equations in Table 9.6.1 by the Gauss-Jordan elimination method. Note that the number of equations is never reduced by each elimination. Had the pivots been taken on the main diagonal in sequence, as in Table 9.7.1, at the end of all eliminations the given matrix becomes a unit matrix and the unit matrix representing the columns of constants becomes the inverse of the given matrix. If the pivots are scattered, then some row interchanges are required to result in a unit matrix in place of the given matrix, as well as in the inverse matrix in place of the unit matrix representing the columns of constants at the beginning.

A computer subroutine[†] can be written to give the inverse of a non-singular matrix wherein there is a full-panel search for the element with the largest absolute value to be the next pivot and wherein the given matrix is replaced by its inverse, thus requiring no additional storage for the unit matrix to the right side of the equals sign, as shown in Table 9.7.1. This kind of subroutine is available in most institutional or commercial computing facilities.

9.8 Matrix Transposition

If a given matrix is rewritten in such a way that the ith row of the original matrix becomes the ith column of the new matrix, the new matrix is defined

---

[†]For instance, see C.-K. Wang, *Matrix Methods of Structural Analysis*, 2d ed., American Publishing Company, Madison, Wis., 1970, Appendix B.

Table 9.7.1 Solution of simultaneous linear equations by Gauss-Jordan elimination

	x_1	x_2	x_3	x_4	=				
Given data	+15	−3	−8	+1	−50	+1.0	0	0	0
	+2	+10	−6	+5	+54	0	+1.0	0	0
	+4	−5	+12	−3	−89	0	0	+1.0	0
	+3	−8	+1	−9	−61	0	0	0	+1.0
1st elimination	+1.0	−0.20000	−0.53333	+0.06666	−3.3333	+0.06666	0	0	0
	0	+10.4	−4.93333	+4.86666	+60.6666	−0.13333	+1	0	0
	0	−4.2	+14.13333	−3.26666	−75.6666	−0.26666	0	+1.0	0
	0	−7.4	+2.6	−9.2	−51	−0.20000	0	0	+1.0
2d elimination	+1.0	0	−0.62820	+0.16026	−2.16666	+0.064103	+0.019231	0	0
	0	+1.0	−0.47356	+0.46795	+5.83333	−0.012820	+0.096154	0	0
	0	0	+12.14102	−1.30128	−51.16666	−0.320513	+0.403846	+1.0	0
	0	0	−0.91026	−5.73718	−7.83333	−0.294872	+0.711538	0	+1.0
3d elimination	+1.0	0	0	+0.092926	−4.814150	+0.047518	+0.040129	+0.051742	0
	0	+1.0	0	+0.417107	+3.834212	−0.025343	+0.111934	+0.039071	0
	0	0	+1.0	−0.107180	−4.214363	−0.026399	+0.033267	+0.082365	0
	0	0	0	−5.834738	−11.669487	−0.318902	+0.741820	+0.074974	+1.0
4th elimination	+1.0	0	0	0	−5.00000	+0.042439	+0.051943	+0.052936	+0.015926
	0	+1.0	0	0	+3.00000	−0.048140	+0.164964	+0.044431	+0.071487
	0	0	+1.0	0	−4.00000	−0.020541	+0.019640	+0.080988	−0.018369
	0	0	0	+1.0	+2.00000	+0.054656	−0.127138	−0.012850	−0.171387

as the *transpose* of the original matrix, and vice versa. The symbol $[A^T]$ is used to denote the transpose of $[A]$. Thus if

$$[A]_{4\times 3} = \begin{bmatrix} +2 & +9 & +14 \\ +6 & +8 & +17 \\ +4 & +11 & +5 \\ +7 & +10 & +12 \end{bmatrix}$$

then

$$[A^T]_{3\times 4} = \begin{bmatrix} +2 & +6 & +4 & +7 \\ +9 & +8 & +11 & +10 \\ +14 & +17 & +5 & +12 \end{bmatrix}$$

Matrix transposition plays an important role in structural analysis; in fact, the principle of virtual work, which will be presented in Chaps. 10, 11, and 12, can be conveniently represented by a matrix transposition relationship.

9.9 Some Useful Properties of Matrices

Some useful properties of matrices are described and proved in this section so that they may be quoted later.

(a) The transpose of the product of matrices $[A]$ and $[B]$ is the product of the transpose of $[B]$ and the transpose of $[A]$, or

$$[AB]^T = [B^T][A^T] \qquad (9.9.1)$$

Proof for Eq. (9.9.1)

Statement

(1) The element at the ith row and jth column of the $[AB]^T$ matrix = the element at the jth row and ith column of the $[AB]$ matrix.

(2) Right side of (1) = sum of inner products of the jth row of the $[A]$ matrix and the ith column of the $[B]$ matrix.

(3) Right side of (2) = sum of inner products of the jth column of the $[A^T]$ matrix and the ith row of the $[B^T]$ matrix.

(4) Left side of (1) = right side of (1) = right side of (2) = right side of (3).

Reason

(1) Definition of matrix transposition.

(2) Rule for matrix multiplication.

(3) Definition of matrix transposition.

(4) Apply steps 1, 2, and 3 above.

(5) Step 4 means

$$[AB]^T = [B^T][A^T]$$

(5) Rule for matrix multiplication.

(b) When a square matrix is premultiplied or postmultiplied by a unit matrix of the same size, the original matrix is unchanged, or

$$[I][A] = [A] \qquad (9.9.2a)$$

$$[A][I] = [A] \qquad (9.9.2b)$$

Equations (9.9.2a and b) can be easily verified when one begins to write the longhand expressions of the square matrix and the unit matrix; thus

$$\begin{bmatrix} 1 & 0 & 0 \\ 0 & 1 & 0 \\ 0 & 0 & 1 \end{bmatrix} \begin{bmatrix} a_{11} & a_{12} & a_{13} \\ a_{21} & a_{22} & a_{23} \\ a_{31} & a_{32} & a_{33} \end{bmatrix} = \begin{bmatrix} a_{11} & a_{12} & a_{13} \\ a_{21} & a_{22} & a_{23} \\ a_{31} & a_{32} & a_{33} \end{bmatrix}$$

and

$$\begin{bmatrix} a_{11} & a_{12} & a_{13} \\ a_{21} & a_{22} & a_{23} \\ a_{31} & a_{32} & a_{33} \end{bmatrix} \begin{bmatrix} 1 & 0 & 0 \\ 0 & 1 & 0 \\ 0 & 0 & 1 \end{bmatrix} = \begin{bmatrix} a_{11} & a_{12} & a_{13} \\ a_{21} & a_{22} & a_{23} \\ a_{31} & a_{32} & a_{33} \end{bmatrix}$$

(c) A symmetric matrix is a square matrix which is identical to its own transpose, or vice versa.

If $[A] = [A^T]$, $[A]$ is a symmetric matrix. (9.9.3a)

If $[A]$ is a symmetric matrix, $[A] = [A^T]$. (9.9.3b)

Proof for Eq. (9.9.3a)

Statement

(1) The element at the ith row and jth column of the $[A]$ matrix = the element at the ith column and jth row of the $[A^T]$ matrix.
(2) Right side of (1) = the element at the ith column and jth row of the $[A]$ matrix.
(3) Left side of (1) = right side of (1) = right side of (2).
(4) Step 3 means $[A]$ is a symmetric matrix.

Reason

(1) Definition of $[A^T]$ of any matrix $[A]$.

(2) Given condition that $[A] = [A^T]$.

(3) Apply steps 1 and 2 above.

(4) Definition of a symmetric matrix.

Proof for Eq. (9.9.3b)

Statement

(1) The element at the ith row and jth column of the $[A]$ matrix = the element at the jth row and ith column of the $[A]$ matrix.
(2) Right side of (1) = the element at the jth column and ith row of the $[A^T]$ matrix.
(3) Left side of (1) = right side of (1) = right side of (2).
(4) Step 3 above means $[A] = [A^T]$.

Reason

(1) Definition of a symmetric matrix.

(2) Definition of $[A^T]$ of any matrix $[A]$.

(3) Apply steps 1 and 2 above.
(4) Two matrices are identical if the respective elements at every identical location are equal.

As an example, the following matrix is symmetric.

$$[A] = [A^T] = \begin{bmatrix} +10 & -5 & +7 \\ -5 & +9 & +3 \\ +7 & +3 & +14 \end{bmatrix}$$

Note that the first row is the same as the first column, the second row is the same as the second column, etc.

(d) The unit matrix is symmetric, by observation, or

$$[I] = [I^T] \tag{9.9.4}$$

(e) The inverse of a symmetric matrix is still symmetric, or

$$\text{If } [A] = [A^T], [A^{-1}] = [A^{-1}]^T. \tag{9.9.5}$$

Proof for Eq. (9.9.5)

Statement

(1) $[A] = [A^T]$
(2) $[A^{-1}A] = [I]$
(3) $[A^{-1}A]^T = [I]^T$ or $[I]$
(4) $[A^T][A^{-1}]^T = [I]$
(5) $[A][A^{-1}]^T = [I]$
(6) $[A^{-1}][A][A^{-1}]^T = [A^{-1}][I]$
(7) $[I][A^{-1}]^T = [A^{-1}][I]$
(8) $[A^{-1}]^T = [A^{-1}]$

Reason

(1) Given condition.
(2) Definition of an inverse matrix.
(3) Take transpose of each side of (2).
(4) Apply Eq. (9.9.1) to left side of (3).
(5) Substitute (1) in left side of (4).
(6) Premultiply each side of (5) by $[A^{-1}]$.
(7) Substitute (2) in left side of (6).
(8) Apply Eqs. (9.9.2a and b).

9.10 Exercises

9.1 Given the $[A]$ and $[B]$ matrices as shown, find $[C] = [A][B]$. If $\{X\} = [A]\{Y\}$, $\{Y\} = [B]\{Z\}$, and the $\{Z\}$ matrix is as given, (a) find $\{X\}$ from $\{X\} = [C]\{Z\}$, and (b) find $\{Y\} = [B]\{Z\}$ and then $\{X\}$ from $\{X\} = [A]\{Y\}$.

$$[A]_{3 \times 2} = \begin{bmatrix} +6 & -17 \\ +11 & -2 \\ -5 & +24 \end{bmatrix} \quad [B]_{2 \times 2} = \begin{bmatrix} +9 & +3 \\ -8 & +15 \end{bmatrix} \quad \{Z\}_{2 \times 1} = \begin{Bmatrix} -3 \\ +5 \end{Bmatrix}$$

9.2 Multiply $[A]$ by $[B]$ if

$$[A] = \begin{bmatrix} a & b & c \\ d & e & f \end{bmatrix} \quad [B] = \begin{bmatrix} +3 & 0 & 0 \\ 0 & +5 & 0 \\ 0 & 0 & +4 \end{bmatrix}$$

9.3 Multiply $[A]$ by $[B]$ if

$$[A] = \begin{bmatrix} +3 & 0 & 0 \\ 0 & +5 & 0 \\ 0 & 0 & +4 \end{bmatrix} \quad [B] = \begin{bmatrix} a & d \\ b & e \\ c & f \end{bmatrix}$$

9.4 Find the inverse of the $[A]$ matrix shown below by forward elimination and backward substitution using pivots on the main diagonal in sequence. Check the solution by computing $[A][A^{-1}]$.

$$[A] = \begin{bmatrix} +5 & +4 & -9 \\ +8 & -2 & +10 \\ -16 & -3 & +18 \end{bmatrix}$$

9.5 Solve Exercise 9.4 but use a full-panel search for the element with the largest absolute value as the next pivot. Check the solution by computing $[A][A^{-1}]$.

9.6 Solve Exercise 9.4 by Gauss-Jordan elimination. Check the solution by computing $[A][A^{-1}]$.

9.7 Solve Exercise 9.5 by Gauss-Jordan elimination. Check the solution by computing $[A][A^{-1}]$.

9.8 Find the inverse of

$$\begin{bmatrix} s_{ii} & s_{ij} \\ s_{ji} & s_{jj} \end{bmatrix}$$

9.9 Find the inverse of

$$\begin{bmatrix} f_{ii} & f_{ij} \\ f_{ji} & f_{jj} \end{bmatrix}$$

CHAPTER
TEN
MATRIX DISPLACEMENT METHOD OF TRUSS ANALYSIS

10.1 Degree of Freedom, Number of Independent Unknown Forces, and Degree of Indeterminacy

The basic concept of the displacement method of truss analysis has been briefly described in Sec. 1.4 by means of the four-bar, pin-connected structure shown in Fig. 10.1.1. It was stated then that Navier (1785–1836) held that this structure is "determinate" if the horizontal and vertical displacements of pin joint E are taken as the unknowns at the beginning. Since these two displacements are the primary unknowns to be determined in advance of all other information about the structure, they are denoted by X_1 horizontally to the right and X_2 vertically upward. Due to any kind of imposition on this structure, whether it be applied forces like W_1 and W_2 in Fig. 10.1.1a, or fabrication errors in the lengths of the bars, or support settlements, these two displacements are the *freedoms* in which the structure can exercise to show its "displeasure," more technically called *deformation response*. Thus the *degree of freedom* of a truss (the pin-connected structure of Fig. 10.1.1 satisfies the definition of a truss) may be defined as the number of unknown displacements of the joints. As another example, the degree of freedom of the truss in Fig. 10.1.2 is 6.

To fulfill the definition of a truss, the loads can be applied only at the joints. Then at the maximum, there can be both horizontal and vertical forces applied to every joint in a truss, although those applied in any direction at a hinged support or applied in a direction perpendicular to the rolling surface of a roller support, will not affect the forces in the bars or the deformed shape of the truss. Thus there can be a nonzero force in the direction of every degree

358 INTERMEDIATE STRUCTURAL ANALYSIS

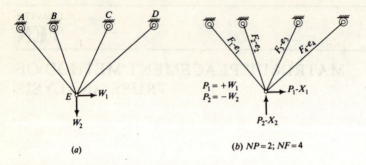

Figure 10.1.1 A four-bar, pin-connected structure.

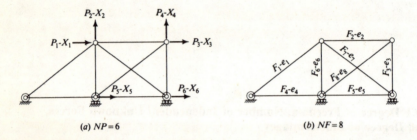

Figure 10.1.2 A two-panel truss.

of freedom, and the external force system can be expressed by a column matrix of which the number of rows is equal to the degree of freedom, hereafter denoted by the symbol NP for number of P values (in fact, NX could have been used for number of X values). Mathematically speaking, there is a one-to-one correspondence between the P's and the X's, hence the notations P_i-X_i for $i = 1$ to NP in Figs. 10.1.1b and 10.1.2a.

The force response of a truss is completely defined by the values of the reactions and the forces in the bars. However, once the forces in the bars are known, every reaction can be determined by the equation of statics along the direction of that reaction; thus the forces in the bars are the *independent unknown forces*, the number of which is denoted by the symbol NF. It can be shown that the degree of indeterminacy of a truss, denoted by the symbol NI, is equal to $NF - NP$, or

$$NI = NF - NP \tag{10.1.1}$$

The total number of unknown forces for a truss is $NF + NR$, wherein NR is the number of reactions, but the total number of equations of statics is $2(NJ)$, wherein NJ is the number of joints. Thus,

$$NI = NF + NR - 2(NJ) \tag{10.1.2}$$

but physically,

$$2(NJ) = NP + NR \tag{10.1.3}$$

Substituting Eq. (10.1.3) into Eq. (10.1.2),

$$NI = (NF + NR) - (NP + NR) = NF - NP$$

which is Eq. (10.1.1). In fact Eq. (10.1.1) is the same as Eq. (1.1.2) except that lowercase symbols were used in Eq. (1.1.2). Symbols denoted by two-letter capitals are used in this chapter for convenience in computer programming.

The unknown forces in the bars, having one-to-one correspondence with the elongations (the e's) in themselves, are numbered consecutively as F_i-e_i for $i = 1$ to NF in Figs. 10.1.1b and 10.1.2b.

10.2 The Deformation Matrix [B]

The deformation matrix [B] of a truss expresses the elongations in the bars in terms of the joint displacements; thus the size of the deformation matrix is $NF \times NP$, and

$$\{e\}_{NF \times 1} = [B]_{NF \times NP}\{X\}_{NP \times 1} \tag{10.2.1}$$

Here it is important to emphasize again the first-order assumption that the length of a bar is affected only by the longitudinal component of any joint displacement and not by the transverse component. Thus in Fig. 10.2.1a, if the

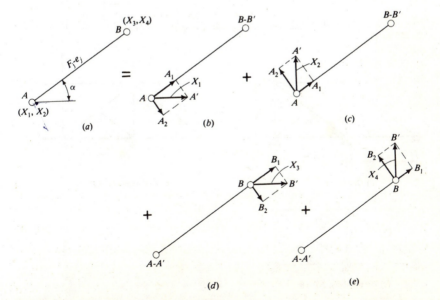

Figure 10.2.1 Bar elongation due to joint displacements.

360 INTERMEDIATE STRUCTURAL ANALYSIS

joint displacements of A are X_1 to the right and X_2 upward and those of B are X_3 to the right and X_4 upward, by applying the first-order assumption to Fig. 10.2.1b to e it may be seen that

$$e_j = -X_1 \cos \alpha - X_2 \sin \alpha + X_3 \cos \alpha + X_4 \sin \alpha \qquad (10.2.2)$$

For a typical truss like the one shown in Fig. 10.2.2, the $[B]$ matrix as shown in Eq. (10.2.3) can best be established one column at a time by visual inspection, using the concept advanced in Fig. 10.2.1.

$[B]_{8 \times 6} =$

X \ e	1	2	3	4	5	6
1	+0.8	+0.6	0.	0.	0.	0.
2	−1.0	0.	+1.0	0.	0.	0.
3	0.	0.	0.	+1.0	0.	0.
4	0.	0.	0.	0.	+1.0	0.
5	0.	0.	0.	0.	−1.0	+1.0
6	0.	+1.0	0.	0.	0.	0.
7	−0.8	+0.6	0.	0.	0.	+0.8
8	0.	0.	+0.8	+0.6	−0.8	0.

(10.2.3)

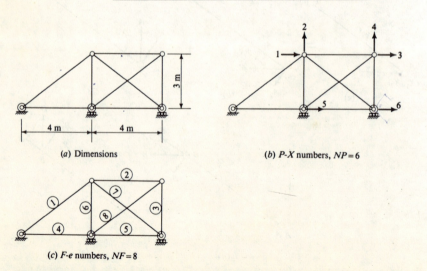

(a) Dimensions

(b) P-X numbers, NP=6

(c) F-e numbers, NF=8

Figure 10.2.2 P-X and F-e numbers for a typical truss.

MATRIX DISPLACEMENT METHOD OF TRUSS ANALYSIS 361

10.3 The Element Stiffness Matrix [S]

The element stiffness matrix $[S]$ of a truss expresses the forces in the bars (or the elements) in terms of the elongations in them; thus this matrix is a square matrix $NF \times NF$. By Hooke's law, the force in any one bar is expressible by the elongation in itself as

$$F_j = S_j e_j \quad \text{for } j = 1 \text{ to } NF \quad (10.3.1a)$$

wherein

$$S_j = \frac{E_j A_j}{L_j} \quad (10.3.1b)$$

L_j and A_j are the length and the area of the jth bar. The symbol S_j (in a departure from using lowercase symbols for the elements within a matrix) is used to denote the stiffness of the bar, which is the force to cause a unit elongation, much like the spring constant of a spring. Thus the element stiffness matrix of a truss is a diagonal matrix so that

$$\{F\}_{NF \times 1} = [S]_{NF \times NF} \{e\}_{NF \times 1} \quad (10.3.2a)$$

in which

$$[S]_{NF \times NF} = \begin{array}{|c|c|c|c|c|} \hline {}_{F}\!\diagdown\!{}^{e} & 1 & 2 & \cdots & NF \\ \hline 1 & \dfrac{EA_1}{L_1} & \cdots & \cdots & \cdots \\ \hline 2 & \cdots & \dfrac{EA_2}{L_2} & \cdots & \cdots \\ \hline \cdot & \cdots & \cdots & \cdots & \cdots \\ \hline NF & \cdots & \cdots & \cdots & \dfrac{EA_{NF}}{L_{NF}} \\ \hline \end{array} \quad (10.3.2b)$$

10.4 The Force-Displacement Matrix [SB]

Substituting Eq. (10.2.1) into Eq. (10.3.2a), one obtains

$$\{F\}_{NF \times 1} = [S]_{NF \times NF} \{e\}_{NF \times 1}$$
$$= [S]_{NF \times NF} [B]_{NF \times NP} \{X\}_{NP \times 1}$$
$$= [SB]_{NF \times NP} \{X\}_{NP \times 1} \quad (10.4.1)$$

The matrix product $[SB]$, of which the diagonal matrix $[S]$ is the premultiplier and the matrix $[B]$ is the postmultiplier, expresses the forces in the bars of a truss in terms of the joint displacements. The force-displacement matrix $[SB]$ is an important matrix, not only as the needed intermediate matrix to obtain the matrix $[ASB]$ defined in Sec. 10.7 *before* the joint displacements are determined, but also as the matrix to be used again in the final stage of the analysis to obtain the forces in the bars *after* the joint displacements have been found.

10.5 The Statics Matrix [A]

The statics matrix $[A]$ of a truss expresses the external joint forces in terms of the forces in the bars; thus the size of the statics matrix is $NP \times NF$, and

$$\{P\}_{NP\times 1} = [A]_{NP\times NF}\{F\}_{NF\times 1} \qquad (10.5.1)$$

For the truss shown in Fig. 10.2.2, the statics matrix $[A]$ is

$[A]_{6\times 8} =$

F / P	1	2	3	4	5	6	7	8
1	+0.8	−1.0	0.	0.	0.	0.	−0.8	0.
2	+0.6	0.	0.	0.	0.	+1.0	+0.6	0.
3	0.	+1.0	0.	0.	0.	0.	0.	+0.8
4	0.	0.	+1.0	0.	0.	0.	0.	+0.6
5	0.	0.	0.	+1.0	−1.0	0.	0.	−0.8
6	0.	0.	0.	0.	+1.0	0.	+0.8	0.

$$(10.5.2)$$

The contents of the $[A]$ matrix can be determined either by rows or by columns. To establish by rows, one needs to sketch a diagram such as Fig. 10.5.1, which shows the free-body diagrams of all the joints. Note that the

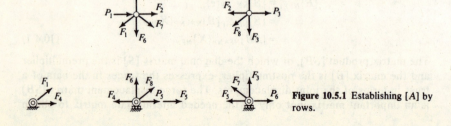

Figure 10.5.1 Establishing $[A]$ by rows.

MATRIX DISPLACEMENT METHOD OF TRUSS ANALYSIS   363

forces in the bars are assumed to be tensile if positive; hence they pull on the joints in Fig. 10.5.1. Applying $\Sigma F_x = 0$ and $\Sigma F_y = 0$ to the upper left joint,

$$P_1 = +0.8F_1 - 1.0F_2 - 0.8F_7$$
$$P_2 = +0.6F_1 + 1.0F_6 + 0.6F_7$$

Applying $\Sigma F_x = 0$ and $\Sigma F_y = 0$ to the upper right joint,

$$P_3 = +1.0F_2 + 0.8F_8$$
$$P_4 = +1.0F_3 + 0.6F_8$$

Applying $\Sigma F_x = 0$ to the lower middle joint,

$$P_5 = +1.0F_4 - 1.0F_5 - 0.8F_8$$

Applying $\Sigma F_x = 0$ to the lower right joint,

$$P_6 = +1.0F_5 + 0.8F_7$$

When the six preceding equations expressing P_1 to P_6 in terms of F_1 to F_8 are arranged as a rectangular matrix, the result is Eq. (10.5.2).

To fill out the [A] matrix by columns, it is convenient to assess what external joint forces each bar force can resist. In Fig. 10.5.2a, let the tensile force in the bar be F_j and the external joint forces be P_1 to P_4. Then the P forces that F_j can resist, from inspection of the free-body diagrams of the two joints, are

$$P_1 = -F_j \cos \alpha$$
$$P_2 = -F_j \sin \alpha$$
$$P_3 = +F_j \cos \alpha$$
$$P_4 = +F_j \sin \alpha$$
\hfill (10.5.3)

Equations (10.5.3) can be visualized even more conveniently by making use of the free-body diagram of the bar itself as shown in Fig. 10.5.2b. The effective components of the two F_j forces pulling on the bar, along the positive

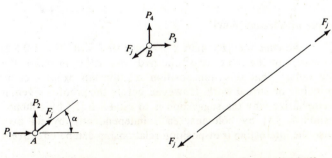

(a) Equilibrium at the joints   (b) Free-body diagram of bar itself

Figure 10.5.2 Establishing [A] by columns.

directions of P_1, P_2, P_3, and P_4 are observed to be those expressed by Eqs. (10.5.3). This latter concept will be used exclusively in the future to establish the $[A]$ matrix by columns. Returning to Fig. 10.2.2, if one visualizes two pulling forces on the ends of bar 1, the lower force has no effective components along any P's, but the upper force has an 80 percent component in the positive direction of P_1 and a 60 percent component in the positive direction of P_2. These values are entered into the first column of the $[A]$ matrix in Eq. (10.5.2). Likewise the pulling force at the left end of bar 2 has a 100 percent component in the negative direction of P_1 and the pulling force at the right end has a 100 percent component in the positive direction of P_3. These values are entered into the second column of the $[A]$ matrix in Eq. (10.5.2). The reader may think through the remaining six columns in Eq. (10.5.2).

It may be noted here that if NF is equal to NP, then, following Eq. (10.1.1), $NI = NF - NP = 0$ and the truss is statically determinate. In such a case the square matrix $[A]_{NP \times NP}$, if it is nonsingular as it has to be for a statically stable truss, can be inverted to yield

$$\{F\}_{NF \times 1} = [A^{-1}]_{NF \times NP} \{P\}_{NP \times 1} \quad \text{for } NF = NP \text{ only} \qquad (10.5.4)$$

Equation (10.5.4) can be used to determine the forces in the bars of any statically determinate and stable truss. A computer program can also be written using the subroutine on matrix inversion.† Furthermore, using the property $[B] = [A^T]$, which will be proved in the next section, Eq. (10.2.1) will become

$$\{e\}_{NF \times 1} = [A^T]_{NF \times NP} \{X\}_{NP \times 1}$$

from which

$$\{X\}_{NP \times 1} = [A^T]^{-1}_{NP \times NF} \{e\}_{NF \times 1} \quad \text{for } NF = NP \text{ only} \qquad (10.5.5)$$

Thus the combined use of Eqs. (10.5.4) and (10.5.5) will produce not only the forces in the bars but also the joint displacements of any statically determinate and stable truss due to a set of joint forces.

10.6 The Principle of Virtual Work

When Eq. (10.2.3) showing the $[B]$ matrix is compared with Eq. (10.5.2) showing the $[A]$ matrix of the same truss, one finds that either matrix is the transpose of the other. That this transposition relationship exists can be proved by the principle of virtual work. However, before the proof is given, it is important to emphasize why the suggestion is to establish $[B]$ by columns and then to establish $[A]$ by columns, each independently by its own definition, because the interesting transposition relationship can be effectively used as a check.

†For instance, see C.-K. Wang, *Matrix Methods of Structural Analysis*, 2d ed., American Publishing Company, Madison, Wis., 1970, Appendix C.

MATRIX DISPLACEMENT METHOD OF TRUSS ANALYSIS 365

The principle of virtual work is axiomatic, because it simply states that the product of a zero force and a nonzero displacement, each existent under a different set of circumstances, is zero. For instance, under one circumstance, in the truss of Fig. 10.6.1a, only F_7 is nonzero while all other F's are zero. This is the situation of the seventh column of the $[A]$ matrix in Eq. (10.5.2), repeated partially below as Eq. (10.6.1).

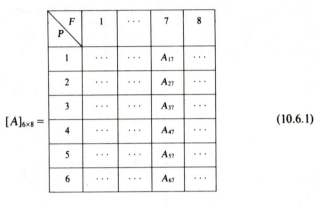

$$[A]_{6\times 8} = \begin{array}{|c|c|c|c|c|} \hline F\backslash P & 1 & \cdots & 7 & 8 \\ \hline 1 & \cdots & \cdots & A_{17} & \cdots \\ \hline 2 & \cdots & \cdots & A_{27} & \cdots \\ \hline 3 & \cdots & \cdots & A_{37} & \cdots \\ \hline 4 & \cdots & \cdots & A_{47} & \cdots \\ \hline 5 & \cdots & \cdots & A_{57} & \cdots \\ \hline 6 & \cdots & \cdots & A_{67} & \cdots \\ \hline \end{array} \quad (10.6.1)$$

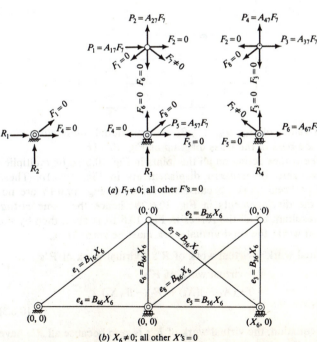

(a) $F_7 \neq 0$; all other F's $= 0$

(b) $X_6 \neq 0$; all other X's $= 0$

Figure 10.6.1 Principle of virtual work.

The actual values of $A_{17}, A_{27}, \ldots, A_{67}$ are immaterial, but these joint forces which $F_7 \neq 0$ can resist are shown on the free-body diagrams of the joints in Fig. 10.6.1a. Since the resultant of each of the five concurrent force systems is zero, the resultant of all the directed forces acting on all the joints in Fig. 10.6.1a is zero.

Consider another entirely different circumstance. The truss happens to take some deformed shape, due to the application of some set of joint forces, in which only X_6 is nonzero while all other X's are zero. The elongations in the bars due to $X_6 \neq 0$ alone are given by the sixth column of the $[B]$ matrix in Eq. (10.2.3), repeated partially below as Eq. (10.6.2).

$$[B]_{8\times 6} = \begin{array}{|c|c|c|c|c|} \hline X \backslash e & 1 & \cdots & 5 & 6 \\ \hline 1 & \cdots & \cdots & \cdots & B_{16} \\ \hline 2 & \cdots & \cdots & \cdots & B_{26} \\ \hline 3 & \cdots & \cdots & \cdots & B_{36} \\ \hline 4 & \cdots & \cdots & \cdots & B_{46} \\ \hline 5 & \cdots & \cdots & \cdots & B_{56} \\ \hline 6 & \cdots & \cdots & \cdots & B_{66} \\ \hline 7 & \cdots & \cdots & \cdots & B_{76} \\ \hline 8 & \cdots & \cdots & \cdots & B_{86} \\ \hline \end{array} \qquad (10.6.2)$$

Again, the actual values of $B_{16}, B_{26}, \ldots, B_{86}$ are immaterial; these elongations are written over the bars in the truss diagram of Fig. 10.6.1b.

Now let all the forces acting on all the joints in Fig. 10.6.1a be multiplied by the respective zero or nonzero displacements in Fig. 10.6.1b. These products cannot be "real work" because the forces in Fig. 10.6.1a are not even related to the displacements in Fig. 10.6.2b; hence the name *virtual work*. Since the resultant of all the forces in Fig. 10.6.1a is zero, then by the principle of virtual work, the total virtual work must be zero. Thus,

Total virtual work = virtual work of R's + virtual work of P's
  + virtual work of F's
$$= 0 + (A_{67}F_7)(X_6) + (F_7)(-B_{76}X_6)$$
$$= 0 \qquad (10.6.3)$$

In the preceding equation, the virtual work of R's is zero because all R's have zero displacement; the virtual work of P's are zero except $P_6 = A_{67}F_7$ which has a displacement of X_6 in the same positive direction; and the virtual work of the two nonzero F_7's toward each other in Fig. 10.6.1a moving apart by the amount of elongation in Fig. 10.6.1b is negative. Canceling F_6 and X_6 in Eq. (10.6.3),

$$A_{67} = B_{76} \qquad (10.6.4)$$

The proof of Eq. (10.6.4), using subscripts of 6 and 7, is perfectly general because the subscripts could be i and j. Since $A_{ij} = B_{ji}$, then

$$[A] = [B^T] \quad \text{or} \quad [B] = [A^T] \qquad (10.6.5)$$

10.7 The Global Stiffness Matrix $[K] = [ASA^T]$

Substituting Eq. (10.4.1) into Eq. (10.5.1), one obtains

$$\{P\}_{NP\times 1} = [A]_{NP\times NF}\{F\}_{NF\times 1} = [A]_{NP\times NF}[SB]_{NF\times NP}\{X\}_{NP\times 1}$$
$$= [ASB]_{NP\times NP}\{X\}_{NP\times 1} = [ASA^T]_{NP\times NP}\{X\}_{NP\times 1}$$
$$= [K]_{NP\times NP}\{X\}_{NP\times 1} \qquad (10.7.1a)$$

wherein

$$[K]_{NP\times NP} = [ASA^T]_{NP\times NP} =$$

P \ X	1	2	$\cdots$	NP
1	K_{11}	$\cdots$	$\cdots$	$K_{1\text{-}NP}$
2	K_{12}	$\cdots$	$\cdots$	$K_{2\text{-}NP}$
.	$\cdots$	$\cdots$	$\cdots$	$\cdots$
NP	$K_{NP\text{-}1}$	$\cdots$	$\cdots$	$K_{NP\text{-}NP}$

$$(10.7.1b)$$

The matrix product $[K] = [ASA^T]$, of which the statics matrix $[A]$ is the premultiplier and the force-displacement matrix $[SB]$ or $[SA^T]$ is the postmultiplier, expresses the joint forces in terms of the joint displacements for the entire truss, therefore the name *global stiffness matrix*. This matrix is always a square matrix; and if the structure is statically stable (refer to Fig. 1.1.5 again), this matrix is nonsingular and has an inverse.

Example 10.7.1 Obtain numerically the global stiffness matrix $[K] = [ASA^T]$ of the truss shown in Fig. 10.7.1a.

SOLUTION (*a*) *Assignment of P-X and F-e numbers.* Joints A and B being hinged, the only freedoms are the horizontal and vertical displacements of joints C and D. Thus the degree of freedom NP is equal to 4 as shown in Fig. 10.7.1b. The bar forces and elongations are assigned the numbers 1 to 5 as shown in Fig. 10.7.1c; NF is equal to 5. Thus, $NI = NF - NP = 5 - 4 = 1$.

(b) *The* [A], [B], *and* [S] *matrices.* These matrices are shown in Table 10.7.1. The [A] and [B] matrices are determined independently of each other by their own definitions, and the transposition relationship is used for the check. The values shown on the diagonal of the [S] matrix are simply the $S = EA/L$ values of the bars.

Table 10.7.1 Global stiffness matrix $[K] = [ASA^T]$ of truss in Example 10.7.1†

[A] =

F \ P	1	2	3	4	5
1	0.	+0.6	−0.6	0.	0.
2	+1.0	+0.8	+0.8	0.	0.
3	0.	0.	0.	+0.78087	−0.78087
4	−1.0	0.	0.	+0.62470	+0.62470

[B] =

X \ e	1	2	3	4
1	0.	+1.0	0.	−1.0
2	+0.6	+0.8	0.	0.
3	−0.6	+0.8	0.	0.
4	0.	0.	+0.78087	+0.62470
5	0.	0.	−0.78087	+0.62470

[S] =

e \ F	1	2	3	4	5
1	+200.				
2		+106.666			
3			+106.666		
4				+222.114	
5					+222.114

$$[SB] = \begin{array}{c|cccc} \diagdown X \\ F \diagdown & 1 & 2 & 3 & 4 \\ \hline 1 & 0. & +200. & 0. & -200. \\ 2 & +64. & +85.33 & 0. & 0. \\ 3 & -64. & +85.33 & 0. & 0. \\ 4 & 0. & 0. & +173.44 & +138.75 \\ 5 & 0. & 0. & -173.44 & +138.75 \end{array}$$

$$[ASB] = \begin{array}{c|cccc} \diagdown X \\ P \diagdown & 1 & 2 & 3 & 4 \\ \hline 1 & +76.8 & 0. & 0. & 0. \\ 2 & 0. & +336.53 & 0. & -200. \\ 3 & 0. & 0. & +270.87 & 0. \\ 4 & 0. & -200. & 0. & +373.35 \end{array}$$

†Dimensional units, where applicable, are in 10^3 kN/m.

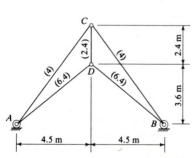

Numbers in parentheses are areas in $10^{-3} m^2$
$E = 200 \times 10^6$ kN/m^2

(a) The given truss

(b) P-X numbers, NP = 4

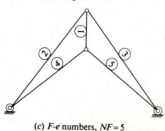

(c) F-e numbers, NF = 5

Figure 10.7.1 Truss of Example 10.7.1.

(c) *The* [SB] *and* [K] = [ASA^T] *matrices.* These two matrices are obtained by matrix multiplication as shown in Table 10.7.1.

(d) *Discussion.* Note that the [K] = [ASA^T] matrix is a square symmetric matrix. The proof of this symmetry will be made in Sec. 10.10.

10.8 The Local Stiffness Matrix

The *local stiffness matrix* is here defined to be the global stiffness matrix of a single bar, as shown in Fig. 10.8.1. Certainly a single bar with four degrees of freedom will not make a stable truss; in other words, the 4 × 4 global stiffness matrix of a single bar would be far from the possibility of having an inverse. Thus the appropriate name for this matrix is local stiffness matrix, and the global stiffness matrix of a whole truss containing many bars should be the assemblage of many local stiffness matrices.

The local stiffness matrix of a typical bar in a truss is shown in Table 10.8.1. The local [B] matrix is taken from Eq. (10.2.2); and the local [A]

Table 10.8.1 Local stiffness matrix of a typical bar in a truss

$[B] =$

X \ e	1	2	3	4
e_j	$-\cos \alpha$	$-\sin \alpha$	$+\cos \alpha$	$+\sin \alpha$

$[S] =$

$e_j \setminus F_j$	j
j	S_j

$[SB] =$

X \ F_j	1	2	3	4
j	$-S_j \cos \alpha$	$-S_j \sin \alpha$	$+S_j \cos \alpha$	$+S_j \sin \alpha$

$[A] =$

F \ P	j
1	$-\cos \alpha$
2	$-\sin \alpha$
3	$+\cos \alpha$
4	$+\sin \alpha$

$[ASB] =$

X \ P	1	2	3	4
1	$+T1$	$+T2$	$-T1$	$-T2$
2	$+T2$	$+T3$	$-T2$	$-T3$
3	$-T1$	$-T2$	$+T1$	$+T2$
4	$-T2$	$-T3$	$+T2$	$+T3$

$T1 = +S_j \cos^2 \alpha \qquad T2 = +S_j \cos \alpha \sin \alpha \qquad T3 = +S_j \sin^2 \alpha$

MATRIX DISPLACEMENT METHOD OF TRUSS ANALYSIS 371

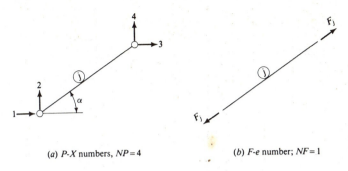

(a) P-X numbers, NP = 4   (b) F-e number; NF = 1

Figure 10.8.1 Local stiffness matrix of a truss element.

matrix, from Eqs. (10.5.3). The local $[SB]$ and the local $[K]$ matrices in Table 10.8.1 are obtained by matrix multiplication.

10.9 The Direct Stiffness Method

There are two convenient ways of building up the global stiffness matrix of any structure, including, of course, that of a truss. The first has been shown in Sec. 10.7 and Example 10.7.1, where the equation $[K] = [ASA^T]$ was used. In this equation, all the matrices—$[A]$, $[S]$, $[B] = [A^T]$, and $[K]$—are global in nature. The second involves the feeding of the local stiffness matrices of all the bars into the appropriate slots in the global stiffness matrix. In so doing it is necessary to note the four global degree-of-freedom numbers which correspond to the four local degrees of freedom of each bar. Then the 16 elements in the local stiffness matrix of the bar can be put in the correct locations in the global stiffness matrix.

As an illustration, the local and the corresponding global degrees of freedom of all the bars in the truss of Fig. 10.9.1a and b are shown in Fig. 10.9.1c. Each bar has a direction indicated by an arrow going from the end with local degrees of freedom $L1$ and $L2$ toward the end with local degrees of freedom $L3$ and $L4$. The angle α for each bar is measured at the end with local degrees of freedom $L1$ and $L2$, in the counterclockwise direction from the positive horizontal direction to the bar direction. For a global restraint, a global degree of freedom number equal to $NP + 1$ is used.

In Table 10.9.1, there are 16 quantities in each local stiffness matrix and there are $NP \times NP$, or 16, slots in the global stiffness matrix. The quantities in all the local stiffness matrices are then fed *directly* into these 16 slots in the global stiffness matrix, ignoring these quantities in any $G5$ row or $G5$ column; hence the name *direct stiffness method*. The direct stiffness method may not

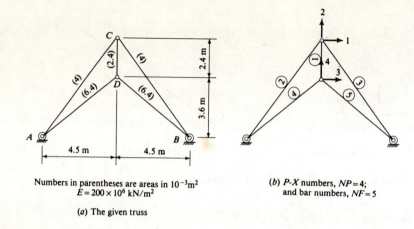

(a) The given truss

Numbers in parentheses are areas in $10^{-3} m^2$
$E = 200 \times 10^6$ kN/m²

(b) P-X numbers, NP = 4; and bar numbers, NF = 5

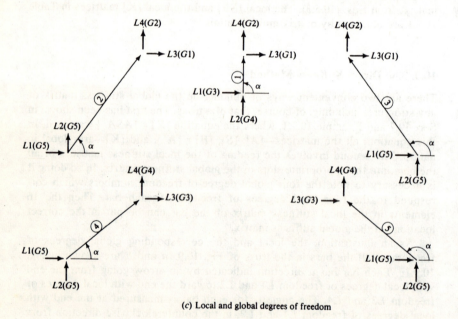

(c) Local and global degrees of freedom

Figure 10.9.1 Direct stiffness method.

MATRIX DISPLACEMENT METHOD OF TRUSS ANALYSIS 373

be so convenient as the global $[K] = [ASA^T]$ method in longhand computations; however, the opposite is true in the computer method, because in a computer program† all the local stiffness matrices can be fed into the global stiffness matrix in a single DO LOOP.

†For instance, see ibid., Appendix J.

Table 10.9.1 Local stiffness matrices of the bars in truss of Example 10.9.1

Bar 1: $S = 200,000$ kN/m; $\alpha = 90°$				
m \ kN	L1(G3)	L2(G4)	L3(G1)	L4(G2)
L1(G3)	0	0	0	0
L2(G4)	0	+200,000	0	−200,000
L3(G1)	0	0	0	0
L4(G2)	0	−200,000	0	+200,000

Bar 2: $S = 106,667$ kN/m; $\alpha = 53.13°$				
m \ kN	L1(G5)	L2(G5)	L3(G1)	L4(G2)
L1(G5)	+38,400	+51,200	−38,400	−51,200
L2(G5)	+51,200	+68,267	−51,200	−68,267
L3(G1)	−38,400	−51,200	+38,400	+51,200
L4(G2)	−51,200	−68,267	+51,200	+68,267

Bar 3: $S = 106,667$ kN/m; $\alpha = 126.87°$				
m \ kN	L1(G5)	L2(G5)	L3(G1)	L4(G2)
L1(G5)	+38,400	−51,200	−38,400	+51,200
L2(G5)	−51,200	+68,267	+51,200	−68,267
L3(G1)	−38,400	+51,200	+38,400	−51,200
L4(G2)	+51,200	−68,267	−51,200	+68,267

Table 10.9.1 (Cont.)

Bar 4: $S = 222{,}114$ kN/m; $\alpha = 38.66°$

kN \ m	L1(G5)	L2(G5)	L3(G3)	L4(G4)
L1(G5)	+135,435	+108,348	−135,435	−108,348
L2(G5)	+108,348	+ 86,679	−108,348	− 86,679
L3(G3)	−135,435	−108,348	+135,435	+108,348
L4(G4)	−108,348	− 86,679	+108,348	+ 86,679

Bar 5: $S = 222{,}114$ kN/m; $\alpha = 141.34°$

kN \ m	L1(G5)	L2(G5)	L3(G3)	L4(G4)
L1(G5)	+135,435	−108,348	−135,435	+108,348
L2(G5)	−108,348	+ 86,679	+108,348	− 86,679
L3(G3)	−135,435	+108,348	+135,435	−108,348
L4(G4)	+108,348	− 86,679	−108,348	+ 86,679

Global $[K] = \Sigma \text{local}\,[K]$

kN \ m	G1	G2	G3	G4
G1	+76,800	0	0	0
G2	0	+336,534	0	−200,000
G3	0	0	+270,870	0
G4	0	−200,000	0	+373,358

10.10 The Law of Reciprocal Forces and the Law of Reciprocal Displacements

The global stiffness matrix $[K] = [ASA^T]$ can be proved to be always symmetric, or

$$K_{ij} = K_{ji} \tag{10.10.1}$$

To show that a square matrix is symmetric, following Eq. (9.9.3a) it is necessary only to prove that its transpose is identical to itself, or

$$[ASA^T]^T = [ASA^T]$$

Proof

Statement	Reason
(1) $[ASA^T]^T = [SA^T]^T[A]^T$	(1) Apply Eq. (9.9.1).
(2) Right side of (1) $= [A^T]^T[S]^T[A^T]$ $= [A][S]^T[A^T]$	(2) Apply Eq. (9.9.1); also, a matrix twice transposed becomes itself again, by definition.
(3) Right side of (2) $= [A][S][A^T]$	(3) The $[S]$ matrix, being a diagonal matrix, is symmetric.
(4) $[ASA^T]^T = [ASA^T]$	(4) Apply steps 1, 2, and 3 above.

The global stiffness matrix $[K] = [ASA^T]$ expresses the joint forces in terms of the joint displacements; or

$$\{P\}_{NP \times 1} = [K]_{NP \times NP}\{X\}_{NP \times 1} \qquad (10.10.2)$$

The inverse of the global stiffness matrix may be called the *global flexibility matrix* $[\delta]$, which expresses the joint displacements in terms of the joint forces, or

$$\{X\}_{NP \times 1} = [K^{-1}]_{NP \times NP}\{P\}_{NP \times 1} = [\delta]_{NP \times NP}\{P\}_{NP \times 1} \qquad (10.10.3)$$

Applying Eq. (9.9.5), which states that the inverse of a symmetric matrix is still symmetric,

$$\delta_{ij} = \delta_{ji} \qquad (10.10.4)$$

Equation (10.10.1) is the law of reciprocal forces, which states that the force in the ith degree of freedom due to a unit displacement in the jth degree of freedom is equal to the force in the jth degree of freedom due to a unit displacement in the ith degree of freedom. Likewise, Eq. (10.10.4) is the law of reciprocal displacements, which states that the displacement in the ith degree of freedom due to a unit force in the jth degree of freedom is equal to the displacement in the jth degree of freedom due to a unit force in the ith degree of freedom. These two reciprocal laws can also be directly derived as corollaries of the general reciprocal virtual-work theorem proved in Sec. 4.3.

10.11 The Joint-Force Matrix $\{P\}$

The imposition on a truss may be (1) forces actually applied on the joints, (2) fabrication errors or temperature changes in the lengths of some bars, or (3) support settlements. Categories 2 and 3 will be treated in Secs. 10.13 and 10.14. So far as category 1 is concerned, that of forces actually applied on the joints, the joint-force matrix $\{P\}$ can be easily filled out from the given data, with the number of rows equal to the degree of freedom and the number of columns equal to the number of loading conditions NLC.

10.12 Numerical Examples

The matrix displacement method of truss analysis, as in all other finite-element methods, involves two distinct major steps; these are as follows:

$$\{X\}_{NP \times NLC} = [K]^{-1}_{NP \times NP} \{P\}_{NP \times NLC} \qquad (10.12.1)$$

and
$$\{F\}_{NF \times NLC} = [SA^T]_{NF \times NP} \{X\}_{NP \times NLC} \qquad (10.12.2)$$

The global stiffness matrix $[K]$ may be built up from the equation $[K] = [ASA^T]$ using matrix multiplication, or alternatively it may be built up directly by compiling the local stiffness matrices of all the bars in the truss. The former method will be used in the longhand numerical examples in this section, mostly because of its educational value, but the latter procedure is more automatic in computer programs. The same is true in applying Eq. (10.12.2); in the first method, all symbols in this equation are global in nature, as the subscripts show, but in the second method Eq. (10.12.2) becomes

$$[F_j]_{1 \times NLC} = [S_j A_j^T]_{1 \times 4} \{X\}_{4 \times NLC} \qquad (10.12.3)$$

so that this equation should be applied locally NF times in the DO LOOP of the computer program.

Example 10.12.1 Analyze the truss shown in Fig. 10.12.1a by the matrix displacement method. Note that this truss is the same as the one in Examples 5.2.1 and 5.3.1.

SOLUTION (a) *The output matrices* $\{X\}$ *and* $\{F\}$. The $[K] = [ASA^T]$ matrix has been worked out in Table 10.7.1.

$$\{X\}_{4 \times 1} = [K]^{-1}_{4 \times 4} \{P\}_{4 \times 1}$$

$$= \begin{bmatrix} +76.8 & 0. & 0. & 0. \\ 0. & +336.53 & 0. & -200. \\ 0. & 0. & +270.87 & 0. \\ 0. & -200. & 0. & +373.35 \end{bmatrix}^{-1} \begin{Bmatrix} 0. \\ -54. \\ 0. \\ 0. \end{Bmatrix}$$

$$= \begin{Bmatrix} 0. \\ -0.2354 \times 10^{-3} \text{ m} \\ 0. \\ -0.1261 \times 10^{-3} \text{ m} \end{Bmatrix}$$

In the Gauss-Jordan eliminations for the solution of the simultaneous equations, all pivots invariably stay on the main diagonal, even if a full-panel search is attempted for the element with the largest absolute value. This is naturally so because each element on the main

MATRIX DISPLACEMENT METHOD OF TRUSS ANALYSIS **377**

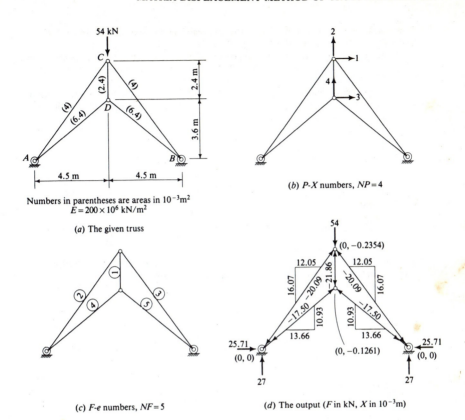

(a) The given truss

(b) P-X numbers, NP=4

(c) F-e numbers, NF=5

(d) The output (F in kN, X in 10^{-3}m)

Numbers in parentheses are areas in $10^{-3} m^2$
$E = 200 \times 10^6$ kN/m²

Figure 10.12.1 Truss of Example 10.12.1.

diagonal must be predominant in its own row since it represents the force and displacement along the same degree of freedom. Using $[SA^T]$ from Table 10.7.1,

$$\{F\}_{5\times 1} = [SA^T]_{5\times 4}\{X\}_{4\times 1} = \begin{bmatrix} 0. & +200. & 0. & -200. \\ +64. & +85.33 & 0. & 0. \\ -64. & +85.33 & 0. & 0. \\ 0. & 0. & +173.44 & +138.75 \\ 0. & 0. & -173.44 & +138.75 \end{bmatrix} \begin{Bmatrix} 0. \\ -0.2354 \\ 0. \\ -0.1261 \end{Bmatrix}$$

$$= \begin{Bmatrix} -21.86 \text{ kN} \\ -20.09 \text{ kN} \\ -20.09 \text{ kN} \\ -17.50 \text{ kN} \\ -17.50 \text{ kN} \end{Bmatrix}$$

(b) *The statics and deformation checks.* The four X values and the five F values from the output are shown on the truss diagram of Fig. 10.12.1d. Arrows indicating pulling or pushing of the bar forces on the joints are added, and so are the horizontal and vertical components of each inclined bar force. The four reactions are next computed from $\Sigma F_x = 0$ and $\Sigma F_y = 0$ at joints A and B.

The number of statics checks is equal to $NP = 4$; they require that the joints are in equilibrium along the degrees of freedom. These four checks can be obtained by inspection of the free-body diagrams of the joints in Fig. 10.12.1d.

The number of deformation checks is equal to $NF = 5$; they require that the change in length of each bar due to the force in it is equal to the change in length due to the joint displacements of the ends as computed by Eq. (3.6.2), which is

$$e = \Delta L = (X_3 - X_1)\cos\alpha + (X_4 - X_2)\sin\alpha$$

For instance, for bar 2,

$$\frac{F_2 L_2}{E A_2} \stackrel{?}{=} (0 - 0)(+0.6) + (-0.2354 \times 10^{-3} - 0)(+0.8)$$

$$\frac{-20.09(7.5)}{(200 \times 10^6)(4 \times 10^{-3})} \stackrel{?}{=} -0.1883 \times 10^{-3} \text{ m}$$

$$-0.1883 \times 10^{-3} \text{ m} = -0.1883 \times 10^{-3} \text{ m} \quad \text{(OK)}$$

When all the statics and deformation checks are obtained, the correctness of the solution is certain.

Example 10.12.2 Analyze the truss shown in Fig. 10.12.2a by the matrix displacement method. Note that this truss is the same as the one in Example 5.2.2.

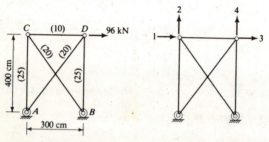

Numbers in parentheses are areas in cm²
$E = 20{,}000$ kN/cm²

(a) The given truss

(b) P-X numbers, $NP = 4$

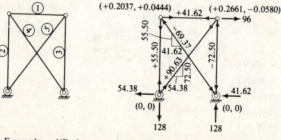

(c) F-e numbers, $NF = 5$

(d) The output (F in kN, X in cm)

Figure 10.12.2 Truss of Example 10.12.2.

Table 10.12.1 Global stiffness matrix $[K] = [ASA^T]$ of truss in Example 10.12.1†

$[A] = $

P \ F	1	2	3	4	5
1	−1.0	0.	0.	−0.6	0.
2	0.	+1.0	0.	+0.8	0.
3	+1.0	0.	0.	0.	+0.6
4	0.	0.	+1.0	0.	+0.8

$[B] = $

e \ X	1	2	3	4
1	−1.0	0.	+1.0	0.
2	0.	+1.0	0.	0.
3	0.	0.	0.	+1.0
4	−0.6	+0.8	0.	0.
5	0.	0.	+0.6	+0.8

$[S] = $

F \ e	1	2	3	4	5
1	+666.66				
2		+1250.			
3			+1250.		
4				+800.	
5					+800.

$[SB] = $

F \ X	1	2	3	4
1	−666.66	0.	+666.66	0.
2	0.	+1250.	0.	0.
3	0.	0.	0.	+1250.
4	−480.	+640.	0.	0.
5	0.	0.	+480.	+640.

Table 10.12.1 (Cont.)

P \ X	1	2	3	4
1	+954.66	−384.	−666.66	0.
2	−384.	+1762.	0.	0.
3	−666.66	0.	+954.66	+384.
4	0.	0.	+384.	+1762.

$[ASB] =$ (shown at left of table)

†Dimensional units, where applicable, are in kN/cm.

SOLUTION (a) *The output matrices $\{X\}$ and $\{F\}$.* The P-X and F-e numbers are shown in Fig. 10.2.2b and c, wherein $NP = 4$, $NF = 5$, and $NI = 1$. The $[A]$, $[B]$, $[S]$, $[SB]$, and $[K] = [ASB]$ matrices are shown in Table 10.12.1.

$$\{X\}_{4\times 1} = [K]^{-1}_{4\times 4}\{P\}_{4\times 1}$$

$$= \begin{bmatrix} +954.66 & -384. & -666.66 & 0. \\ -384. & +1762. & 0. & 0. \\ -666.66 & 0. & +954.66 & +384. \\ 0. & 0. & +384. & +1762. \end{bmatrix}^{-1} \begin{Bmatrix} 0. \\ 0. \\ +96. \\ 0. \end{Bmatrix}$$

$$= \begin{Bmatrix} +0.2037 \text{ cm} \\ +0.0444 \text{ cm} \\ +0.2661 \text{ cm} \\ -0.0580 \text{ cm} \end{Bmatrix}$$

Using $[SA^T]$ from Table 10.12.1,

$$\{F\}_{5\times 1} = [SA^T]_{5\times 4}\{X\}_{4\times 1}$$

$$= \begin{bmatrix} -666.66 & 0. & +666.66 & 0. \\ 0. & +1250. & 0. & 0. \\ 0. & 0. & 0. & +1250. \\ -480. & +640. & 0. & 0. \\ 0. & 0. & +480. & +640. \end{bmatrix} \begin{Bmatrix} +0.2037 \\ +0.0444 \\ +0.2661 \\ -0.0580 \end{Bmatrix}$$

$$= \begin{Bmatrix} +41.62 \text{ kN} \\ +55.50 \text{ kN} \\ -72.50 \text{ kN} \\ -69.37 \text{ kN} \\ +90.63 \text{ kN} \end{Bmatrix}$$

(b) *The statics and deformation checks.* The same discussion in part (b) of Example 10.12.1 applies equally well here.

Example 10.12.3 Analyze the truss shown in Fig. 10.12.3a by the matrix displacement method. Note that this truss is the same as the one in Example 5.3.4.

MATRIX DISPLACEMENT METHOD OF TRUSS ANALYSIS 381

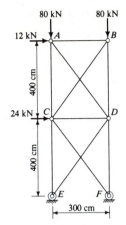
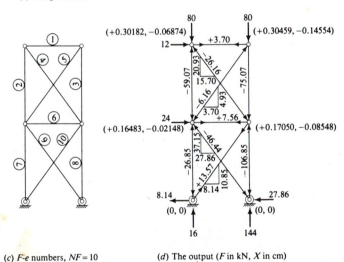

(a) The given truss
Numbers in parentheses are areas in cm²
$E = 20,000$ kN/cm²

(b) P-X numbers, $NP = 8$

(c) F-e numbers, $NF = 10$

(d) The output (F in kN, X in cm)

Figure 10.12.3 Truss of Example 10.12.3.

SOLUTION (a) *The output matrices $\{X\}$ and $\{F\}$.* The P-X and F-e numbers are shown in Fig. 10.12.3b and c, wherein $NP = 8$, $NF = 10$, and $NI = 2$. The $[A]$, $[B]$, $[S]$, $[SB]$, and $[K] = [ASB]$ matrices are shown in Table 10.12.2.

The contents of the $\{P\}$ matrix are $P_1 = +12$ kN, $P_2 = -80$ kN, $P_3 = 0$, $P_4 = -80$ kN, $P_5 = +24$ kN, $P_6 = 0$, $P_7 = 0$, and $P_8 = 0$. From

$$\{X\}_{8\times 1} = [K]^{-1}_{8\times 8}\{P\}_{8\times 1} = [K \text{ in Table } 10.12.2]^{-1}\{P\}$$

Table 10.12.2 Global stiffness matrix $[K] = [ASA^T]$ of truss in Example 10.12.3†

$[A] =$

P \ F	1	2	3	4	5	6	7	8	9	10
1	−1.0			−0.6						
2		+1.0		+0.8						
3	+1.0				+0.6					
4			+1.0		+0.8					
5					−0.6	−1.0			−0.6	
6		−1.0			−0.8		+1.0	+0.8		
7				+0.6		+1.0				+0.6
8			−1.0	−0.8				+1.0		+0.8

$[B] =$

e \ X	1	2	3	4	5	6	7	8
1	−1.0		+1.0					
2		+1.0				−1.0		
3				+1.0				−1.0
4	−0.6	+0.8					+0.6	−0.8
5			+0.6	+0.8	−0.6	−0.6		
6					−1.0		+1.0	
7						+1.0		
8								+1.0
9					−0.6			
10							+0.6	+0.8

$[S] =$

F \ e	1	2	3	4	5	6	7	8	9	10
1	+1333.3									
2		+1250.								
3			+1250.							
4				+400.						
5					+400.					
6						+1333.3				
7							+1250.			
8								+1250.		
9									+400.	
10										+400.

MATRIX DISPLACEMENT METHOD OF TRUSS ANALYSIS 383

$[SB] =$

F \ X	1	2	3	4	5	6	7	8
1	−1333.3		+1333.3					
2		+1250.				−1250.		
3				+1250.				−1250.
4	− 240.	+ 320.					+ 240.	− 320.
5			+ 240.	+ 320.	− 240.	− 320.		
6					−1333.3		+1333.3	
7						+1250.		
8								+1250.
9					− 240.	+ 320.		
10							+ 240.	+ 320.

$[ASB] =$

P \ X	1	2	3	4	5	6	7	8
1	+1477.3	− 192.	−1333.3				− 144.	+ 192.
2	− 192.	+1506.				−1250.	+ 192.	− 256.
3	−1333.3		+1477.3	+ 192.	− 144.	− 192.		
4			+ 192.	+1506.	− 192.	− 256.		−1250.
5			− 144.	− 192.	+1621.3		−1333.3	
6			−1250.	− 192.	− 256.		+3012.	
7	− 144.	+ 192.			−1333.3		+1621.3	
8	+ 192.	− 256.		−1250.				+3012.

†Dimensional units where applicable are in kN/cm.

the contents of the output matrix $\{X\}$ are found to be

$X_1 = +0.30182$ cm $X_2 = -0.06874$ cm
$X_3 = +0.30459$ cm $X_4 = -0.14554$ cm
$X_5 = +0.16483$ cm $X_6 = -0.02148$ cm
$X_7 = +0.17050$ cm $X_8 = -0.08548$ cm

$\{F\}_{10\times 1} = [SA^T]_{10\times 8}\{X\}_{8\times 1}$
$= [SA^T$ in Table 10.12.2]$\{X$ from preceding equation$\}$

the contents of the output matrix $\{F\}$ are found to be

$F_1 = +\ 3.70$ kN $F_2 = -\ 59.07$ kN
$F_3 = -75.07$ kN $F_4 = -\ 26.16$ kN
$F_5 = -\ 6.16$ kN $F_6 = +\ 7.56$ kN
$F_7 = -26.85$ kN $F_8 = -106.85$ kN
$F_9 = -46.44$ kN $F_{10} = +\ 13.57$ kN

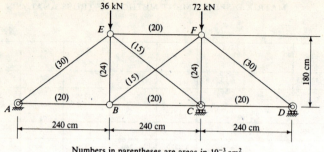

Numbers in parentheses are areas in 10^{-3} cm^2
$E = 20{,}000$ kN/cm^2

(a) The given truss

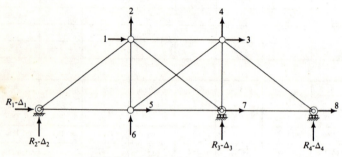

(b) P-X numbers, $NP = 18$
(R-Δ numbers for use in Example 10.14.1)

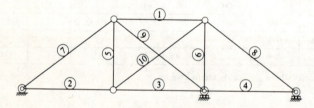

(c) F-e numbers, $NF = 10$

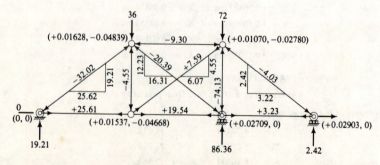

(d) The output (F in kN, X in cm)

Figure 10.12.4 Truss of Example 10.12.4.

MATRIX DISPLACEMENT METHOD OF TRUSS ANALYSIS 385

(b) *The statics and deformation checks.* The same discussion in part (b) of Example 10.12.1 applies equally well here.

Example 10.12.4 Analyze the truss shown in Fig. 10.12.4a by the matrix displacement method. Note that this truss is the same as the one in Example 5.4.1.

SOLUTION (a) *The output matrices* $\{X\}$ *and* $\{F\}$. The P-X and F-e numbers are shown in Fig. 10.12.4b and c, wherein $NP = 8$, $NF = 10$, and $NI = 2$. The $[A]$, $[B]$, $[S]$, $[SB]$, and $[K] = [ASB]$ matrices are shown in Table 10.12.3.

The contents of the $\{P\}$ matrix are $P_1 = 0$, $P_2 = -36$ kN, $P_3 = 0$, $P_4 = -72$ kN, $P_5 = 0$, $P_6 = 0$, $P_7 = 0$, and $P_8 = 0$. From

$$\{X\}_{8\times1} = [K]^{-1}_{8\times8}\{P\}_{8\times1} = [K \text{ in Table } 10.12.3]^{-1}\{P\}$$

Table 10.12.3 Global stiffness matrix $[K] = [ASA^T]$ of truss in Example 10.12.4†

$[A] =$

F \ P	1	2	3	4	5	6	7	8	9	10
1	−1.0						+0.8		−0.8	
2					+1.0		+0.6		+0.6	
3	+1.0							−0.8		+0.8
4						+1.0		+0.6		+0.6
5		+1.0	−1.0						−0.8	
6						−1.0				−0.6
7			+1.0	−1.0					+0.8	
8				+1.0				+0.8		

$[B] =$

X \ e	1	2	3	4	5	6	7	8
1	−1.0		+1.0					
2					+1.0			
3					−1.0		+1.0	
4							−1.0	+1.0
5		+1.0				−1.0		
6				+1.0				
7	+0.8	+0.6						
8			−0.8	+0.6				+0.8
9	−0.8	+0.6					+0.8	
10			+0.8	+0.6	−0.8	+0.6		

Table 10.12.3 (Cont.)

$[S] =$

e\F	1	2	3	4	5	6	7	8	9	10
1	+1666.7									
2		+1666.7								
3			+1666.7							
4				+1666.7						
5					+2666.7					
6						+2666.7				
7							+2000.			
8								+2000.		
9									+1000.	
10										+1000.

$[SB] =$

X\F	1	2	3	4	5	6	7	8
1	−1666.7		+1666.7					
2					+1666.7			
3					−1666.7		+1666.7	
4							−1666.7	+1666.7
5		+2666.7				−2666.7		
6				+2666.7				
7	+1600.	+1200.						
8			−1600.	+1200.				+1600.
9	−800.	+600.					+800.	
10			+800.	+600.	−800.	−600.		

$[ASB] =$

X\P	1	2	3	4	5	6	7	8
1	+3586.7	+480.	−1666.7				−640.	
2	+480.	+3746.7				−2666.7	+480.	
3	−1666.7		+3586.7	−480.	−640.	−480.		−1280.
4			−480.	+3746.7	−480.	−360.		+960.
5			−640.	−480.	+3973.6	+480.	−1666.7	
6		−2666.7	−480.	−360.	+480.	+3026.7		
7	−640.	+480.			−1666.7		+3973.3	−1666.7
8			−1280.	+960.			−1666.7	+2946.7

†Dimensional units, where applicable, are in kN/cm.

the contents of the output matrix $\{X\}$ are found to be

$$X_1 = +0.01628 \text{ cm} \qquad X_2 = -0.04839 \text{ cm}$$
$$X_3 = +0.01070 \text{ cm} \qquad X_4 = -0.02780 \text{ cm}$$
$$X_5 = +0.01537 \text{ cm} \qquad X_6 = -0.04668 \text{ cm}$$
$$X_7 = +0.02709 \text{ cm} \qquad X_8 = +0.02903 \text{ cm}$$

From

$$\{F\}_{10\times 1} = [SA^T]_{10\times 8}\{X\}_{8\times 1}$$
$$= [SA^T \text{ in Table } 10.12.3]\{X \text{ from preceding equation}\}$$

the contents of the output matrix $\{F\}$ are found to be

$$F_1 = -\ 9.30 \text{ kN} \qquad F_2 = +25.61 \text{ kN}$$
$$F_3 = +19.54 \text{ kN} \qquad F_4 = +\ 3.23 \text{ kN}$$
$$F_5 = -\ 4.55 \text{ kN} \qquad F_6 = -74.13 \text{ kN}$$
$$F_7 = -32.02 \text{ kN} \qquad F_8 = -\ 4.03 \text{ kN}$$
$$F_9 = -20.39 \text{ kN} \qquad F_{10} = +\ 7.59 \text{ kN}$$

(b) *The statics and deformation checks.* The same discussion in part (b) of Example 10.12.1 applies equally well here.

10.13 Effect of Fabrication Errors or Temperature Changes

Occasionally design considerations may be such that the effect of fabrication errors or temperature changes in the lengths of some bars needs to be investigated. If the positions of the holes for pins or bolts at the ends of some bars are mislocated, the assembled truss will be out of shape. Likewise when there is sudden rise or fall in the temperature of some bars, the truss will assume a different form. However, if the truss is statically determinate, reactions would not be induced to act on the truss nor would axial forces be developed in the bars. This is obvious when one considers Eq. (10.5.4), $\{F\} = [A^{-1}]\{P\}$, for a statically determinate and stable truss in that if $\{P\}$ is zero, $\{F\}$ must be zero. If the joint displacements are required, Eq. (10.5.5), $\{X\} = [A^T]^{-1}\{e_0\}$, may be used in that $\{e_0\}$ is the column matrix of possible fabrication errors or of length changes due to a rise or fall in temperature.

For statically indeterminate trusses with $NR = 3$, small changes in the lengths of the bars would not induce reactions to act on the truss, but there would be axial forces developed in the bars within the region that involves change in the shape of the truss as well as redundancy of bars. For statically indeterminate trusses with $NR \geq 4$, small changes in the lengths of the bars would usually induce both reactions on the truss and forces in the bars.

To determine by the matrix displacement method the forces in the bars due to small changes in the lengths of the bars, one approach is to derive formulas for the bar forces $\{F\}$ in terms of the properties of the truss defined by the $[A]$, $[B]$, and $[S]$ matrices and the given small changes in bar lengths defined by the $\{e_0\}$ matrix. The statics equation

$$\{P\} = [A]\{F\} \qquad (10.13.1)$$

remains always true, as does the deformation equation

$$\{e\} = [B]\{X\} \qquad (10.13.2)$$

But

$$\{F\} = [S]\{e - e_0\} \qquad (10.13.3)$$

because the force in the bar has to be that corresponding to the additional elongation $e - e_0$. Substituting Eqs. (10.13.2) and (10.13.3) into Eq. (10.13.1),

$$\{P\} = [A]\{F\} = [A][S]\{e - e_0\} = [A][S]\{e\} - [A][S]\{e_0\}$$
$$= [A][S][B]\{X\} - [A][S]\{e_0\} \qquad (10.13.4)$$

Setting $\{P\}$ in Eq. (10.13.4) equal to zero and solving for $\{X\}$,

$$\{X\} = [ASA^T]^{-1}[A][S]\{e_0\} \qquad (10.13.5)$$

Substituting Eq. (10.13.2) into Eq. (10.13.3),

$$\{F\} = [SA^T]\{X\} - [S]\{e_0\} \qquad (10.13.6)$$

Example 10.13.1 Determine the effect of a fabrication error on the truss of Fig. 10.12.4a, in which bar CE is made 0.75 cm too short.

SOLUTION (a) *The output matrices* $\{X\}$ *and* $\{F\}$. Using Eq. (10.13.5) and Table 10.12.3,

$$\{X\}_{8\times 1} = [ASA^T]^{-1}_{8\times 8}[A]_{8\times 10}[S]_{10\times 10}\{e_0\}_{10\times 1}$$
$$= [ASA^T]^{-1}[A][S][0.\ 0.\ 0.\ 0.\ 0.\ 0.\ 0.\ 0.\ -0.750\ 0.]^T$$
$$= [ASA^T]^{-1}[A][0.\ 0.\ 0.\ 0.\ 0.\ 0.\ 0.\ 0.\ -750.\ 0.]^T$$
$$= [ASA^T]^{-1}[+600.\ -450.\ 0.\ 0.\ 0.\ 0.\ -600.\ 0.]^T$$

The X values obtained from the preceding equation are

$X_1 = +0.24630$ cm $X_2 = -0.38136$ cm
$X_3 = +0.07796$ cm $X_4 = -0.02169$ cm
$X_5 = +0.03052$ cm $X_6 = -0.33106$ cm
$X_7 = -0.04627$ cm $X_8 = +0.01476$ cm

Using Eq. (10.13.6), $[SA^T]$ from Table 10.12.3, and $\{X\}$ and $\{Se_0\}$ from the preceding equation,

$$\{F\}_{10\times 1} = [SA^T]_{10\times 8}\{X\}_{8\times 1} - [S]_{10\times 10}\{e_0\}_{10\times 1}$$

The F values obtained from the preceding equation are

$F_1 = -280.56$ kN $F_2 = +\ 50.86$ kN
$F_3 = -127.99$ kN $F_4 = +101.72$ kN
$F_5 = -134.14$ kN $F_6 = -\ 57.85$ kN
$F_7 = -\ 63.57$ kN $F_8 = -127.14$ kN
$F_9 = +287.13$ kN $F_{10} = +223.56$ kN

(b) *The statics and deformation checks.* The eight X values and the ten F values from the output are shown on the truss diagram of Fig. 10.13.1. Arrows indicating pulling or pushing of the bar forces on the joints are added, and so are the horizontal and vertical

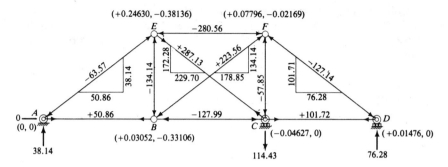

Figure 10.13.1 Output for Example 10.13.1.

components of each inclined bar force. The four reactions are next computed from $\Sigma F_x = 0$ at joint A, and $\Sigma F_y = 0$ at joints A, C, and D.

The $NP = 8$ statics checks require equilibrium of the joints in the directions of the eight degrees of freedom. These are done by inspection using Fig. 10.13.1.

The $NF = 10$ deformation checks require that the change in length of every bar due to the combined effect of the fabrication error and that computed from FL/EA be identical to that computed from the joint-displacement equation, Eq. (3.6.2), using the joint displacements shown in Fig. 10.13.1. Thus, for member CE, the only member with fabrication error,

$$-0.750 + \frac{287.13(300)}{20{,}000(15)} \stackrel{?}{=} (+0.24630 + 0.04627)(-0.8) + (-0.38136 - 0)(+0.6)$$

$$-0.75000 + 0.28713 \stackrel{?}{=} -0.46287$$

$$-0.46287 = -0.46287 \quad \text{(OK)}$$

10.14 Effect of Support Settlements

Sometimes in the design process, in case of long-span bridge trusses in particular, there is need to investigate the effect of uneven settlements of the supports. Usually the analysis is done for one support at a time, and the resulting forces in the bars may then be combined with those due to dead load, live load, and sometimes impact load as an added percentage of the live load. If the number of unknown reactions is equal to three, which is the minimum number required for statical equilibrium, any support settlement will not affect the shape of the truss, even if the truss contains redundant bars within it. However, if the truss has more than two supports exerting vertical reactions and one exerting the horizontal reaction, any nonalignment of the supports will induce a set of vertical reacting forces to act on the truss. In the matrix method of truss analysis, the sequence involves first the determination of the joint displacements using the support displacements and truss properties as input data, then the bar forces, and finally the induced reactions. The needed formulas are derived below.

Consider a typical truss as shown in Fig. 10.14.1. In Fig. 10.14.1b, not only the P-X numbers but also the R-Δ numbers are shown. It should be emphasized that both the R's and Δ's are in the direction of the arrows; thus in ordinary cases of support settlements, the contents of the input $\{\Delta\}_{4\times1}$ matrix would be negative. Define a statics matrix $[A_R]$ to express the reactions in terms of the bar forces; this matrix can be established either by rows or by columns in the same way as the ordinary statics matrix $[A]$, which could have been denoted by $[A_P]$. For the truss in Fig. 10.14.1,

$[A_R]_{4\times8} = $

F \ R	1	2	3	4	5	6	7	8
1	-0.8	0.	0.	-1.0	0.	0.	0.	0.
2	-0.6	0.	0.	0.	0.	0.	0.	0.
3	0.	0.	0.	0.	0.	-1.0	0.	-0.6
4	0.	0.	-1.0	0.	0.	0.	-0.6	0.

From the definition of the $[A_R]$ matrix,

$$\{R\}_{NR\times1} = [A_R]_{NR\times NF}\{F\}_{NF\times1} \qquad (10.14.1)$$

By the principle of virtual work, the elongations in the bars due to support displacements would be equal to $[A_R^T]\{\Delta\}$; thus the elongations in the bars due to both joint displacements and support settlements are

$$\{e\}_{NF\times1} = [A^T]_{NF\times NP}\{X\}_{NP\times1} + [A_R^T]_{NF\times NR}\{\Delta\}_{NR\times1} \qquad (10.14.2)$$

The forces in the bars are, using Eq. (10.14.2),

$$\{F\}_{NF\times1} = [S]_{NF\times NF}\{e\}_{NF\times1} = [S]_{NF\times NF}\{A^TX + A_R^T\Delta\}_{NF\times1}$$
$$= [SA^T]_{NF\times NP}\{X\}_{NP\times1} + [SA_R^T]_{NF\times NR}\{\Delta\}_{NR\times1} \qquad (10.14.3)$$

Then,

$$\{P\}_{NP\times1} = [A]_{NP\times NF}\{F\}_{NF\times1}$$
$$= [ASA^T]_{NP\times NP}\{X\}_{NP\times1} + [ASA_R^T]_{NP\times NR}\{\Delta\}_{NR\times1} \qquad (10.14.4)$$

Setting $\{P\}$ in Eq. (10.14.4) equal to zero and solving for $\{X\}$,

$$\{X\}_{NP\times1} = [ASA^T]^{-1}_{NP\times NP}\{-ASA_R^T\Delta\}_{NP\times1} \qquad (10.14.5)$$

In actually solving a problem involving support settlements, Eqs. (10.14.5), (10.14.3), and (10.14.1) will be used in this order to obtain $\{X\}$, $\{F\}$, and $\{R\}$.

MATRIX DISPLACEMENT METHOD OF TRUSS ANALYSIS 391

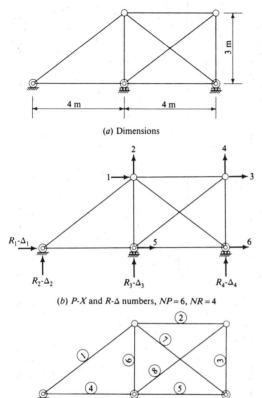

(a) Dimensions

(b) P-X and R-Δ numbers, NP=6, NR=4

(c) F-e numbers, NF=8

Figure 10.14.1 P-X, R-Δ, and F-e numbers for a typical truss.

Example 10.14.1 Determine the effect of a vertical settlement of 0.36 cm of the support at C of the truss shown in Fig. 10.12.4a.

SOLUTION (a) *The output matrices* $\{X\}$ *and* $\{F\}$. Assigning the R-Δ numbers as shown in Fig. 10.12.4b and using the P-X and F-e numbers in Fig. 10.12.4b and c, one can establish the $[A_R]$ and $[A_R^T]$ matrices independently of each other as follows:

	F \ R	1	2	3	4	5	6	7	8	9	10
	1	0.	−1.0	0.	0.	0.	0.	−0.8	0.	0.	0.
$[A_R]_{4 \times 10} =$	2	0.	0.	0.	0.	0.	0.	−0.6	0.	0.	0.
	3	0.	0.	0.	0.	0.	−1.0	0.	0.	−0.6	0.
	4	0.	0.	0.	0.	0.	0.	0.	−0.6	0.	0.

$$[A_R^T]_{10\times 4} =$$

e \ Δ	1	2	3	4
1	0.	0.	0.	0.
2	−1.0	0.	0.	0.
3	0.	0.	0.	0.
4	0.	0.	0.	0.
5	0.	0.	0.	0.
6	0.	0.	−1.0	0.
7	−0.8	−0.6	0.	0.
8	0.	0.	0.	−0.6
9	0.	0.	−0.6	0.
10	0.	0.	0.	0.

Using Eq. (10.14.5), the matrices $[ASA^T]$, $[A]$, and $[S]$ from Table 10.12.3, and $[A_R^T]$ from the preceding equation,

$$\{X\}_{8\times 1} = [ASA^T]^{-1}[-A][S][A_R^T][0.\ \ 0.\ \ -0.360\ \ 0.]^T$$
$$= [ASA^T]^{-1}[-A][S][0.\ \ 0.\ \ 0.\ \ 0.\ \ 0.\ \ +0.360\ \ 0.\ \ 0.\ \ +0.216\ \ 0.]^T$$
$$= [ASA^T]^{-1}[-A][0.\ \ 0.\ \ 0.\ \ 0.\ \ 0.\ \ +960.\ \ 0.\ \ 0.\ \ +216.\ \ 0.]^T$$
$$= [ASA^T]^{-1}[+172.8\ \ -129.6\ \ 0.\ \ -960.\ \ 0.\ \ 0.\ \ -172.8\ \ 0.]^T$$

The X values obtained from the preceding equation are

$X_1 = +0.12283$ cm $X_2 = −0.23605$ cm
$X_3 = +0.05484$ cm $X_4 = −0.31382$ cm
$X_5 = +0.04163$ cm $X_6 = −0.24320$ cm
$X_7 = +0.09853$ cm $X_8 = +0.18179$ cm

Using Eq. (10.14.3), the matrix $[SA^T]$ from Table 10.12.3, and $\{X\}$ and $\{SA_R^T\Delta\}$ from the preceding equations,

$$\{F\}_{10\times 1} = [SA^T]_{10\times 8}\{X\}_{8\times 1} + [SA_R^T]_{10\times 4}\{\Delta\}_{4\times 1}$$

The F values obtained from the preceding equation are

$F_1 = −113.32$ kN $F_2 = + 69.38$ kN
$F_3 = + 94.83$ kN $F_4 = +138.77$ kN
$F_5 = + 19.08$ kN $F_6 = +123.16$ kN
$F_7 = − 86.73$ kN $F_8 = −173.46$ kN
$F_9 = + 54.93$ kN $F_{10} = − 31.80$ kN

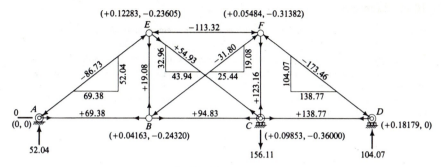

Figure 10.14.2 Output for Example 10.14.1.

(b) *The statics and deformation checks.* The eight X-values and the ten F values from the output are shown on the truss diagram of Fig. 10.14.2. Arrows indicating pulling or pushing of the bar forces on the joints are added, and so are the horizontal and vertical components of each inclined bar force. The reactions could have been computed from $\{R\} = [A_R]\{F\}$ but may as well have been found from the free-body diagrams of the joints A, C, and D in Fig. 10.14.2.

The $NP = 8$ statics checks require equilibrium of the joints in the directions of the eight degrees of freedom. These are done by inspection using Fig. 10.14.2.

The $NF = 10$ deformation checks require that the change in length of every bar computed from FL/EA be identical to that computed from the joint-displacement equation, Eq. (3.6.2), using the support settlements as well as the joint displacements shown in Fig. 10.14.2. Thus, for member CE,

$$\frac{(+54.94)(300)}{20,000(15)} \stackrel{?}{=} (+0.09853 - 0.12283)(+0.8) + (-0.36000 + 0.23605)(-0.6)$$

$$+0.05493 = +0.05493 \quad \text{(OK)}$$

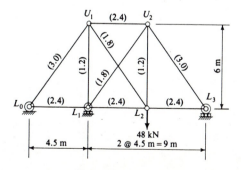

Numbers in parentheses are areas in 10^{-3}m^2
$E = 200 \times 10^6 \text{ kN/m}^2$

Figure 10.15.1 Exercise 10.3.

10.15 Exercises

10.1 Analyze the truss of Exercise 5.1 by the matrix displacement method.
10.2 Analyze the truss of Exercise 5.4 by the matrix displacement method.
10.3 Analyze the truss shown in Fig. 10.15.1 by the matrix displacement method.
10.4 Analyze by the matrix displacement method the truss of Exercise 10.3 for a fabrication error of 0.60 cm too long in the length of bar L_1U_2.
10.5 Analyze by the matrix displacement method the truss of Exercise 10.3 for a 0.45-cm settlement of the support at L_3.

CHAPTER
ELEVEN
MATRIX DISPLACEMENT METHOD OF BEAM ANALYSIS

11.1 Joints and Elements in Beams vs. Degree of Freedom

For a truss, whether it be statically determinate or indeterminate, the number of joints and the number of bars (hereafter called *elements*) are both definite and can be counted. For a beam, one can consider that there is a 180° welded (rigid) joint anywhere along the beam; hence the number of joints to be used in the analysis is up to the analyst. Sometimes even a simple beam may be divided into many elements so that the slopes and deflections at the joints, once they are obtained by means of a computer program, can be used to produce a graphical display of the elastic curve. The simple beam shown in Fig. 11.1.1a is divided into two elements, each 5 m long; thus the number of joints is three. It is convenient to mark the rigid-joint locations by solid circles as shown by Fig. 11.1.1b, in contrast to the open circles used for the pin joints in trusses.

Without regard to whatever imposition may act on the beam, be it applied loads or support settlements, each joint in Fig. 11.1.1b may be examined as to its freedom in rotation and in deflection. The unknown rotations are denoted by X_1, X_2, and X_3; and the unknown deflection, by X_4. Thus the degree of freedom NP is 4, much like that of a truss; the only difference is that for a beam there may be two kinds of joint displacements, rotations as well as deflections. Generally all rotations are numbered in sequence first, followed by deflections. Had this beam actually been subjected to concentrated moments or forces at the joints only, the contents of the $\{P\}$ matrix in the directions of the degree of freedom would be known at the outset. How

396 INTERMEDIATE STRUCTURAL ANALYSIS

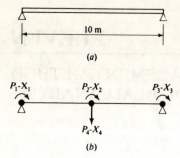

Figure 11.1.1 Joints and elements in a beam.

distributed and concentrated loads acting on the elements themselves can be treated will be discussed in the next section.

11.2 Sending Loads on the Elements to the Joints

The simple beam shown in Fig. 11.2.1a is divided into two elements with $NP = 4$. Suppose that a concentrated load of 54 kN acts on the right element, as shown in Fig. 11.2.1b. Now if the two joints at both ends of this element are completely fixed in rotation and in deflection, the fixed-end moments and

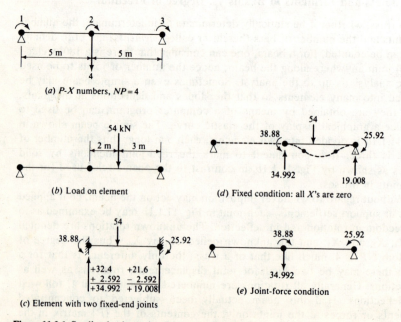

Figure 11.2.1 Sending loads on the element to the joints.

MATRIX DISPLACEMENT METHOD OF BEAM ANALYSIS 397

the fixed-end reactions can be determined and are shown in Fig. 11.2.1c. The complete force system in Fig. 11.2.1c is then transported to the simple beam, as shown in Fig. 11.2.1d, wherein the elastic curve of the right element should be identical to that of the element by itself in Fig. 11.2.1c. In Fig. 11.2.1d, all joint rotations and deflections are zero; hence the name *fixed condition*. Now place the opposites of the moments and forces acting on the joints of the fixed condition on a separate diagram as shown by Fig. 11.2.1e, which is called the *joint-force condition*, noting that the 19.008-kN downward force will be resisted by the right support of the simple beam. The physical superposition of Fig. 11.2.1d and e is obviously the given simple beam of Fig. 11.2.1b.

In the matrix displacement method of beam analysis, the problem to be solved is only that of the joint-force condition, such as in Fig. 11.2.1e, wherein all the loads, be they moments or forces, are applied at the joints, much like those acting on a truss. When the loads acting on the elements themselves are sent to the joints through the fixed condition in the manner just described, the joint displacements in the joint-force condition will indeed be identical to those of the beam to be analyzed.

11.3 Degree of Indeterminacy vs. Independent Unknown Forces

Attention is now focused on the joint-force condition alone, in which there is no load acting on any of the elements. The free-body diagram of a typical element, then, should consist of two moments acting on its two ends and a pair of equal but opposite end reactions. However, only the two end moments are independent unknowns, because the end reactions can be determined by statics from these two moments. For a truss, too, one can consider the axial force at one end as the independent unknown and that at the other end as obtainable by statics. At this point, in order to make the treatment on all matrix displacement methods a *unified* one, the name *generalized force*, or simply *force*, will be construed to mean either a translational force or a rotational force (which is a moment). Consequently the independent unknown forces on the two beam elements in Fig. 11.3.1b are denoted by F_1 to F_4, positive if clockwise. Note the cross marks on the elements in Fig. 11.3.1b; they are there to remind the reader that there is no load on the element in the joint-force condition.

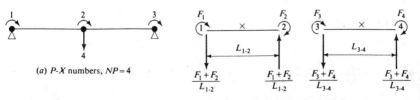

Figure 11.3.1 Free-body diagrams of the elements.

The number of independent unknown forces in a beam, NF, is always equal to twice the number of elements; for the simple beam of Fig. 11.3.1, $NF = 4$. As stated in Eq. (10.1.1), the degree of indeterminacy NI of a structure is always equal to $NF - NP$, or

$$NI = NF - NP \tag{11.3.1}$$

For the simple beam in Fig. 11.3.1, $NI = NF - NP = 4 - 4 = 0$.

11.4 The Statics Matrix [A]

As for a truss, in the joint-force condition of beams, the statics matrix $[A]$ expresses the joint forces in terms of the independent forces in the elements. In Sec. 10.5, suggestion was made that it is more convenient to establish the $[A]$ matrix of a truss by columns by assessing the effective components of the forces as they act on the bars in the directions of the degrees of freedom. The same concept may be applied to beams, although the advantages and disadvantages of establishing the $[A]$ matrix by rows or by columns are about equally divided. In any case it is easier to explain the basis of the $[A]$ matrix by rows, because each row is just an equation of statics. For the beam in Fig. 11.4.1, the $[A]$ matrix is

$$[A]_{4 \times 4} = \begin{array}{|c|c|c|c|c|} \hline \diagdown \begin{array}{c} F \\ P \end{array} & 1 & 2 & 3 & 4 \\ \hline 1 & +1.0 & & & \\ \hline 2 & & +1.0 & +1.0 & \\ \hline 3 & & & & +1.0 \\ \hline 4 & -\dfrac{1}{L_{1\text{-}2}} & -\dfrac{1}{L_{1\text{-}2}} & +\dfrac{1}{L_{3\text{-}4}} & +\dfrac{1}{L_{3\text{-}4}} \\ \hline \end{array} \tag{11.4.1}$$

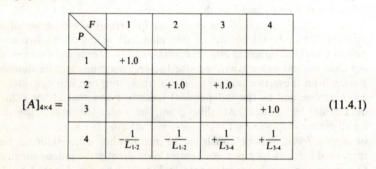

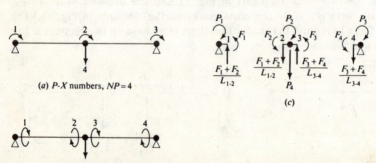

Figure 11.4.1 Free-body diagrams for the statics matrix $[A]$ of a beam.

MATRIX DISPLACEMENT METHOD OF BEAM ANALYSIS

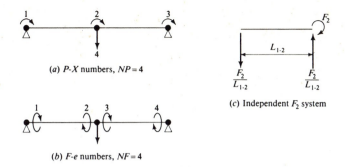

(a) P-X numbers, NP = 4

(b) F-e numbers, NF = 4

(c) Independent F_2 system

Figure 11.4.2 Establishing the [A] matrix by columns.

The four equations in the preceding matrix read as follows:

$$P_1 = +1.0F_1$$
$$P_2 = +1.0F_2 + 1.0F_3$$
$$P_3 = +1.0F_4$$
$$P_4 = -\frac{F_1 + F_2}{L_{1-2}} + \frac{F_3 + F_4}{L_{3-4}}$$

These are exactly the equations of statics which may be observed from the free-body diagrams of the joints shown in Fig. 11.4.1c.

To establish the [A] matrix of Eq. (11.4.1) by columns, for example the second column, one needs to visualize a diagram like that shown in Fig. 11.4.2c, which is the entire F_2 system acting on the element. The effective components are 100 percent of the right-end moment in the positive direction of P_2, and $1/L_{1-2}$ of the right-end reaction in the negative direction of P_4; note that the left-end reaction has no effective component in any degree of freedom.

11.5 The Deformation Matrix [B]

The deformation matrix [B] of a beam expresses the end rotations of the elements, measured clockwise from the straight line joining the ends of the element to the elastic curve tangent, in terms of the joint displacements (which may include both joint rotations and joint deflections). It is always advisable to establish the [B] matrix by columns by considering the effect of one joint displacement at a time, on the NF end rotations. Again, in order to make the treatment of all matrix displacement methods a unified one, the NF end rotations will be denoted by $e_1, e_2, \ldots, e_{NF}$. In trusses, the e is the first letter in the word *elongation*, but for beams it means the end rotations of the element. In fact, the e's here are the same as the ϕ's defined in the slope-deflection method in Chap. 7.

400 INTERMEDIATE STRUCTURAL ANALYSIS

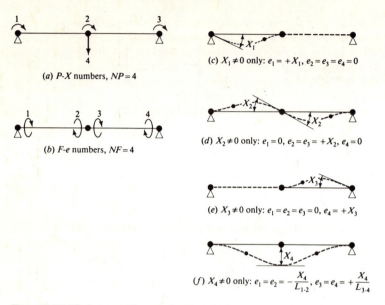

Figure 11.5.1 Displacement diagrams for the deformation matrix $[B]$ of a beam.

For the beam in Fig. 11.5.1, the $[B]$ matrix is

$$[B]_{4\times 4} = \begin{array}{|c|c|c|c|c|} \hline e \backslash X & 1 & 2 & 3 & 4 \\ \hline 1 & +1.0 & & & -\dfrac{1}{L_{1\text{-}2}} \\ \hline 2 & & +1.0 & & -\dfrac{1}{L_{1\text{-}2}} \\ \hline 3 & & +1.0 & & +\dfrac{1}{L_{3\text{-}4}} \\ \hline 4 & & & +1.0 & +\dfrac{1}{L_{3\text{-}4}} \\ \hline \end{array} \qquad (11.5.1)$$

The contents of the four columns in Eq. (11.5.1) may be obtained by observing the four displacement diagrams in Fig. 11.5.1c to f. The elastic curve in each diagram is that resulting from one single joint displacement. The question is not on what forces need to be applied to the beam to give this elastic curve (although these forces are indeed those in the corresponding column of the $[K] = [ASA^T]$ matrix). What is to be observed is the rotation at

MATRIX DISPLACEMENT METHOD OF BEAM ANALYSIS **401**

each end of the element measured clockwise from the straight line joining the ends of the element to the elastic-curve tangent at that end. For the cases in Fig. 11.5.1c to e, the straight line joining the ends of the element is still the original axis of the element, for both the left and right elements. But for the case in Fig. 11.5.1f, the straight line joining the ends of the left element has turned a clockwise angle of X_4/L_{1-2} from the original element axis, but the elastic-curve tangents are still horizontal at both ends; thus the end rotations measured from the inclined straight line to the elastic-curve tangents are counterclockwise, or negative. For the right segment, the end rotations measured from the inclined straight line to the elastic-curve tangents are clockwise, or positive X_4/L_{3-4}.

11.6 The Principle of Virtual Work

When the $[A]$ matrix of Eq. (11.4.1) is compared with the $[B]$ matrix of Eq. (11.5.1) for the same beam, one again observes that $[A] = [B^T]$ and $[B] = [A^T]$. This is, as proved for trusses in Sec. 10.6, the natural consequence of the principle of virtual work. Although the reasoning for the proof remains unchanged, another description will be made below to prove that A_{43} in Eq. (11.4.1) is equal to B_{34} in Eq. (11.5.1).

Now A_{43}, the element on the fourth row and the third column of the $[A]$ matrix for the beam in Fig. 11.6.1a and b, means that the P_4 which F_3 alone

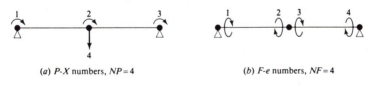

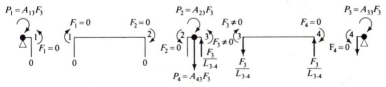

(c) Free-body diagrams of joints and elements for $F_3 \neq 0$ only

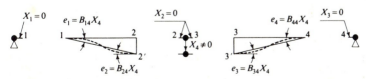

(d) Joint-displacements and element rotations for $X_4 \neq 0$ only

Figure 11.6.1 The transposition relationship between $[A]$ and $[B]$ for beams.

can counteract is equal to $A_{43}F_3$. This P_4 is shown on the free-body diagram of the intermediate joint in Fig. 11.6.1c; in fact, in this figure, the free-body diagrams of all three joints and of both elements are shown. B_{34}, the element on the third row and fourth column of the $[B]$ matrix, means that the end rotation at the left end of the second element due to $X_4 \neq 0$ only is $B_{34}X_4$. This end rotation, $e_3 = B_{34}X_4$ (shown in Fig. 11.6.1d), is measured clockwise from the straight line joining the ends of the second element to the elastic-curve tangent, which is horizontal to maintain X_2 at zero. If fact, the zero rotations of the three joints, the downward displacement X_4 of the intermediate joint, and the rotation of the two element axes are also shown in Fig. 11.6.1d. Now let the five force systems, all in equilibrium, in Fig. 11.6.1c go through the displacements in Fig. 11.6.1d; the total virtual work must be zero.

$$\text{Total virtual work} = +(A_{43}F_3)(X_4) - (F_3)(B_{34}X_4) - \frac{F_3}{L_{3\text{-}4}}(X_4) + \frac{F_3}{L_{3\text{-}4}}(X_4)$$

$$= 0$$

Note that in the preceding equation, whenever the force (or moment) and the displacement (or rotation) are in the same direction, the virtual work is positive; otherwise it is negative. From this equation,

$$A_{43} = B_{34} \tag{11.6.1}$$

11.7 The Element Stiffness Matrix [S]

The element stiffness matrix $[S]$ of a beam element expresses the clockwise moments F_i and F_j acting at the ends of the element in terms of the rotations e_i and e_j at the ends, measured clockwise from the straight line joining the ends of the element to the elastic-curve tangent. Applying conjugate-beam theorem 2 to Fig. 11.7.1,

$$e_i = +\frac{F_i L}{3EI} - \frac{F_j L}{6EI} \tag{11.7.1a}$$

$$e_j = -\frac{F_i L}{6EI} + \frac{F_j L}{3EI} \tag{11.7.1b}$$

When Eqs. (11.7.1a and b) are expressed in matrix form, the element flexibility matrix $[D]$ is obtained; thus

$$[D] = \begin{bmatrix} +f_{ii}\dfrac{L}{EI} & +f_{ij}\dfrac{L}{EI} \\ +f_{ji}\dfrac{L}{EI} & +f_{jj}\dfrac{L}{EI} \end{bmatrix} = \begin{bmatrix} +\dfrac{L}{3EI} & -\dfrac{L}{6EI} \\ -\dfrac{L}{6EI} & +\dfrac{L}{3EI} \end{bmatrix} \tag{11.7.2}$$

Note here the inconsistency in notation; a capital D is used to indicate the element flexibility matrix (because F has been reserved for the generalized force matrix); but the lowercase f_{ii}, $f_{ij} = f_{ji}$, and f_{jj} are used to indicate the

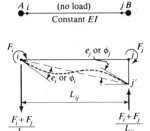

Figure 11.7.1 Rotations at ends of a beam element.

flexibility coefficients. The inverse of the element flexibility matrix becomes the element stiffness matrix; thus

$$[S] = \begin{bmatrix} +s_{ii}\dfrac{EI}{L} & +s_{ij}\dfrac{EI}{L} \\ +s_{ji}\dfrac{EI}{L} & +s_{jj}\dfrac{EI}{L} \end{bmatrix} = \begin{bmatrix} +\dfrac{4EI}{L} & +\dfrac{2EI}{L} \\ +\dfrac{2EI}{L} & +\dfrac{4EI}{L} \end{bmatrix} \quad (11.7.3)$$

For the beam with two elements in Figs. 11.5.1 or 11.6.1, the element stiffness matrix $[S]$ is a 4×4 matrix; thus

$$[S]_{4\times 4} = \begin{array}{|c|c|c|c|c|} \hline \diagdown_F^{\,e} & 1 & 2 & 3 & 4 \\ \hline 1 & +\dfrac{4EI_{1\text{-}2}}{L_{1\text{-}2}} & +\dfrac{2EI_{1\text{-}2}}{L_{1\text{-}2}} & & \\ \hline 2 & +\dfrac{2EI_{1\text{-}2}}{L_{1\text{-}2}} & +\dfrac{4EI_{1\text{-}2}}{L_{1\text{-}2}} & & \\ \hline 3 & & & +\dfrac{4EI_{3\text{-}4}}{L_{3\text{-}4}} & +\dfrac{2EI_{3\text{-}4}}{L_{3\text{-}4}} \\ \hline 4 & & & +\dfrac{2EI_{3\text{-}4}}{L_{3\text{-}4}} & +\dfrac{4EI_{3\text{-}4}}{L_{3\text{-}4}} \\ \hline \end{array} \quad (11.7.4)$$

Note that only the moment of inertia within the element has to be constant to qualify for $s_{ii} = s_{jj} = +4$ and $s_{ij} = s_{ji} = +2$, but the moment of inertia of one element can be different from that of any other element.

11.8 The Force-Displacement Matrix [SB]

The force-displacement matrix $[SB]$ of a beam expresses the generalized member-end forces (i.e., end moments) in terms of the end rotations. The size of this matrix is $NF \times NP$, and it can be obtained by multiplying the $[S]$ matrix by the $[B]$ matrix. Thus,

$$\{F\}_{NF \times 1} = [S]_{NF \times NF}[B]_{NF \times NP}\{X\}_{NP \times 1} = [SB]_{NF \times NP}\{X\}_{NP \times 1} \quad (11.8.1)$$

The end moments given by Eq. (11.8.1) are only those in the joint-force condition. Based on the concept developed in Fig. 11.2.1, the end moments in the original beam with loads on the elements should be the sum of the fixed-end moments in Fig. 11.2.1d plus the end moments in the joint-force condition of Fig. 11.2.1e. Using the notation $\{F^*\}$ to mean the final end moments in the original beam and $\{F_0\}$ to mean the fixed-end moments,

$$\{F^*\}_{NF \times 1} = \{F_0\}_{NF \times 1} + \{F\}_{NF \times 1}$$
$$= \{F_0\}_{NF \times 1} + [SB]_{NF \times NP}\{X\}_{NP \times 1} \quad (11.8.2)$$

Actually, Eq. (11.8.2), when applied locally to one single element, becomes identical to the two slope-deflection equations in Chap. 7. To summarize, one may say that the matrix displacement method of beam analysis is in fact the slope-deflection method in matrix notation.

11.9 The Global Stiffness Matrix $[K] = [ASA^T]$

The global stiffness matrix $[K] = [ASA^T]$ of a beam expresses the joint forces (generalized to include moments) in terms of the joint displacements (generalized to include both rotations and deflections), or

$$\{P\}_{NP \times 1} = [A]_{NP \times NF}\{F\}_{NF \times 1} = [A]_{NP \times NF}[SB]_{NF \times NP}\{X\}_{NP \times 1}$$
$$= [ASA^T]_{NP \times NP}\{X\}_{NP \times 1} = [K]_{NP \times NP}\{X\}_{NP \times 1} \quad (11.9.1)$$

When the rows of the $[A]$ matrix, such as the one shown in Eq. (11.4.1), are examined, one will find that each row represents either a joint-moment equation of equilibrium or a joint-force equation of equilibrium, just as those used in the slope-deflection method of Chap. 7. Thus the contents of the $\{P\}$ matrix can be surmised to be either the opposite of the sum of the fixed-end moments or the opposite of the reacting forces to prevent sidesway displacements. Moreover, the $[ASA^T]\{X\}$ values should represent the joint forces (generalized to include moments) which the deformation response, or the $\{X\}$ matrix, can resist. It was mentioned several times in Chap. 7 that the coefficients on the left sides of the simultaneous equations in the unknown joint displacements are symmetrical with respect to the main diagonal downward to the right. That the $[ASA^T]$ matrix is always symmetric, as demonstrated in Sec. 10.10, furnishes the proof; note that the $[S]$ matrix for the beam elements, although not a single diagonal matrix, is still symmetric.

11.10 The Local Stiffness Matrix

The global stiffness matrix $[K] = [ASA^T]$ can be obtained by matrix multiplication using the $[A]$, $[B]$, and $[S]$ matrices, all in the global dimensions. Alternatively, it can be obtained by compiling the contribution of the local stiffness matrix of every element into the appropriate slots in the global stiffness matrix, once the four global degree-of-freedom numbers corresponding to the four local degree-of-freedom numbers are given to the computer. Although the former method is suitable to explain the numerical examples in the next section, it will be instructive, as well as useful in the event of writing the computer program, to derive the local stiffness martrix of a beam element.

The local stiffness matrix of a beam element can itself be obtained from the formula $[K] = [ASA^T]$, as shown in Table 11.10.1 and Fig. 11.10.1. The local degrees of freedom 1 and 2 are taken positive if clockwise; 3 and 4, positive if downward. The member-end moments F_i and F_j are taken positive if clockwise. The contents of the $[B]$ matrix in Table 11.10.1 are obtained by columns, each described by one of the four displacement diagrams in Fig. 11.10.1c. The contents of the $[A]$ matrix in Table 11.10.1 are obtained by rows, described by the free-body diagrams of the joints in Fig. 11.10.1d.

The local stiffness matrix of a beam element, obtained by the formula $[ASA^T]$, is, of course, symmetric. The four P values in each column, if applied to the joints at both ends of the element, will produce an elastic curve as shown by the respective displacement diagram in Fig. 11.10.1c.

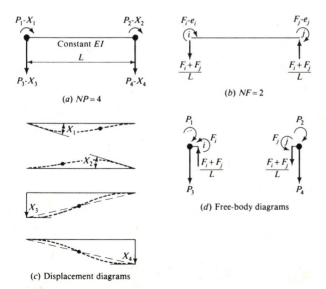

Figure 11.10.1 Local stiffness matrix of a beam element.

Table 11.10.1 Local stiffness matrix of a beam element

$[B] =$

e \ X	1	2	3	4
i	+1.0	0.	$+\dfrac{1}{L}$	$\dfrac{1}{L}$
j	0.	+1.0	$+\dfrac{1}{L}$	$\dfrac{1}{L}$

$[S] =$

F \ e	i	j
i	$+\dfrac{4EI}{L}$	$+\dfrac{2EI}{L}$
j	$+\dfrac{2EI}{L}$	$+\dfrac{4EI}{L}$

$[SB] =$

F \ X	1	2	3	4
i	$+\dfrac{4EI}{L}$	$+\dfrac{2EI}{L}$	$+\dfrac{6EI}{L^2}$	$-\dfrac{6EI}{L^2}$
j	$+\dfrac{2EI}{L}$	$+\dfrac{4EI}{L}$	$+\dfrac{6EI}{L^2}$	$-\dfrac{6EI}{L^2}$

$[A] =$

P \ F	i	j
1	+1.0	0.
2	0.	+1.0
3	$+\dfrac{1}{L}$	$+\dfrac{1}{L}$
4	$-\dfrac{1}{L}$	$-\dfrac{1}{L}$

$[ASB] =$

P \ X	1	2	3	4
1	$+\dfrac{4EI}{L}$	$+\dfrac{2EI}{L}$	$+\dfrac{6EI}{L^2}$	$-\dfrac{6EI}{L^2}$
2	$+\dfrac{2EI}{L}$	$+\dfrac{4EI}{L}$	$+\dfrac{6EI}{L^2}$	$-\dfrac{6EI}{L^2}$
3	$+\dfrac{6EI}{L^2}$	$+\dfrac{6EI}{L^2}$	$+\dfrac{12EI}{L^3}$	$-\dfrac{12EI}{L^3}$
4	$-\dfrac{6EI}{L^2}$	$-\dfrac{6EI}{L^2}$	$-\dfrac{12EI}{L^3}$	$+\dfrac{12EI}{L^3}$

Example 11.10.1 Determine the global stiffness matrix for the simple beam of two elements in Fig. 11.10.2 by the formula $[K] = [ASA^T]$ and also by the direct stiffness method of compiling the local stiffness matrices of the elements.

SOLUTION (a) $[K] = [ASA^T]$. The P-X and F-e numbers are assigned as shown in Fig. 11.10.2b and c, wherein $NP = 4$, $NF = 4$, and $NI = NF - NP = 4 - 4 = 0$. The $[A]$, $[B]$, and $[S]$ matrices shown in Table 11.10.2 are taken from Eqs. (11.4.1), (11.5.1), and (11.7.4) since the same simple beam has been used to define these three input matrices. The matrix multiplications are then performed to obtain the $[SB]$ and $[ASB]$ matrices as shown. The symmetry of the $[ASA^T]$ matrix should be noted as a check.

(b) *Direct stiffness method.* The contents of the two local stiffness matrices shown in Table 11.10.3 are identical in this problem because the two elements are alike. However, the global degree-of-freedom numbers corresponding to the local degree-of-freedom numbers

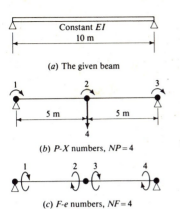

Figure 11.10.2 Beam of Example 11.10.1.

Table 11.10.2 Global stiffness matrix $[K] = [ASA^T]$ of beam in Example 11.10.1

$[B] = $

e \ X	1	2	3	4
1	+1.0			−0.2
2		+1.0		−0.2
3		+1.0		+0.2
4			+1.0	+0.2

$[S] = $

F \ e	1	2	3	4
1	+0.8	+0.4		
2	+0.4	+0.8		
3			+0.8	+0.4
4			+0.4	+0.8

$* EI$

$[SB] = $

F \ X	1	2	3	4
1	+0.8	+0.4	0.	−0.24
2	+0.4	+0.8	0.	−0.24
3	0.	+0.8	+0.4	+0.24
4	0.	+0.4	+0.8	+0.24

$* EI$

$[A] = $

P \ F	1	2	3	4
1	+1.0			
2		+1.0	+1.0	
3				+1.0
4	−0.2	−0.2	+0.2	+0.2

$[ASB] = $

P \ X	1	2	3	4
1	+0.8	+0.4	0.	−0.24
2	+0.4	+1.6	+0.4	0.
3	0.	+0.4	+0.8	+0.24
4	−0.24	0.	+0.24	+0.192

$* EI$

407

shown in the top and left margins are different for the two elements; these are taken from the correspondence shown in Fig. 11.10.3. In the compilation process, all values on the $G5$ row or column are ignored. In a computer program, an additional row and column may be added beyond the size of the $[ASA^T]$ matrix and used as "dumping grounds" for all values on the $G(NP+1)$th rows and the $G(NP+1)$th columns in the local stiffness matrices of the elements.

Table 11.10.3 Direct stiffness method for beam of Example 11.10.1

P \ X	Beam element 1			
	$L1(G1)$	$L2(G2)$	$L3(G5)$	$L4(G4)$
$L1(G1)$	+0.8	+0.4	+0.24	−0.24
$L2(G2)$	+0.4	+0.8	+0.24	−0.24
$L3(G5)$	+0.24	+0.24	+0.096	−0.096
$L4(G4)$	−0.24	−0.24	−0.096	+0.096

∗ EI

P \ X	Beam element 2			
	$L1(G2)$	$L2(G3)$	$L3(G4)$	$L4(G5)$
$L1(G2)$	+0.8	+0.4	+0.24	−0.24
$L2(G3)$	+0.4	+0.8	+0.24	−0.24
$L3(G4)$	+0.24	+0.24	+0.096	−0.096
$L4(G5)$	−0.24	−0.24	−0.096	+0.096

∗ EI

P \ X	Global stiffness matrix by compilation			
	$G1$	$G2$	$G3$	$G4$
$G1$	+0.8	+0.4	0.	−0.24
$G2$	+0.4	+1.6	+0.4	0.
$G3$	0.	0.	+0.8	+0.24
$G4$	−0.24	−0.24	+0.24	+0.192

∗ EI

MATRIX DISPLACEMENT METHOD OF BEAM ANALYSIS **409**

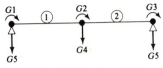

(a) Global P-X numbers, NP = 4

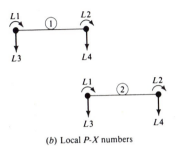

(b) Local P-X numbers

Figure 11.10.3 Global vs. local degree-of-freedom numbers for beam of Example 11.10.1.

11.11 Numerical Examples

As in all matrix displacement methods, the two major steps involved in the matrix displacement method of beam analysis are as follows:

$$\{X\}_{NP \times 1} = [ASA^T]^{-1}_{NP \times NP} \{P\}_{NP \times 1} \tag{11.11.1}$$

and

$$\{F^*\}_{NF \times 1} = \{F_0\}_{NF \times 1} + [SA^T]_{NF \times NP} \{X\}_{NP \times 1} \tag{11.11.2}$$

In checking the output, both statics and deformation checks should be made. First of all, from the $\{F^*\}$ values the end reactions R_i and R_j acting on each element should be computed by adding the simple-beam reactions and the reactions due to the algebraic sum of the F_i^* and F_j^* values, as done in Chaps. 6, 7, and 8. The NR (where NR is the number of reactions on the entire beam) conditions of equilibrium at the joints are used to compute the external reactions. Then the NP statics checks on the equilibrium along all the degrees of freedom may be made.

There should be two deformation checks per element, one at each end. The end rotations, measured clockwise from the straight line joining the ends of the element to the elastic-curve tangent, are computed first by adding the end rotations e_{0i} and e_{0j} due to load on the element if the end moments F_i^* and F_j^* are zero and those due to F_i^* and F_j^*; thus

$$\begin{Bmatrix} e_i \\ e_j \end{Bmatrix} = \begin{Bmatrix} e_{0i} \\ e_{0j} \end{Bmatrix} + \begin{bmatrix} +\dfrac{L}{3EI} & -\dfrac{L}{6EI} \\ -\dfrac{L}{6EI} & +\dfrac{L}{3EI} \end{bmatrix} \begin{Bmatrix} F_i^* \\ F_j^* \end{Bmatrix} \tag{11.11.3}$$

The end rotations e_{0i} and e_{0j} can be conveniently computed by the conjugate-beam method. Then, too, the end rotations e_i and e_j can be computed from the X values in the output by the equations

$$e_i = X_i - R_{ij} \quad (11.11.4a)$$
$$e_j = X_j - R_{ij} \quad (11.11.4b)$$

in which X_i and X_j are the rotations of the joints at the ends of the element, and R_{ij} is the clockwise rotation of the element axis as may be observed from the translational displacements of the joints in the $\{X\}$ matrix. In fact Eqs. (11.11.4a and b) are the rows in the $[B]$ matrix. That the two end rotations computed from Eq. (11.11.3) are equal to those computed from Eqs. (11.11.4a and b) constitutes the two deformation checks for the element.

When all the NP statics checks and the NF deformation checks are obtained, the correctness of the solution is certain.

Example 11.11.1 Analyze the simple beam shown in Fig. 11.11.1 by the matrix displacement method using two 5-m elements.

SOLUTION (a) *The global stiffness matrix* $[K] = [ASA^T]$. This matrix has already been computed in Example 11.10.1.

(b) *The joint-force matrix* $\{P\}$. The treatment for the concentrated load on the right element was described in Sec. 11.2; the reader may review that section. The fixed-end

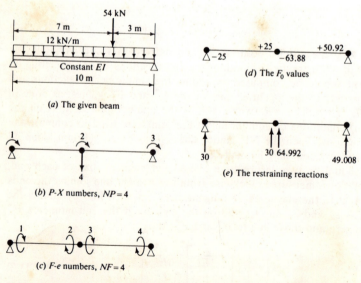

Figure 11.11.1 Beam of Example 11.11.1.

moments are computed as follows:

$$F_{01} = -\frac{12(5)^2}{12} = -25 \text{ kN·m} \qquad F_{02} = +25 \text{ kN·m}$$

$$F_{03} = -\frac{12(5)^2}{12} - \frac{54(2)(3)^2}{5^2} = -25 - 38.88 = -63.88 \text{ kN·m}$$

$$F_{04} = +25 + \frac{54(3)(2)^2}{5^2} = +25 + 25.92 = +50.92 \text{ kN·m}$$

These moments are shown in Fig. 11.11.1d, below the element at the left end and above the element at the right end. (The reason for doing this is that in a complex rigid frame more than two elements may meet at one joint.) The needed reactions at the ends of the elements in the fixed condition are shown in Fig. 11.11.1e. The entries in the $\{P\}$ matrix are the *opposites* of the restraining moments or forces required to hold all X values to zero in the fixed condition; thus

$$P_1 = -(F_{01}) = -(-25) = +25 \text{ kN·m}$$
$$P_2 = -(F_{02} + F_{03}) = -(+25 - 63.88) = +38.88 \text{ kN·m}$$
$$P_3 = -(F_{04}) = -(+50.92) = -50.92 \text{ kN·m}$$
$$P_4 = -(R_{02} + R_{03}) = -(-30 - 64.992) = +94.992 \text{ kN}$$

Note that the positive direction for F_{01} to F_{04} is clockwise acting on the element, and the positive direction for R_{02} and R_{03} is downward acting on the element, both the same as the positive directions for the degree of freedom. It is more straightforward to see physically from Fig. 11.11.1e that the releasing force along the fourth degree of freedom is 94.992 kN downward.

(c) *The output matrix* $\{X\}$. Using $[ASA^T]$ from Table 11.10.2,

$$\{X\} = [ASA^T]^{-1}\{P\} = [ASA^T]^{-1}\begin{bmatrix} +25.00 \\ +38.88 \\ -50.92 \\ +94.992 \end{bmatrix}$$

$$= \left[+\frac{745.7}{EI} \quad +\frac{43.2}{EI} \quad -\frac{821.3}{EI} \quad +\frac{2453.5}{EI} \right]^T$$

In fact, these $\{X\}$ values may be obtained by applying the conjugate-beam method to the given simple beam of Fig. 11.11.1a and substituted into the equation $\{P\} = [ASA^T]\{X\}$ for verification, inasmuch as the purpose here is to demonstrate how the matrix displacement method works.

(d) *The output matrix* $\{F^*\}$. Using $[SA^T]$ from Table 11.10.2 and $\{X\}$ from part (c),

$$\{F^*\} = \{F_0\} + [SA^T]\{X\}$$

$$= \begin{Bmatrix} -25 \\ +25 \\ -63.88 \\ +50.92 \end{Bmatrix} + [SA^T]\{X\}$$

$$= \begin{Bmatrix} -25 \\ +25 \\ -63.88 \\ +50.92 \end{Bmatrix} + \begin{Bmatrix} +25 \\ -256 \\ +294.88 \\ -50.92 \end{Bmatrix} = \begin{Bmatrix} 0. \\ -231. \\ +231. \\ 0. \end{Bmatrix}$$

412 INTERMEDIATE STRUCTURAL ANALYSIS

(e) *The statics checks.* The free-body diagrams of the elements are shown in Fig. 11.11.2a. Note that a positive sign for the F^* value means a clockwise moment acting on the member end. The $NP = 4$ statics checks are that the moments on the left and right supports of the simple beam are both zero, and the moments at the intermediate joint are balanced, plus the fact that the right-end reaction of 16.2 kN downward on the left element is equal to the left-end reaction of 16.2 kN upward on the right element.

(f) *The deformation checks.* The end rotations as dictated by the X values are shown in Fig. 11.11.2b; these are

$$e_1 = X_1 - R_{12} = +\frac{745.7}{EI} - \left(+\frac{2453.5}{EI}\bigg/5\right) = +\frac{255}{EI}$$

$$e_2 = X_2 - R_{12} = +\frac{43.2}{EI} - \left(+\frac{2453.5}{EI}\bigg/5\right) = -\frac{447.5}{EI}$$

$$e_3 = X_2 - R_{34} = +\frac{43.2}{EI} - \left(-\frac{2453.5}{EI}\bigg/5\right) = +\frac{533.9}{EI}$$

$$e_4 = X_3 - R_{34} = -\frac{821.3}{EI} - \left(-\frac{2453.5}{EI}\bigg/5\right) = -\frac{330.6}{EI}$$

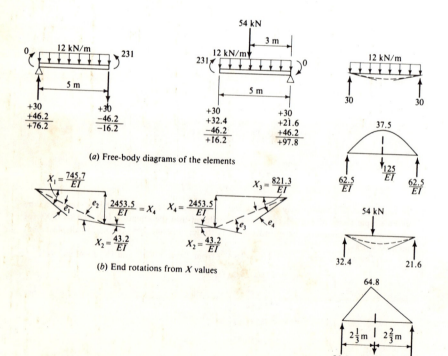

(a) Free-body diagrams of the elements

(b) End rotations from X values

(c) The e_0 values

Figure 11.11.2 Statics and deformation checks for beam of Example 11.11.1.

MATRIX DISPLACEMENT METHOD OF BEAM ANALYSIS 413

Using the e_{01} to e_{04} values from the reactions to the conjugate beams in Fig. 11.11.2c,

$$e_1 = e_{01} + \frac{F_1^*(5)}{3EI} - \frac{F_2^*(5)}{6EI} = +\frac{62.5}{EI} + 0 - \frac{(-231)(5)}{6EI}$$

$$= +\frac{255}{EI} \quad \text{(Check)}$$

$$e_2 = e_{02} - \frac{F_1^*(5)}{6EI} + \frac{F_2^*(5)}{3EI} = -\frac{62.5}{EI} + 0 + \frac{(-231)(5)}{3EI}$$

$$= -\frac{447.5}{EI} \quad \text{(Check)}$$

$$e_3 = e_{03} + \frac{F_3^*(5)}{3EI} - \frac{F_4^*(5)}{6EI} = +\frac{62.5}{EI} + \frac{86.4}{EI} + \frac{(+231)(5)}{3EI} - 0$$

$$= +\frac{533.9}{EI} \quad \text{(Check)}$$

$$e_4 = e_{04} - \frac{F_3^*(5)}{6EI} + \frac{F_4^*(5)}{3EI} = -\frac{62.5}{EI} - \frac{75.6}{EI} - \frac{(+231)(5)}{6EI} - 0$$

$$= -\frac{330.6}{EI} \quad \text{(Check)}$$

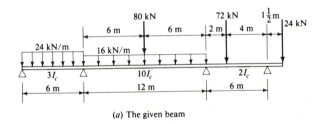

(a) The given beam

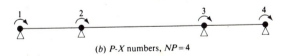

(b) P-X numbers, NP = 4

(c) F-e numbers, NF = 6

(d) The F_0 values

Figure 11.11.3 Beam of Example 11.11.2.

Example 11.11.2 Analyze the beam shown in Fig. 11.11.3a by the matrix displacement method. Note that this beam is the same as the one in Example 7.3.1.

SOLUTION Although there is no upper limit on the number of elements to be used in the analysis of beams by the displacement method, there is a lower limit since the element should not involve any unknown reaction on it. For this beam, the minimum number of elements is three, because the overhang on the right end is a statically known cantilever for which there is a fixed-end moment only on one side. For three elements, the P-X and F-e numbers are shown in Fig. 11.11.3b and c, wherein $NP = 4$, $NF = 6$, and $NI = 2$. The fixed-end moments are shown in Fig. 11.11.3d.

$$[K] = [ASA^T] =$$

$$= \begin{bmatrix} +1.0 & & & & & \\ & +1.0 & +1.0 & & & \\ & & & +1.0 & +1.0 & \\ & & & & & +1.0 \end{bmatrix} \begin{bmatrix} +2. & +1. & & & & \\ +1. & +2. & & & & \\ & & +\frac{10}{3} & +\frac{5}{3} & & \\ & & +\frac{5}{3} & +\frac{10}{3} & & \\ & & & & +\frac{4}{3} & +\frac{2}{3} \\ & & & & +\frac{2}{3} & +\frac{4}{3} \end{bmatrix} * EI_c$$

$$\begin{bmatrix} +1.0 & & & & \\ & +1.0 & & & \\ & +1.0 & & & \\ & & & +1.0 & \\ & & & +1.0 & \\ & & & & +1.0 \end{bmatrix}$$

$$= \begin{bmatrix} +1.0 & & & & & \\ & +1.0 & +1.0 & & & \\ & & & +1.0 & +1.0 & \\ & & & & & +1.0 \end{bmatrix} \begin{bmatrix} +2. & +1. & & & & \\ +1. & +2. & & & & \\ & & +\frac{10}{3} & +\frac{5}{3} & & \\ & & +\frac{5}{3} & +\frac{10}{3} & & \\ & & & & +\frac{4}{3} & +\frac{2}{3} \\ & & & & +\frac{2}{3} & +\frac{4}{3} \end{bmatrix} * EI_c$$

$$= \begin{bmatrix} +2.0 & +1.0 & & \\ +1.0 & +\frac{16}{3} & +\frac{5}{3} & \\ & +\frac{5}{3} & +\frac{14}{3} & +\frac{2}{3} \\ & & +\frac{2}{3} & +\frac{4}{3} \end{bmatrix} * EI_c$$

$$\{X\} = [K]^{-1}\{P\}$$

$$= [K \text{ from preceding equation}]^{-1} \begin{Bmatrix} + 72. \\ +240. \\ -248. \\ + 4. \end{Bmatrix} = \begin{Bmatrix} + 0.20 \\ +71.60 \\ -85.23 \\ +45.62 \end{Bmatrix} * \frac{1}{EI_c}$$

MATRIX DISPLACEMENT METHOD OF BEAM ANALYSIS **415**

$$\{F^*\} = \{F_0\} + [SA^T]\{X\}$$

$$= \begin{Bmatrix} -72. \\ +72. \\ -312. \\ +312. \\ -64. \\ +32. \end{Bmatrix} + [SA^T \text{ from first equation}]\{X \text{ from preceding equation}\}$$

$$= \begin{Bmatrix} -72. \\ +72. \\ -312. \\ +312. \\ -64. \\ +32. \end{Bmatrix} + \begin{Bmatrix} +72. \\ +143.40 \\ +96.62 \\ -164.77 \\ -83.23 \\ +4.01 \end{Bmatrix} = \begin{Bmatrix} 0. \\ +215.40 \\ -215.38 \\ +147.23 \\ -147.23 \\ +36.01 \end{Bmatrix}$$

The above output matrices $\{X\}$ and $\{F^*\}$ are same as those obtained by the slope-deflection method in Example 7.3.1. The $NP = 4$ statics checks are obviously satisfied by looking at the $\{F^*\}$ values. The $NF = 6$ deformation checks can be made in the same way as illustrated in Example 11.11.1. Note that making NF deformation checks in which the X values are taken from the output is equivalent to the making of the NI compatibility checks in Example 7.3.1.

Example 11.11.3 Analyze the beam shown in Fig. 11.11.4a by the matrix displacement method. Note that this beam is the same as the one in Example 7.3.2.

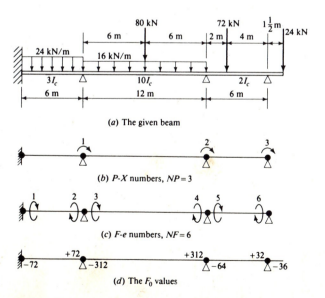

Figure 11.11.4 Beam of Example 11.11.3.

416 INTERMEDIATE STRUCTURAL ANALYSIS

SOLUTION Using the minimum number of elements to simplify the solution, the P-X and F-e numbers are shown in Fig. 11.11.4b and c, wherein $NP = 3$, $NF = 6$, and $NI = 3$. The fixed-end moments are shown in Fig. 11.11.4d.

$[K] = [ASA^T]$

$$= \begin{bmatrix} 0. & +1.0 & +1.0 & 0. & 0. & 0. \\ 0. & 0. & 0. & +1.0 & +1.0 & 0. \\ 0. & 0. & 0. & 0. & 0. & +1.0 \end{bmatrix} \begin{bmatrix} +2. & +1. & & & & \\ +1. & +2. & & & & \\ & & +\tfrac{10}{3} & +\tfrac{5}{3} & & \\ & & +\tfrac{5}{3} & +\tfrac{10}{3} & & \\ & & & & +\tfrac{4}{3} & +\tfrac{2}{3} \\ & & & & +\tfrac{2}{3} & +\tfrac{4}{3} \end{bmatrix} * EI_c$$

$$\begin{bmatrix} 0. & 0. & 0. \\ +1.0 & 0. & 0. \\ +1.0 & 0. & 0. \\ 0. & +1.0 & 0. \\ 0. & +1.0 & 0. \\ 0. & 0. & +1.0 \end{bmatrix}$$

$$= \begin{bmatrix} 0. & +1.0 & +1.0 & 0. & 0. & 0. \\ 0. & 0. & 0. & +1.0 & +1.0 & 0. \\ 0. & 0. & 0. & 0. & 0. & +1.0 \end{bmatrix} \begin{bmatrix} +1.0 & & \\ +2.0 & & \\ +\tfrac{10}{3} & +\tfrac{5}{3} & \\ +\tfrac{5}{3} & +\tfrac{10}{3} & \\ & & +\tfrac{4}{3} & +\tfrac{2}{3} \\ & & +\tfrac{2}{3} & +\tfrac{4}{3} \end{bmatrix} * EI_c$$

$$= \begin{bmatrix} +\tfrac{16}{3} & +\tfrac{5}{3} & \\ +\tfrac{5}{3} & +\tfrac{14}{3} & +\tfrac{2}{3} \\ & +\tfrac{2}{3} & +\tfrac{4}{3} \end{bmatrix} * EI_c$$

$\{X\} = [K]^{-1}\{P\}$

$= [K \text{ from preceding equation}]^{-1} \begin{Bmatrix} +240. \\ -248. \\ + 4. \end{Bmatrix} = \begin{Bmatrix} +71.64 \\ -85.25 \\ +45.63 \end{Bmatrix} * \dfrac{1}{EI_c}$

$\{F^*\} = \{F_0\} + [SA^T]\{X\}$

$= \begin{Bmatrix} -72. \\ +72. \\ -312. \\ +312. \\ -64. \\ +32. \end{Bmatrix} + [SA^T]\{X \text{ from preceding equation}\}$

$= \begin{Bmatrix} -72. \\ +72. \\ -312. \\ +312. \\ -64. \\ +32. \end{Bmatrix} + \begin{Bmatrix} +71.64 \\ +143.28 \\ +96.72 \\ -164.77 \\ -83.52 \\ +4.01 \end{Bmatrix} = \begin{Bmatrix} -0.36 \\ +215.28 \\ -215.28 \\ +147.23 \\ -147.25 \\ +36.01 \end{Bmatrix}$

MATRIX DISPLACEMENT METHOD OF BEAM ANALYSIS 417

The above output matrices $\{X\}$ and $\{F^*\}$ are the same as those obtained by the slope-deflection method in Example 7.3.2. The $NP = 3$ statics checks are obviously satisfied by looking at the $\{F^*\}$ values. The $NF = 6$ deformation checks can be made in the same way as illustrated in Example 11.11.1.

11.12 Effect of Support Settlements

The matrix displacement method can be used in just about the same way as the slope-deflection method to obtain the effect of uneven settlements of supports in continuous beams. First the fixed-end moments are determined to fix all joints against rotation while allowing the uneven settlements to occur. These fixed-end moments, according to Eq. (8.8.1), are

$$F_{0i} = F_{0j} = -\frac{6EIR}{L} \tag{11.12.1}$$

wherein R is the clockwise rotation of the member axis due to the uneven support settlements, and L and EI are the length and flexural rigidity of the member, respectively. Then the opposites of the sum of the fixed-end moments around each rotational degree of freedom are applied as the generalized joint forces (actually joint moments). The two major steps in the matrix displacement method are then performed:

$$\{X\} = [ASA^T]^{-1}\{P\} \tag{11.2.2}$$

and
$$\{F^*\} = \{F_0\} + [SA^T]\{X\} \tag{11.2.3}$$

wherein the $\{F_0\}$ values are those of Eq. (11.12.1).

Example 11.12.1 By the matrix displacement method analyze the beam shown in Fig. 11.12.1a for a 15-mm settlement of the support at B. Note that this problem is the same as Example 7.5.2.

SOLUTION (a) *The $\{F_0\}$ and $\{P\}$ values.* Using the P-X and F-e numbers shown in Fig. 11.12.1b and c, the $\{F_0\}$ and $\{P\}$ matrices are computed below.

$$F_{01} = F_{02} = -\frac{6EIR}{L} = -\frac{6E(3I_c)R}{L}$$

$$= -\frac{6(200)(1200)(+0.015/6)}{6} = -600 \text{ kN·m}$$

$$F_{03} = F_{04} = -\frac{6EIR}{L} = -\frac{6E(10I_c)R}{L}$$

$$= -\frac{6(200)(4000)(-0.015/6)}{12} = +500 \text{ kN·m}$$

$$F_{05} = F_{06} = 0$$

418 INTERMEDIATE STRUCTURAL ANALYSIS

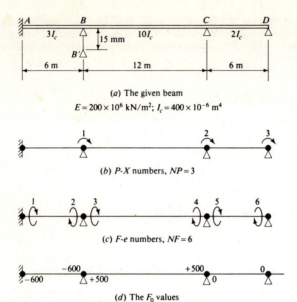

(a) The given beam
$E = 200 \times 10^6 \text{ kN/m}^2$; $I_c = 400 \times 10^{-6} \text{ m}^4$

(b) P-X numbers, NP = 3

(c) F-e numbers, NF = 6

(d) The F_0 values

Figure 11.12.1 Beam of Example 11.12.1.

The above $\{F_0\}$ values are shown in Fig. 11.12.1d. The $\{P\}$ values are

$$P_1 = -(F_{02} + F_{03}) = -(-600 + 500) = +100 \text{ kN·m}$$
$$P_2 = -(F_{04} + F_{05}) = -(+500 + 0) = -500 \text{ kN·m}$$
$$P_3 = -F_{06} = 0$$

(b) *The element stiffness matrix.* It is more logical to use the absolute numerical values rather than the relative values in terms of EI_c for the contents of the element stiffness matrix. For member 1-2,

$$S_{11} = S_{22} = \frac{4EI}{L} = \frac{4E(3I_c)}{L} = \frac{4(200)(1200)}{6} = 160{,}000 \text{ kN·m/rad}$$

$$S_{12} = S_{21} = \frac{2EI}{L} = 80{,}000 \text{ kN·m/rad}$$

For member 3-4,

$$S_{33} = S_{44} = \frac{4EI}{L} = \frac{4E(10I_c)}{L} = \frac{4(200)(4000)}{12} = 266{,}666 \text{ kN·m/rad}$$

$$S_{34} = S_{43} = \frac{2EI}{L} = 133{,}333 \text{ kN·m/rad}$$

For member 5-6,

$$S_{55} = S_{66} = \frac{4EI}{L} = \frac{4E(2I_c)}{L} = \frac{4(200)(800)}{6} = 106{,}666 \text{ kN·m/rad}$$

$$S_{56} = S_{65} = \frac{2EI}{L} = 53{,}333 \text{ kN·m/rad}$$

MATRIX DISPLACEMENT METHOD OF BEAM ANALYSIS 419

(c) *The output matrices* $\{X\}$ *and* $\{F^*\}$.

$$[K] = [ASA^T]$$

$$= \begin{bmatrix} 0. & +1.0 & +1.0 & 0. & 0. & 0. \\ 0. & 0. & 0. & +1.0 & +1.0 & 0. \\ 0. & 0. & 0. & 0. & 0. & +1.0 \end{bmatrix}$$

$$\begin{bmatrix} +160{,}000 & +80{,}000 & & & & \\ +80{,}000 & +160{,}000 & & & & \\ & & +266{,}666 & +133{,}333 & & \\ & & +133{,}333 & +266{,}666 & & \\ & & & & +106{,}666 & +53{,}333 \\ & & & & +53{,}333 & +106{,}666 \end{bmatrix} \begin{bmatrix} 0. & 0. & 0. \\ +1.0 & 0. & 0. \\ +1.0 & 0. & 0. \\ 0. & +1.0 & 0. \\ 0. & +1.0 & 0. \\ 0. & 0. & +1.0 \end{bmatrix}$$

$$= \begin{bmatrix} 0. & +1.0 & +1.0 & 0. & 0. & 0. \\ 0. & 0. & 0. & +1.0 & +1.0 & 0. \\ 0. & 0. & 0. & 0. & 0. & +1.0 \end{bmatrix} \begin{bmatrix} +80{,}000 & & \\ +160{,}000 & & \\ +266{,}666 & +133{,}333 & \\ +133{,}333 & +266{,}666 & \\ & +106{,}666 & +53{,}333 \\ & +53{,}333 & +106{,}666 \end{bmatrix}$$

$$= \begin{bmatrix} +426{,}666 & +133{,}333 & \\ +133{,}333 & +373{,}333 & +53{,}333 \\ & +53{,}333 & +106{,}666 \end{bmatrix}$$

$$\{X\} = [K]^{-1}\{P\}$$

$$= [K]^{-1} \begin{Bmatrix} +100. \\ -500. \\ 0. \end{Bmatrix} = \begin{Bmatrix} +0.7787 \times 10^{-3} \\ -1.7418 \times 10^{-3} \\ +0.8709 \times 10^{-3} \end{Bmatrix}$$

$$\{F^*\} = \{F_0\} + [SA^T]\{X\}$$

$$= \begin{Bmatrix} -600. \\ -600. \\ +500. \\ +500. \\ 0. \\ 0. \end{Bmatrix} + [SA^T]\{X\}$$

$$= \begin{Bmatrix} -600. \\ -600 \\ +500. \\ +500. \\ 0. \\ 0. \end{Bmatrix} + \begin{Bmatrix} +62.30 \\ +124.59 \\ -24.59 \\ -360.65 \\ -139.34 \\ 0. \end{Bmatrix} = \begin{Bmatrix} -537.70 \\ -475.41 \\ +475.41 \\ +139.35 \\ -139.34 \\ 0. \end{Bmatrix}$$

(d) *The statics and deformation checks.* By looking at the $\{F^*\}$ values, the $NP = 3$ statics checks are obviously satisfied because $F_2^* + F_3^* = 0$, $F_4^* + F_5^* = 0$, and $F_6^* = 0$. The two deformation checks for member 3-4 are shown below. Using Eq. (11.11.3),

$$\begin{Bmatrix} e_3 \\ e_4 \end{Bmatrix} = \begin{Bmatrix} e_{03} \\ e_{04} \end{Bmatrix} + \begin{bmatrix} +\dfrac{12}{3(200)(4000)} & -\dfrac{12}{6(200)(4000)} \\ -\dfrac{12}{6(200)(4000)} & +\dfrac{12}{3(200)(4000)} \end{bmatrix} \begin{Bmatrix} +475.41 \\ -139.35 \end{Bmatrix}$$

$$= \begin{Bmatrix} 0. \\ 0. \end{Bmatrix} + \begin{Bmatrix} +2.0287 \times 10^{-3} \\ -0.4918 \times 10^{-3} \end{Bmatrix} = \begin{Bmatrix} +2.0287 \times 10^{-3} \\ -0.4918 \times 10^{-3} \end{Bmatrix}$$

Using Eqs. (11.11.4a and b),

$$e_3 = X_1 - R_{34} = (+0.7787 \times 10^{-3}) - (-0.015/12)$$
$$= +2.0287 \times 10^{-3} \quad \text{(Check)}$$
$$e_4 = X_2 - R_{34} = (-1.7418 \times 10^{-3}) - (-0.015/12)$$
$$= -0.4918 \times 10^{-3} \quad \text{(Check)}$$

When the six deformation checks here are compared with the three compatibility checks made in Example 6.4.2, one will note that in order to ensure the correctness of the solution, one can either make the NI compatibility checks in which no X values are borrowed from the output, or make the NF deformation checks in which the NP values of X are needed.

11.13 Exercises

11.1 By the matrix displacement method analyze the simple beam shown in Fig. 11.13.1 by using two 5-m elements.

11.2 By the matrix displacement method analyze the simple beam of Exercise 11.1 by using a left 4-m element and a right 6-m element.

11.3 By the matrix displacement method analyze the cantilever beam shown in Fig. 11.13.2 by using two 5-m elements.

11.4 By the matrix displacement method analyze the cantilever beam of Exercise 11.3 by using a left 6-m element and a right 4-m element.

11.5 to 11.10 Analyze the continuous beams of Exercises 7.1 to 7.6 by the matrix displacement method. In each case, use the minimum number of elements possible. Note that Exercises 7.3 and 7.4 are for uneven settlement of supports.

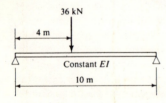

Figure 11.13.1 Exercise 11.1.

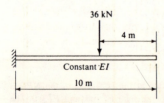

Figure 11.13.2 Exercise 11.3.

CHAPTER
TWELVE

MATRIX DISPLACEMENT METHOD OF RIGID-FRAME ANALYSIS

12.1 Matrix Displacement Method vs. Slope-Deflection Method

Having observed in Chap. 7 the solution of beam problems by the slope-deflection method and in Chap. 11 of those same problems by the matrix displacement method, the reader should by this time be fully aware of the fact that the two methods are identical in substance and different only in format. Once one becomes familiar with matrix notations and operations, the judgment might be that the matrix displacement method is superior to the slope-deflection method for the following reasons:

1. The global stiffness matrix, which constitutes the left side of the simultaneous equations in the slope-deflection method, can be systematically determined by either the formula $[K] = [ASB]$ or the direct stiffness method. When the former approach is used, the $[A]$ and $[B]$ matrices should be independently established, each by its own definition, and the transposition relationship between them can be nicely used for checking.
2. The equilibrium conditions in the slope-deflection method are constructed using the end moments when there are loads acting on the elements. Thus for each loading condition one will need to repeat the task, even though the left side of the simultaneous equations always turns out to be the same. Yet in the matrix displacement method, for more than one loading condition, one needs only to expand the joint-force matrix from a column matrix to a rectangular matrix.
3. For rigid frames with multiple degree of freedom in sidesway, particularly for those with nonrectangular joints, the use of the input matrices $[A]$ and

[B] and the check of the transposition relationship between them, plus the uncoupling of the fixed condition from the joint-force condition, are unsurpassable by any measure of the slope-deflection method.

4. If the rigid frame contains elements with variable moment of inertia, the modification necessary in the matrix displacement method involves only the element stiffness matrices and the fixed-end moments for these elements.

5. A short and concise computer program† can be written for the matrix displacement method, using $[A], [S], \{P\}$, and $\{F_0\}$ as input and giving $\{X\}$ and $\{F^*\}$ as output.

In the remainder of this chapter, the examples which were solved by the slope-deflection method in Chap. 7 will be solved again by the matrix displacement method. In all cases, it is just a matter of showing how the input matrices $[A], [B], [S], \{F_0\}$, and $\{P\}$ are established, because the output matrices $\{X\}$ and $\{F^*\}$ are then computed by the same two major steps

$$\{X\} = [ASB]^{-1}\{P\} \qquad (12.1.1)$$

and
$$\{F^*\} = \{F_0\} + [SB]\{X\} \qquad (12.1.2)$$

Upon finding the output matrices, the correctness of the solution can be checked by making the NP statics checks and the NF deformation checks.

12.2 Analysis of Rigid Frames without Sidesway

When none of the joints in a rigid frame can have an unknown change in position, all the unknown joint displacements are rotations. In such a case, the contents of the $[A]$ and $[B]$ matrices are either 0. or $+1.0$. The sketches for the free-body diagrams of the joints and for the displacement diagrams are shown for the first example only in this section. The contents for the $\{P\}$ matrix are simply

$$P_i = -(\Sigma F_0 \text{ along the } i\text{th degree of freedom}) \qquad (12.2.1)$$

Example 12.2.1 Analyze the rigid frame shown in Fig. 12.2.1a by the matrix displacement method. Note that this rigid frame is the same as the one in Example 7.6.1.

SOLUTION The minimum number of joints permitted is three, at B, C, and D, because BA can be treated as a statically known cantilever. The P-X and F-e numbers are shown in Fig. 12.2.1b and c, wherein $NP = 1$, $NF = 4$, and $NI = 3$. The fixed-end moments are computed and shown in Fig. 12.2.1d, below the element at the left end and above the element at the right end, with a vertical element treated as a horizontal element as viewed from the right side. The only row in the $[A]$ matrix, as may be observed from the free-body diagram of the joint in Fig. 12.2.1e, is $P_1 = F_2 + F_3$. The only column in the $[B]$ matrix, as may be observed from the displacement diagram in Fig. 12.2.1f, is $e_2 = e_3 = +1.0X_1$.

---

†For instance, see C.-K. Wang, *Matrix Methods of Structural Analysis*, 2d ed., American Publishing Company, Madison, Wis., 1970, Appendix E.

MATRIX DISPLACEMENT METHOD OF RIGID-FRAME ANALYSIS 423

$[K] = [A][S][B]$

$$= [0.\ +1.\ +1.\ 0.] \begin{bmatrix} +0.8 & +0.4 & & \\ +0.4 & +0.8 & & \\ & & +1.6 & +0.8 \\ & & +0.8 & +1.6 \end{bmatrix} * EI_c \begin{Bmatrix} 0. \\ +1. \\ +1. \\ 0. \end{Bmatrix}$$

$$= [0.\ +1.\ +1.\ 0.] \begin{Bmatrix} +0.4 \\ +0.8 \\ +1.6 \\ +0.8 \end{Bmatrix} * EI_c$$

$= +2.4 EI_c$

$\{P\} = -(\Sigma F_0) = -(+54. - 135.) = +81.$

$\{X\} = [K]^{-1}\{P\} = [+2.4 EI_c]^{-1}\{+81.\} = +\dfrac{33.75}{EI_c}$

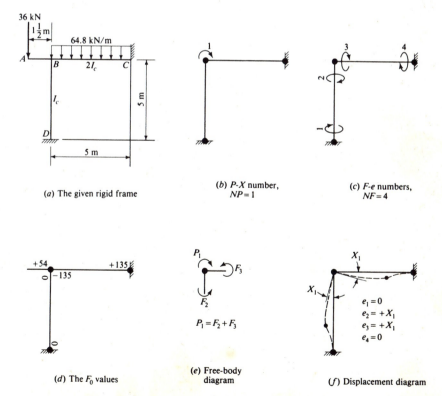

(a) The given rigid frame

(b) P-X number, NP = 1

(c) F-e numbers, NF = 4

(d) The F_0 values

(e) Free-body diagram

$P_1 = F_2 + F_3$

(f) Displacement diagram

$e_1 = 0$
$e_2 = +X_1$
$e_3 = +X_1$
$e_4 = 0$

Figure 12.2.1 Rigid frame of Example 12.2.1.

424 INTERMEDIATE STRUCTURAL ANALYSIS

$$\{F^*\} = \{F_0\} + [SB]\{X\}$$

$$= \begin{Bmatrix} 0. \\ 0. \\ -135. \\ +135. \end{Bmatrix} + \begin{Bmatrix} +0.4 \\ +0.8 \\ +1.6 \\ +0.8 \end{Bmatrix} \{+33.75\}$$

$$= \begin{Bmatrix} 0. \\ 0. \\ -135. \\ +135. \end{Bmatrix} + \begin{Bmatrix} +13.5 \\ +27.0 \\ +54.0 \\ +27.0 \end{Bmatrix} = \begin{Bmatrix} +13.5 \\ +27.0 \\ -81.0 \\ +162.0 \end{Bmatrix}$$

Example 12.2.2 Analyze the rigid frame shown in Fig. 12.2.2a by the matrix displacement method. Note that this rigid frame is the same as the one in Example 7.6.2.

SOLUTION The minimum number of joints being four, the P-X and F-e numbers are shown in Fig. 12.2.2b and c, wherein $NP = 3$, $NF = 6$, and $NI = 3$. The fixed-end moments are shown in Fig. 12.2.2d.

$$K = [A][S][B]$$

$$= \begin{bmatrix} +1. & 0. & 0. & 0. & 0. & 0. \\ 0. & +1. & +1. & 0. & 0. & 0. \\ 0. & 0. & 0. & +1. & +1. & 0. \end{bmatrix}$$

$$\begin{bmatrix} +0.8 & +0.4 & & & & \\ +0.4 & +0.8 & & & & \\ & & +0.8 & +0.4 & & \\ & & +0.4 & +0.8 & & \\ & & & & +1.0 & +0.5 \\ & & & & +0.5 & +1.0 \end{bmatrix} * EI_c \begin{bmatrix} +1. & 0. & 0. \\ 0. & +1. & 0. \\ 0. & +1. & 0. \\ 0. & 0. & +1. \\ 0. & 0. & +1. \\ 0. & 0. & 0. \end{bmatrix}$$

$$= \begin{bmatrix} +1. & 0. & 0. & 0. & 0. & 0. \\ 0. & +1. & +1. & 0. & 0. & 0. \\ 0. & 0. & 0. & +1. & +1. & 0. \end{bmatrix} \begin{bmatrix} +0.8 & +0.4 & \\ +0.4 & +0.8 & \\ +0.8 & +0.4 \\ +0.4 & +0.8 \\ & +1.0 \\ & +0.5 \end{bmatrix} * EI_c$$

$$= \begin{bmatrix} +0.8 & +0.4 & \\ +0.4 & +1.6 & +0.4 \\ & +0.4 & +1.8 \end{bmatrix} * EI_c$$

$$\{X\} = [K]^{-1}\{P\} = [K]^{-1} \begin{Bmatrix} +120.0 \\ +52.8 \\ -115.2 \end{Bmatrix} = \begin{Bmatrix} +142.98 \\ +14.03 \\ -67.12 \end{Bmatrix} * \frac{1}{EI_c}$$

MATRIX DISPLACEMENT METHOD OF RIGID-FRAME ANALYSIS

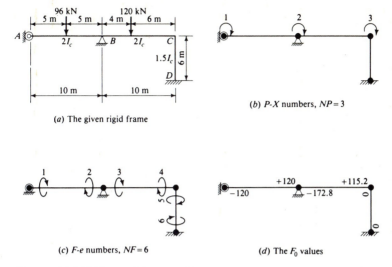

(a) The given rigid frame

(b) P-X numbers, $NP = 3$

(c) F-e numbers, $NF = 6$

(d) The F_0 values

Figure 12.2.2 Rigid frame of Example 12.2.2.

$$\{F^*\} = \{F_0\} * [SB]\{X\}$$

$$= \begin{Bmatrix} -120.0 \\ +120.0 \\ -172.8 \\ +115.2 \\ 0. \\ 0. \end{Bmatrix} + \begin{bmatrix} +0.8 & +0.4 \\ +0.4 & +0.8 \\ +0.8 & +0.4 \\ +0.4 & +0.8 \\ & & +1.0 \\ & & +0.5 \end{bmatrix} \begin{Bmatrix} +142.98 \\ +14.03 \\ -67.12 \end{Bmatrix}$$

$$= \begin{Bmatrix} -120.0 \\ +120.0 \\ -172.8 \\ +115.2 \\ 0. \\ 0. \end{Bmatrix} + \begin{Bmatrix} +120.0 \\ +68.42 \\ -15.62 \\ -48.08 \\ -67.12 \\ -33.56 \end{Bmatrix} = \begin{Bmatrix} +0.00 \\ +188.42 \\ -188.42 \\ +67.12 \\ -67.12 \\ -33.56 \end{Bmatrix}$$

Example 12.2.3 Analyze the rigid frame shown in Fig. 12.2.3a by the matrix displacement method. Note that this rigid frame is the same as the one in Example 7.6.3.

SOLUTION If the loading on the rigid frame were unsymmetrical, the degree of freedom would be $NP = 4$, three in the rotations of joints A, B, and C and one in the sidesway of ABC, in which case with $NF = 10$ the degree of indeterminacy NI is 6. Because of the

426 INTERMEDIATE STRUCTURAL ANALYSIS

symmetry not only in the makeup of the rigid frame but also in the applied loads, there can be no sidesway. To simplify the solution, a fictitious support to prevent the sidesway is added at C so that the degree of freedom becomes $NP = 3$, as shown in Fig. 12.2.3b. The fixed-end moments are shown in Fig. 12.2.3d.

$[A] =$

$P \diagdown F$	1	2	3	4	5	6	7	8	9	10
1	+1.				+1.					
2		+1.	+1.				+1.			
3				+1.					+1.	

$\{P\} =$

$P \diagdown LC$	1
1	+256.
2	0.
3	−256.

$[B] =$

$e \diagdown X$	1	2	3
1	+1.		
2		+1.	
3		+1.	
4			+1.
5	+1.		
6			
7		+1.	
8			
9			+1.
10			

MATRIX DISPLACEMENT METHOD OF RIGID-FRAME ANALYSIS 427

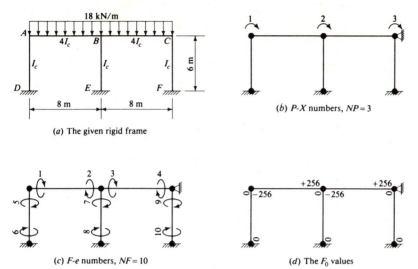

(a) The given rigid frame

(b) P-X numbers, $NP = 3$

(c) F-e numbers, $NF = 10$

(d) The F_0 values

Figure 12.2.3 Rigid frame of Example 12.2.3.

$$[S] = \begin{array}{c} \\ \\ \end{array}$$

F \ e	1	2	3	4	5	6	7	8	9	10
1	+2.	+1.								
2	+1.	+2.								
3			+2.	+1.						
4			+1.	+2.						
5					$+\frac{2}{3}$	$+\frac{1}{3}$				
6					$+\frac{1}{3}$	$+\frac{2}{3}$				
7							$+\frac{2}{3}$	$+\frac{1}{3}$		
8							$+\frac{1}{3}$	$+\frac{2}{3}$		
9									$+\frac{2}{3}$	$+\frac{1}{3}$
10									$+\frac{1}{3}$	$+\frac{2}{3}$

$* EI_c$

$$[SB] = \begin{array}{|c|c|c|c|} \hline \diagdown X \\ F \diagdown & 1 & 2 & 3 \\ \hline 1 & +2. & +1. & \\ \hline 2 & +1. & +2. & \\ \hline 3 & & +2. & +1. \\ \hline 4 & & +1. & +2. \\ \hline 5 & +\frac{2}{3} & & \\ \hline 6 & +\frac{1}{3} & & \\ \hline 7 & & +\frac{2}{3} & \\ \hline 8 & & +\frac{1}{3} & \\ \hline 9 & & & +\frac{2}{3} \\ \hline 10 & & & +\frac{1}{3} \\ \hline \end{array} * EI_c$$

$$[K] = [ASB] = \begin{array}{|c|c|c|c|} \hline \diagdown X \\ P \diagdown & 1 & 2 & 3 \\ \hline 1 & +2\frac{2}{3} & +1. & \\ \hline 2 & +1. & +4\frac{2}{3} & +1. \\ \hline 3 & & +1. & +2\frac{2}{3} \\ \hline \end{array} * EI_c$$

$$\{X\} = [K]^{-1}\{P\} = [K]^{-1} \begin{Bmatrix} +256. \\ 0. \\ -256. \end{Bmatrix} = \begin{Bmatrix} +96. \\ 0. \\ -96. \end{Bmatrix} * \frac{1}{EI_c}$$

$$\{F^*\} = \{F_0\} + [SB]\{X\} = \begin{Bmatrix} -256. \\ +256. \\ -256. \\ +256. \\ 0. \\ 0. \\ 0. \\ 0. \\ 0. \\ 0. \end{Bmatrix} + \begin{Bmatrix} +192. \\ +96. \\ -96. \\ -192. \\ +64. \\ +32. \\ 0. \\ 0. \\ -64. \\ -32. \end{Bmatrix} = \begin{Bmatrix} -64. \\ +352. \\ -352. \\ +64. \\ +64. \\ +32. \\ 0. \\ 0. \\ -64. \\ -32. \end{Bmatrix}$$

12.3 Analysis of Rigid Frames with Sidesway

For some rigid frames, even though the number of joints to be used in the analysis is the minimum permitted, some joints may undergo unknown amounts of translation. In such cases the degrees of freedom will include both rotations and translations. For rigid frames with horizontal and vertical members only, the only translations are those of the horizontal members in the horizontal direction; thus they are called sidesways. The degree of freedom in sidesway can be easily observed by inspection, in almost all cases.

The contents of the [A] matrix can be obtained either by rows from the free-body diagrams, or by columns from the effective components along the degrees of freedom of each F system acting on the member. The contents of the [B] matrix should be obtained by columns from the displacement diagrams. The contents of the {P} matrix can be obtained from the opposites of the restraining moments and forces along the degrees of freedom in the fixed condition.

Example 12.3.1 Analyze the rigid frame shown in Fig. 12.3.1a by the matrix displacement method. Note that this rigid frame is the same as the one in Example 7.7.1.

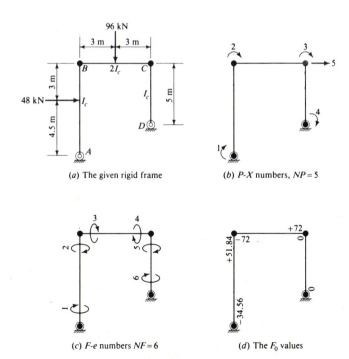

(a) The given rigid frame

(b) P-X numbers, $NP = 5$

(c) F-e numbers $NF = 6$

(d) The F_0 values

Figure 12.3.1 Rigid frame of Example 12.3.1.

430 INTERMEDIATE STRUCTURAL ANALYSIS

SOLUTION (a) *The [A] matrix.* The P-X and F-e diagrams are shown in Fig. 12.3.1b and c. The [A] matrix is

$$[A] = \begin{array}{|c|c|c|c|c|c|c|} \hline {}_P\!\diagdown\!{}^F & 1 & 2 & 3 & 4 & 5 & 6 \\ \hline 1 & +1. & & & & & \\ \hline 2 & & +1. & +1. & & & \\ \hline 3 & & & & +1. & +1. & \\ \hline 4 & & & & & & +1. \\ \hline 5 & -\dfrac{1}{7.5} & -\dfrac{1}{7.5} & & & -\dfrac{1}{5} & -\dfrac{1}{5} \\ \hline \end{array}$$

One can see that each row in this matrix is simply an equation of equilibrium for each of the five free-body diagrams shown in Fig. 12.3.2a. Also the entries in the first column of this matrix are simply the effective components of the generalized forces in the F_1 system shown in Fig. 12.3.2b along the degrees of freedom; so are those in the other five columns.

(b) *The [B] matrix.* The [B] matrix is established by columns from observing the five displacement diagrams shown in Fig. 12.3.3, recalling that the member-end rotations are measured *clockwise* from the straight line joining the ends of the element to the elastic-curve tangent. Thus,

$$[B] = \begin{array}{|c|c|c|c|c|c|} \hline {}_e\!\diagdown\!{}^X & 1 & 2 & 3 & 4 & 5 \\ \hline 1 & +1. & & & & -\dfrac{1}{7.5} \\ \hline 2 & & +1. & & & -\dfrac{1}{7.5} \\ \hline 3 & & +1. & & & \\ \hline 4 & & & +1. & & \\ \hline 5 & & & +1. & & -\dfrac{1}{5} \\ \hline 6 & & & & +1. & -\dfrac{1}{5} \\ \hline \end{array}$$

(c) *The [S] matrix.* The [S] matrix is a 6×6 matrix containing the following three 2×2 submatrices.

MATRIX DISPLACEMENT METHOD OF RIGID-FRAME ANALYSIS 431

$$[S] = \begin{array}{c|cccccc} F \backslash e & 1 & 2 & 3 & 4 & 5 & 6 \\ \hline 1 & +\dfrac{4}{7.5} & +\dfrac{2}{7.5} & & & & \\ 2 & +\dfrac{2}{7.5} & +\dfrac{4}{7.5} & & & & \\ 3 & & & +\dfrac{4}{3} & +\dfrac{2}{3} & & \\ 4 & & & +\dfrac{2}{3} & +\dfrac{4}{3} & & \\ 5 & & & & & +0.8 & +0.4 \\ 6 & & & & & +0.4 & +0.8 \end{array} * EI_c$$

(d) *The* $\{P\}$ *matrix.* From Fig. 12.3.1d,

$$P_1 = -F_{01} = -(-34.56) = +34.56 \text{ kN·m}$$
$$P_2 = -(F_{02} + F_{03}) = -(+51.84 - 72.) = +20.16 \text{ kN·m}$$
$$P_3 = -(F_{04} + F_{05}) = -(+72. + 0.) = -72. \text{ kN·m}$$
$$P_4 = -F_{06} = 0.$$

So far as P_5 is concerned, it should be equal to the opposite of the restraining force. The restraining force itself, as seen from Fig. 12.3.4, is 31.104 kN to the left, or $R_{05} = -31.104$ kN (positive is to the right, as P_5). Thus,

$$P_5 = -R_{05} = -(-31.104) = +31.104 \text{ kN}$$

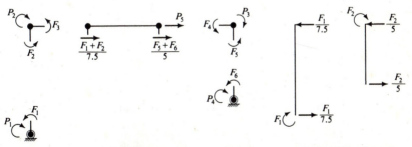

(a) Establishing [A] by rows

(b) Establishing first and second columns of [A]

Figure 12.3.2 Free-body diagrams for the [A] matrix in Example 12.3.1.

(e) *The output matrices $\{X\}$ and $\{F^*\}$.*

$[SB] =$

X / F	1	2	3	4	5
1	+1.6/3	+0.8/3			−0.32/3
2	+0.8/3	+1.6/3			−0.32/3
3		+4/3	+2/3		
4		+2/3	+4/3		
5			+0.8	+0.4	−0.24
6			+0.4	+0.8	−0.24

$* EI_c$

$[K] =$

X / P	1	2	3	4	5
1	+1.6/3	+0.8/3			−0.32/3
2	+0.8/3	+5.6/3	+2/3		−0.32/3
3		+2/3	+6.4/3	+0.4	−0.24
4			+0.4	+0.8	−0.24
5	−0.32/3	−0.32/3	−0.24	−0.24	+1.12/9

$* EI_c$

$$\{X\} = [K]^{-1}\{P\}$$

$$= [K]^{-1} \begin{Bmatrix} +34.56 \\ +20.16 \\ -72.00 \\ 0.00 \\ +31.104 \end{Bmatrix} = \begin{Bmatrix} +336.42 \\ +29.82 \\ +41.40 \\ +409.12 \\ +1432.7 \end{Bmatrix} * \frac{1}{EI_c}$$

$$\{F^*\} = \{F_0\} + [SB]\{X\}$$

$$= \begin{Bmatrix} -34.56 \\ +51.84 \\ -72. \\ +72. \\ 0. \\ 0. \end{Bmatrix} + \begin{Bmatrix} +34.55 \\ -47.21 \\ +67.36 \\ +75.08 \\ -147.08 \\ +0.01 \end{Bmatrix} = \begin{Bmatrix} -0.01 \\ +4.63 \\ -4.64 \\ +147.08 \\ -147.08 \\ +0.01 \end{Bmatrix}$$

MATRIX DISPLACEMENT METHOD OF RIGID-FRAME ANALYSIS 433

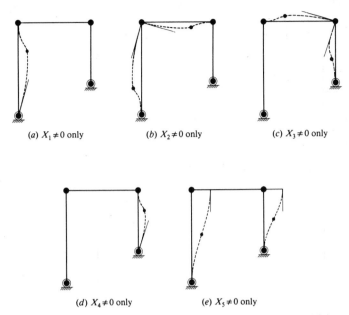

(a) $X_1 \neq 0$ only (b) $X_2 \neq 0$ only (c) $X_3 \neq 0$ only

(d) $X_4 \neq 0$ only (e) $X_5 \neq 0$ only

Figure 12.3.3 Displacement diagrams for the $[B]$ matrix in Example 12.3.1.

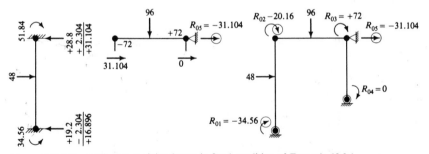

Figure 12.3.4 Generalized restraining forces in fixed condition of Example 12.3.1.

(f) *The statics and deformation checks.* To make complete output checks, it will be necessary to make the $NP = 5$ statics checks and $NF = 6$ deformation checks, as well as to draw the free-body, shear, and moment diagrams of the members, plus the sketch of the elastic curve. Refer to Example 4.6.2 and Fig. 4.6.5.

Example 12.3.2 Analyze the rigid frame shown in Fig. 12.3.5a by the matrix displacement method. Note that this rigid frame is the same as the one in Example 7.7.2.

SOLUTION Since the rigid frame itself and the loading acting on it in this problem are identical to those of the preceding example except that the supports are now fixed, the 3×3 $[K]$ matrix

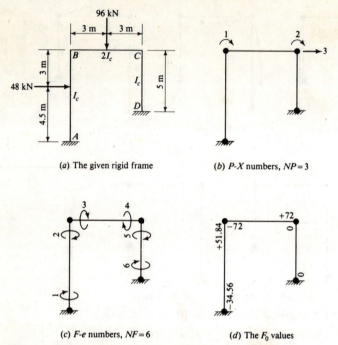

Figure 12.3.5 Rigid frame of Example 12.3.2.

and the 3×1 $\{P\}$ matrix for this problem can be taken from the preceding example by omitting the rows and columns relating to the freedoms to rotate at the hinged supports. Thus,

$$\{X\} = [K]^{-1}\{P\}$$

$$= \begin{bmatrix} +\dfrac{5.6}{3} & +\dfrac{2}{3} & -\dfrac{0.32}{3} \\ +\dfrac{2}{3} & +\dfrac{6.4}{3} & -0.24 \\ -\dfrac{0.32}{3} & -0.24 & +\dfrac{1.12}{9} \end{bmatrix}^{-1} * \dfrac{1}{EI_c} \begin{Bmatrix} +20.16 \\ -72.00 \\ +31.104 \end{Bmatrix} = \begin{Bmatrix} +30.422 \\ -15.586 \\ +245.96 \end{Bmatrix} * \dfrac{1}{EI_c}$$

Similarly, the 6×3 $[SB]$ matrix for this problem can be taken from the 6×5 $[SB]$ matrix in the preceding example; thus

$$\{F^*\} = \{F_0\} + [SB]\{X\}$$

$$= \begin{Bmatrix} -34.56 \\ +51.84 \\ -72.00 \\ +72.00 \\ 0.00 \\ 0.00 \end{Bmatrix} + \begin{bmatrix} +0.8/3 & -0.32/3 & \\ +1.6/3 & -0.32/3 & \\ +4/3 & +2/3 & \\ +2/3 & +4/3 & \\ & +0.8 & -0.24 \\ & +0.4 & -0.24 \end{bmatrix} \begin{Bmatrix} +30.422 \\ -15.586 \\ +245.96 \end{Bmatrix} = \begin{Bmatrix} -52.68 \\ +41.83 \\ -41.83 \\ +71.50 \\ -71.50 \\ -65.26 \end{Bmatrix}$$

MATRIX DISPLACEMENT METHOD OF RIGID-FRAME ANALYSIS

Example 12.3.3 Analyze the rigid frame shown in Fig. 12.3.6a by the matrix displacement method. Note that this rigid frame is the same as the one in Example 7.7.3.

SOLUTION (a) *The input matrices* $[A]$, $[B]$, $[S]$, and $\{P\}$. The P-X and F-e numbers are shown in Fig. 12.3.6b and c, wherein $NP = 4$, $NF = 10$, and $NI = 6$. The fixed-end moments are shown in Fig. 12.3.6d. The input matrices $[A]$, $\{P\}$, $[B]$, and $[S]$ are shown below:

$[A] =$

P \ F	1	2	3	4	5	6	7	8	9	10
1	+1.				+1.					
2		+1.	+1.				+1.			
3				+1.					+1.	
4					$-\frac{1}{10}$	$-\frac{1}{10}$	$-\frac{1}{8}$	$-\frac{1}{8}$	$-\frac{1}{4}$	$-\frac{1}{4}$

$\{P\} =$

P \ LC	1
1	+96.
2	−96.
3	0.
4	0.

$[B] =$

e \ X	1	2	3	4
1	+1.			
2		+1.		
3		+1.		
4			+1.	
5	+1.			$-\frac{1}{10}$
6				$-\frac{1}{10}$
7		+1.		$-\frac{1}{8}$
8				$-\frac{1}{8}$
9			+1.	$-\frac{1}{4}$
10				$-\frac{1}{4}$

436 INTERMEDIATE STRUCTURAL ANALYSIS

$[S] =$

e\F	1	2	3	4	5	6	7	8	9	10
1	1.50	0.75								
2	0.75	1.50								
3			1.50	0.75						
4			0.75	1.50						
5					0.8	0.4				
6					0.4	0.8				
7							1.0	0.5		
8							0.5	1.0		
9									2.0	1.0
10									1.0	2.0

$* EI_c$

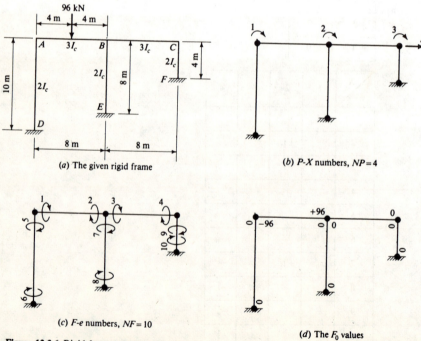

(a) The given rigid frame

(b) P-X numbers, $NP = 4$

(c) F-e numbers, $NF = 10$

(d) The F_0 values

Figure 12.3.6 Rigid frame of Example 12.3.3.

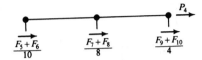

(a) Free-body diagram for fourth row of $[A]$

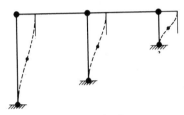

(b) Displacement diagram for fourth column of $[B]$

Figure 12.3.7 Diagrams for the sidesway degree of freedom in Example 12.3.3.

For the $[A]$ matrix, only the free-body diagram for its fourth row is shown in Fig. 12.3.7a; for the $[B]$ matrix, only the displacement diagram for $X_4 \neq 0$ is shown in Fig. 12.3.7b. For the $\{P\}$ matrix, the fixed condition is shown in Fig. 12.3.8, wherein the required restraining moments and forces are

$$R_{01} = -96. \quad R_{02} = +96. \quad R_{03} = 0. \quad R_{04} = 0.$$

Thus,

$$P_1 = -R_{01} = +96. \quad P_2 = -R_{02} = -96. \quad P_3 = -R_{03} = 0. \quad P_4 = -R_{04} = 0.$$

(b) The output matrices $\{X\}$ and $\{F^*\}$.

$[SB] =$

F \ X	1	2	3	4	
1	+1.50	+0.75			
2	+0.75	+1.50			
3		+1.50	+0.75		
4		+0.75	+1.50		
5	+0.8			−0.12	$*EI_c$
6	+0.4			−0.12	
7		+1.0		−0.1875	
8		+0.5		−0.1875	
9			+2.0	−0.75	
10			+1.0	−0.75	

438 INTERMEDIATE STRUCTURAL ANALYSIS

$$[K] = \begin{array}{|c|c|c|c|c|} \hline \diagbox{P}{X} & 1 & 2 & 3 & 4 \\ \hline 1 & +2.30 & +0.75 & & -0.12 \\ \hline 2 & +0.75 & +4.00 & +0.75 & -0.1875 \\ \hline 3 & & +0.75 & +3.50 & -0.75 \\ \hline 4 & -0.12 & -0.1875 & -0.75 & +3.567/8 \\ \hline \end{array} * EI_c$$

$$\{X\} = \begin{array}{|c|c|} \hline \diagbox{X}{LC} & 1 \\ \hline 1 & +54.33 \\ \hline 2 & -35.48 \\ \hline 3 & +11.79 \\ \hline 4 & +19.529 \\ \hline \end{array} * \frac{1}{EI_c}$$

$$\{F^*\} = \{F_0\} + [SB]\{X\}$$

$$= \begin{Bmatrix} -96. \\ +96. \\ 0. \\ 0. \\ 0. \\ 0. \\ 0. \\ 0. \\ 0. \\ 0. \end{Bmatrix} + \begin{Bmatrix} +54.88 \\ -12.47 \\ -44.38 \\ -8.92 \\ +41.12 \\ +19.39 \\ -39.14 \\ -21.40 \\ +8.93 \\ -2.86 \end{Bmatrix} = \begin{Bmatrix} -41.12 \\ +83.53 \\ -44.38 \\ -8.92 \\ +41.12 \\ +19.39 \\ -39.14 \\ -21.40 \\ +8.93 \\ -2.86 \end{Bmatrix}$$

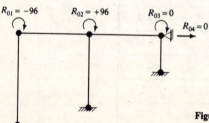

Figure 12.3.8 Generalized restraining forces in fixed condition of Example 12.3.3.

MATRIX DISPLACEMENT METHOD OF RIGID-FRAME ANALYSIS 439

Example 12.3.4 Analyze the rigid frame shown in Fig. 12.3.9a by the matrix displacement method. Note that this rigid frame is the same as the one in Example 7.7.4.

SOLUTION (a) *The input matrices* [A], [B], [S], *and* {P}. The P-X and F-e numbers are shown in Fig. 12.3.9b and c, wherein $NP = 6$, $NF = 12$, and $NI = 6$. There are no loads acting on the elements themselves, so there are no fixed-end moments anywhere. The [A], {P}, [B], and [S] matrices are shown below. The free-body diagrams for the first four rows of the [A] matrix and the displacement diagrams for the first four columns of the [B] matrix are not shown; in fact they really only show which member ends go into which joint. The fifth and sixth rows of [A], however, are established from the free-body diagrams shown in Fig. 12.3.10a. The fifth and sixth columns of [B] are based on the displacement diagrams of Fig. 12.3.10b. With some experience now, one may even omit sketching the elastic curves in Fig. 12.3.10b and merely show the rotations of the member axes. From the equation $e_i = X_i - R_{ij}$, one may see that any clockwise rotation of the member axis means negative end rotation. Due to $X_5 \neq 0$ in Fig. 12.3.10b, the member axes 5-6 and 7-8 rotate clockwise; therefore $e_5 = e_6 = e_7 = e_8 = -X_5/6.4$. Owing to $X_6 \neq 0$ in Fig. 12.3.10b, the member axes 5-6 and 7-8 rotate counterclockwise but member axes 9-10 and 11-12 rotate clockwise; therefore $e_5 = e_6 = e_7 = e_8 = +X_6/6.4$, $e_9 = e_{10} = -X_6/6.4$, *and* $e_{11} = e_{12} = -X_6/3.2$.

$[A] = $

F \\ P	1	2	3	4	5	6	7	8	9	10	11	12
1	+1.				+1.							
2		+1.					+1.					
3			+1.			+1.				+1.		
4				+1.				+1.				+1.
5					$-\dfrac{1}{6.4}$	$-\dfrac{1}{6.4}$	$-\dfrac{1}{6.4}$	$-\dfrac{1}{6.4}$				
6					$+\dfrac{1}{6.4}$	$+\dfrac{1}{6.4}$	$+\dfrac{1}{6.4}$	$+\dfrac{1}{6.4}$	$-\dfrac{1}{6.4}$	$-\dfrac{1}{6.4}$	$-\dfrac{1}{3.2}$	$-\dfrac{1}{3.2}$

$\{P\} = $

LC \\ P	1
1	0.
2	0.
3	0.
4	0.
5	+12.
6	+24.

440 INTERMEDIATE STRUCTURAL ANALYSIS

$[B] =$

e \ X	1	2	3	4	5	6
1	+1.					
2		+1.				
3			+1.			
4				+1.		
5	+1.				−1/6.4	+1/6.4
6		+1.			−1/6.4	+1/6.4
7		+1.			−1/6.4	+1/6.4
8				+1.	−1/6.4	+1/6.4
9			+1.			−1/6.4
10						−1/6.4
11				+1.		−1/3.2
12						−1/3.2

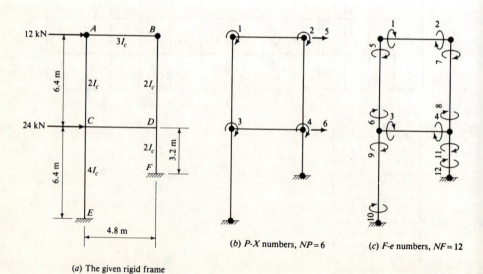

(a) The given rigid frame (b) P-X numbers, NP = 6 (c) F-e numbers, NF = 12

Figure 12.3.9 Rigid frame of Example 12.3.4.

MATRIX DISPLACEMENT METHOD OF RIGID-FRAME ANALYSIS 441

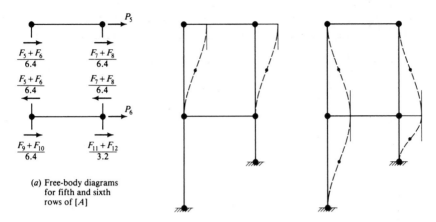

(a) Free-body diagrams for fifth and sixth rows of [A]

(b) Displacement diagrams for fifth and sixth columns of [B]

Figure 12.3.10 Diagrams for the sidesway degrees of freedom in Example 12.3.4.

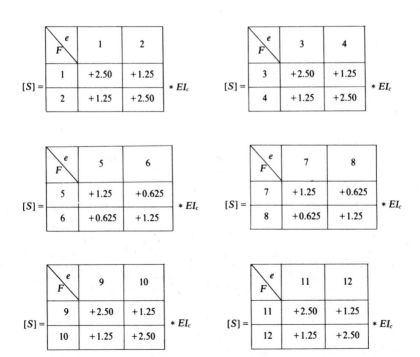

442 INTERMEDIATE STRUCTURAL ANALYSIS

$[SB] =$

F \ X	1	2	3	4	5	6
1	+2.50	+1.25				
2	+1.25	+2.50				
3			+2.50	+1.25		
4			+1.25	+2.50		
5	+1.25		+0.625		−0.29297	+0.29297
6	+0.625		+1.25		−0.29297	+0.29297
7		+1.25		+0.625	−0.29297	+0.29297
8		+0.625		+1.25	−0.29297	+0.29297
9			+2.50			−0.58594
10			+1.25			−0.58594
11				+2.50		−1.17188
12				+1.25		−1.17188

$* EI_c$

$[K] =$

P \ X	1	2	3	4	5	6
1	+3.75	+1.25	+0.625		−0.29297	+0.29297
2	+1.25	+3.75		+0.625	−0.29297	+0.29297
3	+0.625		+0.625	+1.25	−0.29297	−0.29297
4		+0.625	+1.25	+0.625	−0.29297	−0.87891
5	−0.29297	−0.29297	−0.29297	−0.29297	+0.183106	−0.183106
6	+0.29297	+0.29297	−0.29297	−0.87891	−0.183106	+1.098638

$* EI_c$

$$\{X\} = \begin{array}{|c|c|} \hline \diagdown\text{LC} & 1 \\ X \diagdown & \\ \hline 1 & +\ 6.6138 \\ \hline 2 & +\ 4.6592 \\ \hline 3 & +\ 7.9347 \\ \hline 4 & +\ 15.7440 \\ \hline 5 & +186.010 \\ \hline 6 & +\ 64.550 \\ \hline \end{array} * \dfrac{1}{EI_c}$$

$$\{F^*\} = \{F_0\} + [SB]\{X\} = \begin{array}{|c|c|} \hline \diagdown\text{LC} & 1 \\ F^* \diagdown & \\ \hline 1 & +22.36 \\ \hline 2 & +19.92 \\ \hline 3 & +39.52 \\ \hline 4 & +49.28 \\ \hline 5 & -22.36 \\ \hline 6 & -21.53 \\ \hline 7 & -19.92 \\ \hline 8 & -12.99 \\ \hline 9 & -17.99 \\ \hline 10 & -27.90 \\ \hline 11 & -36.28 \\ \hline 12 & -55.96 \\ \hline \end{array}$$

12.4 Analysis of Rigid Frames for Yielding of Supports

Just like the analysis for uneven support settlements in continuous beams, the matrix displacement method can be used to obtain the effects of yielding of supports in rigid frames. First the fixed-end moments are determined to fix all joints against rotation and translation while allowing the yielding of supports to happen. These fixed-end moments are, according to Eqs. (8.11.1) and (8.11.2),

$$M_{0i} = +\frac{4EI}{L}d\theta \qquad M_{0j} = +\frac{2EI}{L}d\theta \qquad \text{for rotational yielding of } d\theta \text{ at } i\text{th end} \qquad (12.4.1)$$

and

$$M_{0i} = -\frac{6EI\Delta}{L^2} \qquad M_{0j} = -\frac{6EI\Delta}{L^2} \qquad \text{for settlement of } \Delta \text{ at } j\text{th end relative to } i\text{th end} \qquad (12.4.2)$$

For these fixed-end moments, there may be reactions acting on the member ends, the opposites of these restraining reactions may contribute to become the active sidesway forces in the $\{P\}$ matrix.

Example 12.4.1 By the matrix displacement method analyze the rigid frame shown in Fig. 12.4.1a for a rotational slip of 0.002 rad clockwise of joint D and a vertical settlement of 15 mm at joint D.

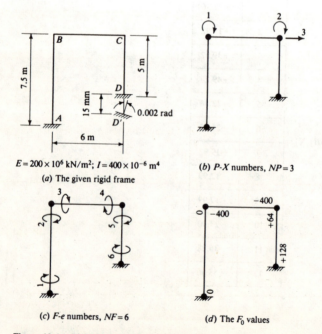

Figure 12.4.1 Rigid frame of Example 12.4.1.

MATRIX DISPLACEMENT METHOD OF RIGID-FRAME ANALYSIS 445

SOLUTION (a) The input matrices [A], [B], [S], and {P}. The P-X and F-e numbers are shown in Fig. 12.4.1b and c, wherein NP = 3, NF = 6, and NI = 3. The [A] and [B] matrices are as follows:

$$[A] = \begin{array}{c|c|c|c|c|c|c|} \diagdown^F_P & 1 & 2 & 3 & 4 & 5 & 6 \\ \hline 1 & & +1. & +1. & & & \\ \hline 2 & & & & +1. & +1. & \\ \hline 3 & -\dfrac{1}{7.5} & -\dfrac{1}{7.5} & & & -\dfrac{1}{5} & -\dfrac{1}{5} \\ \end{array}$$

$$[B] = \begin{array}{c|c|c|c|} \diagdown^X_e & 1 & 2 & 3 \\ \hline 1 & & & -\dfrac{1}{7.5} \\ \hline 2 & +1. & & -\dfrac{1}{7.5} \\ \hline 3 & +1. & & \\ \hline 4 & & +1. & \\ \hline 5 & & +1. & -\dfrac{1}{5} \\ \hline 6 & & & -\dfrac{1}{5} \\ \end{array}$$

The entries in the [S] matrix are computed in absolute dimensional units, in kilonewton-meters per radian; thus

$\diagdown^e_F$	1	2	3	4	5	6
1	+42666.	+21333.				
2	+21333.	+42666.				
3			+106666.	+53333.		
4			+53333.	+106666.		
5					+64000.	+32000.
6					+32000.	+64000.

[S] =

446 INTERMEDIATE STRUCTURAL ANALYSIS

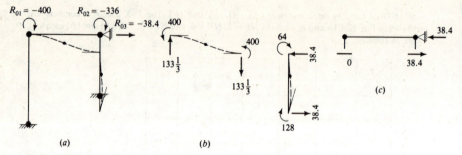

Figure 12.4.2 Generalized restraining forces in fixed condition of Example 12.4.1.

The shape of the rigid frame, with all joints fixed in rotation and translation while allowing the yielding of supports to happen, is shown in Fig. 12.4.2a. The restraining moments and reactions on the member ends are shown in Fig. 12.4.2b. From Fig. 12.4.2c, R_{03} is 38.4 kN to the left, or

$$R_{03} = -38.4 \text{ kN} \quad \text{(positive is to the right)}$$

Thus,

$$P_1 = -(F_{02} + F_{03}) = -(0. - 400.) = +400. \text{ kN} \cdot \text{m}$$
$$P_2 = -(F_{04} + F_{05}) = -(-400. + 60.) = +336. \text{ kN} \cdot \text{m}$$
$$P_3 = -R_{03} = -(-38.4) = +38.4 \text{ kN}$$

(b) *The output matrices* $\{X\}$ *and* $\{F^*\}$.

$[SB] = $

F \ X	1	2	3
1	+ 21333.		− 8533.3
2		+ 42666.	− 8533.3
3	+ 106666.	+ 53333.	
4	+ 53333.	+ 106666.	
5		+ 64000.	− 19200.
6		+ 32000.	− 19200.

$[K] = $

P \ X	1	2	3
1	+ 149333.	+ 53333.	− 8533.3
2	+ 53333.	+ 170666.	− 19200.
3	− 8533.3	− 19200.	+ 9455.5

$$\{X\} = [K]^{-1}\{P\} = \begin{array}{|c|c|} \hline \diagdown LC & \\ X \diagdown & 1 \\ \hline 1 & +2.4253 \times 10^{-3} \text{ rad} \\ \hline 2 & +2.3992 \times 10^{-3} \text{ rad} \\ \hline 3 & +10.563 \times 10^{-3} \text{ m} \\ \hline \end{array}$$

$$\{F^*\} = \{F_0\} + [SB]\{X\} = \begin{Bmatrix} 0. \\ 0. \\ -400. \\ -400. \\ +64. \\ +128. \end{Bmatrix} + \begin{Bmatrix} -38.40 \\ +13.34 \\ +386.66 \\ +385.26 \\ -49.26 \\ -126.04 \end{Bmatrix} = \begin{Bmatrix} -38.40 \\ +13.34 \\ -13.34 \\ -14.74 \\ +14.74 \\ +1.96 \end{Bmatrix}$$

12.5 Analysis of Gable Frames

The basic equations in the matrix displacement method of rigid-frame analysis are no different from those of beam analysis or of truss analysis. In the future, these basic equations will be shown to be equally applicable to rigid-frame analysis when axial and shear deformations are considered, in Chaps. 16 and 21; to the analysis of horizontal grid frames vertically loaded, in Chap. 19; to rigid frames with curved members and with semirigid connections, in Chaps. 18 and 20; and to beam elements supported on elastic foundation, in Chap. 22. In every case, the input matrices are $[A]$, $[B]$, $[S]$, and $\{P\}$, and the output matrices $\{X\}$ and $\{F^*\}$ are obtained in the two major steps

$$\{X\} = [ASB]^{-1}\{P\} \qquad (12.5.1)$$

and
$$\{F^*\} = \{F_0\} + [SB]\{X\} \qquad (12.5.2)$$

This section will describe how the input matrices can be determined for the analysis of rigid frames with nonrectangular joints by neglecting the effect of axial deformation. Single-span gable frames with two fixed supports which have been solved by the slope-deflection and moment-distribution methods in Chaps. 7 and 8 will be used in the examples.

Oftentimes even the lowest degrees of freedom permitted to be used in the analysis of a rigid frame may involve both rotation and sidesway. The degree of freedom in rotation, NPR, is easy to determine because only at the fixed supports is the freedom to rotate suppressed. The degree of freedom in sidesway, NPS, is the number of independent unknown linear displacements of the joints which are necessary to define the deflected positions of all joints. This number of unknowns can best be obtained by actually counting how many independent joint translations are needed to graphically determine the

448 INTERMEDIATE STRUCTURAL ANALYSIS

deflected positions of all the joints, when all member lengths remain unchanged. In addition, an algebraic formula may be derived to give the degree of freedom in sidesway, NPS, of a rigid frame, if axial deformation is to be ignored in its analysis. Let

NFS = number of fixed supports
NHS = number of hinged supports
NRS = number of roller supports
NJ = number of joints
NM = number of members

Then

$$NPS = 2(NJ) - [2(NFS + NHS) + NRS + NM] \quad (12.5.3)$$

Since each member length furnishes one constraint in that the distance between the two joints representing the ends of the member must remain unchanged, the degree of freedom in sidesway is NM less than the number of unknown joint translations. Equation (12.5.3) may be used to double-check the result of the graphical consideration.

Example 12.5.1 By the matrix displacement method analyze the gable frame shown in Fig. 12.5.1a for two separate loading conditions: the first is the uniform load in the vertical direction and the second is the uniform load in the horizontal direction. Note that this problem is a combination of Examples 7.9.1 and 7.9.2.

SOLUTION (a) *Degree of freedom in sidesway.* The minimum number of elements to be used is four, making five the minimum number of joints, as shown in Fig. 12.5.1b and c. From the graphical sketch of the possible joint translations shown in Fig. 12.5.2, there are four unknown joint translations, but only two are independent. Any two of the four—the horizontal deflections of joints B, C, and D and the vertical deflection of joint C—may be chosen as the freedoms in sidesway. The choice is shown in Fig. 12.5.1b. Also, by Eq. (12.5.3),

$$NPS = 2(NJ) - [2(NFS + NHS) + NRS + NM]$$
$$= 2(5) - [2(2 + 0) + 0 + 4)] = 2$$

Thus, $NP = NPR + NPS = 3 + 2 = 5$, $NF = 8$, and $NI = 3$.

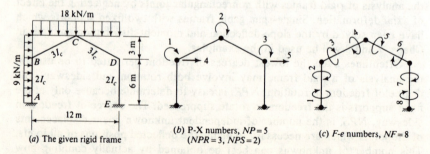

(a) The given rigid frame   (b) P-X numbers, $NP = 5$ (c) F-e numbers, $NF = 8$
(NPR = 3, NPS = 2)

Figure 12.5.1 Gable frame of Example 12.5.1.

MATRIX DISPLACEMENT METHOD OF RIGID-FRAME ANALYSIS 449

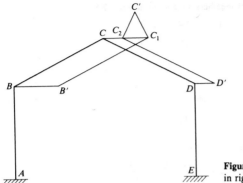

Figure 12.5.2 Possible joint translations in rigid frame of Example 12.5.1.

(*b*) *The* [*A*] *matrix.* For rigid frames with rectangular joints, whether it is more convenient to establish the [*A*] matrix by rows or by columns is open to question and may be decided randomly by the analyst. For rigid frames with nonrectangular joints, invariably it is more convenient to establish the sidesway equations in the [*A*] matrix by rows, because one needs only to draw one set of free-body diagrams in which all end moments are nonzero. From the free-body diagrams of members 1-2 and 7-8 in Fig. 12.5.3,

$$H_2 = \frac{F_1 + F_2}{6} \qquad H_7 = \frac{F_7 + F_8}{6}$$

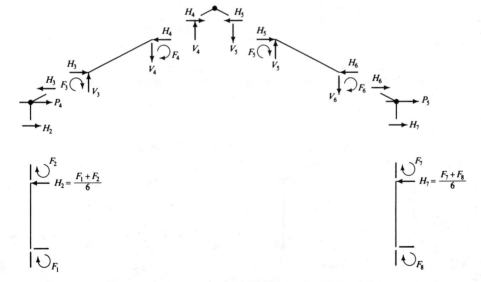

Figure 12.5.3 Free-body diagrams for the [*A*] matrix in Example 12.5.1.

From the free-body diagrams of members 3-4 and 5-6 in Fig. 12.5.3,

$$+3H_4 - 6V_4 = F_3 + F_4 \tag{12.5.4}$$

$$-3H_5 - 6V_5 = F_5 + F_6 \tag{12.5.5}$$

But for equilibrium at joint C,

$$H_4 = H_5 \qquad V_4 = V_5$$

Solving Eqs. (12.5.4) and (12.5.5) simultaneously,

$$H_4 = H_5 = \frac{(F_3 + F_4) - (F_5 + F_6)}{6}$$

which is also H_3 or H_6. For horizontal equilibrium at joint B and C,

$$P_4 = -H_2 + H_3 = -\frac{F_1 + F_2}{6} + \frac{F_3 + F_4}{6} - \frac{F_5 + F_6}{6}$$

and

$$P_5 = -H_6 - H_7 = -\frac{F_3 + F_4}{6} + \frac{F_5 + F_6}{6} - \frac{F_7 + F_8}{6}$$

Thus the $[A]$ matrix is

P \ F	1	2	3	4	5	6	7	8
1		+1.	+1.					
2				+1.	+1.			
3						+1.	+1.	
4	−1/6	−1/6	+1/6	+1/6	−1/6	−1/6		
5			−1/6	−1/6	+1/6	+1/6	−1/6	−1/6

$[A] =$

(c) *The $[B]$ matrix.* It is always advisable to obtain the $[B]$ matrix by columns. The joint-displacement diagrams for $X_4 \neq 0$ only and for $X_5 \neq 0$ only are shown in Fig. 12.5.4a

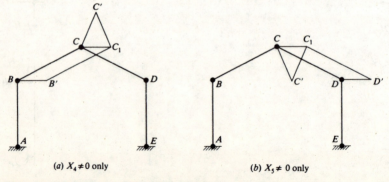

(a) $X_4 \neq 0$ only   (b) $X_5 \neq 0$ only

Figure 12.5.4 Displacement diagrams for the $[B]$ matrix in Example 12.5.1.

MATRIX DISPLACEMENT METHOD OF RIGID-FRAME ANALYSIS **451**

and b, but the elastic curves are not drawn because the member-end rotations can be obtained by the relationships

$$e_i = X_i - R_{ij} = 0 - R_{ij} = -R_{ij}$$
$$e_j = X_j - R_{ij} = 0 - R_{ij} = -R_{ij}$$

For $X_4 \neq 0$ only,

$$e_1 = e_2 = -R_{12} = -\left(+\frac{BB'}{AB}\right) = -\frac{X_4}{6}$$

$$e_3 = e_4 = -R_{34} = -\left(-\frac{C_1 C'}{BC}\right) = \frac{(+\sqrt{5}/2)X_4}{3\sqrt{5}} = +\frac{X_4}{6}$$

$$e_5 = e_6 = -R_{56} = -\left(+\frac{CC'}{CD}\right) = \frac{(-\sqrt{5}/2)X_4}{3\sqrt{5}} = -\frac{X_4}{6}$$

$$e_7 = e_8 = 0$$

For $X_5 \neq 0$ only,

$$e_1 = e_2 = 0$$

$$e_3 = e_4 = -R_{34} = -\left(+\frac{CC'}{BC}\right) = \frac{(-\sqrt{5}/2)X_5}{3\sqrt{5}} = -\frac{X_5}{6}$$

$$e_5 = e_6 = -R_{56} = -\left(-\frac{C_1 C'}{CD}\right) = \frac{(+\sqrt{5}/2)X_5}{3\sqrt{5}} = +\frac{X_5}{6}$$

$$e_7 = e_8 = -R_{78} = -\left(+\frac{DD'}{DE}\right) = -\frac{X_5}{6}$$

Thus the $[B]$ matrix is

$[B] =$

e \ X	1	2	3	4	5
1				−1/6	
2	+1.			−1/6	
3	+1.			+1/6	−1/6
4		+1.		+1/6	−1/6
5		+1.		−1/6	+1/6
6			+1.	−1/6	+1/6
7			+1.		−1/6
8					−1/6

(d) *The $[S]$ matrix.* For members 1-2 and 7-8,

$$\frac{4EI}{L} = \frac{4E(2I_c)}{6} = \frac{4}{3} EI_c \qquad \frac{2EI}{L} = \frac{2}{3} EI_c$$

For members 3-4 and 5-6,

$$\frac{4EI}{L} = \frac{4E(3I_c)}{6.7082} = 1.7888 EI_c \qquad \frac{2EI}{L} = 0.8944 EI_c$$

Thus the [S] matrix is

$[S] =$

e \ F	1	2	3	4	5	6	7	8
1	+1.3333	+0.6667						
2	+0.6667	+1.3333						
3			+1.7888	+0.8944				
4			+0.8944	+1.7888				
5					+1.7888	+0.8944		
6					+0.8944	+1.7888		
7							+1.3333	+0.6667
8							+0.6667	+1.3333

$* EI_c$

(e) *The* {P} *matrix.* The {P} matrix has two columns, one for each separate loading condition. As has been proved in part (a) of Example 7.9.1, the fixed-end moments due to a uniform load on an inclined member are $\frac{1}{12}$ of the intensity of vertical load per unit horizontal distance times the square of the horizontal projection of the member, or $\frac{1}{12}$ of the intensity of horizontal load per unit vertical distance times the square of the vertical projection of the member. Thus for the vertical loading condition in Fig. 12.5.5,

$$F_{03} = -\frac{18(6)^2}{12} = -54 \text{ kN·m} \qquad F_{04} = +54 \text{ kN·m}$$

$$F_{05} = -54 \text{ kN·m} \qquad F_{06} = +54 \text{ kN·m}$$

$$P_1 = -(F_{02} + F_{03}) = -(0 - 54) = +54 \text{ kN·m}$$

$$P_2 = -(F_{04} + F_{05}) = -(+54 - 54) = 0$$

$$P_3 = -(F_{06} + F_{07}) = -(+54 + 0) = -54 \text{ kN·m}$$

and, for the horizontal loading condition in Fig. 12.5.6,

$$F_{01} = -\frac{9(6)^2}{12} = -27 \text{ kN·m} \qquad F_{02} = +27 \text{ kN·m}$$

$$F_{03} = -\frac{9(3)^2}{12} = -6.75 \text{ kN·m} \qquad F_{04} = +6.75 \text{ kN·m}$$

$$P_1 = -(F_{02} + F_{03}) = -(+27 - 6.75) = -20.25 \text{ kN·m}$$

$$P_2 = -(F_{04} + F_{05}) = -(+6.75 + 0) = -6.75 \text{ kN·m}$$

$$P_3 = -(F_{06} + F_{07}) = 0$$

The restraining forces R_{04} and R_{05} (positive if in the direction of the degree of freedom) in each loading case are determined from the free-body diagrams in Figs. 12.5.5 and 12.5.6, respectively.

MATRIX DISPLACEMENT METHOD OF RIGID-FRAME ANALYSIS 453

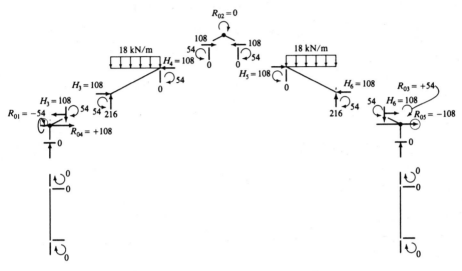

Figure 12.5.5 Fixed condition for vertical loading in Example 12.5.1.

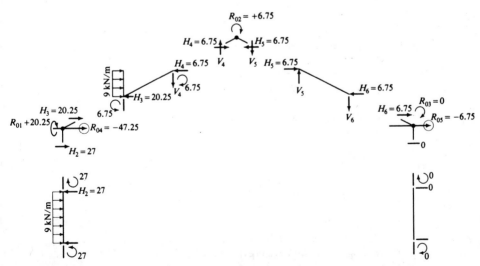

Figure 12.5.6 Fixed condition for horizontal loading in Example 12.5.1.

Vertical loading From the free-body diagrams of members 3-4 and 5-6 in Fig. 12.5.5,

$$H_3 = H_4 = \frac{108(3)}{3} = 108 \text{ kN} \qquad H_5 = H_6 = 108 \text{ kN}$$

From the free-body diagrams of joints B and C in Fig. 12.5.5,

$$R_{04} = H_3 = +108 \text{ kN} \qquad R_{05} = -H_6 = -108 \text{ kN}$$

Thus,

$$P_4 = -R_{04} = -108 \text{ kN} \qquad P_5 = -R_{05} = +108 \text{ kN}$$

Horizontal loading From the free-body diagrams of members 3-4 and 5-6 in Fig. 12.5.6,

$$+3H_4 - 6V_4 = 27(1.5) = 40.5 \qquad (12.5.6)$$

$$+3H_5 + 6V_5 = 0 \qquad (12.5.7)$$

Solving Eqs. (12.5.6) and (12.5.7) simultaneously and noting that $H_4 = H_5$ and $V_4 = V_5$ by virtue of equilibrium at joint C,

$$H_4 = H_5 = +6.75$$

For horizontal equilibrium of members 3-4 and 5-6 in Fig. 12.5.6,

$$H_3 = 27 - H_4 = 27 - 6.75 = +20.25$$
$$H_6 = H_5 = +6.75$$

From the free-body diagrams of joints B and C in Fig. 12.5.6,

$$R_{04} = -H_2 - H_3 = -27 - 20.25 = -47.25$$
$$R_{05} = -H_6 = -6.75$$

Thus,

$$P_4 = -R_{04} = -(-47.25) = +47.25 \text{ kN}$$
$$P_5 = -R_{05} = -(-6.75) = +6.75 \text{ kN}$$

Finally, the $[P]$ matrix can be shown as

$[P] =$

LC \ P	1	2
1	+ 54.	− 20.25
2	0.	− 6.75
3	− 54.	0.
4	− 108.	+ 47.25
5	+ 108.	+ 6.75

(f) *The output matrices $[X]$ and $[F^*]$.*

MATRIX DISPLACEMENT METHOD OF RIGID-FRAME ANALYSIS 455

$[SB] =$

F \ X	1	2	3	4	5
1	+0.6667			−0.3333	
2	+1.3333			−0.3333	
3	+1.7888	+0.8944		+0.4472	−0.4472
4	+0.8944	+1.7888		+0.4472	−0.4472
5		+1.7888	+0.8944	−0.4472	+0.4472
6		+0.8944	+1.7888	−0.4472	+0.4472
7			+1.3333		−0.3333
8			+0.6667		−0.3333

$* EI_c$

$[K] = [ASB] =$

P \ X	1	2	3	4	5
1	+3.1221	+0.8944		+0.1139	−0.4472
2	+0.8944	+3.5776	+0.8944		
3		+0.8944	+3.1221	−0.4472	+0.1139
4	+0.1139		−0.4472	+0.4093	−0.2982
5	−0.4472		+0.1139	−0.2982	+0.4093

$* EI_c$

$[X] = [K]^{-1}[P] =$

X \ LC	1	2
1	+ 52.17	+ 33.56
2	0	− 23.56
3	− 52.17	+ 53.12
4	− 194.04	+409.5
5	+194.04	+336.7

$* \dfrac{1}{EI_c}$

456 INTERMEDIATE STRUCTURAL ANALYSIS

$[F^*] = [F_0] + [SB][X]$

$$= \begin{bmatrix} 0. & -27. \\ 0. & +27. \\ -54. & -6.75 \\ +54. & +6.75 \\ -54. & 0. \\ +54. & 0. \\ 0. & 0. \\ 0. & 0. \end{bmatrix} + \begin{bmatrix} +99.46 & -114.13 \\ +134.24 & -91.75 \\ -80.23 & +71.51 \\ -126.89 & +20.43 \\ +126.89 & -27.19 \\ +80.23 & +41.40 \\ -134.24 & -41.41 \\ -99.46 & -76.82 \end{bmatrix} = \begin{bmatrix} +99.46 & -141.13 \\ +134.24 & -64.75 \\ -134.23 & +64.76 \\ -72.89 & +27.18 \\ +72.89 & -27.19 \\ +134.23 & +41.40 \\ -134.24 & -41.41 \\ -99.46 & -76.82 \end{bmatrix}$$

(g) *Statics and deformation checks.* Refer to Parts (f) and (g) of both Examples 7.9.1 and 7.9.2 for the free-body diagrams and the sketches of the elastic curves, for both loading conditions. So far as statics checks are concerned, there should be $NP = 5$ independent checks. From the results in the output, the facts that $F_2^* + F_3^* = 0$, $F_4^* + F_5^* = 0$, and $F_6^* + F_7^* = 0$ constitute the checks along the first three degrees of freedom. Since the shear and moment diagrams are completely defined by the end moments F^*, the remaining unknowns in the "force response" are the axial forces in the four members. There being six resolution equations of equilibrium at joints B, C, and D, the two extra conditions beyond the unknown axial forces form the two remaining statics checks, regardless of which degrees of freedom in sidesway have been chosen in the solution procedure.

In regard to the "deformation response," if the properties of the elastic curve were determined solely from the moment diagram, as in Examples 7.9.1 and 7.9.2, there should be $NI = 3$ compatibility checks. If the five X values from the output are used, then there should be $NF = NP + NI = 8$ deformation checks, 2 per member. In the latter event, it is necessary to determine all the joint displacements first, from the two independent joint displacements in the output, by using the joint-displacement equation, Eq. (3.6.2). Only then can the member-axis rotations R_{ij} be obtained so that the deformation checks, Eq. (11.11.3) vs. Eqs. (11.11.4a and b), can be made.

12.6 Element Stiffness Matrix of Beam Element with Variable Moment of Inertia

The matrix displacement method of rigid-frame analysis can be easily adapted to cases wherein some elements have variable moment of inertia from one end to the other. The first necessary modification is that the element stiffness matrix of the element with variable moment of inertia becomes

$$[S] = \begin{array}{c|c|c} {}_F \diagdown {}^e & i & j \\ \hline i & s_{ii} \dfrac{EI_c}{L} & s_{ij} \dfrac{EI_c}{L} \\ \hline j & s_{ji} \dfrac{EI_c}{L} & s_{jj} \dfrac{EI_c}{L} \end{array} \quad (12.6.1)$$

in which s_{ii}, $s_{ij} = s_{ji}$, and s_{jj} are the stiffness coefficients in Sec. 7.10. The second necessary modification is that if there is load on the element with variable moment of inertia, the fixed-end moments F_{0i} and F_{0j} would be different from those had the moment of inertia been constant along the length of the element. These modified fixed-end moments can be computed using the method described in Sec. 7.10 or by the column-analogy method to be described in Chap. 15. In fact, the stiffness coefficients themselves can be independently obtained by the column-analogy method as well. The third necessary modification is that in doing the two deformation checks for the element with variable moment of inertia, the modified flexibility coefficients f_{ii}, $f_{ij} = f_{ji}$, and f_{jj} should be used instead of $+1/3$, $-1/6$, and $+1/3$, and the rotations e_{0i} and e_{0j} at the ends due to load on the element should be computed by the conjugate-beam method taking the variation in moment of inertia into account.

12.7 Exercises

Exercises 7.7 through 7.35 in Chap. 7 can all be solved using matrix form. For convenience of reference, they are listed in the following categories:
  (a) Rigid frames without sidesway, Exercises 7.7 to 7.13.
  (b) Rigid frames with sidesway, Exercises 7.14 to 7.25.
  (c) Rigid frames with yielding of supports, Exercises 7.26 and 7.27.
  (d) Gable frames, Exercises 7.28 to 7.31. Note that these four exercises can be considered as four loading conditions of the same gable frame. Variations of the exercises may be attempted by choosing different independent degrees of freedom in sidesway.
  (e) Continuous beams containing spans with variable moment of inertia, Exercises 7.32 to 7.35.

CHAPTER
THIRTEEN

INFLUENCE LINES AND MOVING LOADS

13.1 Definition of Influence Lines

So far the treatment of the determination of the force and deformation response of beams, trusses, and rigid frames, both statically determinate and indeterminate, has been for a fixed load pattern. The distributed loads are usually the weight of the structural element itself or the weight of something which rests directly on the element. Of course, if only to satisfy the definition of a truss, all loads are concentrated at its joints. The concentrated loads may be the weights of some equipment which hangs on the structure, or more often they are the opposites of the reactions to another structure which runs perpendicular to the structure being considered.

The loads on a structure, whether distributed or concentrated, however, are seldom in a fixed pattern for an indefinite period of time. The equipment may hang on different locations at different times. Wind and snow may come at the same or at different times. People or stored goods may move or be moved around. Traffic on a bridge may pass back and forth, setting a variable pattern with time. The load that is permanently there is called *dead load*, which should always include the weight of the structure itself and perhaps the weight of a fixed wall partition or some such similar fixture resting on the structure. The load that may or may not be there is called *live load*, which may include snow, wind, people, or stored goods for a building structure, or traffic such as trucks on a highway bridge and locomotives with trains on a railway bridge. In such cases, a certain percentage of the live load is added on as *impact load* as well.

Since the live load may or may not be there and since it may take any pattern on the structure, the question arises as to where to place the live load so that the effect on the structure is most critical. Of course, no such problem exists for dead load because it is always there. The key to the answer is to consider first the live load involving only a single concentrated load of 1 unit measure of weight, such as 1.0 kN. Then the effect of this unit concentrated load moving about the structure is investigated. For example, the effect on the shear at a chosen point in a beam or the effect on one of the reactions to the beam may be investigated. If the magnitude of this effect is plotted right at the position of the moving unit concentrated load, the result is an *influence line*. Thus, an influence line is the graphical display showing the effect on a chosen function of a moving unit concentrated load. It will be shown that the influence line is the essential device by which the critical position of a more complex live-load system to exert the maximum effect on a structure can be determined.

13.2 Influence Lines for Statically Determinate Beams

The influence lines for reactions, shears, and bending moments on statically determinate beams are always made of linear segments, because such functions due to a moving unit concentrated load taking any position at a distance x from a reference point is always a linear function of x. Consider the overhanging beam ABC shown in Fig. 13.2.1. Assume that the problem is to

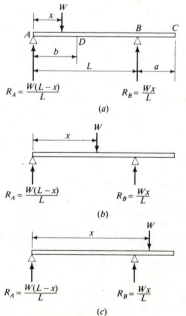

Figure 13.2.1 Reaction, shear, and bending moment as function of load position.

derive expressions for R_A, V_D, and M_D as functions of the position x of a moving concentrated load W. So far as R_A is concerned,

$$R_A = +\frac{W(L-x)}{L} \quad \text{for } 0 \leq x \leq L+a \quad (13.2.1)$$

But for V_D, it makes a difference whether W is to the left or right of D; thus

$$V_D = R_A - W = +\frac{W(L-x)}{L} - W \quad \text{for } 0 \leq x < b \quad (13.2.2a)$$

$$V_D = R_A = +\frac{W(L-x)}{L} \quad \text{for } b < x \leq L+a \quad (13.2.2b)$$

There is no distinct value for V_D when W is just at D because both Eqs. (13.2.2a and b) are not valid for $x = b$. For M_D, there is again a difference depending on whether W is to the left or right of D; thus

$$M_D = R_A b - W(b-x)$$
$$= \frac{W(L-x)}{L} b - W(b-x) \quad \text{for } 0 \leq x \leq b \quad (13.2.3a)$$

$$M_D = R_A b = \frac{W(L-x)}{L} b \quad \text{for } b \leq x \leq L+a \quad (13.2.3b)$$

In this case, Eqs. (13.2.3a and b) are both valid for $x = b$.

From the discussion in the preceding paragraph, it may be generally stated that the complete influence lines of statically determinate beams consist only of linear segments, which may be discontinuous at critical sections such as point D in Fig. 13.2.1. Although equations such as Eqs. (13.2.1) to (13.2.3) may be derived first and then various parts of the whole influence line plotted from them, in practice it is more convenient to compute the influence values at the critical sections only and then obtain the whole influence line by joining the critical points by straight segments, as will be illustrated in the following example.

Example 13.2.1 Show graphically the influence lines for R_A, V_D, and M_D of the overhanging beam shown in Fig. 13.2.2.

SOLUTION The three influence lines shown in Fig. 13.2.2b to d may be seen to be the graphs of the linear equations in Eqs. (13.2.1) to (13.2.3). However, it is not suggested that this is the most convenient way to obtain the answers. The easier way is to plot the critical values on the influence lines by the heavy dots in Fig. 13.2.2b to d and then join all adjacent dots by straight lines. To accomplish this, the four free-body diagrams of the beam under a unit load at each of the four critical positions A, D, B, and C are shown in Fig. 13.2.2e to h. For instance, the five heavy dots in Fig. 13.2.2c are answers to the following questions:

1. For a unit load at A, how much is V_D?
2. For a unit load just to left of D, how much is V_D?
3. For a unit load just to right of D, how much is V_D?

INFLUENCE LINES AND MOVING LOADS **461**

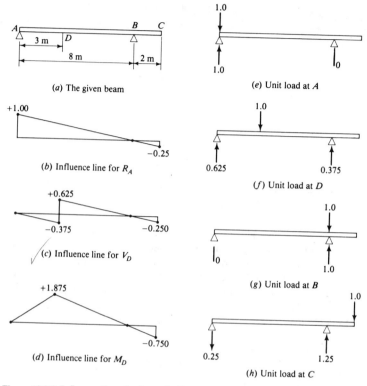

Figure 13.2.2 Influence lines for beam in Example 13.2.1.

4. For a unit load at B, how much is V_D?
5. For a unit load at C, how much is V_D?

The answers to these questions are, of course, provided by making observations of the free-body diagrams in Fig. 13.2.2e to h.

To reiterate the definition of an influence line: an influence line is the graphical display showing the effect on a chosen function of a moving unit concentrated load. As shown by Fig. 13.2.3a, since y_1, y_2, and y_3 are the values of the function due to a unit load at the respective positions where y_1, y_2, and y_3 are measured, due to the presence of W_1, W_2, and W_3 at those positions,

$$\text{Value of function} = \Sigma\, Wy \qquad (13.2.4)$$

On the other hand, if there is a uniform load of finite length on the beam as shown in Fig. 13.2.3b,

$$\text{Value of function} = \int_{x_1}^{x_2} (q\, dx)(y) = q \text{ times area } A \qquad (13.2.5)$$

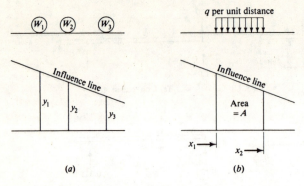

Figure 13.2.3 Use of influence line to compute value of function.

That Eqs. (13.2.4) and (13.2.5) hold true can be used not only to determine where to place the live load for the most critical effect but also to obtain the maximum positive and negative values of the function due to live load. Of course, once the position of the live load is known, the value of the function can be computed by the usual free-body method as well. This last step should be done to check the results obtained by using Eqs. (13.2.4) and (13.2.5).

Example 13.2.2 For the overhanging beam in Fig. 13.2.4, determine the maximum positive and negative R_A due to the three concentrated loads as shown, which may come on the beam in either direction.

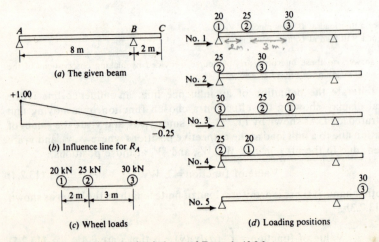

Figure 13.2.4 Maximum reaction in beam of Example 13.2.2.

SOLUTION (a) *Maximum positive* R_A. Any one of the first four loading positions may be critical. For position 1,

$$R_A = \Sigma Wy = 20(1.0) + 25(1.0)(\tfrac{6}{8}) + 30(1.0)(\tfrac{3}{8}) = +50 \text{ kN}$$

$$R_A \text{ (from free body)} = \frac{20(8) + 25(6) + 30(3)}{8} = +50 \text{ kN}$$

For position 2,

$$R_A = \Sigma Wy = 25(1.0) + 30(1.0)(\tfrac{5}{8}) = +43.75 \text{ kN}$$

$$R_A \text{ (from free body)} = \frac{25(8) + 30(5)}{8} = +43.75 \text{ kN}$$

For position 3,

$$R_A = \Sigma Wy = 30(1.0) + 25(1.0)(\tfrac{5}{8}) + 20(1.0)(\tfrac{3}{8}) = +53.125 \text{ kN} \quad \text{(maximum)}$$

$$R_A \text{ (from free body)} = \frac{30(8) + 25(5) + 20(3)}{8} = +53.125 \text{ kN}$$

For position 4,

$$R_A = \Sigma Wy = 25(1.0) + 20(1.0)(\tfrac{6}{8}) = +40 \text{ kN}$$

$$R_A \text{ (from free body)} = \frac{25(8) + 20(6)}{8} = +40 \text{ kN}$$

(b) *Maximum negative* R_A. For position 5,

$$R_A = \Sigma Wy = 30(-0.25) = -7.5 \text{ kN}$$

$$R_A \text{ (from free body)} = -\frac{30(2)}{8} = -7.5 \text{ kN}$$

(c) *Discussion.* When there is a long series of moving concentrated loads, many loading positions may have to be tried. In Sec. 13.3, the method of obtaining the most critical loading position will be further treated for simple beams.

Example 13.2.3 For the overhanging beam in Fig. 13.2.5, determine the maximum positive and negative R_A due to a uniform load of 6 kN/m that may come to act on any portion or portions of the beam.

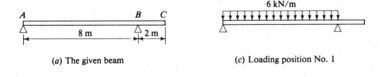

(a) The given beam

(c) Loading position No. 1

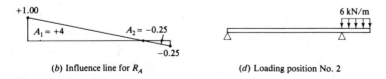

(b) Influence line for R_A

(d) Loading position No. 2

Figure 13.2.5 Maximum reaction in beam of Example 13.2.3.

SOLUTION (a) Maximum positive R_A. For position 1,

$$R_A = q \text{ times } A_1 = 6(+4) = +24 \text{ kN}$$
$$R_A \text{ (from free body)} = \tfrac{1}{2}(6)(8) = +24 \text{ kN}$$

(b) Maximum negative R_A. For position 2,

$$R_A = q \text{ times } A_2 = 6(-0.25) = -1.5 \text{ kN}$$
$$R_A \text{ (from free body)} = -\frac{6(2)^2/2}{8} = -1.5 \text{ kN}$$

Example 13.2.4 For the overhanging beam in Fig. 13.2.6, determine the maximum positive and negative V_D due to the three concentrated loads as shown, which may come on the beam in either direction.

SOLUTION (a) Maximum positive V_D. For position 1,

$$V_D = \Sigma \, Wy = 20(+0.625) + 25(+0.625)(\tfrac{3}{5}) = +21.875 \text{ kN}$$
$$V_D \text{ (from free body)} = \frac{20(5) + 25(3)}{8} = +21.875 \text{ kN}$$

For position 2,

$$V_D = \Sigma \, Wy = 30(+0.625) + 25(+0.625)(\tfrac{2}{5}) = +25 \text{ kN} \qquad \text{(maximum)}$$
$$V_D \text{ (from free body)} = \frac{30(5) + 25(2)}{8} = +25 \text{ kN}$$

(b) Maximum negative V_D. For position 3,

$$V_D = \Sigma \, Wy = 20(-0.375) + 25(-0.375)(\tfrac{1}{3}) = -10.625 \text{ kN}$$
$$V_D \text{ (from free body)} = \frac{25(7) + 20(5)}{8} - (25 + 20) = -10.625 \text{ kN}$$

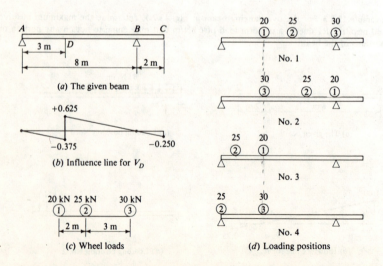

Figure 13.2.6 Maximum positive and negative shears in beam of Example 13.2.4.

For position 4,
$$V_D = \Sigma \, Wy = 30(-0.375) = -11.25 \text{ kN} \quad \text{(numerically maximum)}$$
$$V_D \text{ (from free body)} = \frac{30(5) + 25(8)}{8} - (25 + 30) = -11.25 \text{ kN}$$

(c) *Discussion.* The result of position 1 should be compared with that of placing wheel 2 at D; and the result of position 2, also with that of placing wheel 2 at D. The alternate positions give lower values in both cases.

Example 13.2.5 For the overhanging beam in Fig. 13.2.7, determine the maximum positive and negative V_D due to a uniform load of 6 kN/m that may come to act on any portion or portions of the beam.

SOLUTION (a) *Maximum positive V_D.* For position 1,
$$V_D = q \text{ times } A_2 = 6(+1.5625) = +9.375 \text{ kN}$$
$$V_D \text{ (from free body)} = R_A = \frac{6(5)^2/2}{8} = +9.375 \text{ kN}$$

(b) *Maximum negative V_D.* For position 2,
$$V_D = q \text{ times } A_1 \text{ and } A_3 = 6(-0.5625 - 0.25) = -4.875 \text{ kN}$$
$$V_D \text{ (from free body)} = R_A - 6(3) = \frac{18(6.5) - 12(1.0)}{8} - 18$$
$$= -4.875 \text{ kN}$$

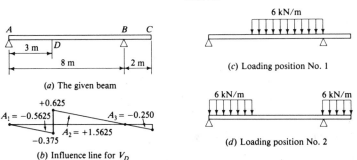

(a) The given beam

(b) Influence line for V_D

(c) Loading position No. 1

(d) Loading position No. 2

Figure 13.2.7 Maximum positive and negative shears in beam of Example 13.2.5.

Example 13.2.6 For the overhanging beam in Fig. 13.2.8, determine the maximum positive and negative M_D due to the three concentrated loads as shown, which may come on the beam in either direction.

SOLUTION (a) *Maximum positive M_D.* It is difficult to see whether the three concentrated loads should straddle across the peak point of $+1.875$ on the influence line. From the theory to be expounded in Sec. 13.4, one of the concentrated loads has to be right at D. In fact, the theory will also reveal which of the concentrated loads should be placed at D to produce the maximum M_D. Try position 1:
$$M_D = \Sigma \, Wy = 20(1.875)(\tfrac{1}{3}) + 25(1.875) + 30(1.875)(\tfrac{2}{3})$$
$$= +81.875 \text{ kN·m} \quad \text{(maximum)}$$
$$M_D \text{ (from free body)} = R_A(3) - 20(2) = 40.625(3) - 40$$
$$= +81.875 \text{ kN·m}$$

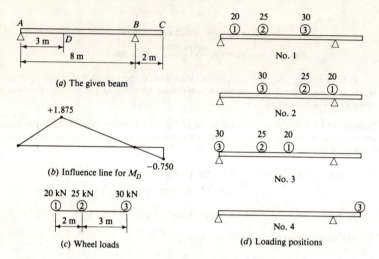

Figure 13.2.8 Maximum positive and negative bending moments in beam of Example 13.2.6.

For position 2,

$$M_D = \Sigma Wy = 30(1.875) + 25(1.875)(\tfrac{2}{3}) = +75 \text{ kN·m}$$
$$M_D \text{ (from free body)} = R_A(3) = 25(3) = +75 \text{ kN·m}$$

For position 3,

$$M_D = \Sigma Wy = 25(1.875) + 20(1.875)(\tfrac{2}{3}) = +69.375 \text{ kN·m}$$
$$M_D \text{ (from free body)} = R_A(3) - 30(3) = 53.125(3) - 90$$
$$= +69.375 \text{ kN·m}$$

(b) *Maximum negative* M_D. For position 4,

$$M_D = \Sigma Wy = 30(-0.75) = -22.5 \text{ kN·m}$$
$$M_D \text{ (from free body)} = -R_A(3) = -\frac{30(2)}{8}(3) = -22.5 \text{ kN·m}$$

Example 13.2.7 For the overhanging beam in Fig. 13.2.9, determine the maximum positive and negative M_D due to a uniform load of 6 kN/m that may come to act on any portion or portions of the beam.

SOLUTION (a) *Maximum positive* M_D. For position 1,

$$M_D = q \text{ times } A_1 = 6(+7.5) = +45 \text{ kN·m}$$
$$M_D \text{ (from free body)} = R_A(3) - 18(1.5) = 24(3) - 27 = +45 \text{ kN·m}$$

(b) *Maximum negative* M_D. For position 2,

$$M_D = q \text{ times } A_2 = 6(-0.75) = -4.5 \text{ kN·m}$$
$$M_D \text{ (from free body)} = -R_A(3) = -1.5(3) = -4.5 \text{ kN·m}$$

INFLUENCE LINES AND MOVING LOADS 467

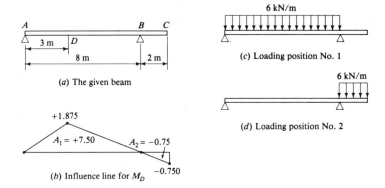

Figure 13.2.9 Maximum positive and negative bending moments in beam of Example 13.2.7.

13.3 Criterion for Maximum Reaction or Shear in Simple Beams

From the preceding section, it can be seen that so far as uniform live load is concerned, the influence line is infallible as a means of telling which portions of the beam should be so loaded for the most critical effect. It is not quite so for a series of moving concentrated loads, especially if there are more than three or four concentrated loads at relatively short spacings. Moreover, in the case of a locomotive with trains on a railway bridge, after the engine wheels come the trains carrying uniform load. In such cases, some guesswork is required and the critical value is taken from the results of several trials.

It will be shown in this section that a method can be devised to obtain the critical position of the loads for giving the maximum reaction or shear in simple beams.

Consider the effect on the shear at C of the simple beam AB in Fig. 13.3.1a due to several concentrated loads which may move in either direction. The load position for maximum positive shear at C might be one of those shown in Fig. 13.3.1c; and that for maximum negative shear at C, in Fig. 13.3.1d. When the load positions of Fig. 13.3.1d are compared with those of Fig. 13.3.1e, one sees that the maximum negative shear at C should be identical to the maximum positive shear at C', which is a point symmetrically located with respect to C. Thus, as a general problem, it is necessary only to derive a criterion for determining which of the two successive load positions in Fig. 13.3.1c will give a larger value for positive V_C.

Returning to the comparison of the two load positions in Fig. 13.3.1c, the gain on V_C by moving W_2 to C is Gs_{1-2}/L, in which G is the total load on the beam which moves up along the influence line. In fact, Gs_{1-2}/L is the increase in the reaction at A. The loss in V_C is W_1 because W_1 is inside AC in the second position. Thus, the answer to whether W_1 at C or W_2 at C will give

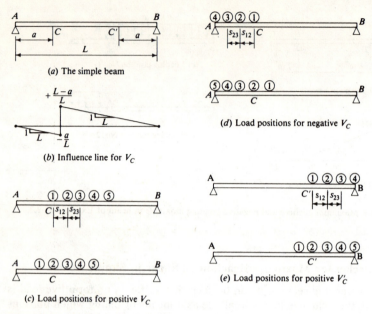

Figure 13.3.1 Maximum shear in simple beams due to moving concentrated loads.

larger V_C is:

$$\text{Gain} = Gs_{1\text{-}2}/L \qquad \text{Loss} = W_1$$

$$\text{If gain} < \text{loss, place } W_1 \text{ at } C \qquad (13.3.1a)$$

$$\text{If gain} > \text{loss, place } W_2 \text{ at } C \text{ and compare with } W_3 \text{ at } C \qquad (13.3.1b)$$

In common cases, the first wheel load might be light so that placing W_2 at C could be more critical, but the need of moving W_3 to C seldom arises.

For maximum reaction at A, Eqs. (13.3.1a and b) can be used as well, except that the G in that equation should not include W_1 because W_1 is moving off the span and there is no increase of its moment arm about point B in computing the reaction at A.

Example 13.3.1 For the simple beam in Fig. 13.3.2a, determine the maximum reaction at A due to the passage in either direction of the five wheel loads followed by an indefinite length of uniform load as shown in Fig. 13.3.2c.

SOLUTION (a) Position. Comparing W_1 at A with W_2 at A,

$$G \; (W_1 \text{ at } A, \text{ not including } W_1) = 160 + 16(18 - 8) = 320$$

$$G \; (W_2 \text{ at } A, \text{ not including } W_1) = 160 + 16(18 - 6) = 352$$

$$\text{Gain} = Gs_{1\text{-}2}/L = (\text{between 320 and 352})(2)/18$$

$$= (\text{between 35.6 and 39.1})$$

$$\text{Loss} = W_1 = 20$$

INFLUENCE LINES AND MOVING LOADS **469**

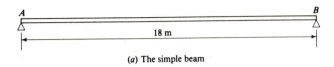

(a) The simple beam

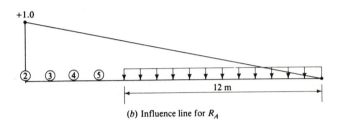

(b) Influence line for R_A

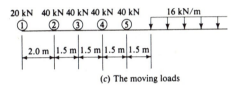

(c) The moving loads

Figure 13.3.2 Maximum reaction in beam of Example 13.3.1.

Gain > loss; W_2 at A will cause the larger R_A than W_1 at A.
Comparing W_2 at A with W_3 at A,

G (W_2 at A, not including W_1 and W_2) = 120 + 16(18 − 6) = 312

G (W_3 at A, not including W_1 and W_2) = 120 + 16(18 − 4.5) = 336

Gain = Gs_{2-3}/L = (between 312 and 336)(1.5)/18

= (between 26 and 28)

Loss = W_2 = 40

Gain < loss; W_2 at A will cause the larger R_A than W_3 at A.

(b) *Maximum* R_A. Referring to Fig. 13.3.2 again, with W_2 at A,

R_A (influence-line method) = $40(1) + 40\left(\dfrac{16.5}{18}\right) + 40\left(\dfrac{15}{18}\right) + 40\left(\dfrac{13.5}{18}\right) + 16\left(\dfrac{1}{2}\right)\left(\dfrac{12}{18}\right)(12)$

= 204 kN

R_A (free-body method) = $\dfrac{160(15.75) + 16(12)(6)}{18}$ = 204 kN

Example 13.3.2 For the simple beam in Fig. 13.3.3a, determine the maximum positive and negative shear at C due to the passage in either direction of the five wheel loads followed by an indefinite length of uniform load as shown in Fig. 13.3.3d.

470 INTERMEDIATE STRUCTURAL ANALYSIS

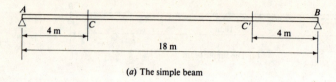

(a) The simple beam

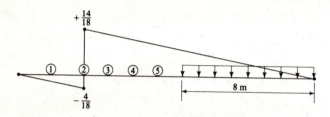

(b) Influence line for V_C

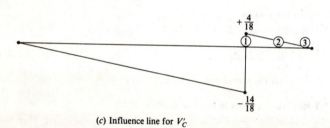

(c) Influence line for V'_C

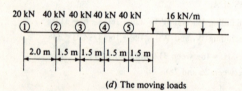

(d) The moving loads

Figure 13.3.3 Maximum positive and negative shears in beam of Example 13.3.2.

SOLUTION (a) *Maximum positive shear at C.* Comparing W_1 at C with W_2 at C,

$$G \ (W_1 \text{ at } C) = 180 + 16(14 - 8) = 276$$
$$G \ (W_2 \text{ at } C) = 180 + 16(14 - 6) = 308$$

Gain = Gs_{1-2}/L = (between 276 and 308)(2)/18

= (between 30.7 and 34.2)

Loss = $W_1 = 20$

Gain > loss; W_2 at C will cause the larger V_C than W_1 at C.
Comparing W_2 at C with W_3 at C,

$$G\ (W_2\ \text{at}\ C) = 180 + 16(14-6) = 308$$
$$G\ (W_3\ \text{at}\ C) = 180 + 16(14-4.5) = 332$$
$$\text{Gain} = Gs_{2\text{-}3}/L = (\text{between } 308 \text{ and } 332)(1.5)/18$$
$$= (\text{between } 25.7 \text{ and } 27.7)$$
$$\text{Loss} = W_2 = 40$$

Gain < loss; W_2 at C will cause the larger V_C than W_3 at C.
With W_2 at C,

$$V_C\ (\text{influence-line method}) = 20\left(-\frac{4}{18}\right)\left(\frac{2}{4}\right) + 160\left(\frac{14}{18}\right)\left(\frac{11.75}{14}\right) + 16\left(\frac{1}{2}\right)\left(\frac{14}{18}\right)\left(\frac{8}{14}\right)(8)$$
$$= +130.66\ \text{kN}$$

$$V_C\ (\text{free-body method}) = \frac{20(16) + 160(11.75) + 128(4)}{18} - 20$$
$$= +130.66\ \text{kN}$$

(b) *Maximum positive shear at C'*. Comparing W_1 at C' with W_2 at C',

$$G\ (W_1\ \text{at}\ C') = 100 \qquad G\ (W_2\ \text{at}\ C') = 140$$
$$\text{Gain} = Gs_{1\text{-}2}/L = (\text{between } 100 \text{ and } 140)(2)/18$$
$$= (\text{between } 11.1 \text{ and } 15.6)$$
$$\text{Loss} = W_1 = 20$$

Gain < loss; W_1 at C' will cause the larger $V_{C'}$ than W_2 at C'.
With W_1 at C',

$$V_{C'}\ (\text{influence-line method}) = 20\left(\frac{4}{18}\right) + 40\left(\frac{4}{18}\right)\left(\frac{2}{4}\right) + 40\left(\frac{4}{18}\right)\left(\frac{0.5}{4}\right) = +10\ \text{kN}$$

$$V_{C'}\ (\text{free-body method}) = R_A = \frac{20(4) + 40(2) + 40(0.5)}{18} = +10\ \text{kN}$$

(c) *Discussion.* So far as the shear at C is concerned, because of the passage of the live load in either direction, its value may vary between -10 kN and $+130.66$ kN.

13.4 Criterion for Maximum Bending Moment in Simple Beams

When there is a long series of concentrated loads which may pass back and forth on a simple beam, such as in the several longitudinal beams carrying a highway bridge slab, or in the two main girders on each side of the tracks of a railway bridge, the problem arises as to which position of the live load will cause the largest bending moment at a designated point along the span. Whatever this position, it is certain that the bending-moment diagram consists only of linear segments under the concentrated loads, in which case a maximum can occur only at the corners of the bending-moment diagram. Thus it can be concluded that one of the concentrated loads must act right at the designated point. The same conclusion will be reached by the purely mathematical derivation which follows.

Consider some basic position of the seven wheel loads on the simple

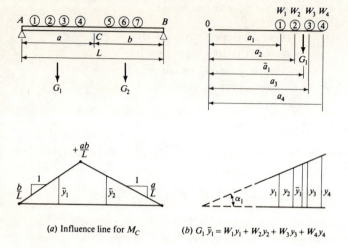

(a) Influence line for M_C (b) $G_1 \bar{y}_1 = W_1 y_1 + W_2 y_2 + W_3 y_3 + W_4 y_4$

Figure 13.4.1 Maximum bending moment in simple beams due to moving concentrated loads.

beam AB of Fig. 13.4.1. The bending moment at C is

$$M_C = \sum_{i=1}^{i=7} W_i y_i = G_1 y_1 + G_2 y_2 \tag{13.4.1}$$

It can be shown that, so long as the influence line is a linear segment, the value of the function ΣWy for concentrated loads may be obtained as well from the product of the resultant load on the line segment and the influence ordinate under the resultant. Referring to Fig. 13.4.1b, taking moments of the four W forces about point O and equating the sum to the moment of G_1,

$$G_1 \bar{a}_1 = W_1 a_1 + W_2 a_2 + W_3 a_3 + W_4 a_4$$

Multiplying every term in the above equation by $\tan \alpha_1$,

$$G_1 \bar{a}_1 \tan \alpha_1 = W_1 a_1 \tan \alpha_1 + W_2 a_2 \tan \alpha_1 + W_3 a_3 \tan \alpha_1 + W_4 a_4 \tan \alpha_1$$

from which

$$G_1 \bar{y}_1 = W_1 y_1 + W_2 y_2 + W_3 y_3 + W_4 y_4$$

Return to the basic position of Eq. (13.4.1); if the load system advances a small distance dx from the right toward the left, G_1 slides down along the influence line and G_2 climbs up along the influence line. The net increase is

$$dM_C = -G_1 \left(\frac{ab/L}{a}\right) dx + G_2 \left(\frac{ab/L}{b}\right) dx$$

$$= -\frac{G_1 b}{L} dx + \frac{G_2 a}{L} dx \tag{13.4.2}$$

Letting $G = G_1 + G_2$, the above equation becomes

$$dM_C = -\frac{G_1 b}{L} dx + \frac{(G-G_1)a}{L} dx = \frac{Ga}{L} dx - G_1 dx$$

from which

$$\frac{dM_C}{dx} = \frac{Ga}{L} - G_1 \qquad (13.4.3)$$

From observing Eq. (13.4.3), it can be stated that, for the maximum bending moment to occur at C, the loading position must be such that the load G_1 inside AC is equal to Ga/L. This equality can rarely happen exactly; however, if a load is placed right at C, any fraction of it can be considered to be inside AC, thus providing a chance of satisfying the condition

$$\frac{dM_C}{dx} = 0 \qquad (13.4.4)$$

Practically speaking, then, to determine the loading position for which the bending moment at C is maximum, do the following:

1. Place a load at C, determine the total load G on the span, and compute Ga/L.
2. There are two possible values of the load G_1 inside AC: a smaller value not including the load at C and a larger value including the load at C.
3. If the smaller value of G_1, the value of Ga/L, and the larger value of G_1 are in sequence, the correct load at C has been obtained to give the maximum bending moment at C.

In cases where complications may arise—such as when there are loads simultaneously at A, C, and B—one can always go back to the fundamental equation, Eq. (13.4.2), which can be restated as follows:

$$\text{Gain} = \frac{G_2 a}{L} dx \qquad \text{Loss} = \frac{G_1 b}{L} dx \qquad \text{for } dx \text{ toward the left}$$
$$(13.4.5a)$$

$$\text{Loss} = \frac{G_2 a}{L} dx \qquad \text{Gain} = \frac{G_1 b}{L} dx \qquad \text{for } dx \text{ toward the right}$$
$$(13.4.5b)$$

If a net loss results by moving the position being investigated either dx toward the left or dx toward the right, the peak, or optimum, position has been obtained.

Sometimes the criterion as stated in Eq. (13.4.4) may be satisfied by two or more successive wheels placed at C; then the actual values of the bending moment at C for all such positions should be computed and the largest is the answer being sought.

For the same designated point C, there is a maximum when traffic goes in one way, and another maximum in the opposite way. Rather than draw another sketch of the load system heading in the opposite direction, a more convenient approach is to determine the maximum bending moment at a point C', symmetrically located on the span with respect to point C, for the same traffic direction as for point C. The value so obtained at point C' for traffic heading toward the left is identical to that at point C for traffic heading toward the right.

Example 13.4.1 For the simple beam in Fig. 13.4.2a, determine the maximum bending moment at C due to the passage in either direction of the five wheel loads followed by an indefinite length of uniform load as shown in Fig. 13.4.2d.

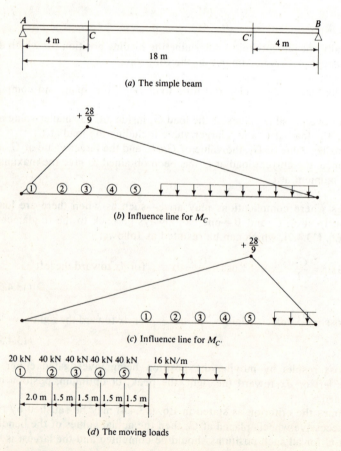

(a) The simple beam

(b) Influence line for M_C

(c) Influence line for $M_{C'}$

(d) The moving loads

Figure 13.4.2 Maximum bending moment in beam of Example 13.4.1.

INFLUENCE LINES AND MOVING LOADS **475**

SOLUTION (a) *Maximum bending moment at C*. The trial for the wheel to be placed at C so that the smaller value of G_1, the value of Ga/L, and the larger value of G_1 are in sequence is shown in the accompanying table:

Load at C	G_1	$\dfrac{Ga}{L} = \dfrac{4}{18}G$	G_1	Criterion satisfied?
W_2	20	$\dfrac{4}{18}(308) = 68.4$	60	No
W_3	60	$\dfrac{4}{18}(322) = 73.8$	100	Yes
W_4	80	$\dfrac{4}{18}(336) = 74.7$	120	No

With W_3 at C,

$$M_C \text{ (free-body method)} = \frac{16(9.5)^2/2 + 160(13.25) + 20(17.5)}{18}(4) - 40(1.5) - 20(3.5)$$

$$= 579.33 \text{ kN} \cdot \text{m}$$

$$M_C \text{ (influence-line method)} = 120\left(\frac{28}{9}\right)\left(\frac{12.5}{14}\right) + 40\left(\frac{28}{9}\right)\left(\frac{2.5}{4}\right) + 20\left(\frac{28}{9}\right)\left(\frac{0.5}{4}\right)$$

$$+ 16\left(\frac{1}{2}\right)\left(\frac{28}{9}\right)\left(\frac{9.5}{14}\right)(9.5)$$

$$= 579.33 \text{ kN} \cdot \text{m}$$

(b) *Maximum bending moment at C'*.

Load at C'	G_1	$\dfrac{Ga}{L} = \dfrac{14}{18}G$	G_1	Criterion satisfied?
W_4	100	$\dfrac{14}{18}(196) = 152.4$	140	No
W_5	140	$\dfrac{14}{18}(200) = 171.1$	180	Yes

With W_5 at C',

$$M_{C'} \text{ (free-body method)} = \frac{16(2.5)^2/2 + 160(6.25) + 20(10.5)}{18}(14) - 120(3) - 20(6.5)$$

$$= 490 \text{ kN} \cdot \text{m}$$

$$M_{C'} \text{ (influence-line method)} = 160\left(\frac{28}{9}\right)\left(\frac{11.75}{14}\right) + 20\left(\frac{28}{9}\right)\left(\frac{7.5}{14}\right) + 16\left(\frac{1}{2}\right)\left(\frac{28}{9}\right)\left(\frac{2.5}{4}\right)(2.5)$$

$$= 490 \text{ kN} \cdot \text{m}$$

(c) *Conclusion.* The largest bending moment at C for traffic heading toward the left is 579.33 kN·m, and the largest bending moment at C', also for traffic heading toward the left, is 490 kN·m. Thus the maximum bending moment at C is the larger of the values 579.33 and 490, or 579.33 kN·m.

13.5 Absolute Maximum Bending Moment in a Simple Beam

For short and small-sized simple beams of a constant cross section, there are no designated points along the span at which maximum bending moments need to be determined, but the question remains as to what may be the largest bending moment that may ever happen to the beam. This largest bending moment is called the *absolute maximum bending moment*. The absolute maximum bending moment, wherever it may be on the span, must occur somewhere near the middle of the span under a concentrated load. This is so because the moment diagram for concentrated loads can consist only of linear segments, and it is inconceivable that the largest bending moment can occur in the end portions of the beam. Thus the usual procedure is to determine what load placed at the midspan would cause the maximum bending moment there first. Then the position of this load becomes the unknown, which is solved on the condition that the bending moment at this load is to be maximum. This procedure is a general one in that it can be used in the case where the wheel loads are followed by an indefinite length of uniform load.

In the special case where the moving loads consist of concentrated loads only, a formula may be derived for the position x, in Fig. 13.5.1, such that the bending moment at, say, W_3 is maximum. From the free-body diagram,

$$M \text{ at } W_3 = R_A x - (\text{moment of } W_1 \text{ and } W_2 \text{ about } W_3)$$
$$= \frac{G(L-c-x)}{L}x - (\text{moment of } W_1 \text{ and } W_2 \text{ about } W_3)$$

(13.5.1)

Taking the derivative of the above expression with respect to x and setting it equal to zero,

$$\frac{G}{L}(L - x - 2x) = 0$$

Solving the above equation for x,

$$x = \frac{L-c}{2}$$

(13.5.2)

The distance of G from the right end is

$$L - c - x = L - c - \frac{L-c}{2} = \frac{L-c}{2}$$

(13.5.3)

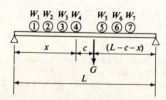

Figure 13.5.1 Absolute maximum bending moment at W_3.

Comparing Eq. (13.5.2) and Eq. (13.5.3), it can be said that W_3 and G should be placed at equal distances from the left and right ends of the beam, respectively, so that the bending moment at W_3 is maximum.

Example 13.5.1 Determine the maximum bending moment at midspan of the simple beam in Fig. 13.5.2a and then the absolute maximum bending moment due to the passage in either direction of the five wheel loads followed by an indefinite length of uniform load as shown in Fig. 13.5.2d.

SOLUTION (a) *Maximum bending moment at midspan.*

Load at midspan	G_l	$\dfrac{Ga}{L} = \tfrac{1}{2}G$	G_l	Criterion satisfied?
W_3	60	$\tfrac{1}{2}(252) = 126$	100	No
W_4	100	$\tfrac{1}{2}(276) = 138$	140	Yes
W_5	140	$\tfrac{1}{2}(300) = 150$	180	Yes

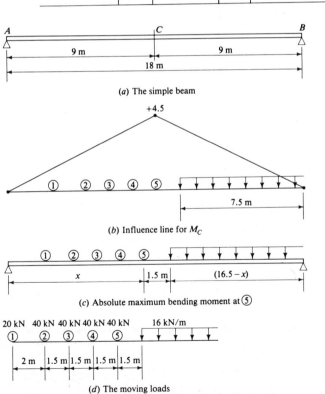

(a) The simple beam

(b) Influence line for M_C

(c) Absolute maximum bending moment at ⑤

(d) The moving loads

Figure 13.5.2 Absolute maximum bending moment in beam of Example 13.5.1.

With W_4 at midspan, by both free-body and influence-line methods,

$$M \text{ at midspan} = 784 \text{ kN·m}$$

With W_5 at midspan, by both free-body and influence-line methods,

$$M \text{ at midspan} = 790 \text{ kN·m}$$

Thus the maximum bending moment at midspan is 790 kN·m.

(b) *Absolute maximum bending moment.* Determine x in Fig. 13.5.2c so that the bending moment at W_5 is maximum.

$$M \text{ at } W_5 = R_A x - (\text{moment of } W_1 \text{ to } W_4 \text{ about } W_5)$$

$$= \frac{16(16.5-x)^2/2 + 160(18-x+2.25) + 20(18-x+6.5)}{18} x$$

$$- (\text{moment of } W_1 \text{ to } W_4 \text{ about } W_5)$$

$$\frac{d(M \text{ at } W_5)}{dx} = 328.22 - 49.33x + 1.3333x^2 = 0$$

$$x = 8.698 \text{ m}$$

$$M \text{ at } W_5 = \frac{16(7.802)^2/2 + 160(11.552) + 20(15.802)}{18}(8.698) - 490$$

$$= 791.2 \text{ kN·m}$$

(c) *Discussion.* By comparing Fig. 13.5.2b with Fig. 13.5.2c, it can be seen that by bringing $16(0.302) = 4.83$ kN more of uniform load on the beam at the right end, the bending moment at W_5 is raised from 790 to only 791.2 kN·m. For such a case, it is not worth the effort to obtain the absolute maximum bending moment; just the maximum bending moment at midspan would suffice.

Example 13.5.2 Determine the maximum bending moment at midspan of the simple beam in Fig. 13.5.3a and then the absolute maximum bending moment due to the passage in either direction of the three concentrated loads as shown in Fig. 13.5.3d.

SOLUTION (a) *Maximum bending moment at midspan.*

Load at midspan	G_l	$\frac{Ga}{L} = \frac{1}{2}G$	G_r	Criterion satisfied?
W_1	0	$\frac{1}{2}(60) = 30$	20	No
W_2	20	$\frac{1}{2}(100) = 50$	60	Yes
W_3	40	$\frac{1}{2}(80) = 40$	80	Yes

Even though both positions, W_2 or W_3 at midspan, satisfy the criterion, it is obvious that W_2 at midspan should give the larger value, because W_1 is off the beam when W_3 is placed at midspan. With W_2 at midspan,

$$M \text{ at midspan} = 250 \text{ kN·m}$$

computed by either the free-body or the influence-line method.

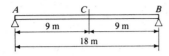

(a) The simple beam

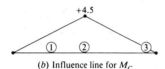

(b) Influence line for M_C

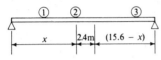

(c) Absolute maximum bending moment at ②

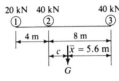

(d) The moving loads

Figure 13.5.3 Absolute maximum bending moment in beam of Example 13.5.2.

(b) *Absolute maximum bending moment.* From Fig. 13.5.3d, the distance $\bar{x}$ of G from W_3 is

$$\bar{x} = \frac{40(8) + 20(12)}{100} = 5.6 \text{ m}$$

and the distance c is

$$c = 8 - 5.6 = 2.4 \text{ m}$$

Using Eq. (13.5.2),

$$x = \frac{L-c}{2} = \frac{18-2.4}{2} = 7.8 \text{ m}$$

From the free-body diagram in Fig. 13.5.3c,

$$M \text{ at } W_2 = R_A(7.8) - 20(4) = \frac{100(7.8)}{18}(7.8) - 80 = 258 \text{ kN·m}$$

(c) *Discussion.* In this case, the absolute maximum bending moment is 258 kN·m, while the maximum bending moment at midspan is 250 kN·m.

13.6 Müller-Breslau Influence Theorem for Statically Determinate Beams

In 1886 and 1887 Müller-Breslau came up with an ingenious way of visualizing and obtaining the influence lines of statically determinate and indeterminate structures.† This section is limited to a discussion of statically determinate beams.

According to Müller-Breslau, the influence line for the reaction, such as on the simple beam of Fig. 13.6.1a, can be obtained by lifting the beam off the support a unit distance, as shown in Fig. 13.6.1b. The influence line for shear at C (Fig. 13.6.1c) can be obtained by cutting the beam at C and separating the two parts until C_1 and C_2 are a unit distance apart while keeping AC_1 parallel to C_2B. The influence line for bending moment at C (Fig. 13.6.1d) can

†S. P. Timoshenko, *History of Strength of Materials*, McGraw-Hill Book Company, New York, 1953, p. 310.

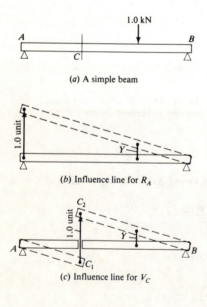

(a) A simple beam

(b) Influence line for R_A

(c) Influence line for V_C

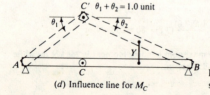

(d) Influence line for M_C

Figure 13.6.1 Müller-Breslau influence theorem for statically determinate beams.

INFLUENCE LINES AND MOVING LOADS 481

be obtained by installing a hinge at C and lifting the hinge up until $\theta_1 + \theta_2 = 1$ rad. Common in the three cases are: (1) for reaction, lift the beam off the support; for shear, cut the shear resistance; for bending moment, remove the moment resistance; then (2) introduce a unit displacement in the same direction as R_A; introduce a unit separation at the cut points in the directions as V_C acting on CA and V_C acting on CB, as shown in Fig. 13.6.2b; introduce a unit change in direction $\theta_1 + \theta_2$ such that the directions of θ_1 and θ_2 are the same as M_C acting on CA and M_C acting on CB, as shown in Fig. 13.6.2b. Thus the Müller-Breslau influence theorem for statically determinate beams may be stated as follows:

The influence line for an assigned function on a statically determinate beam may be obtained by removing the restraint offered by that function and introducing a directly related generalized unit displacement at the location and in the direction of the function.

This theorem may be proved by showing that the Y values in Fig. 13.6.1b to d are equal to R_A, V_C, and M_C in Fig. 13.6.2b, respectively. With regard to the reaction, let the entire force system in Fig. 13.6.2a go through the rigid-body motion of the beam in Fig. 13.6.1b. The total virtual work must be zero since the resultant of the force system is zero.

$$W = R_A(+1.0) - 1.0(Y) + R_3(0) = 0$$

from which

$$Y = R_A \qquad (13.6.1)$$

With regard to V_C, let the two force systems acting on the left and right parts of Fig. 13.6.2b go through the rigid-body motions of AC and CB in Fig. 13.6.1c; then

$$W = R_A(0) + V_C(CC_1) - M\left(\frac{CC_1}{AC}\right) + R_B(0) - 1.0(Y) + V_C(CC_2) + M\left(\frac{CC_2}{CB}\right) = 0$$

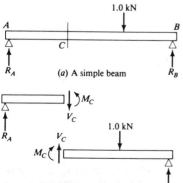

(a) A simple beam

(b) Two parts of the simple beam

Figure 13.6.2 Free-body diagrams for two parts of a simple beam under unit load.

482 INTERMEDIATE STRUCTURAL ANALYSIS

from which
$$Y = V_C \tag{13.6.2}$$

With regard to M_C, again let the two force systems in Fig. 13.6.2b go through the rigid-body motions of AC and CB in Fig. 13.6.1d; then

$$W = R_A(0) - V(CC') + M_C(\theta_1) + R_B(0) - 1.0(Y) + V(CC') + M(\theta_2) = 0$$

from which
$$Y = M_C \tag{13.6.3}$$

The usefulness of the Müller-Breslau influence theorem lies in the fact that the influence lines as obtained by the "heavy dots" method described in Sec. 13.2 may be visually reviewed by applying this theorem.

Example 13.6.1 For the statically determinate beam shown in Fig. 13.6.3a, sketch the influence lines for R_A, R_B, R_C, and M_C by applying the Müller-Breslau influence theorem.

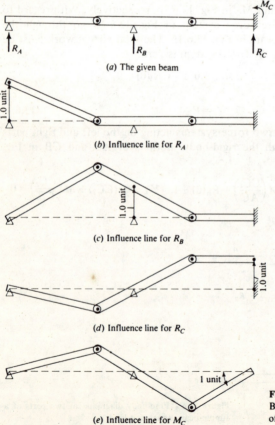

(a) The given beam

(b) Influence line for R_A

(c) Influence line for R_B

(d) Influence line for R_C

(e) Influence line for M_C

Figure 13.6.3 Use of Müller-Breslau influence theorem in beam of Example 13.6.1.

SOLUTION The results are shown in Fig. 13.6.3b to e. In each case, the restraint offered by the function is removed, a directly related displacement of a unit value is introduced in the direction of the function, and the rigid-body motions of the three pieces are permitted so long as the boundary conditions are still satisfied. Note that in Fig. 13.6.3d, the fixed support itself is lifted by 1 unit. In Fig. 13.3.3e, the fixed support is replaced by a hinged support and the unit rotation is all on one side.

13.7 Influence Lines for Statically Determinate Trusses

For roof trusses, the concentrated loads may come directly to the panel points, such as from hanging a piece of equipment under the ceiling. For bridge trusses, the main trusses on each side support the transverse floor beams at the panel points. The floor beams in turn support the stringers which run parallel to the traffic. In case of a railway bridge such as that shown in Fig. 13.7.1, the stringers support the ties and the ties support the rails. Thus the wheel loads and the uniform load following them do not come to act on the loaded chord of the truss at all; they are transmitted to the panel points through the stringers and the floor beams.

For a bridge truss such as the one shown in Fig. 13.7.2, the influence values for the force in bar, say U_2L_3, due to loads applied directly at the panel points can be easily computed and are plotted as heavy dots in that figure. The question remains as to what is the force in this bar due to the pair of unit loads acting on the stringers at a distance x to right of L_4. Through the transmission, the floor beam at L_4 carries a pair of equal loads of $(1.0)(p-x)/p$ each; and that at L_5, of $(1.0)(x/p)$ each. In turn, the loads at L_4 and L_5 of each truss are $(1.0)(p-x)/p$ and $(1.0)(x/p)$, respectively. Thus, referring

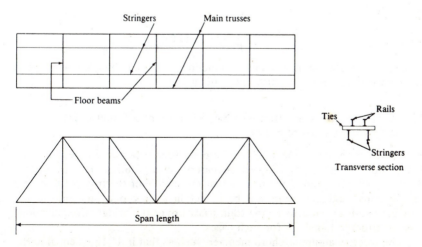

Figure 13.7.1 A typical layout of a railway bridge.

484 INTERMEDIATE STRUCTURAL ANALYSIS

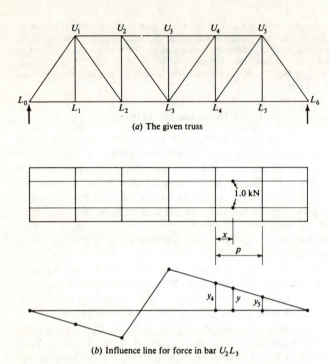

(a) The given truss

(b) Influence line for force in bar U_2L_3

Figure 13.7.2 Influence line between adjacent panel points of a truss.

to Fig. 13.7.2b,

$$y = (1.0)\left(\frac{p-x}{p}\right)(y_4) + (1.0)\left(\frac{x}{p}\right)(y_5) \quad (13.7.1)$$

Since the above value of y is a linear function of x, the influence line between adjacent panel points of a truss has to be linear.

13.8 Criterion for Maximum Bending Moment at a Panel Point on the Loaded Chord of a Truss

For the truss shown in Fig. 13.8.1a, there are four upper-chord members, six lower-chord members, and 11 web members. In large bridge trusses, double diagonals may be used in the panels near the middle of the truss so that only one diagonal taking tension will be in action at any one time due to the variable live-load patterns as the train passes over the bridge. Complications arise in structural analysis for such cases.

The force in an upper-chord member, such as that in U_1U_2, is equal to the bending moment at panel point L_2, divided by the length U_2L_2 and multiplied

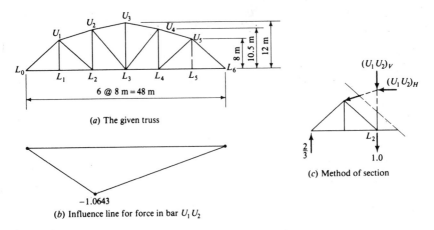

Figure 13.8.1 Influence line for force in upper-chord member of a truss.

by sec α, where α is the angle of inclination of U_1U_2, as may be verified by observing the method of sections shown in Fig. 13.8.1c. Thus the influence lines for the four upper-chord members resemble those for bending moments at the panel points on the loaded chord. The criterion for determining the load position which causes the largest compressive force in any upper-chord member is, therefore, the same as for the maximum bending moment at the panel point opposite to that chord member.

13.9 Criterion for Maximum Bending Moment at a Panel Point on the Unloaded Chord of a Truss

The force in a lower-chord member of a truss, such as that in L_2L_3 of the truss in Fig. 13.8.1a, is equal to the bending moment at the panel point U_2, divided by the length U_2L_2. So long as the upper-chord panel points are vertically above the lower-chord panel points, the shape of the influence line, for the bending moment or for the force in the lower chord, still takes the form of a simple triangle, and the method for determining the critical loading position of the moving loads is the same as for the maximum bending moment at designated points on a simple beam.

When, however, the panel points on the unloaded chord are not vertically above those on the loaded chord, the general shape of the influence line is as shown in Fig. 13.9.1. A common case is that the unloaded panel point is halfway between the two adjacent loaded panel points. This is the case to be treated here.

The influence line shown in Fig. 13.9.1 is for the bending moment at a panel point on the unloaded chord, with the left adjacent panel point on the loaded chord at m panels from the left support in a truss of n panels. As a

486 INTERMEDIATE STRUCTURAL ANALYSIS

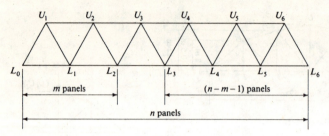

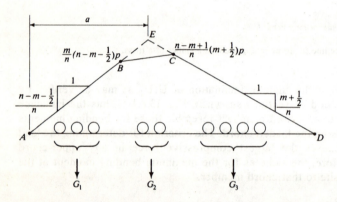

Figure 13.9.1 Influence line for bending moment at a panel point on the unloaded chord.

matter of interest, it can be proved that the prolongations of *AB* and *DC* intersect at a point *E* that is directly under the panel point U_3. The derivation of the criterion for the critical loading position to cause the maximum bending moment at U_3 will now be derived.

Consider the basic position in Fig. 13.9.1 of a series of concentrated loads, of which G_1 is to the left of the left adjacent panel point on the loaded chord, G_2 is within the panel opposite the panel point on the unloaded chord, and G_3 is to the right of the right adjacent panel point on the loaded chord. If the loads are advanced an infinitesimal distance dx farther to the left, G_1 goes down along *BA* of the influence line, G_2 goes down along *CB*, and G_3 comes up along *DC*. Thus,

$$dM = -G_1\left(\frac{n-m-\tfrac{1}{2}}{n}\right)dx + G_2\left[\frac{n-m-1}{n}(m+\tfrac{1}{2}) - \frac{m}{n}(n-m-\tfrac{1}{2})\right]dx$$
$$+ G_3\left(\frac{m+\tfrac{1}{2}}{n}\right)dx$$

Letting $G_3 = G - G_1 - G_2$ in the preceding equation and simplifying,

$$\frac{dM}{dx} = G\left(\frac{m+\frac{1}{2}}{n}\right) - (G_1 + \tfrac{1}{2}G_2) = \frac{Ga}{L} - (G_1 + \tfrac{1}{2}G_2) \qquad (13.9.1)$$

For the critical position, dM/dx in Eq. (13.9.1) must change from positive to negative; this is possible only when a load is placed either at the left adjacent or the right adjacent panel point on the loaded chord. Both of these possibilities must be investigated, and the larger of the two results is the final maximum bending moment being sought.

Example 13.9.1 For the truss in Fig. 13.9.2a, determine the maximum bending moment at U_2 due to the passage in either direction of the five wheel loads followed by an indefinite length of uniform load as shown in Fig. 13.9.2d.

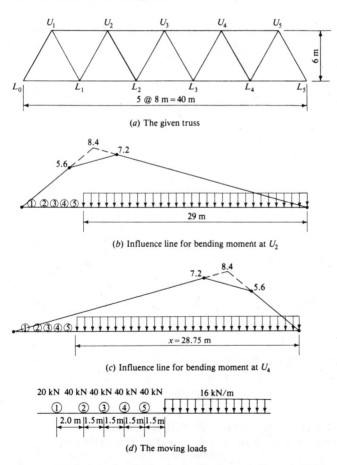

(a) The given truss

(b) Influence line for bending moment at U_2

(c) Influence line for bending moment at U_4

(d) The moving loads

Figure 13.9.2 Maximum bending moment at U_2 of truss in Example 13.9.1.

SOLUTION (a) Maximum bending moment at U_2.

Load at L_1	$G_1 + \tfrac{1}{2}G_2$	$\dfrac{Ga}{L} = \dfrac{1.5}{5}G$	$G_1 + \tfrac{1}{2}G_2$	Criterion satisfied?
W_3	$G_1 = 60 \quad G_2 = 176$ $G_1 + \tfrac{1}{2}G_2 = 148$	$0.3G = 0.3(620) = 186$	$G_1 = 100 \quad G_2 = 136$ $G_1 + \tfrac{1}{2}G_2 = 168$	No
W_4	$G_1 = 100 \quad G_2 = 160$ $G_1 + \tfrac{1}{2}G_2 = 180$	$0.3G = 0.3(644) = 193.2$	$G_1 = 140 \quad G_2 = 120$ $G_1 + \tfrac{1}{2}G_2 = 200$	Yes
W_5	$G_1 = 140 \quad G_2 = 144$ $G_1 + \tfrac{1}{2}G_2 = 212$	$0.3G = 0.3(668) = 200.4$	$G_1 = 180 \quad G_2 = 104$ $G_1 + \tfrac{1}{2}G_2 = 232$	No

The bending moment at U_2 can be computed in three ways: (1) by taking the average of the bending moments at L_1 and L_2; (2) by taking moment about U_2 directly, in which case the loads within the panel L_1L_2 will have to be attributed to the panel points L_1 and L_2; and (3) by making use of the influence line in Fig. 13.9.2. With W_4 at L_1,

M at $L_1 = R_0(8) - 40(1.5) - 40(3) - 20(5) = 2261.6$ kN·m

M at $L_2 = R_0(8) - 16(5)^2/2 - 160(8.75) - 20(13) = 3223.2$ kN·m

M at $U_2 = \tfrac{1}{2}(2261.6 + 3223.2) = 2742.4$ kN·m

M at U_2 (directly) $= R_0(12) - 120(5.5) - 20(9) - \dfrac{16(5)^2/2 + 40(6.5)}{8}$ (4)

$\qquad = 2742.4$ kN·m

M at U_2 (influence-line method) $= 20(2.1) + 40(3.5 + 4.55 + 5.6 + 5.9) + 16(33.5 + 86.4)$

$\qquad = 2742.4$ kN·m

In this problem, with W_4 at L_1 the uniform load has already extended into the panel L_1L_2; therefore there is no longer a concentrated load which can be placed at L_2 to satisfy the criterion of Eq. (13.9.1).

(b) Maximum bending moment at U_4. Referring to the loading position in Fig. 13.9.2c,

$$G_1 = 180 + 16(x - 16) \qquad G_2 = 16(8) = 128 \qquad G = 180 + 16x$$

Substituting the above values into Eq. (13.9.1) and setting it equal to zero,

$$\dfrac{Ga}{L} - (G_1 + \tfrac{1}{2}G_2) = \dfrac{(180 + 16x)(3.5)}{5} - [180 + 16(x - 16) + \tfrac{1}{2}(128)] = 0$$

from which

$$x = 28.75 \text{ m}$$

For this position, $R_6 = 326.3125$ kN, and

M at $L_4 = R_6(8) - 16(8)^2/2 = 2098.5$ kN·m

M at $L_3 = R_6(16) - 16(16)^2/2 = 3173.0$ kN·m

M at $U_4 = \tfrac{1}{2}(2098.5 + 3173.0) = 2635.75$ kN·m

M at U_4 (directly) $= R_6(12) - 16(8)(8) - \tfrac{1}{4}(16)(8)(4) = 2635.75$ kN·m

M at U_4 (influence-line method) $= 20(0.975) + 40(1.575 + 2.025 + 2.475 + 2.925)$
$\qquad + 16(141.015625)$

$\qquad = 2635.75$ kN·m.

(c) *Discussion.* If this bridge can carry only one-way traffic from right toward the left, then the maximum bending moment at U_2 is 2742.4 kN·m, and that at U_4 is 2635.75 kN·m. If, however, there is to be two-way traffic, then the maximum bending moment at U_4 is also 2742.4 kN·m.

13.10 Criterion for Maximum Force in the Web Member of a Truss

Although many large trusses are more complex than the one shown in Fig. 13.10.1, for purpose of explanation, the six-panel, curved upper-chord truss may be regarded as typical. Of the 11 web members, the influence line for the force in the end post L_0U_1 (or U_5L_6) is in the form of a simple triangle with its apex on the lower side, such as shown in Fig. 13.10.1b. The influence line for the force in the hanger U_1L_1 (or U_5L_5) is as shown in Fig. 13.10.1c. If counter diagonals† are not used, there is no chance for the center vertical U_3L_3 to receive any force due to live load. The influence lines for the remaining two pairs of diagonals and one pair of verticals are shown, typically, in Fig. 13.10.1d and e.

It is interesting to note, as will be substantiated in the numerical examples, that the first and third straight segments of the influence lines in Fig. 13.10.1d and e intersect at a point vertically under the point of intersection of the upper and lower chords, which with the web member in question are the three members to be cut so as to separate the entire truss into two parts. This fact can be used to check the correctness of the critical influence ordinates computed by the method of sections.

When the influence area has two parts with different signs, even for a two-way bridge it is necessary to consider only the live load coming to act on the right part of the truss, for *every* member. Then, what may happen to, say, U_1L_2 when the left part of the truss is loaded is the same as what may happen to L_4U_5 when the right part of the truss is loaded.

So far as the criterion for the maximum force in a web member of a truss is concerned, the same criterion

$$\frac{dF}{dx} = \frac{Ga}{L} - G_1 \qquad (13.10.1)$$

that was derived for bending moment in simple beams in Sec. 13.4 can be used. For the maximum tensile force in bar U_2L_3, of which the influence line is shown in Fig. 13.10.1e, the distances a and L to be used in Eq. (13.10.1) are as shown in that figure. However, G_1 should include all the loads before the panel point L_3, even if there may be one or two loads beyond the point of zero influence; and so should G. There is little possibility that any load will extend beyond point L_2, in which case Eq. (13.10.1) is no longer valid.

---

†For more detailed discussion on use of counters, see C.-K. Wang and C. L. Eckel, *Elementary Theory of Structures*, McGraw-Hill Book Company, New York, 1957, p. 220.

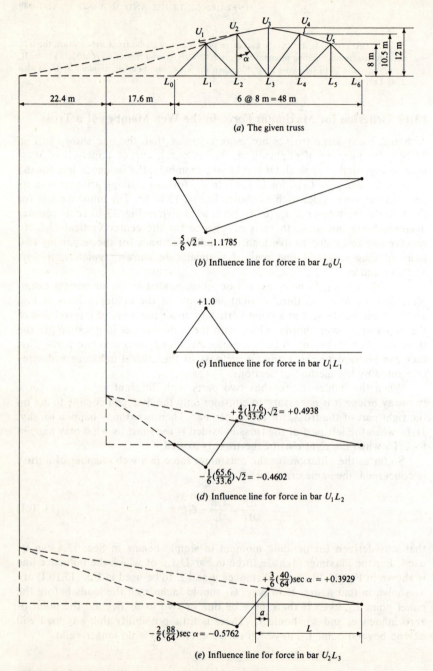

Figure 13.10.1 Influence lines for forces in the web members of a truss.

INFLUENCE LINES AND MOVING LOADS 491

Example 13.10.1 Determine the maximum tensile force and the maximum compressive force in bar U_2L_3 of the bridge truss of Fig. 13.10.2a due to the passage in either direction of the five wheel loads followed by an indefinite length of uniform load as shown in Fig. 13.10.2d.

SOLUTION (a) *Maximum tensile force in* U_2L_3. The influence line for the force in bar U_2L_3 shown in Fig. 13.10.2b is the same as that of Fig. 13.10.1e, of which the critical ordinates are obtained by the method of sections, taking moment about the point of intersection of the extensions of bars U_2U_3 and L_2L_3. A further check is made by noting the fact that the prolongations of the two major line segments of the influence line intersect at a point under the moment center used in the method of sections. The critical loading position on the basis of the triangle on the right side of the influence line is obtained from the accompanying table.

Load at L_3	G_1 in L_2L_3	$\dfrac{Ga}{L} = G\dfrac{3.243}{27.243} = 0.1190G$	G_1 in L_2L_3	Criterion satisfied?
W_1	0	0.1190(436) = 51.9	20	No
W_2	20	0.1190(468) = 55.7	60	Yes
W_3	60	0.1190(492) = 58.5	100	No

With W_2 at L_3,

$$R_0 = \frac{16(18)^2/2 + 160(24 - 2.25) + 20(26)}{48} = 137.33$$

$$\text{Panel-point load at } L_2 = \frac{W_1 \text{ times 2 m}}{8 \text{ m}} = 5$$

Taking moments of forces on left section about the point of intersection of U_2U_3 and L_2L_3,

$$137.33(40) - 5(56) - U_2L_3 \cos \alpha (64) = 0$$

$$U_2L_3 = 102.40 \text{ kN} \quad \text{(tension)}$$

$$U_2L_3 \text{ (influence-line method)} = \frac{20(0.1198) + 160(0.2832) + 16(\tfrac{1}{2})(0.234375)(18)}{\cos \alpha}$$

$$= 102.40 \text{ kN} \quad \text{(tension)}$$

(b) *Maximum compressive force in* L_3U_4. The influence line for the force in bar L_3U_4 shown in Fig. 13.10.2c can be obtained from the opposite image of Fig. 13.10.2b. The critical loading position on the basis of the triangle on the right side of this influence line is obtained from the accompanying table.

Load at L_4	G_1 in L_3L_4	$\dfrac{Ga}{L} = G\dfrac{4.757}{20.757} = 0.2292G$	G_1 in L_3L_4	Criterion satisfied?
W_2	20	0.2292(340) = 77.9	60	No
W_3	60	0.2292(364) = 83.4	100	Yes
W_4	100	0.2292(388) = 88.9	140	No

492 INTERMEDIATE STRUCTURAL ANALYSIS

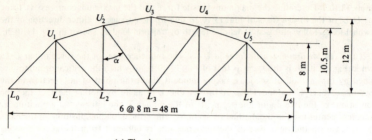

(a) The given truss

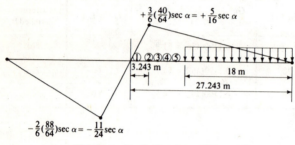

(b) Influence line for force in bar U_2L_3

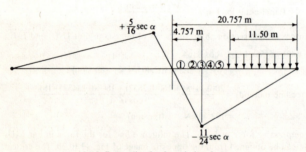

(c) Influence line for force in bar L_3U_4

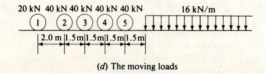

(d) The moving loads

Figure 13.10.2 Maximum tension and compression in bar U_2L_3 of truss in Example 13.10.1.

With W_3 at L_4,

$$R_0 = \frac{16(11.5)^2/2 + 160(16 - 0.75) + 20(16 + 3.5)}{48} = 81$$

$$\text{Panel-point load at } L_3 = \frac{40(1.5) + 20(3.5)}{8} = 16.25$$

Taking moments of forces on the left section about the point of intersection of U_3U_4 and L_3L_4,

$$81(88) - 16.25(64) + L_3U_4 \cos\alpha(64) = 0$$

$$L_3U_4 = -119.58 \text{ kN} \quad \text{(negative means compression)}$$

L_3U_4 (influence-line method)

$$= \frac{(-0.12111)(20) + (-0.31381)(40) + (-0.41536)(120) + 16(\tfrac{1}{2})(-0.32942)(11.5)}{\cos\alpha}$$

$$= -119.58 \text{ kN}$$

(c) *Discussion.* For a two-way bridge, the force in bar U_2L_3, or in bar L_3U_4, may vary from a compression of 119.58 kN to a tension of 102.40 kN, so far as the effect of the moving concentrated wheel loads is concerned.

13.11 Müller-Breslau Influence Theorem for Statically Determinate Trusses

In Sec. 13.6, and especially in Example 13.6.1, it has been shown how the Müller-Breslau influence theorem is useful as a means of obtaining visually the influence lines of statically determinate beams. This theorem, in fact, is so general that it applies to all structures: beams, trusses, and rigid frames, whether they be statically determinate or indeterminate. In this section the validity of the theorem as applied to statically determinate trusses will be shown. One might reach the conclusion after studying this section that the straightforward method of connecting the heavy dots of the function at the panel-point locations, as described in Sec. 13.7, is shorter than the influence-theorem method, so far as statically determinate trusses are concerned.

Following the general tone of the theorem as applied to statically determinate beams, one can surmise that the Müller-Breslau influence theorem for statically determinate trusses may be stated as follows:

The influence line for the force in a bar of a statically determinate truss is the same as the plot of the vertical positions of the panel points on the loaded chord, obtained by cutting the bar in question and pulling the cut ends with a pair of equal forces until there is an overlap of 1 unit.

As an illustration, the influence line for the force in bar L_2L_3 of the truss in Fig. 13.11.1a can be obtained by cutting the bar and applying the pair of forces F and F until the overlap reaches 1 unit. The two parts of the truss, one to the left of U_2L_2 and the other to the right of U_2L_3, would be squeezed together so that the lower chord can arch up only around the point U_2, with

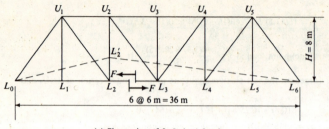

(a) Shortening of L_2L_3 by 1.0 unit

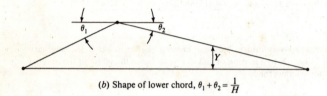

(b) Shape of lower chord, $\theta_1 + \theta_2 = \frac{1}{H}$

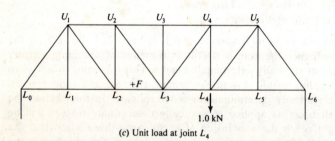

(c) Unit load at joint L_4

Figure 13.11.1 Müller-Breslau influence theorem for statically determinate trusses.

$\theta_1 + \theta_2$ in Fig. 13.11.2b equal to 1.0 unit divided by the height H of the truss. That Fig. 13.11.1b is the influence line for the force in bar L_2L_3 becomes obvious when it is compared with the influence line for bending moment at U_2 as described by Fig. 13.6.1d, because the force in bar L_2L_3 is indeed equal to $1/H$ times the bending moment at U_2.

That Y in Fig. 13.11.1b is equal to F in Fig. 13.11.1c can be further proved by separating the equilibrium state of Fig. 13.11.1c into 12 free bodies, one for each joint. Applying the principle of virtual work, the sum of products of all the forces acting on the above-mentioned 12 free bodies and their corresponding displacements in the compatible state of Fig. 13.11.1a must be zero; thus

$$-(1.0 \text{ force})(Y) + F(1.0 \text{ displacement}) = 0$$

or, Y in Fig. 13.11.1a is equal to F in Fig. 13.11.1c.

INFLUENCE LINES AND MOVING LOADS 495

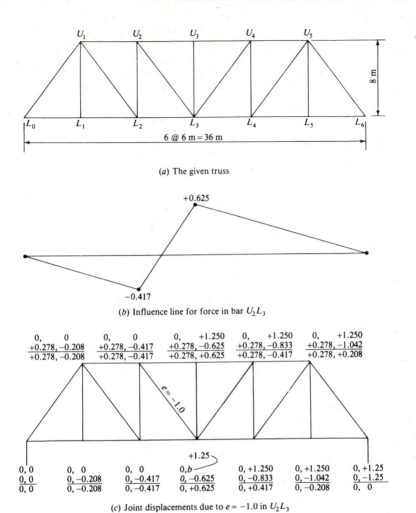

Figure 13.11.2 Influence line for force in bar U_2L_3 by Müller-Breslau influence theorem.

Example 13.11.1 For the statically determinate truss shown in Fig. 13.11.2a, first obtain the influence line for the force in bar U_2L_3 by its own definition, and then by determining the vertical deflections of all lower-chord panel points due to a contraction of 1.0 unit in the length of bar U_2L_3.

SOLUTION The influence line for the force in bar U_2L_3, worked out by its own definition, is shown in Fig. 13.11.2b. The vertical deflections of all lower-chord panel points, due to an elongation of -1.0 unit in bar U_2L_3 only, are worked out by the geometric method using L_0U_1 as the reference member and applying the joint-displacement equation, Eq. (3.6.2). The results as shown in Fig. 13.11.2c are identical to the influence ordinates in Fig. 13.11.2b.

13.12 Influence Lines for Statically Indeterminate Beams

Unlike for statically determinate beams, the influence lines for statically indeterminate beams or girders do not consist of "long stretches" of straight lines, but of "major curves." Why the terms *long stretches* and *major curves* are used will be explained subsequently.

First, there is no real difference between a beam and a girder when each is by itself. Relatively speaking, a girder is larger than a beam; in particular, a girder generally gives support to beams which run perpendicular to it. For smaller highway bridges, there may be only longitudinal beams, parallel to the traffic, which in turn support the transverse slab. Alternately, in larger bridges, there may be two main longitudinal griders, one on each side of the bridge, which support the floor beams equally spaced along the main span. The transverse floor beams in turn support longitudinal beams, often called stringers, which in turn support the bridge deck. In the former type, the influence lines for statically indeterminate cases will be curved continuously along the length of the *beam* (called a beam because there is no other comparison), as shown by the solid curve in Fig. 13.12.1b. In the latter type, the short segments between the panel points where the floor beams come in should be straight, as shown by the dashed line in Fig. 13.12.1b; thus the influence lines do not have "long stretches" of straight lines, but they are made of short, straight segments in the shape of "major curves" along the length of the *girder* (called a girder because there are floor beams running into it).

The distinction between the influence lines of beams and those of girders having been made, the next problem is to describe the ways by which the influence ordinates at selected points on the span can be determined. Certainly, for the two-span continuous beam shown in Fig. 13.12.1, the influence ordinates at points 1, 2, 4, 5, and 6 can be obtained by analyzing the beam for five loading conditions, each containing a single unit load at points 1, 2, 4, 5, and 6, respectively. This analysis can be performed by one of the common methods, such as the three-moment equation, slope-deflection, moment-distribution, or matrix displacement method. Only in the last method can all the

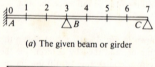

(*a*) The given beam or girder

— if unit load comes directly on beam
--- if unit load comes to girder only at panel points

(*b*) Influence line for bending moment at *B*

Figure 13.12.1 Influence lines for statically indeterminate beams or girders.

five loading conditions be treated altogether by using a rectangular $[P]$ matrix. One may call this straightforward method the *influence-line-definition* method.

Example 13.12.1 For the two-span continuous beam shown in Fig. 13.12.2a, compute the influence ordinates at 2-m intervals and sketch the influence lines for R_A, R_B, R_C, M_A, M_B, and M_5. Assume that the moving load may come directly on the beam.

SOLUTION The given beam is analyzed for five loading conditions, each containing a unit load at points 1, 2, 4, 5, and 6, by the matrix displacement method. The analysis itself is not shown; only the results are tabulated in Table 13.12.1. When the load is placed at A, B, or C, it goes directly into the support, causing a +1.0 for the reaction under the load but nothing else within the beam itself. The influence lines are plotted in Fig. 13.12.1b to g.

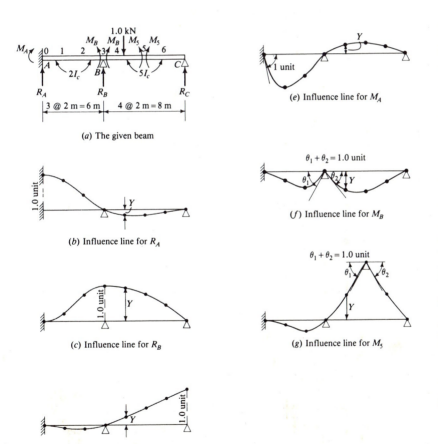

Figure 13.12.2 Influence lines for beam in Example 13.12.1.

Table 13.12.1 Influence ordinates for beam in Example 13.12.1

Unit load at	1	2	4	5	6
R_A	+0.78692	+0.35161	−0.13636	−0.15584	−0.09740
R_B	+0.24555	+0.71333	+0.95454	+0.73376	+0.39610
R_C	−0.03247	−0.06494	+0.18182	+0.42208	+0.70130
M_A	−0.98124	−0.62915	+0.27273	+0.31169	+0.19480
M_B	−0.25974	−0.51949	−0.54545	−0.62338	−0.38961
M_5	−0.12988	−0.25976	+0.72728	+1.68832	+0.80520

13.13 Müller-Breslau Influence Theorem for Statically Indeterminate Beams

Perhaps the most important practical application of the Müller-Breslau influence theorem is in the experimental determination of the influence lines for the reactions on statically indeterminate beams. For instance, the influence line for R_A on the continuous beam of Fig. 13.12.1a can be very inexpensively obtained in the laboratory by clamping a piece of plastic (with the correct relative moments of inertia) at the left end, holding the points B and C between nails, and then lifting the clamp upward by 1 unit. Because the beam itself is statically indeterminate only to the second degree, once the three influence lines for R_A, R_B, and R_C such as shown in Fig. 13.12.2b to d are obtained from measurements in the laboratory, their mutual agreement can be reviewed.

So far as influence lines for bending moments at A, B, and point 5 are concerned, such as shown in Fig. 13.12.2e to g, the Müller-Breslau method is the same. A hinge is installed at the point, then a pair M and M (and a pair of V and V for point 5) are applied until $\theta_1 + \theta_2$ is equal to 1.0 unit. Thus the Müller-Breslau influence theorem for statically indeterminate beams can be stated exactly as it is for statically determinate beams. As for the proof, however, it is no longer possible to apply the principle of virtual work to one or two finite rigid bodies. The equilibrium state of Fig. 13.12.2a has to be divided into an infinite number of free bodies of infinitesimal length. It is the virtual work done by all the generalized forces acting on all these free bodies, in going through the corresponding displacements in the compatible state of, say, Fig. 13.12.2c, that is equated to zero. Thus,

$$+(R_B)(+1.0 \text{ displacement}) - (1.0 \text{ load})(Y) = 0$$

or

$$Y \text{ in Fig. } 13.12.2c = R_B \text{ in Fig. } 13.12.2a \qquad (13.13.1)$$

The virtual work done by all other generalized forces is zero, because these always occur in pairs at the adjacent faces of two consecutive free bodies.

The proof for Eq. (13.13.1) can also be made by applying the reciprocal virtual-work theorem as already proved in Sec. 4.3, by calling Fig. 13.12.2a the P system and Fig. 13.12.2c the Q system, both applied to the beam without the support at B, although the deflection at B of the beam in Fig. 13.12.2a is still zero. Thus,

$$P * \Delta Q = Q * \Delta P$$
$$-(1.0)(Y) + R_B(1.0) = 0$$

Note that the virtual work done by the forces in the Q system in going through the displacements in the P system is zero. The same proof applies between the P system of Fig. 13.12.2a and the Q system of Fig. 13.12.2g, both applied to the beam with a hinge installed at point 5, although the tangent at point 5 in Fig. 13.12.2a is still continuous.

The ordinates on each of the six elastic curves in Fig. 13.12.2 can be obtained by either the force method or the displacement method. If the force method is used to determine the elastic curve of Fig. 13.12.2f, it will be necessary to obtain M_B in Fig. 13.13.1a so that the discontinuity of the tangents at B is 1.0 rad. If the displacement method is used, the problem is to find the final elastic curve starting with one of the two fixed conditions in Fig. 13.13.1b. If the force method is used to determine the elastic curve of Fig. 13.12.2g, M_5 and V_5 in Fig. 13.13.2a ought to be solved from the conditions that there is slope discontinuity of 1.0 rad but deflection continuity at point 5. If the matrix displacement method is used, the P matrix has to be established from one of the two fixed conditions in Fig. 13.13.2b.

While the foregoing discussion relating to the algebraic solution of the six elastic curves in Fig. 13.12.2 by the force and the displacement methods may be of academic interest and provide good exercises for applying the material treated in Chaps. 4, 6, 7, 8, and 11, the fact of the matter is that for a beam statically indeterminate to the second degree, only two influence lines, usually

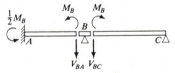

(a) Force method: M_B to cause 1.0-rad discontinuity

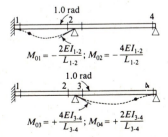

(b) Displacement method: Two fixed conditions

Figure 13.13.1 Alternate methods of solution for unit rotational discontinuity at a support.

500 INTERMEDIATE STRUCTURAL ANALYSIS

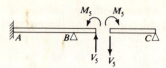

(a) Force method: M_5 and V_5 to cause 1.0-rad discontinuity

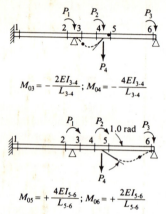

(b) Displacement method: Two fixed conditions

Figure 13.13.2 Alternate methods of solution for unit rotational discontinuity at any point.

of any two reactions, need to be first found from the Müller-Breslau influence theorem, either analytically or experimentally, and the other influence lines can be obtained by applying statics to each of the loading conditions.

13.14 Influence Lines for Statically Indeterminate Trusses vs. Müller-Breslau Influence Theorem

The Müller-Breslau influence theorem for statically indeterminate trusses, so far as the force in a bar is concerned, can be stated exactly as it is for statically determinate trusses. The advantage of using this theorem to obtain the influence line for the force in any one bar, over that of the influence-line-definition method, is not great, particularly when there remains a truss with redundant reactions or redundant bars even after the bar for which the influence line is to be found has been cut.

In cases where the truss is only externally indeterminate, it would be convenient to obtain the influence lines, just for the redundant reactions, by means of the Müller-Breslau influence theorem. Any other influence line can then be obtained by applying statics to each of the loading conditions.

Example 13.14.1 Outline a procedure by which the influence line for R_3 of the statically indeterminate truss shown in Fig. 13.14.1a may be obtained.

INFLUENCE LINES AND MOVING LOADS 501

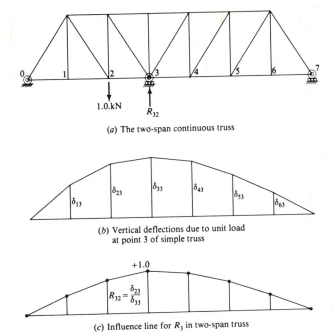

(a) The two-span continuous truss

(b) Vertical deflections due to unit load at point 3 of simple truss

(c) Influence line for R_3 in two-span truss

Figure 13.14.1 Influence line for reaction on a two-span continuous truss.

SOLUTION Apply a unit upward force at panel point 3 of the statically determinate truss without the intermediate support. The vertical deflections of all lower-chord panel points are as shown in Fig. 13.14.1b, in which δ_{ij} is the deflection at the ith panel point due to a unit load at the jth panel point of the simple truss. Then obtain the influence line for R_3 in Fig. 13.14.1c by dividing all ordinates in Fig. 13.14.1b by δ_{33}.

13.15 Exercises

13.1 to 13.4 For the statically determinate beams shown in Figs. 13.15.1 to 13.15.4, draw the influence lines for the functions as listed. Then compute the range of variation in the function if there is a dead load of 6 kN/m on the beam and a live load of 18 kN/m that may be on any portion or portions of the beam. Obtain these values by both the influence-line and the free-body methods.

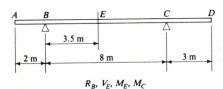

R_B, V_E, M_E, M_C

Figure 13.15.1 Exercise 13.1.

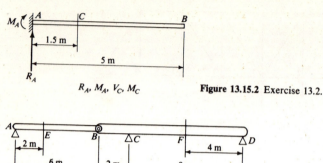

R_A, M_A, V_C, M_C **Figure 13.15.2** Exercise 13.2.

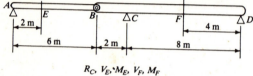

R_C, V_E, M_E, V_F, M_F **Figure 13.15.3** Exercise 13.3.

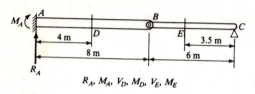

R_A, M_A, V_D, M_D, V_E, M_E **Figure 13.15.4** Exercise 13.4.

13.5 to 13.8 For the simple beams shown in Figs. 13.15.5 to 13.15.8, determine the range of variation in the values of R_A, V_C, and M_C due to the passage in either direction of the five wheel loads followed by an indefinite length of uniform load as shown in Fig. 13.3.1c. Obtain these values by both the influence-line and the free-body methods.

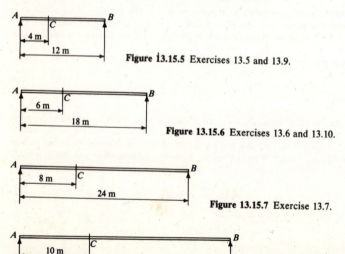

Figure 13.15.5 Exercises 13.5 and 13.9.

Figure 13.15.6 Exercises 13.6 and 13.10.

Figure 13.15.7 Exercise 13.7.

Figure 13.15.8 Exercise 13.8.

INFLUENCE LINES AND MOVING LOADS   503

13.9 and 13.10 For the simple beams in Exercises 13.5 and 13.6, determine the maximum bending moment at the center of the span and also the absolute maximum bending moment due to the passage of the same moving loads as used in Exercises 13.5 and 13.6.

13.11 and 13.12 Determine the absolute maximum bending moment in a simple beam of 12 m (Exercise 13.11) and 18 m (Exercise 13.12) in length, respectively, due to the passage of four equal concentrated loads of 40 kN each at three equal spacings of 3 m each.

13.13 to 13.16 Obtain the influence lines for the statically determinate beams in Exercises 13.1 to 13.4 by the Müller-Breslau influence theorem.

13.17 and 13.18 Obtain the maximum bending moment at U_1 and U_5 (Exercise 13.17) and at U_3 (Exercise 13.18) for the truss in Example 13.9.1 (Fig. 13.9.2) for the same loading in that example. Obtain this bending moment by both the influence-line and the free-body methods.

13.19 and 13.20 Obtain the maximum tensile and the maximum compressive forces which may come into the bar L_0U_1 (Exercise 13.19) and the bar U_1L_2 (Exercise 13.20) for the truss in Example 13.10.1 (Figs. 13.10.1 and 13.10.2) for the same loading in that example. Obtain the bar force in every case by both the influence-line and the free-body methods.

13.21 and 13.22 Obtain the influence lines for the force in bar U_1L_2 (Exercise 13.21) and in bar U_2L_2 (Exercise 13.22) of the truss in Example 13.11.1 (Fig. 13.11.2), first by their own definitions and then by the Müller-Breslau influence theorem.

13.23 to 13.28 Obtain each of the six influence lines in Fig. 13.12.2 (or in the influence table of Table 13.12.1) by the Müller-Breslau influence theorem using the force method.

13.29 to 13.34 Using the displacement method, obtain each of the six influence lines in Fig. 13.12.2 (or in Table 13.12.1) by the Müller-Breslau influence theorem.

CHAPTER
FOURTEEN
APPROXIMATE METHODS
OF MULTISTORY-FRAME ANALYSIS

14.1 Vertical- and Lateral-Load Analysis of Multistory Frames

Large, tall buildings, often called skyscrapers, have been analyzed, designed, and built since the early decades of the twentieth century, long before the extensive use of electronic computers in structural analysis from the 1960s onward. These high-rise building frames may be six to eight bays in width and 30, 40, or more stories in height. Chosen for illustration is a three-bay, two-story building frame such as that shown in Fig. 14.1.1.

So far as vertical loading is concerned, horizontal beams could be designed as simple beams supported at columns, which would then be subjected to axial forces only. But for lateral loads, usually due to wind, there have to be shears and bending moments in the columns, thus requiring moment-resisting connections between beams and columns. In theory, therefore, the joints should be taken as rigid, or at least partially rigid, joints in structural analysis. In reinforced concrete construction, the joints, being monolithic, are indeed rigid; thus the beam and column ends meeting at the same joint should rotate the same amount. In steel construction, the beam-to-column connections can be designed to carry the moments resulting from an analysis assuming either perfect or partial rigidity. Analysis of rigid frames with semirigid connections is to be treated in Chap. 20. Of course, once the joints are treated as perfectly rigid or as partially rigid, the same basis will have to be used for both vertical- and lateral-load analysis.

In regard to the general case of taking all joints in a multistory frame as perfectly rigid, its analysis can be made by the slope-deflection, moment-distribution, or matrix displacement method. Because of the convenience of the electronic computer, the matrix displacement method is in favor, although

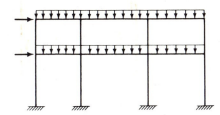

Figure 14.1.1 A three-bay, two-story building frame.

the moment-distribution method can also be programmed on the computer. For many years, however, the above-mentioned methods, called *elastic analysis by the exact method*, were much too time-consuming to perform, except for final review of the structure after several trials in the design process or for unusual or irregular cases. Nevertheless there have been approximate methods developed and used successfully, particularly for lateral-load analysis. It is the object of this chapter to describe these approximate methods so that they may continue to be useful in preliminary design; they may also help the analyst and the designer to understand and appreciate more fully the voluminous output sheets from an elastic analysis on the computer.

14.2 Degree of Indeterminacy vs. Number of Assumptions

The degree of indeterminacy of a structure is one of its characteristics; it does not depend on any assumption used in analysis, such as that of inextensible member length. From the viewpoint of the displacement method, the degree of indeterminacy NI of a rigid frame is equal to $NF - NP$, where NF is equal to twice the number of members and NP is the sum of unknown joint rotations and sidesway. The basis for this statement has been demonstrated in Eq. (11.3.1).

To recapitulate, for a rigid frame like the one shown in Fig. 14.1.1, the independent unknowns are the moments at the ends of all beams and columns, because the axial forces in them can be obtained by the resolution equations of equilibrium at the joints. Of the 16 resolution equations of equilibrium provided at the eight joints, 2 have already been used up as sidesway conditions in the displacement method of solution, leaving 14 equations for the 14 unknown axial forces. Thus the number of independent unknowns is the same as the number of end moments, or 28. The number of unknown generalized joint displacements is 10, of which 8 are rotations and 2 are sidesways. The degree of indeterminacy NI is $NF - NP = 28 - 10 = 18$. Indeed, a general formula for NI may be derived for a multistory frame in terms of its number of bays and number of stories.

From the viewpoint of the force method, the degree of indeterminacy is the number of generalized redundant forces acting with the applied loads on the basic determinate structure, which is derived from the given indeterminate structure by cuts or releases at the locations of the redundants. For the rigid frame of Fig. 14.1.1, repeated in Fig. 14.2.1a, one possible choice for the basic determinate structure involves the four vertical cantilever structures of Fig. 14.2.1b with cuts somewhere in the middle of all beams. At each cut there are three pairs of unknowns, for which the compatibility conditions are equal slope, and horizontal and vertical deflections at both sides of the cut. Consequently, the degree of indeterminacy is simply 3 times the number of beams, or 18. The column-analogy method to be described in Chap. 15 is essentially a force method, suitable for the analysis of closed frames with one single cell. It will be shown that a closed cell or ring structure is statically indeterminate to the third degree. If the three imaginary members with infinite flexural rigidity are added between the fixed supports in Fig. 14.2.1a, one observes six closed cells in that figure; then again the degree of indeterminacy is 3 times the number of cells, or 18.

Let NB be the number of bays and NS be the number of stories; then the degree of indeterminacy NI of a regular rectangular multistory frame is

$$NI = 3 * NB * NS \qquad (14.2.1)$$

This formula follows directly from the force method of analysis. For the displacement method,

$$NI = NF - NP$$
$$= 2[NB * NS + NS * (NB + 1)] - [(NB + 1) * NS + NS] = 3 * NB * NS$$

just as before.

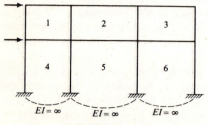

(a) The given structure

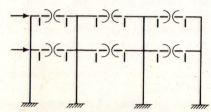

(b) Basic determinate structure under applied loads and 18 pairs of redundants

Figure 14.2.1 Degree of indeterminacy from viewpoint of force method.

14.3 Assumptions for Vertical-Load Analysis

For a regular multistory frame such as that shown in Fig. 14.3.1a, it has been ascertained in the preceding section that the degree of indeterminacy is equal to 3 times the number of beams. If this same number of assumptions is made regarding the "force response" of the structure, then the solution of the problem can be completed by statics alone. So far as vertical uniform loadings on all beams are concerned, the bending-moment diagram for any beam should take the form of Fig. 14.3.1b; only the moments at the ends are the unknowns. For typical interior spans, the two end moments should be nearly equal, and, if the beam is of constant cross section in steel, an economical distribution of moments along the span would be that of equal positive and negative moments in the amount of $\frac{1}{16} wL^2$. In this case, the points of inflection are at $0.146L$ from the columns. For exterior spans, the point of inflection near the exterior column may be somewhat less than $0.146L$ from it, and that near the interior column may be slightly more than $0.146L$ from it. If the analyst, using intuition or past experience, would spot arbitrarily the locations of two points of inflection on each beam span, the degree of indeterminacy of the structure is reduced by 2 times the number of beams.

Under vertical loading the horizontal reactions at the column bases are usually small so that the axial forces in the beams are also small. If the further assumption is made that the axial force in every beam is zero, the degree of

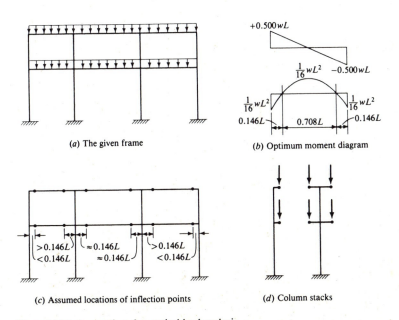

(a) The given frame

(b) Optimum moment diagram

(c) Assumed locations of inflection points

(d) Column stacks

Figure 14.3.1 Assumptions for vertical-load analysis.

508 INTERMEDIATE STRUCTURAL ANALYSIS

indeterminacy of the structure is accordingly further reduced to zero, because the total number of assumptions is now 3 times the number of beams.

The foregoing assumptions having been made, the portions of beams between points of inflection can be designed as simple beams without end moments, and the columns stacks can be designed as cantilever structures as shown in Fig. 14.3.1d. In this way, a preliminary design is obtained so that relative moments of inertia of all beams and columns become available as input in the more rigorous analysis. In fact, the locations of the points of inflection, as well as the magnitudes of the axial forces, as determined from the rigorous analysis, can be compared with those used in the approximate analysis. This kind of information can be compiled and kept on file in a design office for future use.

14.4 Assumptions for Lateral-Load Analysis

The force and deformation response of a regular multistory building frame under lateral loading can be described in a qualitative way, as shown in Fig. 14.4.1. The following observations can be made from the sketches:

1. The amount of sidesway increases rather fast from the lower toward the upper stories; all joint rotations are clockwise for lateral loading coming from the left.

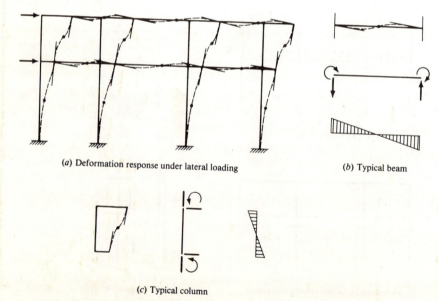

(a) Deformation response under lateral loading

(b) Typical beam

(c) Typical column

Figure 14.4.1 Multistory frame under lateral loading.

2. All column-end moments are counterclockwise, producing resisting horizontal shears from right to left at lower ends; there is one point of inflection in each column.
3. All beam-end moments are clockwise, producing resisting vertical shears downward at the left end and upward at the right end; there is one point of inflection in each beam.

For the frame shown in Fig. 14.4.1, there are 28 independently unknown end moments but only 10 equations of statics, leaving 18 as the degree of indeterminacy. The 10 equations of statics are as follows:

1. At each of the eight joints, the numerical sum of the counterclockwise column-end moments must be equal to the numerical sum of the clockwise beam-end moments (the counterclockwise or clockwise sense refers to the action on the member end);
2. The sum of the resisting horizontal shears at the lower ends of all columns in each of the two stories must be equal to the total lateral load above the level of the resisting horizontal shears.

In order that a lateral-load analysis be made by statics alone for the frame in Fig. 14.4.1, short of the rigor provided by the force or displacement methods, there have to be 18 assumptions, equal in number to the degree of indeterminacy. Fourteen of these assumptions are that the points of inflection are at the midpoints of all columns and beams. The remaining four assumptions, two per each story, pertain to the relative magnitudes of the vertical shear forces in the beams. There being three beams in each story, the relative magnitudes between the shears in them can count for only two independent assumptions.

In general terms, letting NB be the number of bays and NS be the number of stories, the number of beams is $NB * NS$, the number of columns is $(NB + 1) * NS$, and the number of specified points of inflection (PI) is

$$\text{Number of PIs} = (NB * NS) + (NB + 1) * NS \qquad (14.4.1)$$

The relative magnitudes of beam shears in each story furnish $(NB - 1)$ independent assumptions per story; thus the number of additional assumptions is

$$\text{Number of assumptions for relative beam shears} = (NB - 1) * NS$$
$$(14.4.2)$$

Adding Eqs. (14.4.1) and (14.4.2),

$$\text{Total number of assumptions made} = 3 * (NB * NS) \qquad (14.4.3)$$

which is the same as the degree of indeterminacy NI expressed by Eq. (14.2.1).

There have been two well-known approximate methods for lateral-load analysis of multistory frames: the portal method and the cantilever method.

The portal method was proposed by Albert Smith in his paper "Wind Stresses in the Frames of Office Buildings" in the *Journal of the Western Society of Engineers* (April 1915). The cantilever method was proposed by A. C. Wilson in his paper "Wind Bracing with Knee Braces or Gusset Plates," *Engineering Record* (September 5, 1908). The assumptions locating the points of inflection at the midpoints of all beams and columns were made in both of these methods; the difference was in the ways of assigning relative magnitudes to the shears in the beams. In the following two sections, each method will be described with an explanation as to why each is so named.

14.5 Portal Method

In the portal method, the shear forces in the beams of the same story are assumed to be all identical.

The same three-bay and two-story frame which has been used for illustration previously is again shown in Fig. 14.5.1a. Here the column lines are denoted by A, B, C, and D, and the beam lines by 1 and 2. The free-body diagrams of the frame from the top to the midheight of the columns in the second and first stories are shown, respectively, in Figs. 14.5.1b and c. From the basic assumption of constant shear along each beam line, further inferences can be drawn:

1. There are tensile and compressive axial forces of equal amount in the exterior windward and leeward columns, but no axial forces in the interior columns.
2. The beam-end moments, being the products of the vertical shear and one-half of the respective beam spans, are proportional to the beam spans, such as $L_1/L_2/L_3$.
3. The column-end moments are proportional to L_1, $L_1 + L_2$, $L_2 + L_3$, and L_4, or to $L_1/2$, $(L_1 + L_2)/2$, $(L_2 + L_3)/2$, and $L_4/2$, just as the ratios of horizontal dimensions tributary to each column.
4. The resisting horizontal shear forces at the lower ends of the columns in the same story are also proportional to $L_1/2$, $(L_1 + L_2)/2$, $(L_2 + L_3)/2$, and $L_4/2$, which are the ratios of horizontal dimensions tributary to each column.

The fourth inference is of particular significance in that it can be shown as in Fig. 14.5.2, where the total lateral load is divided into four parts in the ratios of $L_1/2$, $(L_1 + L_2)/2$, $(L_2 + L_3)/2$, and $L_4/2$. If the beam spans are equal, the shear in the interior column is twice that in the exterior column. The same observation can be made if the entire structure is thought of as the superposition of three portal frames shown in the lower part of Fig. 14.5.2 because the horizontal reactions at the base of two adjacent portals will be combined for an interior column. The name *portal method* came about in this way.

The application of the portal method in a numerical problem is quite

APPROXIMATE METHODS OF MULTISTORY-FRAME ANALYSIS **511**

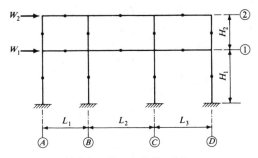

(a) Column lines A, B, C, and D; beam lines 1 and 2

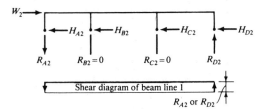

(b) Free-body diagram to inflection points of columns in second story

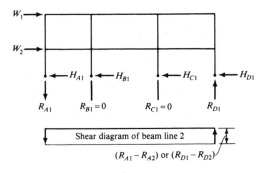

(c) Free-body diagram to inflection points of columns in first story

Figure 14.5.1 Assumptions in the portal method.

simple if the beam spans are constant. If the beam spans are different, the following steps in reference to Fig. 14.5.1 may be followed:

1. Solve for the axial forces in the exterior columns of each story by applying a moment equation to each of the free bodies such as shown in Fig. 14.5.1b and c.

512 INTERMEDIATE STRUCTURAL ANALYSIS

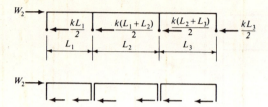

Figure 14.5.2 The portal concept.

2. Solve for the clockwise beam-end moments along each beam line by multiplying the constant shear by one-half of the respective beam spans.
3. Solve for the counterclockwise column-end moments by using the condition that the numerical sum of clockwise beam-end moments must be equal to the numerical sum of counterclockwise column-end moments for the same joint.
4. Solve for the horizontal shear forces in all the columns from the column-end moments and check that the sum of resisting horizontal shears in the columns of the same story is equal to the total lateral load above that level.

The alternate procedure would be to reverse the four steps by determining the horizontal shears in the columns first and using the moment equations mentioned in the first step for checking.

Example 14.5.1 For the multistory frame shown in Fig. 14.5.3a, determine by the portal method all column-end and beam-end moments due to the lateral loads as shown.

SOLUTION (a) *Axial forces in exterior columns.* Referring to Fig. 14.5.3b,

$$(R_{A2} \text{ or } R_{D2})(L_1 + L_2 + L_3) = W_2 H_2/2$$

$$(R_{A2} \text{ or } R_{D2})(18) = 16(1.8) \qquad R_{A2} = R_{D2} = 1.60 \text{ kN}$$

Referring to Fig. 14.5.3c,

$$(R_{A1} \text{ or } R_{D1})(L_1 + L_2 + L_3) = W_2\left(H_2 + \frac{H_1}{2}\right) + W_1 H_1/2$$

$$(R_{A1} \text{ or } R_{D1})(18) = 16(6.3) + 40(2.7) \qquad R_{A1} = R_{D1} = 11.60 \text{ kN}$$

(b) *Shears and moments in beams.* The shear along beam line 2 is $-R_{A2} = -1.60$; the shear along beam line 1 is $-(R_{A1} - R_{A2}) = -(11.6 - 1.6) = -10.0$. For these shears, the moment diagrams are shown in Fig. 14.5.4. The moment values along beam line 2 are $1.60(L_1/2)$, $1.60(L_2/2)$, $1.60(L_3/2)$; or 3.84, 5.76, 4.80. The moment values along beam line 1 are $10.0(L_1/2)$, $10.0(L_2/2)$, $10.0(L_3/2)$; or 24, 36, 30.

(c) *Column-end moments.* The clockwise beam-end moments determined in part (b) are first written down in Fig. 14.5.5a. Then the counterclockwise column-end moments are worked out around the joints on beam lines 2 and 1, in succession, using the condition that the sum of clockwise beam-end moments must be equal to the sum of the counterclockwise column-end moments, at the same joint.

(d) *Column-end shears.* The resisting shear forces at the lower ends of all the eight columns are computed by dividing the column-end moments by one-half of the respective story

APPROXIMATE METHODS OF MULTISTORY-FRAME ANALYSIS **513**

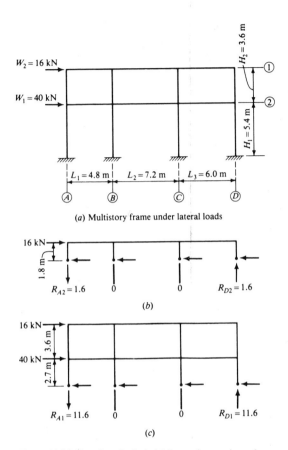

Figure 14.5.3 Portal method: Axial forces in exterior columns.

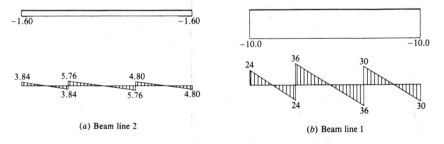

Figure 14.5.4 Portal method: Shears and moments in beams.

514 INTERMEDIATE STRUCTURAL ANALYSIS

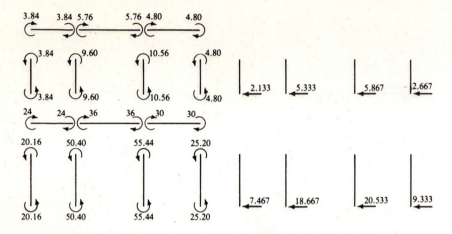

(a) From beam moments to column moments

(b) From column moments to column shears

Figure 14.5.5 Portal method: Shears and moments in columns.

height. These values are written down in Fig. 14.5.5b. Note the checks that

$$2.133 + 5.333 + 5.667 + 2.667 = W_2 = 16.000$$

and

$$7.467 + 18.667 + 20.533 + 9.333 = W_2 + W_1 = 56.000$$

14.6 Cantilever Method

In the cantilever method, the axial forces in the columns are assumed to be proportional to their respective distances from the centroid of column areas, tensile on one side and compressive on the other, taking all column areas to be equal.

The same three-bay, two-story frame used in the presentation of the portal method is shown in Fig. 14.6.1a. The assumption for the linear variation of axial forces in the columns means that the relative magnitudes of the axial forces in the columns are the same for all stories, such as shown in Fig. 14.6.1b and c. In turn, the relative magnitudes of the shears in the beam spans are the same for all stories, also shown in Fig. 14.6.1b and c. These shear values are numerically largest in the center span or spans of the beam, in contrast to the constant shear for all spans in the portal method. Consequently, the beam-end moments, as well as the column-end moments near the middle of the building width, are relatively larger than those in the portal method.

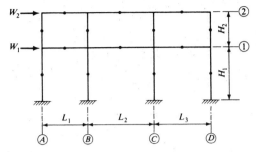

(a) Column lines A, B, C, and D; beam lines 1 and 2

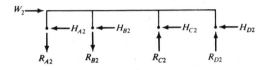

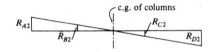

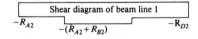

(b) Free-body diagram to inflection points of columns in second story

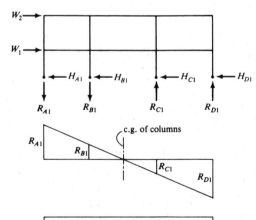

(c) Free-body diagram to inflection points of columns in first story

Figure 14.6.1 Assumptions in the cantilever method.

516 INTERMEDIATE STRUCTURAL ANALYSIS

The application of the cantilever method in a numerical problem can be made in the following steps:

1. Determine the location of the centroid of the columns assuming that all columns are of equal areas.
2. Solve for the axial forces in the columns of each story by applying a moment equation to the free body from the top of the frame to the inflection points of the columns in that story.
3. Draw the shear diagrams for the beams in each story and from them compute the clockwise beam-end moments of that story.
4. Solve for the counterclockwise column-end moments using the conditions that the sum of the counterclockwise column-end moments must be equal to the sum of the clockwise beam-end moments at each joint.
5. Solve for the horizontal resisting shears at the lower ends of all columns in each story and check the total value against the total lateral load above that level.

The alternate procedure would be to reverse steps 2 to 5 above, in which case one still needs to go through steps 2 to 5, first using only relative magnitudes in order to arrive at the distribution ratios by which the total lateral load can be divided into resisting shears in the columns.

A cantilever beam is one which is fixed at one end and free at the other; in addition, the distribution of stress in the beam is linear, changing from tension to compression at the centroid of the cross section. As shown in Fig. 14.6.2a, a tall building frame under lateral loading can be regarded as a vertical cantilever; and, considering the equilibrium of the free body shown in Fig. 14.6.2b, the axial force in each column, by its resemblance to a solid cantilever beam, may be assumed to be proportional to its distance from the

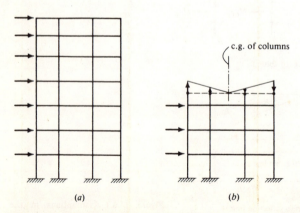

Figure 14.6.2 The cantilever concept.

APPROXIMATE METHODS OF MULTISTORY-FRAME ANALYSIS **517**

centroid of all column areas. The name *cantilever method* comes from this concept.

Example 14.6.1 For the multistory frame shown in Fig. 14.6.3a, determine by the cantilever method all column-end and beam-end moments due to the lateral loads as shown.

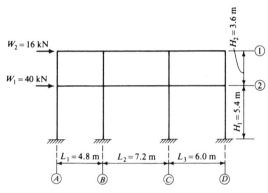

(a) Multistory frame under lateral loads

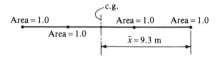

(b) Centroid of column areas

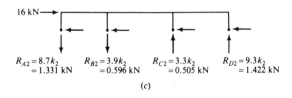

(c)

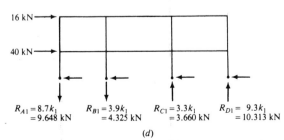

(d)

Figure 14.6.3 Cantilever method: Axial forces in columns.

SOLUTION (a) *Centroid of column.* In reference to Fig. 14.6.3b, the distance $\bar{x}$ of the centroid of columns from the column line D is

$$\bar{x} = \frac{1.0(6.0) + 1.0(13.2) + 1.0(18.0)}{4.0} = 9.3 \text{ m}$$

(b) *Axial forces in columns of each story.* The axial forces are tensile in the windward columns and compressive in the leeward columns, each equal to a constant times its distance from the centroid of the columns. Using the free body in Fig. 14.6.3c and taking moments about the centroid of the columns,

$$16(1.8) = k_2(8.7^2 + 3.9^2 + 3.3^2 + 9.3^2)$$

$$k_2 = \frac{16(1.8)}{188.28} = 0.15296$$

Then,

$$R_{A2} = 8.7k_2 = 1.331 \text{ kN} \qquad R_{B2} = 3.9k_2 = 0.596 \text{ kN}$$
$$R_{C2} = 3.3k_2 = 0.505 \text{ kN} \qquad R_{D2} = 9.3k_2 = 1.422 \text{ kN}$$

Using the free body in Fig. 14.6.3d and taking moments about the centroid of the columns,

$$16(6.3) + 40(2.7) = k_1(8.7^2 + 3.9^2 + 3.3^2 + 9.3^2)$$

$$k_1 = \frac{16(6.3) + 40(2.7)}{188.28} = 1.1090$$

Then,

$$R_{A1} = 8.7k_1 = 9.648 \text{ kN} \qquad R_{B1} = 3.9k_1 = 4.325 \text{ kN}$$
$$R_{C1} = 3.3k_1 = 3.660 \text{ kN} \qquad R_{D2} = 9.3k_1 = 10.314 \text{ kN}$$

(c) *Shears and moments in beams.* Using the axial forces in the columns obtained in part (b) and shown in Fig. 14.6.3c and d, the shears and then the moments along beam lines 2 and 1 are computed and shown in Fig. 14.6.4.

(d) *Column-end moments.* The counterclockwise column-end moments shown in Fig. 14.6.5a are obtained from the conditions that the sum of counterclockwise column-end moments must be equal to the sum of clockwise beam-end moments for the same joint.

(e) *Column-end shears.* The lower-end column shears shown in Fig. 14.6.5b are obtained by dividing each of the counterclockwise column-end moments in Fig. 14.6.5a by one-half of

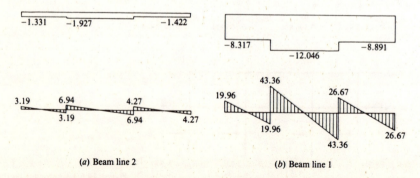

(a) Beam line 2   (b) Beam line 1

Figure 14.6.4 Cantilever method: Shears and moments in beams.

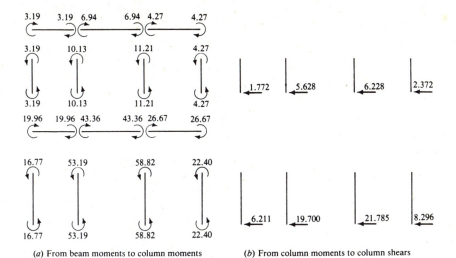

(a) From beam moments to column moments   (b) From column moments to column shears

Figure 14.6.5 Cantilever method: Shears and moments in columns.

the respective column height. The sum of the four lower-end column shears in the second story is $1.772 + 5.628 + 6.228 + 2.372 = 16.000 = W_2$, and the sum of those in the first story is $6.211 + 19.700 + 21.785 + 8.296 = 55.992 \approx W_2 + W_1$.

14.7 Alternate Moment and Shear Distribution

In Sec. 8.10, the application of the moment-distribution method to the analysis of statically indeterminate rigid frames with unknown joint translation was described. In this method, sometimes called the *indirect method*, there needs to be one moment distribution for the fixed-end moments due to applied loads when joint translations are prohibited, and then one additional moment distribution for each arbitrary amount of unknown joint translation. It was also mentioned in Sec. 8.10 that Morris has proposed a method of alternate moment and shear distribution in which the releasing of joint rotation while sidesways are prevented alternates with locking of joint rotation while sidesways are permitted. In this section, it will be shown that this method of alternate moment and shear distribution is extremely suitable to the analysis of multistory frames under lateral loads.

The multistory frame under lateral loads as shown in Fig. 14.7.1a is statically indeterminate to the eighteenth degree and as such can be analyzed to satisfy both equilibrium and compatibility by an "elastic analysis." (The approximate methods described in Secs. 14.5 and 14.6 do not satisfy compatibility.) To make the elastic analysis, it is necessary to have the relative magnitudes of the moments of inertia of the beams and columns in the frame, as indicated in Fig. 14.7.1a.

520 INTERMEDIATE STRUCTURAL ANALYSIS

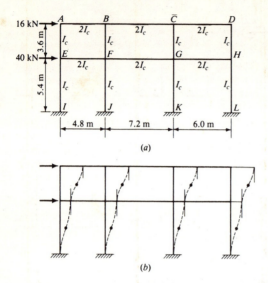

Figure 14.7.1 Concept of alternate moment and shear distribution.

The method of alternate moment and shear distribution should begin with a moment distribution if there are fixed-end moments due to loads acting on the members themselves; if not, with a shear distribution. Then the moment and shear distributions will alternate, and the process continues until the changes in the moment values in two successive cycles are all within the desired tolerance. In general, the process should stop at the end of a moment distribution.

Example 14.7.1 Analyze the multistory frame of Fig. 14.7.1a by the alternate moment- and shear-distribution method.

SOLUTION (a) *Head of distribution table.* The complete distribution table is shown in Table 14.7.1. The joints in Fig. 14.7.1a having been designated from A to L, lines 1 to 4 in this table are made up as in any moment distribution.

(b) *First shear distribution, line 5.* If all eight joints are locked against rotation but the beams are permitted to sway to the right, as shown in Fig. 14.7.1b, the relative magnitudes of the column-end moments should be same as those of $6(EI)_{col}\Delta/H^2$, or simply $(EI)_{col}$ for constant Δ and H. The sum of the four pairs of column-end moments in the second story should be $W_2H_2 = 16(3.6) = 57.60$ kN·m, which if divided by 8 gives 7.20. The sum of the four pairs of column-end moments in the first story should be $(W_1 + W_2)(H_2) = (16 + 40)(5.4) = 302.4$, which if divided by 8 gives 37.80. These values are entered in line 5. Note that if the flexural rigidity values $(EI)_{col}$ are not the same, the total shear force in the story must be divided among the columns in the ratio of their EI values.

(c) *First moment distribution, lines 6 and 7.* The joints are permitted to rotate; they are balanced and carry-overs are made.

(d) *Second shear distribution, line 8.* Since the shear was balanced for the second story on line 5, the undesirable column-end moments added on are $(+2.88 + 7.10) + (14.21 + 1.44) + (2.06 + 5.40) + (10.80 + 1.03) + (2.25 + 5.82) + (11.64 + 1.12) + (3.27 + 7.85) + (15.70 + 1.64) = +94.21$, which if divided by 8 and reversed in sign gives -11.78. For the first story the undesirable column-end moments on lines 6 and 7 are $(9.47+0)+(0+4.74)+(7.20+0)+(0+3.60)+(7.76+0)+(0+3.88)+(10.46+0)+(0+5.23)=+52.34$, which if divided by 8 and reversed in sign gives -6.54. The values of -11.78 and -6.54 are placed at the respective column ends on line 8.

(e) *Alternate moment and shear distribution, lines 9 to 11, 12 to 14, 15 to 17, and 18 to 20.* The processes done on lines 6 to 8 are repeated in these four alternate moment and shear distributions.

(f) *Stop at moment distribution, line 21.* The balancing moments on this line are deemed within the tolerance of 0.02 or 0.01.

(g) *Final end moments, line 22.* The values on this line are the sums of those from lines 5 through 21.

(h) *Fixed-end moments due to sidesway, line 23.* The values on this line are the sums of those on lines 5, 8, 11, 14, 17, and 20.

(i) *Check on moment distribution, lines 24 to 27.* The usual check procedure for moment distribution is applied to obtain the joint rotations.

(j) *Results from computer output, line 28.* These values are taken from the computer output using a computer program tailored to tall building–frame analysis by a marching technique in which the largest inversion is no more than the number of joints in each story plus 1.† The complete solution is shown in Fig. 14.7.2.

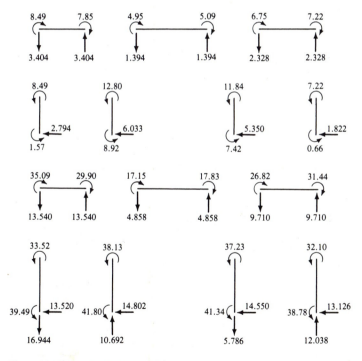

Figure 14.7.2 Results of elastic analysis.

†C.-K. Wang, *Matrix Methods of Structural Analysis*, 2d ed., American Publishing Company, Madison, Wis., 1970, Chap. 19 and Appendix P.

Table 14.7.1a Alternate moment and shear distribution (part 1)

1	2	3	4	5	6	7	8	9	10	11	12	13	14
		F			G				H		I	J	K	L
	Joint Member	FE	FG	FJ	GC	GF	GH	GK	HD	HG	HL	IE	JF	KG	LH
3	$EI/L; (EI_t)$	0.4167	0.2778	0.1852	0.2778	0.2778	0.3333	0.1852	0.2778	0.3333	0.1852	0.1852	0.1852	0.1852	0.1852
4	DF	0.3600	0.2400	0.1600	0.2586	0.2586	0.3103	0.1724	0.3489	0.4186	0.2325				
5	BAL H			−37.80	−7.20			−37.80	−7.20		−37.80	−37.80	−37.80	−37.80	−37.80
6	BAL M	+16.20	+10.80	+ 7.20	+11.64	+11.64	+13.96	+ 7.76	+15.70	+18.84	+10.46				
7	CO	+10.66	+ 5.82		+ 1.12	+ 5.40	+ 9.42		+ 1.64	+ 6.98		+ 4.74	+ 3.60	+ 3.88	+ 5.23
8	BAL H			− 6.54	−11.78			− 6.54	−11.78		− 6.54	− 6.54	− 6.54	− 6.54	− 6.54
9	BAL M	+ 0.29	+ 0.19	+ 0.14	+ 0.62	+ 0.62	+ 0.74	+ 0.40	+ 3.38	+ 4.06	+ 2.26				
10	CO	+ 2.08	+ 0.31		+ 0.46	+ 0.10	+ 2.03		+ 0.58	+ 0.37		+ 0.92	+ 0.07	+ 0.20	+ 1.13
11	BAL H			− 0.87	− 2.10			− 0.87	− 2.10		− 0.87	− 0.87	− 0.87	− 0.87	− 0.87
12	BAL M	+ 0.05	+ 0.04	+ 0.02	+ 0.10	+ 0.10	+ 0.12	+ 0.06	+ 0.70	+ 0.85	+ 0.47				
13	CO	+ 0.52	+ 0.05		+ 0.10	+ 0.02	+ 0.42		− 0.04	+ 0.06		+ 0.23	+ 0.01	+ 0.03	+ 0.24
14	BAL H			− 0.19	− 0.34			− 0.19	− 0.34		− 0.19	− 0.19	− 0.19	− 0.19	− 0.19
15	BAL M	− 0.04	− 0.03	− 0.02		− 0.01	− 0.01		− 0.18	+ 0.21	− 0.12				
16	CO	+ 0.12			+ 0.04	− 0.01	+ 0.10		− 0.03			+ 0.05	− 0.01	+ 0.03	+ 0.06
17	BAL H			− 0.04	− 0.06			− 0.04	− 0.06		− 0.04	− 0.04	− 0.04	− 0.04	− 0.04
18	BAL M	− 0.02	− 0.01	− 0.01	− 0.01	− 0.01	− 0.01		+ 0.05	+ 0.05	+ 0.03				
19	CO	+ 0.03			+ 0.01		+ 0.02		− 0.01			+ 0.01			+ 0.01
20	BAL H			− 0.01	− 0.01			− 0.01	− 0.01		− 0.01	− 0.01	− 0.01	− 0.01	− 0.01
21	BAL M	− 0.01								+ 0.01	+ 0.01				
22	Total M	+29.88	+17.17	−38.12	− 7.41	+17.86	+26.78	−37.23	+ 0.67	+31.43	−32.10	−39.50	−41.78	−41.34	−38.78
23	Σ BAL H			−45.45	−21.49			−45.45	−21.49		−45.45	−45.45	−45.45	−45.45	−45.45
24	Change	+29.88	+17.17	+ 7.33	+14.08	+17.86	+26.78	+ 8.22	+22.16	+31.43	+13.35	+ 5.95	+ 3.67	+ 4.11	+ 6.67
25	−⅓(change)	−17.54	− 8.93	− 1.84	− 4.84	− 8.58	−15.72	− 2.06	− 7.14	−13.39	− 3.34	− 5.96	− 3.66	− 4.11	− 6.68
26	Sum	+12.34	+ 8.24	+ 5.49	+ 9.24	+ 9.28	+11.06	+ 6.16	+15.02	+18.04	+10.01	− 0.01	+ 0.01	0.00	0.00
27	$\theta = \text{sum}/(3EI/L)$	+ 9.87	+ 9.89	+ 9.88	+11.09	+11.13	+11.06	+11.09	+18.02	+18.04	+18.01	OK	OK	OK	OK
28	M (computer)	+29.90	+17.15	−38.13	− 7.42	+17.83	+26.82	−37.23	+ 0.66	+31.44	−32.10	−39.49	−41.80	−41.34	−38.78

522

Table 14.7.1b Alternate moment and shear distribution (part 2)

1	Joint	A		B			C			D			E		
2	Member	AB	AE	BA	BC	BF	CB	CD	CG	DC	DH	EA	EF	EI	FB
3	$EI/L; (EI_i)$	0.4167	0.2778	0.4167	0.2778	0.2778	0.2778	0.3333	0.2778	0.3333	0.2778	0.2778	0.4167	0.1852	0.2778
4	DF	0.6000	0.4000	0.4286	0.2857	0.2857	0.3125	0.3750	0.3125	0.5454	0.4546	0.3158	0.4737	0.2105	0.2400
5	BAL H	+4.32	− 7.20	+3.08	+2.06	− 7.20	+2.25	+2.70	− 7.20	+3.93	− 7.20	− 7.20	+21.32	−37.80	− 7.20
6	BAL M		+ 2.88			+ 2.06			+ 2.25		+ 3.27	+14.21		+ 9.47	+10.80
7	CO	+1.54	+ 7.10	+2.16	+1.12	+ 5.40	+1.03	+1.96	+ 5.82	+1.35	+ 7.85	+ 1.44	+ 8.10	− 6.54	+ 1.03
8	BAL H		−11.78			−11.78			−11.78		−11.78	−11.78			−11.78
9	BAL M	+1.88	+ 1.26	+1.34	+0.88	+ 0.88	+0.93	+1.11	+ 0.93	+1.41	+ 1.17	+ 2.77	+ 4.16	+ 1.85	+ 0.19
10	CO	+0.67	+ 1.38	+0.94	+0.46	+ 0.10	+0.44	+0.70	+ 0.31	+0.56	+ 1.69	+ 0.63	+ 0.14	− 0.87	+ 0.44
11	BAL H		− 2.10			− 2.10			− 2.10		− 2.10	− 2.10			− 2.10
12	BAL M	+0.03	+ 0.02	+0.26	+0.17	+ 0.17	+0.20	+0.25	+ 0.20	−0.08	− 0.97	+ 0.70	+ 1.04	+ 0.46	+ 0.03
13	CO	+0.13	+ 0.35	+0.01	+0.10	+ 0.01	+0.08	−0.04	+ 0.05	+0.12	+ 0.35	+ 0.01	+ 0.02	+ 0.19	+ 0.08
14	BAL H		− 0.34			− 0.34			− 0.34		− 0.34	− 0.34			− 0.34
15	BAL M	−0.08	− 0.06	+0.10	+0.06	+ 0.06	+0.08	+0.09	+ 0.08	−0.07	− 0.06	+ 0.16	+ 0.24	+ 0.10	− 0.03
16	CO	+0.05	+ 0.08	−0.04	+0.04	− 0.01	−0.03	−0.04	− 0.06	+0.04	+ 0.09	− 0.03	− 0.02	− 0.04	+ 0.03
17	BAL H		− 0.06			− 0.06			− 0.06		− 0.06	− 0.06			− 0.06
18	BAL M	−0.04	− 0.03	+0.03	+0.02	+ 0.02	−0.02	+0.03	+ 0.02	−0.04	− 0.03	− 0.05	+ 0.07	+ 0.03	− 0.01
19	CO	+0.1	+ 0.02	−0.02	+0.01		+0.01	−0.02		+0.01	+ 0.02	− 0.01	− 0.01	− 0.01	+ 0.01
20	BAL H		− 0.01			− 0.01			− 0.01		− 0.01	− 0.01			− 0.01
21	BAL M	−0.01	− 0.01	+0.02		0.01	+0.01		+ 0.01	−0.01	− 0.01	− 0.01	+ 0.02	+ 0.01	− 0.01
22	Total M	+8.50	− 8.50	+7.88	+4.92	−12.80	+5.08	+6.74	−11.82	+7.22	− 7.22	− 1.55	+35.08	−33.53	− 8.93
23	Σ BAL H		−21.49			−21.49			−21.49		−21.49	−21.49			−21.49
24	Change	+8.50	+12.99	+7.88	+4.92	+ 8.69	+5.08	+6.74	+ 9.67	+7.22	+14.27	+19.94	+35.08	+11.92	+12.56
25	−⅓(change)	−3.94	− 9.97	−4.25	−2.54	− 6.28	−2.46	−3.61	− 7.04	−3.37	−11.08	− 6.50	−14.94	− 2.98	− 4.34
26	Sum	+4.56	+ 3.02	+3.63	+2.38	+ 2.41	+2.62	+3.13	+ 2.63	+3.85	+ 3.83	+13.44	+20.14	+ 8.94	+ 8.22
27	$\theta = sum/(3EI/L)$	+3.65	+ 3.62	+2.90	+2.86	+ 2.89	+3.14	+3.13	+ 3.16	+3.85		+16.13	+16.11	+16.09	+ 9.86
28	M (computer)	+8.49	− 8.49	+7.85	+4.95	−12.80	+5.09	+6.75	−11.84	+7.22	− 7.22	− 1.57	+35.09	−33.52	− 8.92

523

14.8 Comparison of Methods

The three-bay, two-story building frame as described in the preceding sections is statically indeterminate to the eighteenth degree; its elastic analysis requires as input the relative magnitudes of the moments of inertia of the beams and columns. In light of the results of the elastic analysis made in Sec. 14.7, the 18 assumptions which have been used in the approximate methods may be examined with regard to their divergence from the solution which satisfies compatibility. So far as the 14 assumed locations of the inflection points at the midpoints of all beams and columns are concerned, the following observations may be made from the end-moment values shown in Fig. 14.7.2:

1. The points of inflection in the center beam spans are very close to the midpoint, slightly nearer to the side with less restraint.
2. The points of inflection in the side spans are farther from the exterior than the interior columns, at from 0.52 to 0.54 of the span.
3. The points of inflection in the columns of the second story are much closer to the intermediate floor than the roof, at 0.16, 0.41, and 0.39 of the height for the first three columns from the left, and, in fact, there is no point of inflection in the extreme right column.

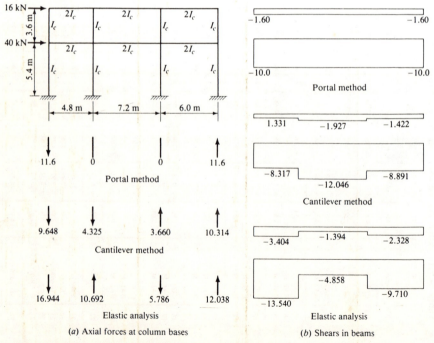

Figure 14.8.1 Comparison of column axial forces and beam shears.

4. The points of inflection in the columns of the first story are farther from the base than the upper floor level, at 0.54, 0.52, 0.53, and 0.55 of the height.

Comparison of the relative magnitudes of the beam shears may be observed from Fig. 14.8.1. In the portal method, the beam shears are constant along the same floor level, which means that the axial forces in the exterior windward and leeward columns are numerically equal but are zero for the interior columns. In the cantilever method, the beam shears are numerically larger in the center spans, owing to the assumption of linear variation of the axial forces in the columns with respect to the centroid of all the column areas. From the elastic analysis, however, the beam shears are much lower in the center spans, because the axial forces in the columns are alternately tensile and compressive starting from the left. At a glance, one would think that the assumption of linear variation of axial forces in the cantilever method is quite reasonable, and the exact analysis of the skeleton structure ought to confirm the conjecture, but it did not. The author has made remarks such as "Common sense comes from an accumulation of knowledge"; perhaps this is a case in point.

14.9 Exercises

14.1 to 14.27 Analyze the rectangular building frames shown in Figs. 14.9.1 to 14.9.3 by the method as indicated in Table 14.9.1.

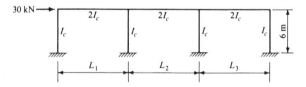

Figure 14.9.1 Exercises 14.1 to 14.9.

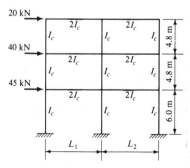

Figure 14.9.2 Exercises 14.10 to 14.15.

Table 14.9.1 Input data for Exercises 14.1 to 14.27

L_1, m	L_2, m	L_3, m	Portal method	Cantilever method	Morris method	
10	10	10	Exercise 14.1	Exercise 14.2	Exercise 14.3	
9	12	9	Exercise 14.4	Exercise 14.5	Exercise 14.6	Fig. 14.9.1
12	6	12	Exercise 14.7	Exercise 14.8	Exercise 14.9	
8	8	...	Exercise 14.10	Exercise 14.11	Exercise 14.12	Fig. 14.9.2
6	10	...	Exercise 14.13	Exercise 14.14	Exercise 14.15	
10	10	10	Exercise 14.16	Exercise 14.17	Exercise 14.18	
9	12	9	Exercise 14.19	Exercise 14.20	Exercise 14.21	Fig. 14.9.3
12	6	12	Exercise 14.22	Exercise 14.23	Exercise 14.24	
8	10	12	Exercise 14.25	Exercise 14.26	Exercise 14.27	

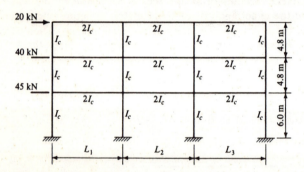

Figure 14.9.3 Exercises 14.16 to 14.27.

CHAPTER FIFTEEN
THE COLUMN-ANALOGY METHOD

15.1 General Introduction

Two important contributions on structural analysis made by Prof. Hardy Cross of the University of Illinois are the moment distribution and the column analogy. It is the object of this chapter to fully describe the column-analogy method† gradually from the simple to the more complex cases, rather than derive the general theorem and apply it to all cases. The general theorem is in fact shown in the last two sections of this chapter.

First, the column-analogy method is useful in determining the fixed-end moments, as well as the stiffness and carry-over factors, for a beam element with constant or variable moment of inertia. Second, it is useful in the complete analysis of symmetrical or unsymmetrical rigid frames, either with two fixed supports or with one closed cell. Although the examples and exercises used in this chapter may refer to a beam element with only a small number of abrupt changes in moment of inertia along its length, the advantage of the method lies in applying the same procedure to a beam element with many changes in moment of inertia for very small divisions along the length, such as the haunched beam element shown in Fig. 15.1.1a. Likewise, the same procedure used for the analysis of the quadrangular, or box, frames in Fig. 15.1.1b and c can be applied to the rigid-frame bridge or the closed-cell sanitary structure, for which the moment of inertia is taken constant only for each small division along the curved center line of the frame. The procedure

---

†Hardy Cross, "The Column Analogy," University of Illinois Engineering Experiment Station, Bulletin 215, 1930; also, Hardy Cross and Newlin D. Morgan, *Continuous Frames of Reinforced Concrete*, John Wiley & Sons, Inc., New York, 1932.

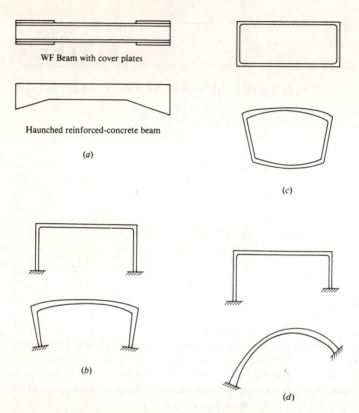

Figure 15.1.1 Structures suitable for analysis by column analogy.

used for the analysis of the unsymmetrical quadrangular frame of Fig. 15.1.1d can be applied to the unsymmetrical arch, except that again the unsymmetrical arch should be divided into 10 or more segments, each with a different moment of inertia.

The elastic-center method† has long been recognized as the convenient method of analyzing fixed arches, bridges in particular. It will be shown in Chap. 18 that the elastic-center method and the column-analogy method are in fact nearly identical; only toward the end of the computational procedure do the two methods diverge. Of course, the same free-body diagram should result as the solution of the problem.

---

†For instance, see C. B. McCullough and E. S. Thayer, *Elastic Arch Bridges*, John Wiley & Sons, Inc., New York, 1931.

15.2 Fixed-End Moments for a Beam Element with Constant Moment of Inertia

The use of the column-analogy method to determine the fixed-end moments for a beam element with constant moment of inertia due to loads on the element will now be developed.

Let it be required to find the fixed-end moments M_A and M_B acting on the ends of beam AB, which is fixed at both ends and subjected to the applied loading as shown in Fig. 15.2.1a. The moment diagram of the given fixed-end beam is the sum of the moment diagram due to the applied loading, if acting on a simple beam AB, Fig. 15.2.1b, and that due to the end moments, Fig. 15.2.1c. The compatibility conditions, from which the redundants M_A and M_B can be solved, are as follows:

1. Change of slope between A and $B = 0$; or sum of moment areas between A and $B = 0$ (because EI is constant); or area of moment diagram in Fig. 15.2.1b = area of moment diagram in Fig. 15.2.1c;
2. Deflection of B from tangent at $A = 0$; or sum of moments of moment areas between A and B about $B = 0$; or moment of moment diagram in Fig. 15.2.1b about B = moment of moment diagram in Fig. 15.2.1c about B.

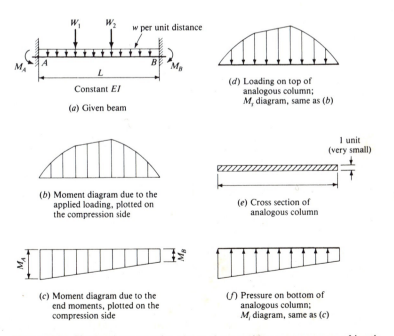

(a) Given beam — Constant EI

(b) Moment diagram due to the applied loading, plotted on the compression side

(c) Moment diagram due to the end moments, plotted on the compression side

(d) Loading on top of analogous column; M_s diagram, same as (b)

(e) Cross section of analogous column

(f) Pressure on bottom of analogous column; M_i diagram, same as (c)

Figure 15.2.1 Fixed-end moments for a beam element with constant moment of inertia.

530 INTERMEDIATE STRUCTURAL ANALYSIS

Now if an imaginary short column with a cross section as shown in Fig. 15.2.1e is so visualized that the loading on the top of the column is the moment diagram of Fig. 15.2.1b and the pressure acting on the bottom is the moment diagram of Fig. 15.2.1c, it is obvious that the column is in equilibrium by reason of the two conditions stated previously, which are (1) total load on the top is equal to the total pressure at the bottom, and (2) moment of load about B is equal to the moment of pressure about B. Thus if the loading diagram of Fig. 15.2.1d is known, the pressure diagram of Fig. 15.2.1f can be determined.

It is necessary to establish a sign convention to be followed in subsequent work. In reference to Fig. 15.2.2, this sign convention includes the following:

1. Loading on top of column is downward if M_s (statical moment, or moment due to the applied loading on simple beam AB as determined by the laws of statics) is positive, which means that it causes compression on the outside.
2. Upward pressure on bottom of column, M_i (indeterminate moment, or redundants to be obtained to satisfy the compatibility conditions) is positive.
3. Moment at any point of the given fixed-end beam is equal to $M = M_s - M_i$, which is positive if it causes compression on the outside.

If, in applying the redundant-force method of analysis, the cantilever beam AB fixed at A and free at B is chosen as the basic determinate beam, the moment diagram of the given beam will be the sum of the moment diagrams shown in Fig. 15.2.3b and c. However, the same two compatibility conditions—(1) change of slope between A and B = 0, and (2) deflection of B from tangent at A = 0—can be applied to the moment diagrams of Fig. 15.2.3b and c as previously; consequently, once the rules for sign convention are strictly followed, the column analogy as described in Fig. 15.2.3d to f is still valid.

Thus, to determine the fixed-end moments for a beam element with constant moment of inertia due to loads on the element, it is necessary only to determine the pressure, or M_i, at the two ends when the analogous column is loaded with the M_s diagram; M_A and M_B are then equal to $M = M_s - M_i$.

Example 15.2.1 Determine the fixed-end moments for the beam shown in Fig. 15.2.4a by the column-analogy method.

SOLUTION (a) M_s diagram from simple beam. As shown in Fig. 15.2.4b, the simple-beam moment diagram due to the uniform load is applied as a *downward* loading on the top of the

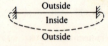

Figure 15.2.2 Sign convention in column analogy.

THE COLUMN-ANALOGY METHOD **531**

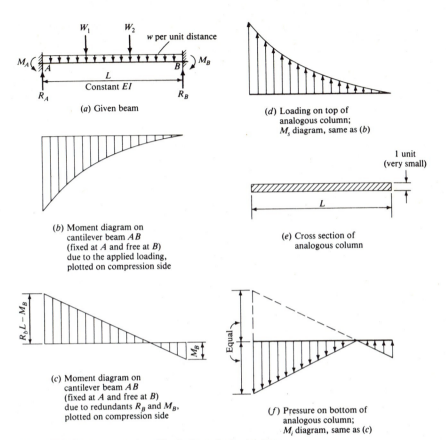

Figure 15.2.3 Alternate choice of basic determinate structure.

analogous column because this moment causes *compression* on the *outside* (or top) of the beam. The pressure along the base of the column is constant in this case and is equal to the total load divided by the area of the column.

$$M_i = \text{pressure} = \frac{wL^3/12}{L} = \frac{wL^2}{12}$$

Thus, at A or B,

$$M_s = 0 \qquad M_i = +\frac{wL^2}{12} \qquad M_A = M_B = M_s - M_i = 0 - \frac{wL^2}{12} = -\frac{wL^2}{12}$$

The negative sign for M_A (or M_B) indicates that M_A (or M_B) is in such a direction as to cause compression on the inside (or bottom) of the beam at A (or B).

(b) M_s *diagram from cantilever beam.* As shown in Fig. 15.2.4c, the moment diagram due to the uniform load acting on a cantilever beam fixed at A and free at B is applied as an *upward* loading on the top of the analogous column because this moment causes compression on the *inside* (or bottom) of the beam. The pressures at A and B can be found by

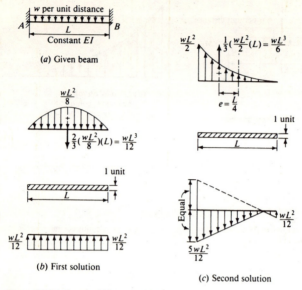

Figure 15.2.4 Beam of Example 15.2.1.

the formula

$$M_i = \text{pressure} = \frac{P}{A} \pm \frac{Mc}{I}$$

For point A,

$$M_s = -\frac{wL^2}{2}$$

$$M_i = -\frac{wL^3/6}{(1)(L)} - \frac{(wL^3/6)(L/4)}{(1)(L)^2/6} = -\frac{wL^2}{6} - \frac{wL^2}{4} = -\frac{5wL^2}{12}$$

$$M_A = M_s - M_i = -\frac{wL^2}{2} - \left(-\frac{5wL^2}{12}\right) = -\frac{wL^2}{12}$$

For point B,

$$M_s = 0$$

$$M_i = -\frac{wL^3/6}{(1)(L)} + \frac{(wL^3/6)(L/4)}{(1)(L)^2/6} = -\frac{wL^2}{6} + \frac{wL^2}{4} = +\frac{wL^2}{12}$$

$$M_B = M_s - M_i = 0 - \left(+\frac{wL^2}{12}\right) = -\frac{wL^2}{12}$$

Note that in determining the pressure M_i at A or B, the upward load when acting at the centroid of the column causes negative pressure, while the clockwise overturning moment causes negative pressure at A and positive pressure at B.

Example 15.2.2 Determine the fixed-end moments for the beam shown in Fig. 15.2.5a by the column-analogy method.

THE COLUMN-ANALOGY METHOD

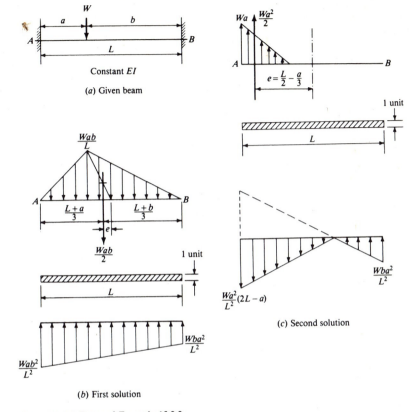

Figure 15.2.5 Beam of Example 15.2.2.

SOLUTION (a) M_s *diagram from simple beam.* As shown in Fig. 15.2.5b,

$$\text{Total downward load on column} = \text{simple-beam moment diagram} = \frac{Wab}{2}$$

$$\text{Eccentricity } e = \frac{1}{3}\left(\frac{L}{2} - a\right) \qquad \frac{P}{A} = \frac{Wab/2}{(1)(L)} = \frac{Wab}{2L}$$

$$\frac{Mc}{I} = \frac{(Wab/2)[\frac{1}{3}(L/2 - a)]}{(1)(L)^2/6} = \frac{Wab(L/2 - a)}{L^2}$$

For point A,

$$M_s = 0$$

$$M_i = \frac{P}{A} + \frac{Mc}{I} = \frac{Wab}{2L} + \frac{Wab(L/2 - a)}{L^2} = \frac{Wab}{2L^2}(L + L - 2a) = +\frac{Wab^2}{L^2}$$

$$M_A = M_s - M_i = 0 - \left(+\frac{Wab^2}{L^2}\right) = -\frac{Wab^2}{L^2}$$

For point B,

$$M_s = 0$$

$$M_i = \frac{P}{A} - \frac{Mc}{I} = \frac{Wab}{2L} - \frac{Wab(L/2-a)}{L^2} = \frac{Wab}{2L^2}(L - L + 2a) = +\frac{Wba^2}{L^2}$$

$$M_B = M_s - M_i = 0 - \left(+\frac{Wba^2}{L^2}\right) = -\frac{Wba^2}{L^2}$$

The negative sign for M_A (or M_B) indicates that M_A (or M_B) is in such a direction as to cause compression on the inside (or bottom) of the beam at A (or B).

(b) M_s *diagram from cantilever beam.* As shown in Fig. 15.2.5c,

$$\text{Total upward load on column} = \text{cantilever-beam moment diagram} = \frac{Wa^2}{2}$$

$$\text{Eccentricity } e = \frac{L}{2} - \frac{a}{3} \qquad \frac{P}{A} = -\frac{Wa^2/2}{(1)(L)} = -\frac{Wa^2}{2L}$$

$$\frac{Mc}{I} = \frac{(Wa^2/2L)(L/2-a/3)}{(1)(L)^2/6} = \frac{Wa^2}{2L^2}(3L - 2a)$$

For point A,

$$M_s = -Wa$$

$$M_i = \frac{P}{A} - \frac{Mc}{I} = -\frac{Wa^2}{2L} - \frac{Wa^2}{2L^2}(3L - 2a) = -\frac{Wa^2}{L^2}(2L - a)$$

$$M_A = M_s - M_i = -Wa - \left[-\frac{Wa^2}{L^2}(2L - a)\right] = -\frac{Wa^2}{L^2}(L^2 - 2La + a^2) = -\frac{Wab^2}{L^2}$$

For point B,

$$M_s = 0$$

$$M_i = \frac{P}{A} + \frac{Mc}{I} = -\frac{Wa^2}{2L} + \frac{Wa^2}{2L^2}(3L - 2a) = +\frac{Wa^2}{2L^2}(-L + 3L - 2a) = +\frac{Wba^2}{L^2}$$

$$M_B = M_s - M_i = 0 - \left(+\frac{Wba^2}{L^2}\right) = -\frac{Wba^2}{L^2}$$

15.3 Stiffness and Carry-Over Factors for a Beam Element with Constant Moment of Inertia

The stiffness factor S_A at A of a beam element AB has been so defined that if a clockwise moment M_A equal to $S_A\phi_A$ is applied at A, the clockwise rotation of the tangent at A is ϕ_A when the end B is fixed. The carry-over factor (COF) from A to B is the ratio of the moment at the fixed end B to the moment applied at A under the above conditions. This beam element with constant moment of inertia is shown in Fig. 15.3.1a to c, where the M/EI diagrams make up the loads on the conjugate beam and ϕ_A and zero are the reactions. Now if the reactions to the conjugate beam are considered as the loads (actually there is only one load ϕ_A) on the analogous column and the M_A/EI and M_B/EI diagrams are considered as positive and negative pressures on the bottom of the column, the column is obviously still in equilibrium. In order that the pressures at A and B may be positive M_A and negative M_B directly, as shown in Fig. 15.3.1d to f, it is convenient to call the width of the

THE COLUMN-ANALOGY METHOD **535**

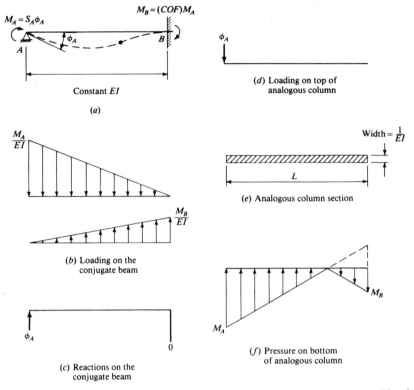

Figure 15.3.1 Stiffness and carry-over factors for a beam element with constant moment of inertia.

analogous column $1/EI$ instead of unity. Thus M_A and M_B can be found directly as the positive and negative pressures at A and B. So,

Area of analogous column section $= \left(\dfrac{1}{EI}\right)(L) = \dfrac{L}{EI}$

Downward load on column $= \phi_A$ at A

Eccentricity $e = L/2$

$$M_A = \dfrac{P}{A} + \dfrac{Mc}{I} = \dfrac{\phi_A}{L/EI} + \dfrac{\phi_A(L/2)}{(1/EI)(L^2/6)} = \dfrac{4EI}{L}\phi_A = S_A \phi_A$$

Stiffness factor S_A at $A = \dfrac{4EI}{L}$

$$M_B = \dfrac{P}{A} - \dfrac{Mc}{I} = \dfrac{\phi_A}{L/EI} - \dfrac{\phi_A(L/2)}{(1/EI)(L^2/6)} = -\dfrac{2EI}{L}\phi_A = -\tfrac{1}{2}M_A$$

The negative sign for M_B means that M_B and M_A have opposite signs according to the sign convention used in the column-analogy method; or M_B and

M_A should have the same sign according to the sign convention used in the slope-deflection, moment-distribution, and matrix displacement methods. Thus,

$$\text{COF} = \frac{M_B}{M_A} = +\tfrac{1}{2}$$

15.4 Fixed-End Moments for a Beam Element with Variable Moment of Inertia

The column-analogy method can be used, quite conveniently, to determine the fixed-end moments for a beam element with variable moment of inertia due to loads applied on the element.

Let it be required to find the fixed-end moments M_A and M_B acting on the ends of the beam element AB with variable moment of inertia due to the applied loading as shown in Fig. 15.4.1a. When the redundant-force method is

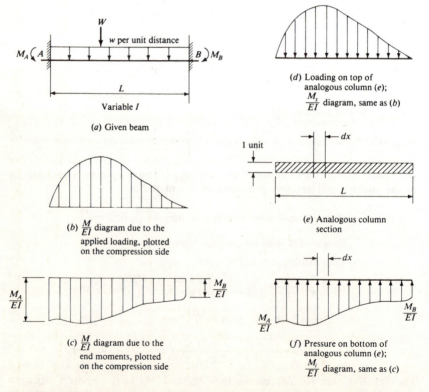

Figure 15.4.1 Fixed-end moments for a beam element with variable moment of inertia.

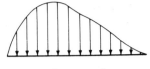

(g) Loading on top of analogous column (h); $\frac{M_s}{EI}$ diagram, same as (b)

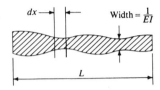

(h) Analogous column section

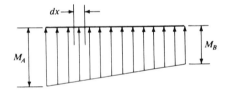

(i) Pressure on bottom of analogous column (h); M_i diagram

Figure 15.4.1 (*continued*).

used and if a simple beam AB is chosen as the basic determinate beam, the compatibility conditions from which M_A and M_B can be determined are as follows:

1. Change of slope between A and $B = 0$, or area of Fig. 15.4.1b = area of Fig. 15.4.1c;
2. Deflection of B from tangent at $A = 0$, or moment of area of Fig. 15.4.1b about B = moment of area of Fig. 15.4.1c about B.

If an imaginary short column with a cross section as shown in Fig. 15.4.1e is subjected to the loading on top and the pressure on bottom as shown in Fig. 15.4.1d and f, respectively, it is obvious that this column is in equilibrium by reason of the two compatibility conditions stated previously. It will be shown that the "loading-section-pressure" diagrams of Fig. 15.4.1d to f can be replaced by those of Fig. 15.4.1g to i. The loading diagrams of Fig. 15.4.1d and g being the same, it remains necessary to show that the upward reactions over any division dx at the base of the respective columns are also the same, which is obviously true because

$$\left[\left(\frac{M_i}{EI}\right)(dx)(1 \text{ unit})\right]_{\text{in Fig. 15.4.1}e \text{ and } f} = \left[(M_i)(dx)\left(\frac{1}{EI}\right)\right]_{\text{in Fig. 15.4.1}h \text{ and } i}$$

The loading-section-pressure system of Fig. 15.4.1g to i will be used to find the fixed-end moments M_A and M_B.

The same sign convention as stated in Sec. 15.2 should again be strictly followed. Furthermore, the discussion in that section regarding the use of the moment diagram due to action of the applied loads on a cantilever beam fixed at A and free at B as the M_s diagram applies equally well in the present problem of a fixed-end beam with variable moment of inertia.

Example 15.4.1 Determine the fixed-end moments for the beam shown in Fig. 15.4.2a by the column-analogy method.

SOLUTION (a) *Properties of the analogous column section.* The length of the analogous column section is the same as that of the span of the given beam, and the width is equal to $1/EI$. Since values of EI change along the span, the width of the analogous column section

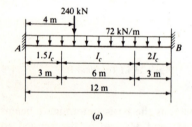

(a)

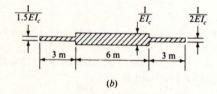

(b)

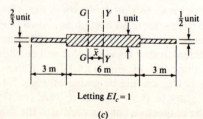

Letting $EI_c = 1$

(c)

Figure 15.4.2 Analogous column section in Example 15.4.1.

THE COLUMN-ANALOGY METHOD 539

changes accordingly, as shown in Fig. 15.4.2b. It is convenient to let $EI_c = 1$, so that the column section shown in Fig. 15.4.2c can be used.

Area of column section = $\tfrac{2}{3}(3) + 1(6) + \tfrac{1}{2}(3) = 9.5$

$$\bar{x} = \frac{2(-4.5) + 6(0) + 1.5(+4.5)}{9.5} = -0.2368$$

$$I_G = I_y - A\bar{x}^2$$

$$= \frac{2(3)^2}{12} + 2(4.5)^2 + \frac{6(6)^2}{12} + \frac{1.5(3)^2}{12} + 1.5(4.5)^2 - 9.5(0.2368)^2$$

$$= 90.967$$

or

$$I_G = \frac{2(3)^2}{12} + 2(4.5 - 0.2368)^2 + \frac{6(6)^2}{12} + 6(0.2368)^2 + \frac{1.5(3)^2}{12} + 1.5(4.5 + 0.2368)^2$$

$$= 90.967$$

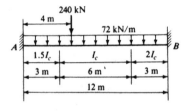

(a) Basic determinate beam

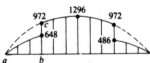

(b) $\dfrac{M_s}{EI}$ diagrams, plotted on the compression side ($EI_c = 1$)

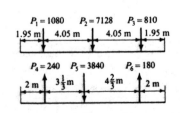

(c) Loads on top of analogous column

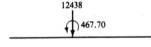

(d) Loads on top of analogous column

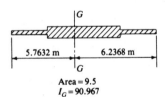

Area = 9.5
$I_G = 90.967$

(e) Properties of the column section

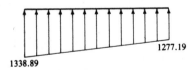

(f) Pressure on bottom of analogous column

Figure 15.4.3 Simple beam as basic determinate beam in Example 15.4.1.

540 INTERMEDIATE STRUCTURAL ANALYSIS

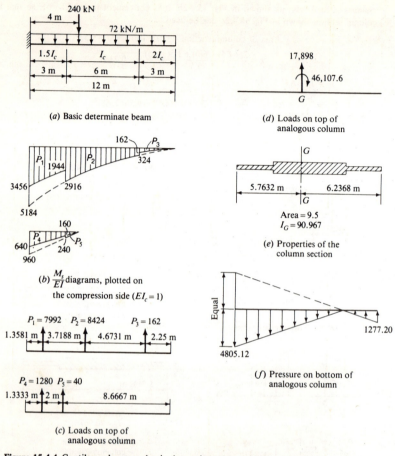

Figure 15.4.4 Cantilever beam as basic determinate beam in Example 15.4.1.

The properties of the analogous column section determined above are summarized in Figs. 15.4.3e and 15.4.4e for convenient use.

(b) M_s *diagram from simple beam.* The simple beam AB is chosen as the basic determinate beam. The M_s/EI diagrams are shown by the solid lines in Fig. 15.4.3b. Note again that EI_c is taken to be equal to unity. The loads on the top of the analogous column are the areas of the M_s/EI diagrams shown in Fig. 15.4.3b.

Let P_1, P_2, and P_3 be the areas of the M_s/EI diagrams due to the uniform load over the 3-m, 6-m, and 3-m sections, respectively.

$$\text{Area of } abc = \int_0^3 (432x - 36x^2)\, dx = 1620$$

$$P_1 = \tfrac{2}{3}(abc) = 1080 \qquad P_3 = \tfrac{1}{2}(abc) = 810$$

THE COLUMN-ANALOGY METHOD 541

Distance between centroid of area abc and the left support $= \dfrac{\int_0^3 (432x - 36x^2)(x)\,dx}{1620} = \dfrac{3159}{1620} = 1.95$

$P_2 = \tfrac{2}{3}(1296)(12) - 2(abc) = 10{,}368 - 2(1620) = 7128$

Let the M_s/EI diagram due to the concentrated load be the area P_5 of the triangle with base equal to 12 m and altitude equal to 640 minus the two small triangles indicated by P_4 and P_6 in Fig. 15.4.3b.

$$P_5 = \tfrac{1}{2}(12)(640) = 3840$$

Location of P_5 from left support $= \dfrac{12+4}{3} = 5\tfrac{1}{3}$

$P_4 = \tfrac{1}{2}(3)(160) = 240 \qquad P_6 = \tfrac{1}{2}(3)(120) = 180$

As shown in Fig. 15.4.3c and d,

Total load on the column $= P_1 + P_2 + P_3 - P_4 + P_5 - P_6$

$= 1080 + 7128 + 810 - 240 + 3840 - 180$

$= 12{,}438$ downward

Total moment of the loads about centroid G $= 1080(5.7632 - 1.95) - 7128(0.2368) - 810(6.2368 - 1.95)$

$- 240(3.7632) + 3840(5.7632 - 5.3333) + 180(4.2368)$

$= 467.70$ counterclockwise

For point A,

$$M_s = 0$$

$$M_i = +\dfrac{12{,}438}{9.5} + \dfrac{467.70(5.7632)}{90.967} = +1309.26 + 29.63 = +1338.89$$

$M_A = M_s - M_i = 0 - (+1338.89) = -1338.89$ kN·m

For point B,

$$M_s = 0$$

$$M_i = +\dfrac{12{,}438}{9.5} - \dfrac{467.70(6.2368)}{90.967} = +1309.26 - 32.07 = +1277.19$$

$M_B = M_s - M_i = 0 - (+1277.19) = -1277.19$ kN·m

The above results check completely with those in Example 7.10.1.

(c) M_s *diagram from cantilever beam.* The cantilever beam AB fixed at A and free at B is chosen as the basic determinate beam. The M_s/EI diagrams are shown by the solid lines in Fig. 15.4.4b. Let P_1, P_2, P_3, P_4, and P_5 be the loads on the analogous column.

$$P_1 = \tfrac{2}{3}\int_9^{12} 36x^2 \, dx = 7992$$

Distance of P_1 from $B = \dfrac{\tfrac{2}{3}\int_9^{12}(3x^2)(x)\,dx}{7992} = 10.6419$

$$P_2 = \int_3^9 36x^2 = 8424$$

Distance of P_2 from $B = \dfrac{\int_3^9 (3x^2)(x)\,dx}{8424} = 6.9231$

$P_3 = \tfrac{1}{3}(162)(3) = 162 \qquad$ Distance of P_3 from $B = \tfrac{3}{4}(3) = 2.25$

$P_4 = \tfrac{1}{2}(640)(4) = 1280 \qquad P_5 = \tfrac{1}{2}(80)(1) = 40$

As shown in Fig. 15.4.4c and d,

$$\text{Total load on the column} = P_1 + P_2 + P_3 + P_4 + P_5$$
$$= 7992 + 8424 + 162 + 1280 + 40$$
$$= 17{,}898 \text{ upward}$$

$$\text{Total moment of the loads about centroid } G = 7992(4.4051) + 8424(0.6862) - 162(3.9868) + 1280(4.4298)$$
$$+ 40(2.4298)$$
$$= 46{,}107.6 \text{ clockwise}$$

For point A,

$$M_s = -6144$$

$$M_i = -\frac{17{,}898}{9.5} + \frac{46{,}107.6(6.2368)}{90.967} = -1884 - 2921.12 = -4805.12$$

$$M_A = M_s - M_i = -6144 - (-4805.12) = -1338.88 \text{ kN·m}$$

For point B,

$$M_s = 0$$

$$M_i = -\frac{17{,}898}{9.5} - \frac{46{,}107.6(6.2368)}{90.967} = -1884 + 3161.20 = +1277.20$$

$$M_B = M_s - M_i = 0 - (+1277.20) = -1277.20 \text{ kN·m}$$

(d) *Discussion.* It may be noted that there are other choices for the basic determinate beam, such as a cantilever beam fixed at B and free at A, or even more conveniently, a fixed-end beam with two internal hinges inserted where the moments of inertia change abruptly.

15.5 Stiffness and Carry-Over Factors for a Beam Element with Variable Moment of Inertia

The stiffness and carry-over factors for a beam element with variable moment of inertia can be determined in the same manner as in the case of constant moment of inertia, except that the width of the analogous column section is variable along the length instead of being equal to the constant of $1/EI$. In this case, it is better not to let EI_c equal 1, but to keep it within the properties of the analogous column section, as will be shown in the following example.

Example 15.5.1 Determine the stiffness factors at A and B and the carry-over factors from A to B and from B to A for the beam element with variable moment of inertia as shown in Fig. 15.5.1a.

SOLUTION (a) *Properties of the analogous column section.* From part (a) of Example 15.4.1,

$$\text{Area of analogous column section} = \frac{9.5}{EI_c}$$

Distance of centroid from $A = 5.7632$

Distance of centroid from $B = 6.2368$

$$\text{Moment of inertia about centroid} = \frac{90.967}{EI_c}$$

THE COLUMN-ANALOGY METHOD

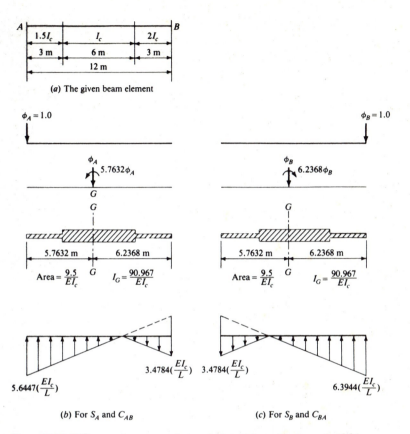

Figure 15.5.1 Stiffness and carry-over factors for beam element in Example 15.5.1.

(b) *Stiffness factor S_A and carry-over factor C_{AB} from A to B.* In Fig. 15.5.1b, a load of ϕ_A is applied at the left edge of the analogous column.

$$M_A = S_A \phi_A = \text{pressure at } A$$
$$= \frac{\phi_A}{9.5/EI_c} + \frac{(5.7632\phi_A)(5.7632)}{90.967/EI_c}$$
$$= \left[\frac{12}{9.5} + \frac{12(5.7632)^2}{90.967}\right]\frac{EI_c}{L}\phi_A = (1.2632 + 4.3815)\frac{EI_c}{L}\phi_A$$
$$= 5.6447\frac{EI_c}{L}\phi_A$$

Thus,

$$S_A = 5.6447\frac{EI_c}{L}$$

M_B = pressure at B

$$= \frac{\phi_A}{9.5/EI_c} - \frac{(5.7632\phi_A)(6.2368)}{90.967/EI_c}$$

$$= \left[\frac{12}{9.5} - \frac{12(5.7632)(6.2368)}{90.967}\right]\frac{EI_c}{L}\phi_A$$

$$= (1.2632 - 4.7616)\frac{EI_c}{L}\phi_A$$

$$= -3.4784\frac{EI_c}{L}$$

Taking clockwise moments acting on the beam element as positive for both ends,

$$C_{AB} = +\frac{3.4784}{5.6447} = +0.61622$$

(c) *Stiffness factor S_B and carry-over factor C_{BA} from B to A.* In Fig. 15.5.1c, a load of ϕ_B is applied at the right edge of the analogous column.

$M_B = S_B \phi_B$ = pressure at B

$$= \frac{\phi_B}{9.5/EI_c} + \frac{(6.2368\phi_B)(6.2368)}{90.967/EI_c}$$

$$= \left[\frac{12}{9.5} + \frac{12(6.2368)^2}{90.967}\right]\frac{EI_c}{L}\phi_B = (1.2632 + 5.1312)\frac{EI_c}{L}\phi_B$$

$$= 6.3944\frac{EI_c}{L}\phi_B$$

Thus,

$$S_B = 6.3944\frac{EI_c}{L}$$

M_A = pressure at A

$$= \frac{\phi_B}{9.5/EI_c} - \frac{(6.2368\phi_B)(5.7632)}{90.967/EI_c}$$

$$= \left[\frac{12}{9.5} - \frac{12(6.2368)(5.7632)}{90.967}\right]\frac{EI_c}{L}\phi_B$$

$$= -3.4784\frac{EI_c}{L}\phi_B$$

Again using the sign convention as in the moment-distribution method,

$$C_{BA} = +\frac{3.4784}{6.3944} = +0.54398$$

15.6 Moments in Quadrangular Frames with One Axis of Symmetry

The column-analogy method can be used to analyze quadrangular frames with one axis of symmetry when such frames have two fixed supports and are subjected to some applied loading. A typical quadrangular frame fulfilling these requirements is shown in Fig. 15.6.1. Note that the axis of symmetry refers only to the properties of the frame and not to the applied loads. The analysis of this frame by the redundant-force method requires the solution of three simultaneous equations. By the column-analogy method, however, the moment at any point on the frame can be obtained by a *direct* procedure.

THE COLUMN-ANALOGY METHOD

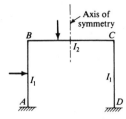

Figure 15.6.1 A quadrangular frame with one axis of symmetry.

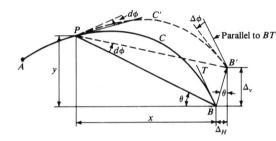

Figure 15.6.2 Generalized displacements at the ends of a curved member.

A simple geometric problem will be discussed first. If the curved line $APCB$ in Fig. 15.6.2 is "opened up" at the point P by a small angle $d\phi$, it is apparent that the area PCB will rotate about point P through an angle $d\phi$ in the counterclockwise direction, thus taking the new position $P'C'B'$. If the angle $d\phi$ is very small, then by the first-order assumption BB' can be assumed to be perpendicular to PB. Let Δ_ϕ, Δ_H, and Δ_V be the rotation of the tangent and the horizontal and vertical deflections at B. It can be seen that

$$\Delta_\phi \text{ at } B = d\phi \text{ counterclockwise} \tag{15.6.1a}$$

$$\Delta_H \text{ at } B = BB' \sin\theta = (PB\, d\phi)\sin\theta = (PB \sin\theta)\, d\phi$$
$$= (+y)\, d\phi = +y\, d\phi \quad \text{to the right} \tag{15.6.1b}$$

$$\Delta_V \text{ at } B = BB' \cos\theta = (PB\, d\phi)\cos\theta = (PB \cos\theta)\, d\phi$$
$$= (-x)\, d\phi = -x\, d\phi \quad \text{(positive for upward)} \tag{15.6.1c}$$

Note that x and y are the coordinates of P in reference to B as origin. Thus in Fig. 15.6.2, x is negative and y is positive; so $-x$, which is then a positive quantity, should be substituted for $PB \cos\theta$ in Eq. (15.6.1c).

Now let the curved line in Fig. 15.6.2 be actually a curved member which is rigid (cannot be deformed) except for an infinitesimal segment ds at P on which there is a moment M acting. M is considered positive if it causes compression on the outside (or the convex side). The curved member, being compressed on the outside at P over the elastic segment ds, will be "opened up." The rotation of the tangent at P, $d\phi$, is equal to $(M\, ds)/EI$, according to Eq. (2.12.1) derived in Chap. 2. By substituting $d\phi = (M\, ds)/EI$ into Eqs.

(15.6.1a to c), the three generalized displacements at B due to the action of M over ds at P are

$$\Delta_\phi \text{ at } B = +\frac{M\,ds}{EI} \quad \text{(positive means counterclockwise)} \quad (15.6.2a)$$

$$\Delta_H \text{ at } B = +\frac{My\,ds}{EI} \quad \text{(positive means to the right)} \quad (15.6.2b)$$

$$\Delta_V \text{ at } B = -\frac{Mx\,ds}{EI} \quad \text{(positive means upward)} \quad (15.6.2c)$$

in which x and y are the coordinates of P in reference to B as origin. Equations (15.6.2a to c), which are derived for the general case of a curved member, can, of course, be applied to a quadrangular member such as $ABCD$ in Fig. 15.6.3a.

The analysis of the quadrangular frame shown in Fig. 15.6.3a by the column-analogy method will now be considered. As discussed previously in the redundant-force method of analysis, the given frame of Fig. 15.6.3a may be considered equivalent to the sum of the two cantilever frames shown in Fig. 15.6.3b and c. The moment at any point P in the frame of Fig. 15.6.3a is equal to the sum of the moments at P in the frames of Fig. 15.6.3b and c. Let the moment at P in the frame of Fig. 15.6.3b be M_s, or the statical moment in the determinate cantilever structure. Note again that all moments are considered positive if they cause compression on the outside. Thus, for the frame of Fig. 15.6.3a and at the point P,

$$M = M_s + (M_D + H_D y - V_D x) \quad (15.6.3)$$

in which x and y are the coordinates of P in reference to D as origin. Note that, in this present case, x in Eq. (15.6.3) is a negative quantity for all positions of P on the frame. Substituting Eq. (15.6.3) into Eqs. (15.6.2a to c) and integrating throughout the entire frame $ABCD$, the total generalized

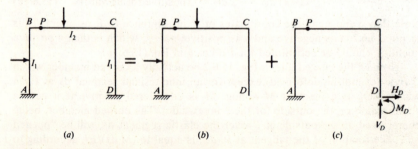

Figure 15.6.3 Redundant-force method of analysis.

displacements at point D are

$$\Delta_\phi \text{ at } D = +\int \frac{(M_s + M_D + H_D y - V_D x)\, ds}{EI} \quad (15.6.4a)$$
(positive means counterclockwise)

$$\Delta_H \text{ at } D = +\int \frac{(M_s + M_D + H_D y - V_D x) y\, ds}{EI} \quad (15.6.4b)$$
(positive means to the right)

$$\Delta_V \text{ at } D = -\int \frac{(M_s + M_D + H_D y - V_D x) x\, ds}{EI} \quad (15.6.4c)$$
(positive means upward)

in which x and y are the coordinates in reference to D as origin. In fact, the three compatibility conditions by which the redundants M_D, H_D, and V_D can be solved are those coming from setting Eqs. (15.6.4a to c) to zero.

The elastic center, point O in Fig. 15.6.4, is defined as the centroid of the analogous column section, which has the same shape as the given structure but with a thickness at any point equal to $1/EI$. However, the so-called thickness here is used only as the means of indicating the "density" of the "line area"; mathematically the thickness of a line is zero.

Let A, I_x, I_y, and I_{xy} be the area, the two moments of inertia, and the product of inertia of the analogous column section about the x and y axes through the centroid. Then, by definition,

$$A = \int \frac{ds}{EI} \quad A\bar{x} = \int \frac{x\, ds}{EI} = 0 \quad A\bar{y} = \int \frac{y\, ds}{EI} = 0$$

$$I_{xy} = \int \frac{xy\, ds}{EI} = 0 \quad \begin{array}{l}\text{(true only if } y \text{ axis is an}\\ \text{axis of symmetry)}\end{array} \quad (15.6.5)$$

$$I_x = \int \frac{y^2\, ds}{EI} \quad I_y = \int \frac{x^2\, ds}{EI}$$

Now if to the given structure of Fig. 15.6.5a a rigid arm OD joining the elastic center O and the fixed support D is attached, the equivalent structure of Fig. 15.6.5b is obtained. Since arm OD is rigid (for which $EI = $ infinity) and

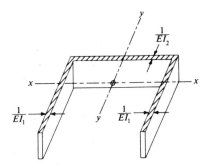

Figure 15.6.4 The analogous column section.

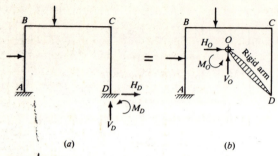

(a)　　　　　　　　　　　　　　(b)

Figure 15.6.5 The equivalent structure with rigid arm.

cannot be deformed, the three generalized displacements at point O are also equal to zero. Let M_O, H_O, and V_O be the three unknown redundants to act at the elastic center. If the letter O is substituted for the letter D in Eqs. (15.6.4a to c) and then the resulting expressions are set equal to zero,

$$\Delta_\phi \text{ at } O = +\int \frac{M_s\,ds}{EI} + M_O\int \frac{ds}{EI} + H_O\int \frac{y\,ds}{EI} - V_O\int \frac{x\,ds}{EI} = 0 \quad (15.6.6a)$$

$$\Delta_H \text{ at } O = +\int \frac{M_s y\,ds}{EI} + M_O\int \frac{y\,ds}{EI} + H_O\int \frac{y^2\,ds}{EI} - V_O\int \frac{xy\,ds}{EI} = 0 \quad (15.6.6b)$$

$$\Delta_V \text{ at } O = -\int \frac{M_s x\,ds}{EI} - M_O\int \frac{x\,ds}{EI} - H_O\int \frac{xy\,ds}{EI} + V_O\int \frac{x^2\,ds}{EI} = 0 \quad (15.6.6c)$$

in which x and y are the coordinates in reference to the elastic center O as origin and the integrals should be through the frame $ABCD$ as well as the rigid arm DO. However, EI of the rigid arm being infinity, all integrals in Eqs. (15.6.6a to c) need in effect only to go over the frame $ABCD$.

The integrals following the redundants M_O, H_O, and V_O in Eqs. (15.6.6a to c) are identical to those defined as the properties of the analogous column section in Eq. (15.6.5). Thus Eqs. (15.6.6a to c) become

$$M_O = -\frac{\int \frac{M_s\,ds}{EI}}{A} \quad (15.6.7a)$$

$$H_O = -\frac{\int \frac{M_s y\,ds}{EI}}{I_x} \quad \text{(true only if } y \text{ axis is an axis of symmetry)} \quad (15.6.7b)$$

$$V_O = +\frac{\int \frac{M_s x\,ds}{EI}}{I_y} \quad \text{(true only if } y \text{ axis is an axis of symmetry)} \quad (15.6.7c)$$

Now examine the numerators in Eqs. (15.6.7a to c). The numerator in Eq. (15.6.7a) is the M_s/EI diagram over the entire structure; it can be regarded as the load on top of the analogous column. The numerators in Eqs. (15.6.7b and c) are the moments of the M_s/EI diagram about the x and y axes, respectively. Let, then, P be the load on the analogous column, and M_x and M_y be the moments of this load about the x and y axes. Equations (15.6.7a to c) become

$$M_O = -\frac{P}{A} \tag{15.6.8a}$$

$$H_O = -\frac{M_x}{I_x} \quad \text{(true only if y axis is an axis of symmetry)} \tag{15.6.8b}$$

$$V_O = +\frac{M_y}{I_y} \quad \text{(true only if y axis is an axis of symmetry)} \tag{15.6.8c}$$

Applying Eq. (15.6.3) shows that the moment M at any point of the equivalent structure in Fig. 15.6.5b, or of the given structure in Fig. 15.6.5a, is equal to

$$M = M_s + (M_O + H_O y - V_O x) \tag{15.6.9}$$

Substituting Eqs. (15.6.8a to c) into Eq. (15.6.9),

$$M = M_s + \left(-\frac{P}{A} - \frac{M_x y}{I_x} - \frac{M_y x}{I_y}\right) = M_s - M_i \tag{15.6.10a}$$

in which

M_i = pressure at bottom of analogous column

$$= \frac{P}{A} + \frac{M_x y}{I_x} + \frac{M_y x}{I_y} \tag{15.6.10b}$$

The column analogy is clearly shown by Eqs. (15.6.10a and b). By this analogy, the moment at as many chosen points on the structure as desired may be obtained directly, once the properties of and the loading on the analogous column have been determined.

It has been pointed out in Chap. 4 that in the application of the redundant-force method of analysis there are in general several ways of choosing the basic determinate structure. Another way of deriving a basic determinate structure from the quadrangular frame with two fixed supports is shown in Fig. 15.6.6a to c; in such a case the redundants are M_A, M_D, and H_D and the compatibility conditions are $\theta_A = 0$, $\theta_D = 0$, and Δ_H at $D = 0$. When a rigid arm joining the elastic center and the fixed support D is attached, it is seen that Fig. 15.6.6d is equivalent to Fig. 15.6.6a, Fig. 15.6.6e to Fig. 15.6.6b, and Fig. 15.6.6f to Fig. 15.6.6c because there is certainly a set of values for M_O, H_O, and V_O which can be found such that the moment at any point in the frame of Fig. 15.6.6f is the same as that in Fig. 15.6.6c. Consequently the moment M at any point in the frame of Fig. 15.6.6d is the sum of the M_s in Fig. 15.6.6e and

550 INTERMEDIATE STRUCTURAL ANALYSIS

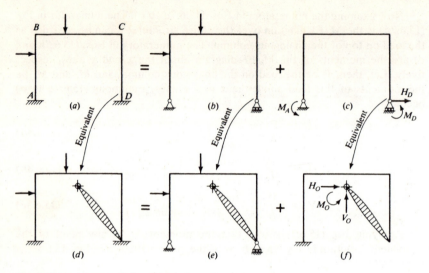

Figure 15.6.6 Alternate choice of basic determinate structure.

the $(M_O + H_O y - V_O x)$ in Fig. 15.6.6f, and thus the column-analogy method still applies.

To summarize, the column-analogy method of analyzing quadrangular frames with one axis of symmetry and two fixed supports involves the following steps:

1. The loading on top of the analogous column is equal to the M_s/EI diagram, where M_s is the statical moment in *any* basic determinate structure derived from the given frame. The loading is in the downward direction if M_s is positive, which means that it causes compression on the outside.
2. The cross section of the analogous column consists of a line area, the shape of which is the same as that of the given frame and the density at any point is equal to $1/EI$.
3. The moment at any point of the given frame is equal to $M = M_s - M_i$, where M_i is the pressure on the bottom of the analogous column at the point under consideration.

Example 15.6.1 Analyze the quadrangular frame shown in Fig. 15.6.7a by the column-analogy method.

SOLUTION (*a*) *Properties of the analogous column section.* The analogous column section by considering EI_c to be unity is shown in Fig. 15.6.7b. Note that the moment of inertia of a line area about its own centroidal axis in the direction of the "line" is equal to zero, because

THE COLUMN-ANALOGY METHOD 551

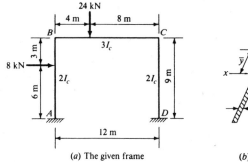
(a) The given frame

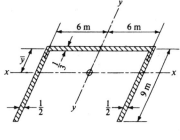
(b) The analogous column section

Figure 15.6.7 Quadrangular frame of Example 15.6.1.

theoretically a line has no lateral dimension.

$$A = \tfrac{1}{3}(12) + \tfrac{1}{2}(9) + \tfrac{1}{2}(9) = 13$$

$$\bar{y} = \frac{4(0) + 2(4.5)(4.5)}{13} = 3.115$$

$$I_x = 4(3.115)^2 + 2\left[\frac{4.5(9)^2}{12} + 4.5(1.385)^2\right] = 116.83 \quad \text{(Use)}$$

$$I_x = I_{BC} - A\bar{y}^2 = 0 + 2(4.5)\frac{(9)^2}{3} - 13(3.115)^2 = 116.86 \quad \text{(Check)}$$

$$I_y = \frac{4(12)^2}{12} + 2(4.5)(6)^2 = 372$$

(b) M_s *diagrams from three different basic determinate structures.* For purpose of illustration, the complete solution, using three different basic determinate structures as shown in Fig. 15.6.8, will be presented.

For Case 1,

$$P = 64 + 432 + 72 = 568 \text{ upward}$$

$$M_x = 432(1.385) + 72(3.885) - 64(3.115)$$

$$= 687.7 \text{ clockwise} \quad \text{(viewed from right)}$$

$$M_y = 64(4.667) + (432 + 72)(6)$$

$$= 3322.7 \text{ clockwise} \quad \text{(viewed from front)}$$

The values of M at points A, B, C, and D are computed in Table 15.6.1. In this table, the signs for the values in the P/A, M_xy/I_x, and M_yx/I_y columns are determined by inspection. For instance, the total load P is upward; thus it causes tension, which is considered negative, at all points in the analogous column. M_x is observed to be clockwise when viewed from the right side; thus it causes negative pressures at A and D and positive pressures at B and C. M_y is clockwise when viewed from front; thus it causes negative pressures at A and B and positive pressures at C and D.

For Case 2,

$$P = 18 + 96 + 256 + 486 + 324 = 1180 \text{ upward}$$

$$M_x = 18(2.115) + (96 + 256)(3.115) + 486(0.115) - 324(2.885)$$

$$= 255.70 \text{ counterclockwise} \quad \text{(viewed from right)}$$

552 INTERMEDIATE STRUCTURAL ANALYSIS

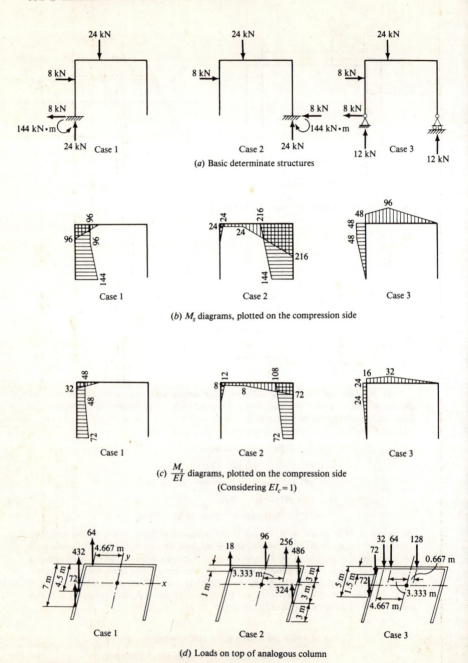

Figure 15.6.8 Three basic determinate structures for Example 15.6.1.

THE COLUMN-ANALOGY METHOD 553

Table 15.6.1 Moments at points A, B, C, and D of quadrangular frame in Example 15.6.1: Case 1

Point	M_s	$\dfrac{P}{A}$	$\dfrac{M_x y}{I_x}$	$\dfrac{M_y x}{I_y}$	M_i	M
A	−144	$-\dfrac{568}{13} = -43.69$	$-\dfrac{678.7}{116.83}(5.885) = -34.19$	$-\dfrac{3322.7}{372}(6) = -53.59$	−131.47	−12.53
B	−96	−43.69	+ (3.115) = +18.10	−53.59	−79.18	−16.82
C	0	−43.69	+ (3.115) = +18.10	+53.59	+28.00	−28.00
D	0	−43.69	− (5.885) = −34.19	+53.59	−24.29	+24.29

Table 15.6.2. Moments at points A, B, C, and D of quadrangular frame in Example 15.6.1: Case 2

Point	M_s	$\dfrac{P}{A}$	$\dfrac{M_x y}{I_x}$	$\dfrac{M_y x}{I_y}$	M_i	M
A	0	$-\dfrac{1180}{13} = -90.77$	$+\dfrac{255.70}{116.83}(5.885) = +12.88$	$+\dfrac{5605.3}{372}(6) = +90.41$	+12.52	−12.52
B	−24	−90.77	− (3.115) = −6.82	+90.41	−7.18	−16.82
C	−216	−90.77	− (3.115) = −6.82	−90.41	−188.00	−28.00
D	−144	−90.77	+ (5.885) = +12.88	−90.41	−168.30	+24.30

$$M_y = 256(3.333) + (486 + 324)(6) - 18(6)$$
$$= 5605.3 \text{ counterclockwise} \quad \text{(viewed from front)}$$

The values of M at points A, B, C, and D are computed in Table 15.6.2.
For Case 3,
$$P = 72 + 72 + 32 + 64 + 128 = 368 \text{ downward}$$
$$M_x = 72(1.615) + (32 + 64 + 128)(3.115) - 72(1.885)$$
$$= 678.32 \text{ clockwise} \quad \text{(viewed from right)}$$
$$M_y = (72 + 72)(6) + 32(4.667) + 64(3.333) - 128(0.667)$$
$$= 1141.3 \text{ counterclockwise} \quad \text{(viewed from front)}$$

The values of M at points A, B, C, and D are computed in Table 15.6.3.

Table 15.6.3 Moments at points A, B, C, and D of quadrangular frame in Example 15.6.1: Case 3

Point	M_s	$\dfrac{P}{A}$	$\dfrac{M_x y}{I_x}$	$\dfrac{M_y x}{I_y}$	M_i	M
A	0	$+\dfrac{368}{13} = +28.31$	$-\dfrac{678.32}{116.83}(5.885) = -34.17$	$+\dfrac{1141.3}{372}(6) = +18.41$	+12.55	−12.55
B	+48	+28.31	+ (3.115) = +18.09	+18.41	+64.81	−16.81
C	0	+28.31	+ (3.115) = +18.09	−18.41	+27.99	−27.99
D	0	+28.31	− (5.885) = −34.17	−18.41	−24.27	+24.27

The results obtained from the three different choices of the basic determinate structure check reasonably well within the limits of accuracy. The answers are

$M_A = -12.53$ kN·m (compression inside)

$M_B = -16.82$ kN·m (compression inside)

$M_C = -28.00$ kN·m (compression inside)

$M_D = +24.29$ kN·m (compression outside)

After the free-body diagrams, the moment diagram, and the generalized joint displacements at points A, B, C, and D of the elastic curve are all computed and drawn as

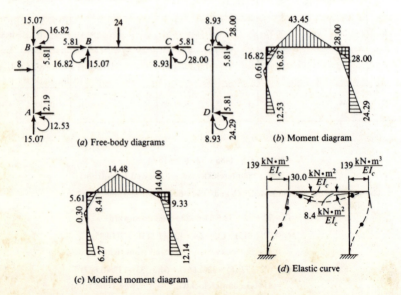

(a) Free-body diagrams
(b) Moment diagram
(c) Modified moment diagram
(d) Elastic curve

Figure 15.6.9 Solution for quadrangular frame in Example 15.6.1.

shown in Fig. 15.6.9, the usual statics checks and compatibility checks are observed, thus ensuring the correctness of the solution.

(c) *Discussion.* It is to be noted that a basic determinate structure can also be derived by placing three hinges (external or internal as the case may be) at any three of the four locations A, B, C, or D.

15.7 Moments in Closed Frames with One Axis of Symmetry

The column-analogy method can be used to analyze closed frames with one axis of symmetry when such frames are subjected to self-balancing loads or may be supported by external reactions. Two typical closed frames fulfilling these requirements are shown in Fig. 15.7.1. One should note again that the axis of symmetry refers only to the properties of the frame and not to the applied loads, although in Fig. 15.7.1b the self-balancing loads happen to be symmetrical as well with respect to the axis of symmetry of the frame itself. By the column-analogy method the moment at any point on the frame can be obtained by a *direct* procedure.

Consider the closed frame of Fig. 15.7.2a. If this frame is cut at any point A and the points on each side of the cut be called A_1 and A_2, respectively, the thrust, shear, and moment acting on point A_1 and A_2 are as shown in Fig. 15.7.2b, wherein, by reason of statics,

$$M_{A1} = M_{A2} \qquad H_{A1} = H_{A2} \qquad V_{A1} = V_{A2}$$

Now if the position of the point A_1 and the direction of the tangent at A_1 are held fixed (after the loads are applied), it is obvious that the Δ_ϕ, Δ_H, and Δ_V at A_2 relative to A_1 should all be zero. Thus, when the redundant-force method of analysis is used, the redundants M_{A2}, H_{A2}, and V_{A2} can be solved from the three compatibility conditions by setting equations similar to Eqs. (15.6.4a to c) to zero, except that point D in Eqs. (15.6.4a to c) is now replaced by point A_2 and M_s is now the statical moment due to the applied loads acting on the cantilever frame fixed at A_1 and free at A_2. If a rigid arm joining the point A_2 and the elastic center O is attached to the frame of Fig. 15.7.2b, the equivalent frame of Fig. 15.7.2c is obtained. Since OA_2 is a rigid arm and cannot be deformed, the Δ_ϕ, Δ_H, and Δ_V of point O relative to point A_1 are also all zero. The redundants M_O, H_O, and V_O can now be solved from

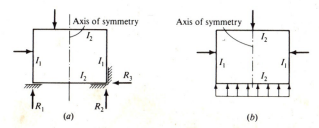

Figure 15.7.1 Typical closed frames with one axis of symmetry.

556 INTERMEDIATE STRUCTURAL ANALYSIS

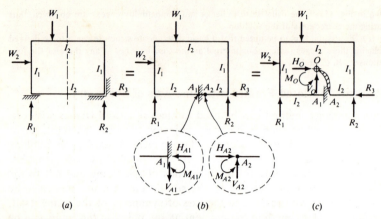

Figure 15.7.2 Redundant-force method of analysis.

equations similar to Eqs. (15.6.6a to c), and each of the three equations contains only one unknown. Then it follows that Eqs. (15.6.7a to c), (15.6.8a to c), and (15.6.10a and b) all apply to the present case of a closed frame. Thus the column-analogy method as described in the preceding section for quadrangular frames can be equally applied to the analysis of closed frames with one axis of symmetry. In fact, when a closed frame has two axes of symmetry, the position of the elastic center is known by inspection.

The discussion relating to the sign convention and the possibility of choosing different basic determinate structures as presented earlier are applicable to closed frames in a similar manner.

Example 15.7.1 Analyze the closed rectangular frame shown in Fig. 15.7.3a by the column-analogy method.

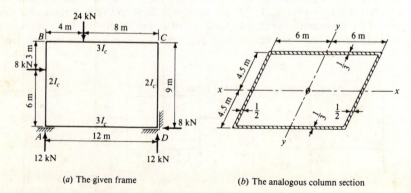

(a) The given frame (b) The analogous column section

Figure 15.7.3 Closed frame of Example 15.7.1.

THE COLUMN-ANALOGY METHOD

SOLUTION (a) *Properties of the analogous column section.* The analogous column section by considering EI_c to be unity is shown in Fig. 15.7.3b.

$$A = 2(\tfrac{1}{3})(12) + 2(\tfrac{1}{2})(9) = 17$$

$$I_x = 2(4)(4.5)^2 + 2(4.5)\frac{(9)^2}{12} = 222.75$$

$$I_y = 2(4.5)(6)^2 + 2(4)\frac{(12)^2}{12} = 420$$

(b) M_s *diagrams from two different basic determinate structures.* For purpose of illustration, the complete solution, using two different basic determinate structures shown in Fig. 15.7.4, will be presented.

For Case 1,

$$P = 72 + 432 + 64 + 288 = 856 \text{ upward}$$

$$M_x = 288(4.5) + 72(2.5) - 64(4.5)$$

$$= 1188 \text{ clockwise} \quad \text{(viewed from right)}$$

$$M_y = 64(4.667) + 288(2) + (432 + 72)(6)$$

$$= 3898.7 \text{ clockwise} \quad \text{(viewed from front)}$$

The values of M at points A, B, C, and D are computed in Table 15.7.1.

For Case 2,

$$P = 18 + 48 + 144 + 162 - 128 = 244 \text{ upward}$$

$$M_x = 18(3.5) + 162(1.5) + (144 + 48 - 128)(4.5)$$

$$= 594 \text{ counterclockwise} \quad \text{(viewed from right)}$$

$$M_y = 162(6) + 144(2) + 128(0.667) - 48(2) - 18(6)$$

$$= 1141.4 \text{ counterclockwise} \quad \text{(viewed from front)}$$

The values of M at points A, B, C, and D are computed in Table 15.7.2. The answers are

$$M_A = -13.95 \text{ kN·m} \quad \text{(compression inside)}$$
$$M_B = -13.95 \text{ kN·m} \quad \text{(compression inside)}$$
$$M_C = -29.35 \text{ kN·m} \quad \text{(compression inside)}$$
$$M_D = +18.65 \text{ kN·m} \quad \text{(compression outside)}$$

Table 15.7.1 Moments at points A, B, C, and D of closed frame in Example 15.7.1: Case 1

Point	M_s	$\dfrac{P}{A}$	$\dfrac{M_x y}{I_x}$	$\dfrac{M_y x}{I_y}$	M_i	M
A	−144	$-\dfrac{856}{17} = -50.35$	$-\dfrac{(1188)(4.5)}{222.75} = -24.00$	$-\dfrac{(3898.7)(6)}{420} = -55.70$	−130.05	−13.95
B	−96	−50.35	+24.00	−55.70	−82.05	−13.95
C	0	−50.35	+24.00	+55.70	+29.35	−29.35
D	0	−50.35	−24.00	+55.70	−18.65	+18.65

558 INTERMEDIATE STRUCTURAL ANALYSIS

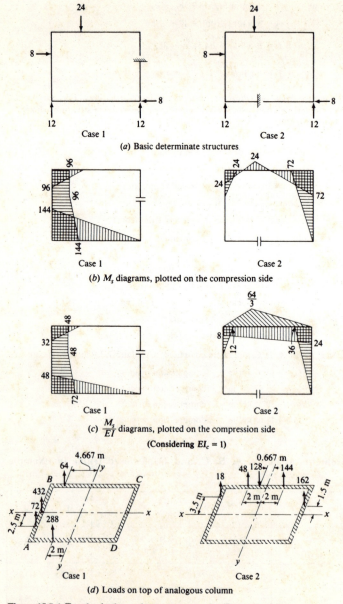

Figure 15.7.4 Two basic determinate structures for Example 15.7.1.

THE COLUMN-ANALOGY METHOD **559**

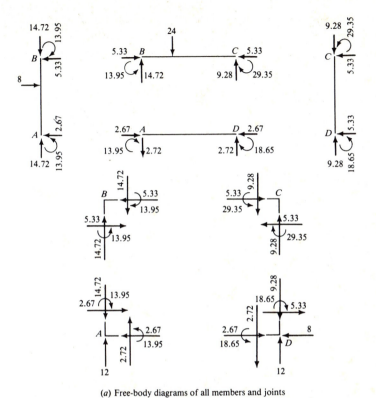

(a) Free-body diagrams of all members and joints

(b) Moment diagram   (c) Elastic curve

Figure 15.7.5 Solution for closed frame in Example 15.7.1.

Table 15.7.2 Moments at points A, B, C, and D of closed frame in Example 15.7.1: Case 2

Point	M_s	$\dfrac{P}{A}$	$\dfrac{M_x y}{I_x}$	$\dfrac{M_y x}{I_y}$	M_i	M
A	0	$-\dfrac{244}{17} = -14.35$	$+\dfrac{(594)(4.5)}{222.75} = +12.00$	$+\dfrac{(1141.4)(6)}{420} = +16.31$	$+13.96$	-13.96
B	-24	-14.35	-12.00	$+16.31$	-10.04	-13.96
C	-72	-14.35	-12.00	-16.31	-42.66	-29.34
D	0	-14.35	$+12.00$	-16.31	-18.66	$+18.66$

After the free-body diagrams, the moment diagram, and the generalized joint displacements at points A, B, C, and D of the elastic curve are all computed and drawn as shown in Fig. 15.7.5, the usual statics checks and compatibility checks are observed in order to ensure the correctness of the solution.

15.8 Moments in Gable Frames with One Axis of Symmetry

A gable frame with one axis of symmetry such as the one shown in Fig. 15.8.1 can be analyzed by the column-analogy method in the same manner as the quadrangular frame with two fixed supports. In determining the moments of inertia of the analogous column section for the gable frame, it is necessary to find the moment of inertia of a line area about a centroidal axis which is at an angle θ with the direction of the line. From Fig. 15.8.2a,

$$I_x = 2\int_0^{L/2} (b\,ds)(s\,\sin\theta)^2 = \left[\dfrac{2}{3}bs^3\right]_0^{L/2} \sin^2\theta$$

$$= \dfrac{(bL)(L\sin\theta)^2}{12} = \dfrac{AL_y^2}{12} \tag{15.8.1a}$$

in which A is the total line area. Similarly,

$$I_y = \dfrac{AL_x^2}{12} \tag{15.8.1b}$$

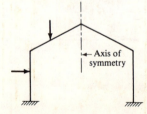

Figure 15.8.1 A gable frame with one axis of symmetry.

THE COLUMN-ANALOGY METHOD

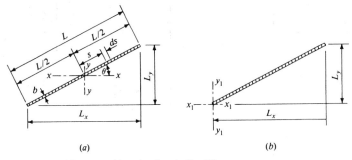

(a)   (b)

Figure 15.8.2 Moments of inertia of an inclined line area.

By the parallel-axis theorem, in Fig. 15.8.2b,

$$I_{x1} = \frac{AL_y^2}{3} \qquad I_{y1} = \frac{AL_x^2}{3} \qquad (15.8.2a \text{ and } b)$$

Example 15.8.1 Analyze the gable frame shown in Fig. 15.8.3a by the column-analogy method. Note that this same gable frame has been solved previously by the slope-deflection, moment-distribution, and matrix displacement methods.

SOLUTION (a) *Properties of the analogous column section*. The analogous column section by considering EI_c to be unity is shown in Fig. 15.8.3b.

$$A = 2(\tfrac{1}{2})(6) + 2(\tfrac{1}{3})(6.7082) = 10.472$$

$$\bar{y} = \frac{2(3)(3) - 2(2.236)(1.5)}{10.472} = 1.0783 \text{ m}$$

$$I_x = \frac{2(3)(6)^2}{12} + 2(3)(1.9217)^2 + 2(2.236)\frac{(3)^2}{12} + 2(2.236)(2.5783)^2 = 73.24$$

or

$$I_x = I_{BD} - A\bar{y}^2 = \frac{2(3)(6)^2}{3} + 2(2.236)\frac{(3)^2}{3} - 10.472(1.0783)^2 = 73.24 \quad \text{(Check)}$$

$$I_y = 2(3)(6)^2 + 2(2.236)\frac{(6)^2}{3} = 269.66$$

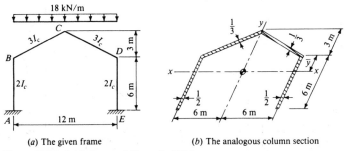

(a) The given frame   (b) The analogous column section

Figure 15.8.3 Gable frame of Example 15.8.1.

(b) *Moments at A, B, C, D, and E.* The choice of the basic determinate structure shown in Fig. 15.8.4a results in a symmetrical M_s diagram and thus symmetrical loads on the analogous column. A basic determinate structure obtained by installing internal hinges at B, C, and D will also result in symmetrical loads on the analogous column. Referring back to Fig. 15.8.4,

$$P_1 = P_2 = \tfrac{2}{3}(108)(6.7082) = 482.99 \qquad P = P_1 + P_2 = 965.98 \text{ downward}$$

$$M_x = 2(482.99)(1.8750 + 1.0783)$$

$$= 2852.8 \text{ clockwise} \qquad \text{(viewed from right)}$$

$$M_y = 0$$

The values of M at points A, B, C, D, and E are computed in Table 15.8.1. The signs in front of all numbers in the P/A and $M_x y/I_x$ columns of this table are obtained by visualizing the action of the downward load, or of the clockwise moment as viewed from the right. The answers are

$$M_A = M_E = +\ 99.47 \text{ kN·m} \qquad \text{(compression outside)}$$

$$M_B = M_D = -134.24 \text{ kN·m} \qquad \text{(compression inside)}$$

$$M_C = +72.90 \text{ kN·m} \qquad \text{(compression outside)}$$

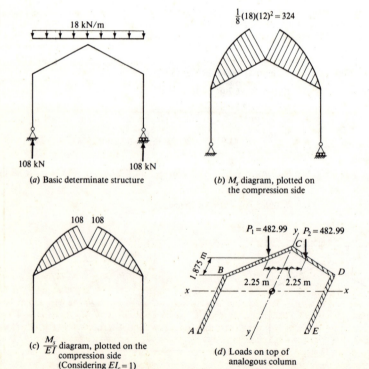

(a) Basic determinate structure

(b) M_s diagram, plotted on the compression side

(c) $\dfrac{M_s}{EI}$ diagram, plotted on the compression side (Considering $EI_c = 1$)

(d) Loads on top of analogous column

Figure 15.8.4 Loads on analogous column of gable frame in Example 15.8.1.

Table 15.8.1 Moments at points A, B, C, D, and E of gable frame in Example 15.8.1

Point	M_s	$\dfrac{P}{A}$	$\dfrac{M_x y}{I_x}$		M_i	M
A	0	$+\dfrac{965.98}{10.472} = +92.24$	$-\dfrac{2852.8}{73.24}(4.9217) = -191.71$		-99.47	$+99.47$
B	0	$+92.24$	$+$	$(1.0783) = +42.00$	$+134.24$	-134.24
C	$+324$	$+92.24$	$+$	$(4.0783) = +158.86$	$+251.10$	$+72.90$
D	0	$+92.24$	$+$	$(1.0783) = +42.00$	$+134.24$	-134.24
E	0	$+92.24$	$-$	$(4.9217) = -191.71$	-99.47	$+99.47$

The free-body diagrams, the moment diagram, and the elastic curve have been worked out in Example 7.9.1.

Example 15.8.2 Analyze the gable frame shown in Fig. 15.8.5 by the column-analogy method. Note that this problem has been solved previously by the slope-deflection, moment-distribution, and matrix displacement methods.

SOLUTION (a) *Properties of the analogous column section.* The analogous column section is the same as that of Example 15.8.1.

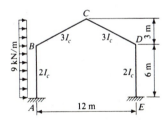

Figure 15.8.5 Gable frame of Example 15.8.2.

(b) *Moments at A, B, C, D, and E.* The basic determinate structure chosen for the M_s diagram is a cantilever structure fixed at A and free at E, as shown in Fig. 15.8.6. The moment diagram on BC has a tangent at C which coincides with BC; thus its area and centroid are conveniently known. The moment diagram on BA is taken as the difference between the total extended diagram with dotted lines and the small parabola above the level B, as shown in Fig. 15.8.6c. Thus,

$$P_1 = A_1 = \tfrac{1}{3}(182.25)(9) - \tfrac{1}{3}(20.25)(3) = 546.75 - 20.25 = 526.5$$
$$P_2 = A_2 = \tfrac{1}{3}(13.5)(6.7082) = 30.19$$

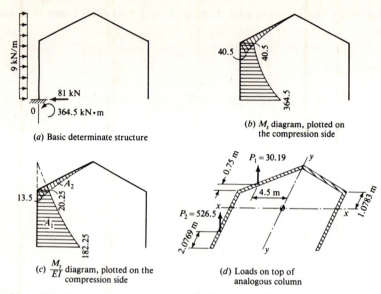

Figure 15.8.6 Loads on analogous column of gable frame in Example 15.8.2.

$$\text{Distance from centroid of } A_1 \text{ to the left support} = \frac{546.75(2.25) - 20.25(6.75)}{526.5} = 2.0769 \text{ m}$$

$P = P_1 + P_2 = 526.5 + 30.19 = 556.69$ upward

$M_x = 526.5(3.9231 - 1.0783) - 30.19(1.0783 + 0.7500)$
$\quad = 1442.6$ clockwise   (viewed from right)

$M_y = 526.5(6) + 30.19(4.5) = 3294.8$ clockwise   (viewed from front)

The values of M at points A, B, C, D, and E are computed in Table 15.8.2.

Table 15.8.2 Moments at points A, B, C, D, and E of gable frame in Example 15.8.2

Point	M_s	$\dfrac{P}{A}$	$\dfrac{M_x y}{I_x}$		$\dfrac{M_y x}{I_y}$		M_i	M
A	−364.5	$-\dfrac{556.69}{10.472} = -53.16$	$-\dfrac{1442.6}{73.24}(4.9217) = -96.94$		$-\dfrac{3294.8}{269.66}(6) = -73.31$		−223.41	−141.09
B	−40.5	−53.16	+	(1.0783) = +21.24	−	(6) = −73.31	−105.23	+64.73
C	0	−53.16	+	(4.0783) = +80.33		0	+27.17	−27.17
D	0	−53.16	+	(1.0783) = +21.24	+	(6) = +73.31	+41.39	−41.39
E	0	−53.16	−	(4.9217) = −96.94	+	(6) = +73.31	−76.79	+76.79

The answers are

$M_A = -141.09$ kN·m (compression inside)
$M_B = +64.73$ kN·m (compression outside)
$M_C = -27.17$ kN·m (compression inside)
$M_D = -41.39$ kN·m (compression inside)
$M_E = +76.79$ kN·m (compression outside)

The free-body diagrams, the moment diagram, and the elastic curve have been worked out in Example 7.9.2.

15.9 Moments in Unsymmetrical Quadrangular Frames

The column-analogy method can be used to analyze unsymmetrical quadrangular frames when such frames have two fixed supports and are subjected to some applied loading. The analysis of an unsymmetrical quadrangular frame as shown in Fig. 15.9.1 by the redundant-force method requires the solution of three simultaneous equations. By the column-analogy method, however, the moment at any point on the frame can be obtained by a *direct* procedure.

The problem of determining the pressure at any point in a column with an unsymmetrical cross section due to the combined action of a direct load at the centroid and two bending moments about a pair of mutually perpendicular axes through the centroid will be discussed first. Consider the short column shown in Fig. 15.9.2a. It is subjected to the downward loads P_1, P_2, P_3, etc. at the points (x_1, y_1), (x_2, y_2), (x_3, y_3), etc. By the principles of statics, the loads P_1, P_2, P_3, etc. shown in Fig. 15.9.2a can be replaced by P, M_x, and M_y shown in Fig. 15.9.2b. Thus,

$$P = P_1 + P_2 + P_3 + \cdots = \Sigma P \quad \text{downward} \tag{15.9.1a}$$

$$M_x = P_1 y_1 + P_2 y_2 + P_3 y_3 + \cdots = \Sigma Py \quad \text{clockwise} \tag{15.9.1b}$$
(viewed from right)

$$M_y = P_1 x_1 + P_2 x_2 + P_3 x_3 + \cdots = \Sigma Px \quad \text{clockwise} \tag{15.9.1c}$$
(viewed from front)

in which P_1, P_2, P_3, etc. are considered positive when acting downward and (x_1, y_1), (x_2, y_2), (x_3, y_3), etc. are coordinates of the points of application of P_1, P_2, P_3, etc. in reference to the centroid as origin.

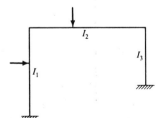

Figure 15.9.1 An unsymmetrical quadrangular frame.

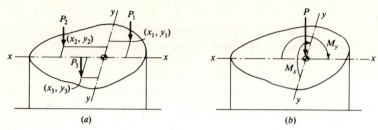

Figure 15.9.2 An unsymmetrical cross section subjected to direct load and biaxial bending.

From the assumption of planar distribution of pressure over the cross section, the pressure p at any point (x, y) can be expressed by

$$p = a + bx + cy \tag{15.9.2}$$

where the constants a, b, and c can be determined from the three equations of statics, namely,

$$\int_0^A p\, dA = P \qquad \int_0^A (p\, dA)(y) = M_x \qquad \int_0^A (p\, dA)(x) = M_y \tag{15.9.3}$$

By substituting Eq. (15.9.2) into Eqs. (15.9.3), and noting that

$$\int_0^A dA = A \qquad \int_0^A y\, dA = 0 \qquad \int_0^A x\, dA = 0$$
$$\int_0^A y^2\, dA = I_x \qquad \int_0^A xy\, dA = I_{xy} \qquad \int_0^A x^2\, dA = I_y \tag{15.9.4}$$

the following equations are obtained:

$$P = \int_0^A p\, dA = \int_0^A (a + bx + cy)\, dA = aA \tag{15.9.5a}$$

$$M_x = \int_0^A (p\, dA)(y) = \int_0^A (ay + bxy + cy^2)\, dA$$
$$= bI_{xy} + cI_x \tag{15.9.5b}$$

$$M_y = \int_0^A (p\, dA)(x) = \int_0^A (ax + bx^2 + cxy)\, dA$$
$$= bI_y + cI_{xy} \tag{15.9.5c}$$

Solving Eqs. (15.9.5a to c) for a, b, and c,

$$a = \frac{P}{A} \qquad b = \frac{M_y'}{I_y'} \qquad c = \frac{M_x'}{I_x'} \tag{15.9.6a}$$

THE COLUMN-ANALOGY METHOD **567**

in which

$$M'_x = M_x - M_y\left(\frac{I_{xy}}{I_y}\right)$$

$$M'_y = M_y - M_x\left(\frac{I_{xy}}{I_x}\right)$$

$$I'_x = I_x\left(1 - \frac{I_{xy}^2}{I_x I_y}\right)$$ (15.9.6b)

$$I'_y = I_y\left(1 - \frac{I_{xy}^2}{I_x I_y}\right)$$

Substituting Eqs. (15.9.6a and b) into Eq. (15.9.2),

$$p = a + bx + cy = \frac{P}{A} + \frac{M'_y x}{I'_y} + \frac{M'_x y}{I'_x} \qquad (15.9.7)$$

Equation (15.9.7) gives the pressure at any point in a column with an unsymmetrical cross section under the combined action of a direct load P and two bending moments M_x and M_y about two reference axes through the centroid.

The analysis by the column-analogy method of the unsymmetrical quadrangular frame shown in Fig. 15.9.3a will now be taken up.

The discussion relating to quadrangular frames with one axis of symmetry in Sec. 15.6 applies equally well to the present problem of unsymmetrical quadrangular frames except that I_{xy} is now *not* equal to zero. Restating Eqs. (15.6.6a) and referring to Fig. 15.9.3b,

$$\Delta_\phi \text{ at } O = +P + M_O A + H_O(0) - V_O(0) = 0 \qquad (15.9.8a)$$

$$\Delta_H \text{ at } O = +M_x + M_O(0) + H_O I_x - V_O I_{xy} = 0 \qquad (15.9.8b)$$

$$\Delta_V \text{ at } O = -M_y - M_O(0) - H_O I_{xy} + V_O I_y = 0 \qquad (15.9.8c)$$

Solving the above equations for M_O, H_O, and V_O,

$$M_O = -\frac{P}{A} \qquad H_O = -\frac{M'_x}{I'_x} \qquad V_O = +\frac{M'_y}{I'_y} \qquad (15.9.9)$$

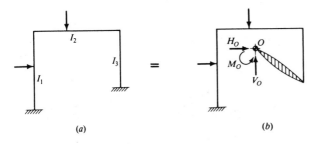

Figure 15.9.3 The equivalent structure with rigid arm.

in which M'_x, M'_y, I'_x, and I'_y are as defined in Eq. (15.9.6b). Substituting Eq. (15.9.9) into Eq. (15.6.9),

$$M = M_s + (M_O + H_O y - V_O x) = M_s - \left(\frac{P}{A} + \frac{M'_x y}{I'_x} + \frac{M'_y x}{I'_y}\right) \quad (15.9.10)$$

The sum of the three terms in the parentheses of Eq. (15.9.10) is seen to be identical to the formula in Eq. (15.9.7) for the pressure in a column when subjected to P, M_x, and M_y. Thus, letting the pressure in the analogous column be M_i, Eq. (15.9.10) becomes

$$M = M_s - M_i \quad (15.9.11)$$

To summarize, the column-analogy method of analyzing unsymmetrical quadrangular frames with two fixed supports involves the following steps:

1. The loading on the top of the analogous column is equal to the M_s/EI diagram. The loading is in the downward direction if M_s is positive, which means that it causes compression on the outside.
2. The properties of the analogous column section may be found by referring to any pair of convenient x and y axes passing through the centroid (or the elastic center). The properties are A, I_x, I_{xy}, I_y, I'_x and I'_y, wherein

$$I'_x = I_x\left(1 - \frac{I_{xy}^2}{I_x I_y}\right) \quad I'_y = I_y\left(1 - \frac{I_{xy}^2}{I_x I_y}\right)$$

3. The moment at any point of the given frame is equal to $M = M_s - M_i$, where

$$M'_x = M_x - M_y\left(\frac{I_{xy}}{I_y}\right)$$

$$M'_y = M_y - M_x\left(\frac{I_{xy}}{I_x}\right)$$

$$M_i = \frac{P}{A} + \frac{M'_x y}{I'_x} + \frac{M'_y x}{I'_y}$$

In applying all formulas shown above, care must be taken in adhering to the correct signs for P, x, y, and I_{xy}, noting that $M_x = Py$ and $M_y = Px$.

Example 15.9.1 Analyze the unsymmetrical quadrangular frame shown in Fig. 15.9.4a by the column-analogy method.

SOLUTION (a) *Properties of the analogous column section.* The analogous column section by considering EI_c to be unity is shown in Fig. 15.9.4b. The moment of inertia I_x and I_y, as well as the product of inertia I_{xy}, are each computed by two ways: first by transferring from the centroidal axes of each straight line area to the centroidal axes of the whole cross section, and second by transferring from the centroidal axes of each straight line area to a pair of convenient axes and then transferring back to the centroidal axes of the whole cross

THE COLUMN-ANALOGY METHOD

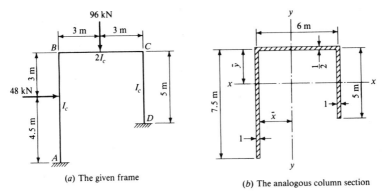

(a) The given frame   (b) The analogous column section

Figure 15.9.4 Quadrangular frame of Example 15.9.1.

section. This double computation helps to ensure the correctness of the results.

$A = 7.5 + 3 + 5 = 15.5$

$\bar{x} = \dfrac{7.5(0) + 3(3) + 5(6)}{15.5} = 2.516 \text{ m}$

$\bar{y} = \dfrac{7.5(3.75) + 3(0) + 5(2.5)}{15.5} = 2.621 \text{ m}$

$I_x = \dfrac{7.5(7.5)^2}{12} + 7.5(1.129)^2 + 3(2.621)^2 + \dfrac{5(5)^2}{12} + 5(0.121)^2$

$= 75.815$ (Use)

$I_x = I_{BC} - 15.5(2.621)^2 = \dfrac{7.5(7.5)^2}{3} + \dfrac{5(5)^2}{3} - 15.5(2.621)^2$

$= 75.812$ (Check)

$I_y = 7.5(2.516)^2 + \dfrac{3(6)^2}{12} + 3(0.484)^2 + 5(3.484)^2 = 117.87$ (Use)

$I_y = I_{AB} - 15.5(2.516)^2 = \dfrac{3(6)^2}{3} + 5(6)^2 - 15.5(2.516)^2 = 117.88$ (Check)

$I_{xy} = 0 + 7.5(+2.516)(+1.129) + 0 + 3(-0.484)(-2.621) + 0 + 5(-3.484)(-0.121)$

$= +27.218$ (Use)

$I_{xy} = I_{AB-BC} - 15.5(-2.516)(+2.621)$

$= 5(-6)(+2.5) - 15.5(-2.516)(+2.621)$

$= +27.214$ (Check)

$1 - \dfrac{I_{xy}^2}{I_x I_y} = 1 - \dfrac{(+27.218)^2}{(75.815)(117.87)} = 0.91710$

$I'_x = 0.91710 I_x = 0.91710(75.815) = 69.53$

$I'_y = 0.91710 I_y = 0.91710(117.87) = 108.10$

(b) M_s *diagrams from two different basic determinate structures.* For purpose of illustration, the complete solution, using two different basic determinate structures as shown

570 INTERMEDIATE STRUCTURAL ANALYSIS

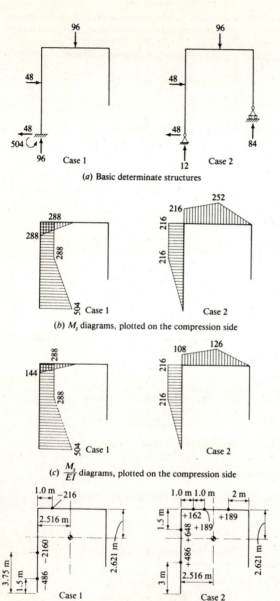

Figure 15.9.5 Two basic determinate structures for Example 15.9.1.

in Fig. 15.9.5, will be presented. In the case of quadrangular frames with one axis of symmetry, emphasis has been placed on physically visualizing whether the action of the direct load and the two bending moments are to cause positive or negative pressures in the analogous column. For unsymmetrical cases, it is more convenient to adhere strictly to the correct signs of P, of the coordinates x and y, and of I_{xy}, $M_x = Py$, and $M_y = Px$.

For Case 1, the loads on the analogous column are as follows:

P	x	y	$M_y = Px$	$M_x = Py$
-486	-2.516	-3.379	$+1222.8$	$+1642.2$
-2160	-2.516	-1.129	$+5434.6$	$+2438.6$
-216	-1.516	$+2.621$	$+327.4$	-566.1
-2862			$+6984.8$	$+3514.7$

$$M'_x = M_x - M_y \frac{I_{xy}}{I_y} = +3514.7 - (+6984.8)\left(\frac{+27.218}{117.87}\right) = +1901.8$$

$$M'_y = M_y - M_x \frac{I_{xy}}{I_x} = +6984.8 - (+3514.7)\left(\frac{+27.218}{75.815}\right) = +5723.0$$

The values of M at points A, B, C, and D are computed in Table 15.9.1.
For Case 2, the loads on the analogous column are as follows:

P	x	y	$M_y = Px$	$M_x = Py$
$+486$	-2.516	-1.879	-1222.8	-913.2
$+648$	-2.516	$+1.121$	-1630.4	$+726.4$
$+162$	-1.516	$+2.621$	-245.6	$+424.6$
$+189$	-0.516	$+2.621$	-97.5	$+495.4$
$+189$	$+1.484$	$+2.621$	$+280.5$	$+495.4$
$+1674$			-2915.8	$+1228.6$

$$M'_x = M_x - M_y \frac{I_{xy}}{I_y} = +1228.6 - (-2915.8)\left(\frac{+27.218}{117.87}\right) = +1901.9$$

$$M'_y = M_y - M_x \frac{I_{xy}}{I_x} = -2915.8 - (+1228.6)\left(\frac{+27.218}{75.815}\right) = -3356.9$$

The values of M at points A, B, C, and D are computed in Table 15.9.2.
The answers, after averaging out the errors, are

$M_A = -52.68$ kN·m   (compression inside)
$M_B = -41.84$ kN·m   (compression inside)
$M_C = -71.50$ kN·m   (compression inside)
$M_D = +65.26$ kN·m   (compression outside)

The free-body diagrams, the moment diagram, and the elastic curve have been worked out in Example 4.6.2.

Table 15.9.1 Moments at points A, B, C, and D of unsymmetrical quadrangular frame in Example 15.9.1: Case 1

Point	x	y	M_s	$\dfrac{P}{A}$	$\dfrac{M'_y x}{I'_y}$	$\dfrac{M'_x y}{I'_x}$	M_i	M
A	-2.516	-4.879	-504	$\dfrac{2862}{15.5} = -184.64$	$\dfrac{+5723.0}{108.10}(-2.516) = -133.20$	$\dfrac{+1901.8}{69.53}(-4.879) = -133.45$	-451.29	-52.71
B	-2.516	$+2.621$	-288	-184.64	$(-2.516) = -133.20$	$(+2.621) = +71.69$	-246.15	-41.85
C	$+3.484$	$+2.621$	0	-184.64	$(+3.484) = +184.45$	$(+2.621) = +71.69$	$+71.50$	-71.50
D	$+3.484$	-2.379	0	-184.64	$(+3.484) = +184.45$	$(-2.379) = -65.07$	-65.26	$+65.26$

Table 15.9.2 Moments at points A, B, C, and D of unsymmetrical quadrangular frame in Example 15.9.1: Case 2

Point	x	y	M_s	$\dfrac{P}{A}$	$\dfrac{M'_y x}{I'_y}$	$\dfrac{M'_x y}{I'_x}$	M_i	M
A	-2.516	-4.879	0	$+\dfrac{1674}{15.5}=+108$	$\dfrac{-3356.9}{108.10}(-2.516)=+79.13$	$\dfrac{+1901.9}{69.53}(-4.879)=-133.46$	$+52.67$	-52.67
B	-2.516	$+2.621$	$+216$	$+108$	$(-2.516)=+78.13$	$(+2.621)=+71.69$	$+257.82$	-41.82
C	$+3.484$	$+2.621$	0	$+108$	$(+3.484)=-108.19$	$(+2.621)=+71.69$	$+71.50$	-71.50
D	$+3.484$	-2.379	0	$+108$	$(+3.484)=-108.19$	$(-2.379)=-65.07$	-65.26	$+65.26$

15.10 Moments in Unsymmetrical Closed Frames

The column-analogy method can be used to analyze unsymmetrical closed frames when such frames are subjected to some applied loading. It has been shown in Sec. 15.7 that the procedure for analyzing closed frames with one axis of symmetry is similar to that for quadrangular frames with one axis of symmetry and two fixed supports. Thus the procedure for analyzing unsymmetrical quadrangular frames with two fixed supports, as described in Sec. 15.9, can be applied equally well to unsymmetrical closed frames.

Example 15.10.1 Analyze the unsymmetrical closed frame shown in Fig. 15.10.1a by the column-analogy method.

SOLUTION (a) *Properties of the analogous column section.* The analogous column section by considering EI_c to be unity is shown in Fig. 15.10.1b. Note that the I_{xy} of an inclined line area about its own centroidal axes is numerically equal to the product of the line area itself and the horizontal and vertical projections of the line divided by 12, positive if it lies in the first and third quadrants and negative if it lies in the second and fourth quadrants. In using the parallel-axis theorem for the product of inertia, the transferred distances may be directed either from the centroidal axes to the reference axes or vice versa.

$$A = \tfrac{1}{3}(12) + \tfrac{1}{3}(12.369) + 1(6) + \tfrac{1}{2}(9) = 4 + 4.123 + 6 + 4.5 = 18.623$$

$$\bar{x} = \frac{6(0) + 4(6) + 4.123(6) + 4.5(12)}{18.623} = 5.5167 \text{ m}$$

$$\bar{y} = \frac{4(0) + 6(3) + 4.5(4.5) + 4.123(7.5)}{18.623} = 3.7144 \text{ m}$$

$$I_x = 4(3.7144)^2 + \frac{6(6)^2}{12} + 6(0.7144)^2 + \frac{4.5(9)^2}{12} + 4.5(0.7856)^2$$

$$+ \frac{4.123(3)^2}{12} + 4.123(3.7856)^2$$

$$= 171.58 \qquad (\text{Use})$$

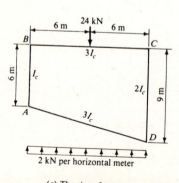

(a) The given frame

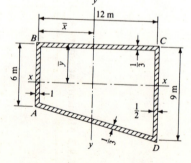

(b) The analogous column section

Figure 15.10.1 Closed frame of Example 15.10.1.

THE COLUMN-ANALOGY METHOD

$I_x = I_{BC} - 18.623(3.7144)^2$

$= \dfrac{6(6)^2}{3} + \dfrac{4.5(9)^2}{3} + \dfrac{4.123(3)^2}{12} + 4.123(7.5)^2 - 18.623(3.7144)^2$

$= 171.57 \quad \text{(Check)}$

$I_y = 6(5.5167)^2 + 4.5(6.4833)^2 + \dfrac{4(12)^2}{12} + 4(0.4833)^2$

$\quad + \dfrac{4.123(12)^2}{12} + 4.123(0.4833)^2$

$= 471.13 \quad \text{(Use)}$

$I_y = I_{AB} - 18.623(5.5167)^2$

$= 4.5(12)^2 + \dfrac{4(12)^2}{3} + \dfrac{4.123(12)^2}{3} - 18.623(5.5167)^2$

$= 471.13 \quad \text{(Check)}$

$I_{xy} = +6(+5.5167)(-0.7144) + 4(-0.4833)(-3.7144)$

$\quad + 4.5(-6.4833)(+0.7856) - \dfrac{4.123(3)(12)}{12} + 4.123(-0.4833)(+3.7856)$

$= -59.298 \quad \text{(Use)}$

$I_{xy} = I_{AB-BC} - 18.623(-5.5167)(+3.7144)$

$= 4.5(-12)(+4.5) - \dfrac{4.123(3)(12)}{12} + 4.123(-6)(+7.5)$

$\quad - 18.623(-5.5167)(+3.7144)$

$= -59.296 \quad \text{(Check)}$

$1 - \dfrac{I_{xy}^2}{I_x I_y} = 1 - \dfrac{(-59.298)^2}{(171.58)(471.13)} = 0.95650$

$I_x' = 0.95650 I_x = 0.95650(171.58) = 164.12$

$I_y' = 0.95650 I_y = 0.95650(471.13) = 450.64$

(b) M_s diagrams from two different basic determinate structures. For purpose of illustration, the complete solution, using two different basic determinate structures as shown in Fig. 15.10.2, will be presented.

For Case 1,

P	x	y	$M_x = Py$	$M_y = Px$
− 144	−3.5167	+3.7144	− 534.87	+ 506.40
− 864	−5.5167	+0.7144	− 617.24	+4766.43
− 197.90	−2.5167	−3.0356	+ 600.75	+ 498.05
− 1205.90			− 551.36	+5770.88

$M_x' = M_x - M_y \dfrac{I_{xy}}{I_y} = -551.36 - (+5770.88)\left(\dfrac{-59.298}{471.13}\right) = -174.98$

$M_y' = M_y - M_x \dfrac{I_{xy}}{I_x} = +5770.88 - (-551.36)\left(\dfrac{-59.298}{171.58}\right) = +5580.3$

The values of M at points A, B, C, and D are computed in Table 15.10.1.

576 INTERMEDIATE STRUCTURAL ANALYSIS

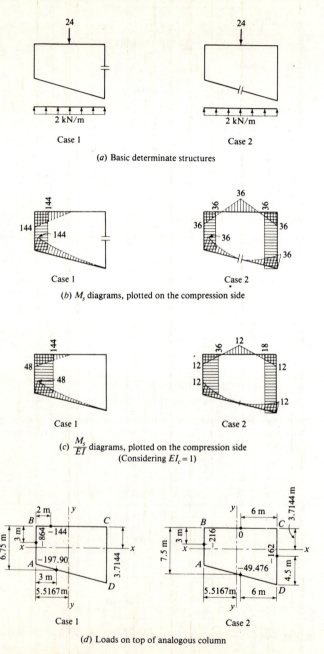

Figure 15.10.2 Two basic determinate structures for Example 15.10.1.

THE COLUMN-ANALOGY METHOD 577

Table 15.10.1 Moments at points A, B, C, and D of unsymmetrical closed frame in Example 15.10.1: Case 1

Point	x	y	M_s	$\dfrac{P}{A}$	$\dfrac{M'_y x}{I'_y}$	$\dfrac{M'_x y}{I'_x}$	M_i	M
A	-5.5167	-2.2856	-144	$\dfrac{-1205.90}{18.623} = -64.75$	$\dfrac{+5580.3}{450.64}(-5.5167) = -68.31$	$\dfrac{+174.98}{164.12}(-2.2856) = -2.44$	-135.50	-8.50
B	-5.5167	$+3.7144$	-144	-64.75	$(-5.5167) = -68.31$	$(+3.7144) = +3.96$	-129.10	-14.90
C	$+6.4833$	$+3.7144$	0	-64.75	$(+6.4833) = +80.28$	$(+3.7144) = +3.96$	$+19.49$	-19.49
D	$+6.4833$	-5.2856	0	-64.75	$(+6.4833) = +80.28$	$(-5.2856) = -5.64$	$+9.89$	-9.89

578 INTERMEDIATE STRUCTURAL ANALYSIS

Table 15.10.2 Moments at points A, B, C, and D of unsymmetrical closed frame in Example 15.10.1: Case 2

Point	x	y	M_s	$\dfrac{P}{A}$	$\dfrac{M'_y x}{I'_y}$	$\dfrac{M'_x y}{I'_x}$	M_i	M
A	-5.5167	-2.2856	-36	$\dfrac{-427.48}{18.623} = -22.95$	$\dfrac{+172.79}{450.64}(-5.5167) = -2.11$	$\dfrac{+175.04}{164.12}(-2.2856) = -2.44$	-27.50	-8.50
B	-5.5167	$+3.7144$	-36	$= -22.95$	$(-5.5167) = -2.11$	$(+3.7144) = +3.96$	-21.10	-14.90
C	$+6.4833$	$+3.7144$	-36	$= -22.95$	$(+6.4833) = +2.48$	$(+3.7144) = +3.96$	-16.51	-19.49
D	$+6.4833$	-5.2856	-36	$= -22.95$	$(+6.4833) = +2.48$	$(-5.2856) = -5.64$	-26.11	-9.89

THE COLUMN-ANALOGY METHOD

For Case 2,

P	x	y	$M_x = Py$	$M_y = Px$
−216	−5.5167	+0.7144	−154.31	+1191.61
−49.48	+0.4833	−3.7856	+187.30	−23.91
−162	+6.4833	−0.7856	+127.27	−1050.29
−427.48			+160.26	+117.41

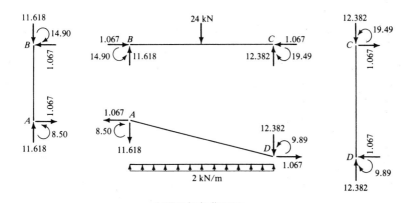

(a) Free-body diagrams

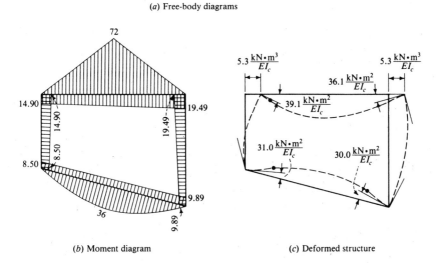

(b) Moment diagram　　　　　　　　(c) Deformed structure

Figure 15.10.3 Solution for closed frame in Example 15.10.1.

$$M'_x = M_x - M_y\frac{I_{xy}}{I_y} = +160.26 - (+117.41)\left(\frac{-59.298}{471.13}\right) = +175.04$$

$$M'_y = M_y - M_x\frac{I_{xy}}{I_x} = +117.41 - (+160.26)\left(\frac{-59.298}{171.58}\right) = +172.79$$

The values of M at points A, B, C, and D are computed in Table 15.10.2. The answers are

$M_A = -\ 8.50$ kN·m   (compression inside)

$M_B = -14.90$ kN·m   (compression inside)

$M_C = -19.49$ kN·m   (compression inside)

$M_D = -\ 9.89$ kN·m   (compression inside)

After the free-body diagrams, the moment diagram, and the generalized joint displacements at points A, B, C, and D of the elastic curve are all computed and drawn as shown in Fig. 15.10.3, the usual statics checks and compatibility checks are observed in order to ensure the correctness of the solution.

15.11 Exercises

Twenty-one exercises are suggested. Except for Exercises 15.8 and 15.9, there are alternate solutions for each problem by using different basic determinate structures. Exercises 15.6 to 15.9 and 15.15 to 15.20 have been used in earlier chapters.

15.1 to 15.5 Determine the fixed-end moments for a beam element with constant moment of inertia (Figs. 15.11.1 to 15.11.5).

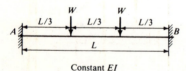

Figure 15.11.1 Exercise 15.1.

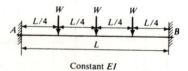

Figure 15.11.2 Exercise 15.2.

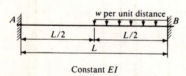

Figure 15.11.3 Exercise 15.3.

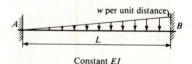

Figure 15.11.4 Exercise 15.4.

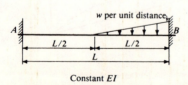

Figure 15.11.5 Exercise 15.5.

THE COLUMN-ANALOGY METHOD 581

15.6 and 15.7 Determine the fixed-end moments for a beam element with variable moment of inertia (Figs. 15.11.6 and 15.11.7).

15.8 and 15.9 Determine the stiffness and carry-over factors for a beam element with variable moment of inertia (Figs. 15.11.8 and 15.11.9).

15.10 to 15.12 Analyze the quadrangular frames with one axis of symmetry (Figs. 15.11.10 to 15.11.12).

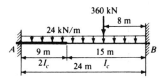

Figure 15.11.6 Exercise 15.6.

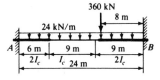

Figure 15.11.7 Exercise 15.7.

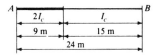

Figure 15.11.8 Exercise 15.8.

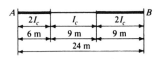

Figure 15.11.9 Exercise 15.9.

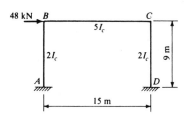

Figure 15.11.10 Exercise 15.10.

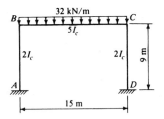

Figure 15.11.11 Exercise 15.11.

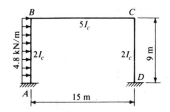

Figure 15.11.12 Exercise 15.12.

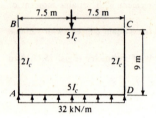
Figure 15.11.13 Exercise 15.13.

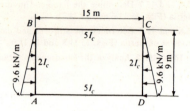
Figure 15.11.14 Exercise 15.14.

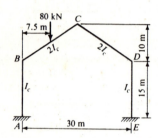
Figure 15.11.15 Exercise 15.15.

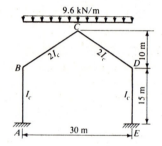
Figure 15.11.16 Exercise 15.16.

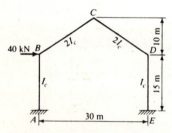
Figure 15.11.17 Exercise 15.17.

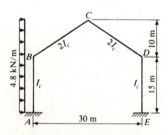
Figure 15.11.18 Exercise 15.18.

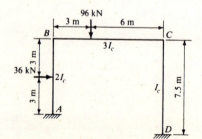
Figure 15.11.19 Exercise 15.19.

THE COLUMN-ANALOGY METHOD 583

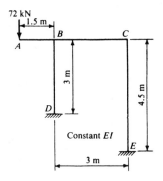

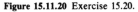

Figure 15.11.20 Exercise 15.20.

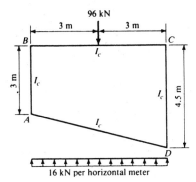

Figure 15.11.21 Exercise 15.21.

15.13 and **15.14** Analyze the closed frames with one axis of symmetry (Figs. 15.11.13 and 15.11.14).

15.15 to **15.18** Analyze the gable frames with one axis of symmetry (Figs. 15.11.15 to 15.11.18).

15.19 and **15.20** Analyze the unsymmetrical quadrangular frames (Figs. 15.11.19 and 15.11.20).

15.21 Analyze the unsymmetrical closed frame (Figs. 15.11.21).

CHAPTER
SIXTEEN
COMPOSITE STRUCTURES AND RIGID FRAMES WITH AXIAL DEFORMATION

16.1 General Description

Up to this point, three types of framed structures, each distinct by itself, have been considered: trusses, beams, and rigid frames. In a truss, each bar is either in axial tension or in axial compression; it becomes longer or shorter due to the presence of axial force. In a beam, every span, or a short segment within the span, is subjected to transverse shear and bending moment. Shear deformation will not be considered until Chap. 21; bending deformation causes a change in slope equal to $(M\,dx)/EI$ in a length equal to dx. In a rigid frame, every bar, or a short segment in a straight bar, may be under the combined action of axial force, transverse shear, and bending moment. So far deformation due to both shear and axial force has been neglected; only bending deformation has been considered. One must note, however, that although axial deformation is ignored in obtaining the "deformation response" of a rigid frame, axial force must be considered in the "force response," and the strength of the members must be designed for combined axial force and bending.

If in rigid-frame analysis axial deformation is to be considered, the results should be more "accurate" than those obtained when axial deformation is neglected, but if they are not much different the question is whether the more elaborate analysis is worth the effort. The results are indeed not much different in most cases, except possibly in the case of a very tall building of irregular story heights, in which the unequal axial deformation in adjacent lower columns where story heights change suddenly would cause rotations of the girder axes. However, the more elaborate analysis—taking axial deformation into consideration—may actually take less time by making the input

COMPOSITE STRUCTURES AND RIGID FRAMES WITH AXIAL DEFORMATION 585

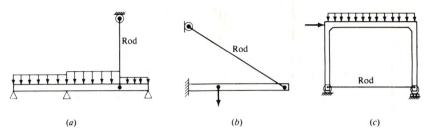

Figure 16.1.1 Typical composite structures.

data more systematic for a computer program. As computing costs become lower and lower, considering axial deformation in rigid-frame analysis may become even more economical. Within the input data, the absolute values of the longitudinal rigidity EA and the flexural rigidity EI must be used, or at least the two kinds of rigidities must be in their correct proportional relationship. This requirement may be a deterrent to considering axial deformation, because otherwise only relative values of the flexural rigidity of the constituent members would be needed.

So far as structural elements are concerned, there are three kinds: one subjected to axial force only, called a *truss element*; one subjected to shear and bending only, called a *beam element*; and one subjected to combined axial force and bending, called a *combined element* (or combined truss and beam element). When a framed structure contains more than one kind of element, it cannot be called a truss, a beam, or a rigid frame; so the name *composite structure* is used. The structure in Fig. 16.1.1a has one truss element and three beam elements; in Fig. 16.1.1b, one truss element and one combined element; and in Fig. 16.1.1c, one truss element and three combined elements. In each case, the number of elements refers only to the smallest number of elements which have to be used in the matrix displacement method of analysis.

For the truss elements in a composite structure, axial deformation should be considered, as is commonly done in truss analysis; for combined elements it is up to the judgment of the analyst whether axial deformation in some or all of them is to be neglected or considered. Because the analysis of composite structures, by the force or displacement method, can be equally applied to the analysis of rigid frames with axial deformation, the two subjects are treated together in this chapter.

16.2 Composite Structure with Truss and Beam Elements—Force Method

As noted earlier, the force method, or the redundant-force method, should rightfully refer to statically indeterminate structures. The compatibility conditions in the force method can be obtained either by physical considerations,

586 INTERMEDIATE STRUCTURAL ANALYSIS

or by applying the theorem of least work. In the following example (which, even though the structure is statically indeterminate only to the first degree, is appropriate for illustration), two alternate choices of the redundant are made; in each case, the compatibility condition is established by physical considerations and by applying the theorem of least work.

Example 16.2.1 By the force method analyze the composite structure shown in Fig. 16.2.1a using first the tension in the rod and then the reaction at E as the redundant.

SOLUTION (a) *T as redundant, theorem of least work.* Applying statics to the beam $ABCDE$ in the structure of Fig. 16.2.1b,

$$R_A = \frac{10.5}{15}(6) + \frac{3}{15}(2.4) - \frac{6}{15}T = 4.68 - 0.4T$$

$$R_E = \frac{4.5}{15}(6) + \frac{12}{15}(2.4) - \frac{9}{15}T = 3.72 - 0.6T$$

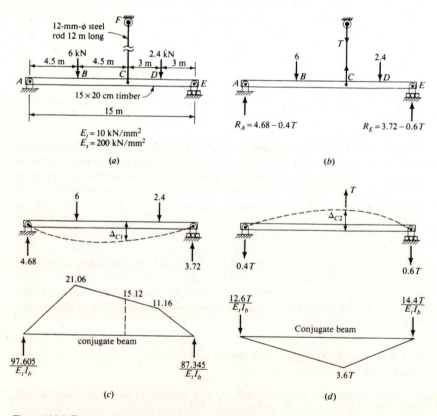

Figure 16.2.1 Force method, T as redundant, Example 16.2.1.

COMPOSITE STRUCTURES AND RIGID FRAMES WITH AXIAL DEFORMATION 587

The total strain energy in the beam and rod is

$$W = \frac{1}{2}\frac{T^2 12}{E_s A_s} + \frac{1}{2}\int \frac{M^2\,dx}{E_t I_b}$$

$$= \frac{1}{2}\frac{T^2 12}{E_s A_s} + \frac{1}{2E_t I_b}\left\{\int_0^{4.5}[(4.68-0.4T)x]^2\,dx\right.$$

$$+ \int_{4.5}^9 [(4.68-0.4T)x - 6(x-4.5)]^2\,dx$$

$$+ \int_0^3 [(3.72-0.6T)x]^2\,dx$$

$$\left.+ \int_3^6 [(3.72-0.6T)x - 2.4(x-3)]^2\,dx\right\}$$

Taking the partial derivative of W with respect to T and setting it equal to zero,

$$\frac{\partial W}{\partial T} = \frac{T(12)}{E_s A_s} + \frac{1}{E_t I_b}\left\{\int_0^{4.5}[(4.68-0.4T)x](-0.4x)\,dx\right.$$

$$+ \int_{4.5}^9 [(4.68-0.4T)x - 6(x-4.5)](-0.4x)\,dx$$

$$+ \int_0^3 [(3.72-0.6T)x](-0.6x)\,dx$$

$$\left.+ \int_3^6 [(3.72-0.6T)x - 2.4(x-3)](-0.6x)\,dx\right\} = 0$$

The compatibility condition is obtained by simplifying the above equation; thus

$$\frac{T(12)}{E_s A_s} - \frac{400.95}{E_t I_b} + \frac{64.8T}{E_t I_b} = 0$$

(b) T as redundant, physical compatibility. If the rod is removed, the downward deflection Δ_{C1} of point C due to the two loads applied on the beam $ABCDE$ is, from applying the conjugate-beam theorem to the moment diagram of Fig. 16.2.1c,

$$\Delta_{C1} = \frac{97.605}{E_t I_b}(9) - \frac{1}{2}\frac{21.06}{E_t I_b}(4.5)(6) - \frac{1}{2}\frac{21.06}{E_t I_b}(4.5)(3) - \frac{1}{2}\frac{15.12}{E_t I_b}(4.5)(1.5)$$

$$= \frac{400.95}{E_t I_b}$$

The upward deflection Δ_{C2} of point C due to the tensile force T in the rod is, from Fig. 16.2.1d,

$$\Delta_{C2} = \frac{12.6T}{E_t I_b}(9) - \frac{1}{2}\frac{3.6T}{E_t I_b}(9)(3) = \frac{64.8T}{E_t I_b}$$

The net downward deflection $\Delta_{C1} - \Delta_{C2}$ must be equal to the elongation of the rod; thus

$$\Delta_{C1} - \Delta_{C2} = \frac{T(12)}{E_s A_s}$$

or
$$\frac{400.95}{E_t I_b} - \frac{64.8T}{E_t I_b} = \frac{T(12)}{E_s A_s}$$

which is identical to the equation for T obtained in part (a).

(c) R_E as redundant, theorem of least work. Applying statics to the beam $ABCDE$ in the structure of Fig. 16.2.2a,

$$R_A = \frac{6R_E + 6(4.5) - 2.4(3)}{9} = \tfrac{2}{3}R_E + 2.20$$

$$T = \frac{6(4.5) + 2.4(12) - 15R_E}{9} = 6.20 - \tfrac{5}{3}R_E$$

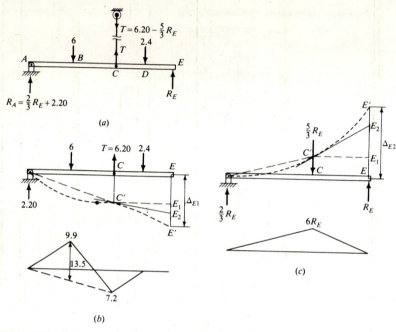

Figure 16.2.2 Force method, R_E as redundant, Example 16.2.1.

The total strain energy in the beam and rod is

$$W = \frac{1}{2}\frac{T^2 L}{E_s A_s} + \frac{1}{2}\int \frac{M^2\,dx}{E_t I_b}$$

$$= \frac{1}{2}\frac{(6.20 - \frac{5}{3}R_E)^2(12)}{E_s A_s} + \frac{1}{2E_t I_b}\left\{\int_0^{4.5}[(\tfrac{2}{3}R_E + 2.20)x]^2\,dx \right.$$

$$+ \int_{4.5}^{9}[(\tfrac{2}{3}R_E + 2.20)x - 6(x-4.5)]^2\,dx$$

$$\left. + \int_0^3 (R_E x)^2\,dx + \int_3^6 [R_E x - 2.4(x-3)]^2\,dx\right\}$$

Setting the partial derivative of W with respect to R_E equal to zero,

$$\frac{\partial W}{\partial R_E} = \frac{(6.20 - \tfrac{5}{3}R_E)(-\tfrac{5}{3})(12)}{E_s A_s} + \frac{1}{E_t I_b}\left\{\int_0^{4.5}[(\tfrac{2}{3}R_E + 2.20)x](\tfrac{2}{3}x)\,dx \right.$$

$$+ \int_{4.5}^{9}[(\tfrac{2}{3}R_E + 2.20)x - 6(x-4.5)](\tfrac{2}{3}x)\,dx$$

$$\left. + \int_0^3 (R_E x)(x)\,dx + \int_3^6 [R_E x - 2.4(x-3)](x)\,dx\right\}$$

$$= 0$$

Simplifying the above equation,

$$-\frac{124}{E_s A_s} + \frac{100 R_E}{3 E_s A_s} - \frac{1.35}{E_t I_b} + \frac{180 R_E}{E_t I_b} = 0$$

COMPOSITE STRUCTURES AND RIGID FRAMES WITH AXIAL DEFORMATION

(d) R_E as redundant, physical compatibility. The physical compatibility requires

$$\Delta_{E1} \text{ in Fig. 16.2.2b} = \Delta_{E2} \text{ in Fig. 16.2.2c}$$

In Fig. 16.2.2b,

$$EE_1 = CC' = \text{elongation of tie rod} = \frac{6.20(12)}{E_s A_s} = \frac{74.4}{E_s A_s}$$

$$\theta_C = \frac{CC'}{9} + \frac{1}{2}\frac{7.2}{E_t I_b}(9)\left(\frac{2}{3}\right) - \frac{1}{2}\frac{13.5}{E_t I_b}(9)\left(\frac{1}{2}\right) = \frac{74.4}{9E_s A_s} - \frac{8.775}{E_t I_b} \quad \text{(clockwise)}$$

$$\Delta_{E1} = EE_1 + E_1 E_2 + E_2 E' = \frac{74.4}{E_s A_s} + 6\theta_C + \frac{1}{2}\frac{7.2}{E_t I_b}(3)(5)$$

$$= \frac{124}{E_s A_s} + \frac{1.35}{E_t I_b}$$

In Fig. 16.2.2c,

$$EE_1 = CC' = \text{shortening of tie rod} = \frac{\frac{5}{3}R_E(12)}{E_s A_s} = \frac{20R_E}{E_s A_s}$$

$$\theta_C = \frac{CC'}{9} + \frac{1}{2}\frac{6R_E}{E_t I_b}(9)\left(\frac{2}{3}\right) = \frac{20R_E}{9E_s A_s} + \frac{18R_E}{E_t I_b}$$

$$\Delta_{E2} = EE_1 + E_1 E_2 + E_2 E' = \frac{20R_E}{E_s A_s} + 6\theta_C + \frac{1}{2}\frac{6R_E}{E_t I_b}(6)(4)$$

$$= \frac{100R_E}{3E_s A_s} + \frac{180R_E}{E_t I_b}$$

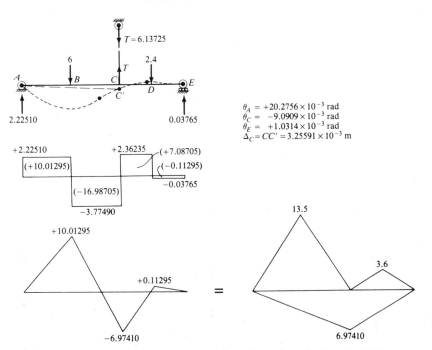

$\theta_A = +20.2756 \times 10^{-3}$ rad
$\theta_C = -9.0909 \times 10^{-3}$ rad
$\theta_E = +1.0314 \times 10^{-3}$ rad
$\Delta_C = CC' = 3.25591 \times 10^{-3}$ m

Figure 16.2.3 Statics and deformation of suspended continuous beam in Example 16.2.1.

Equating Δ_{E1} to Δ_{E2},

$$\frac{124}{E_s A_s} + \frac{1.35}{E_t I_b} = \frac{100 R_E}{3 E_s A_s} + \frac{180 R_E}{E_t I_b}$$

which is identical to the equation for R_E obtained in part (c).

(e) *Statics and compatibility checks.* The expression for T, from parts (a) and (b), is

$$T = \frac{400.95/E_t I_b}{12/E_s A_s + 64.8/E_t I_b} = \frac{400.95 \times 10^{-3}}{(0.530516 + 64.8) \times 10^{-3}} = 6.13725 \text{ kN}$$

The expression for R_E, from parts (c) and (d), is

$$R_E = \frac{124/E_s A_s + 1.35/E_t I_b}{100/3 E_s A_s + 180/E_t I_b} = \frac{(5.48200 + 1.35) \times 10^{-3}}{(1.47366 + 180) \times 10^{-3}} = 0.03765 \text{ kN}$$

The shear and moment diagrams for the beam are shown in Fig. 16.2.3. Because this structure is statically indeterminate to the first degree, there is to be one check on compatibility of deformation. This check may be obtained either (1) by computing the deflection at C of a simple beam $ABCDE$ and seeing that this deflection is equal to the elongation of the tie rod, or (2) by using the elongation of the tie rod as the deflection at C and seeing that the slope at C obtained from the moment diagram on AC is equal to that obtained from the moment diagram on CE. The reader should verify the joint rotations and deflections given in Fig. 16.2.3.

16.3 Composite Structure with Truss and Beam Elements—Displacement Method

The fundamental equations in the displacement method are twofold; they are

$$\{X\} = [ASB]^{-1}\{P\} \quad (16.3.1)$$

and

$$\{F^*\} = \{F_0\} + [SB]\{X\} \quad (16.3.2)$$

In applying the displacement method, a composite structure is first separated into a finite number of elements connected at the joints (or nodes); each element must not have unknown forces acting on itself. The degree of freedom is the sum of the number of unknown joint rotations and of the number of unknown horizontal and vertical joint deflections; in other words it is the total number of unknown generalized joint displacements. The number of independent unknown generalized internal forces is equal to 1 times the number of truss elements, 2 times the number of beam elements, and 2 or 3 times the number of combined elements depending on whether axial deformation of the combined element in question is to be neglected or considered.

When axial deformation in any combined element is to be neglected in the analysis, the axial force in this element is no longer an independent unknown internal force and as such will not be in the $\{F\}$ matrix. Consequently the number of unknown horizontal and vertical deflections will be 1 less than otherwise for each such change from an independent unknown to a dependent

COMPOSITE STRUCTURES AND RIGID FRAMES WITH AXIAL DEFORMATION 591

unknown. Care then must be taken to assign the unknown joint displacements only to the correct locations. In fact, the discussion here pertains more to Sec. 16.5 than to this section, since the following example contains only truss and beam elements.

Example 16.3.1 Analyze the composite structure in Example 16.2.1 by the displacement method, considering axial deformation in the steel rod.

SOLUTION (a) *P-X and F-e numbers*. From observation of the given structure shown again in Fig. 16.3.1a, the smallest possible number of joints to be taken is four, at A, C, E, and F. AC and CE are beam elements and CF is a truss element. The P-X and F-e numbers are shown in Fig. 16.3.1b and c; thus $NI = NF - NP = 5 - 4 = 1$. Note that F_5 is the tensile force in the steel rod, and there are no axial forces in elements 1-2 and 3-4.

(b) *The input matrices* $[A]$, $[B]$, $[S]$, *and* $\{P\}$. The statics matrix $[A]$ is established by rows from the free-body diagrams of the joints in Fig. 16.3.1d; the deformation matrix $[B]$ by columns from the displacement diagrams of Fig. 16.3.1e. Note, however, $X_4 \neq 0$ not only causes end rotations $e_1 = e_2 = -X_4/9$ and $e_3 = e_4 = +X_4/6$, but also elongation $e_5 = +X_4$. The contents of the $[S]$ matrix are

$$S_{11} = S_{22} = \frac{4E_t I_b}{9} = \frac{4(10)(150)(200)^3}{9(12)} \times 10^{-6} = 444.4444 \text{ kN·m}$$

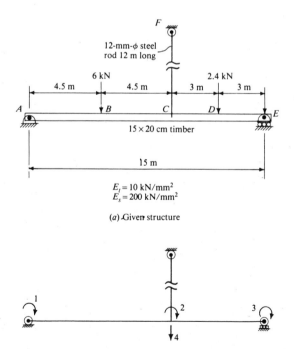

Figure 16.3.1 Displacement method, suspended continuous beam, Example 16.3.1.

592 INTERMEDIATE STRUCTURAL ANALYSIS

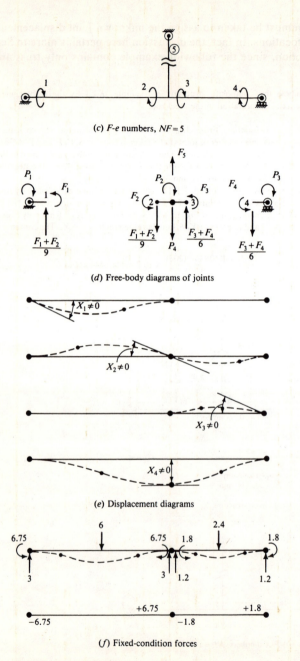

(c) F-e numbers, $NF = 5$

(d) Free-body diagrams of joints

(e) Displacement diagrams

(f) Fixed-condition forces

Figure 16.3.1 (*continued*).

COMPOSITE STRUCTURES AND RIGID FRAMES WITH AXIAL DEFORMATION

$$S_{12} = S_{21} = \frac{2E_tI_b}{9} = 222.2222 \text{ kN·m}$$

$$S_{33} = S_{44} = \frac{4E_tI_b}{6} = \frac{4(10)(150)(200)^3}{6(12)} \times 10^{-6} = 666.6666 \text{ kN·m}$$

$$S_{34} = S_{43} = \frac{2E_tI_b}{6} = 333.3333 \text{ kN·m}$$

$$S_{55} = \frac{E_sA_s}{12} = \frac{200\pi(6)^2}{12} = 1884.9556 \text{ kN/m}$$

The contents of the $\{P\}$ matrix are taken from the fixed-condition forces in Fig. 16.3.1f.

$[A]_{4\times 5} = $

F \\ P	1	2	3	4	5
1	+1.				
2		+1.	+1.		
3				+1.	
4	−1/9	−1/9	+1/6	+1/6	+1.

$\{P\}_{4\times 1} = $

LC \\ P	1
1	+6.75
2	−4.95
3	−1.80
4	+4.20

$[B]_{5\times 4} = $

X \\ e	1	2	3	4
1	+1.			−1/9
2		+1.		−1/9
3		+1.		+1/6
4			+1.	+1/6
5				+1.

$[S]_{5\times 5} = $

e \\ F	1	2	3	4	5
1	S_{11}	S_{12}			
2	S_{21}	S_{22}			
3			S_{33}	S_{34}	
4			S_{43}	S_{44}	
5					S_{55}

594 INTERMEDIATE STRUCTURAL ANALYSIS

(c) *The output matrices $\{X\}$ and $\{F^*\}$*. The $\{X\}$ matrix is taken from Fig. 16.2.3 of Example 16.2.1, and the equation $\{P\} = [ASB]\{X\}$ is verified. The $\{F^*\}$ matrix obtained from $\{F^*\} = \{F_0\} + [SB]\{X\}$ checks exactly with the results of the force method in Example 16.2.1.

$[SB]_{5\times 4} =$

F \ X	1	2	3	4
1	+444.44	+222.22		− 74.073
2	+222.22	+444.44		− 74.073
3		+666.66	+333.33	+ 166.666
4		+333.33	+666.66	+ 166.666
5				+1884.96

$[ASB]_{4\times 4} =$

P \ X	1	2	3	4
1	+444.44	+ 222.22		− 74.073
2	+222.22	+1111.11	+333.33	+ 92.593
3		+ 333.33	+666.66	+ 166.66
4	− 74.073	+ 92.593	+166.66	+1956.98

$\{X\}_{4\times 1} =$

X \ LC	1
1	$+20.2756 \times 10^{-3}$
2	-9.0909×10^{-3}
3	$+1.0314 \times 10^{-3}$
4	$+3.25591 \times 10^{-3}$

$\{F^*\}_{5\times 1} =$

F* \ LC	1
1	0.0000
2	+6.9741
3	−6.9741
4	0.0000
5	+6.13725

16.4 Composite Structure with Truss and Combined Elements—Force Method

Three examples will be shown in this section. The first example involves a cantilever beam with an inclined rod connecting the outer end of the beam

COMPOSITE STRUCTURES AND RIGID FRAMES WITH AXIAL DEFORMATION 595

and a point directly above the support. For this composite structure, statically indeterminate to the first degree, two alternate solutions are considered, one using the tensile force in the rod as the redundant and the other using the reacting moment at the fixed support as the redundant. The second example is the commonly known king-post truss, wherein the carrying capacity of a timber beam can be increased almost fourfold by adding a short timber post under the midpoint of the beam, with two steel rods connecting the lower end of the post to the ends of the timber beam. For this problem, it seems natural just to take the axial force in the king post as the redundant. The third example involves a quadrangular rigid frame, wherein the lower end of the left column is hinged, the lower end of the right column is supported on rollers, and the lower ends of the two columns are tied together by a steel rod.

For all three examples, closed-form expressions for the redundant are obtained by both physical compatibility and the theorem of least work, for the purpose of showing that the theorem of least work is just a mathematical way of stating the physical requirement for compatibility of deformation. It is the nature of the force method, as is evident from the expressions for the redundant, that it is a simple matter to ignore the axial deformation of a truss or a combined element by substituting the value of infinity for the longitudinal stiffness of that element. Because the longitudinal stiffness always appears in the denominator in the compatibility equation, the quotient (which is the longitudinal flexibility of the element) vanishes. Thus the effect of considering or neglecting the axial deformation in any or all of the elements can be easily determined.

Example 16.4.1 By the force method analyze the tied-back cantilever beam shown in Fig. 16.4.1a, using first the tension in the tie rod and then the reacting moment at C as the redundant.

SOLUTION (a) T *as redundant, theorem of least work*. Applying statics to the beam ABC in Fig. 16.4.1b,

$$H_C = 0.8T \qquad V_C = 24 - 0.6T$$
$$M_C = 24(2.5) - 0.6T(4) = 60 - 2.4T$$

The total strain energy in the beam and rod is

$$W = \frac{1}{2}\frac{T^2(5)}{E_s A_s} + \frac{1}{2}\frac{(-0.8T)^2(4)}{E_t A_b}$$
$$+ \frac{1}{2E_t I_b}\left\{\int_0^{1.5}(0.6Tx)^2\,dx + \int_{1.5}^{4}[0.6Tx - 24(x-1.5)]^2\,dx\right\}$$

Setting the partial derivative of W with respect to T equal to zero,

$$\frac{\partial W}{\partial T} = \frac{T(5)}{E_s A_s} + \frac{(-0.8T)(-0.8)(4)}{E_t A_b}$$
$$+ \frac{1}{E_t I_b}\left\{\int_0^{1.5}(0.6Tx)(0.6)\,dx + \int_{1.5}^{4}[0.6Tx - 24(x-1.5)](0.6)\,dx\right\}$$
$$= 0$$

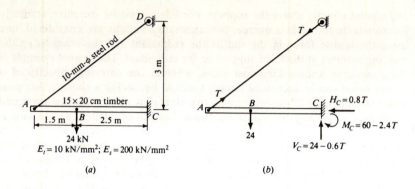

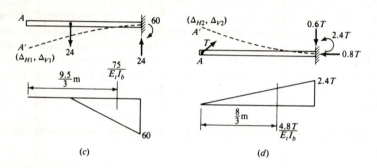

Figure 16.4.1 Force method, T as redundant, Example 16.4.1.

Simplifying the above expression,

$$\frac{5T}{E_s A_s} + \frac{2.56}{E_t A_b} + \frac{7.68T}{E_t I_b} - \frac{142.5}{E_t I_b} = 0$$

(b) *T as redundant, physical compatibility*. Without the tie rod, let the deflection AA' in Fig. 16.4.1c be Δ_{H1} to the right and Δ_{V1} upward; then,

$$\Delta_{H1} = 0 \qquad \Delta_{V1} = -\frac{75}{E_t I_b}\left(\frac{9.5}{3}\right) = -\frac{237.5}{E_t I_b}$$

Owing to force T acting on the beam, the deflection AA' in Fig. 16.4.1d is upward to the right, or

$$\Delta_{H2} = +\frac{0.8T(4)}{E_t A_b} \qquad \Delta_{V2} = \frac{4.8T}{E_t I_b}\left(\frac{8}{3}\right) = +\frac{12.8T}{E_t I_b}$$

The physical condition for compatibility is that the deflection component of point A in the extended direction of the tie rod must be equal to the elongation of the tie rod; thus

$$-(\Delta_{H1} + \Delta_{H2})(0.8) - (\Delta_{V1} + \Delta_{V2})(0.6) = +\frac{T(5)}{E_s A_s}$$

$$-\left(0 + \frac{3.2T}{E_t A_b}\right)(0.8) - \left(-\frac{237.5}{E_t I_b} + \frac{12.8T}{E_t I_b}\right)(0.6) = +\frac{T(5)}{E_s A_s}$$

COMPOSITE STRUCTURES AND RIGID FRAMES WITH AXIAL DEFORMATION

Simplifying the above equation,

$$\frac{5T}{E_sA_s} + \frac{2.56T}{E_tA_b} + \frac{7.68T}{E_tI_b} - \frac{142.5}{E_tI_b} = 0$$

which is identical to the equation for T obtained in part (a).

(c) M_C as redundant, theorem of least work. The three equations of statics are applied to the structure of Fig. 16.4.2a to solve for T, H_C, and V_C in terms of the redundant M_C. For ΣM about $C = 0$,

$$0.6T(4) + M_C = 24(2.5) \qquad T = 25 - \frac{5}{12}M_C$$

For $\Sigma F_x = 0$,

$$H_C = 0.8T = 0.8(25 - \tfrac{5}{12}M_C) = 20 - \tfrac{1}{3}M_C$$

For $\Sigma F_y = 0$,

$$V_C = 24 - 0.6T = 24 - 0.6(25 - \tfrac{5}{12}M_C) = 9 + \tfrac{1}{4}M_C$$

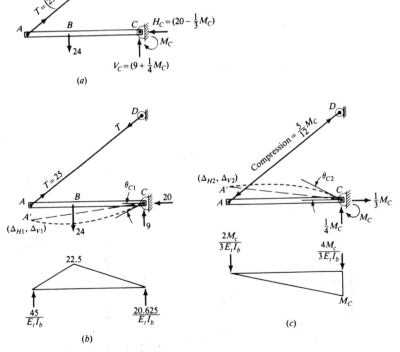

Figure 16.4.2 Force method, M_C as redundant, Example 16.4.1.

The total strain energy in the beam and rod is

$$W = \frac{1}{2}\frac{(25-\frac{5}{12}M_C)^2(5)}{E_sA_s} + \frac{1}{2}\frac{(-20+\frac{1}{3}M_C)^2(4)}{E_tA_b}$$
$$+ \frac{1}{2E_tI_b}\left\{\int_0^{1.5}[0.6(25-\tfrac{5}{12}M_C)x]^2\,dx + \int_0^{2.5}[-M_C+(9+\tfrac{1}{4}M_C)x]^2\,dx\right\}$$

Setting the partial derivative of W with respect to M_C equal to zero,

$$\frac{\partial W}{\partial M_C} = \frac{(25-\tfrac{5}{12}M_C)(-\tfrac{5}{12})(5)}{E_sA_s} + \frac{(-20+\tfrac{1}{3}M_C)(+\tfrac{1}{3})(4)}{E_tA_b}$$
$$+ \frac{1}{E_tI_b}\left\{\int_0^{1.5}[0.6(25-\tfrac{5}{12}M_C)x](-\tfrac{1}{4}x)\,dx\right.$$
$$\left.+ \int_0^{2.5}[-M_C+(9+\tfrac{1}{4}M_C)x](-1+\tfrac{1}{4}x)\,dx\right\}$$

Simplifying the above equation,

$$-\frac{625}{12E_sA_s} + \frac{125M_C}{144E_sA_s} - \frac{80}{3E_tA_b} + \frac{4M_C}{9E_tA_b} - \frac{20.625}{E_tI_b} + \frac{4M_C}{3E_tI_b} = 0$$

(d) M_C **as redundant, physical compatibility.** The basic determinate structure is obtained by changing the fixed support at C into a hinged support. The compatibility condition is that the counterclockwise rotation θ_{C1} in Fig. 16.4.2b should be equal to the clockwise rotation θ_{C2} in Fig. 16.4.2c. In Fig. 16.4.2b,

$$\Delta_{H1}\text{ to right} = \text{shortening of }ABC = \frac{20(4)}{E_tA_b} = \frac{80}{E_tA_b}$$

$$(\Delta_{V1}\text{ downward})(0.6) - (\Delta_{H1}\text{ to right})(0.8) = \text{elongation of }AD = \frac{25(5)}{E_sA_s}$$

$$(\Delta_{V1}\text{ downward})(0.6) - \left(\frac{80}{E_tA_b}\right)(0.8) = \frac{25(5)}{E_sA_s}$$

$$(\Delta_{V1}\text{ downward}) = \frac{625}{3E_sA_s} + \frac{320}{3E_tA_b}$$

$$\theta_{C1} = \frac{\Delta_{V1}\text{ downward}}{4} + \begin{pmatrix}\text{Right reaction to conjugate}\\ \text{beam in Fig. 16.4.2b}\end{pmatrix}$$
$$= \frac{1}{4}\left(\frac{625}{3E_sA_s} + \frac{320}{3E_tA_b}\right) + \frac{20.625}{E_tI_b} = \frac{625}{12E_sA_s} + \frac{80}{3E_tA_b} + \frac{20.625}{E_tI_b}$$

In Fig. 16.4.2c,

$$\Delta_{H2}\text{ to left} = \text{elongation of }ABC = \frac{(\tfrac{1}{3}M_C)(4)}{E_tI_b}$$

$$(\Delta_{V2}\text{ upward})(0.6) - (\Delta_{H2}\text{ to left})(0.8) = \text{shortening of }AD = \frac{(\tfrac{5}{12}M_C)(5)}{E_sA_s}$$

$$(\Delta_{V2}\text{ upward})(0.6) - \frac{4M_C}{3E_tI_b}(0.8) = \frac{25M_C}{12E_sA_s}$$

$$(\Delta_{V2}\text{ upward}) = \frac{125M_C}{36E_sA_s} + \frac{16M_C}{9E_tA_b}$$

$$\theta_{C2} = \frac{\Delta_{V2}\text{ upward}}{4} + \begin{pmatrix}\text{Downward reaction at }C\text{ to}\\ \text{conjugate beam in Fig. 16.4.2c}\end{pmatrix}$$
$$= \frac{1}{4}\left(\frac{125M_C}{36E_sA_s} + \frac{16M_C}{9E_tA_b}\right) + \frac{4M_C}{3E_tI_b} = \frac{125M_C}{144E_sA_s} + \frac{4M_C}{9E_tA_b} + \frac{4M_C}{3E_tI_b}$$

COMPOSITE STRUCTURES AND RIGID FRAMES WITH AXIAL DEFORMATION 599

Equating θ_{C1} to θ_{C2}, the compatibility equation is

$$\frac{625}{12E_sA_s} + \frac{80}{3E_tA_b} + \frac{20.625}{E_tI_b} = \frac{125M_C}{144E_sA_s} + \frac{4M_C}{9E_tA_b} + \frac{4M_C}{3E_tI_b}$$

which is identical to the equation for M_C obtained in part (c).

(e) *Statics and compatibility checks.* The expression for T, from parts (a) and (b), is

$$T = \frac{142.5/E_tI_b}{5/E_sA_s + 2.56/E_tA_b + 7.68/E_tI_b}$$

$$= \frac{142.5 \times 10^{-3}}{(0.31831 + 0.00853 + 7.68000) \times 10^{-3}}$$

$$= 17.7973 \text{ kN}$$

The expression for M_C, from parts (c) and (d), is

$$M_C = \frac{\dfrac{625}{12E_sA_s} + \dfrac{80}{3E_tA_b} + \dfrac{20.625}{E_tI_b}}{\dfrac{125}{144E_sA_s} + \dfrac{4}{9E_tA_b} + \dfrac{4}{3E_tI_b}}$$

$$= \frac{(3.31573 + 0.08889 + 20.625) \times 10^{-3}}{(0.055262 + 0.001481 + 1.333333) \times 10^{-3}}$$

$$= 17.2866 \text{ kN·m}$$

The equilibrium and compatibility checks are then made using the above results, as shown in Fig. 16.4.3.

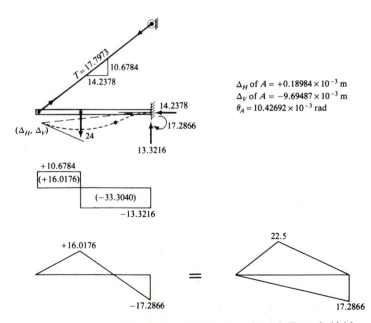

Figure 16.4.3 Statics and deformation of tied-back cantilever in Example 16.4.1.

600 INTERMEDIATE STRUCTURAL ANALYSIS

Example 16.4.2 By the force method analyze the king-post structure of Fig. 16.4.4a, using the compression in the king post as the redundant.

SOLUTION (a) *Using theorem of least work.* By letting S be the compression in the king post, the free-body diagram of the given structure becomes that shown in Fig. 16.4.4b, on which the axial forces in the members are noted. The total strain energy in the structure is

$$W = \frac{1}{2}\frac{(-S)^2(3)}{E_t A_p} + 2\left(\frac{1}{2}\right)\frac{[-S/(2\tan\alpha)]^2(5)}{E_t A_b} + 2\left(\frac{1}{2}\right)\frac{[+S/(2\sin\alpha)]^2\sqrt{34}}{E_s A_s}$$

$$+ \frac{1}{2E_t I_b}\int_0^5 \left[\left(3 - \frac{S}{2}\right)x\right]^2 dx + \frac{1}{2E_t I_b}\int_0^3 \left[\left(7 - \frac{S}{2}\right)x\right]^2 dx$$

$$+ \frac{1}{2E_t I_b}\int_3^5 \left[\left(7 - \frac{S}{2}\right)x - 10(x-3)\right]^2 dx$$

Setting the partial derivative of W with respect to S equal to zero,

$$\frac{\partial W}{\partial S} = \frac{(-S)(3)(-1.0)}{E_t A_p} + 2\frac{[-S/(2\tan\alpha)][-1/(2\tan\alpha)](5)}{E_t A_b} + 2\frac{[+S/(2\sin\alpha)][+1/(2\sin\alpha)]\sqrt{34}}{E_s A_s}$$

$$+ \frac{1}{E_t I_b}\int_0^5 \left[\left(3 - \frac{S}{2}\right)x\right](-\tfrac{1}{2}x)\, dx + \frac{1}{E_t I_b}\int_0^3 \left[\left(7 - \frac{S}{2}\right)x\right](-\tfrac{1}{2}x)\, dx$$

$$+ \frac{1}{E_t I_b}\int_3^5 \left[\left(7 - \frac{S}{2}\right)x - 10(x-3)\right](-\tfrac{1}{2}x)\, dx$$

$$= 0$$

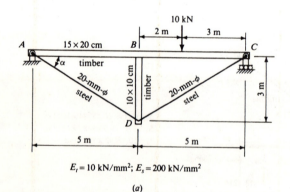

(a)

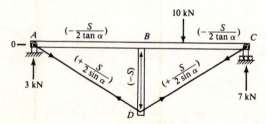

Values in parentheses are axial forces

(b)

Figure 16.4.4 Force method, king-post structure, Example 16.4.2.

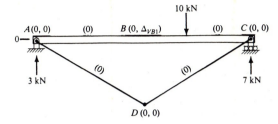

Axial forces and joint displacements

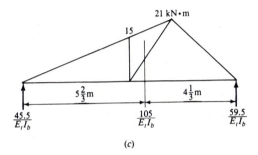

(c)

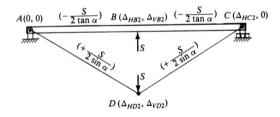

Axial forces and joint displacements

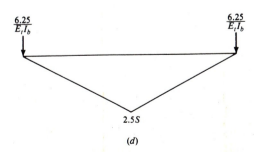

(d)

Figure 16.4.4 (*continued*).

601

Simplifying the above expression,

$$\frac{3S}{E_tA_p} + \frac{125S}{18E_tA_b} + \frac{17\sqrt{34}S}{9E_sA_s} - \frac{165}{E_tI_b} + \frac{125S}{6E_tI_b} = 0$$

(b) *Using physical compatibility.* The king post is removed in Fig. 16.4.4c and its action on the structure is replaced by a pair of S forces in Fig. 16.4.4d. The compatibility condition is that the decrease in distance between B and D in Fig. 16.4.4c minus the increase in distance between B and D in Fig. 16.4.4d must be equal to the shortening of the king post. The axial forces and joint deflections are all zero in Fig. 16.4.4c except for the vertical deflection of point B, which is computed by the conjugate-beam method as

$$\Delta_{VB1} \text{ downward} = \frac{45.5}{E_tI_b}(5) - \frac{1}{2}\frac{15}{E_tI_b}(5)\left(\frac{5}{3}\right) = \frac{165}{E_tI_b}$$

Since point B moves downward and point D does not move, the decrease in distance between B and D in Fig. 16.4.4c is $165/E_tI_b$.

In reference to Fig. 16.4.4d, let Δ_{HB2}, Δ_{HC2}, and Δ_{HD2} be the horizontal deflections to the *right* of points B, C, and D; and Δ_{VB2} and Δ_{VD2} be the vertical deflections *upward* of points B and D.

$$\Delta_{HB2} = -(\text{shortening of } AB) = -\frac{[S/(2\tan\alpha)](5)}{E_tA_b}$$

$$\Delta_{HC2} = -(\text{shortening of } ABC) = -\frac{[S/(2\tan\alpha)](10)}{E_tA_b}$$

$$\Delta_{VB2} = \text{upward deflection at } B \text{ of simple beam } ABC$$
$$= \frac{6.25}{E_tI_b}(5) - \frac{1}{2}\frac{2.5S}{E_tI_b}(5)\left(\frac{5}{3}\right) = \frac{125S}{6E_tI_b}$$

Applying the joint-displacement equation (3.6.2) to bars AD and CD, respectively,

$$+\frac{[S/(2\sin\alpha)]\sqrt{34}}{E_sA_s} = (\Delta_{HD2} - 0)(+\cos\alpha) + (\Delta_{VD2} - 0)(-\sin\alpha)$$

$$+\frac{[S/(2\sin\alpha)]\sqrt{34}}{E_sA_s} = (\Delta_{HD2} - \Delta_{HC2})(-\cos\alpha) + (\Delta_{VD2} - 0)(-\sin\alpha)$$

Solving the above two equations,

$$\Delta_{HD2} = \tfrac{1}{2}(\Delta_{HC2})$$

$$\Delta_{VD2} = -\frac{125S}{18E_tA_b} - \frac{17\sqrt{34}S}{9E_sA_s} \quad \text{(negative means downward)}$$

The increase in distance between B and D in Fig. 16.4.4d is the sum of the upward deflection of B and the downward deflection of D.

The compatibility condition is now

$$\binom{\text{Decrease in distance between}}{B \text{ and } D \text{ in Fig. 16.4.4c}} - \binom{\text{increase in distance between}}{B \text{ and } D \text{ in Fig. 16.4.4d}} = \binom{\text{shortening of}}{\text{king post}}$$

$$\frac{165}{E_tI_b} - \left(\frac{125S}{6E_tI_b} + \frac{125S}{18E_tA_b} + \frac{17\sqrt{34}S}{9E_sA_s}\right) = \frac{S(3)}{E_tA_p}$$

which is identical to the equation for S obtained in part (a).

(c) *Statics and compatibility checks.* The expression for S, from parts (a) and (b), is

$$S = \frac{165/E_tI_b}{\dfrac{3}{E_tA_p} + \dfrac{125}{18E_tA_b} + \dfrac{17\sqrt{34}}{9E_sA_s} + \dfrac{125}{6E_tI_b}}$$

$$= \frac{165 \times 10^{-3}}{(0.03000 + 0.02315 + 0.17529 + 20.83333) \times 10^{-3}}$$

$$= 7.83410 \text{ kN}$$

COMPOSITE STRUCTURES AND RIGID FRAMES WITH AXIAL DEFORMATION 603

The axial forces in all members, the transverse forces acting on ABC, and the shear and moment diagrams for ABC are all shown in Fig. 16.4.5. The compatibility check may be made either (1) by determining all joint deflections without using the shortening of the king post and seeing at last that the shortening of the king post is consistent with the vertical deflections of points B and

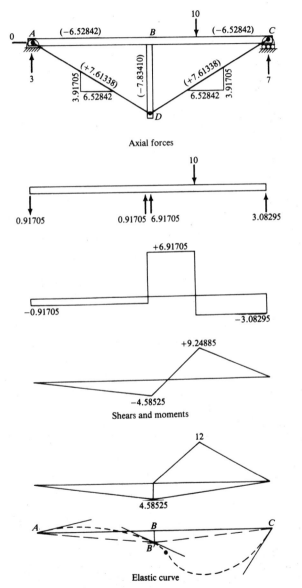

$\theta_A = -3.46311 \times 10^{-3}$ rad
$\theta_B = +8.00000 \times 10^{-3}$ rad
$\theta_C = -10.53689 \times 10^{-3}$ rad
Δ_H of $B = -0.10881 \times 10^{-3}$ m
Δ_V of $B = -1.78964 \times 10^{-3}$ m
Δ_H of $C = -0.21762 \times 10^{-3}$ m
Δ_H of $D = -0.10881 \times 10^{-3}$ m
Δ_V of $D = -1.55461 \times 10^{-3}$ m

Figure 16.4.5 Statics and deformation of king-post structure in Example 16.4.2.

604 INTERMEDIATE STRUCTURAL ANALYSIS

D, or (2) by determining all joint deflections without using the moment diagram on ABC and seeing at last that the slope at B for span BA is equal to the slope at B for span BC. The slopes at A, B, and C, as well as the final positions of points B, C, and D, are shown in Fig. 16.4.5.

Example 16.4.3 By the force method analyze the rigid frame with tie rod as shown in Fig. 16.4.6a, using the tension in the tie rod as the redundant.

SOLUTION (a) *Using theorem of least work.* By letting T be the tension in the tie rod, the free-body diagram of the given structure is that of Fig. 16.4.6b, on which the axial forces in the

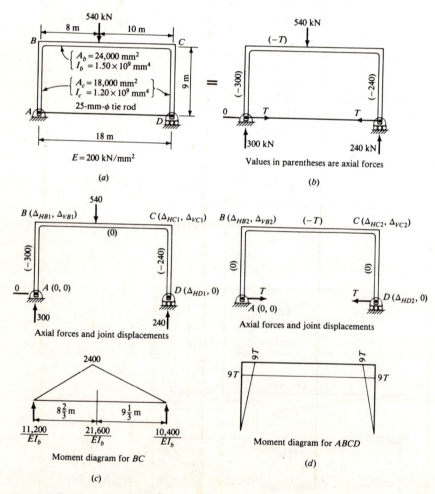

Figure 16.4.6 Force method, rigid frame with tie rod, Example 16.4.3.

COMPOSITE STRUCTURES AND RIGID FRAMES WITH AXIAL DEFORMATION

members are noted. The total strain energy in the structure is

$$W = \frac{1}{2}\frac{(+T)^2(18)}{EA_r} + \frac{1}{2}\frac{(-300)^2(9)}{EA_c} + \frac{1}{2}\frac{(-240)^2(9)}{EA_c} + \frac{1}{2}\frac{(-T)^2(18)}{EA_b}$$

$$+ \frac{1}{2EI_b}\int_0^8 (300x - 9T)^2\, dx + \frac{1}{2EI_b}\int_0^{10}(240x - 9T)^2\, dx + 2\left(\frac{1}{2EI_c}\right)\int_0^9 (Tx)^2\, dx$$

Setting the partial derivative of W with respect to T equal to zero,

$$\frac{\partial W}{\partial T} = \frac{(+T)(1.0)(18)}{EA_r} + \frac{(-T)(-1.0)(18)}{EA_b} + \frac{1}{EI_b}\int_0^8 (300x - 9T)(-9)\, dx$$

$$+ \frac{1}{EI_b}\int_0^{10} (240x - 9T)(-9)\, dx + \frac{2}{EI_c}\int_0^9 (Tx)(x)\, dx$$

$$= 0$$

Simplifying the above equation,

$$\frac{18T}{EA_r} + \frac{18T}{EA_b} - \frac{194{,}400}{EI_b} + \frac{1458T}{EI_b} + \frac{486}{EI_c} = 0$$

(b) *Using physical compatibility.* The tie rod is removed in Fig. 16.4.6c and its action on the structure is replaced by a pair of T forces in Fig. 16.4.6d. The compatibility condition is

$$\begin{pmatrix}\Delta_{HD1}\text{ to right} \\ \text{in Fig. 16.4.6c}\end{pmatrix} - \begin{pmatrix}\Delta_{HD2}\text{ to left} \\ \text{in Fig. 16.4.6d}\end{pmatrix} = \text{elongation of tie rod}$$

In Fig. 16.4.6c,

$$\Delta_{VB1} = \frac{300(9)}{EA_c}\text{ downward} \qquad \Delta_{VC1} = \frac{240(9)}{EA_c}\text{ downward}$$

$$\text{Slope of chord } B'C' = -\frac{2700/EA_c - 2160/EA_c}{18}$$

$$= -\frac{30}{EA_c} \quad \text{(negative means counterclockwise)}$$

From the moment diagram for BC,

$$\theta_B = +\frac{11{,}200}{EI_b} - \frac{30}{EA_c} \qquad \theta_C = -\frac{10{,}400}{EI_b} - \frac{30}{EA_c}$$

$$\Delta_{HB1} = 9\theta_B = 9\left(\frac{11{,}200}{EI_b} - \frac{30}{EA_c}\right) = \frac{100{,}800}{EI_b} - \frac{270}{EA_c} = \Delta_{HC1}$$

$$\Delta_{HD1} = \Delta_{HC1} + 9(\text{counterclockwise } \theta_C)$$

$$= \left(\frac{100{,}800}{EI_b} - \frac{270}{EA_c}\right) + 9\left(\frac{10{,}400}{EI_b} + \frac{30}{EA_c}\right)$$

$$= \frac{194{,}400}{EI_b}$$

In Fig. 16.4.6d,

$$\Delta_{VB2} = \Delta_{VC2} = 0$$

From the moment diagram for BC,

$$\theta_B = -\frac{1}{2}\frac{9T}{EI_b}(18) = -\frac{81T}{EI_b}$$

$$\theta_C = +\frac{81T}{EI_b} \quad \text{(positive means clockwise)}$$

606 INTERMEDIATE STRUCTURAL ANALYSIS

Δ_{HB2} to left = 9(counterclockwise θ_B) + (moment of M/EI area on BA about A)

$$= 9\left(\frac{81T}{EI_b}\right) + \frac{1}{2}\frac{9T}{EI_c}(9)(6) = \frac{729T}{EI_b} + \frac{243T}{EI_c}$$

Δ_{HC2} to left = (Δ_{HB2} to left) + (shortening of BC)

$$= \frac{729T}{EI_b} + \frac{243T}{EI_c} + \frac{T(18)}{EA_b}$$

Δ_{HD2} to left = (Δ_{HC2} to left) + 9(clockwise θ_C) + (moment of M/EI area on CD about D)

$$= \frac{729T}{EI_b} + \frac{243T}{EI_c} + \frac{T(18)}{EA_b} + 9\left(\frac{81T}{EI_b}\right) + \frac{1}{2}\frac{9T}{EI_c}(9)(6)$$

$$= \frac{1458T}{EI_b} + \frac{486T}{EI_c} + \frac{18T}{EA_b}$$

Substituting the expressions just found for Δ_{HD1} and Δ_{HD2} into the compatibility condition,

$$\frac{194,400}{EI_b} - \left(\frac{1458T}{EI_b} + \frac{486T}{EI_c} + \frac{18T}{EA_b}\right) = \frac{T(18)}{EA_r}$$

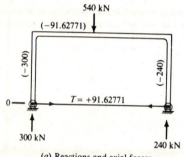

$\theta_A = -2.87666 \times 10^{-3}$ rad
$\theta_B = +12.58552 \times 10^{-3}$ rad
$\theta_C = -9.93552 \times 10^{-3}$ rad
$\theta_D = +5.52666 \times 10^{-3}$ rad
$\Delta_{HB} = +20.49661 \times 10^{-3}$ m
$\Delta_{VB} = -0.75000 \times 10^{-3}$ m
$\Delta_{HC} = +20.15301 \times 10^{-3}$ m
$\Delta_{VC} = -0.60000 \times 10^{-3}$ m
$\Delta_{HD} = +16.79962 \times 10^{-3}$ m

(a) Reactions and axial forces

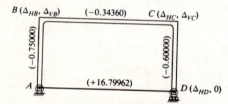

(b) Elongations and joint displacements in 10^{-3} m

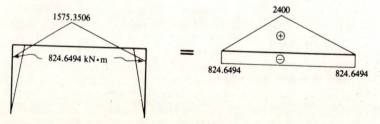

Figure 16.4.7 Statics and deformation of rigid frame with tie rod in Example 16.4.3.

COMPOSITE STRUCTURES AND RIGID FRAMES WITH AXIAL DEFORMATION

which is identical to the equation for T obtained in part (a). Had Δ_{HD1} in Fig. 16.4.6c and Δ_{HD2} in Fig. 16.4.6d been determined by the unit-load method instead of the geometric method using conjugate-beam and moment-area theorems, one would obtain the same integrals that appear in the equation $\partial W/\partial T = 0$ when applying the theorem of least work. Of course, the working formula for the unit-load method, when axial deformation is to be considered, should be the combination of Eqs. (2.6.4) and (3.4.3), or

$$\Delta = \sum \frac{FuL}{EA} + \int_0^L \frac{Mm\,dx}{EI}$$

(c) *Statics and compatibility checks.* The expression for T, from parts (a) and (b), is

$$T = \frac{194{,}400/EI_b}{(18/EA_r) + (18/EA_b) + (486/EA_c) + (1458/EI_b)}$$

$$= \frac{648}{0.183346 + 0.00375 + 2.025 + 4.86}$$

$$= 91.62771 \text{ kN}$$

The reactions and axial forces are shown in Fig. 16.4.7a, and the elongations with joint displacements in Fig. 16.4.7b. The geometric method is used to find the slopes and deflections at A, B, C, and D without using the elongation of the tie rod. The compatibility check is observed by the equality of the horizontal deflection at the roller support to the elongation of the tie rod.

16.5 Composite Structure with Truss and Combined Elements—Displacement Method

As the reader must have observed by this time, the matrix displacement method is quite straightforward in that the twofold formulas of

$$\{X\} = [ASB]^{-1}\{P\} \tag{16.5.1}$$

and $\quad\quad\quad\quad\{F^*\} = \{F_0\} + [SB]\{X\} \tag{16.5.2}$

are to be universally applied. This is why the method is so adaptable to solution by computer. The computer program using the global $[A]$, $[S]$, and $\{P\}$ matrices as input and producing the global $\{X\}$ and $\{F^*\}$ matrices as output is more basic and therefore more instructive to the beginner, and sometimes is more useful in special and irregular problems. The computer program which builds up the global stiffness matrix by feeding into it the individual stiffness matrix of each truss element, beam element, or combined element is more automatic and therefore more desirable in large-scale industrial applications. This latter approach as applied to the general problem of two-dimensional-frame analysis considering axial deformation is further discussed in Chap. 17. Some simplified versions of both types of the computer programs described above are available.†

In this section, the same three examples that have been solved by the force method in Sec. 16.4 will be solved by the displacement method. Again, for purpose of illustration, the equations $\{P\} = [ASB]\{X\}$ and $\{F^*\} = \{F_0\} +$

---

†For instance, see C.-K. Wang, *Matrix Methods of Structural Analysis*, 2d ed., American Publishing Company, Madison, Wis., 1970.

[SB]{X} are verified by using the {X} values provided in the statics and compatibility checks in the solutions for the same problems by the force method. However, there will be further discussion as to what modifications should be made in the solution procedure if the axial deformation in some of the stouter members is to be neglected.

Example 16.5.1 Analyze the tied-back cantilever beam shown in Fig. 16.5.1a by the matrix displacement method, considering the axial deformation in both the timber beam and the steel rod.

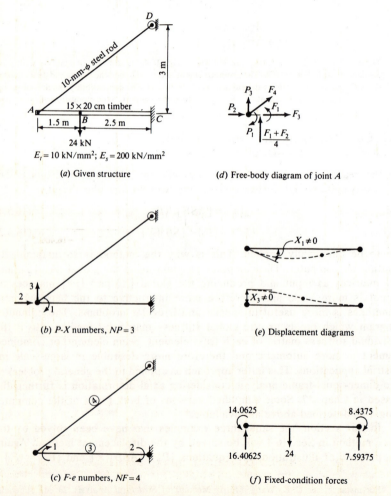

Figure 16.5.1 Displacement method, tied-back cantilever, Example 16.5.1.

COMPOSITE STRUCTURES AND RIGID FRAMES WITH AXIAL DEFORMATION 609

SOLUTION (a) *P-X and F-e numbers.* Using three joints and two elements (one truss element and one combined element), the *P-X* and *F-e* numbers are assigned as shown in Fig. 16.5.1b and c, wherein $NI = NF - NP = 4 - 3 = 1$. Note that a pair of successive odd and even numbers is used for the end moments of the same element, followed by consecutive numbers for the axial forces.

(b) *The input matrices* [A], [B], [S], *and* {P}. The three equilibrium equations in the [A] matrix can be observed from the free-body diagram of joint A in Fig. 16.5.1d. The effect of joint displacements on end rotations e_1 and e_2 can be determined from the displacement diagrams in Fig. 16.5.1e, with the effect on member elongations e_3 and e_4 obtained by inspection. The contents of the {P} matrix are obtained from the fixed-condition forces in Fig. 16.5.1f. The contents of the [S] matrix are

$$S_{11} = S_{22} = \frac{4E_t I_b}{4} = \frac{4(10)(150)(200)^3}{4(12)} \times 10^{-6} = 1000 \text{ kN·m}$$

$$S_{12} = S_{21} = \frac{2E_t I_b}{4} = 500 \text{ kN·m}$$

$$S_{33} = \frac{E_t A_b}{L} = \frac{10(150)(200)}{4} = 75{,}000 \text{ kN/m}$$

$$S_{44} = \frac{E_s A_s}{L} = \frac{200\pi(5)^2}{5} = 3141.59 \text{ kN/m}$$

$[A]_{3 \times 4} =$

P \ F	1	2	3	4
1	+1.			
2			−1.0	−0.8
3	−1/4	−1/4		−0.6

$\{P\}_{3 \times 1} =$

P \ LC	1
1	+14.0625
2	0.
3	−16.40625

$[B]_{4 \times 3} =$

e \ X	1	2	3
1	+1.		−1/4
2			−1/4
3		−1.0	
4		−0.8	−0.6

$[S]_{4 \times 4} =$

F \ e	1	2	3	4
1	S_{11}	S_{12}		
2	S_{21}	S_{22}		
3			S_{33}	
4				S_{44}

(c) *The output matrices* {X} *and* {F*}. The {X} matrix is taken from Fig. 16.4.3 of Example 16.2.1, and the equation $\{P\} = [ASB]\{X\}$ is verified. The {F*} matrix obtained from $\{F^*\} = \{F_0\} + [SB]\{X\}$ checks exactly with the results of the force method in Example 16.4.1.

$[SB]_{4\times 3} =$

F \ X	1	2	3
1	+1000.		−375.
2	+500.		−375.
3		−75000.	
4		−2513.27	−1884.95

$\{F^*\}_{4\times 1} =$

F* \ LC	1
1	0.
2	+17.2866
3	−14.2378
4	+17.7973

$[ASB]_{3\times 3} =$

P \ X	1	2	3
1	+1000.		−375.
2		+77010.62	+1507.96
3	−375.	+1507.96	+1318.47

$\{X\}_{3\times 1} =$

X \ LC	1
1	$+10.42692 \times 10^{-3}$
2	$+0.18984 \times 10^{-3}$
3	-9.69487×10^{-3}

(d) *Discussion.* When axial deformation is to be ignored in any one element, it is necessary to let only the longitudinal stiffness EA of that element be equal to infinity if the force method is being used. If the displacement method is used, especially in a computer solution, some analysts will use a very large EA value (near infinity), as large as the computer will allow, for any element whose axial deformation is to be neglected. This technique sometimes works, but it may lead to a trivial solution for $\{X\}$ or even to grossly erroneous results. For instance, in this present problem, a very large value of S_{33} will cause a very large K_{22} value in the $[K] = [ASB]$ matrix, yielding a near-zero X_2 value and nearly correct X_1 and X_3 values. However, a very large S_{44} value will cause very large K_{22}, K_{23}, K_{32}, and K_{33} values, which may lead to wrong results. The reason is that if axial deformation in the timber beam is to be neglected, one needs to suppress the degree-of-freedom number 2. Yet if axial deformation in the steel rod is to be neglected (an unlikely choice), the horizontal and vertical deflections of point A must bear the ratio of 3 to right and 4 downward; thus only one of the two can be assigned a degree of freedom and the method of using only independent linear degrees of freedom discussed in Chap. 12 will have to be used.

Example 16.5.2 Analyze the king-post structure shown in Fig. 16.5.2a by the displacement method, considering the axial deformation in all members.

COMPOSITE STRUCTURES AND RIGID FRAMES WITH AXIAL DEFORMATION 611

SOLUTION (a) *P-X and F-e numbers.* Using four joints and five elements (three truss elements and two combined elements), the *P-X* and *F-e* numbers are assigned as shown in Fig. 16.5.2b and c, wherein $NI = NF - NP = 9 - 8 = 1$.

(b) *The input matrices* $[A], [B], [S]$, and $\{P\}$. The three moment and five force equations of equilibrium in the $[A]$ matrix are established from the free-body diagrams of the joints in Fig. 16.5.2d. While the effects of joint deflections X_4 to X_8 on member elongations are easily

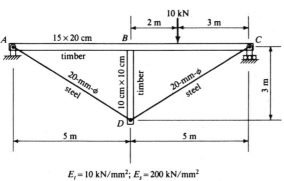

$E_t = 10$ kN/mm^2; $E_s = 200$ kN/mm^2

(a) Given structure

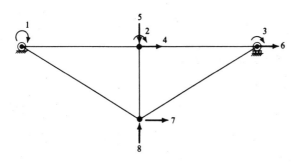

(b) *P-X* numbers, $NP = 8$

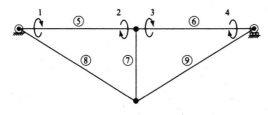

(c) *F-e* numbers, $NF = 9$

Figure 16.5.2 Displacement method, king-post structure, Example 16.5.2.

612 INTERMEDIATE STRUCTURAL ANALYSIS

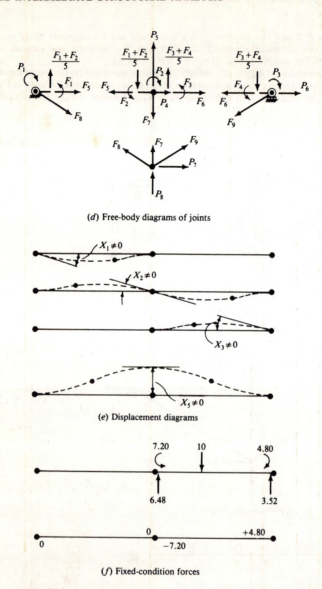

(d) Free-body diagrams of joints

(e) Displacement diagrams

(f) Fixed-condition forces

Figure 16.5.2 (*continued*).

observed, the effects of X_1 to X_3 and of X_5 on member-end rotations e_1 to e_4 can be determined from the joint-displacement diagrams in Fig. 16.5.2e. The contents of the $\{P\}$ matrix are obtained from the fixed-condition forces in Fig. 16.5.2f. The contents of the $[S]$ matrix are

$$S_{11} = S_{22} = S_{33} = S_{44} = \frac{4E_sI_b}{L} = \frac{4(10)(150)(200)^3}{5(12)} \times 10^{-6} = 800 \text{ kN} \cdot \text{m}$$

COMPOSITE STRUCTURES AND RIGID FRAMES WITH AXIAL DEFORMATION

$$S_{12} = S_{21} = S_{34} = S_{43} = \frac{2E_t I_b}{L} = 400 \text{ kN·m}$$

$$S_{55} = S_{66} = \frac{E_t A_b}{5} = \frac{10(150)(200)}{5} = 60,000 \text{ kN/m}$$

$$S_{77} = \frac{E_t A_p}{3} = \frac{10(100)(100)}{3} = 33,333.33 \text{ kN/m}$$

$$S_{88} = S_{99} = \frac{E_s A_s}{L} = \frac{200\pi(10)^2}{\sqrt{34}} = 10,775.37 \text{ kN/m}$$

$[A]_{8\times 9} =$

P \ F	1	2	3	4	5	6	7	8	9
1	+1.								
2		+1.	+1.						
3				+1.					
4					+1.	−1.			
5	+1/5	+1/5	−1/5	−1/5		+1.			
6						+1.			+5/√34
7							+5/√34		−5/√34
8							−1.	−3/√34	−3/√34

$\{P\}_{8\times 1} =$

P \ LC	1
1	0.
2	+7.20
3	−4.80
4	0.
5	−6.48
6	0.
7	0.
8	0.

614 INTERMEDIATE STRUCTURAL ANALYSIS

$[B]_{9\times 8} =$

X \ e	1	2	3	4	5	6	7	8
1	+1.				+1/5			
2		+1.			+1/5			
3		+1.			−1/5			
4			+1.		−1/5			
5				+1.				
6				−1.		+1.		
7					+1.			−1.
8							+5/√34	−3/√34
9						+5/√34	−5/√34	−3/√34

$[S]_{9\times 9} =$

F \ e	1	2	3	4	5	6	7	8	9
1	S_{11}	S_{12}							
2	S_{21}	S_{22}							
3			S_{33}	S_{34}					
4			S_{43}	S_{44}					
5					S_{55}				
6						S_{66}			
7							S_{77}		
8								S_{88}	
9									S_{99}

(c) *The output matrices* $\{X\}$ *and* $\{F^*\}$. The $\{X\}$ matrix is taken from Fig. 16.4.5 of Example 16.4.2, and the equation $\{P\} = [ASB]\{X\}$ is verified. The $\{F^*\}$ matrix obtained from $\{F^*\} = \{F_0\} + [SB]\{X\}$ checks exactly with the results of the force method in Example 16.4.2. Only the $\{X\}$ and $\{F^*\}$ matrices are shown.

COMPOSITE STRUCTURES AND RIGID FRAMES WITH AXIAL DEFORMATION 615

$\{X\}_{8 \times 1} =$

LC X	1
1	$- 3.46311 \times 10^{-3}$
2	$+ 8.00000 \times 10^{-3}$
3	$- 10.53689 \times 10^{-3}$
4	$- 0.10881 \times 10^{-3}$
5	$- 1.78964 \times 10^{-3}$
6	$- 0.21762 \times 10^{-3}$
7	$- 0.10881 \times 10^{-3}$
8	$- 1.55461 \times 10^{-3}$

$\{F^*\}_{9 \times 1} =$

LC F^*	1
1	0.
2	$+4.58525$
3	-4.58525
4	0.
5	-6.52842
6	-6.52842
7	-7.83410
8	$+7.61338$
9	$+7.61338$

(d) *Discussion.* When axial deformation in all members is to be neglected, the degree of freedom is 3 (the joint rotations at A, B, and C) and the independent unknown internal forces are only the end moments on AB and BC, making $NI = NF - NP = 4 - 3 = 1$. If axial deformation in the steel rods only is to be considered, the degree of freedom is 5 (the three joint rotations at A, B, and C, the horizontal deflection of D, and the vertical deflection of B or D) and the number of independent unknown internal forces is 6 (the four end moments on AB and BC and the two axial forces in the steel rods), again making $NI = NF - NP = 6 - 5 = 1$. The degree of indeterminacy of a structure is its intrinsic character; it does not change with the assumptions used in the analysis.

Example 16.5.3 Analyze the rigid frame with tie rod shown in Fig. 16.5.3a by the displacement method, considering the axial deformation in all members.

SOLUTION (a) *P-X and F-e numbers.* Using four joints and four elements (one truss element and three combined elements), the P-X and F-e numbers are assigned as shown in Fig. 16.5.3b and c, wherein $NI = NF - NP = 10 - 9 = 1$.

(b) *The input matrices* $[A]$, $[B]$, $[S]$, *and* $\{P\}$. The four moment and five force equations of equilibrium in the $[A]$ matrix are established from the free-body diagrams of the joints in Fig. 16.5.3d. Two typical displacement diagrams showing the effect of joint rotation or deflection on member-end rotations are shown in Fig. 16.5.3e; note that each column in the $[B]$ matrix is established by taking one X value to be nonzero while keeping all other X values zero. The contents of the $\{P\}$ matrix are obtained from the fixed-condition forces in Fig. 16.5.3f. The contents of the $[S]$ matrix are

$$S_{11} = S_{22} = S_{55} = S_{66} = \frac{4EI_c}{9} = \frac{4(200)(1.20 \times 10^9) \times 10^{-6}}{9} = 106{,}666 \text{ kN·m}$$

$$S_{12} = S_{21} = S_{56} = S_{65} = \frac{2EI_c}{9} = 53{,}333 \text{ kN·m}$$

$$S_{33} = S_{44} = \frac{4EI_b}{18} = \frac{4(200)(1.50 \times 10^9) \times 10^{-6}}{18} = 66{,}666 \text{ kN·m}$$

616 INTERMEDIATE STRUCTURAL ANALYSIS

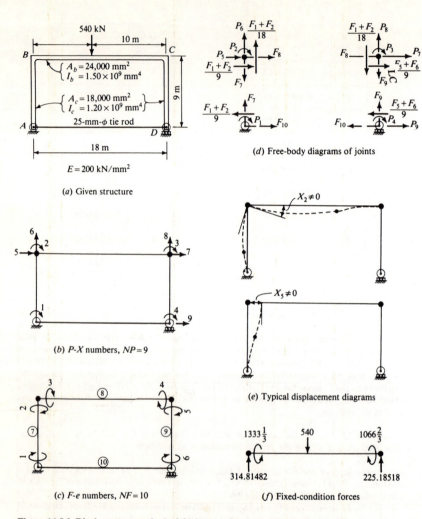

Figure 16.5.3 Displacement method, rigid frame with tie rod, Example 16.5.3.

$$S_{34} = S_{43} = \frac{2EI_b}{18} = 33{,}333 \text{ kN·m}$$

$$S_{77} = S_{99} = \frac{EA_c}{9} = \frac{200(18{,}000)}{9} = 400{,}000 \text{ kN/m}$$

$$S_{88} = \frac{EA_b}{9} = \frac{200(24{,}000)}{18} = 266{,}666 \text{ kN/m}$$

$$S_{1010} = \frac{EA_r}{18} = \frac{200\pi(12.5)^2}{18} = 5454.154 \text{ kN/m}$$

COMPOSITE STRUCTURES AND RIGID FRAMES WITH AXIAL DEFORMATION

$[A]_{9\times 10} =$

F\P	1	2	3	4	5	6	7	8	9	10
1	+1.									
2		+1.	+1.							
3				+1.	+1.					
4						+1.				
5	−1/9	−1/9						−1.		
6			−1/18	−1/18			+1.			
7					−1/9	−1/9			+1.	
8			+1/18	+1/18						+1.
9						+1/9	+1/9			+1.

$[B]_{10\times 9} =$

X\e	1	2	3	4	5	6	7	8	9
1	+1.				−1/9				
2		+1.			−1/9				
3		+1.				−1/18		+1/18	
4			+1.			−1/18		+1/18	
5			+1.				−1/9		+1/9
6				+1.			−1/9		+1/9
7						+1.			
8					−1.		+1.		
9								+1.	
10								+1.	

618 INTERMEDIATE STRUCTURAL ANALYSIS

$[S]_{10 \times 10} =$

e\F	1	2	3	4	5	6	7	8	9	10
1	S_{11}	S_{12}								
2	S_{21}	S_{22}								
3			S_{33}	S_{34}						
4			S_{43}	S_{44}						
5					S_{55}	S_{56}				
6					S_{65}	S_{66}				
7							S_{77}			
8								S_{88}		
9									S_{99}	
10										S_{1010}

$\{P\}_{9 \times 1} =$

LC\P	1
1	0.
2	+1333.33333
3	−1066.66666
4	0.
5	0.
6	−314.81482
7	0.
8	−225.18518
9	0.

COMPOSITE STRUCTURES AND RIGID FRAMES WITH AXIAL DEFORMATION 619

(c) *The output matrices* $\{X\}$ *and* $\{F^*\}$. The $\{X\}$ matrix is taken from Fig. 16.4.7 of Example 16.4.3, and the equation $\{P\} = [ASB]\{X\}$ is verified. The $\{F^*\}$ matrix obtained from $\{F^*\} = \{F_0\} + [SB]\{X\}$ checks exactly with the results of the force method in Example 16.4.3. Only the $\{X\}$ and $\{F^*\}$ matrices are shown.

$\{X\}_{9\times 1} = $

X \ LC	1
1	-2.87666×10^{-3}
2	$+12.58552 \times 10^{-3}$
3	-9.93552×10^{-3}
4	$+5.52666 \times 10^{-3}$
5	$+20.49661 \times 10^{-3}$
6	-0.75000×10^{-3}
7	$+20.15301 \times 10^{-3}$
8	-0.60000×10^{-3}
9	$+16.79962 \times 10^{-3}$

$\{F^*\}_{10\times 1} = $

F* \ LC	1
1	0.
2	$+824.6494$
3	-824.6494
4	$+824.6494$
5	-824.6494
6	0.
7	$-300.$
8	-91.62771
9	-240
10	$+91.62771$

(d) *Discussion.* For this problem, it would not be logical to neglect the axial deformation in the tie rod, because then the tie rod would be infinitely rigid and the horizontal deflection at the roller support would be equal to zero, making it equivalent to a hinged support. If axial deformation in the beam and the columns is to be neglected, the degree of freedom would be reduced to 6 (four joint rotations, the horizontal deflection of B or C, and the horizontal deflection at the roller support) and the number of independent unknown internal forces would be 7 (six end moments and the tension in the tie rod); thus, $NI = NF - NP = 7 - 6 = 1$.

16.6 Exercises

In solving Exercises 16.1 to 16.6 it is more instructive to derive the closed-form expression for the redundant in terms of the longitudinal and flexural rigidities and check it against the closed-form results provided in the examples. Also, such expressions should be obtained by using both the theorem of least work and physical compatibility (this same suggestion applies to Exercises 16.7, 16.8, and 16.9).

16.1 By the force method, solve Example 16.2.1 again using the reaction at A as the redundant.
16.2 By the force method, solve Example 16.2.1 again using the bending moment at C as the redundant.

16.3 By the force method, solve Example 16.4.1 again using the vertical reaction at C as the redundant. In this case, the degenerate structure would have the right end of the beam encased in a large block that may slide freely along a vertical surface.

16.4 By the force method, solve Example 16.4.2 again using the tension in the left steel rod as the redundant.

16.5 By the force method, solve Example 16.4.2 again using the bending moment at B as the redundant. In this case, the degenerate structure would have a pinned joint at B connecting three timber elements.

16.6 By the force method, solve Example 16.4.3 again using the bending moment in the beam at the location of the load as the redundant.

16.7 By the force method, analyze the queen-post structure shown in Fig. 16.6.1 using the compression in the left post as the redundant.

16.8 Solve Exercise 16.7 again using the tension in the horizontal steel rod as the redundant.

16.9 Solve Exercise 16.8 again using the bending moment in the beam at top of the left post as the redundant.

16.10 The timber beam shown in Fig. 16.6.2 rests on three elastic springs which may deform at the rate of 0.5 kN/mm. Analyze by the displacement method, treating each spring as a truss element and using the minimum possible number of joints. Obtain the $\{X\}$ matrix by solving the simultaneous equations on the computer or by solving the problem first by the force method.

16.11 Solve Exercise 16.10 again by the displacement method using five joints, four beam elements, and three truss elements. Do this problem only if a computer program is available.

16.12 By the displacement method, solve Example 16.4.1 again, but consider the axial deformation in the steel rod. Check the results with those in Example 16.4.1 upon making EA of the timber beam equal to infinity.

16.13 By the displacement method, solve Example 16.4.2 again, but consider the axial deformation in the two steel rods. Check the results with those in Example 16.4.2 upon making EA of the timber beam and post equal to infinity.

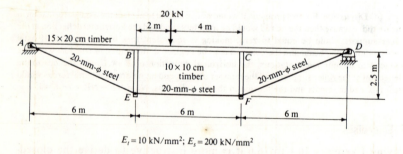

$E_t = 10$ kN/mm^2; $E_s = 200$ kN/mm^2

Figure 16.6.1 Exercises 16.7, 16.8, and 16.9.

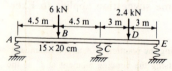

$E_t = 10$ kN/mm^2

Figure 16.6.2 Exercise 16.10.

COMPOSITE STRUCTURES AND RIGID FRAMES WITH AXIAL DEFORMATION

16.14 By the displacement method, solve Example 16.4.3 again, but consider only the axial deformation in the tie rod. Check the results with those in Example 16.4.3 upon making EA of the beam and the columns equal to infinity.

16.15 By the displacement method, solve Exercise 16.7, considering the axial deformation in all members. Obtain the $\{X\}$ matrix by solving the simultaneous equations on the computer or by using the results of the force method in Exercises 16.7, 16.8, or 16.9.

CHAPTER
SEVENTEEN
SECONDARY MOMENTS IN TRUSSES WITH RIGID JOINTS

17.1 General Description

The usual steel truss is a structure composed of individual members so joined together as to form a series of triangles. The joints may be pin-connected, bolted, or welded. In the first stage of structural analysis, however, the joints are assumed to act as smooth hinges. If the truss were indeed so built, it follows that the members would be subjected to axial force of tension or compression only and not to bending. The axial force in each member thus computed, either for a statically determinate or indeterminate truss, is called the *primary axial force.*

When the lengths of members change owing to the presence of primary axial forces in them, the joints (the pins) must move accordingly to some new position in order to accommodate themselves to the new member lengths. These movements, called joint displacements, or deflections, must be accompanied by changes in the angles between members. This may easily happen if the *straight* members can freely rotate around the joints, or, in other words, if the joints can act as perfectly smooth hinges. Even for pin-connected trusses, the members may not be free to turn around the pins because of the friction which may develop. In bolted trusses, the members are almost impossible to rotate at the joints except for a certain amount of yielding, or "play," owing to the possible loss of friction in bearing type of bolted connections. In welded trusses, or in reinforced concrete trusses, the joints are considered rigid; that is, the angles between members cannot change at all. In all cases (except that of smooth hinges), when the angles between member directions tend to change but are restrained from attaining the full amounts as in the case of semirigid joints, or are entirely prevented from doing so as in the case of rigid joints, the members themselves will have to bend. Thus, bending moments, commonly called *secondary moments* to contrast with primary axial forces, will be developed in the members.

SECONDARY MOMENTS IN TRUSSES WITH RIGID JOINTS

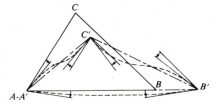

Figure 17.1.1 A triangular truss with rigid joints.

The evaluation of the secondary moments in a truss with rigid joints will be treated in this chapter. When the joints are semirigid, the method presented in Chap. 20 will have to be used.

To illustrate further, consider the triangular truss ABC of Fig. 17.1.1. Owing to the changes in the lengths of the three members, the joints A, B, C are displaced to take the positions A', B', C' (A and A' happen to coincide and BB' happens to be horizontal on account of the support conditions). If the three members $A'B'$, $B'C'$, and $C'A'$ are to remain straight, apparently angles A' and B' should be respectively smaller than the original angles A and B, while angle C' should be larger than the original angle C. If, however, the joints are rigid, then the sizes of the angles cannot change. Thus the members $A'B'$, $B'C'$, and $C'A'$ will have to bend as shown by the dashed lines so that the angles between the tangents to the elastic curves at the member ends remain in their original sizes. Note that each joint may rotate as a whole so long as the angle between tangents to the elastic curves meeting at that joint does not change. In Fig. 17.1.1, the amount of rotation of each joint as a whole is shown by a short arrow starting from the original direction of the member and ending at the tangent to the elastic curve of that member.

17.2 Methods of Analysis

In general, there are, at least theoretically, three different methods by which a truss with rigid joints (as opposed to a truss with pin joints) may be analyzed.

The first is the iteration method, which follows the general description of the preceding section as to how secondary bending moments actually arise in the event that the joints are rigid instead of being pinned. Consequently, in the first step of the iteration process, the joints are assumed to be pinned, and the *primary axial forces* with the corresponding joint displacements are determined. At this point, if the member directions are not allowed to change, fixed-end moments as shown in Fig. 17.2.1c must be applied. In the second step of the iteration process, the structure is analyzed as a rigid frame without sidesway and without axial deformation, but only subjected to the releasing action of the fixed-end moments. For this operation, the moment-distribution method is used most often, although the slope-deflection or matrix displacement method can also be used. The balanced moments acting at the ends of the members are the so-called secondary moments. These end moments in turn

624 INTERMEDIATE STRUCTURAL ANALYSIS

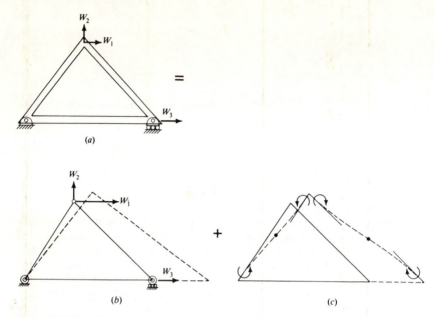

Figure 17.2.1 Iteration method.

require transverse shear forces acting at the member ends to keep the members in equilibrium. In the third step of the iteration process, the opposites of the transverse shear forces on the member ends are to act on the joints of the truss again with pinned connections, giving rise to what may be called the *third axial forces*. As long as the original load system, such as W_1, W_2, W_3 in Fig. 17.2.1a, involves only forces acting on the joints themselves, the third axial forces are usually very small compared with the primary axial forces so that there is no need to go any further in the iteration process.

The alternative to the iteration method is to treat the truss with rigid joints as a "composite structure" or a "rigid frame with axial deformation," as it has been treated in Chap. 16. When this approach is used, the results of the analysis are exact in the sense that they include the totals of an infinite number of iterations in the iteration method. Furthermore, when the load system involves forces other than linear forces acting on the joints—such as the weight of heavy and long members, or heavy equipment hanging on the members, or moments acting on the joints due to eccentricity of the member connection to the gusset plate—this approach may be the only logical method to use because the iteration method may not converge quickly.

The truss with rigid joints, as a composite structure, is statically indeterminate. Therefore, the second and third methods of analysis are the force and displacement methods of composite-structure analysis. The force

SECONDARY MOMENTS IN TRUSSES WITH RIGID JOINTS 625

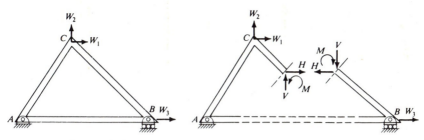

Figure 17.2.2 Force method.

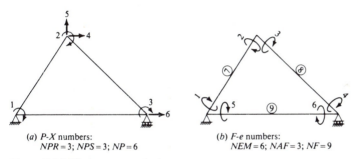

(a) P-X numbers:
NPR = 3; NPS = 3; NP = 6

(b) F-e numbers:
NEM = 6; NAF = 3; NF = 9

Figure 17.2.3 Displacement method.

method is conceptually elegant, but it is difficult to apply. For instance, the structure in Fig. 17.2.2 is statically indeterminate to the third degree. It can be made statically determinate by a complete cut at any point along, say, BC. Three pairs of equal and opposite generalized redundant forces H, V, and M are applied to the sides of the cut. The compatibility conditions are that the relative rotation, horizontal displacement, and vertical displacement at the cut must be zero; that is, if one side of the cut is considered fixed, the three generalized displacements at the other side of the cut must be zero. This method will not be further illustrated in this chapter.

The displacement method, when formulated in matrix notation, is a convenient method to use, especially with the aid of a digital computer. The application of this method follows the procedure outlined in Chap. 16. For example, the P-X and F-e numbers for the triangular truss with rigid joints are shown in Fig. 17.2.3.

In the next few sections, the displacement and iteration methods for the analysis of trusses with rigid joints will be described in detail.

17.3 Solution as a General Two-Dimensional-Frame Problem

The truss with rigid joints is in fact a rigid frame, and as such it may be analyzed by the matrix displacement method. There is no point in analyzing this rigid frame by neglecting axial deformation, because if all the applied loads were at the joints only, there would be no fixed-end moments due to applied loads. Besides there can be no unknown joint deflection because all bars are assumed not to change in length. Thus it is inconceivable that a truss with rigid joints could be analyzed as a rigid frame without axial deformation.

When the truss with rigid joints is analyzed as a rigid frame with axial deformation, then the solution procedure falls into the pattern for a general two-dimensional-frame problem. When this method is used, the loads may act not only at the joints but on the bars themselves as well. The matrix displacement method for such a solution always involves the twofold equations

$$\{X\} = [ASA^T]^{-1}\{P\} \qquad (17.3.1)$$

and
$$\{F^*\} = \{F_0\} + [SA^T]\{X\} \qquad (17.3.2)$$

When a computer program is written to carry out the operations in Eqs. (17.3.1) and (17.3.2), two approaches may be taken.

In the first approach, the input matrices are the global $[A]$ matrix (which has been checked against the global $[B]$ matrix by the relationship $[B] = [A^T]$), the global $[S]$ matrix, and the global $\{F_0\}$ and $\{P\}$ matrices. The $[S]$ matrix consists of a number of 2×2 submatrices for the flexure portion and a diagonal matrix for the axial portion; therefore some short cuts may be devised within the computer software.†

In the second approach, the expressions for the contents in the $[ASA^T]$ matrix are derived for a single inclined member in a rigid frame when that member is under the combined action of axial force and bending. The $[A]$, $[B]$, $[S]$, $[SB]$, and $[ASB]$ matrices of such a member as shown in Fig. 17.3.1 are

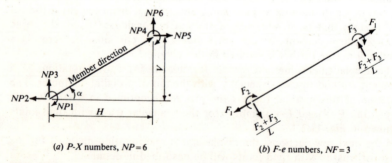

(a) P-X numbers, $NP = 6$ \qquad (b) F-e numbers, $NF = 3$

Figure 17.3.1 A typical member with axial deformation in a rigid frame.

---

†One such computer program is provided in C.-K. Wang, *Matrix Methods of Structural Analysis*, 2d ed., American Publishing Company, Madison, Wis., 1970, Appendix H.

$$[A] = \begin{array}{|c|c|c|c|} \hline \diagdown\!\!\begin{array}{c}F\\P\end{array} & 1 & 2 & 3 \\ \hline 1 & 0 & +1 & 0 \\ \hline 2 & -\cos\alpha & (+\sin\alpha)/L & (+\sin\alpha)/L \\ \hline 3 & -\sin\alpha & (-\cos\alpha)/L & (-\cos\alpha)/L \\ \hline 4 & 0 & 0 & +1 \\ \hline 5 & +\cos\alpha & (-\sin\alpha)/L & (-\sin\alpha)/L \\ \hline 6 & +\sin\alpha & (+\cos\alpha)/L & (+\cos\alpha)/L \\ \hline \end{array} \qquad (17.3.3)$$

$$[B] = \begin{array}{|c|c|c|c|c|c|c|} \hline \diagdown\!\!\begin{array}{c}X\\e\end{array} & 1 & 2 & 3 & 4 & 5 & 6 \\ \hline 1 & 0 & -\cos\alpha & -\sin\alpha & 0 & +\cos\alpha & +\sin\alpha \\ \hline 2 & +1 & (+\sin\alpha)/L & (-\cos\alpha)/L & 0 & (-\sin\alpha)/L & (+\cos\alpha)/L \\ \hline 3 & 0 & (+\sin\alpha)/L & (-\cos\alpha)/L & +1 & (-\sin\alpha)/L & (+\cos\alpha)/L \\ \hline \end{array}$$

$$(17.3.4)$$

$$[S] = \begin{array}{|c|c|c|c|} \hline \diagdown\!\!\begin{array}{c}e\\F\end{array} & 1 & 2 & 3 \\ \hline 1 & +EA/L & 0 & 0 \\ \hline 2 & 0 & +4EI/L & +2EI/L \\ \hline 3 & 0 & +2EI/L & +4EI/L \\ \hline \end{array} \qquad (17.3.5)$$

$[SB] =$

$\diagdown\!\!\begin{array}{c}X\\F\end{array}$	1	2	3	4	5	6
1	0	$-\dfrac{EA}{L}\cos\alpha$	$-\dfrac{EA}{L}\sin\alpha$	0	$+\dfrac{EA}{L}\cos\alpha$	$+\dfrac{EA}{L}\sin\alpha$
2	$+\dfrac{4EI}{L}$	$+\dfrac{6EI}{L^2}\sin\alpha$	$-\dfrac{6EI}{L^2}\cos\alpha$	$+\dfrac{2EI}{L}$	$-\dfrac{6EI}{L^2}\sin\alpha$	$+\dfrac{6EI}{L^2}\cos\alpha$
3	$+\dfrac{2EI}{L}$	$+\dfrac{6EI}{L^2}\sin\alpha$	$-\dfrac{6EI}{L^2}\cos\alpha$	$+\dfrac{4EI}{L}$	$-\dfrac{6EI}{L^2}\sin\alpha$	$+\dfrac{6EI}{L^2}\cos\alpha$

$$(17.3.6)$$

$[ASB] =$

X \ P	1	2	3	4	5	6
1	$+4EI/L$	$+T5$	$-T4$	$+2EI/L$	$-T5$	$+T4$
2	$+T5$	$+(T1+T8)$	$+(T2-T7)$	$+T5$	$-(T1+T8)$	$-(T2-T7)$
3	$-T4$	$+(T2-T7)$	$+(T3+T6)$	$-T4$	$-(T2-T7)$	$-(T3+T6)$
4	$+2EI/L$	$+T5$	$-T4$	$+4EI/L$	$-T5$	$+T4$
5	$-T5$	$-(T1+T8)$	$-(T2-T7)$	$-T5$	$+(T1+T8)$	$+(T2-T7)$
6	$+T4$	$-(T2-T7)$	$-(T3+T6)$	$+T4$	$+(T2-T7)$	$+(T3+T6)$

(17.3.7)

$$T1 = \frac{EA}{L}\cos^2\alpha \qquad T2 = \frac{EA}{L}\sin\alpha\cos\alpha \qquad T3 = \frac{EA}{L}\sin^2\alpha$$

$$T4 = \frac{6EI}{L^2}\cos\alpha \qquad T5 = \frac{6EI}{L^2}\sin\alpha$$

$$T6 = \frac{12EI}{L^3}\cos^2\alpha \qquad T7 = \frac{12EI}{L^3}\sin\alpha\cos\alpha \qquad T8 = \frac{12EI}{L^3}\sin^2\alpha$$

In this way the required input data for building up the global stiffness matrix will be only, for each member (1) the six global degree-of-freedom numbers corresponding to the six local degree-of-freedom numbers, (2) the directed distances H and V (positive to the right and upward) as shown in Fig. 17.3.1a, and (3) the member properties EA and EI. By a single DO loop, with the proper algorithm, the local $[ASA^T]$ matrix of each member is fed into the appropriate slots in the global $[ASA^T]$ matrix. Then Eq. (17.3.1) is used to produce the displacement matrix $\{X\}$. After that, Eq. (17.3.2), or Eq. (17.3.6) in this case, is applied locally to yield the axial force and the end moments for each member.†

Equation (17.3.7) may be called the local stiffness matrix of a combined truss and beam element, in reference to a pair of coordinate axes. Another way of obtaining this matrix is by use of two transformation matrices and a local stiffness matrix with one coordinate axis along the direction of the element. In reference to Fig. 17.3.2a, the forces P_{E1} to P_{E6} (the E in the subscript means "element") may be expressed in terms of X_{E1} to X_{E6} by the following:

---

†A computer program of this nature is provided in ibid., Appendix K.

SECONDARY MOMENTS IN TRUSSES WITH RIGID JOINTS

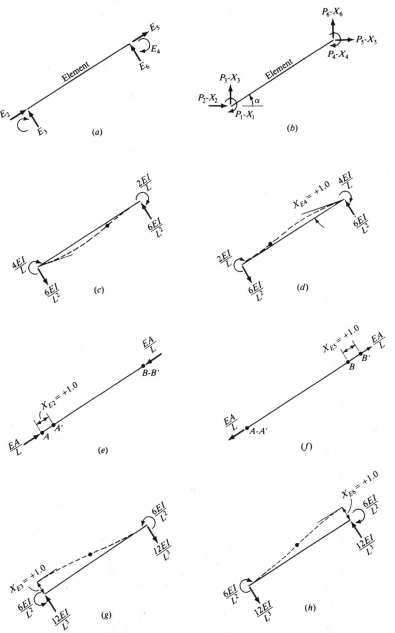

Figure 17.3.2 Use of transformation matrices to derive the local [K] of a combined truss and beam element.

$$\begin{Bmatrix} P_{E1} \\ P_{E2} \\ P_{E3} \\ P_{E4} \\ P_{E5} \\ P_{E6} \end{Bmatrix} = \begin{bmatrix} +\dfrac{4EI}{L} & 0 & -\dfrac{6EI}{L^2} & +\dfrac{2EI}{L} & 0 & +\dfrac{6EI}{L^2} \\ 0 & +\dfrac{EA}{L} & 0 & 0 & -\dfrac{EA}{L} & 0 \\ -\dfrac{6EI}{L^2} & 0 & +\dfrac{12EI}{L^3} & -\dfrac{6EI}{L^2} & 0 & -\dfrac{12EI}{L^3} \\ +\dfrac{2EI}{L} & 0 & -\dfrac{6EI}{L^2} & +\dfrac{4EI}{L} & 0 & +\dfrac{6EI}{L^2} \\ 0 & -\dfrac{EA}{L} & 0 & 0 & +\dfrac{EA}{L} & 0 \\ +\dfrac{6EI}{L^2} & 0 & -\dfrac{12EI}{L^3} & +\dfrac{6EI}{L^2} & 0 & +\dfrac{12EI}{L^3} \end{bmatrix} \begin{Bmatrix} X_{E1} \\ X_{E2} \\ X_{E3} \\ X_{E4} \\ X_{E5} \\ X_{E6} \end{Bmatrix} \quad (17.3.8)$$

The above equation may be obtained by observing the six free-body diagrams in Fig. 17.3.2c to h. Then the forces P_1 to P_6 in Fig. 17.3.2b may be expressed in terms of the forces P_{E1} to P_{E6} by

$$\begin{Bmatrix} P_1 \\ P_2 \\ P_3 \\ P_4 \\ P_5 \\ P_6 \end{Bmatrix} = \begin{bmatrix} +1 & 0 & 0 & 0 & 0 & 0 \\ 0 & +\cos\alpha & -\sin\alpha & 0 & 0 & 0 \\ 0 & +\sin\alpha & +\cos\alpha & 0 & 0 & 0 \\ 0 & 0 & 0 & +1 & 0 & 0 \\ 0 & 0 & 0 & 0 & +\cos\alpha & -\sin\alpha \\ 0 & 0 & 0 & 0 & +\sin\alpha & +\cos\alpha \end{bmatrix} \begin{Bmatrix} P_{E1} \\ P_{E2} \\ P_{E3} \\ P_{E4} \\ P_{E5} \\ P_{E6} \end{Bmatrix} \quad (17.3.9)$$

and the displacements X_{E1} to X_{E6} may be expressed in terms of the displacements X_1 to X_6 by

$$\begin{Bmatrix} X_{E1} \\ X_{E2} \\ X_{E3} \\ X_{E4} \\ X_{E5} \\ X_{E6} \end{Bmatrix} = \begin{bmatrix} +1 & 0 & 0 & 0 & 0 & 0 \\ 0 & +\cos\alpha & +\sin\alpha & 0 & 0 & 0 \\ 0 & -\sin\alpha & +\cos\alpha & 0 & 0 & 0 \\ 0 & 0 & 0 & +1 & 0 & 0 \\ 0 & 0 & 0 & 0 & +\cos\alpha & +\sin\alpha \\ 0 & 0 & 0 & 0 & -\sin\alpha & +\cos\alpha \end{bmatrix} \begin{Bmatrix} X_1 \\ X_2 \\ X_3 \\ X_4 \\ X_5 \\ X_6 \end{Bmatrix} \quad (17.3.10)$$

Calling Eq. (17.3.8) by

$$\{P_E\} = [K_L]\{X_E\}$$

Eq. (17.3.9) by

$$\{P\} = [H]\{P_E\}$$

and Eq. (17.3.10) by

$$\{X_E\} = [H^T]\{X\}$$

SECONDARY MOMENTS IN TRUSSES WITH RIGID JOINTS 631

and then combining the three equations,

$$\{P\} = [H]\{P_E\} = [H][K_L]\{X_E\} = [H][K_L][H^T]\{X\} \quad (17.3.11)$$

The matrix $[H]$ transforms the $\{P_E\}$ forces into $\{P\}$ forces, and the matrix $[H^T]$ transforms the $\{X\}$ displacements into $\{X_E\}$ displacements; hence they are called the transformation matrices. The transposition relationship between the two transformation matrices is the natural consequence of the principle of virtual work; the proof is the same as the proof that $[B] = [A^T]$.

Comparing Eq. (17.3.11) with Eq. (17.3.7),

$$[A][S][B] = [H][K_L][H^T] \quad (17.3.12)$$

In the following two examples, the required input data for the analysis of a four-panel Pratt truss with rigid joints are shown for both approaches (using global $[A]$ matrix versus compiling local stiffness matrix of each member) in the application of the displacement method. The output is shown only once.

Example 17.3.1 Show the input matrices $[A]$, $[B]$, $[S]$, and $\{P\}$ for the analysis of the four-panel Pratt truss with rigid joints shown in Fig. 17.3.3.

SOLUTION (a) *The P-X and F-e numbers.* The P-X and F-e numbers are shown in Fig. 17.3.4. The joint rotations are numbered from 1 to 8; the joint deflections, from 9 to 21; thus $NP = NPR + NPS = 8 + 13 = 21$. The member-end moments are numbered from 1 to 26; the member axial forces, from 27 to 39; thus $NF = NEM + NAF = 26 + 13 = 39$. Note that the same sequence is used for numbering the end moments and the axial forces: the upper chord, the lower chord, the diagonals, and the verticals, all from left to right. Although not needed in the displacement method, as a matter of interest the degree of indeterminacy is $NI = NF - NP = 39 - 21 = 15$.

(b) *The input $[A]$ matrix.* The nonzero elements in each column of the $[A]$ matrix can be tabulated and then fed into the computer. For instance, the *effective* components along

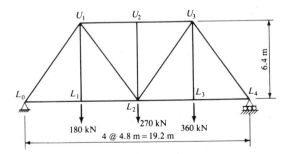

For U_1U_2 and U_2U_3: $A = 90$ cm^2; $I = 3600$ cm^4
For L_0L_1, L_1L_2, L_2L_3, and L_3L_4: $A = 45$ cm^2; $I = 1215$ cm^4
For L_0U_1 and U_3L_4: $A = 112.5$ cm^2; $I = 7200$ cm^4
For U_1L_2 and L_2U_3: $A = 56.25$ cm^2; $I = 1440$ cm^4
For U_1L_1 and U_3L_3: $A = 72$ cm^2; $I = 3240$ cm^4
For U_2L_2: $A = 18$ cm^2; $I = 135$ cm^4
$E = 20,000$ kN/cm^2

Figure 17.3.3 The four-panel Pratt truss with rigid joints in Example 17.3.1.

632 INTERMEDIATE STRUCTURAL ANALYSIS

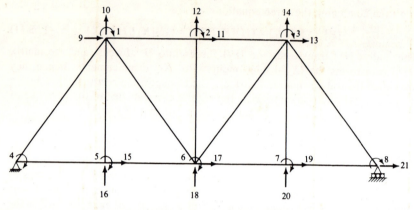

(a) P-X numbers; $NPR = 8$, $NPS = 13$, $NP = 21$

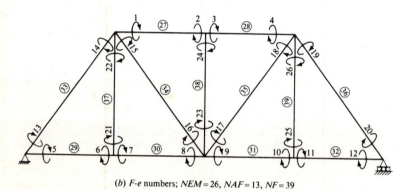

(b) F-e numbers; $NEM = 26$, $NAF = 13$, $NF = 39$

Figure 17.3.4 The P-X and F-e numbers for truss in Example 17.3.1.

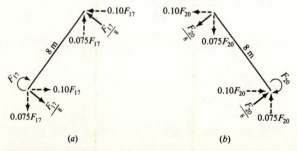

(a)   (b)

Figure 17.3.5 Nonzero elements in seventeenth and twentieth columns of $[A]$ for truss in Example 17.3.1.

SECONDARY MOMENTS IN TRUSSES WITH RIGID JOINTS   633

Table 17.3.1 Nonzero elements in seventeenth and twentieth columns of [A], Example 17.3.1

i	j	A(i, j)	Check with B(j, i)
6	17	+1.	See B(17, 6)
13	17	−0.100	See B(17, 13)
14	17	+0.075	
17	17	+0.100	
18	17	−0.075	
8	20	+1.	
13	20	−0.100	See B(20, 13)
14	20	−0.075	
21	20	+0.100	

the positive directions of the degree of freedom, of the F_{17} and F_{20} internal force systems shown in Fig. 17.3.5, are shown in Table 17.3.1. Note that there is no vertical degree of freedom at the end where F_{20} is applied.

(c) *The input [B] matrix.* The nonzero elements in each column of the [B] matrix can be carefully observed by inspection and then checked off with the listings in the tabulation of the [A] matrix. For instance, the nonzero elements in the sixth and thirteenth columns of the [B] matrix are shown in Table 17.3.2. A joint rotation like X_6 will cause rotations equal to X_6 at all member ends entering into that joint. A joint deflection like X_{13} will cause three member-axis rotations as shown in Fig. 17.3.6. Since $e = X - R$, clockwise R's mean negative e's. Thus $e_{17} = e_{18} = -0.8X_{13}/8 = -0.100X_{13}$, $e_{19} = e_{20} = -0.8X_{13}/8 = -0.100X_{13}$, and $e_{25} = e_{26} = -X_{13}/6.4$. X_{13} will also cause member elongations of $e_{28} = +X_{13}$, $e_{35} = +0.6X_{13}$, and $e_{36} = -0.6X_{13}$. Note again that in actual practice Table 17.3.2 need not be made; by noting that $B(i, j) = A(j, i)$ check marks may be placed in the check column of the tabulation of the [A] matrix. After the analyst thinks through all columns of [B], the check column of the [A] tabulation should then be completely filled with check marks.

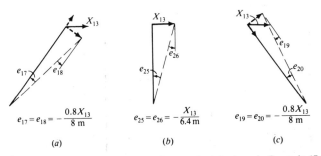

Figure 17.3.6 Effect of X_{13} on member rotations for truss in Example 17.3.1.

Table 17.3.2 Nonzero elements in sixth and thirteenth columns of [B], Example 17.3.1

i	j	B(i, j)	
8	6	+1.	
9	6	+1.	
16	6	+1.	
17	6	+1.	Go to A(6, 17)
23	6	+1.	
17	13	−0.100	Go to A(13, 17)
18	13	−0.100	
19	13	−0.100	
20	13	−0.100	Go to A(13, 20)
25	13	−1/6.4	
26	13	−1/6.4	
28	13	+1.	
35	13	+0.6	
36	13	−0.6	

(d) *The input [S] matrix.* The contents of the [S] matrix are $+4EI/L$ and $+2EI/L$ in flexure and $+EA/L$ in axial elongation, as shown below typically for end moments 17 and 18 and axial forces 35, 36, and 37.

	e F	17	18	35	36	37
	17	144000.	72000.			
	18	72000.	144000.			
[S] =	35			1406.25		
	36				2813.5	
	37					2250.

SECONDARY MOMENTS IN TRUSSES WITH RIGID JOINTS

$$S_{17-17} = S_{18-18} = \frac{4EI}{L} = \frac{4(20,000)(1440)}{800} = 144,000 \text{ kN·cm/rad}$$

$$S_{17-18} = S_{18-17} = \frac{2EI}{L} = 72,000 \text{ kN·cm/rad}$$

$$S_{35} = \frac{EA}{L} = \frac{20,000(56.25)}{800} = 1406.25 \text{ kN/cm}$$

$$S_{36} = \frac{EA}{L} = \frac{20,000(112.5)}{800} = 2812.5 \text{ kN/cm}$$

$$S_{37} = \frac{EA}{L} = \frac{20,000(72)}{640} = 2250 \text{ kN/cm}$$

(e) *The input* $\{P\}$ *matrix.* Although this general method can accommodate loads applied on the members themselves as well, in this particular problem of a "truss with rigid joints," there are no fixed-end moments to begin with (zero $\{F_0\}$ matrix) and the $\{P\}$ matrix includes

$$P_6 = -180 \quad P_{18} = -270 \quad P_{20} = -360$$

(f) *The output* $\{X\}$ *matrix.* The elements in the $\{X\}$ matrix as given out by the computer are

$X_1 = +0.60077 \times 10^{-3}$ rad $\quad X_2 = +0.15320 \times 10^{-3}$ rad
$X_3 = -0.59243 \times 10^{-3}$ rad $\quad X_4 = +0.99698 \times 10^{-3}$ rad
$X_5 = +0.60661 \times 10^{-3}$ rad $\quad X_6 = +0.19052 \times 10^{-3}$ rad
$X_7 = -0.65862 \times 10^{-3}$ rad $\quad X_8 = -1.28890 \times 10^{-3}$ rad
$X_9 = +0.39717$ cm $\quad X_{10} = -0.49762$ cm
$X_{11} = +0.28929$ cm $\quad X_{12} = -0.77637$ cm
$X_{13} = +0.18140$ cm $\quad X_{14} = -0.59921$ cm
$X_{15} = +0.14385$ cm $\quad X_{16} = -0.57741$ cm
$X_{17} = +0.28790$ cm $\quad X_{18} = -0.77872$ cm
$X_{19} = +0.46782$ cm $\quad X_{20} = -0.75876$ cm
$X_{21} = +0.64756$ cm

(g) *The output* $\{F\} = \{F^*\}$ *matrix.* The elements in the $\{F\} = \{F^*\}$ matrix as given out by the computer are

Member U_1U_2: $F_1 = -116.23$ kN·cm $\quad F_2 = -250.50$ kN·cm $\quad F_{27} = -404.57$ kN
Member U_2U_3: $F_3 = +246.36$ kN·cm $\quad F_4 = + 22.67$ kN·cm $\quad F_{28} = -404.58$ kN
Member L_0L_1: $F_5 = -102.08$ kN·cm $\quad F_6 = -141.61$ kN·cm $\quad F_{29} = +269.71$ kN
Member L_1L_2: $F_7 = + 14.73$ kN·cm $\quad F_8 = - 27.39$ kN·cm $\quad F_{30} = +270.11$ kN
Member L_2L_3: $F_9 = - 15.47$ kN·cm $\quad F_{10} = -101.45$ kN·cm $\quad F_{31} = +337.34$ kN
Member L_3L_4: $F_{11} = +216.28$ kN·cm $\quad F_{12} = +152.47$ kN·cm $\quad F_{32} = +337.01$ kN
Member L_0U_1: $F_{13} = +102.08$ kN·cm $\quad F_{14} = - 40.55$ kN·cm $\quad F_{33} = -449.42$ kN
Member U_1L_2: $F_{15} = + 31.09$ kN·cm $\quad F_{16} = + 1.55$ kN·cm $\quad F_{34} = +224.04$ kN
Member L_2U_3: $F_{17} = + 36.87$ kN·cm $\quad F_{18} = - 19.51$ kN·cm $\quad F_{35} = +112.09$ kN
Member U_3L_4: $F_{19} = + 98.26$ kN·cm $\quad F_{20} = -152.47$ kN·cm $\quad F_{36} = -561.59$ kN
Member U_1L_1: $F_{21} = +126.87$ kN·cm $\quad F_{22} = +125.69$ kN·cm $\quad F_{37} = +179.52$ kN
Member U_2L_2: $F_{23} = + 4.45$ kN·cm $\quad F_{24} = + 4.14$ kN·cm $\quad F_{38} = + 1.32$ kN
Member U_3L_3: $F_{25} = -114.83$ kN·cm $\quad F_{26} = -101.43$ kN·cm $\quad F_{39} = +358.99$ kN

636 INTERMEDIATE STRUCTURAL ANALYSIS

(h) *The statics and deformation checks.* The 21 statics checks and 39 deformation checks, although not shown, should be made. In this case, the statics checks are essential. The deformation checks, because of the absence of loads acting between member ends, are dependent only on the correctness of the [S] matrix.

Example 17.3.2 For the analysis of the truss with rigid joints in the preceding example by the displacement method, show the member information that is needed as input to a computer program in which the global stiffness matrix is compiled by feeding the contribution of the local stiffness of each member.

SOLUTION In accordance with the member numbers and directions shown in Fig. 17.3.7, the global degree-of-freedom numbers in Fig. 17.3.4a, the local degree-of-freedom numbers in Fig. 17.3.1a, and the member dimensions and properties in Fig. 17.3.3, the information for the 13 members to be used as input is shown in Table 17.3.3. Note that where there is restraint, a global degree of freedom equal to 1 larger than the degree of freedom of the total structure is used; in this case, 22. In this way the twenty-second row and the twenty-second column reserved for the global stiffness matrix are used as the "dump" for the unwanted elements in the local stiffness matrix of any member.

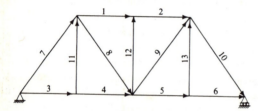

Figure 17.3.7 Member directions for truss in Example 17.3.2.

Table 17.3.3 Member information to be used as input in the direct stiffness method: Example 17.3.2

Member number	NP1	NP2	NP3	NP4	NP5	NP6	H, cm	V, cm	A, cm²	I, cm⁴
1	1	9	10	2	11	12	+480.	0.	90.	3600.
2	2	11	12	3	13	14	+480.	0.	90.	3600.
3	4	22	22	5	15	16	+480.	0.	45.	1215.
4	5	15	16	6	17	18	+480.	0.	45.	1215.
5	6	17	18	7	19	20	+480.	0.	45.	1215.
6	7	19	20	8	21	22	+480.	0.	45.	1215.
7	4	22	22	1	9	10	+480.	+640.	112.5	7200.
8	1	9	10	6	17	18	+480.	−640.	56.25	1440.
9	6	17	18	3	13	14	+480.	+640.	56.25	1440.
10	3	13	14	8	21	22	+480.	−640.	112.5	7200.
11	5	15	16	1	9	10	0.	+640.	72.	3240.
12	6	17	18	2	11	12	0.	+640.	18.	135.
13	7	19	20	3	13	14	0.	+640.	72.	3240.

17.4 Iteration Method—From Primary Axial Forces to Secondary Bending Moments

If the truss with rigid joints becomes a statically determinate truss once the rigid joints are replaced by pin joints, the primary axial forces and the corresponding joint displacements can be determined by the various methods presented in Chap. 3. As shown in Fig. 17.2.1c, the two fixed-end moments required to hold the tangents to the elastic curve at the member ends fixed while permitting the member axis to rotate (because of joint displacements) are functions of the clockwise member-axis rotation R, the length of the member L, and the flexural rigidity of the member EI; thus

$$M_{0i} = M_{0j} = -\frac{6EIR}{L} \quad (17.4.1)$$

The above formula is identical to Eq. (8.8.1) in Chap. 8.

The member-axis rotation R can be computed from the displacements of its ends. For the typical member shown in Fig. 17.4.1, on the basis of first-order assumption,

$$R = R_1 - R_2 = +\frac{(X_3 - X_1)\sin\alpha}{L} - \frac{(X_4 - X_2)\cos\alpha}{L} \quad (17.4.2)$$

Equation (17.4.2) can be either written into a computer program or used as such in hand computation. In hand computation, the analyst can more easily visualize the graphical picture of each member with its ends displaced, and then write out the numerical expression for R, in its correct sign.

The joint displacements used to compute the member-axis rotations need not be the true displacements satisfying the external boundary conditions,

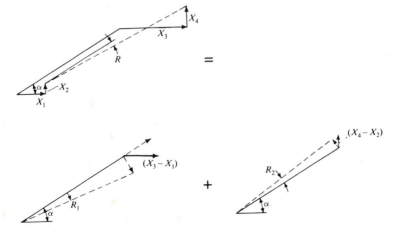

Figure 17.4.1 Member-axis rotation due to joint displacements.

638 INTERMEDIATE STRUCTURAL ANALYSIS

such as zero displacement for the hinge or zero vertical displacement at a roller support. Instead, the joint displacements corresponding to a zero member-axis rotation for any randomly chosen reference member may be used. The reason is that secondary bending moments are due to the rigidity of joints so that the displaced truss joints may be placed in any position relative to the original shape of the truss in obtaining a set of consistent fixed-end moments. In all cases the same secondary moments ought to be obtained. In another way of thinking, if the rigid-joint truss is rotated as a whole through the same angle of rotation, the fixed-end moments will be proportional to the EI/L values of the members, and they will all be relieved in the first cycle of moment distribution, resulting in no secondary moments. Only when the members are forced to have different member-axis rotations will the members have to bend themselves to maintain the same angle between the member directions (or tangents to the elastic curve) meeting at the same joint. A member axis is the straight line joining the ends of the member, while the member direction refers to the tangent to the elastic curve at a member end.

Following the argument in the preceding paragraph, the member-axis rotations R in the primary condition may be obtained by any one of the following three methods, *after* choosing at random any one member to be the reference member with zero member-axis rotation. Of course, as an alternative, the true member-axis rotations satisfying the boundary conditions may be used. The three methods are as follows:

1. Using Eqs. (3.5.1a to c) in the angle-weights method, find the changes in the sizes of all angles in all the triangles of the truss. Then, as shown in Fig. 17.4.2a, if BE is chosen as the reference member, the R values for the other two sides of triangle number 2 may be obtained by going along the

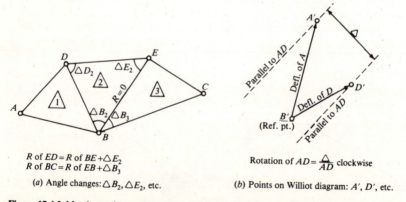

R of $ED = R$ of $BE + \triangle E_2$
R of $BC = R$ of $EB + \triangle B_3$

(a) Angle changes: $\triangle B_2, \triangle E_2$, etc.

Rotation of $AD = \dfrac{\triangle}{AD}$ clockwise

(b) Points on Williot diagram: A', D', etc.

Figure 17.4.2 Member-axis rotations from angle changes or from Williot diagram.

sides of the triangle in the *counterclockwise* direction. Thus,

R of $BE = 0$
R of $ED = (R$ of $BE) +$ (increase in size of angle at E, ΔE_2)
R of $DB = (R$ of $ED) +$ (increase in size of angle at D, ΔD_2)
R of $BE = (R$ of $DB) +$ (increase in size of angle at B, $\Delta B_2) = 0$

In Fig. 17.4.2a, with the R of BD known, going along the sides of triangle number 1 in the counterclockwise direction will give the member-axis rotations of DA and AB. Similarly, going along the sides of triangle number 3 in the counterclockwise direction gives the member-axis rotations of BC and CE.

2. Using one end of the chosen reference member as the reference point, obtain the joint displacements of all joints by successive use of the joint-displacement equation, as illustrated in Example 3.6.1 and Fig. 3.6.2b. With these displacements the member-axis rotations can then be computed using the concept of Fig. 17.4.1. Note that translation and rotation, in order to satisfy the actual external boundary conditions as in Fig. 3.6.2d, are not required.

3. Using only the Williot part of the Williot-Mohr diagram, obtain the relative transverse displacement between the member ends and divide by the original member length to arrive at the member-axis rotation. In this case the sign of the member-axis rotation R must be determined by inspection. For instance, as shown in Fig. 17.4.2b, the relative transverse displacement of member ends A and D is equal to Δ; by locating A' and D' on Fig. 17.4.2a from their deflections in Fig. 17.4.2b, the R of AD is observed to be clockwise, or positive.

Once the R values of all members have been obtained, the fixed-end moments are computed using Eq. (17.4.1). The moment distribution, along with the checks to obtain the joint rotations, is then carried out in a moment-distribution table. The balanced moments are the secondary bending moments.

Example 17.4.1 Compute the secondary bending moments in all members of the four-panel Pratt truss with rigid joints in Examples 17.3.1 and 17.3.2.

SOLUTION (a) *Member-axis rotations from joint displacements.* The joint displacements resulting from using U_1 as the reference point and U_1L_2 as the reference member are taken from Fig. 3.6.2b of Example 3.6.1 and shown again in Fig. 17.4.3a. The member-axis rotations, computed by applying the concept of Fig. 17.4.1, are written on the members in Fig. 17.4.3a. The sample computation for L_2U_3, using the sketches in Fig. 17.4.3b, is

$$R \text{ of } L_2U_3 = -\frac{3.12(0.8)}{8000} - \frac{3.34(0.6)}{8000} = -0.5625 \times 10^{-3}$$

The negative sign in the above expressions are obtained by inspecting Fig. 17.4.3b.

640 INTERMEDIATE STRUCTURAL ANALYSIS

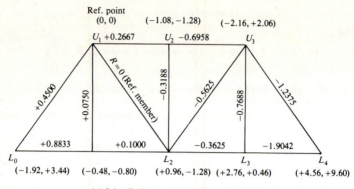

(a) Joint displacements (mm) from Fig. 3.6.2b; member-axis rotations in 10^{-3} rad

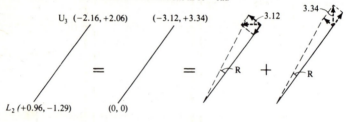

(b) Computation of R for member L_2U_3

Figure 17.4.3 Member-axis rotations from joint displacements of truss in Example 17.4.1.

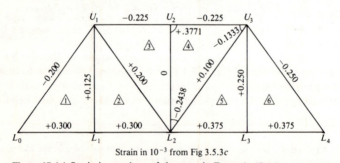

Strain in 10^{-3} from Fig 3.5.3c

Figure 17.4.4 Strain in members of the truss in Example 17.4.1.

(b) *Member-axis rotations from angle changes.* The unit strain values shown in Fig. 17.4.4 are taken from Fig. 3.5.3c in Example 3.5.1. The angle changes in the three interior angles of the six triangles designated by numbers 1 to 6 are computed using Eqs. (3.5.1a to c). The sample computations for triangle number 4 are as follows:

$$\text{Angle change at } L_2 = (-0.225 - 0)(0) + (-0.225 - 0.100)(\tfrac{3}{4}) = -0.2438$$
$$\text{Angle change at } U_3 = (0 + 0.225)(0) + (0 - 0.100)(\tfrac{4}{3}) = -0.1333$$
$$\text{Angle change at } U_2 = (+0.100 - 0)(\tfrac{4}{3}) + (+0.100 + 0.225)(\tfrac{3}{4}) = +0.3771$$

SECONDARY MOMENTS IN TRUSSES WITH RIGID JOINTS 641

Note that the angle changes are in 10^{-3} rad and the sum of the three changes of the same triangle is zero. For clarity, only the angle changes for triangle number 4 are shown in Fig. 17.4.4. With these angle changes, the R values can be computed starting from triangle number 3 by setting R of U_1L_2 equal to zero, then triangle numbers 4, 5, and 6 toward the right. Then starting from triangle number 2, one goes toward the left, completing triangle number 1. The same R values as shown in Fig. 17.4.3a are obtained.

(c) *Fixed-end moments due to member-axis rotations.* Equation (17.4.1) gives the two fixed-end moments at the ends of a member for which there has been a rotation of the member axis, but the tangents to the elastic curve at the ends remain in the direction of the original member axis. Clockwise member-axis rotations are taken as positive, and clockwise moments acting on the member ends are positive; therefore there is a negative sign in Eq. (17.4.1). Using this equation, the fixed-end moments at the ends of each member when U_1L_2 is taken as the reference member are computed and shown in Table 17.4.1.

(d) *Moment distribution.* The moment distribution is carried out in Table 17.4.2. In the check process, the rotations of the joints (when the member axis of U_1L_2 does not rotate) are found. Actually the member axis of U_1L_2 does rotate, in order to satisfy the boundary conditions of the truss, in the amount of (referring to Fig. 17.4.3a)

$$\frac{9.60 - 3.44}{19{,}200} = 0.3208 \times 10^{-3} \text{ rad}$$

in the clockwise direction. By adding this rotation to those in the check process of Table 17.4.2, the true rotations of the rigid joints in the truss are as follows:

θ at $U_1 = +0.605 \times 10^{-3}$ rad θ at $L_0 = +0.997 \times 10^{-3}$ rad
θ at $U_2 = +0.154 \times 10^{-3}$ rad θ at $L_1 = +0.603 \times 10^{-3}$ rad
θ at $U_3 = -0.595 \times 10^{-3}$ rad θ at $L_2 = +0.184 \times 10^{-3}$ rad
θ at $L_3 = -0.663 \times 10^{-3}$ rad
θ at $L_4 = -1.285 \times 10^{-3}$ rad

These rotations are approximately equal to those in part (f) of Example 17.3.1.

(e) *Alternate moment distribution.* An alternate moment distribution may be made by using any other member as the reference member, such as U_2L_2, or by using the true

Table 17.4.1 Fixed-end moments due to member-axis rotation taking U_1L_2 as reference member; Example 17.4.1

Member	L, cm	R, 10^{-3} rad	E, kN/cm^2	I, cm^4	$-6EIR/L$, kN·cm
U_1U_2	480	+0.2667	20,000	3600	− 240
U_2U_3	480	−0.6958	20,000	3600	+ 626
L_0L_1	480	+0.8833	20,000	1215	− 268
L_1L_2	480	+0.1000	20,000	1215	− 30
L_2L_3	480	−0.3625	20,000	1215	+ 110
L_3L_4	480	−1.9042	20,000	1215	+ 578
L_0U_1	800	+0.4500	20,000	7200	− 486
U_1L_2	800	0.0000	20,000	1440	0
L_2U_3	800	−0.5625	20,000	1440	+ 122
U_3L_4	800	−1.2375	20,000	7200	+1336
U_1L_1	640	+0.0750	20,000	3240	− 46
U_2L_2	640	−0.3188	20,000	135	+ 8
U_3L_3	640	−0.7688	20,000	3240	+ 467

Table 17.4.2 Moment Distribution for Example 17.4.1

Joint		U_1				U_2			U_3				L_0	
Member		U_1L_0	U_1L_1	U_1L_2	U_1U_2	U_2U_1	U_2L_2	U_2U_3	U_3U_2	U_3L_3	U_3L_3	U_3L_4	L_0U_1	L_0L_1
EI/L, 10^3 kN·cm		180	101	36	150	150	4	150	150	36	101	180	180	51
Cycle	DF	0.385	0.217	0.077	0.321	0.493	0.014	0.493	0.321	0.077	0.217	0.385	0.780	0.220
1	FEM	−486	−46	0	−240	−240	+8	+626	+626	+122	+467	+1336	−486	−268
	BAL	+298	+167	+59	+248	−194	−6	−194	−819	−196	−553	−983	+588	+166
2	CO	+294	+86	−21	−97	+124	−2	−410	−97	−22	−289	−747	+149	+43
	BAL	−101	−57	−20	−84	+142	+4	+142	+371	+89	+250	+445	−150	−42
3	CO	−75	−34	+18	+71	−42	+2	+186	+71	+18	+129	+248	−50	−17
	BAL	+8	+4	+2	+6	−72	−2	−72	−150	−36	−101	−179	+52	+15
4	CO	+26	+6	8	−36	+3	−1	−75	−36	−8	−55	−112	+4	+3
	BAL	+5	+2	1	+4	+36	+1	+36	+68	+16	+46	+81	−5	−2
5	CO	−2	0	+4	+18	+2	0	+34	+18	+4	+24	+46	+2	0
	BAL	−8	−4	−2	−6	−18	0	−18	−30	−7	−20	−35	−2	0
6	CO	−1	−1	−2	−9	+3	0	−15	−9	−2	−10	−20	−4	0
	BAL	+5	+3	+1	+4	+9	0	+9	+13	+3	+9	+16	+3	+1
7	CO	+1	+1	+1	+4	+2	0	+6	+4	+1	+5	+10	+2	0
	BAL	−3	−1	−1	−2	−4	0	−4	−6	−2	−4	−8	−1	0
8	CO	−1	0	0	−2	−1	0	−3	−2	0	−2	−4	−1	0
	BAL	+1	+1	0	+1	+2	0	+2	+2	+1	+2	+3	+1	0
9	CO	0	0	0	+1	0	0	+1	+1	0	0	0	0	0
	BAL	−1	0	0	0	0	0	−1	0	0	0	−1	0	0
Total M		−40	+127	+32	−119	−254	+4	+250	+25	−19	−102	+96	+101	−101
Check:														
Change		+446	+173	+32	+121	−14	−4	−376	−601	−141	−569	−1240	+587	+167
−½(change)		−294	−86	0	+7	−60	+2	+300	+188	+44	+292	+745	−223	−64
Sum		+152	+87	+32†	+128	−74	−2	−76	−413	−97	−277	−495	+364	+103
Sum/(3EI/L)		+0.282	+0.286	+0.296†	+0.284	−0.164	−0.158†	−0.169	−0.918	−0.898†	−0.912	−0.917	+0.674	+0.678
θ (average)		+0.284				−0.167			−0.916				+0.676	

†Not used in computing average θ.

Table 17.4.2 Moment distribution for Example 17.4.1 (continued)

Joint		L_1				L_2					L_3			L_4	
Member		L_1L_0	L_1U_1	L_1L_2	L_2L_1	L_2U_1	L_2U_2	L_2U_3	L_2L_3	L_3L_2	L_3U_3	L_3L_4	L_3L_3	L_4L_3	L_4U_3
EI/L, 10^3 kN·cm		51	101	51	51	36	4	36	51	51	101	51	51	51	180
Cycle	DF	0.250	0.500	0.250	0.285	0.203	0.024	0.203	0.285	0.250	0.500	0.250	0.220	0.780
1	FEM	−268	−46	−30	−30	0	+8	+122	+110	+110	+467	+578	+578	+1336
	BAL	+86	+172	+86	−60	−42	−5	−43	−60	−288	−578	−289	−420	−1494
2	CO	+83	+84	−30	+43	+30	−3	−98	−144	−30	−276	−210	−144	−492
	BAL	−34	−68	−35	+49	+35	+4	+35	+49	+129	+258	+129	+140	+496
3	CO	−21	−28	+24	−18	−10	+2	+44	+64	+24	+125	+70	+64	+222
	BAL	+6	+12	+7	−23	−17	−2	−17	−23	−54	−110	−55	−63	−223
4	CO	+8	+2	−12	+4	+1	−1	−18	−27	−12	−50	−32	−28	−90
	BAL	+1	+1	0	+12	+8	+1	+8	+12	+23	+47	+24	+26	+92
5	CO	−1	+1	+6	0	0	0	+8	+12	+6	+23	+13	+12	+40
	BAL	−1	+3	−2	−6	−4	0	−4	−6	−10	−21	−11	−11	−41
6	CO	0	−2	−3	−1	−1	0	+4	+5	−3	−10	+6	+6	+18
	BAL	+1	+2	+2	+3	+2	+1	+2	+3	+4	+10	+5	+5	+19
7	CO	0	+1	+1	+1	0	0	+1	+2	+1	+4	+2	+2	+8
	BAL	0	−1	−1	−1	−1	0	−1	−1	−1	−4	−2	−1	−9
8	CO	0	0	0	0	0	0	−1	0	0	−2	0	0	−4
	BAL	0	0	0	+1	0	0	0	0	+1	+1	0	−1	+4
9	CO	0	0	0	0	0	0	0	0	0	+1	0	0	−1
	BAL	0	0	0	0	0	0	0	0	0	−1	0	0	−1
Total M		−140	+127	+13	−26	+1	+5	+34	−14	−100	−116	+216	+154	−154
Check:															
Change		+128	+173	+43	+4	+1	−3	−88	−124	−211	−583	−362	−424	−1490
−$\frac{1}{2}$(change)		−84	−86	−2	−22	−16	+2	+70	+150	+62	+284	+212	+181	+620
Sum		+44	+87	+41	−18	−15	−1	−18	−19	−149	−299	−150	−243	−870
Sum/(3EI/L)		+0.290	+0.286	+0.270	−0.118	−0.139	−0.079†	−0.167	−0.125	−0.981	−0.984	−0.988	−1.600	−1.611
θ(average)			+0.282			−0.137					−0.984			−1.606	

†Not used in computing average θ.

member-axis rotations as determined from the true joint displacements in Fig. 3.6.2d of Example 3.6.1. In all cases the same balanced moments should be obtained, as well as the same true joint rotations, provided the true rotation of the reference member is added to the joint rotations in the check process of the moment distribution.

17.5 Iteration Method—From Secondary Bending Moments to Third Axial Forces

By moment distribution, the rigid frame in the shape of the truss with rigid joints has been analyzed for unequal member-axis rotations that are forced into the rigid frame on the basis of infinite longitudinal rigidity EA, that is, neglecting axial deformation. The balanced moments acting at the member ends require transverse shear forces, also acting at the member ends, for equilibrium, such as shown in Fig. 17.5.1a. The reversals of these shear forces on member ends will in turn act on the joints of the truss, such as shown in Fig. 17.5.1b. If the reversals of all the transverse shear forces acting at all member ends are assembled at the joints, a set of external forces acting on the joints can be obtained. Since each pair of shear forces acting on the joints is statically equivalent to the sum of the two moments acting on the ends of the member, and since the sum of all end moments acting on all members is zero, it follows that the resultant of the set of joint forces replacing the shear forces must be zero. Consequently there will be no modification of the external reactions on the truss with rigid joints.

At this point, the joints of the truss are considered to be pins again. The axial forces in the members due to the set of joint forces arising from the transverse shear forces of the secondary bending moments are called the *third axial forces*. Invariably, if the primary joint displacements are due to external forces applied at the joints only, the third axial forces are but a small fraction of the primary axial forces. This shows that the iteration process may stop when the secondary bending moments are obtained. In the following example, the third axial forces are computed to demonstrate their almost negligible magnitudes.

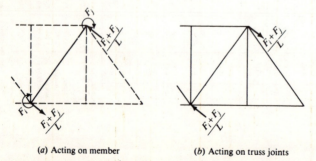

(a) Acting on member   (b) Acting on truss joints

Figure 17.5.1 Transverse shear forces due to secondary bending moments.

SECONDARY MOMENTS IN TRUSSES WITH RIGID JOINTS

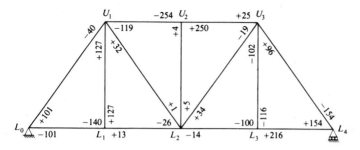

(a) Secondary bending moments in kN·cm from Example 17.4.1

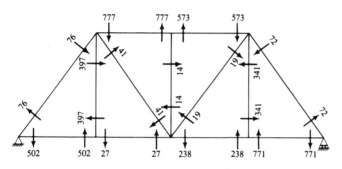

(b) Transverse shear forces in 10^{-3} kN acting on joints

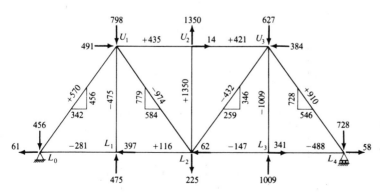

(c) Equivalent joint forces and third axial forces in 10^{-3} kN

Figure 17.5.2 Third axial forces for truss in Example 17.5.1.

Example 17.5.1 Compute the third axial forces from the secondary bending moments already obtained in Example 17.4.1.

SOLUTION (a) *Transverse shear forces.* The secondary bending moments, positive if acting clockwise at the member end, are taken from the moment-distribution table in Example 17.4.1 and shown in Fig. 17.5.2a. The transverse shear forces as they act on the joints are shown in Fig. 17.5.2b, by considering the free-body diagram of each member. The forces shown in Fig. 17.5.2b, when transformed into horizontal and vertical forces, are shown in Fig. 17.5.2c. As a check, the horizontal, vertical, and moment summations of all the joint forces in Fig. 17.5.2c should be zero; thus

$\Sigma F_x = 0$: $+904 - 904 = 0$

$\Sigma F_y = 0$: $+2834 - 2834 = 0$

ΣM about $L_0 = 0$: $+16,302 - 16,302 = 0$

(b) *Third axial forces.* By the method of joints, the axial forces (the third axial forces) due to the set of external joint forces in Fig. 17.5.2c are computed, with the answers written on the respective members in Fig. 17.5.2c.

17.6 Comparison of Methods

A simple four-panel Pratt truss with rigid joints has been analyzed for three unequal vertical loads applied at the three lower-chord joints. First the rigid-joint truss is treated as a composite structure, that is, a rigid frame with axial deformation. The matrix displacement method is used and the results are obtained through a computer program since there are 21 simultaneous equations involved. Next the iteration method is used by first ignoring the rigidity of the joints but considering only the axial deformation, then by ignoring the axial deformation but considering the rigidity of the joints, and finally by ignoring the rigidity of the joints again. It can be shown that the iteration method gives almost identical results as the displacement method of composite-structure analysis. This is so because the *primary* action is a *truss* action, and the bending moments are *secondary* owing to the deformations in the primary action. When the load system involves forces other than linear forces acting on the joints, then the primary action is no longer a truss action and the iteration method may not converge. For the present example, results obtained by the two methods are nearly identical; in fact the differences might be due to the fact that a smaller number of significant figures is used in the iteration method than in the computer displacement method. One should note that the iteration method itself can also be written to a computer program.

Example 17.6.1 Compare the sum of the primary and the third axial forces in each member of the truss in Example 17.5.1 with the axial force of the displacement method in Example 17.4.1.

SOLUTION The comparison is shown in Table 17.4.2.

17.7 Exercises

17.1 Using U_2L_2 as the reference member, compute the secondary bending moments in all members of the truss with rigid joints in Example 17.4.1.

SECONDARY MOMENTS IN TRUSSES WITH RIGID JOINTS

Table 17.4.2 Comparison of axial forces, Example 17.6.1

Member	Iteration method			Displacement method
	Primary	Third	Sum	
U_1U_2	−405	+0.44	−404.56	−404.57
U_2U_3	−405	+0.42	−404.58	−404.58
L_0L_1	+270	−0.28	+269.72	+269.71
L_1L_2	+270	+0.12	+270.12	+270.11
L_2L_3	+337.5	−0.15	+337.35	+337.34
L_3L_4	+337.5	−0.49	+337.01	+337.01
L_0U_1	−450	+0.57	−449.43	−449.42
U_1L_2	+225	−0.97	+224.03	+224.04
L_2U_3	+112.5	−0.43	+112.07	+112.09
U_3L_4	−562.5	+0.91	−561.59	−561.59
U_1L_1	+180	−0.48	+179.52	+179.52
U_2L_2	0	+1.35	+1.35	+ 1.32
U_3L_3	+360	−1.01	+358.99	+358.99

17.2 Using L_0U_1 as the reference member, compute the secondary bending moments in all members of the truss with rigid joints in Example 17.4.1.

17.3 Using U_2L_2 as the reference member, compute the secondary bending moments in all members of the truss shown in Fig. 3.8.1 except that all joints are now rigid. Use the following for the moments of inertia: 2600 cm⁴ for the upper chord, 1800 cm⁴ for the lower chord, 3600 cm⁴ for the end diagonals, 1200 cm⁴ for the intermediate diagonals, 800 cm⁴ for the side verticals, and 600 cm⁴ for the center vertical.

17.4 Solve Exercise 17.3, but use L_0U_1 as the reference member.

17.5 Analyze the truss with rigid joints shown in Fig. 17.7.1 as a rigid frame with axial deformation by the slope-deflection method. To reduce the number of unknowns, analyze an

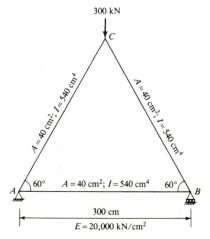

Figure 17.7.1 Exercises 17.5 to 17.7.

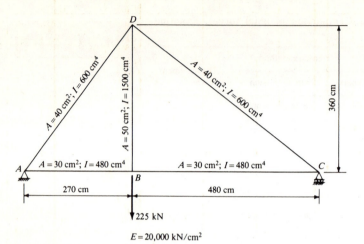

Figure 17.7.2 Exercises 17.8 to 17.10.

equivalent structure having a roller support with a horizontal reaction at C and two roller supports at A and B, each with a vertical reaction. The three unknown displacements are the vertical displacement of C, and the absolute magnitudes of the horizontal displacement and the joint rotation at A or B.

17.6 Using three rotational and three linear degrees of freedom, obtain numerically the matrices $[A]$, $[B]$, $[S]$, $\{P\}$, $[SB]$, and $[ASB]$ in the analysis of the structure in Exercise 17.5 by the matrix displacement method. Making use of the answers to Exercise 17.5, verify the equations $\{P\} = [ASB]\{X\}$ and $\{F^*\} = \{F_0\} + [SB]\{X\}$.

17.7 Analyze the truss with rigid joints in Exercises 17.5 and 17.6 by the iteration method.

17.8 Prepare numerically the input matrices $[A]$, $[B]$, $[S]$, and $\{P\}$ for the analysis of the truss with rigid joints shown in Fig. 17.7.2.

17.9 Show the member information that is needed as input for the analysis of the structure in Fig. 17.7.2 by a computer program in which the global stiffness matrix is built by the direct stiffness method.

17.10 Analyze the truss with rigid joints in Exercises 17.8 and 17.9 by the iteration method.

CHAPTER
EIGHTEEN
RIGID FRAMES WITH CURVED MEMBERS

18.1 General Description

Although the general name *curved members* has been chosen for the title of this chapter, perhaps the word *arches* is a more common name for most curved members by themselves. The general name is used here because the methods to be described are equally useful in the analysis of assemblies including curved members in machine parts, automobiles, airplanes, or even the human skeleton.

In construction, arches have been used to advantage for bridges that can be supported on rock formations at the ends of the span over a deep valley, in situations where very little support settlements are expected during the life of the structure. Support settlements, because of the flexibility that exists due to the rise of the arch, usually induce sizable internal forces in the arch relative to those due to dead and live loads. A typical fixed-arch bridge is shown in Fig. 18.1.1a, wherein the loads on the bridge deck are carried to the top of each of the two main arches at the panel points only. A reinforced concrete roof arch, of constant dimensions in the longitudinal direction of a building, may be connected to heavy footings at the ends of the transverse span by means of dowels so that the arch itself may be considered to have two fixed supports. In the analysis of such an arch as shown in Fig. 18.1.1b, only a typical slice of the arch of a unit width in the longitudinal direction needs to be considered. On the other hand, steel arches may be used at regular intervals in the longitudinal direction of a building carrying closely spaced purlins to which corrugated sheets are attached to form the roof surface, as shown in Fig. 18.1.1c. The continuous-arch bridge on elastic piers shown in Fig. 18.1.1d and the two-span building frame of Fig. 18.1.1e are indeed rigid frames with curved members. In Fig. 18.1.1d, there are three curved members and four straight members; in Fig. 18.1.1e, one curved member and four straight members.

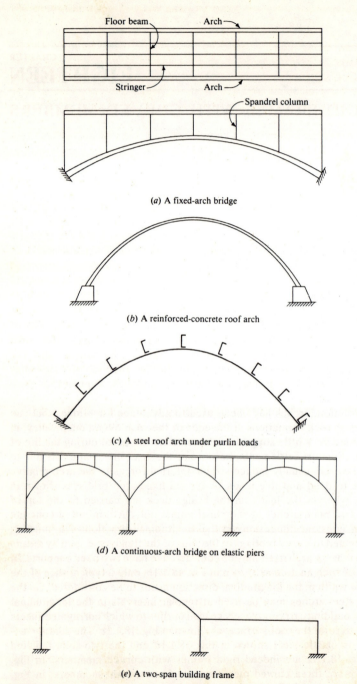

Figure 18.1.1 Typical arches for bridges and roofs.

RIGID FRAMES WITH CURVED MEMBERS

Single-span arches can be statically determinate, if, in addition to two hinged supports at the extreme ends, there is a third hinge, commonly called the *crown hinge*, installed at the highest point of the arch. Such arches usually contain only truss elements and are used for roof structures over a large space. A three-hinged arch truss can be analyzed by the method presented in Chap. 3.

Ribbed arches, as arches containing solid cross sections along the length are commonly called, can be divided into a number of combined truss and beam elements if axial deformation is to be considered in the analysis. Single-span ribbed arches are rarely statically determinate because such an arch, with one hinged support and one roller support, would have too large deflections. A tie rod may be used to connect the end where the roller support is to the hinged end; in this case the structure is statically indeterminate to the first degree with one curved element and one truss element. If both supports are hinged, without the tie rod, the structure is still statically indeterminate to the first degree.

The objective of this chapter is to provide the basis for the general matrix displacement method of analyzing rigid frames with curved members, such as shown in Fig. 18.1.1d and e. The fixed-condition forces for a curved member are actually the reactions on a commonly called fixed arch. The stiffness matrix of the curved member is a 6×6 matrix which expresses the six internal forces (one moment, one horizontal force, and one vertical force at each end of the curved member) in terms of the six corresponding displacements (one rotation, one horizontal displacement, and one vertical displacement at each end of the curved member). It will be shown that the contents of the stiffness matrix are actually the induced reactions acting on a fixed arch when there is yielding in rotation, in horizontal displacement, and in vertical displacement at each support of the fixed arch.

18.2 Fixed-Condition Forces for a Curved Member

Whether a single-span fixed arch is to be analyzed by itself, or the curved member is only one element in a large rigid frame, the fixed-condition forces (or, more simply, the reactions) due to a particular loading (a single load in influence-line computations) must first be determined. Sometimes the shape of the curved member is made up by circular arcs of several different radii or parts of ellipses and parabolas. However, more often the moment of inertia varies along the arch axis not following any mathematical formula, as in the case of a steel arch with a variable-depth web plate and cover plates of different lengths, or a reinforced concrete arch with variable concrete outline as well as reinforcement.. Thus, aside from most regular cases, say of constant cross section on an arch axis of a circle or parabola, closed-form formulas for the reactions are not obtainable. The procedures to be described in this chapter are applicable when the shape of the arch axis and the variation in moment of inertia have both been numerically defined from point to point.

If the displacement method is used, the curved members shown in Fig. 18.2.1a are replaced by rigid frames with straight segments wherein the corners lie on the true arch axis. Since each segment is taken to have a constant cross section, the results are more accurate when more segments are used. The resulting rigid frames, as shown by Fig. 18.2.1b, may be analyzed by the method of Chap. 12 neglecting axial deformation, or by the method of Chap. 16 considering axial deformation. The method of Chap. 12 is not desirable because of the difficulty in obtaining the [A] and [B] matrices for a rigid frame with nonrectangular joints yet with a large degree of freedom in sidesway. The method of Chap. 16 is suitable in cases where a computer program using the direct stiffness method is available, because then the input data can be conveniently prepared, as illustrated in Example 17.3.2 or Table 17.3.3 in connection with secondary moments in trusses with rigid joints.

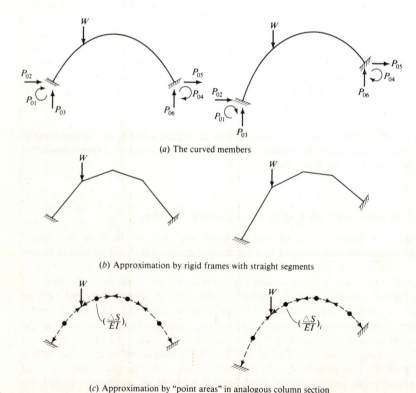

(a) The curved members

(b) Approximation by rigid frames with straight segments

(c) Approximation by "point areas" in analogous column section

Figure 18.2.1 Fixed-condition forces for curved members.

RIGID FRAMES WITH CURVED MEMBERS 653

When the first approach is used, only the relative values of the moments of inertia for the straight segments are required; but if axial deformation is considered, the absolute values of both the cross-sectional area and the moment of inertia, or at least the correct numerical relationship between EA and EI, will have to be within the input data.

In all practical situations of design, it is usually only the relative values of the moments of inertia that can be conveniently estimated before analysis begins. In addition, consideration of axial deformation does not affect the fixed-condition forces to any degree of significance, because by nature the deformation is almost all due to bending moment. If this is the guiding policy, the column-analogy method presented in Chap. 15 lends itself very nicely to the analysis of the rigid frames with two fixed supports shown in Fig. 18.2.1b. Inasmuch as the rigid frame has a fair number of short segments, it seems unnecessary, in depicting the analogous column section, to use a line area of ΔS in length and $1/EI$ in width and then include the moment of inertia of the line area about its own centroidal axes. Hence the entire contribution of the segment to the analogous column may be represented by a "point area," which is equal to $\Delta S/EI$, as shown in Fig. 18.2.1c. The detailed procedure will be further explained in the subsequent sections. It should be emphasized again that the column-analogy method is a force method using compatibility conditions to solve for the redundants and wherein axial-deformation effects are not considered.

18.3 Analysis of Fixed Arches

A curved member with two fixed supports, or the fixed arch, is completely analyzed when the thrust, shear, and moment at any section perpendicular to the arch axis are known. The thrust N, usually in the form of a push, is the total force acting perpendicular to the section at its centroid. The shear V is the total force acting parallel to the section. The moment M is the total moment about the point of application of N in the cross section. In regard to the signs for N, V, and M, the directions as shown in Fig. 18.3.1 are positive. Note that thrust (or compression) has been taken as positive in this particular situation, common in masonry and reinforced concrete design.

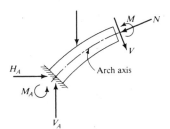

Figure 18.3.1 Sign convention for N, V, and M in a fixed arch.

It is apparent that the thrust, shear, and moment at any section of a fixed arch can be readily determined by the laws of statics if the six reactions at the two fixed supports are known. Since, by considering the whole arch as a free body, only three independent equations of statics are available, the fixed arch is statically indeterminate to the third degree. The column-analogy method, and its variation, the elastic-center method, are the most convenient methods of analyzing the fixed arch.

As with any statically indeterminate structure, the properties of the fixed arch must be completely known before it can be analyzed. Thus, before an arch can be analyzed, a design must be assumed. From the results of the ensuing analysis, the assumed design may be modified, and, if necessary, reanalyzed. This process should be repeated until the last adopted design can sustain at all sections the thrust, shear, and moment as found in the final analysis.

While a fixed arch may be used for roofs or bridges, its analysis can best be performed by constructing influence lines for the reactions and for the moments at certain evenly spaced sections. Usually the position of live load for maximum combined effect at any section is taken as that satisfying the criterion for maximum moment, and the corresponding shear and thrust are then found for this loading position. In this chapter the discussion is limited to the determination of the influence lines for reactions and moments in a fixed arch when its properties are given. In fact, for this purpose, only the shape of the arch axis and the relative values of the moments of inertia at different sections along the arch axis are required. Then the effect of temperature, shrinkage, rib shortening, and foundation yielding will be discussed. In the latter case, the absolute sizes of the moments of inertia are required.

18.4 Elastic-Center Method vs. Column-Analogy Method

As discussed in Sec. 18.2, usually the shape of the arch axis and the way in which the moments of inertia vary along the arch axis do not fit into any mathematical equations. Thus the curved member has to be considered as consisting of a finite number of straight segments, each having a constant moment of inertia. The more divisions that are made in the arch, the more accurate will be the results; usually eight to ten divisions are sufficient for arches with moderate rise. The segments should have approximately equal length along the arch axis, although in the numerical examples the segments are taken to have equal horizontal projections. This is done simply for purpose of illustration so that the reader may check readily on the arithmetical aspects of the examples. In this connection, a quadrangular frame with two fixed supports can be considered a "simplified arch" with only three finite members; and a single-span gable frame, with four finite members. Thus the column-analogy method presented in Chap. 15 can be applied as well to the analysis of fixed arches.

It is sometimes said that there are two methods of analyzing the fixed arch: namely, the elastic-center method and the column-analogy method. In reality, the two methods are *identical* in every detail of arithmetic and only slightly different in the final step when the moment at any section is found. In the elastic-center method, the redundant forces M_O, H_O, and V_O as shown in Fig. 18.4.1a are first found, and the moment M at any section is determined as the bending moment in the cantilever structure $ABCDEFGO$ fixed at A and free at the elastic center O; or

$$M = M_s + M_O + H_O y - V_O x \tag{18.4.1}$$

in which M_s is the bending moment due to the applied loads (usually negative) and x and y are the coordinates of the point on the arch axis at which M is desired, in reference to the elastic center as origin. In the column-analogy method, there is neither mention of the rigid arm nor use of the redundant forces M_O, H_O, and V_O; the moment M at any section is directly determined from the equation

$$M = M_s - M_i \tag{18.4.2}$$

where M_s is as defined previously and M_i is the compressive stress in the analogous column.

A mixed procedure is suggested for use in hand computations wherein (1) values of M_O, H_O, and V_O are computed by formulas as provided by the elastic-center method; (2) values of the reactions at the left support are computed by taking the whole cantilever structure of Fig. 18.4.1a as a free body; (3) values of the reactions at the right support are computed by taking the rigid arm itself as a free body; (4) the three equations of equilibrium are numerically checked out by taking the whole arch as a free body without the rigid arm; (5) values of the bending moments required at chosen points on the arch axis are computed by taking either the left or the right portion of the arch as the free body; and (6) values of the bending moments already obtained in step 5, as well as those at the left and right supports already

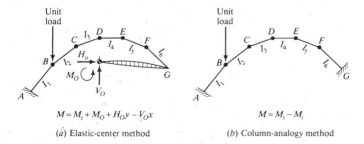

(a) Elastic-center method $\quad\quad\quad\quad\quad$ (b) Column-analogy method

Figure 18.4.1 Elastic-center method vs. column-analogy method.

obtained in steps 2 and 3, are recomputed from the equation $M = M_s - M_i$ in the column-analogy method. Naturally, when a computer program is to be written, duplicate calculations are wasteful; only in the sample problem which is used to debug the program is extreme care in hand computations essential.

For convenience, the formulas already derived and used in Chap. 15 are repeated below:

For both symmetrical and unsymmetrical arches,

$$A = \sum \frac{\Delta s}{EI} \qquad I_x = \sum \frac{y^2 \Delta s}{EI} \qquad I_{xy} = \sum \frac{xy \, \Delta s}{EI} \qquad I_y = \sum \frac{x^2 \Delta s}{EI}$$

$$P = \sum \frac{M_s \Delta s}{EI} \qquad M_x = \sum \frac{M_s y \, \Delta s}{EI} \qquad M_y = \sum \frac{M_s x \, \Delta s}{EI}$$

where (x, y) = coordinates of point areas in reference to the elastic center as origin.

For symmetrical arches,

$$M_O = -\frac{P}{A} \qquad H_O = -\frac{M_x}{I_x} \qquad V_O = +\frac{M_y}{I_y}$$

$$M = M_s + M_O + H_O y - V_O x \quad \text{(elastic-center method)}$$

or $\qquad M = M_s - M_i \quad$ (column-analogy method)

in which

$$M_i = \frac{P}{A} + \frac{M_x y}{I_x} + \frac{M_y x}{I_y}$$

For unsymmetrical arches,

$$M'_x = M_x - M_y \left(\frac{I_{xy}}{I_y}\right) \qquad M'_y = M_y - M_x \left(\frac{I_{xy}}{I_x}\right)$$

$$I'_x = I_x \left(1 - \frac{I_{xy}^2}{I_x I_y}\right) \qquad I'_y = I_y \left(1 - \frac{I_{xy}^2}{I_x I_y}\right)$$

$$M_O = -\frac{P}{A} \qquad H_O = -\frac{M'_x}{I'_x} \qquad V_O = +\frac{M'_y}{I'_y}$$

$$M = M_s + M_O + H_O y - V_O x \text{ (elastic-center method)}$$

or $\qquad M = M_s - M_i \quad$ (column-analogy method)

in which

$$M_i = \frac{P}{A} + \frac{M'_x y}{I'_x} + \frac{M'_y x}{I'_y}$$

18.5 Influence Lines for a Symmetrical Fixed Arch

The influence lines for some reactions, shears, and moments in a typical symmetrical fixed arch are computed and shown in Example 18.5.1. From the results it is important to note that, in general, (1) loading the left three-eighths and the right five-eighths of the span causes negative and positive bending moment, respectively, at the left support; (2) this same loading position causes positive and negative bending moments, respectively, at the quarter point of the span; (3) loading the two outer three-eighths and the center one-fourth of the span causes negative and positive bending moments, respectively, at the crown. These three conclusions can hardly be guessed at by physically visualizing the response behavior preceding the analysis, although one can rationalize and feel agreeable to the results of the analysis. This knowledge, then, becomes useful as a common-sense way to prescribe, for instance, which portion of the roof should be loaded with snow to cause the most severe effects at various points on the arch.

Example 18.5.1 For the symmetrical parabolic fixed arch shown in Fig. 18.5.1a, draw influence lines for (1) the horizontal reaction at the left support, (2) the vertical reaction at the left support, (3) the vertical shear at the crown, (4) the moment at the left support, (5) the moment at the quarter point of the horizontal span, and (6) the moment at the crown.

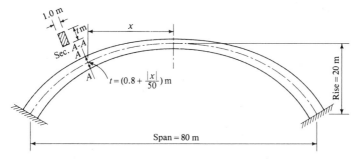

(a) The given fixed arch

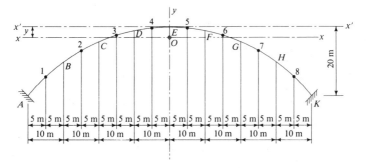

(b) The divided segments

Figure 18.5.1 Symmetrical fixed arch in Example 18.5.1.

658 INTERMEDIATE STRUCTURAL ANALYSIS

Table 18.5.1 Properties of analogous column section for symmetrical fixed arch in Example 18.5.1

Segment	Centroid	Δs, m	t, m	n ($I = nI_c$)	$\dfrac{\Delta s}{EI} = \dfrac{\Delta s}{E(nI_c)} = \dfrac{\Delta s}{n}$ (let $EI_c = 1$)	y'	$y'\dfrac{\Delta s}{EI}$	x	y	$x^2 \dfrac{\Delta s}{EI}$	$y^2 \dfrac{\Delta s}{EI}$
AB	1	13.29	1.50	6.592	2.016	−15.31	−30.86	−35	−11.17	2470	251.5
BC	2	11.79	1.30	4.291	2.748	− 7.81	−21.46	−25	− 3.67	1718	37.0
CD	3	10.68	1.10	2.600	4.108	− 2.81	−11.54	−15	+ 1.33	924	7.3
DE	4	10.08	0.90	1.424	7.079	− 0.31	− 2.19	− 5	+ 3.83	177	103.8
					15.951	$\bar{y}' = -4.14$	−66.05			5289	399.6
					$A = 31.902$					$I_y = 10{,}578$	$I_x = 799.2$

SOLUTION (a) *Properties of the analogous column section.* The arch rib is arbitrarily replaced by eight segments of equal horizontal projection, as represented by *AB, BC, CD*, etc. in Fig. 18.5.1*b*. As noted earlier, it might be better to divide the arch rib into segments of equal length along its own axis, but for the sake of convenience for the reader in reviewing the arithmetic, the division is made horizontally. Likewise, it might be better to locate the centroid at the midpoint of the true length of each segment than at the midpoint of its horizontal projection, which is done again for sake of convenient arithmetical computation. These centroids are noted as points 1, 2, 3, etc. in Fig. 18.5.1*b*. The moment of inertia of the section through the crown, point *E*, is called I_c, and the thicknesses of the arch rib at points 1, 2, 3, etc. are used to compute the moments of inertia $n_1 I_c$, $n_2 I_c$, $n_3 I_c$, etc. for the segments. Each point area in the analogous column section is equal to $(\Delta S)_i/(n_i E I_c)$, $i = 1$ to 8; and it is located at points 1, 2, 3, etc. which lie on the arch axis. The length $(\Delta S)_i$ of each segment is taken as the straight-line distance of *AB, BC, CD*, etc. while points *A, B, C, D*, etc. lie on the arch axis. With these stipulations, the properties of the analogous column section are computed as shown in Table 18.5.1 (sample calculation for segment *CD* is shown below). Note that the quantity EI_c should appear as the denominator in the properties of the analogous column section such as A, I_x, and I_y.

Sample calculation for segment CD (see Fig. 18.5.2):

y' of point $3 = -(\tfrac{15}{40})^2(20) = -(\tfrac{3}{8})^2(20) = -2.81$ m

$\Delta s = \sqrt{(10)^2 + (3.75)^2} = 10.68$ m

$t = 0.8 + \tfrac{15}{50} = 1.10$ m

$n = \dfrac{I}{I_c} = \dfrac{(1)(t)^3/12}{(1)(0.8)^3/12} = \left(\dfrac{t}{0.8}\right)^3 = \left(\dfrac{1.10}{0.8}\right)^3 = 2.600$

$\bar{y}' = -\dfrac{66.05}{15.951} = -4.14$ m

y of point $3 = 4.14 - 2.81 = +1.33$ m

(*b*) *Computations for* M_O, H_O, *and* V_O. The points on the required influence lines are to be determined as the unit load takes successively the positions *B, C, D, E, F, G*, and *H*. On account of symmetry, considerations for loading positions *F, G*, and *H* are not necessary; for instance, the reactions at the right support when the unit load is at *C* will be the same as those at the left support when the unit load is at *G*, with due regard to symmetry (mirror-image) requirements. For loading positions *B, C, D*, and *E*, the basic determinate structure from which the M_s values are obtained is the cantilever structure with a fixed support at *A* and free at the elastic center, as shown in Fig. 18.5.3. Since the $\Delta S/(nEI_c)$'s are point areas, M_s times them will be negative (upward) concentrated loads on top of the analogous column. In the top half of Table 18.5.2, the formulas

$$M_O = -\dfrac{P}{A} \qquad H_O = -\dfrac{M_x}{I_y} \qquad V_O = +\dfrac{M_y}{I_y}$$

are used to compute the redundant forces acting at the elastic center. Note that positive M_O is counterclockwise, positive H_O is to the right, and positive V_O is upward.

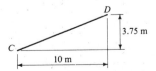

Figure 18.5.2 Geometry of straight line *CD*.

660 INTERMEDIATE STRUCTURAL ANALYSIS

Table 18.5.2 Computations for redundants at the elastic center and moments at three points on the symmetrical fixed arch in Example 18.5.1

Properties of analogous-column section						Load at B				Load at C		
Segment	Centroid	x	y	$\dfrac{\Delta s}{EI}$	M_s	$\dfrac{M_s \Delta s}{EI}$	$\dfrac{M_s x \Delta s}{EI}$	$\dfrac{M_s y \Delta s}{EI}$	M_s	$\dfrac{M_s \Delta s}{EI}$	$\dfrac{M_s x \Delta s}{EI}$	$\dfrac{M_s y \Delta s}{EI}$
AB	1	−35	−11.17	2.016	−5	−10.08	+352.8	+112.6	−15	−30.24	+1058.4	+337.7
BC	2	−25	− 3.67	2.748					− 5	−13.74	+ 343.5	+ 50.4
CD	3	−15	+ 1.33	4.108								
DE	4	− 5	+ 3.83	7.079							+1401.9	+388.1
	$A = 31.902$					−10.08	+352.8	+112.6		−43.98	+1401.9	+388.1
	$I_x = 799.2$					$P = -10.08$	$M_y = +352.8$	$M_x = +112.6$		$P = -43.98$	$M_y = +1401.9$	$M_x = +388.1$
	$I_y = 10,578$											

$M_O = -\dfrac{P}{A} = -\dfrac{-10.08}{31.902} = +0.316$

$H_O = -\dfrac{M_x}{I_x} = -\dfrac{+112.6}{799.2} = -0.1409$

$V_O = +\dfrac{M_y}{I_y} = +\dfrac{+352.8}{10,578} = +0.0334$

$M_O = -\dfrac{P}{A} = -\dfrac{-43.98}{31.902} = +1.378$

$H_O = -\dfrac{M_x}{I_x} = -\dfrac{+388.1}{799.2} = -0.4856$

$V_O = +\dfrac{M_y}{I_y} = +\dfrac{+1401.9}{10,578} = +0.1325$

RIGID FRAMES WITH CURVED MEMBERS

Load at B

Point	x	y	M_s	$\dfrac{P}{A}$	$\dfrac{M_s y}{I_x}$	$\dfrac{M_s x}{I_y}$	M_i	M
A	−40	−15.86	−10	$\dfrac{-10.08}{31.902} = -0.316$	$\dfrac{+112.6}{799.2}(-15.86) = -2.234$	$\dfrac{+352.8}{10,578}(-40) = -1.334$	−3.884	−6.116
E	0	+4.14	0	−0.316	$(+4.14) = +0.583$	$(0) = 0$	+0.267	−0.267
K	+40	−15.86	0	−0.316	$(-15.86) = -2.234$	$(+40) = +1.334$	−1.216	+1.216

Load at C

Point	x	y	M_s	$\dfrac{P}{A}$	$\dfrac{M_s y}{I_x}$	$\dfrac{M_s x}{I_y}$	M_i	M
A	−40	−15.86	−20	$\dfrac{-43.98}{31.902} = -1.378$	$\dfrac{+388.1}{799.2}(-15.86) = -7.704$	$\dfrac{+1401.9}{10,578}(-40) = -5.301$	−14.383	−5.617
E	0	+4.14	0	−1.378	$(+4.14) = +2.011$	$(0) = 0$	+0.633	−0.633
K	+40	−15.86	0	−1.378	$(-15.86) = -7.704$	$(+40) = +5.301$	−3.781	+3.781

Table 18.5.2 Computations for redundants at the elastic center and moments at three points on the symmetrical fixed arch in Example 18.5.1 (*continued*)

Properties of analogous-column section					Load at D					Load at E		
Segment	Centroid	x	y	$\dfrac{\Delta s}{EI}$	M_s	$\dfrac{M_s \Delta s}{EI}$	$\dfrac{M_s x \Delta s}{EI}$	$\dfrac{M_s y \Delta s}{EI}$	M_s	$\dfrac{M_s \Delta s}{EI}$	$\dfrac{M_s x \Delta s}{EI}$	$\dfrac{M_s y \Delta s}{EI}$
AB	1	-35	-11.17	2.016	-25	-50.40	$+1764.0$	$+563.0$	-35	-70.56	$+2469.6$	$+788.2$
BC	2	-25	-3.67	2.748	-15	-41.22	$+1030.5$	$+151.3$	-25	-68.70	$+1717.5$	$+252.1$
CD	3	-15	$+1.33$	4.108	-5	-20.54	$+308.1$	-27.3	-15	-61.62	$+924.3$	-82.0
DE	4	-5	$+3.83$	7.079					-5	-35.40	$+177.0$	-135.6
						-112.16	$+3102.6$	$+687.0$		-236.28	$+5288.4$	$+822.7$
$A = 31.902$					$P = -112.16$		$M_y = +3102.6$	$M_x = +687.0$	$P = -236.28$		$M_y = +5288.4$	$M_x = +822.7$
$I_x = 799.2$												
$I_y = 10,578$												

$$M_O = -\dfrac{P}{A} = -\dfrac{-112.16}{31.902} = +3.516 \qquad M_O = -\dfrac{P}{A} = -\dfrac{-236.28}{31.902} = +7.406$$

$$H_O = -\dfrac{M_x}{I_x} = -\dfrac{+687.0}{799.2} = -0.8596 \qquad H_O = -\dfrac{M_x}{I_x} = -\dfrac{+822.7}{799.2} = -1.0294$$

$$V_O = +\dfrac{M_y}{I_y} = +\dfrac{+3102.6}{10,578} = +0.2933 \qquad V_O = +\dfrac{M_y}{I_y} = +\dfrac{+5288.4}{10,578} = +0.5000$$

Load at D

Point	x	y	M_s	$\dfrac{P}{A}$	$\dfrac{M_s y}{I_x}$	$\dfrac{M_s x}{I_y}$	M_i	M
A	−40	−15.86	−30	$\dfrac{-112.16}{31.902} = -3.516$	$\dfrac{+687.0}{799.2}(-15.86) = -13.633$	$\dfrac{+3102.6}{10,578}(-40) = -11.732$	−28.881	−1.119
E	0	+4.14	0	−3.516	$(+4.14) = +3.559$	$(0) = 0$	+0.043	−0.043
K	+40	−15.86	0	−3.516	$(-15.86) = -13.633$	$(+40) = +11.732$	−5.417	+5.417

Load at E

Point	x	y	M_s	$\dfrac{P}{A}$	$\dfrac{M_s y}{I_x}$	$\dfrac{M_s x}{I_y}$	M_i	M
A	−40	−15.86	−40	$\dfrac{-236.28}{31.902} = -7.406$	$\dfrac{+822.7}{799.2}(-15.86) = -16.326$	$\dfrac{+5288.4}{10,578}(-40) = -20.000$	−43.732	+3.732
E	0	+4.14	0	−7.406	$(+4.14) = +4.262$	$(0) = 0$	−3.144	+3.144
K	+40	−15.86	0	−7.406	$(-15.86) = -16.326$	$(+40) = +20.000$	−3.732	+3.732

664 INTERMEDIATE STRUCTURAL ANALYSIS

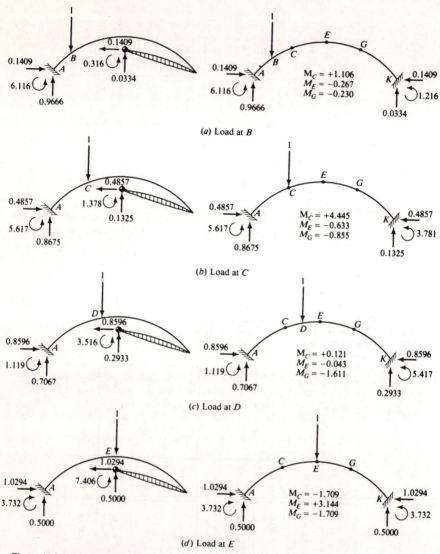

Figure 18.5.3 Free-body diagrams for various load positions in Example 18.5.1.

(c) *Computations for the reactions at the left and right supports.* The results of part (b) are used to complete the free-body diagrams shown on the left side of Fig. 18.5.3. By the application of $\Sigma F_x = 0$, $\Sigma F_y = 0$, and $\Sigma M = 0$ to these free bodies, the reactions at the left support are computed. The reactions at the right support, shown on the free-body diagrams on the right side of Fig. 18.5.3, can be obtained by using the rigid arm as the free body. However, separate free-body diagrams for the rigid arms need not be drawn; it is necessary

RIGID FRAMES WITH CURVED MEMBERS 665

only to transfer the forces acting at the elastic center to the right support. The horizontal and vertical forces acting at the elastic center remain to act at the right support with the same magnitude and direction; the bending moment of M_O, H_O, and V_O about the point B will act on the arch at point B. Then finally the three equations of equilibrium $\Sigma F_x = 0$, $\Sigma F_y = 0$, and $\Sigma M = 0$ are checked out using the free-body diagrams shown on the right side of Fig. 18.5.3.

(d) *Computations for bending moments at the left quarter point, the crown, and the right quarter point.* The bending moments at these three locations are computed using the free-body diagrams shown on the right side of Fig. 18.5.3. The results are shown there as well.

(e) *Application of the column-analogy formula* $M = M_s - M_i$. Without using the values of M_O, H_O, and V_O, the bending moment at any point on the arch axis can be computed directly using the formulas

$$M = M_s - M_i \qquad M_i = \frac{P}{A} + \frac{M_x y}{I_x} + \frac{M_y x}{I_y}$$

This is done for points A, E, and K as tabulated in the lower half of Table 18.5.2. The final tally of the bending moments computed here with those of part (d), already entered on the right half of Fig. 18.5.3, ensures the correctness of the statics portion, although not yet the compatibility portion, of the solution.

(f) *Compatibility checks.* To ensure that the compatibility conditions are satisfied, it is necessary to compute the bending moments at points 1 to 8 and to apply the three generalized moment-area theorems, as Eqs. (15.6.2a to c) may be called, between the points A and K. Thus,

$$\sum_1^8 \frac{M_i \Delta S_i}{n_i EI_c} = 0 \qquad \sum_1^8 \frac{M_i y_i \Delta S_i}{n_i EI_c} = 0 \qquad \sum_1^8 \frac{M_i x_i \Delta S_i}{n_i EI_c} = 0$$

in which x_i and y_i are the coordinates of point i in reference to K as origin. These computations, although not shown here, can be conveniently carried out in tabular form.

(g) *Influence table and influence lines.* Before plotting the influence lines it is advisable to make an influence table as shown in Table 18.5.3. Note how the influence values for unit loads at F, G, and H can be picked up by inspection from the free-body diagrams shown on the right side of Fig. 18.5.3. The required influence lines are then drawn as in Fig. 18.5.4.

Table 18.5.3 Influence table for symmetrical fixed arch in Example 18.5.1

Load at	Distance from A	H_A	V_A	V_E	M_A	M_C	M_E
A	0	0	1.000	0	0	0	0
B	10	0.141	0.967	−0.033	−6.12	+1.11	−0.27
C	20	0.486	0.868	−0.132	−5.62	+4.44	−0.63
D	30	0.860	0.707	−0.293	−1.12	+0.12	−0.04
E	40	1.029	0.500	−0.500	+3.73	−1.71	+3.14
				+0.500			
F	50	0.860	0.293	+0.293	+5.42	−1.61	−0.04
G	60	0.486	0.132	+0.132	+3.78	−0.86	−0.63
H	70	0.141	0.033	+0.033	+1.22	−0.23	−0.27
K	80	0	0	0	0	0	0

666 INTERMEDIATE STRUCTURAL ANALYSIS

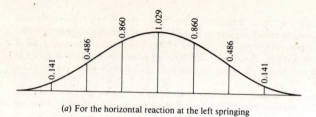

(a) For the horizontal reaction at the left springing

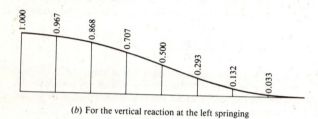

(b) For the vertical reaction at the left springing

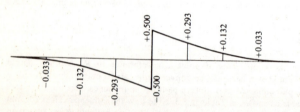

(c) For the vertical shear at the crown

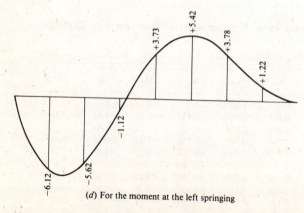

(d) For the moment at the left springing

Figure 18.5.4 Influence lines for symmetrical fixed arch in Example 18.5.1.

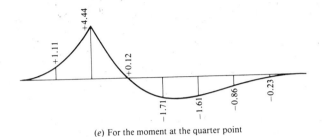

(e) For the moment at the quarter point

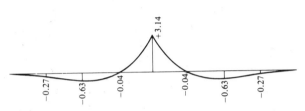

(f) For the moment at the crown

Figure 18.5.4 (*continued*).

18.6 Influence Lines for an Unsymmetrical Fixed Arch

Much that has been said about influence lines for a symmetrical fixed arch applies as well to the unsymmetrical fixed arch, except that this time the unit load must move across the entire span instead of half of the span, and that the formulas listed under the unsymmetrical case wherein I_{xy} is no longer zero must be used for obtaining M_O, H_O, and V_O in the elastic-center method and M_i in the column-analogy method.

When the unit load is closer to the left support than the right support, it is sensible to obtain the M_s values from a cantilever structure fixed at the left and free at the right. Likewise, if the unit load is closer to the right support than the left support, one can make use of a cantilever structure fixed at the right and free at the left to obtain the M_s values at the point areas to right of the unit load. In such cases, although there is no special attention needed in using the formula $M = M_s - M_i$ in the column-analogy method, the redundant forces M_O, H_O, and V_O obtained from the formulas in the elastic-center method plus the forces in the entire M_s system (unit load and reacting force and moment at the right support) must still be applied to the cantilever structure fixed at the left end and free at the elastic center. This is due to the fact that the formulas for M_O, H_O, and V_O have been derived by using such a structure. The same argument applies even when the M_s values are obtained from a basic determinate structure with a left hinged support and a right roller support.

Example 18.6.1 For the unsymmetrical parabolic fixed arch shown in Fig. 18.6.1a, draw influence lines for (1) the horizontal reaction at the left support, (2) the vertical reaction at the left support, (3) the vertical shear at the crown, (4) the moment at the left support, (5) the moment at the quarter point of the horizontal span, (6) the moment at the crown, and (7) the moment at the right support.

SOLUTION (a) *Properties of the analogous column section.* The arch rib is arbitrarily replaced by eight segments of equal horizontal projection, as represented by AB, BC, CD, etc. in Fig. 18.6.1b. In the solution to Example 18.5.1, the discussion in part (a) regarding the

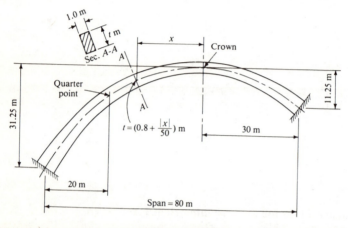

(a) The given fixed arch

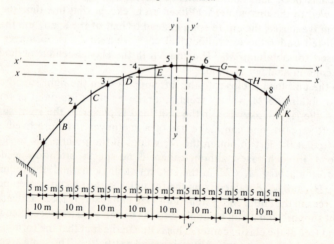

(b) The divided segments

Figure 18.6.1 Unsymmetrical fixed arch in Example 18.6.1.

Table 18.6.1 Properties of analogous column section for unsymmetrical fixed arch in Example 18.6.1

Segment	Centroid	Δs, m	t, m	n ($I = nI_c$)	$\dfrac{\Delta s}{EI} = \dfrac{\Delta s}{n}$ ($EI_c = 1$)	x'	y'	$x'\dfrac{\Delta s}{EI}$	$y'\dfrac{\Delta s}{EI}$	x	y	$x^2\dfrac{\Delta s}{EI}$	$xy\dfrac{\Delta s}{EI}$	$y^2\dfrac{\Delta s}{EI}$
AB	1	15.052	1.70	9.5957	1.5686	−45	−25.3125	−70.587	−39.7052	−40.5128	−20.8283	2574.52	+1323.605	680.488
BC	2	13.288	1.50	6.5918	2.0158	−35	−15.3125	−70.553	−30.8669	−30.5128	−10.8283	1876.77	+ 666.024	236.357
CD	3	11.792	1.30	4.2910	2.7481	−25	− 7.8125	−68.702	−21.4695	−20.5128	− 3.3283	1156.33	+ 187.620	30.442
DE	4	10.680	1.10	2.5996	4.1083	−15	− 2.8125	−61.625	−11.5546	−10.5128	+ 1.6717	454.04	− 72.200	11.481
EF	5	10.078	0.90	1.4238	7.0782	− 5	− 0.3125	−35.391	− 2.2119	− 0.5128	+ 4.1717	1.86	− 15.142	123.182
FG	6	10.078	0.90	1.4238	7.0782	+ 5	− 0.3125	+35.391	− 2.2119	+ 9.4872	+ 4.1717	637.09	+ 280.139	123.182
GH	7	10.680	1.10	2.5996	4.1083	+15	− 2.8125	+61.625	−11.5546	+19.4872	+ 1.6717	1560.13	+ 133.835	11.481
HK	8	11.792	1.30	4.2910	2.7481	+25	− 7.8125	+68.702	−21.4695	+29.4872	− 3.3283	2389.46	− 269.705	30.442
					31.4536	−4.4872	−4.4842	−141.140	−141.0441			10,650.20	+2234.176	1247.055
					A	$\bar{x}'$	$\bar{y}'$					I_y	I_{xy}	I_x

$A = 31.4536$

$$I'_x = I_x \left(1 - \dfrac{I_{xy}^2}{I_x I_y}\right) = (1247.055)\left[1 - \dfrac{(+2234.176)^2}{(1247.055)(10{,}650.20)}\right] = (1247.055)(0.62417) = 778.374$$

$$I'_y = I_y \left(1 - \dfrac{I_{xy}^2}{I_x I_y}\right) = (10{,}650.20)(0.62417) = 6647.54$$

preferred practice of using segments of equal length along the arch axis is even more applicable in this case because the segment AB with the same horizontal projection is much longer than the segment, say EF, near the crown. To repeat, it is only for the ease in which the reader may follow the logic behind the arithmetic that division is made on the basis of equal horizontal projection. Again I_c is the moment of inertia at the crown, while $n_i I_c$ is the moment of inertia of the ith segment based on the thickness of the arch at the location of the ith point area. The expression EI_c should appear as the denominator in the properties of the analogous column section, for which computations are shown in Table 18.6.1. One may note that in this example all computations are carried out to an unusual degree of accuracy. This is done so that a good check can be obtained at the end showing that the reactions and moments due to the simultaneous action of all unit loads are indeed the sums of those respective values due to the separately applied unit loads. In an actual design problem, the usual practice of using three or four significant figures in all numbers is recommended. When the computer is utilized, of course, accuracy of seven, eight, or nine significant figures becomes automatic.

(b) *Computations for M_O, H_O, and V_O.* The fixed arch is analyzed for eight loading conditions: unit loads applied singly at B, C, D, E, F, G, or H and then unit loads applied jointly at all seven points. As shown on the left side of Fig. 18.6.2, for unit loads applied singly at B, C, D, or E, the M_s values are taken from a basic structure fixed at A and free at K; but for unit loads applied singly at F, G, or H, the M_s values are taken from a basic structure fixed at K and free at A. For unit loads applied jointly at all seven locations, the basic structure has the hinged support at A and the roller support at B. In the top half of Tables 18.6.2 to 18.6.5, the formulas

$$M_O = -\frac{P}{A} \qquad H_O = -\frac{M'_x}{I'_x} \qquad V_O = +\frac{M'_y}{I'_y}$$

are used to compute the redundant forces acting at the elastic center. These redundant forces in their actual directions are placed on the free-body diagrams shown on the left side of Fig. 18.6.2. Note particularly here the set of self-balancing forces in the entire M_s system in Fig. 18.6.2e to g, and the inclusion of the vertical reaction at the roller support in the free-body diagram of Fig. 18.6.2h.

(c) *Computations for the reactions at the left and right supports.* The reactions at the left support are obtained by applying the equations $\Sigma F_x = 0$, $\Sigma F_y = 0$, and $\Sigma M = 0$ to the free-body diagrams shown on the left side of Fig. 18.6.2. The reactions at the right support are computed by using the summation of horizontal forces, vertical forces, and moments about point B of all the forces acting on the rigid arm between the elastic center and point B as shown on the right side of Fig. 18.6.2. Then the checks are made by applying $\Sigma F_x = 0$, $\Sigma F_y = 0$, and $\Sigma M = 0$ to the whole arch as a free body without the rigid arm.

(d) *Computations for bending moments at the left quarter point and the crown.* Since the problem statement requires only the influence line for moment at the left quarter point, unlike the symmetrical arch, the moment at the right quarter point need not be computed because the unit loads are in fact applied at all locations from one end to the other. The bending moments at the left quarter point and the crown are computed using the free-body diagrams shown on the right side of Fig. 18.6.2. The results are shown there as well.

(e) *Application of the column-analogy formula $M = M_s - M_i$.* Without using the values of M_O, H_O, and V_O, the bending moment at any point on the arch axis can be computed directly using the formulas

$$M = M_s - M_i \qquad M_i = \frac{P}{A} + \frac{M'_x y}{I'_x} + \frac{M'_y x}{I'_y}$$

This is done for points A, C, F, and K as tabulated in the lower half of Tables 18.6.2 to 18.6.5. The final tally of the bending moments computed here with those of part (d), already entered on the right half of Fig. 18.6.2, ensures the correctness of the statics portion, although not yet the compatibility portion, of the solution.

Table 18.6.2 Computations for redundants at the elastic center and moments at four points for a unit load at B and then at C in Example 18.6.1

	Properties of analogous column section					Load at B					Load at C			
Segment	Centroid	x	y	$\frac{\Delta s}{EI}$	M_s	$\frac{M_s \Delta s}{EI}$	$\frac{M_s x \Delta s}{EI}$	$\frac{M_s y \Delta s}{EI}$		M_s	$\frac{M_s \Delta s}{EI}$	$\frac{M_s x \Delta s}{EI}$	$\frac{M_s y \Delta s}{EI}$	
AB	1	−40.5128	−20.8283	1.5686	−5	−7.8430	+317.742	+163.356		−15	−23.529	+953.226	+490.069	
BC	2	−30.5128	−10.8283	2.0158						−5	−10.079	+307.538	+109.138	
CD	3	−20.5128	−3.3283	2.7481										
DE	4	−10.5128	+1.6717	4.1083										
EF	5	−0.5128	+4.1717	7.0782										
FG	6	+9.4872	+4.1717	7.0782										
GH	7	+19.4872	+1.6717	4.1083										
HK	8	+29.4872	−3.3283	2.7481										
						−7.8430	+317.742	+163.356			−33.608	+1260.764	+599.207	
						P	M_y	M_x			P	M_y	M_x	

$A = 31.4536$
$I'_x = 778.374$
$I'_y = 6647.54$
$\frac{I_{xy}}{I_x} = +1.79156$
$\frac{I_{xy}}{I_y} = +0.20978$

$M'_x = +163.356 - (+317.742)(0.20978) = +96.700$　　$M'_x = +599.207 - (+1260.764)(0.20978) = +334.72$
$M'_y = +317.742 - (+163.356)(1.79156) = +25.080$　　$M'_y = +1260.764 - (+599.207)(1.79156) = +187.25$

$M_O = -\frac{P}{A} = -\frac{-7.8430}{31.4536} = +0.249$　　　$M_O = -\frac{P}{A} = -\frac{-33.608}{31.4536} = +1.068$

$H_O = -\frac{M'_x}{I'_x} = -\frac{+96.700}{778.374} = -0.1242$　　$H_O = -\frac{M'_x}{I'_x} = -\frac{+334.72}{778.374} = -0.4300$

$V_O = +\frac{M'_y}{I'_y} = +\frac{+25.080}{6647.54} = +0.0038$　　$V_O = +\frac{M'_y}{I'_y} = +\frac{+187.25}{6647.54} = +0.0281$

Table 18.6.2 Computations for redundants at the elastic center and moments at four points for a unit load at B and then at C in Example 18.6.1 (*continued*)

Load at B

Point	x	y	M_s	$\dfrac{P}{A}$	$\dfrac{M'_x y}{I_x}$	$\dfrac{M'_y x}{I_y}$	M_i	M
A	−45.5128	−26.7658	−10	$\dfrac{-7.8430}{31.4536} = -0.249$	$\dfrac{+96.700}{778.374}(-26.7658) = -3.325$	$\dfrac{+25,080}{6647.54}(-45.5128) = -0.171$	−3.745	−6.255
C	−25.5128	−6.7658	0	−0.249	(−6.7658) = −0.840	(−25.5128) = −0.096	−1.185	+1.185
F	+4.4872	+4.4842	0	−0.249	(+4.4842) = +0.557	(+4.4872) = +0.017	+0.325	−0.325
K	+34.4872	−6.7658	0	−0.249	(−6.7658) = −0.840	(+34.4872) = +0.130	−0.959	+0.959

Load at C

Point	x	y	M_s	$\dfrac{P}{A}$	$\dfrac{M'_x y}{I_x}$	$\dfrac{M'_y x}{I_y}$	M_i	M
A	−45.5128	−26.7658	−20	$\dfrac{+33.608}{31.4536} = -1.068$	$\dfrac{+334.72}{778.374}(-26.7658) = -11.510$	$\dfrac{+187.25}{6647.54}(-45.5128) = -1.279$	−13.857	−6.143
C	−25.5128	−6.7658	0	−1.068	(−6.7658) = −2.909	(−25.5128) = −0.717	−4.694	+4.694
F	+4.4872	+4.4842	0	−1.068	(+4.4842) = +1.928	(+4.4872) = +0.126	+0.986	−0.986
K	+34.4872	−6.7658	0	−1.068	(−6.7658) = −2.909	(+34.4872) = +0.970	−3.007	+3.007

Table 18.6.4 Computations for redundants at the elastic center and moments at four points for a unit load at F and then at G in Example 18.6.1

Properties of analogous column section					Load at F					Load at G			
Segment	Centroid	x	y	$\frac{\Delta s}{EI}$	M_s	$\frac{M_s \Delta s}{EI}$	$\frac{M_s x \Delta s}{EI}$	$\frac{M_s y \Delta s}{EI}$	M_s	$\frac{M_s \Delta s}{EI}$	$\frac{M_s x \Delta s}{EI}$	$\frac{M_s y \Delta s}{EI}$
AB	1	−40.5128	−20.8283	1.5686								
BC	2	−30.5128	−10.8283	2.0158								
CD	3	−20.5128	−3.3283	2.7481								
DE	4	−10.5128	+1.6717	4.1083								
EF	5	−0.5128	+4.1717	7.0782	−5	−35.391						
FG	6	+9.4872	+4.1717	7.0782	−15	−61.624	−335.76	−147.641				
GH	7	+19.4872	+1.6717	4.1083	−25	−68.702	−1200.88	−103.017	−5	−20.542	−400.31	−34.340
HK	8	+29.4872	−3.3283	2.7481			−2025.83	+228.661	−15	−41.222	−1215.52	+137.199
						−165.717	−3562.47	−21.997		−61.764	−1615.83	+102.859
						P	M_y	M_x		P	M_y	M_x

$A = 31.4536$
$I'_x = 778.374$
$I'_y = 6647.54$
$\frac{I_{xy}}{I_x} = +1.79156$
$\frac{I_{xy}}{I_y} = +0.20978$

$M'_x = -21.997 - (-3562.47)(0.20978) = +725.34$
$M'_y = -3562.47 - (-21.997)(1.79156) = -3523.05$

$M_O = -\frac{P}{A} = -\frac{-165.717}{31.4536} = +5.269$

$H_O = -\frac{M'_x}{I'_x} = -\frac{+725.34}{778.374} = -0.9319$

$V_O = +\frac{M'_y}{I'_y} = +\frac{-3523.05}{6647.54} = -0.5300$

$M'_x = +102.859 - (-1615.83)(0.20978) = +441.83$
$M'_y = -1615.83 - (+102.859)(1.79156) = -1800.11$

$M_O = -\frac{P}{A} = -\frac{-61.764}{31.4536} = +1.964$

$H_O = -\frac{M'_x}{I'_x} = -\frac{+441.83}{778.374} = -0.5676$

$V_O = +\frac{M'_y}{I'_y} = +\frac{-1800.11}{6647.54} = -0.2708$

Table 18.6.4 Computations for redundants at the elastic center and moments at four points for a unit load at F and then at G in Example 18.6.1 (continued)

Load at F

Point	x	y	M_s	$\dfrac{P}{A}$	$\dfrac{M'_x y}{I'_x}$	$\dfrac{M'_y x}{I'_y}$	M_i	M
A	−45.5128	−26.7658	0	$\dfrac{-165.717}{31.4536} = -5.269$	$\dfrac{+725.34}{778.374}(-26.7658) = -24.942$	$\dfrac{-3523.05}{6647.54}(-45.5128) = +24.121$	−6.090	+6.090
C	−25.5128	−6.7658	0	−5.269	$(-6.7658) = -6.305$	$(-25.5128) = +13.521$	+1.947	−1.947
F	+4.4872	+4.4842	0	−5.269	$(+4.4842) = +4.179$	$(+4.4872) = -2.378$	−3.468	+3.468
K	+34.4872	−6.7658	−30	−5.269	$(-6.7658) = -6.305$	$(+34.4872) = -18.277$	−29.851	−0.149

Load at G

Point	x	y	M_s	$\dfrac{P}{A}$	$\dfrac{M'_x y}{I'_x}$	$\dfrac{M'_y x}{I'_y}$	M_i	M
A	−45.5128	−26.7658	0	$\dfrac{-61.764}{31.4536} = -1.964$	$\dfrac{441.83}{778.374}(-26.7658) = -15.193$	$\dfrac{-1800.11}{6647.54}(-45.5128) = +12.325$	−4.832	+4.832
C	−25.5128	−6.7658	0	−1.964	$(-6.7658) = -3.840$	$(-25.5128) = +6.909$	+1.105	−1.105
F	+4.4872	+4.4842	0	−1.964	$(+4.4842) = +2.545$	$(+4.4872) = -1.215$	−0.634	+0.634
K	+34.4872	−6.7658	−20	−1.964	$(-6.7658) = -3.840$	$(+34.4872) = -9.339$	−15.143	−4.857

Table 18.6.5 Computations for redundants at the elastic center and moments at four points for a unit load at H and then unit loads at all seven locations in Example 18.6.1

Properties of analogous column section					Load at H					Loads at $B, C, D, E, F, G,$ and H			
Segment	Centroid	x	y	$\dfrac{\Delta s}{EI}$	M_s	$\dfrac{M_s \Delta s}{EI}$	$\dfrac{M_s x \Delta s}{EI}$	$\dfrac{M_s y \Delta s}{EI}$	M_s	$\dfrac{M_s \Delta s}{EI}$	$\dfrac{M_s x \Delta s}{EI}$	$\dfrac{M_s y \Delta s}{EI}$
AB	1	-40.5128	-20.8283	1.5686					$+17.5$	$+27.450$	-1112.076	-571.737
BC	2	-30.5128	-10.8283	2.0158					$+47.5$	$+95.750$	-2921.601	-1036.810
CD	3	-20.5128	-3.3283	2.7481					$+67.5$	$+185.497$	-3805.063	-617.390
DE	4	-10.5128	$+1.6717$	4.1083					$+77.5$	$+318.393$	-3347.202	$+532.258$
EF	5	-0.5128	$+4.1717$	7.0782					$+77.5$	$+548.561$	-281.302	$+2288.432$
FG	6	$+9.4872$	$+4.1717$	7.0782					$+67.5$	$+477.778$	$+4532.775$	$+1993.146$
GH	7	$+19.4872$	$+1.6717$	4.1083					$+47.5$	$+195.144$	$+3802.810$	$+326.222$
HK	8	$+29.4872$	-3.3283	2.7481	-5	-13.740	-405.169	$+45.732$	$+17.5$	$+48.092$	$+1418.098$	-160.065
						-13.740	-405.169	$+45.732$		$+1896.665$	-1713.561	$+2754.056$
						P	M_y	M_x		P	M_y	M_x

$A = 31.4536$
$I'_x = 778.374$
$I'_y = 6647.54$
$\dfrac{I_{xy}}{I_x} = +1.79156$
$\dfrac{I_{xy}}{I_y} = +0.20978$

$M'_x = +45.732 - (-405.169)(0.20978) = +130.728$
$M'_y = -405.169 - (+45.732)(1.79156) = -487.10$

$M_O = -\dfrac{P}{A} = -\dfrac{-13.740}{31.4536} = +0.437$

$H_O = -\dfrac{M'_x}{I'_x} = -\dfrac{+130.728}{778.374} = -0.1680$

$V_O = +\dfrac{M'_y}{I'_y} = +\dfrac{-487.10}{6647.54} = -0.0733$

$M'_x = +2754.056 - (-1713.561)(0.20978) = +3113.527$
$M'_y = -1713.561 - (+2754.056)(1.79156) = -6647.62$

$M_O = -\dfrac{P}{A} = -\dfrac{+1896.665}{31.4536} = -60.300$

$H_O = -\dfrac{M'_x}{I'_x} = -\dfrac{+3113.527}{778.374} = -4.0000$

$V_O = +\dfrac{M'_y}{I'_y} = +\dfrac{-6647.62}{6647.52} = -1.0000$

Table 18.6.5 Computations for redundants at the elastic center and moments at four points for a unit load at H and then unit loads at all seven locations in Example 18.6.1 (*continued*)

Load at H

Point	x	y	M_s	$\dfrac{P}{A}$	$\dfrac{M_s'y}{I_x'}$	$\dfrac{M_s'x}{I_y'}$	M_i	M
A	−45.5128	−26.7658	0	$\dfrac{-13.740}{31.4536} = -0.437$	$\dfrac{130.728}{778.374}(-26.7658) = -4.495$	$\dfrac{-487.10}{6647.54}(-45.5128) = +3.335$	−1.597	+1.597
C	−25.5128	− 6.7658	0	−0.437	$(- 6.7658) = -1.136$	$(-25.5128) = +1.870$	+0.297	−0.297
F	+ 4.4872	+ 4.4842	0	−0.437	$(+ 4.4842) = +0.753$	$(+ 4.4872) = -0.329$	−0.013	+0.013
K	+34.4872	− 6.7658	−10	−0.437	$(- 6.7658) = -1.136$	$(+34.4872) = -2.527$	−4.100	−5.900

Loads at $B, C, D, E, F, G,$ and H

Point	x	y	M_s	$\dfrac{P}{A}$	$\dfrac{M_s'y}{I_x'}$	$\dfrac{M_s'x}{I_y'}$	M_i	M
A	−45.5128	−26.7658	0	$\dfrac{1896.665}{31.4536} = +60.300$	$\dfrac{3113.527}{778.374}(-26.7658) = -107.064$	$\dfrac{-6647.62}{6647.54}(-45.5128) = +45.513$	− 1.251	+1.251
C	−25.5128	− 6.7658	+60	+60.300	$(- 6.7658) = - 27.063$	$(-25.5128) = +25.513$	+58.750	+1.250
F	+ 4.4872	+ 4.4842	+75	+60.300	$(+ 4.4842) = + 17.937$	$(+ 4.4872) = - 4.487$	+73.750	+1.250
K	+34.4872	− 6.7658	0	+60.300	$(- 6.7658) = - 27.063$	$(+34.4872) = -34.487$	− 1.250	+1.250

RIGID FRAMES WITH CURVED MEMBERS **679**

(f) *Compatibility checks.* To ensure that the compatibility conditions are satisfied, it is necessary that the bending moments at points 1 to 8 be computed and the three generalized moment-area theorems [Eqs. (15.6.2a to c)] be applied between the points A and K. Thus,

$$\sum_1^8 \frac{M_i \Delta S_i}{n_i EI_c} = 0 \qquad \sum_1^8 \frac{M_i y_i \Delta S_i}{n_i EI_c} = 0 \qquad \sum_1^8 \frac{M_i x_i \Delta S_i}{n_i EI_c} = 0$$

in which x_i and y_i are the coordinates of point i in reference to K as origin. These computations, although not shown here, can be conveniently carried out in tabular form.

(g) *Superposition check.* The total effects on the fixed arch due to the singly applied loads at B, C, D, E, F, G, or H must be the same as the effect due to the jointly applied

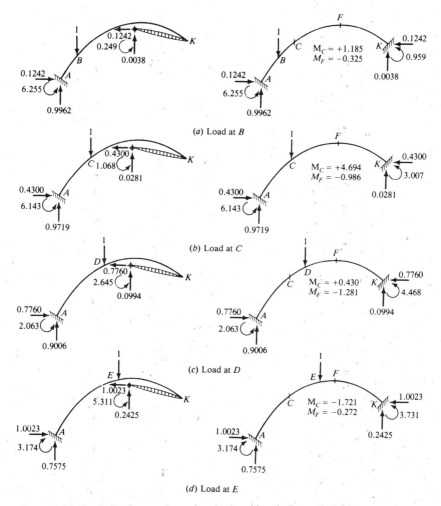

(a) Load at B

(b) Load at C

(c) Load at D

(d) Load at E

Figure 18.6.2 Free-body diagrams for various load positions in Example 18.6.1.

680 INTERMEDIATE STRUCTURAL ANALYSIS

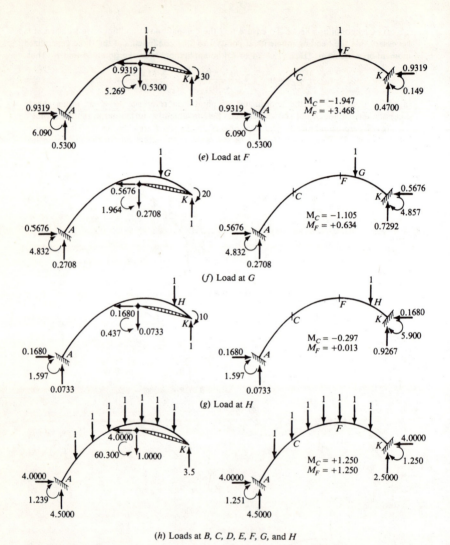

(h) Loads at B, C, D, E, F, G, and H

Figure 18.6.2 (continued).

loads at all seven locations. The verification of this effect serves as another check on the correctness of the solution. The comparisons for H_A, V_A, M_A, M_C, M_F, and M_K are listed in Table 18.6.6. It was found that the properties of the analogous column section, particularly the location of the elastic center, need to be computed to six or seven significant figures in order that a satisfactorily close check may be obtained.

(h) *Influence table and influence lines.* The influence table in which all necessary values for plotting the required influence lines are recorded is shown in Table 18.6.7. The required influence lines are drawn in Fig. 18.6.3.

RIGID FRAMES WITH CURVED MEMBERS 681

Table 18.6.6 Superposition check for unsymmetrical fixed arch in Example 18.6.1

	Sum of values due to unit loads applied singly at $B, C, D, E, F, G,$ or H	Combined value due to unit loads applied jointly at $B, C, D, E, F, G,$ and H
H_A	4.0000	4.0000
V_A	4.5003	4.5000
M_A	1.232	1.251
M_C	1.239	1.250
M_F	1.251	1.250
M_K	1.259	1.250

Table 18.6.7 Influence table for unsymmetrical fixed arch in Example 18.6.1

Load at	Distance from A	H_A	V_A	V_F	M_A	M_C	M_F	M_K
A	0	0	1.000	0	0	0	0	0
B	10	0.124	0.996	−0.004	−6.26	+1.18	−0.32	+0.96
C	20	0.430	0.972	−0.028	−6.14	+4.69	−0.99	+3.01
D	30	0.776	0.901	−0.099	−2.06	+0.43	−1.28	+4.47
E	40	1.002	0.758	−0.242	+3.17	−1.72	−0.27	+3.73
F	50	0.932	0.530	−0.470 +0.530	+6.09	−1.95	+3.47	−0.15
G	60	0.568	0.271	+0.271	+4.83	−1.10	+0.63	−4.86
H	70	0.168	0.073	+0.073	+1.60	−0.30	+0.01	−5.90
K	80	0	0	0	0	0	0	0

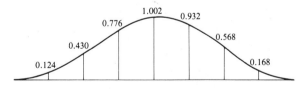

(a) For the horizontal reaction at the left springing

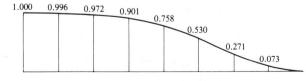

(b) For the vertical reaction at the left springing

Figure 18.6.3 Influence lines for unsymmetrical fixed arch in Example 18.6.1.

682 INTERMEDIATE STRUCTURAL ANALYSIS

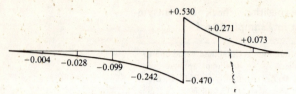

(c) For the vertical shear at the crown

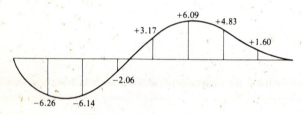

(d) For the moment at the left springing

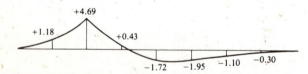

(e) For the moment at the quarter point

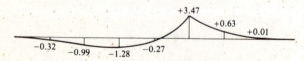

(f) For the moment at the crown

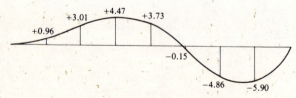

(g) For the moment at the right springing

Figure 18.6.3 (*continued*).

18.7 Stiffness Matrix of a Curved Member

Suppose that the curved member shown in Fig. 18.7.1 is a component in a large rigid frame. The ends A and B of this curved member may have rotations of X_1 and X_4, horizontal displacements of X_2 and X_5, and vertical displacements of X_3 and X_6, all shown in their positive directions in Fig. 18.7.1. It is required to derive a stiffness matrix $[K]$ (neglecting axial deformation) for this curved member satisfying the equation

$$\{P\}_{6\times1} = [K]_{6\times6}\{X\}_{6\times1} \qquad (18.7.1)$$

The elastic-center method can be conveniently used to obtain the contents of the $[K]$ matrix, one column at a time.

Consider the statically determinate cantilever structure shown in Fig. 18.7.2a. Due to the three generalized forces M_O, H_O, and V_O acting at the free end (identical to the elastic center), the moment M at $P(x, y)$ is, by noting the sign convention in Fig. 18.7.2b,

$$M = M_O - H_O y + V_O x \qquad (18.7.2)$$

Owing to a clockwise rotation of X_1, a horizontal displacement of X_2 to the right, and a vertical displacement of X_3 upward, all occurring at point A, the three generalized displacements at the elastic center are

$$\theta_O = +X_1 \quad \text{(clockwise)} \qquad (18.7.3a)$$
$$\Delta_{HO} = -y_A X_1 + X_2 \quad \text{(to right)} \qquad (18.7.3b)$$
$$\Delta_{VO} = +x_A X_1 + X_3 \quad \text{(upward)} \qquad (18.7.3c)$$

where (x_A, y_A) are the coordinates of point A in reference to the elastic center as origin. The actual displacement being OO' due to X_1 as shown in Fig. 18.7.2c, the correct results, both in sign and in magnitude, can be obtained by the negative sign before $y_A X_1$ in Eq. (18.7.3b) and by the positive sign before $x_A X_1$ in Eq. (18.7.3c). Due to X_2 and X_3, of course, the rigid arm just has a simple translation in each case.

Applying the unit loads as shown in Fig. 18.7.2d to f, the moments at point P are

$$m_\theta = +1.0 \qquad m_H = -1.0y \qquad m_V = +1.0x \qquad (18.7.4)$$

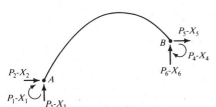

Figure 18.7.1 A curved member in a rigid frame.

684 INTERMEDIATE STRUCTURAL ANALYSIS

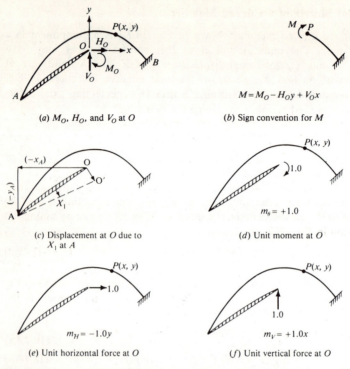

(a) M_O, H_O, and V_O at O

(b) Sign convention for M

(c) Displacement at O due to X_1 at A

(d) Unit moment at O

(e) Unit horizontal force at O

(f) Unit vertical force at O

Figure 18.7.2 A curved member with a fixed support at right end only.

Using the unit-load method for computing displacements, and the properties A, I_x, I_{xy}, I_y, I'_x, and I'_y of the analogous column section as defined in Chap. 15,

$$\theta_O = X_1 = \sum \frac{Mm_\theta \, \Delta S}{EI} = \sum \frac{(M_O - H_O y + V_O x)(+1.0) \, \Delta S}{EI}$$

$$X_1 = M_O A - H_O(0) + V_O(0) = M_O A \quad (18.7.5a)$$

$$\Delta_{HO} = -y_A X_1 + X_2 = \sum \frac{Mm_H \, \Delta S}{EI}$$

$$= \sum \frac{(M_O - H_O y + V_O x)(-1.0y) \, \Delta S}{EI}$$

$$-y_A X_1 + X_2 = -M_O(0) + H_O I_x - V_O I_{xy} = H_O I_x - V_O I_{xy} \quad (18.7.5b)$$

$$\Delta_{VO} = +x_A X_1 + X_3 = \sum \frac{Mm_V \, \Delta S}{EI}$$

$$= \sum \frac{(M_O - H_O y + V_O x)(+1.0x) \, \Delta S}{EI}$$

$$+x_A X_1 + X_3 = +M_O(0) - H_O I_{xy} + V_O I_y = -H_O I_{xy} + V_O I_y \quad (18.7.5c)$$

RIGID FRAMES WITH CURVED MEMBERS **685**

Solving the three simultaneous equations (18.7.5a to c),

$$M_O = +\frac{X_1}{A} \tag{18.7.6a}$$

$$H_O = \frac{X_1[-y_A + x_A(I_{xy}/I_y)] + X_2 + X_3(I_{xy}/I_y)}{I'_x} \tag{18.7.6b}$$

$$V_O = \frac{X_1[x_A - y_A(I_{xy}/I_x)] + X_2(I_{xy}/I_x) + X_3}{I'_y} \tag{18.7.6c}$$

The first three columns of the [K] matrix in Eq. (18.7.1) can be established by letting X_1, X_2, and X_3 each equal 1 unit in Eqs. (18.7.6a to c), computing the M_O, H_O, and V_O values, applying them to the free-body diagrams of Fig. 18.7.2a, and then determining the reactions at A and B.

Next consider the statically determinate cantilever structure shown in Fig. 18.7.3a. Due to the three generalized forces M_O, H_O, and V_O acting at the free

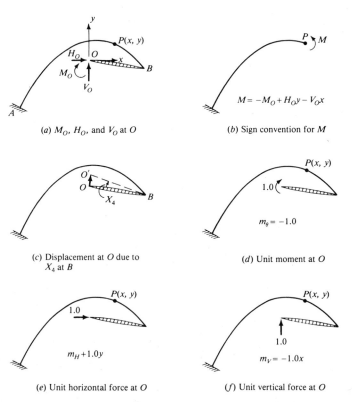

(a) M_O, H_O, and V_O at O

(b) Sign convention for M
$M = -M_O + H_O y - V_O x$

(c) Displacement at O due to X_4 at B

(d) Unit moment at O
$m_\theta = -1.0$

(e) Unit horizontal force at O
$m_H + 1.0 y$

(f) Unit vertical force at O
$m_V = -1.0 x$

Figure 18.7.3 A curved member with a fixed support at left end only.

end (identical to the elastic center), the moment at $P(x, y)$ is, by noting the sign convention in Fig. 18.7.3b,

$$M = -M_O + H_O y - V_O x \qquad (18.7.7)$$

Owing to a clockwise rotation of X_4, a horizontal displacement of X_5 to the right, and a vertical displacement of X_6 upward, all occurring at point B, the three generalized displacements at the elastic center are

$$\theta_O = +X_4 \quad \text{(clockwise)} \qquad (18.7.8a)$$
$$\Delta_{HO} = -y_B X_4 + X_5 \quad \text{(to right)} \qquad (18.7.8b)$$
$$\Delta_{VO} = -x_B X_4 + X_6 \quad \text{(upward)} \qquad (18.7.8c)$$

where (x_B, y_B) are the coordinates of point B in reference to the elastic center as origin. The actual displacement being OO' due to X_4 as shown in Fig. 18.7.3c, the currect results, both in sign and in magnitude, can be obtained by the negative sign before $y_B X_4$ in Eq. (18.7.8b) and by the positive sign before $x_B X_4$ in Eq. (18.7.8c). Due to X_5 and X_6, of course, the rigid arm just has a simple translation in each case.

Applying the unit loads as shown in Fig. 18.7.3d to f, the moments at point P are

$$m_\theta = -1.0 \qquad m_H = +1.0y \qquad m_V = -1.0x \qquad (18.7.9)$$

Using the unit-load method for computing displacements and the properties of the analogous column section,

$$\theta_O = X_4 = \sum \frac{M m_\theta \, \Delta S}{EI} = \sum \frac{(-M_O + H_O y - V_O x)(-1.0) \, \Delta S}{EI}$$

$$X_4 = M_O A - H_O(0) + V_O(0) = M_O A \qquad (18.7.10a)$$

$$\Delta_{HO} = -y_B X_4 + X_5 = \sum \frac{M m_H \, \Delta S}{EI}$$

$$= \sum \frac{(-M_O + H_O y - V_O x)(+1.0y) \, \Delta S}{EI}$$

$$-y_B X_4 + X_5 = -M_O(0) + H_O I_x - V_O I_{xy} = H_O I_x - V_O I_{xy} \qquad (18.7.10b)$$

$$\Delta_{VO} = +x_B X_4 + X_6 = \sum \frac{M m_V \, \Delta S}{EI}$$

$$= \sum \frac{(-M_O + H_O y - V_O x)(-1.0x) \, \Delta S}{EI}$$

$$+x_B X_4 + X_5 = +M_O(0) - H_O I_{xy} + V_O I_y = -H_O I_{xy} + V_O I_y \qquad (18.7.10c)$$

Equations (18.7.10a to c) are nearly identical to Eqs. (18.7.5a to c); solving them,

$$M_O = +\frac{X_4}{A} \qquad (18.7.11a)$$

$$H_O = \frac{X_4[-y_B + x_B(I_{xy}/I_y)] + X_5 + X_6(I_{xy}/I_y)}{I'_x} \qquad (18.7.11b)$$

$$V_O = \frac{X_4[+x_B - y_B(I_{xy}/I_x)] + X_5(I_{xy}/I_x) + X_6}{I'_y} \qquad (18.7.11c)$$

The last three columns of the $[K]$ matrix in Eq. (18.7.1) can be established by letting X_4, X_5, and X_6 each equal 1 unit in Eqs. (18.7.11a to c), computing the M_O, H_O, and V_O values, applying them to the free-body diagrams of Fig. 18.7.3a, and then determining the reactions at A and B.

Example 18.7.1 Establish numerically the 6×6 stiffness matrix of a curved member which has identical properties to the unsymmetrical parabolic fixed arch in Example 18.6.1.

SOLUTION (a) *Properties of the analogous column section.* The location of the elastic center shown in Fig. 18.7.4a and Fig. 18.7.5a is taken from Example 18.6.1. The other

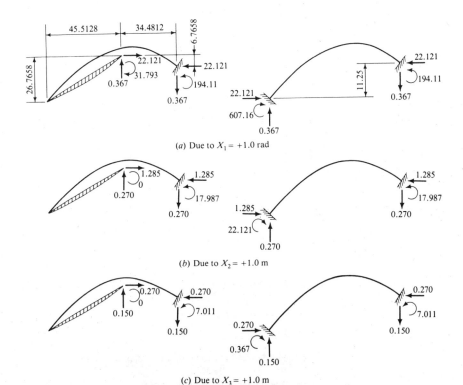

Figure 18.7.4 First three columns of the stiffness matrix in Example 18.7.1.

688 INTERMEDIATE STRUCTURAL ANALYSIS

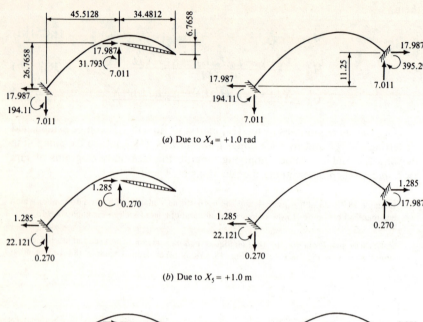

(a) Due to $X_4 = +1.0$ rad

(b) Due to $X_5 = +1.0$ m

(c) Due to $X_6 = +1.0$ m

Figure 18.7.5 Last three columns of the stiffness matrix in Example 18.7.1.

properties of the analogous column section taken from Example 18.6.1 are as follows:

$$A = \frac{31.4536 \text{ m}}{EI_c} \quad I'_x = \frac{778.374 \text{ m}^3}{EI_c} \quad I'_y = \frac{6647.54 \text{ m}^3}{EI_c}$$

$$\frac{I_{xy}}{I_x} = 1.79156 \quad \frac{I_{xy}}{I_y} = 0.20978$$

(b) *First three columns of the stiffness matrix.* Using Eqs. (18.7.6 a to c),

$$M_O = +\frac{X_1}{A} = +\frac{X_1}{31.4536/EI_c} = +31.793 X_1 \times 10^{-3} EI_c$$

$$H_O = \frac{X_1(-y_A + x_A I_{xy}/I_y) + X_2 + X_3(I_{xy}/I_y)}{I'_x}$$

$$= \frac{X_1[+26.7658 - 45.5128(0.20978)] + X_2 + X_3(0.20978)}{778.374/EI_c}$$

$$= (22.121 X_1 + 1.285 X_2 + 0.270 X_3) \times 10^{-3} EI_c$$

RIGID FRAMES WITH CURVED MEMBERS **689**

$$V_O = \frac{X_1(x_A - y_A I_{xy}/I_x) + X_2(I_{xy}/I_x) + X_3}{I'_y}$$

$$= \frac{X_1[-45.5128 + 26.7658(1.79156)] + X_2(1.79156) + X_3}{6647.54/EI_c}$$

$$= (0.367X_1 + 0.270X_2 + 0.150X_3) \times 10^{-3} EI_c$$

The actual directions and magnitudes of M_O, H_O, and V_O obtained from the above three equations due to $X_1 = +1.0$, $X_2 = +1.0$, and $X_3 = +1.0$ separately are shown on the left side of Fig. 18.7.4. The reactions at A and B are shown on the right side of Fig. 18.7.4, from which the first three columns of the stiffness matrix shown in Table 18.7.1 are assembled.

(c) *Last three columns of the stiffness matrix.* Using Eqs. (18.7.11a to c),

$$M_O = +\frac{X_4}{A} = +\frac{X_4}{31.4536/EI_c} = +31.793 X_4 \times 10^{-3} EI_c$$

$$H_O = \frac{X_4(-y_B + x_B I_{xy}/I_y) + X_5 + X_6(I_{xy}/I_y)}{I'_x}$$

$$= \frac{X_4[+6.7658 + 34.4872(0.20978)] + X_5 + X_6(0.20978)}{778.374/EI_c}$$

$$= (17.987 X_4 + 1.285 X_5 + 0.270 X_6) \times 10^{-3} EI_c$$

$$V_O = \frac{X_4(x_B - y_B I_{xy}/I_x) + X_5(I_{xy}/I_x) + X_6}{I'_y}$$

$$= \frac{X_4[+34.4872 + 6.7658(1.79156)] + X_5(1.79156) + X_6}{6647.54/EI_c}$$

$$= (7.011 X_4 + 0.270 X_5 + 0.150 X_6) \times 10^{-3} EI_c$$

The actual directions and magnitudes of M_O, H_O, and V_O obtained from the above three equations due to $X_4 = +1.0$, $X_5 = +1.0$, and $X_6 = +1.0$ separately are shown on the left side of Fig. 18.7.5. The reactions at A and B are shown on the right side of Fig. 18.7.5, from which the last three columns of the stiffness matrix in Table 18.7.1 are assembled.

Table 18.7.1 Stiffness matrix of the curved member in Example 18.7.1

P \ X	1	2	3	4	5	6
1	+607.16	+22.121	+0.367	−194.11	−22.121	−0.367
2	+ 22.121	+ 1.285	+0.270	− 17.987	− 1.285	−0.270
3	+ 0.367	+ 0.270	+0.150	− 7.011	− 0.270	−0.150
4	−194.11	− 17.987	−7.011	+395.29	+17.987	+7.011
5	− 22.121	− 1.285	−0.270	+ 17.987	+ 1.285	+0.270
6	− 0.367	− 0.270	−0.150	+ 7.011	+ 0.270	+0.150

$[K] =$ ... $* EI_c$

690 INTERMEDIATE STRUCTURAL ANALYSIS

(d) *Symmetry property of the stiffness matrix.* It can be seen that the [K] matrix is symmetric, as it must be by the law of reciprocal forces mentioned in Sec. 10.10. Besides, in the absence of forces acting between the ends of the curved element, there are three dependent equations of statics within the [K] matrix. Thus the rank of the 6×6 matrix is only 3, and the stiffness matrix of the element itself has no inverse.

18.8 Flexibility Matrix of a Curved Member with a Fixed Support

In the preceding section the stiffness matrix of a curved member is in fact obtained by inverting two 3×3 flexibility matrices. First Eqs. (18.7.5a to c) are solved simultaneously, and then Eqs. (18.7.10a to c) are solved simultaneously. By use of the elastic center, the answer for M_O comes directly, and the answers for H_O and V_O come out of the solution of two simultaneous equations; however, for a symmetrical curved element, I_{xy} is equal to zero and the equations for H_O and V_O are uncoupled. This feature marks the essence in the advantage of using the elastic center.

With the convenience provided by the electronic hand calculator, not to say computers of greater capacity, the solution of three simultaneous equations is no longer much of a chore. Consider the curved element with a fixed support at the right end, free at the left end, and without the rigid arm, as shown in Fig. 18.8.1. Let X_1, X_2, and X_3 be the displacements at the left end in the directions of P_1, P_2, and P_3. Applying the unit-load method,

$$X_1 = \sum \frac{Mm_1 \Delta S}{EI} = \sum \frac{(P_1 - P_2 y + P_3 x)(+1.0) \Delta S}{EI}$$
$$= +P_1 A - P_2 A \bar{y} + P_3 A \bar{x} \qquad (18.8.1a)$$

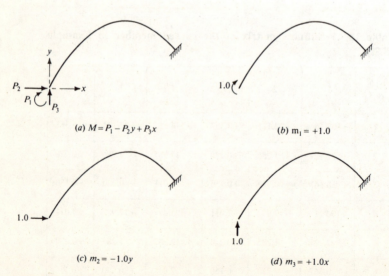

Figure 18.8.1 Flexibility matrix of a curved member with a fixed support at right end only.

RIGID FRAMES WITH CURVED MEMBERS

$$X_2 = \sum \frac{Mm_2 \Delta S}{EI} = \sum \frac{(P_1 - P_2 y + P_3 x)(-1.0 y) \Delta S}{EI}$$
$$= -P_1 A\bar{y} + P_2 I_x - P_3 I_{xy} \tag{18.8.1b}$$

$$X_3 = \sum \frac{Mm_3 \Delta S}{EI} = \sum \frac{(P_1 - P_2 y + P_3 x)(+1.0 x) \Delta S}{EI}$$
$$= +P_1 A\bar{x} - P_2 I_{xy} + P_3 I_y \tag{18.8.1c}$$

in which $(\bar{x}, \bar{y})$ are the coordinates of the elastic center in reference to A as origin, and (I_x, I_y, I_{xy}) all refer to a pair of coordinate axes through point A. Thus the 3×3 flexibility matrix for the displacements at the left end of a curved element is the matrix form of Eqs. (18.8.1a to c), or

$$[D] = \begin{array}{|c|c|c|c|} \hline {}_X\!\diagdown\!{}^P & 1 & 2 & 3 \\ \hline 1 & +A & -A\bar{y} & +A\bar{x} \\ \hline 2 & -A\bar{y} & +I_x & -I_{xy} \\ \hline 3 & +A\bar{x} & -I_{xy} & +I_y \\ \hline \end{array} \tag{18.8.2}$$

and the inverse of Eq. (18.8.2) is

$$[K] = \begin{array}{|c|c|c|c|} \hline {}_P\!\diagdown\!{}^X & 1 & 2 & 3 \\ \hline 1 & K_{11} & K_{12} & K_{13} \\ \hline 2 & K_{21} & K_{22} & K_{23} \\ \hline 3 & K_{31} & K_{32} & K_{33} \\ \hline \end{array} \tag{18.8.3}$$

which is a submatrix of the 6×6 matrix of Eq. (18.7.1).

Similarly, referring to Fig. 18.8.2,

$$X_4 = \sum \frac{Mm_4 \Delta S}{EI} = \sum \frac{(-P_4 + P_5 y - P_6 x)(-1.0) \Delta S}{EI}$$
$$= +P_4 A - P_5 A\bar{y} + P_6 A\bar{x} \tag{18.8.4a}$$

$$X_5 = \sum \frac{Mm_5 \Delta S}{EI} = \sum \frac{(-P_4 + P_5 y - P_6 x)(+1.0 y) \Delta S}{EI}$$
$$= -P_4 A\bar{y} + P_5 I_x - P_6 I_{xy} \tag{18.8.4b}$$

$$X_6 = \sum \frac{Mm_6 \Delta S}{EI} = \sum \frac{(-P_4 + P_5 y - P_6 x)(-1.0 x) \Delta S}{EI}$$
$$= +P_4 A\bar{x} - P_5 I_{xy} + P_6 I_y \tag{18.8.4c}$$

692 INTERMEDIATE STRUCTURAL ANALYSIS

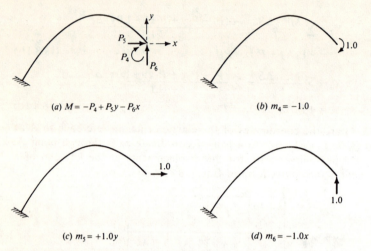

(a) $M = -P_4 + P_5 y - P_6 x$

(b) $m_4 = -1.0$

(c) $m_5 = +1.0y$

(d) $m_6 = -1.0x$

Figure 18.8.2 Flexibility matrix of a curved member with a fixed support at left end only.

in which $(\bar{x}, \bar{y})$ are the coordinates of the elastic center in reference to B as origin, and (I_x, I_y, I_{xy}) all refer to a pair of coordinate axes through point B. Thus the 3×3 flexibility matrix for the displacements at the right end of a curved element is the matrix form of Eqs. (18.8.4a to c), or

$$[D] = \begin{array}{c|ccc} P \backslash X & 4 & 5 & 6 \\ \hline 4 & +A & -A\bar{y} & +A\bar{x} \\ 5 & -A\bar{y} & +I_x & -I_{xy} \\ 6 & +A\bar{x} & -I_{xy} & +I_y \end{array} \quad (18.8.5)$$

and the inverse of Eq. (18.8.3) is

$$[K] = \begin{array}{c|ccc} X \backslash P & 4 & 5 & 6 \\ \hline 4 & K_{44} & K_{45} & K_{46} \\ 5 & K_{54} & K_{55} & K_{56} \\ 6 & K_{64} & K_{65} & K_{66} \end{array} \quad (18.8.6)$$

which is a submatrix of the 6×6 matrix of Eq. (18.7.1).

RIGID FRAMES WITH CURVED MEMBERS **693**

If this method is used, however, the 3×3 submatrix at the lower left and at the upper right within the 6×6 stiffness matrix of the curved element will have to be computed by applying the three equations of equilibrium to the free body represented by Figs. 18.8.1a and 18.8.2a.

Example 18.8.1 Obtain the 3×3 flexibility matrix for the displacements at the left and right ends, respectively, of the curved member represented by the unsymmetrical parabolic fixed arch in Example 18.6.1.

SOLUTION (a) *Flexibility matrix for displacements at left end.* The point areas and the coordinates of their locations in reference to point A as origin are

	A	x	y
1	$1.5688/EI_c$	$+5$	$+5.9375$
2	$2.0158/EI_c$	$+15$	$+15.9375$
3	$2.7481/EI_c$	$+25$	$+23.4375$
4	$4.1083/EI_c$	$+35$	$+28.4375$
5	$7.0782/EI_c$	$+45$	$+30.9375$
6	$7.0782/EI_c$	$+55$	$+30.9375$
7	$4.1083/EI_c$	$+65$	$+28.4375$
8	$2.7481/EI_c$	$+75$	$+23.4375$

The summations are

$$A = 31.4536/EI_c \qquad A\bar{x} = +1431.54/EI_c \qquad A\bar{y} = +841.88/EI_c$$
$$I_x = 23{,}780.7/EI_c \qquad I_y = 75{,}803.5/EI_c \qquad I_{xy} = +40{,}550.5/EI_c$$

Substituting the above values into Eq. (18.8.2) yields the required flexibility matrix

$$[D] = \begin{array}{c|ccc} P \backslash X & 1 & 2 & 3 \\ \hline 1 & +31.4536 & -841.88 & +1431.54 \\ 2 & -841.88 & +23{,}780.7 & -40{,}550.5 \\ 3 & +1431.54 & -40{,}550.5 & +75{,}803.5 \end{array} * \frac{1}{EI_c}$$

The inverse of the above matrix is found to be identical to the results of Example 18.7.1.

(b) *Flexibility matrix for displacements at right end.* The point areas and the coordinates of their locations in reference to point B as origin are

	A	x	y
1	$1.5686/EI_c$	-75	$+ 3.4375$
2	$2.0158/EI_c$	-65	$+ 8.4375$
3	$2.7481/EI_c$	-55	$+10.9375$
4	$4.1083/EI_c$	-45	$+10.9375$
5	$7.0782/EI_c$	-35	$+ 8.4375$
6	$7.0782/EI_c$	-25	$+ 3.4375$
7	$4.1083/EI_c$	-15	$- 4.0625$
8	$2.7481/EI_c$	$- 5$	-14.0625

The summations are

$$A = 31.4536/EI_c \quad A\bar{x} = -1084.75/EI_c \quad A\bar{y} = +212.809/EI_c$$
$$I_x = 2686.88/EI_c \quad I_y = 48{,}060.2/EI_c \quad I_{xy} = -5105.01/EI_c$$

Substituting the above values into Eq. (18.8.5) yields the required flexibility matrix

X \ P	4	5	6	
4	$+\ 31.4536$	$-\ 212.809$	$-\ 1084.75$	
$[D] =$ 5	$-\ 212.809$	$+2686.88$	$+\ 5105.01$	$* \dfrac{1}{EI_c}$
6	$-\ 1084.75$	$+5105.01$	$+48{,}060.2$	

The inverse of the above matrix is found to be identical to the results of Example 18.7.1.

18.9 Flexibility Matrix of a Curved Member with One Hinged Support and One Roller Support

Toward the end of Example 18.7.1, in part (d), it was mentioned that the rank of the 6×6 stiffness matrix of a curved element is 3; thus one can surmise that there can be only three independent unknown internal forces in a curved element. In fact, the curved element can be imagined to be represented by the dashed straight element as shown in Fig. 18.9.1, in which case the local stiffness matrix $[K]$ can be obtained by

$$[K] = [A][S][B] \tag{18.9.1}$$

RIGID FRAMES WITH CURVED MEMBERS

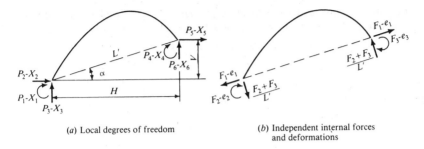

(a) Local degrees of freedom

(b) Independent internal forces and deformations

Figure 18.9.1 Local P-X and F-e numbers of a curved member.

The matrices $[A]$ and $[B]$ are identical to those defined in Sec. 17.3; only the $[S]$ matrix of a curved element needs to be determined. This 3×3 element stiffness matrix is the inverse of its flexibility matrix.

The flexibility matrix of the curved element shown in Fig. 18.9.1b can be obtained by providing a hinged support at the left end and a roller support at the right end, as shown in Fig. 18.9.2. The increase in distance e_1 between the left and right ends, the clockwise angle e_2 from the original element direction to the tangent to the elastic curve at the left end (excluding rigid-body rotation of the entire curved element), and the similarly defined angle e_3 at the right end, corresponding to the generalized internal forces F_1, F_2, and F_3, can be obtained by the unit-load method.

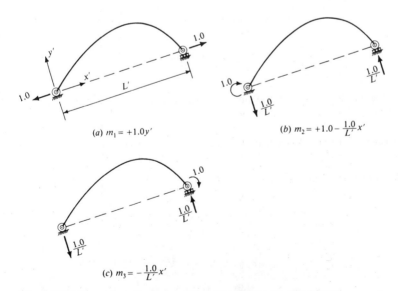

(a) $m_1 = +1.0 y'$

(b) $m_2 = +1.0 - \dfrac{1.0}{L'} x'$

(c) $m_3 = -\dfrac{1.0}{L'} x'$

Figure 18.9.2 Expressions for m_1, m_2, and m_3 due to unit generalized forces.

Let m_1, m_2, and m_3 be the bending moments at any point on the curved element due to $F_1 = +1.0$, $F_2 = +1.0$, and $F_3 = +1.0$, respectively. Then,

$$e_1 = \sum \frac{(m_1 F_1 + m_2 F_2 + m_3 F_3)(m_1) \Delta S}{EI}$$

$$= F_1 \sum \frac{m_1^2 \Delta S}{EI} + F_2 \sum \frac{m_1 m_2 \Delta S}{EI} + F_3 \sum \frac{m_1 m_3 \Delta S}{EI} \quad (18.9.1a)$$

$$e_2 = \sum \frac{(m_1 F_1 + m_2 F_2 + m_3 F_3)(m_2) \Delta S}{EI}$$

$$= F_1 \sum \frac{m_2 m_1 \Delta S}{EI} + F_2 \sum \frac{m_2^2 \Delta S}{EI} + F_3 \sum \frac{m_2 m_3 \Delta S}{EI} \quad (18.9.1b)$$

$$e_3 = \sum \frac{(m_1 F_1 + m_2 F_2 + m_3 F_3)(m_3) \Delta S}{EI}$$

$$= F_1 \sum \frac{m_3 m_1 \Delta S}{EI} + F_2 \sum \frac{m_3 m_2 \Delta S}{EI} + F_3 \sum \frac{m_3^2 \Delta S}{EI} \quad (18.9.1c)$$

The flexibility matrix of the imaginary straight member, equivalent to the curved member, is the matrix form of Eqs. (18.9.1a to c), or

	F e	1	2	3
[D] =	1	$\sum \frac{m_1^2 \Delta S}{EI}$	$\sum \frac{m_1 m_2 \Delta S}{EI}$	$\sum \frac{m_1 m_3 \Delta S}{EI}$
	2	$\sum \frac{m_2 m_1 \Delta S}{EI}$	$\sum \frac{m_2^2 \Delta S}{EI}$	$\sum \frac{m_2 m_3 \Delta S}{EI}$
	3	$\sum \frac{m_3 m_1 \Delta S}{EI}$	$\sum \frac{m_3 m_2 \Delta S}{EI}$	$\sum \frac{m_3^2 \Delta S}{EI}$

(18.9.2)

Indeed, Eq. (18.9.2) conveys a general method for determining the flexibility matrix of any kind of discrete element.

Example 18.9.1 Re-solve Example 18.7.1 by using $[K] = [A][S][B]$ in which the $[S]$ matrix is that of an equivalent straight member joining the ends of the curved member.

SOLUTION Referring to Figs. 18.9.2 and 18.9.3,

$$\sin \alpha = 1/\sqrt{17} \qquad \cos \alpha = 4/\sqrt{17}$$

$$x' = x \cos \alpha + y \sin \alpha \qquad y' = -x \sin \alpha + y \cos \alpha$$

$$m_1 = +1.0 y' = +\frac{1}{\sqrt{17}}(4y - x)$$

$$m_2 = +1.0 - \frac{1.0 x'}{L'} = 1.0 - \frac{4x + y}{340}$$

$$m_3 = -\frac{1.0 x'}{L'} = -\frac{4x + y}{340}$$

RIGID FRAMES WITH CURVED MEMBERS 697

Using the $\Delta S/EI$ and (x, y) values shown in part (a) of the solution to Example 18.8.1,

$$\sum \frac{m_1^2 \Delta S}{EI} = \frac{+7758.251}{EI_c} \quad \sum \frac{m_2^2 \Delta S}{EI} = \frac{+6.321950}{EI_c} \quad \sum \frac{m_3^2 \Delta S}{EI} = \frac{+13.50732}{EI_c}$$

$$\sum \frac{m_1 m_2 \Delta S}{EI} = \frac{+184.0920}{EI_c} \quad \sum \frac{m_1 m_3 \Delta S}{EI} = \frac{-285.4528}{EI_c} \quad \sum \frac{m_2 m_3 \Delta S}{EI} = \frac{-5.81397}{EI_c}$$

The flexibility matrix of the equivalent straight member is

$[D] =$

F \ e	1	2	3
1	+7758.251	+184.0920	−285.4528
2	+184.0920	+6.321950	−5.81397
3	−285.4528	−5.81397	+13.50372

$* \dfrac{1}{EI_c}$

of which the inverse is

$[S] =$

e \ F	1	2	3
1	+1.34491	−21.5510	+19.1512
2	−21.5510	+607.196	−194.141
3	+19.1512	−194.141	+395.304

$* EI_c$

Using the $[A]$ matrix as defined in Sec. 17.3,

$[A] =$

P \ F	1	2	3
1	0.	+1.0	0.
1	$-4/\sqrt{17}$	+1/340	+1/340
3	$-1/\sqrt{17}$	−1/85	−1/85
4	0.	0.	+1.0
5	$+4/\sqrt{17}$	−1/340	−1/340
6	$+1/\sqrt{17}$	+1/85	+1/85

the same matrix $[K] = [A][S][A^T]$ as in Example 18.7.1 is obtained.

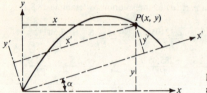

Figure 18.9.3 New coordinates due to rotation of axes.

18.10 Effects of Temperature, Shrinkage, Rib Shortening, and Foundation Yielding on Fixed Arches

Any structure, whether statically determinate or indeterminate, will have internal forces developed in it when loads are applied. An unloaded structure, if it is statically determinate, is not subjected to internal forces as a result of change in temperature or yielding of supports, but a statically indeterminate structure is stressed by such impositions. Inasmuch as the fixed arch is statically indeterminate to the third degree, it will be subjected to self-balancing reactions due to a change in temperature or yielding of supports, which may be a rotational slip, horizontal displacement, or vertical displacement.

If the fixed arch is built of reinforced concrete, it is necessary to investigate the effect of shrinkage. It is conceivable that the effect of shrinkage is equivalent to that of a temperature drop. Usually the amount of shrinkage is considered to be the same as that due to a drop of 8 to 10°C, depending on the properties of the concrete mix. The induced reactions at the left and right supports of a fixed arch due to displacements (rotation, horizontal displacement, or vertical displacement) of the supports can be obtained by multiplying the appropriate columns in the stiffness matrix of the curved member representing the arch by the known or estimated amount of the support displacement. In the case of a temperature rise under the restraint of the fixed supports, one can first visualize the picture of Fig. 18.10.1a, where the arch has no restraints and therefore is free to expand. From Fig. 18.10.1a

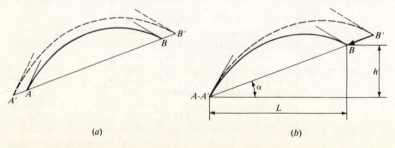

Figure 18.10.1 Expansion of arch rib due to temperature rise.

to b, the boundary condition at the fixed left support is satisfied, but not that at the right support. Thus the effect of a temperature rise is equivalent to a forced displacement at the right support of $BB'\cos\alpha$ to the left and $BB'\sin\alpha$ downward, while BB' is equal to the product of the length of straight line AB times the temperature rise times the coefficient of expansion.

One final note: throughout this chapter, axial deformation of the arch rib has not been considered. As a matter of expediency some analysts tend to approximate the effect of an average compressive force in the arch rib equivalent to that of a temperature drop. This rib-shortening effect can be accomplished by using a temperature drop such that its product with the coefficient of contraction is equal to f_c/E_c, wherein f_c is the estimated average compressive stress in the arch rib and E_c is the modulus of elasticity of concrete.

18.11 Exercises

18.1. For the symmetrical, circular fixed arch shown in Fig. 18.11.1, draw influence lines for (*a*) the horizontal reaction at the left support, (*b*) the vertical reaction at the left support, (*c*) the vertical shear at the crown, (*d*) the moment at the left support, (*e*) the moment at the quarter point of the horizontal span, and (*f*) the moment at the crown. Divide the arch rib into eight segments of equal horizontal projection and place the point area $\Delta S/EI$ on the arch rib midway horizontally from the boundary points of each segment.

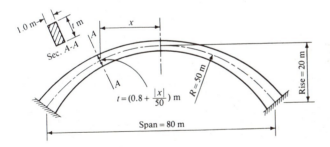

Figure 18.11.1 Exercises 18.1, 18.3, 18.4, 18.5.

18.2 For the unsymmetrical, circular fixed arch shown in Fig. 18.11.2, draw influence lines for (*a*) the horizontal reaction at the left support, (*b*) the vertical reaction at the left support, (*c*) the vertical shear at the crown, (*d*) the moment at the left support, (*e*) the moment at the quarter point of the horizontal span, (*f*) the moment at the crown, and (*g*) the moment at the right support. Divide the arch rib into eight segments of equal horizontal projection and place the point area $\Delta S/EI$ on the arch rib midway horizontally from the boundary points of each segment.

18.3 Using the elastic-center method, obtain numerically the 6×6 stiffness matrix of a curved member identical in shape and in variation of moment of inertia to the symmetrical, circular fixed arch of Exercise 18.1.

18.4 Solve Exercise 18.3 by finding the inverse of the flexibility matrix for the displacements at each end of the curved member while the other end is fixed.

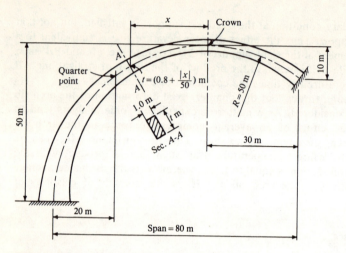

Figure 18.11.2 Exercises 18.2, 18.6, 18.7, 18.8.

18.5 Solve Exercise 18.3 by replacing the curved member by an imaginary straight member and using the equation $[K] = [A][S][B]$. Note that $[S]$ is the inverse of a flexibility matrix for the independent deformations when one end of the curved member is hinged and the other end is on rollers.

18.6 Solve Exercise 18.3, but let the curved member be represented by an unsymmetrical, circular fixed arch in Fig. 18.11.2.

18.7 Solve Exercise 18.4, but let the curved member be represented by the unsymmetrical, circular fixed arch in Fig. 18.11.2.

18.8 Solve Exercise 18.5, but let the curved member be represented by the unsymmetrical, circular fixed arch in Fig. 18.11.2.

18.9 For the unsymmetrical fixed arch shown in Fig. 18.11.3 compute by the elastic-center method the reactions at A and E due to a 1.0-kN vertical load applied at C. Check by a different basic determinate structure. The point area are $A_1 = 3$ m/EI_c, $A_2 = 8$ m/EI_c, $A_3 = 5$ m/EI_c, and $A_4 = 4$ m/EI_c.

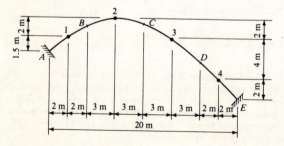

Figure 18.11.3 Exercise 18.9.

18.10 Obtain the first and second columns of the element stiffness matrix of the right-angled member as shown in Fig. 18.11.4, first by any conventional method (for instance, moment-distribution method), and then check by the unsymmetrical elastic-center method.

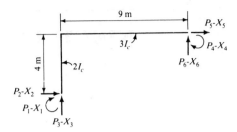

Figure 18.11.4 Exercise 18.10.

18.11 Analyze by the moment-distribution method the rigid frame shown in Fig. 18.11.5 for a rotational slip of $1000 \, \text{kN} \cdot \text{m}^3/EI$ clockwise of support A. Verify the results by the unsymmetrical elastic-center method.

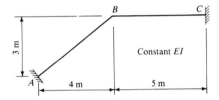

Figure 18.11.5 Exercise 18.11.

18.12 Using the elastic-center method and treating $ABCD$ of Fig. 18.11.6 as one structural member, obtain the 6×6 stiffness matrix of the bent-shaped element.

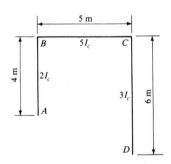

Figure 18.11.6 Exercise 18.12.

CHAPTER
NINETEEN
DISPLACEMENT METHOD OF HORIZONTAL GRID-FRAME ANALYSIS

19.1 Definition of Horizontal Grid Frames

A horizontal grid frame subjected to vertical loading usually consists of two sets of parallel beams, with one set perpendicular to the other. Each beam may be simply supported at the ends, or fixed in rotation about the transverse axis as well, or sometimes also fixed against rotation about the longitudinal axis. If one set of beams sits directly above the other set, then there are only vertical interactions between the two sets at the points of intersection. For example, the three east-west beams in Fig. 19.1.1a are directly above the three north-south beams, making a horizontal (i.e., in the horizontal plane) grid that consists of 12 pedestal supports and 24 beam elements.

On the other hand, if the two sets of beams are all at the same elevation and if the intersecting joints are rigid (welded steel or monolithic reinforced concrete), such as shown in Fig. 19.1.1b, each of the 24 elements is capable of resisting torsion as well as bending moment and shear. An element which may take up torsion and bending, by virtue of its end connections, is called a combined beam and torsion element. The 12 supports in Fig. 19.1.1b may be designed and constructed to take vertical reaction only, or vertical reaction with torsion resistance, or with moment resistance, or more likely with both. A specially designed support is needed in order for a support to take torsion and vertical reaction without having bending-moment resistance.

The horizontal grid frame of Fig. 19.1.1c is a cellular structure with 40 combined beam and torsion elements, supported on four pedestals at its four corners. Such a grate structure may be quite small, cast in steel and assigned to take very heavy loads, or it may be very large, built as a glued-wood roof structure over a garden. One can think of many practical and important

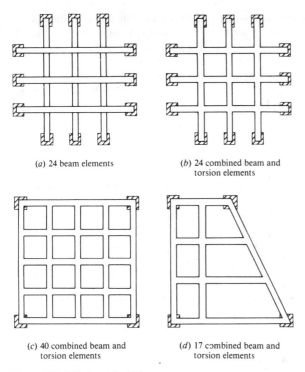

(a) 24 beam elements
(b) 24 combined beam and torsion elements
(c) 40 combined beam and torsion elements
(d) 17 combined beam and torsion elements

Figure 19.1.1 Horizontal grid frames.

structures which should be analyzed as vertically loaded horizontal grid frames, containing either beam elements or combined beam and torsion elements.

Actually, the intersecting elements need not always be at right angles to qualify for the name *grid frame*. For example, the frame shown in Fig. 19.1.1d might be used as a roof frame for an odd-shaped piece of land.

19.2 Methods of Analysis

Before the advent of the electronic computer, horizontal grid frames with beam elements only, without torsion, were analyzed by the force method, that is, using redundants as the primary unknowns. The grid of Fig. 19.1.1a is made of two sets of simple beams; one could use the nine vertical interactions at the points of intersection as the redundants, whereas the compatibility conditions are that the vertical deflections at the intersecting points resulting from either set of beams must be the same. When torsion resistance may be present and should be included in the analysis, the degree of indeterminacy becomes very high and the compatibility conditions are more cumbersome to establish. Thus the common assumption is to neglect torsion, a practice which yields reasonable answers for bending and deflection but gives no indication as to the amount of torsion. In such cases, a nominal amount of torsional moment is usually provided for in the design process.

With the convenience in the application of the matrix displacement method on the computer, analysis of horizontal grid frames, with or without torsion, becomes rather a simple matter so far as making up the input is concerned. The standard twofold equations

$$\{X\} = [ASB]^{-1}\{P\} \tag{19.2.1}$$

and $$\{F^*\} = \{F_0\} + [SB]\{X\} \tag{19.2.2}$$

can be easily programmed by using global $[A]$, $[S]$, $\{F_0\}$ and $\{P\}$ matrices, or by feeding the local $[ASB]$ matrix of each element into the global $[ASB]$ matrix. Both versions of the computer program are available.† In this intermediate text, the first approach will be described and illustrated by several examples; since the second approach is more of a software exercise.

The important part of the displacement method is the assignment of the P-X and F-e numbers. So far as the degree of freedom is concerned, each node may have slopes in two reference directions (rotations about east-west and north-south axes, respectively) and vertical deflection, making a total of three unknown generalized displacements at the most. With three orthogonal axes—x axis positive toward the east, y axis positive toward the north, and z axis positive downward—the three degrees of freedom at a typical node are shown in Fig. 19.2.1a, wherein the "right-hand screw" vector notation in double arrowheads is used for rotations. With the same vector notation, the free-body diagram of a typical combined beam and torsion element in the east-west direction is shown in Fig. 19.2.1b; and in the north-south direction, in Fig. 19.2.1c. Thus, for a beam element the number of independent unknown internal forces is two (F_i and F_j); for a combined beam and torsion element, it is three (F_i, F_j, and F_k).

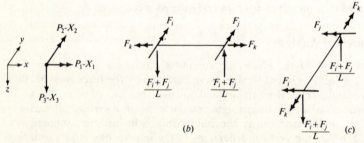

Figure 19.2.1 Typical joint and elements in a horizontal grid.

†See C.-K. Wang, *Matrix Methods of Structural Analysis*, 2d ed., American Publishing Company, Madison, Wis., 1970, Appendixes H and L.

For the grid frame of Fig. 19.1.1a, there are 30 unknown slopes and 9 unknown deflections, making the degree of freedom NP equal to 39; the 24 beam elements require 48 independent unknown internal forces, making NF equal to 48; thus the degree of indeterminacy $NI = NF - NP = 48 - 39 = 9$, as previously stated in connection with the conventional force method. Assuming completely fixed supports, the degree of freedom of the grid frame in Fig. 19.1.1b is 3 times the 9 points of intersection, or $NP = 27$, while NF is equal to 72 for the 24 combined beam and torsion elements; thus $NI = 72 - 27 = 45$. The closed grid frame of Fig. 19.1.1c has four vertical restraints only; thus the degree of freedom is $[3(25) - 4] = 71$; with $NF = 3(40) = 120$, and $NI = 120 - 71 = 49$. Finally, the degree of freedom of the grid frame in Fig. 19.1.1d is $NP = [3(12) - 4] = 32$; $NF = 3(17) = 51$; and $NI = 51 - 32 = 19$. Of course, in the displacement method, the degree of indeterminacy NI is only a matter of interest; it is only in the force method that NI redundants have to be chosen at the outset.

19.3 Grid Frames with Beam Elements Only

When a grid frame contains beam elements only, the force method might still be a viable method, because the vertical interactions at the points of intersection can be conveniently used as the redundants. Even in the displacement-method category, without resorting to matrix notations, the slope-deflection method may be used by writing out the two slope-deflection equations for each beam element and processing them through to obtain the simultaneous equations, from which slopes and deflections are obtained. The matrix displacement method simply puts the slope-deflection method in matrix format, as illustrated in the example which follows.

Example 19.3.1 By the force method and then the displacement method, analyze the horizontal grid frame shown in Fig. 19.3.1a, wherein beam AB is directly above beam CD, and the supports A, B, C, and D are fixed against rotation.

SOLUTION (a) *Force method.* For a beam with both ends fixed, many handbooks give the expression for the deflection at any point on the beam for one concentrated load. For convenience, such an expression is repeated in Fig. 19.3.1c and used in solving this problem. Even though this frame is statically indeterminate to the fifth degree, four redundants have already been removed by the formula for the fixed-end beam in Fig. 19.3.1c. Consequently, only the interaction R between the beams is left as the redundant. The compatibility condition is that the vertical deflection at E on beam AB should be equal to that on beam CD, as shown by Fig. 19.3.1b. Using the formula in Fig. 19.3.1c,

$$\Delta_E \text{ in beam } AB = \frac{3.0(6)^2(7.5)^2}{6(600)(17.5)^3}[3(11.5)(17.5) - 3(11.5)(7.5) - 6(7.5)]$$

$$- \frac{R(7.5)^3(10)^3}{3(600)(17.5)^3}$$

$$= 0.0944606 - 0.0437318R$$

$$\Delta_E \text{ in beam } CD = \frac{R(8)^3(5)^3}{3(250)(13)^3} = 0.0388408R$$

706 INTERMEDIATE STRUCTURAL ANALYSIS

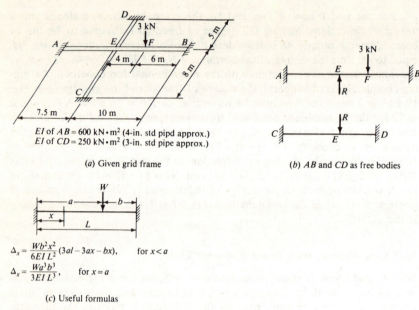

EI of $AB = 600$ kN·m² (4-in. std pipd approx.)
EI of $CD = 250$ kN·m² (3-in. std pipe approx.)

(a) Given grid frame

(b) AB and CD as free bodies

$\Delta_x = \dfrac{Wb^2x^2}{6EIL^2}(3al - 3ax - bx)$, for $x < a$

$\Delta_x = \dfrac{Wa^3b^3}{3EIL^3}$, for $x = a$

(c) Useful formulas

Figure 19.3.1 Force method, grid frame of Example 19.3.1.

Equating the Δ_E's,

$$0.0944606 - 0.0437318R = 0.0388408R$$

$$R = 1.14397 \text{ kN}$$

Then, using the well-used formulas for fixed-end moments and following the designer's sign convention,

$$M_A = -\frac{3(11.5)(6)^2}{(17.5)^2} + \frac{1.14397(7.5)(10)^2}{(17.5)^2} = -1.25395 \text{ kN·m}$$

$$M_B = -\frac{3(6)(11.5)^2}{(17.5)^2} + \frac{1.14397(10)(7.5)^2}{(17.5)^2} = -5.67189 \text{ kN·m}$$

$$M_C = -\frac{1.14397(8)(5)^2}{(13)^2} = -1.35381 \text{ KN·m}$$

$$M_D = -\frac{1.14397(5)(8)^2}{(13)^2} = -2.16610 \text{ kN·m}$$

(b) *Displacement method.* Since there can be no unknown forces acting on any element, the minimum number of elements that have to be used is four. The P-X numbers are shown in Fig. 19.3.2b, wherein X_1 is the counterclockwise slope at E of beam CD when viewed from right toward left, X_2 is the clockwise slope at E of beam AB when viewed from front toward back, and X_3 is the downward deflection of E. The F-e numbers are shown in Fig. 19.3.2c, wherein all end moments are clockwise when viewed from right toward left or from front toward back.

The three equations of equilibrium at joint E, as observed from its free-body diagram

DISPLACEMENT METHOD OF HORIZONTAL GRID–FRAME ANALYSIS

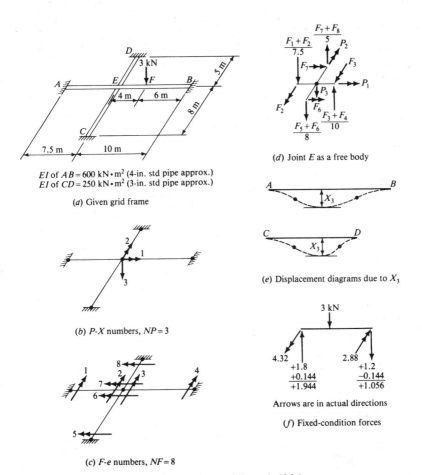

EI of $AB = 600$ kN·m² (4-in. std pipe approx.)
EI of $CD = 250$ kN·m² (3-in. std pipe approx.)

(a) Given grid frame

(b) P-X numbers, $NP = 3$

(c) F-e numbers, $NF = 8$

(d) Joint E as a free body

(e) Displacement diagrams due to X_3

Arrows are in actual directions

(f) Fixed-condition forces

Figure 19.3.2 Displacement method, grid frame of Example 19.3.1.

in Fig. 19.3.1d, are expressed in the rows of the [A] matrix shown below:

F P	1	2	3	4	5	6	7	8
1						-1.0	-1.0	
[A]$_{3\times8}$ = 2	$+1.0$	$+1.0$						
3	$-\dfrac{1}{7.5}$	$-\dfrac{1}{7.5}$	$+\dfrac{1}{10}$	$+\dfrac{1}{10}$	$-\dfrac{1}{8}$	$-\dfrac{1}{8}$	$+\dfrac{1}{5}$	$+\dfrac{1}{5}$

708 INTERMEDIATE STRUCTURAL ANALYSIS

So far as the contents in the $[B]$ matrix are concerned, due to X_1 counterclockwise viewed from right toward left, $e_6 = -X_1$ and $e_7 = -X_1$; due to X_2 clockwise viewed from front toward back, $e_2 = +X_2$ and $e_3 = +X_2$. The displacement diagrams due to X_3 are shown in Fig. 19.3.2e, wherein the e's are the clockwise rotations from the displaced member axis to the elastic-curve tangent; thus

$$e_1 = e_2 = -\frac{X_3}{7.5} \qquad e_3 = e_4 = +\frac{X_4}{10}$$

$$e_5 = e_6 = -\frac{X_3}{8} \qquad e_7 = e_8 = +\frac{X_3}{5}$$

Displayed in matrix form, the $[B]$ matrix is

$[B]_{3\times 8} =$

X \ e	1	2	3
1			$-1/7.5$
2		$+1.0$	$-1/7.5$
3		$+1.0$	$+1/10$
4			$+1/10$
5			$-1/8$
6	-1.0		$-1/8$
7	-1.0		$+1/5$
8			$+1/5$

The contents of the $[S]$ matrix are $4EI/L$ and $2EI/L$ in kilonewton-meters per radian; thus

$[S]_{8\times 8} =$

F \ e	1	2	3	4	5	6	7	8
1	320	160						
2	160	320						
3			240	120				
4			120	240				
5					125	62.5		
6					62.5	125		
7							200	100
8							100	200

The fixed-condition forces are, from Fig. 19.3.2f,

$$F_{03} = -4.32 \text{ kN·m} \quad F_{04} = +2.88 \text{ kN·m}$$
$$R_{03} = -1.944 \text{ kN} \quad R_{04} = -1.056 \text{ kN}$$

Thus the contents of the $\{P\}$ matrix are

$$P_1 = 0 \quad P_2 = -(F_{02} + F_{03}) = +4.32 \quad P_3 = -(R_{02} + R_{03}) = +1.944$$

From the twofold equations, repeated in this chapter as Eqs. (19.2.1) and (19.2.2), the output matrices $\{X\}$ and $\{F^*\}$ are

$\{X\}_{3\times 1} =$

LC X	1
1	+0.0049987 rad
2	+0.0099359 rad
3	+0.044433 m

$\{F^*\}_{8\times 1} =$

LC F^*	1
1	−1.25395
2	+0.33579
3	−0.33580
4	+5.67189
5	−1.35381
6	−1.66623
7	+1.66623
8	+2.16610

To check the output, there should be three statics checks and eight deformation checks. The deformation checks are to follow Eqs. (11.13.3) and (11.11.4a and b), such that

$$\begin{Bmatrix} X_i - R_{ij} \\ X_j - R_{ij} \end{Bmatrix} = \begin{Bmatrix} e_{0i} \\ e_{0j} \end{Bmatrix} + \begin{bmatrix} +\dfrac{L}{3EI} & -\dfrac{L}{6EI} \\ -\dfrac{L}{6EI} & +\dfrac{L}{3EI} \end{bmatrix} \begin{Bmatrix} F_i^* \\ F_j^* \end{Bmatrix} \qquad (19.2.3)$$

Although not shown, these checks have been made.

19.4 Grid Frames with Combined Beam and Torsion Elements

When the joints in a grid frame are rigid, such as in cases where several steel pipes coming into the same joint are welded together or where the complete frame is cast monolithically in reinforced concrete, each element between joints can take torsion as well as bending. If the element itself is unloaded except at the ends, its free-body diagram would be as shown in Fig. 19.2.1b. If the element is subjected to a concentrated torsional moment along its length, as is common in machine parts, the reacting torsional moments in the fixed condition will be as shown in Fig. 19.4.1, provided the torsional rigidity is constant. Even a simple grid frame with torsion action has a very high degree of indeterminacy, and the compatibility conditions in the force method usually require the determination of many deformation quantities due to the applied loads and the redundants. Although the number of simultaneous equations that need to be solved in the displacement method may be as great or sometimes greater than in the force method, these equations themselves can be systematically written out or processed on the computer following a unified scheme in matrix notation. Almost universally now, grid frames with torsion should be analyzed by the matrix displacement method.

Suppose that the four steel pipe sections in Fig. 19.4.2 are welded together at the intersection point E and the resulting frame with four fixed supports is subjected to the load W. The given structure can be separated into four

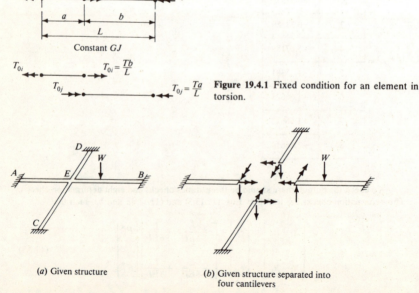

Figure 19.4.1 Fixed condition for an element in torsion.

(a) Given structure

(b) Given structure separated into four cantilevers

Figure 19.4.2 Analysis of simple grid frame by force method.

DISPLACEMENT METHOD OF HORIZONTAL GRID–FRAME ANALYSIS   711

cantilevers, each statically determinate. At the end of each cantilever, there are three unknowns: a torsional moment, a bending moment, and a vertical force. Since the 12 unknowns must satisfy the three equations of equilibrium at joint E, there are nine independent unknown redundants. The nine compatibility equations are that the rotations of the cross section at E about the x and y axes, as well as the vertical deflection there obtained from each cantilever, must be identical. The foregoing is an illustration of the force method of analysis, which requires the analyst to think of a tailor-made method of attack on any one specific problem. The analysis of the frame in Fig. 19.4.2 by the displacement method will now be shown.

Example 19.4.1 By the displacement method, analyze the horizontal grid frame shown in Fig. 19.4.3a, wherein four pipe sections are welded at point E, and four fixed supports are provided at A, B, C, and D.

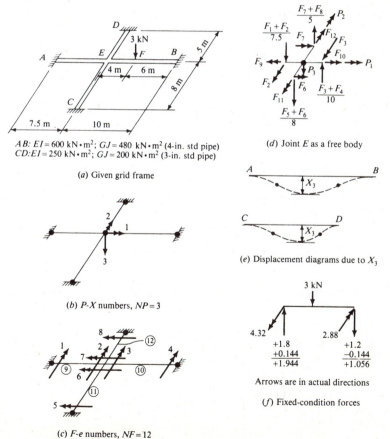

Figure 19.4.3 Grid frame with torsion, Example 19.4.1.

712 INTERMEDIATE STRUCTURAL ANALYSIS

SOLUTION (a) *P-X and F-e numbers.* At joint E the unknown rotations of the cross section are X_1 and X_2 and the unknown vertical deflection is X_3. In reality, point E is within the hollow pipes and the cross sections through E are purely theoretical. The P-X numbers in Fig. 19.4.3b appear identical to those of Fig. 19.3.2b, but in that example X_1 and X_2 are uncoupled because each is in a separate beam. The first eight F-e numbers in Fig. 19.4.3c are identical to those in Fig. 19.3.2c; each of F_9 to F_{12} involves a pair of torsional moments whose vectors are similar to those of a pair of tensile forces. Thus, for this problem, $NP = 3$, $NF = 12$, $NI = 9$.

(b) *The input matrices* $[A], [B], [S],$ *and* $\{P\}$. The $[A]$ matrix for this problem should have four additional columns than that of Example 19.3.1. From Fig. 19.4.3d, the P forces which F_9 through F_{12} can counteract are

$$P_1 = +F_9 - F_{10} \qquad P_2 = +F_{11} - F_{12} \qquad P_3 = 0$$

which when assembled in the $[A]$ matrix become

$[A]_{3 \times 12} =$

F P	9	10	11	12
1	+1.0	−1.0		
2			+1.0	−1.0
3				

The effect of X_1 would be a positive e_9 and a negative e_{10}; the effect of X_2, a positive e_{11} and a negative e_{12}. Thus the additional four rows in the $[B]$ matrix for this problem are

$[B]_{12 \times 3} =$

X e	1	2	3
9	+1.0		
10	−1.0		
11		+1.0	
12		−1.0	

The $[S]$ matrix will now be 12×12, with an additional 4×4 diagonal matrix having GJ/L as its element, or

$[S]_{12 \times 12} =$

e F	9	10	11	12
9	64			
10		48		
11			25	
12				40

The $\{P\}$ matrix is the same as that in Example 19.3.1:

$$[P]_{3\times 1} = \begin{array}{|c|c|} \hline \text{LC} \diagdown P & 1 \\ \hline 1 & 0. \\ \hline 2 & +4.32 \\ \hline 3 & +1.944 \\ \hline \end{array}$$

(c) *The output matrices $\{X\}$ and $\{F^*\}$.* The output matrices $\{X\}$ and $\{F^*\}$ are

$$\{X\}_{3\times 1} = \begin{array}{|c|c|} \hline \text{LC} \diagdown X & 1 \\ \hline 1 & +0.0035903 \text{ rad} \\ \hline 2 & +0.0088344 \text{ rad} \\ \hline 3 & +0.042912 \text{ rad} \\ \hline \end{array}$$

$$\{F^*\}_{12\times 1} = \begin{array}{|c|c|} \hline \text{LC} \diagdown F^* & 1 \\ \hline 1 & -1.33284 \\ \hline 2 & +0.08066 \\ \hline 3 & -0.65493 \\ \hline 4 & +5.48494 \\ \hline 5 & -1.23013 \\ \hline 6 & -1.45453 \\ \hline 7 & +1.85664 \\ \hline 8 & +2.21567 \\ \hline 9 & +0.22978 \\ \hline 10 & -0.17233 \\ \hline 11 & +0.22086 \\ \hline 12 & -0.35338 \\ \hline \end{array}$$

(d) *Statics and deformation checks.* For this problem, there would be three statics checks and 12 deformation checks. Although not shown, these checks have been made.

714 INTERMEDIATE STRUCTURAL ANALYSIS

(e) *Discussion.* The pipe sections have relatively much more torsional resistance than other open sections since the GJ values are 80 percent of the EI values when Poisson's ratio μ is 0.25, and $J = 2I$. Even so, the deflection at point E only decreases from 0.044433 to 0.042912 m when torsional resistance is considered. The bending moments are generally on the same order of magnitude, and the torsional moments are fairly small. This is the reason why many analysts ignore torsion in the analysis of monolithically built horizontal grid frames so far as bending is concerned and then provide nominal torsional strength in addition. This practice, common in reinforced concrete design, must of course be tempered with judgment; in cases where there are indeed externally applied torsional moments, such as a concentrated load applied on a bracket on one side of a beam, the full analysis using combined beam and torsion elements should be executed.

Example 19.4.2 By the displacement method, analyze the closed grid frame shown in Fig. 19.4.4a, which is supported on three columns and subjected to a uniform load on top of 0.6 kN/m.

SOLUTION (a) *P-X and F-e numbers.* There are four elements and four joints. The P-X and F-e numbers are assigned as shown in Fig. 19.4.4b and c, wherein $NP = 9$, $NF = 12$, and $NI = 3$.

(b) *The input matrix [A].* It has been emphasized that each row in the $[A]$ matrix is an equation of statics, which is established from observing the free-body diagram of the joint, in order to convey the concept as simply as possible. However, the simplest way to get the correct $[A]$ matrix is to establish it by columns. Consider the independent F_7 system in Fig. 19.4.4d. The reverses of these three generalized forces will act on the joints at the ends of this element; these reverses are capable of balancing some P forces. Consequently the entries in the seventh column of the $[A]$ matrix will simply be the *effective* components of the three generalized forces in the F_7 system of Fig. 19.4.4d in the directions of the degree of freedom. The components of F_7 itself are $+0.8F_7$ along P_5, $+0.6F_7$ along P_6; and the component of $F_7/10$ down along P_9 is $\frac{1}{10}F_7$. Thus the nonzero elements in the seventh column of $[A]$ are $A_{57} = +0.8$, $A_{67} = +0.6$, and $A_{97} = +\frac{1}{10}$. The components of the two F_{12}'s in Fig. 19.4.4d along the degrees of freedom are $+0.6F_{12}$ along P_3, $-0.8F_{12}$ along P_4, $-0.6F_{12}$ along P_5, and $+0.8F_{12}$ along P_6. Thus the nonzero elements in the twelfth column of $[A]$ are $A_{3\text{-}12} = +0.6$, $A_{4\text{-}12} = -0.8$, $A_{5\text{-}12} = -0.6$, and $A_{6\text{-}12} = +0.8$. Generally it is not even necessary to draw out diagrams like those shown in Fig. 19.4.4d; one can make up those pictures mentally and fill out the $[A]$ matrix by columns. The nonzero elements in the $[A]$ matrix are

$A_{21} = +1.$ $A_{42} = +1.$ $A_{83} = +1.$ $A_{93} = -\frac{1}{4}$

$A_{64} = +1.$ $A_{94} = -\frac{1}{4}$ $A_{15} = -1.$ $A_{76} = -1.$

$A_{57} = +0.8$ $A_{67} = +0.6$ $A_{97} = +\frac{1}{10}$ $A_{38} = +0.8$

$A_{48} = +0.6$ $A_{98} = +\frac{1}{10}$ $A_{19} = -1.$ $A_{39} = +1.$

$A_{5\text{-}10} = +1.$ $A_{7\text{-}10} = -1.$ $A_{2\text{-}11} = -1.$ $A_{8\text{-}11} = +1.$

$A_{3\text{-}12} = +0.6$ $A_{4\text{-}12} = -0.8$ $A_{5\text{-}12} = -0.6$ $A_{6\text{-}12} = +0.8$

(c) *The input matrix [B].* The input matrix $[B]$ can also be filled out by columns. If the X's are rotations in double arrowheads, it is imply necessary to list the components of each X along the e directions, which are also in double arrowheads. For instance, the components of X_4 are $+1.0X_4$ along e_2, $+0.6X_4$ along e_8, and $-0.8X_4$ along e_{12}. Thus the nonzero elements in the fourth column of $[B]$ are $B_{24} = +1.0$, $B_{84} = +0.6$, and $B_{12\text{-}4} = -0.8$. So far as the effects of X_9 are concerned, it tends to rotate the element 3-4 axis in the clockwise direction, thus reducing the clockwise bending rotations e_3 and e_4 by $X_9/4$; it also tends to rotate the element 7-8 axis in the counterclockwise direction (viewed toward northeast), thus increasing the clockwise bending rotations (also viewed toward northeast) e_7 and e_8 by $X_9/10$. Of course, a check should always be made to make sure that $[B] = [A^T]$.

DISPLACEMENT METHOD OF HORIZONTAL GRID–FRAME ANALYSIS 715

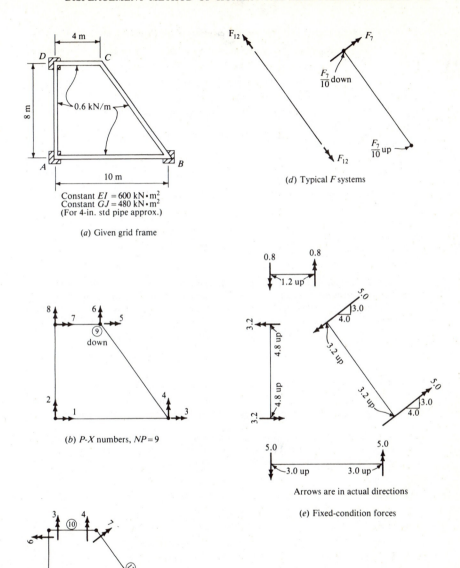

Figure 19.4.4 Displacement method, closed grid frame, Example 19.4.2.

(d) *The input matrix* [S]. The nonzero elements in the [S] matrix are:

$S_{11} = 240$ $S_{12} = 120$ $S_{21} = 120$ $S_{22} = 240$
$S_{33} = 600$ $S_{34} = 300$ $S_{43} = 300$ $S_{44} = 600$
$S_{55} = 300$ $S_{56} = 150$ $S_{65} = 150$ $S_{66} = 300$
$S_{77} = 240$ $S_{78} = 120$ $S_{87} = 120$ $S_{88} = 240$
$S_{99} = 48$ $S_{10\text{-}10} = 120$ $S_{11\text{-}11} = 60$ $S_{12\text{-}12} = 48$

(e) *The input matrix* {P}. The fixed-condition forces are shown in Fig. 19.4.4e. The elements in the {F_0} and {P} matrices are

$F_{01} = -5$ $F_{02} = +5$ $F_{03} = -0.8$ $F_{04} = +0.8$
$F_{05} = -3.2$ $F_{06} = +3.2$ $F_{07} = -5.0$ $F_{08} = +5.0$
$F_{09} = 0.$ $F_{0\text{-}10} = 0.$ $F_{0\text{-}11} = 0.$ $F_{0\text{-}12} = 0.$

and

$P_1 = -3.2$ $P_2 = +5.0$ $P_3 = -4.0$ $P_4 = -8.0$
$P_5 = +4.0$ $P_6 = +2.2$ $P_7 = +3.2$ $P_8 = +0.8$
$P_9 = +4.2$

(f) *The output matrices* {X} *and* {F*}. The output matrices are obtained from the computer program mentioned in Sec. 19.2. The purpose of this example is to show how to make the input matrices and how to check the output matrices. The elements in the {X} and {F*} matrices are listed below and are shown in their actual directions in Fig. 19.4.5a and b.

$X_1 = -0.026466$ $X_2 = +0.058875$ $X_3 = -0.102066$ $X_4 = -0.041989$
$X_5 = -0.033826$ $X_6 = +0.107301$ $X_7 = +0.007407$ $X_8 = +0.127064$
$X_9 = +0.496535$

$F_1^* = +4.0913$ $F_2^* = +1.9877$ $F_3^* = -4.0914$ $F_4^* = -8.4202$
$F_5^* = +3.6288$ $F_6^* = +4.9479$ $F_7^* = +9.0105$ $F_8^* = +1.7105$
$F_9^* = -3.6288$ $F_{10}^* = -4.9479$ $F_{11}^* = +4.0913$ $F_{12}^* = +3.7674$

(g) *Statics checks.* In this problem, there should be nine statics checks along the nine degrees of freedom. The free-body diagrams for the elements with regard to bending are shown in Fig. 19.4.5c. The nine checks are as follows:

(1) Along P_1, $+3.6288 - 3.6288 = 0$
(2) Along P_2, $+4.0913 - 4.0913 = 0$
(3) Along P_3, $+1.7105(0.8) + 3.7674(0.6) - 3.6288 = 0$
(4) Along P_4, $+1.7105(0.6) + 1.9877 - 3.7674(0.8) = 0$
(5) Along P_5, $+9.0105(0.8) - 3.7674(0.6) - 4.9479 = 0$
(6) Along P_6, $+9.0105(0.6) + 3.7674(0.8) - 8.4202 = 0$
(7) Along P_7, $+4.9479 - 4.9479 = 0$
(8) Along P_8, $+4.0913 - 4.0914 \approx 0$
(9) Along P_9, $-1.9279 + 1.9279 = 0$

From Fig. 19.4.5c, the reactions at A, B, and D are

$R_A = 1.3279 + 2.3921 = 3.72$ kN
$R_B = 3.6079 + 4.0721 = 7.68$ kN
$R_D = 4.3279 + 3.4721 = 7.80$ kN

DISPLACEMENT METHOD OF HORIZONTAL GRID–FRAME ANALYSIS

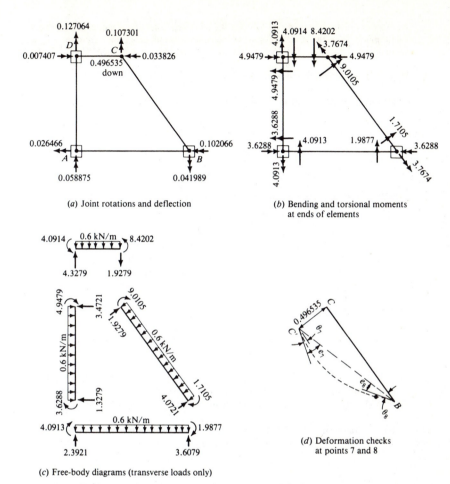

Figure 19.4.5 Statics and deformation of closed grid frame in Example 19.4.2.

These three reactions are last obtained in the displacement method as shown here. In fact, for this problem, they can be determined at the outset by taking the entire grid frame as a free body and considering the equilibrium of a noncoplanar-force system. So the indeterminacy is all inside the closed frame.

(h) *Deformation checks.* There should be three deformation checks per element, or a total of twelve deformation checks. The three deformation checks for element 7-8 when viewed toward the northeast will be shown. For torsion, the angle of twist at point 7 with vector outward from the element plus that at point 8 also with vector outward from the element must be equal to the total twist in the length 7-8; thus from Fig. 19.4.5a and b,

$$[0.107301(0.8) + 0.033826(0.6)] + [0.041989(0.8) - 0.102066(0.6)] \stackrel{?}{=} \frac{3.7674(10)}{480}$$

$$0.078488 = 0.078488 \quad \text{(OK)}$$

The other two deformation checks are that the clockwise rotations e_7 and e_8 in Fig. 19.4.5d as determined from the X values must be equal to those as determined from the simple-beam end rotations due to the applied uniform load and the end moments F_7^* and F_8^*; thus

(X vector toward northeast at point 7) $- R_{78} \stackrel{?}{=} e_{07} + \dfrac{F_7^* L}{3EI} - \dfrac{F_8^* L}{6EI}$

$[0.107301(0.6) - 0.033826(0.8)] - \left(-\dfrac{0.496535}{10}\right)$

$\stackrel{?}{=} + \dfrac{0.6(10)^3}{24(600)} + \dfrac{(+9.0105)(10)}{3(600)} - \dfrac{(+1.7105)(10)}{6(600)}$

$0.086973 \approx 0.086974$ (OK)

and

(X vector toward northeast at point 8) $- R_{78} \stackrel{?}{=} e_{08} - \dfrac{F_7^* L}{6EI} + \dfrac{F_8^* L}{3EI}$

$[-0.102066(0.8) - 0.041989(0.6)] - \left(-\dfrac{0.496535}{10}\right)$

$\stackrel{?}{=} - \dfrac{0.6(10)^3}{24(600)} - \dfrac{(+9.0105)(10)}{6(600)} + \dfrac{(+1.7105)(10)}{3(600)}$

$-0.057193 = -0.057193$ (OK)

19.5 Exercises

19.1 In the grid frame shown in Fig. 19.5.1, the two north-south beams are directly under the two east-west beams. Using the interaction at one of the intersection points as redundant, analyze by the force method for equal uniform load per horizontal distance on all the beams. Assume that the eight supports offer vertical restraints only and all beams are of the same size.

19.2 Analyze the grid frame of Exercise 19.1 by the displacement method. Because of double symmetry, it is possible to consider only one-fourth of the entire structure, with five nodes and four elements. When a node is installed at the line of symmetry, the slope there, although not the deflection, is zero.

19.3 In the grid frame shown in Fig. 19.5.2, the three north-south beams are directly under the three east-west beams. Using four redundant interactions (one at center, one at the corner, and

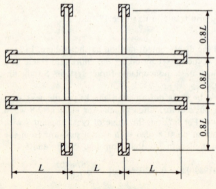

Constant EI throughout

Figure 19.5.1 Exercises 19.1 and 19.2.

DISPLACEMENT METHOD OF HORIZONTAL GRID–FRAME ANALYSIS 719

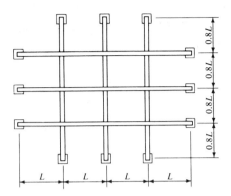

Constant EI throughout

Figure 19.5.2 Exercises 19.3 and 19.4.

one each at the middle of the side beams in the north-south and east-west directions), analyze by the force method for equal uniform load per unit distance on all the beams. Assume that the 12 supports offer vertical restraints only and all beams are of the same size.

19.4 Analayze the grid frame of Exercise 19.3 by the displacement method. Because of double symmetry, it is possible to consider only one-fourth of the entire structure, with eight nodes and eight elements. The beam on the line of symmetry would have one-half of the moment of inertia and be subjected to one-half of the uniform load. The corner node on the quarter frame (center of whole frame) would have zero rotation about both axes and zero deflection. The node on a line of symmetry would have zero rotation about an axis perpendicular to the direction of the beam.

19.5 The closed grid frame shown in Fig. 19.5.3 has constant flexural rigidity EI and torsional rigidity GJ, and is supported at the four corners with vertical restraint only. Using the bending moment at the end of one of the elements as the redundant, analyze by the force method for the same uniform load acting on all four elements.

19.6 Analyze the closed grid frame of Exercise 19.5 by the displacement method, using four nodes and four elements. (The degree of freedom being 8, it might be more instructive to verify the correctness of the input matrices by substituting the answers to Exercise 19.5 into the twofold equations of the displacement method.)

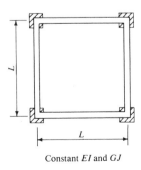

Constant EI and GJ

Figure 19.5.3 Exercises 19.5 and 19.6.

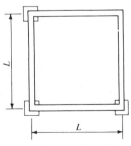

Constant EI and GJ

Figure 19.5.4 Exercises 19.7 and 19.8.

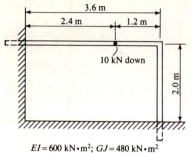

Figure 19.5.5 Exercises 19.9 and 19.10.   **Figure 19.5.6** Exercises 19.11 and 19.12.

19.7 The closed grid frame as shown in Fig. 19.5.4 has constant flexural rigidity EI and torsional rigidity GJ, and is supported at three corners with vertical restraint only. By cutting the frame at the unsupported corner and using the interacting biaxial moments and the vertical interaction as the three redundants, analyze by the force method for the same uniform load acting on all four elements.

19.8 Analyze the closed grid frame of Exercise 19.7 by the displacement method, using four nodes and four elements. (The degree of freedom being 9, it might be more instructive to verify the correctness of the input matrices by substituting the answers to Exercise 19.7 into the twofold equations of the displacement method.)

19.9 Analyze the L-shaped frame shown in Fig. 19.5.5 by the force method for the single concentrated load. The pipe is completely encased in the adjacent walls so that the basic determinate structure involves two cantilevers by cutting the pipe at the corner.

19.10 Analyze the L-shaped frame of Exercise 19.9 by the displacement method.

19.11 Analyze the bent-angle-shaped frame shown in Fig. 19.5.6 by the force method for the single concentrated load. The pipe is completely encased in the wall so that the basic determinate structure involves two cantilevers by cutting the pipe at the corner.

19.12 Analyze the bent-angle-shaped frame of Exercise 19.11 by the displacement method.

CHAPTER
TWENTY
RIGID FRAMES WITH SEMIRIGID CONNECTIONS

20.1 General Description

In building frames of steel, beam-to-column connections may be designed as being rigid, simple, or semirigid. A rigid connection holds unchanged the original angles between intersecting members; a simple connection allows the beam end to rotate freely under gravity load; a semirigid connection possesses a moment capacity intermediate between the simple and the rigid.

One typical model of a beam with semirigid end connections is shown in Fig. 20.1.1, where the ends are bolted to two plates which are fixed in translation and in rotation. Assume that the bolts do not fit tightly in the holes; consequently a moment M_i or M_j acting on the connection may cause a loss of friction and thus a rotational slip in the direction of the beam end equal to ψ_i or ψ_j. One can expect that, for the same semirigid connection, a larger acting moment M_i should cause a larger slip angle ψ_i. Although the absolute relationship can be obtained only from laboratory tests, a simplified linear variation may be expressed in the form

$$M_i = R_i \frac{EI}{L} \psi_i \qquad (20.1.1a)$$

$$M_j = R_j \frac{EI}{L} \psi_j \qquad (20.1.1b)$$

Using the length L and the flexural rigidity EI of the member itself in Eqs. (20.1.1a and b) is only a way of obtaining a nondimensional quantity R_i or R_j, which is a measure of the degree of rigidity of the connection; hence R_i or R_j will be called the *rigidity index*. For a simple connection, R_i or R_j is zero; for a rigid connection, $R_i \propto R_j$ is infinity.

721

722 INTERMEDIATE STRUCTURAL ANALYSIS

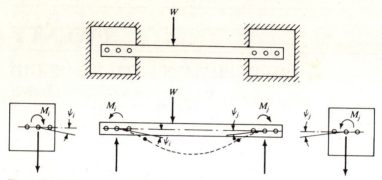

Figure 20.1.1 A beam span with semirigid end connections.

This chapter deals with the mathematical equations and procedures by which a rigid frame containing semirigid member-end connections may be analyzed.

20.2 The Modified Member Flexibility Matrix

The member AB shown in Fig. 20.2.1 is connected to a plate at A with a rigidity index R_i and to a plate at B with rigidity index R_j. Let the clockwise moments acting on the member ends be called M_i and M_j. Let the plates at A and B rotate through clockwise angles θ_A and θ_B and the relative transverse displacements of the two plates be such that the member axis rotates from

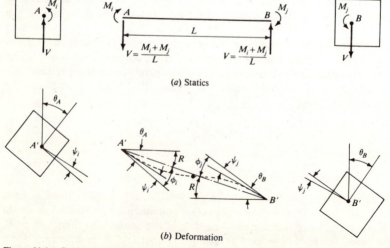

(*a*) Statics

(*b*) Deformation

Figure 20.2.1 Statics and deformation of a member with semirigid end connections.

AB to $A'B'$ by a clockwise angle R. Let the clockwise angles measured from the new member axis $A'B'$ to the new directions of the plates at A and B be called ϕ_i and ϕ_j; then

$$\phi_i = \theta_A - R \tag{20.2.1a}$$

$$\phi_j = \theta_B - R \tag{20.2.1b}$$

The slip angles ψ_i and ψ_j, as expressed by Eqs. (20.1.1a and b), are the counterclockwise angles measured from the new directions of the plates at A and B to the directions of the elastic curve at A and B, respectively. Note that ψ_i and ψ_j should be measured counterclockwise from the plate directions because the moments M_i and M_j act counterclockwise on the plates.

The modified member flexibility matrix $[D]$ for a member with semirigid connections expresses the member-end rotations ϕ_i and ϕ_j, as they are defined in Fig. 20.2.1, in terms of the member-end moments M_i and M_j; then

$$\begin{Bmatrix} \phi_i \\ \phi_j \end{Bmatrix} = [D] \begin{Bmatrix} M_i \\ M_j \end{Bmatrix} \tag{20.2.2}$$

Expressions for the elements in the $[D]$ matrix will now be derived.

By the conjugate-beam theorem, which states that the angles between the new member axis and the directions of the elastic curve at the beam ends are equal to the reactions on a simple beam loaded by the M/EI diagram,

$$\phi_i - \psi_i = +\frac{M_i L}{3EI} - \frac{M_j L}{6EI} \tag{20.2.3a}$$

$$\phi_j - \psi_j = -\frac{M_i L}{6EI} + \frac{M_j L}{3EI} \tag{20.2.3b}$$

Substituting the expressions for ψ_i and ψ_j in Eqs. (20.1.1a and b) into Eqs. (20.2.3a and b),

$$\phi_i - \frac{M_i L}{R_i EI} = +\frac{M_i L}{3EI} - \frac{M_j L}{6EI}$$

$$\phi_j - \frac{M_j L}{R_j EI} = -\frac{M_i L}{6EI} + \frac{M_j L}{3EI}$$

Rearranging,

$$\phi_i = \left(\frac{1}{3} + \frac{1}{R_i}\right)\left(\frac{M_i L}{EI}\right) + \left(-\frac{1}{6}\right)\left(\frac{M_j L}{6EI}\right) \tag{20.2.4a}$$

$$\phi_j = \left(-\frac{1}{6}\right)\left(\frac{M_i L}{EI}\right) + \left(\frac{1}{3} + \frac{1}{R_j}\right)\left(\frac{M_j L}{EI}\right) \tag{20.2.4b}$$

From Eqs. (20.2.4a and b), the modified member flexibility matrix $[D]$ as defined by Eq. (20.2.2) becomes

$$[D] = \frac{L}{EI}\begin{bmatrix} f_{ii} & f_{ij} \\ f_{ji} & f_{jj} \end{bmatrix} = \frac{L}{EI}\begin{bmatrix} +\frac{1}{3p_i} & -\frac{1}{6} \\ -\frac{1}{6} & +\frac{1}{3p_j} \end{bmatrix} \quad (20.2.5a)$$

in which

$$p_i = \frac{1}{1 + 3/R_i} \qquad p_j = \frac{1}{1 + 3/R_j} \quad (20.2.5b)$$

and, conversely,

$$R_i = \frac{3p_i}{1 - p_i} \qquad R_j = \frac{3p_j}{1 - p_j} \quad (20.2.5c)$$

p_i and p_j may be called the *fixity factors*. For a hinged connection, both the rigidity index and the fixity factor are zero; but for a rigid (e.g., welded) connection, the rigidity index is infinity and the fixity factor is 100 percent. Since the fixity factor varies from 0 to 100 percent, it is more convenient for a designer to use.

20.3 The Modified Member Stiffness Matrix

The modified member stiffness matrix $[S]$ for a member with semirigid end connections expresses the member-end moments M_i and M_j in terms of the member-end rotations ϕ_i and ϕ_j. Note that ϕ_i and ϕ_j, as shown in Fig. 20.2.1, are the clockwise angles measured from the new member axis $A'B'$ to the new plate directions, not to the directions of the elastic curve. Symbolically,

$$\begin{Bmatrix} M_i \\ M_j \end{Bmatrix} = [S]\begin{Bmatrix} \phi_i \\ \phi_j \end{Bmatrix} = \frac{EI}{L}\begin{bmatrix} s_{ii} & s_{ij} \\ s_{ji} & s_{jj} \end{bmatrix}\begin{Bmatrix} \phi_i \\ \phi_j \end{Bmatrix} \quad (20.3.1a)$$

Expressions for s_{ii}, $s_{ij} = s_{ji}$, and s_{jj} may be obtained by inverting the $[D]$ matrix of Eq. (20.2.5a); thus

$$s_{ii} = \frac{f_{jj}}{f_{ii}f_{jj} - f_{ij}^2} = \frac{1/(3p_j)}{1/(9p_ip_j) - 1/36} = \frac{12/p_j}{4/(p_ip_j) - 1} \quad (20.3.1b)$$

$$s_{ij} = s_{ji} = \frac{-f_{ij}}{f_{ii}f_{jj} - f_{ij}^2} = \frac{1/6}{1/(9p_ip_j) - 1/36} = \frac{6}{4/(p_ip_j) - 1} \quad (20.3.1c)$$

$$s_{jj} = \frac{f_{ii}}{f_{ii}f_{jj} - f_{ij}^2} = \frac{1/(3p_i)}{1/(9p_ip_j) - 1/36} = \frac{12/p_i}{4/(p_ip_j) - 1} \quad (20.3.1d)$$

The modified member stiffness matrix $[S]$, as expressed by Eqs. (20.3.1a to d) will be needed in the displacement method of analyzing rigid frames in which there are semirigid member-end connections.

20.4 The Modified Fixed-End Moments

The external stiffness matrix $[ASA^T]$ of a rigid frame containing semirigid connections may be obtained in the usual manner; only the formulas for the elements of the modified member stiffness matrix as expressed by Eqs. (20.3.1a to d) are to be used for those members with semirigid end connections. There remains the problem of deriving the formulas for the modified fixed-end moments M_{0i}^{SR} and M_{0j}^{SR}, as shown by Fig. 20.4.1b, in terms of the usual fixed-end moments M_{0i} and M_{0j}, as shown by Fig. 20.4.1a. The transformation may be made in terms of either the fixity factors p_i and p_j or the

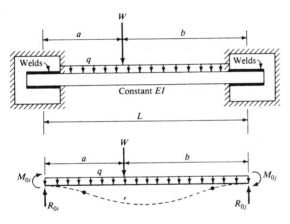

(a) Fixed-end moments M_{0i} and M_{0j} as commonly used

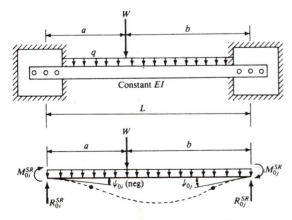

(b) Modified fixed-end moments M_{0i}^{SR} and M_{0j}^{SR}

Figure 20.4.1 Free-body diagram of a member connected to two fixed plates at its ends.

rigidity indexes R_i and R_j since they are mutually expressible by Eqs. (20.2.5b and c).

Typical expressions for the usual fixed-end moments M_{0i} and M_{0j} for the member in Fig. 20.4.1a are

$$M_{0i} = -\frac{1}{12}qL^2 - \frac{Wab^2}{L^2} \qquad (20.4.1a)$$

$$M_{0j} = +\frac{1}{12}qL^2 + \frac{Wba^2}{L^2} \qquad (20.4.1b)$$

If this member were hinged at both ends, the clockwise rotations ϕ_{0i} and ϕ_{0j} of its ends would be

$$\phi_{0i} = -\frac{L}{3EI}M_{0i} + \frac{L}{6EI}M_{0j} \qquad (20.4.2a)$$

$$\phi_{0j} = +\frac{L}{6EI}M_{0i} - \frac{L}{3EI}M_{0j} \qquad (20.4.2b)$$

Since, by definition, M_{0i} and M_{0j} have the capacity to force ϕ_{0i} and ϕ_{0j} back to zero, the reverses of M_{0i} and M_{0j} would result in the rotations ϕ_{0i} and ϕ_{0j}.

The modified fixed-end moments, M_{0i}^{SR} and M_{0j}^{SR} in Fig. 20.4.1a, should be such that they can nullify the hinged-end rotations ϕ_{0i} and ϕ_{0j} back to zero; thus,

$$\begin{Bmatrix} M_{0i}^{SR} \\ M_{0j}^{SR} \end{Bmatrix} = [S \text{ (modified)}] \begin{Bmatrix} -\phi_{0i} \\ -\phi_{0j} \end{Bmatrix} \qquad (20.4.3)$$

Substituting Eqs. (20.4.2a and b) into Eq. (20.4.3),

$$\begin{Bmatrix} M_{0i}^{SR} \\ M_{0j}^{SR} \end{Bmatrix} = \begin{bmatrix} s_{ii} & s_{ij} \\ s_{ji} & s_{jj} \end{bmatrix} \begin{Bmatrix} +\frac{1}{3}M_{0i} - \frac{1}{6}M_{0j} \\ -\frac{1}{6}M_{0i} + \frac{1}{3}M_{0j} \end{Bmatrix}$$

from which

$$M_{0i}^{SR} = (+\tfrac{1}{3}s_{ii} - \tfrac{1}{6}s_{ij})M_{0i} + (-\tfrac{1}{6}s_{ii} + \tfrac{1}{3}s_{ij})M_{0j} \qquad (20.4.4a)$$

$$M_{0j}^{SR} = (-\tfrac{1}{6}s_{jj} + \tfrac{1}{3}s_{ji})M_{0i} + (+\tfrac{1}{3}s_{jj} - \tfrac{1}{6}s_{ji})M_{0j} \qquad (20.4.4b)$$

Example 20.4.1 Determine the modified fixed-end moments M_{0i}^{SR} and M_{0j}^{SR} for the member with semirigid end connections as shown in Fig. 20.4.2, using the fixity factors $p_i = 0.40$ and $p_j = 0.50$. Obtain the slip angles ψ_i and ψ_j from the linear relationship between the rigidity index and the end moment. Check these values against the end slopes of the elastic curve as computed from the moment diagram by the conjugate-beam method.

SOLUTION (a) Compute M_{0i} and M_{0j} if $p_i = p_j = 1.0$.

$$M_{0i} = -\frac{12(12)(6)^2}{18^2} = -16.0 \text{ kN·m}$$

$$M_{0j} = +\frac{12(6)(12)^2}{18^2} = +32.0 \text{ kN·m}$$

RIGID FRAMES WITH SEMIRIGID CONNECTIONS 727

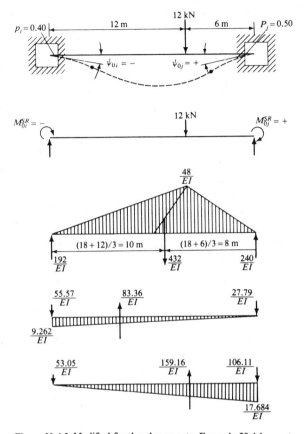

Figure 20.4.2 Modified fixed-end moments, Example 20.4.1.

(b) Compute s_{ii}, $s_{ij} = s_{ji}$, and s_{jj}.

$$s_{ii} = \frac{12/0.50}{4/[(0.40)(0.50)] - 1} = \frac{24}{19} = 1.2632$$

$$s_{ij} = s_{ji} = \frac{6}{4/[(0.40)(0.50)] - 1} = \frac{6}{19} = 0.3158$$

$$s_{jj} = \frac{12/0.40}{4/[(0.40)(0.50)] - 1} = \frac{30}{19} = 1.5789$$

(c) Compute M_{0i}^{SR} and M_{0j}^{SR}:

$$M_{0i}^{SR} = (+\tfrac{1}{3}s_{ii} - \tfrac{1}{6}s_{ij})M_{0i} + (-\tfrac{1}{6}s_{ii} + \tfrac{1}{3}s_{ij})M_{0j}$$

$$= \left(+\frac{1.2632}{3} - \frac{0.3158}{6}\right)(-16.0) + \left(-\frac{1.2632}{6} + \frac{0.3158}{3}\right)(+32.0)$$

$$= -9.262 \text{ kN} \cdot \dot{\text{m}}$$

$$M_{0j}^{SR} = (-\tfrac{1}{6}s_{ij} + \tfrac{1}{3}s_{ji})M_{0i} + (\tfrac{1}{3}s_{jj} - \tfrac{1}{6}s_{ji})M_{0j}$$

$$= \left(-\frac{1.5789}{6} + \frac{0.3158}{3}\right)(-16.0) + \left(+\frac{1.5789}{3} - \frac{0.3158}{6}\right)(+32.0)$$

$$= +17.684 \text{ kN·m}$$

Note that, because of the semirigid end connections, the absolute magnitudes of the fixed-end moments come down from -16.0 and $+32.0$ to -9.262 and $+17.684$, respectively.

(d) Compute ψ_i and ψ_j.

$$R_i = \frac{3p_i}{1 - p_i} = \frac{3(0.40)}{1 - 0.40} = 2.0$$

$$\psi_i = \frac{M_i}{R_i EI/L} = -\frac{9.262}{2.0 EI/18} = -\frac{83.36}{EI} \text{ kN·m}^2$$

$$R_j = \frac{3p_j}{1 - p_j} = \frac{3(0.50)}{1 - 0.50} = 3.0$$

$$\psi_j = \frac{M_j}{R_j EI/L} = +\frac{17.684}{3.0 EI/18} = +\frac{106.10}{EI} \text{ kN·m}^2$$

Note that the sign convention for ψ_i and ψ_j is that they are positive if the sense of rotation from the plate direction (horizontal in this case) to the elastic curve is *counterclockwise*.

(e) Compute the slopes θ_i and θ_j. Applying the conjugate-beam method to the three M/EI diagrams shown in Fig. 20.4.2,

$$\theta_A = +\frac{192}{EI} - \frac{55.57}{EI} - \frac{53.05}{EI} = +\frac{83.38}{EI} \text{ kN·m}^2 \quad \text{(Check)}$$

$$\theta_B = -\frac{240}{EI} + \frac{27.79}{EI} + \frac{106.11}{EI} = -\frac{106.10}{EI} \text{ kN·m}^2 \quad \text{(Check)}$$

Note that the sign convention used here for θ_A and θ_B is that they are positive if the sense of rotation from the horizontal direction to the elastic curve is *clockwise*.

20.5 The Displacement Method

In the displacement-method analysis of rigid frames with semirigid joints, the usual equations $\{X\} = [ASA^T]^{-1}\{P\}$ and $\{F^*\} = \{F_0\} + [SA^T]\{X\}$ are to be used; however, for those members with semirigid end connections, the modified fixed-end moments F_{0i}^{SR} and F_{0j}^{SR} as given by Eqs. (20.4.4a and b) (note that F and M are synonymous), and the modified member stiffness matrix $[S]$ as given by Eqs. (20.3.1a to d), should be used.

In hand computations, after the modified fixed-end moments have been computed, it is desirable to check their correctness by computing the slip angles ψ_i and ψ_j from the linear relationship between the rigidity index and the end moment. Then these slip angles may be checked with the end slopes of the elastic curve as obtained by applying the conjugate-beam theorem to the moment diagram.

For the final checks after the $\{X\}$ and $\{F^*\}$ values have been obtained, the usual statics checks equal in number to the degree of freedom should be made first.

In regard to the deformation checks, there ought to be two checks per member. In ordinary cases of rigid end connections, these checks are rather redundant, since they are solely dependent on whether the usual fixed-end moments and the usual member stiffness matrix have been entered correctly in the computer input. However, for members with semirigid end connections, the deformation checks are more significant because mistakes can be made in the process (of course, this process can itself be built into the computer program) of hand-computing the modified fixed-end moments and the elements of the modified member stiffness matrix.

The two deformation checks for each member may be made by comparing the angles ϕ_{mi} and ϕ_{mj} as computed from Eqs. (20.5.1a) and (20.5.2a), with these same angles as computed from Eqs. (20.5.1b) and (20.5.2b); these equations are

$$\phi_{mi} = \theta_i - R - \psi_i = \theta_i - R - \frac{M_i^*}{R_i EI/L} \quad (20.5.1a)$$

$$\phi_{mi} = \phi_{0i} + \frac{M_i^* L}{3EI} - \frac{M_j^* L}{6EI} \quad (20.5.1b)$$

$$\phi_{mj} = \theta_j - R - \psi_j = \theta_j - R - \frac{M_j^*}{R_j EI/L} \quad (20.5.2a)$$

$$\phi_{mj} = \phi_{0j} - \frac{M_i^* L}{6EI} + \frac{M_j^* L}{3EI} \quad (20.5.2b)$$

As shown by Fig. 20.5.1, ϕ_{mi} and ϕ_{mj} are the clockwise rotations from the new member axis to the tangents to the elastic curve at the ith and jth ends, respectively. θ_i and θ_j are the clockwise plate rotations, and, as such, they are taken from the output $\{X\}$ values. R is the clockwise rotation from the original member axis to the new member axis; this, too, can be obtained from the sidesway values in the $\{X\}$ matrix. The slip angles ψ_i and ψ_j are the

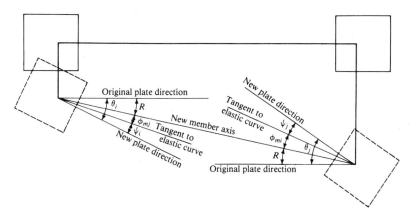

Figure 20.5.1 Geometry of deformation checks.

730 INTERMEDIATE STRUCTURAL ANALYSIS

counterclockwise rotations from the new plate directions to the tangents to the elastic curve; they can be computed from the final values of the end moments and the rigidity indexes. Equations (20.5.1a) and (20.5.2a) can be verified by physical inspection of the geometry shown in Fig. 20.5.1.

So far as Eqs. (20.5.1b) and (20.5.2b) are concerned, they are simple applications of the conjugate-beam method, wherein ϕ_{0i} and ϕ_{0j} are the clockwise end rotations due to the applied loads on the member had it been hinged at both ends. The remaining terms in these equations are simply the clockwise end rotations due to the final end moments M_i^* and M_j^*, which are elements in the output $\{F^*\}$ matrix.

Example 20.5.1 By the displacement method make a complete analysis of the rigid frame shown in Fig. 20.5.2a, wherein the fixity factors at member-end connections numbers 1 to 6 are as shown in Fig. 20.5.2c. Draw the final shear and moment diagrams. Also show the slip angles at all semirigid end connections on the final elastic curve.

SOLUTION (a) *Member 1-2.* In reference to parts (b) and (c) of the solution for Example 20.4.1, the modified stiffness coefficients for this member are $s_{11} = 1.2632$, $s_{12} = s_{21} = 0.3158$, and $s_{22} = 1.5789$; the modified fixed-end moments are

$$F_{01}^{SR} = -9.262 \text{ kN} \cdot \text{m} \qquad F_{02}^{SR} = +17.684 \text{ kN} \cdot \text{m}$$

The member stiffness matrix becomes

$$[S] = \begin{bmatrix} 1.2632 & 0.3158 \\ 0.3158 & 1.5789 \end{bmatrix} \frac{E(3I_c)}{18} = EI_c * \begin{bmatrix} 0.2105 & 0.0526 \\ 0.0526 & 0.2632 \end{bmatrix}$$

Also, the rigidity indexes are $R_1 = 2.0$ and $R_2 = 3.0$.

(b) *Member 3-4.* $p_3 = 0.60$; $p_4 = 0.40$.

$$s_{33} = \frac{12/0.40}{4/[(0.60)(0.40)] - 1} = \frac{30}{47/3} = 1.9149$$

$$s_{34} = s_{43} = \frac{6}{47/3} = 0.3830$$

$$s_{44} = \frac{12/0.60}{47/3} = \frac{20}{47/3} = 1.2766$$

$$[S] = \begin{bmatrix} 1.9149 & 0.3830 \\ 0.3830 & 1.2766 \end{bmatrix} \frac{E(5I_c)}{20} = EI_c * \begin{bmatrix} 0.4787 & 0.0958 \\ 0.0958 & 0.3192 \end{bmatrix}$$

$$F_{03} = -\frac{2.4(20)^2}{12} = -80 \text{ kN} \cdot \text{m} \qquad F_{04} = +80 \text{ kN} \cdot \text{m}$$

$$F_{03}^{SR} = \left(+\frac{1.9149}{3} - \frac{0.3830}{6}\right)(-80) + \left(-\frac{1.9149}{6} + \frac{0.3830}{3}\right)(+80)$$

$$= -45.960 - 15.320 = -61.280 \text{ kN} \cdot \text{m}$$

$$F_{04}^{SR} = \left(-\frac{1.2766}{6} + \frac{0.3830}{3}\right)(-80) + \left(+\frac{1.2766}{3} - \frac{0.3830}{6}\right)(+80)$$

$$= +6.808 + 28.936 = +35.744 \text{ kN} \cdot \text{m}$$

$$R_3 = \frac{3(0.60)}{1 - 0.60} = 4.5 \qquad R_4 = \frac{3(0.40)}{1 - 0.40} = \frac{1.2}{0.6} = 2$$

A check on the modified fixed-end moments may be made by comparing the slip angles (positive if counterclockwise from the fixed plate, or horizontal direction, to the elastic

RIGID FRAMES WITH SEMIRIGID CONNECTIONS   731

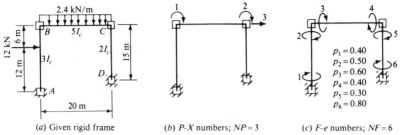

(a) Given rigid frame   (b) P-X numbers; NP = 3   (c) F-e numbers; NF = 6

Figure 20.5.2 A rigid frame with semirigid joints, Example 20.5.1.

curve at the ends) with the end slopes (positive if clockwise from the member axis to the elastic curve at the ends) obtained from the moment diagram.

$$\psi_3 = \frac{F_{03}^{SR}}{R_3 EI/20} = \frac{-61.280}{4.5 EI/20} = -\frac{272.36}{EI}$$

$$\psi_4 = \frac{F_{04}^{SR}}{R_4 EI/20} = \frac{+35.744}{2 EI/20} = +\frac{357.44}{EI}$$

$$\phi_{m3} = \phi_{03} + \frac{F_{03}^{SR} L}{3EI} - \frac{F_{04}^{SR} L}{6EI}$$

$$= +\frac{800}{EI} - \frac{408.53}{EI} - \frac{119.15}{EI} = +\frac{272.32}{EI} \quad \text{(Check)}$$

$$\phi_{m4} = \phi_{04} - \frac{F_{03}^{SR} L}{6EI} + \frac{F_{04}^{SR} L}{3EI}$$

$$= -\frac{800}{EI} + \frac{204.27}{EI} + \frac{238.29}{EI} = -\frac{357.44}{EI} \quad \text{(Check)}$$

(c) Member 5-6. $p_5 = 0.30$; $p_6 = 0.80$.

$$S_{55} = \frac{12/0.80}{4/[(0.30)(0.80)] - 1} = \frac{15}{47/3} = 0.9574$$

$$S_{56} = S_{65} = \frac{6}{47/3} = 0.3830$$

$$S_{66} = \frac{12/0.30}{47/3} = \frac{40}{47/3} = 2.5532$$

$$[S] = \begin{bmatrix} 0.9574 & 0.3830 \\ 0.3830 & 2.5532 \end{bmatrix} \frac{E(2I_c)}{15} = EI_c * \begin{bmatrix} 0.1276 & 0.0511 \\ 0.0511 & 0.3404 \end{bmatrix}$$

$$R_5 = \frac{3(0.30)}{1 - 0.30} = 1.2857 \qquad R_6 = \frac{3(0.80)}{1 - 0.80} = 12$$

(d) The input matrices [A], [B], [S], and {P}. The input matrices are as follows:

F\P	1	2	3	4	5	6
1	+1.	+1.				
2			+1.	+1.		
3	$-\frac{1}{18}$	$-\frac{1}{18}$			$-\frac{1}{15}$	$-\frac{1}{15}$

$[A]_{3 \times 6} = $

732 INTERMEDIATE STRUCTURAL ANALYSIS

$[B]_{6\times 3} =$

e \ X	1	2	3
1			$-\frac{1}{18}$
2	+1.		$-\frac{1}{18}$
3	+1.		
4		+1.	
5		+1.	$-\frac{1}{15}$
6			$-\frac{1}{15}$

$[S]_{6\times 6} =$

F \ e	1	2	3	4	5	6
1	0.2105	0.0526				
2	0.0526	0.2632				
3			0.4787	0.0958		
4			0.0958	0.3192		
5					0.1276	0.0511
6					0.0511	0.3404

$* EI_c$

$\{P\}_{3\times 1} =$

P \ LC	1
1	+43.596
2	−35.744
3	+ 8.4679

Note, once again, that the $[A]$ and $[B]$ matrices should be determined independently of each other and the transposition relationship $[B] = [A^T]$ verified. The $\{P\}$ matrix is obtained from

$$P_1 = -(F_{02}^{SR} + F_{03}^{SR}) = -(+17.684 - 61.280) = +43.596 \text{ kN·m}$$

$$P_2 = -(F_{04}^{SR} + F_{05}^{SR}) = -(+35.744 + 0) = -35.744 \text{ kN·m}$$

P_3 = (leftward reaction at point 2) + (leftward reaction at point 5)
$= 8 + (17.684 - 9.262)/18 = +8.4679$ kN

RIGID FRAMES WITH SEMIRIGID CONNECTIONS 733

(e) *The output matrices* $\{X\}$ *and* $\{F^*\}$. The output matrices $\{X\}$ and $\{F^*\}$ may be obtained by using a typical computer program, or by longhand operations $\{X\} = [ASA^T]^{-1}\{P\}$ and $\{F^*\} = \{F_0\} + [SA^T]\{X\}$. They are

$$X_1 = +119.35 \text{ kN·m}^2/EI_c$$
$$X_2 = -43.62 \text{ kN·m}^2/EI_c$$
$$X_3 = +2324.1 \text{ kN·m}^3/EI_c$$

and

$$F_1^* = -9.262 - 27.691 = -36.953 \text{ kN·m}$$
$$F_2^* = +17.684 - 9.360 = +8.324 \text{ kN·m}$$
$$F_3^* = -61.280 + 52.955 = -8.325 \text{ kN·m}$$
$$F_4^* = +35.744 - 2.490 = +33.254 \text{ kN·m}$$
$$F_5^* = +0. - 33.255 = -33.255 \text{ kN·m}$$
$$F_6^* = +0. - 62.888 = -62.888 \text{ kN·m}$$

(f) *Statics checks.* The $NP = 3$ statics checks may be made by observing the equilibrium along the three degrees of freedom; they are (1) $F_2^* + F_3^* = 0.0$, (2) $F_4^* + F_5^* = 0.0$, and (3) ΣF_x at joint $C = 0.0$. For the third check note in Fig. 20.5.3 that the 6.4095-kN end compression at point 4 as determined from member 1-2 and then member 3-4 is equal to the 6.4095-kN shear force at point 5 of member 5-6.

(g) *Deformation checks.* The deformation checks may be made by referring to Fig. 20.5.3 and using Eqs. (20.5.1a and b) and (20.5.2a and b). For member 1-2,

$$\phi_{m1} = \theta_1 - R_{12} - \psi_1 = 0. - \frac{2324.1}{18EI_c} - \frac{-36.953}{2.0E(3I_c)/18}$$

$$= -\frac{129.12}{EI_c} + \frac{110.86}{EI_c} = -\frac{18.26}{EI_c}$$

$$\phi_{m1} = \phi_{01} + \frac{F_1^* L}{3EI} - \frac{F_2^* L}{6EI} = +\frac{64}{EI_c} + \frac{(-36.953)(18)}{3E(3I_c)} - \frac{(+8.324)(18)}{6E(3I_c)}$$

$$= +\frac{64}{EI_c} - \frac{73.91}{EI_c} - \frac{8.32}{EI_c} = -\frac{18.23}{EI_c} \quad \text{(Check)}$$

$$\phi_{m2} = \theta_2 - R_{12} - \psi_2 = +\frac{119.35}{EI_c} - \frac{2324.1}{18EI_c} - \frac{+8.324}{3E(3I_c)/18}$$

$$= +\frac{119.35}{EI_c} - \frac{129.12}{18EI_c} - \frac{16.65}{EI_c} = -\frac{26.42}{EI_c}$$

$$\phi_{m2} = \phi_{02} - \frac{F_1^* L}{6EI} + \frac{F_2^* L}{3EI} = -\frac{80}{EI_c} - \frac{(-36.953)(18)}{6E(3I_c)} + \frac{(+8.324)(18)}{3E(3I_c)}$$

$$= -\frac{80}{EI_c} + \frac{36.95}{EI_c} + \frac{16.64}{EI_c} = -\frac{26.41}{EI_c} \quad \text{(Check)}$$

For member 3-4,

$$\phi_{m3} = \theta_3 - R_{34} - \psi_3 = +\frac{119.35}{EI_c} - 0. - \frac{-8.325}{4.5E(5I_c)/20}$$

$$= +\frac{119.35}{EI_c} + \frac{7.40}{EI_c} = +\frac{126.75}{EI_c}$$

$$\phi_{m3} = \phi_{03} + \frac{F_3^* L}{3EI} - \frac{F_4^* L}{6EI} = +\frac{160}{EI_c} + \frac{(-8.325)(20)}{3E(5I_c)} - \frac{(+33.254)(20)}{6E(5I_c)}$$

$$= +\frac{160}{EI_c} - \frac{11.10}{EI_c} - \frac{22.17}{EI_c} = +\frac{126.73}{EI_c} \quad \text{(Check)}$$

734 INTERMEDIATE STRUCTURAL ANALYSIS

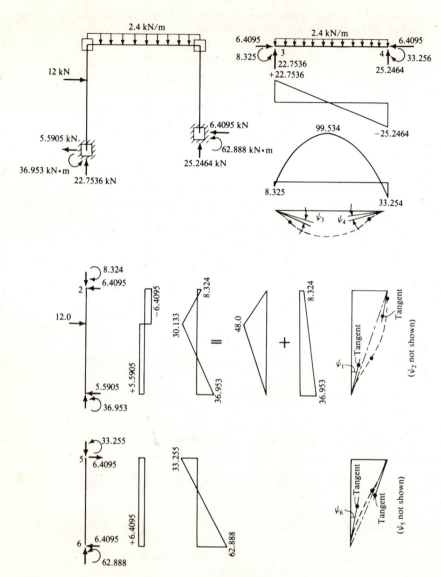

Figure 20.5.3 Statics and deformation checks, Example 20.5.1.

$$\phi_{m4} = \theta_4 - R_{34} - \psi_4 = -\frac{43.62}{EI_c} - 0 - \frac{+33.254}{2E(5I_c)/20}$$

$$= -\frac{43.62}{EI_c} - \frac{66.51}{EI_c} = -\frac{110.13}{EI_c}$$

$$\phi_{m4} = \phi_{04} - \frac{F_i^* L}{6EI} + \frac{F_j^* L}{3EI} = -\frac{160}{EI_c} - \frac{(-8.325)(20)}{6E(5I_c)} + \frac{(+33.254)(20)}{3E(5I_c)}$$

$$= -\frac{160}{EI_c} + \frac{5.55}{EI_c} + \frac{44.34}{EI_c} = -\frac{110.11}{EI_c} \quad \text{(Check)}$$

For member 5-6,

$$\phi_{m5} = \theta_5 - R_{56} - \psi_5 = -\frac{43.62}{EI_c} - \frac{2324.1}{EI_c(15)} - \frac{-33.255}{1.2857E(2I_c)/15}$$

$$= -\frac{43.62}{EI_c} - \frac{154.94}{EI_c} + \frac{193.99}{EI_c} = -\frac{4.57}{EI_c}$$

$$\phi_{m5} = \phi_{05} + \frac{F_i^* L}{3EI} - \frac{F_j^* L}{6EI} = 0 + \frac{(-33.255)(15)}{3E(2I_c)} - \frac{(-62.888)(15)}{6E(2I_c)}$$

$$= -\frac{83.14}{EI_c} + \frac{78.61}{EI_c} = -\frac{4.53}{EI_c} \quad \text{(Check)}$$

$$\phi_{m6} = \theta_6 - R_{56} - \psi_6 = 0 - \frac{2324.1}{EI_c(15)} - \frac{-62.888}{12E(2I_c)/15}$$

$$= -\frac{154.94}{EI_c} + \frac{39.30}{EI_c} = -\frac{115.64}{EI_c}$$

$$\phi_{m6} = \phi_{06} - \frac{F_i^* L}{6EI} + \frac{F_j^* L}{3EI} = 0 - \frac{(-33.255)(15)}{6E(2I_c)} + \frac{(-62.888)(15)}{3E(2I_c)}$$

$$= +\frac{41.57}{EI_c} - \frac{157.22}{EI_c} = -\frac{115.65}{EI_c} \quad \text{(Check)}$$

20.6 Treatment of Semirigid Connections as Rotational Springs

When a semirigid connection exhibits a linear relationship between the applied moment and the slip angle, as shown by Eqs. (20.1:1a and b) repeated here,

$$M_i = R_i \frac{EI}{L} \psi_i \qquad M_j = R_j \frac{EI}{L} \psi_j$$

its behavior becomes analogous to that of a rotational spring, on which the applied moment is equal to the product of its stiffness and the angle of rotation. It is conceivable, then, that for purpose of analysis, one such spring be considered to exist in place of each semirigid connection. It will be shown that the degree of freedom of the structure is increased by the number of rotational springs thus added, but this is not a great disadvantage if a computer is used. The advantage, of course, lies in the fact that no longer is it necessary to compute the modified fixed-end moments and the modified member stiffness matrices.

Shown in Fig. 20.6.1 is the behavior analogy between a linear spring, a

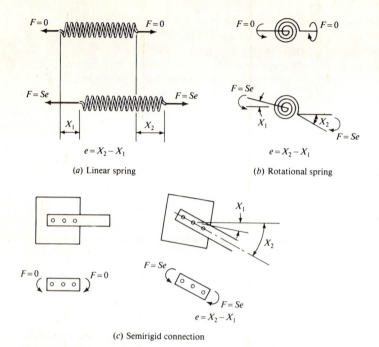

Figure 20.6.1 Behavior analogy between linear spring, rotational spring, and semirigid connection.

rotational spring, and a semirigid connection, on which the applied moment is a linear function of the slip angle. As shown in Fig. 20.6.1a, a linear spring behaves exactly like a member in a truss; the elongation e is equal to $X_2 - X_1$, wherein X_1 and X_2 are the displacements of the ends. For the rotational spring shown in Fig. 20.6.1b, the internal forces F and F are actually moments (generalized forces); the internal deformation e is again $X_2 - X_1$, but X_1 and X_2 are the clockwise rotations at the extremities of the coil. In the semirigid connection shown in Fig. 20.6.1c, X_1 is the new clockwise slope of the plate axis, X_2 is the new clockwise slope of the member axis, and $e = X_2 - X_1$ is the relative clockwise rotation of the member axis over that of the plate. This latter definition is important in order to indicate a clear-cut sign convention for the pairs of moments acting on the equivalent rotational spring.

The rigid frame in Fig. 20.5.2 contains three members and six semirigid connections. If the approach involving modified fixed-end moments and modified member stiffness matrices is used, the degree of freedom NP is 3 and the number of internal forces NF is 6. Now if six rotational springs are added in place of the six semirigid connections, the degree of freedom will be raised by 6 to $NP = 9$ and the number of internal forces will be raised also by 6 to $NF = 12$. While the six additional internal forces are obviously the moments acting on either side of each semirigid connection, the six additional

RIGID FRAMES WITH SEMIRIGID CONNECTIONS 737

joint displacements should be observed to be the true clockwise slopes of the member axis at each end of the three members. The entire calculation may be carried out longhand as shown in the following example, or by use of a typical computer program†, which can accommodate two kinds of internal forces, one with a 2×2 member stiffness matrix and the other with a single-element member stiffness.

Example 20.6.1 Reanalyze the rigid frame of Example 20.5.1 by replacing each semirigid connection with an equivalent rotational spring. The given data are repeated in Fig. 20.6.2a and b.

SOLUTION (a) *The rigidity indexes.* The rigidity indexes shown in Fig. 20.6.2b have been computed from the fixity factors by the formula

$$R_i = \frac{3p_i}{1 - p_i}$$

(b) *The P-X and F-e numbers.* In order that the joint displacements to be obtained in this problem may be conveniently compared with those already found in Example 20.5.1, the first three P-X numbers in Fig. 20.6.2c are assigned to correspond with those in Fig. 20.5.2. One additional nodal point is assumed to exist between each rotational spring and its associated member end; the slopes at these nodes are assigned the P-X numbers of 4 to 9 in Fig. 20.6.2c. In a similar manner, the member-end moments are numbered from 1 to 6, as shown by the F-e numbers in Fig. 20.6.2d; the pairs of moments on the rotational springs are numbered from 7 to 12. Note that the pair of positive moments on each rotational spring is to cause a relative clockwise rotation of the member axis over that of the adjoining plate.

(c) *The [A] and [B] matrices.* The [A] and [B] matrices shown below are obtained from their respective definitions. Note that the transposition relationship $[A] = [B^T]$ should be used as a check.

	F P	1	2	3	4	5	6	7	8	9	10	11	12
	1								−1.	−1.			
	2											−1.	−1.
	3	$-\frac{1}{18}$	$-\frac{1}{18}$			$-\frac{1}{15}$	$-\frac{1}{15}$						
	4	+1.					+1.						
[A] =	5		+1.					+1.					
	6			+1.					+1.				
	7				+1.					+1.			
	8					+1.					+1.		
	9						+1.						+1.

†See C.-K. Wang, *Matrix Methods of Structural Analysis*, 2d ed., American Publishing Company, Madison, Wis., 1970, Appendix H.

$[B] =$

e \ X	1	2	3	4	5	6	7	8	9
1			$-\frac{1}{18}$	$+1.$					
2			$-\frac{1}{18}$		$+1.$				
3						$+1.$			
4							$+1.$		
5		$-\frac{1}{15}$						$+1.$	
6		$-\frac{1}{15}$							$+1.$
7				$+1.$					
8	$-1.$					$+1.$			
9	$-1.$						$+1.$		
10		$-1.$						$+1.$	
11		$-1.$							$+1.$
12									$+1.$

(d) *The [S] matrix.* The usual flexural stiffness coefficients of 4 and 2 are to be used for members 1-2, 3-4, and 5-6. For each rotational spring the single element stiffness is simply $R_i EI/L$.

$[S] =$

F \ e	1	2
1	$2EI_c/3$	$EI_c/3$
2	$EI_c/3$	$2EI_c/3$

F \ e	3	4
3	EI_c	$EI_c/2$
4	$EI_c/2$	EI_c

F \ e	5	6
5	$8EI_c/15$	$4EI_c/15$
6	$4EI_c/15$	$8EI_c/15$

$$S_7 = \frac{2.0E(3I_c)}{18} = \frac{EI_c}{3} \qquad S_8 = \frac{3.0E(3I_c)}{18} = \frac{EI_c}{2}$$

$$S_9 = \frac{4.5E(5I_c)}{20} = 1.125 EI_c \qquad S_{10} = \frac{2.0E(5I_c)}{20} = \frac{EI_c}{2}$$

$$S_{11} = \frac{1.2857E(2I_c)}{15} = 0.17143 EI_c \qquad S_{12} = \frac{12.0E(2I_c)}{15} = 1.6 EI_c$$

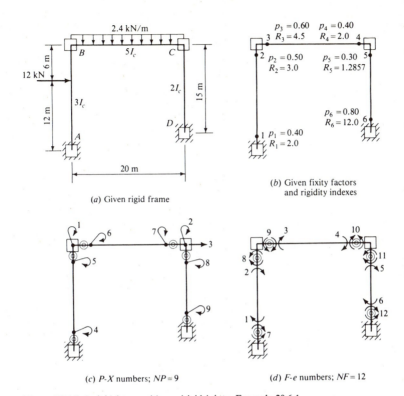

Figure 20.6.2 A rigid frame with semirigid joints, Example 20.6.1.

(e) *The $\{P\}$ matrix.* The elements in the $\{P\}$ matrix are

$P_1 = 0.$ $P_2 = 0.$ $P_3 = 8.8889$ kN

$P_4 = +16.0$ kN·m $P_5 = -32.0$ kN·m $P_6 = +80.0$ kN·m

$P_7 = -80.0$ kN·m $P_8 = 0.$ $P_9 = 0.$

(f) *The output matrices $\{X\}$, $\{F\}$, and $\{F^*\}$.* Using the equations $\{X\} = [ASA^T]^{-1}\{P\}$, $\{F\} = [SA^T]\{X\}$, and $\{F^*\} = \{F_0\} + \{F\}$, the output matrices are shown below. Note that the same solution as in Example 20.5.1 has been obtained.

$X_1 = +119.35/EI_c$ $X_2 = -43.62/EI_c$ $X_3 = +2324.1/EI_c$
$X_4 = +110.86/EI_c$ $X_5 = +102.70/EI_c$ $X_6 = +126.74/EI_c$
$X_7 = -110.12/EI_c$ $X_8 = +150.39/EI_c$ $X_9 = +39.30/EI_c$

$F_1^* = F_{01} + F_1 = -16.0 - 20.95 = -36.95$
$F_2^* = F_{02} + F_2 = +32.0 - 23.65 = +8.32$
$F_3^* = F_{03} + F_3 = -80.0 + 71.68 = -8.32$
$F_4^* = F_{04} + F_4 = +80.0 - 46.75 = +33.25$
$F_5^* = F_{05} + F_5 = 0.0 - 33.25 = -33.25$
$F_6^* = F_{06} + F_6 = 0.0 - 62.89 = -62.89$

$$F_7 = +36.95 \quad F_8 = -8.32 \quad F_9 = +8.32$$
$$F_{10} = -33.25 \quad F_{11} = +33.25 \quad F_{12} = +62.89$$

(g) *The statics and deformation checks.* In this case, the $NP = 9$ statics checks and $NF = 12$ deformation checks are much more convenient to make than these same checks in Example 20.5.1 because the X_4 to X_9 values are the true slopes of the elastic curve at the six member ends.

20.7 Exercises

20.1 For the beam connected to two fixed plates shown in Fig. 20.7.1, compute the modified fixed-end moments and make the two deformation checks to show that the results are correct.

20.2 Solve Exercise 20.1 by the matrix displacement method after installing a rotational spring for the partially fixed connection.

20.3 The 12.5-m-long member shown in Fig. 20.7.2 is subjected to a 40-kN concentrated load and connected by three bolts at each end to a plate which is fixed to a rigid wall. Assume that the

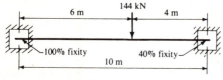

Figure 20.7.1 Exercises 20.1 and 20.2.

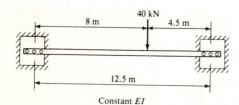

Figure 20.7.2 Exercises 20.3 and 20.4.

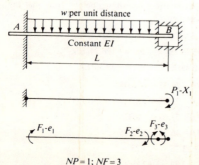

Figure 20.7.3 Exercise 20.5.

RIGID FRAMES WITH SEMIRIGID CONNECTIONS   741

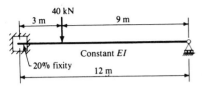

Figure 20.7.4 Exercise 20.6.

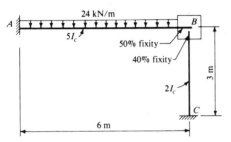

Figure 20.7.5 Exercise 20.7.

connection is so designed that the fixity factor may be taken as 50 percent at the left end and 20 percent at the right end. Using the method of modified fixed-end moments, determine the shear and moment diagrams for the member, and show the slopes of the elastic curve at the ends.

20.4 Solve Exercise 20.3 by the matrix displacement method, after the installation of two equivalent rotational springs, each between the fixed plate and the member end.

20.5 Given a fixity factor of 0.50 at B of the beam shown in Fig. 20.7.3, obtain the bending moments at A and B by using the formulas for modified fixed-end moments. Solve also by using the P-X and F-e numbers as shown in the matrix displacement method.

20.6 Analyze the beam shown in Fig. 20.7.4 by (a) assuming that the right end is connected to a fixed plate with a zero fixity factor, (b) assuming that the right end is connected with a 100 percent fixity factor to a plate which itself is free to rotate on a roller support ($NP = 1$, $NF = 2$), and (c) installing a rotational spring at the left end ($NP = 2$, $NF = 3$).

20.7 Analyze the L-shaped frame shown in Fig. 20.7.5 by using the rotation of the plate at B as the unknown in the matrix displacement method.

CHAPTER
TWENTY ONE
EFFECTS OF SHEAR DEFORMATIONS

21.1 General Introduction

The commonly accepted definition for the elastic curve of a beam is that the rotations of the cross sections and the deflections are due to bending moment alone. On this basis, the relative rotation $d\theta$ between two cross sections at a distance dx apart (Fig. 21.1.1a) has been shown to be

$$d\theta = \frac{M\,dx}{EI} \qquad (21.1.1)$$

in which M is the bending moment and EI is the flexural rigidity. However, the shearing stresses within the cross sections should also cause some externally appreciable change in the position of the beam axis. Owing to shear alone, the adjacent cross sections should slide vertically (if the original beam axis is horizontal) relative to each other, *without rotations*. On this basis, the shear deflection Δ_s (Fig. 21.1.1b) under the load would be equal to $\gamma_1 a$ or $\gamma_2 b$, in which γ_1 and γ_2 are the relative vertical displacements between two adjacent cross sections at a unit distance apart. Indeed, in ordinary cases wherein the span-to-depth ratio is 10 or more, the shear deflection Δ_s is insignificant in relation to the bending deflection Δ_b. But when both deep beams and beams of ordinary proportions act together in a complete assembly, the inclusion of shear-deformation effects in the analysis processes might make a difference worthy of investigation.

It is the purpose of this chapter to present the theoretical methods by which the effects of shear deformations may be included in the analysis of continuous beams and rigid frames.

EFFECTS OF SHEAR DEFORMATIONS 743

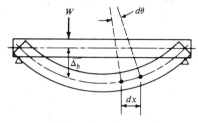

(a) Effect of bending moment

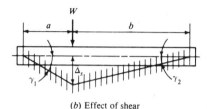

(b) Effect of shear   Figure 21.1.1 Deflection of a beam.

21.2 Shape Factors

The relative displacement $d\Delta_s$ (Fig. 21.2.1a) between two adjacent cross sections at a distance dx apart is a function of the shear force V, the cross-sectional area A, the shear modulus of elasticity G, and a shape factor α which is a function of the shear-stress distribution in the cross section; it can be shown that this relationship may be expressed by

$$d\Delta_s = \gamma\, dx \qquad (21.2.1a)$$

in which
$$\gamma = \frac{\alpha V}{GA} \qquad (21.2.1b)$$

The more rigorous analysis of a beam as an elastic solid by the theory of elasticity is a complex and difficult subject; as a matter of fact, complete satisfaction of the boundary conditions and the stress-strain relationships is not possible. The flexure formula and the shear-stress formula as presented in elementary texts on mechanics of materials are only approximations within the accuracy requirements of design and construction. Nevertheless, once a shear-stress distribution is accepted, the shape factor α as defined in Eqs. (21.2.1a and b) can be obtained by using the principle of equality between the external work and the internal strain energy.

If the shear-stress distribution over the cross section is uniform (Fig. 21.2.1b), the shape factor α is equal to 1.0. The external work W_{ext} done on the body $ABCD$ (Fig. 21.2.1) is

$$W_{ext} = \tfrac{1}{2} V\, d\Delta_s \qquad (21.2.2)$$

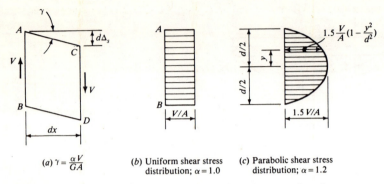

Figure 21.2.1 Shear deflection $d\Delta_s$ between two adjacent cross sections dx apart.

For uniform shear-stress distribution the internal strain energy W_{int} is

$$W_{int} = \frac{1}{2}\frac{(\text{unit shear stress})^2}{G}(\text{volume})$$

$$= \frac{1}{2}\frac{(V/A)^2}{G}(A\,dx) = \frac{V^2}{2GA}\,dx \qquad (21.2.3)$$

Equating the external work of Eq. (21.2.2) to the internal strain energy of Eq. (21.2.3),

$$d\Delta_s = \frac{V}{GA}\,dx$$

When the above equation is compared with Eq. (21.2.1), one sees that the shape factor α is 1.0.

For a rectangular cross section of width b and depth d under a parabolic shear-stress distribution as shown in Fig. 21.2.1c, the shape factor α is equal to 1.2. In this situation, the external work is again given by Eq. (21.2.2), but the internal strain energy W_{int} is

$$W_{int} = \frac{1}{2}G\int_{-d/2}^{+d/2}\left[1.5\frac{V}{A}\left(1-\frac{y^2}{d^2}\right)\right]^2 b\,dy\,dx$$

$$= \frac{1}{G}\int_{0}^{+d/2}\left[1.5\frac{V}{A}\left(1-\frac{y^2}{d^2}\right)\right]^2 b\,dy\,dx$$

which, upon being integrated with respect to y, becomes

$$W_{int} = \frac{0.6V^2}{GA}\,dx \qquad (21.2.4)$$

Equating the external work of Eq. (21.2.2) to the internal strain energy of Eq. (21.2.4),

$$d\Delta_s = \frac{1.2V}{GA}\,dx$$

EFFECTS OF SHEAR DEFORMATIONS

When the above equation is compared with Eq. (21.2.1), one sees that the shape factor α is 1.2.

21.3 Shear Deflections of Statically Determinate Beams

The shear deflections of statically determinate beams can be obtained either from physical visualization of the deformation geometry or by means of the unit-load method. While there is to be no general formula for the physical approach, the latter method may be expressed by the formula

$$1.0 * \Delta_s = \int V \frac{\alpha v}{GA} dx \tag{21.3.1}$$

in which Δ_s is the required shear deflection at a section, V and v are the shear forces in the real compatible state and the dummy unit-load equilibrium state, α is the shape factor, and GA is the shear rigidity. The derivation of Eq. (21.3.1) is to be based on the reciprocal virtual-work theorem (see Sec. 4.3). The product of the external forces in Fig. 21.3.1b and the corresponding displacements in Fig. 21.3.1a is

$$W_{\text{ext}} = 1.0 * \Delta_s \tag{21.3.2}$$

The product of the internal shear forces V in Fig. 21.3.1a and the corresponding displacements in Fig. 21.3.1b is

$$W_{\text{int}} = V \frac{\alpha v}{GA} dx \tag{21.3.3}$$

Equation (21.3.1) is obtained from equating Eq. (21.3.2) to Eq. (21.3.3).

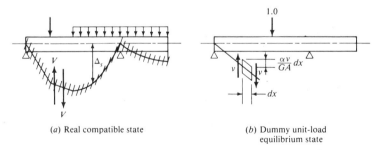

(a) Real compatible state

(b) Dummy unit-load equilibrium state

Figure 21.3.1 The unit-load method for shear deflection.

Example 21.3.1 For the simple beam shown in Fig. 21.3.2a, compute the shear deflection at a section 11 m from the left support.

SOLUTION (a) *Physical approach.* Since the relative deflection of two adjacent cross

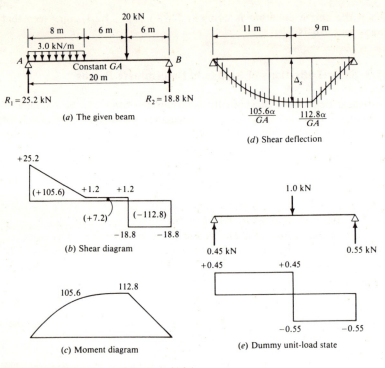

Figure 21.3.2 Simple beam of Example 21.3.1.

sections at a distance dx apart is

$$d\Delta_s = \gamma \, dx = \frac{\alpha V}{GA} dx$$

the total relative deflection between any two sections is equal to

$$\text{Relative } \Delta_s = \frac{\alpha}{GA}\int_1^2 V \, dx = \frac{\alpha}{GA}(M_2 - M_1)$$

which is α/GA times the area of the shear diagram (or the change in the bending moment) between these two sections. Note that if $M_2 - M_1$ is positive, $\Delta_{s2} - \Delta_{s1}$ is also positive, or point 2 is lower than point 1. For a simple beam the shear-deflection diagram is geometrically similar to the moment diagram, except for the constant multiplier α/GA. Thus, from Fig. 21.3.2d, the shear deflection at a section 11 m from the left support is

$$\Delta_s = \frac{\alpha}{GA}\binom{\text{bending moment}}{\text{at the section}} = \frac{\alpha}{GA}[105.6 + 1.2(3)] = \frac{109.2\alpha}{GA} \text{ kN·m}$$

(b) *Unit-load method.* If α/GA is constant for the entire beam, the unit-load-method formula becomes

$$1.0 * \Delta_s = \frac{\alpha}{GA}\int Vv \, dx$$

Using the V and v diagrams of Fig. 21.3.2b and e,

$$\Delta_s = \frac{\alpha}{GA}\left[\int_0^8 (+25.2 - 3x)(+0.45)\,dx + \int_0^3 (+1.2)(+0.45)\,dx\right.$$
$$\left. + \int_0^3 (+1.2)(-0.55)\,dx + \int_0^6 (-18.8)(-0.55)\,dx\right]$$

$$= \frac{\alpha}{GA}(+47.52 + 1.62 - 1.98 + 62.04)$$

$$= \frac{109.2\alpha}{GA}\text{ kN·m} \quad \text{(Check)}$$

Example 21.3.2 Compute the shear deflection at the free end of the cantilever beam shown in Fig. 21.3.3a.

SOLUTION (a) *Physical approach.* Since the relative deflection between any two sections is equal to α/GA times the shear area between these two sections, and since the shear deflection is zero at the fixed end of the cantilever, it is necessary only to integrate the shear area beginning with the fixed end in order to obtain the shear deflection at any section. The variation of shear deflection obtained in this way is shown in Fig. 21.3.3d, wherein the shear deflection Δ_s at the free end is

$$\Delta_s = \frac{466\alpha}{GA}\text{ kN·m}$$

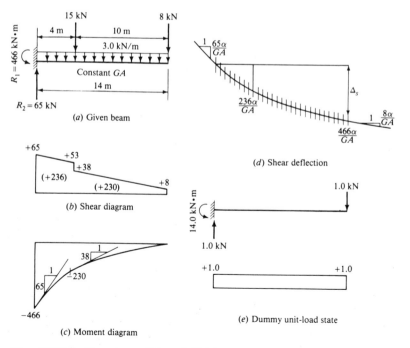

Figure 21.3.3 Cantilever beam of Example 21.3.2.

748 INTERMEDIATE STRUCTURAL ANALYSIS

(b) *Unit-load method.* Using the V and v diagrams of Fig. 21.3.3b and e,

$$\Delta_s = \int \frac{V\alpha v}{GA} dx = \frac{\alpha}{GA}\left[\int_0^4 (+65 - 3x)(1.0)\, dx + \int_0^{10}(+38 - 3x)(+1.0)\, dx\right]$$

$$= \frac{\alpha}{GA}(236 + 230) = \frac{466\alpha}{GA} \text{ kN·m} \quad \text{(Check)}$$

Example 21.3.3 Compute the shear rotation (rotation of the cross section) and deflection at points B, C, and E of the overhanging beam shown in Fig. 21.3.4a.

SOLUTION (a) *Physical approach.* On the basis of the shear and moment diagrams of Fig. 21.3.4b and c for the given overhanging beam, the shear deflection, part 1, of Fig. 21.3.4d is obtained by making it geometrically similar to the moment diagram, except for the constant multiplier α/GA. In order to satisfy the boundary condition that the deflection at D must be zero, the rigid-body rotation of Fig. 21.3.4e, or shear deflection, part 2, is to be added. The resulting shear deflection is shown in Fig. 21.3.4f, from which

$$\theta_s \text{ at } B, C, \text{ or } E = \frac{2.7\alpha}{GA}$$

and

$$\Delta_s \text{ at } B = \frac{112.8\alpha}{GA} \quad \Delta_s \text{ at } C = \frac{105.6\alpha}{GA} \quad \Delta_s \text{ at } E = 70.2\frac{\alpha}{GA}$$

(b) *Unit-load method.* Using the V and v diagrams of Fig. 21.3.4b and g,

$$\theta_s \text{ at } B = \int \frac{V\alpha v}{GA} dx = \frac{\alpha}{GA}\left[\int_0^6 (+16.1)\left(-\frac{1}{20}\right)dx + \int_0^6 (-3.9)\left(-\frac{1}{20}\right)dx\right.$$
$$\left.+ \int_0^8 (-3.9 - 3x)\left(-\frac{1}{20}\right)dx\right]$$

$$= \frac{\alpha}{GA}(-4.83 + 1.17 + 6.36) = +\frac{2.7\alpha}{GA} \quad \text{(clockwise)}$$

Whether the unit moment is applied at B or any other point, the shear diagram of Fig. 21.3.4g holds true. Thus the rotation of the cross section is constant along the beam.

Using the V and v diagrams of Fig. 21.3.4b and h, then of Fig. 21.3.4b and i, and finally of Fig. 21.3.4b and j, the shear deflections at B, C, and E are found to be

$$\Delta_s \text{ at } B = \frac{\alpha}{GA}\left[\int_0^6 (+16.1)(+0.7)\, dx + \int_0^6 (-3.9)(-0.3)\, dx + \int_0^8 (-3.9 - 3x)(-0.3)\, dx\right]$$

$$= \frac{\alpha}{GA}(67.62 + 7.02 + 38.16) = \frac{112.8\alpha}{GA} \quad \text{(Check)}$$

$$\Delta_s \text{ at } C = \frac{\alpha}{GA}\left[\int_0^6 (+16.1)(+0.4)\, dx + \int_0^6 (-3.9)(+0.4)\, dx + \int_0^8 (-3.9 - 3x)(-0.6)\, dx\right]$$

$$= \frac{\alpha}{GA}(38.64 - 9.36 + 76.32) = \frac{105.6\alpha}{GA} \quad \text{(Check)}$$

$$\Delta_s \text{ at } E = \frac{\alpha}{GA}\left[\int_0^6 (+16.1)(-0.3)\, dx + \int_0^6 (-3.9)(-0.3)\, dx + \int_0^8 (-3.9 - 3x)(-0.3)\, dx\right.$$
$$\left.+ \int_0^6 (+3x)(+1.0)\, dx\right]$$

$$= \frac{\alpha}{GA}(-28.98 + 7.02 + 38.16 + 54) = \frac{70.2\alpha}{GA} \quad \text{(Check)}$$

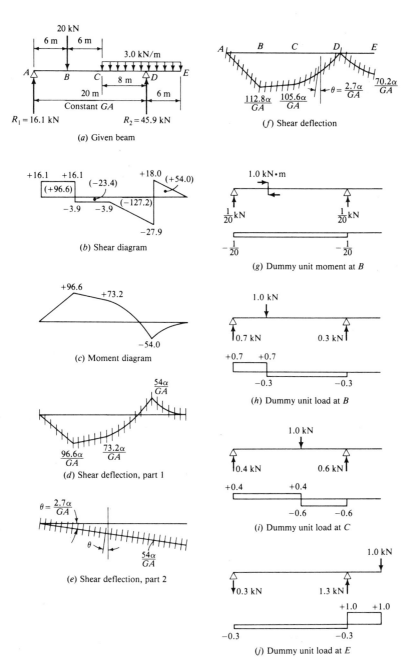

Figure 21.3.4 Overhanging beam of Example 21.3.3.

21.4 Relative Significance of Shear Deflections to Bending-Moment Deflections

The basis and method of determining deflections and rotations due to shear having been established, it is desirable to estimate the relative significance of shear deflections to bending-moment deflections. It can be shown that the ratio of shear deflection Δ_s to bending deflection Δ_b is in the range of 2 to 3 percent when the ratio of depth d to span L is about one-tenth. This ratio of Δ_s to Δ_b becomes larger as the ratio of d to L increases. Table 21.4.1 shows the ratio of Δ_s to Δ_b at the midspan of a simple beam with a rectangular cross section for which $\alpha = 1.2$, under the assumption that $G = 0.4E$.

At the midspan of a simple rectangular beam of depth d and span L, under a total uniform load of W,

$$\Delta_b = \frac{5WL^3}{384EI} \quad \Delta_s = \frac{WL\alpha}{8GA}$$

$$\frac{\Delta_s}{\Delta_b} = \left(\frac{WL\alpha}{8GA}\right)\left(\frac{384EI}{5WL^3}\right) = \frac{384}{40}\alpha\left(\frac{E}{G}\right)\left(\frac{I}{A}\right)\left(\frac{1}{L^2}\right)$$

$$= \frac{384}{40}(1.2)\left(\frac{1}{0.4}\right)\left(\frac{d^2}{12}\right)\left(\frac{1}{L^2}\right) = 2.4\left(\frac{d}{L}\right)^2 \tag{21.4.1}$$

For the same simple beam, but under a concentrated load W at midspan,

$$\Delta_b = \frac{WL^3}{48EI} \quad \Delta_s = \frac{WL\alpha}{4GA}$$

$$\frac{\Delta_s}{\Delta_b} = \left(\frac{WL\alpha}{4GA}\right)\left(\frac{48EI}{WL^3}\right) = \frac{48}{4}\alpha\left(\frac{E}{G}\right)\left(\frac{I}{A}\right)\left(\frac{1}{L^2}\right)$$

$$= \frac{48}{4}(1.2)\left(\frac{1}{0.4}\right)\left(\frac{d^2}{12}\right)\left(\frac{1}{L^2}\right) = 3\left(\frac{d}{L}\right)^2 \tag{21.4.2}$$

Values in Table 21.4.1 are obtained by applying Eqs. (21.4.1) and (21.4.2).

Table 21.4.1 Ratio of shear deflection Δ_s to bending deflection Δ_b

Ratio of depth d to span L	Ratio of Δ_s to Δ_b at midspan, $\alpha = 1.2$, $G = 0.4E$	
	Uniform load	Concentrated load at midspan
$\frac{1}{12}$	0.01666	0.02083
$\frac{1}{10}$	0.02400	0.03000
$\frac{1}{8}$	0.03750	0.04688
$\frac{1}{6}$	0.06666	0.08333
$\frac{1}{4}$	0.15000	0.18750

21.5 Force Method

For statically indeterminate beams and rigid frames, the inclusion of shear-deformation effects will cause some changes in the magnitudes of the redundant forces or moments, which in turn would affect the magnitudes of all internal forces and joint displacements. If the displacement method were to be used, one could surmise that new expressions for the fixed-end moments and the member stiffness matrix should be derived by including shear-deformation effects. It is believed, however, that the treatment of the force method will enhance the understanding of the displacement method to be described in the next section.

In the example which follows, a continuous beam resting on three supports will be analyzed. Certainly any one of the three reactions, or the bending moment at the intermediate support, may be used as the redundant. The solution using the center reaction as the redundant will be shown in detail.

After the value of the redundant is obtained from the compatibility condition, the equilibrium of the given indeterminate beam becomes completely known; the closure of the shear and moment diagrams ensures the statics checks. Then the rotations of the cross sections at the left and right ends of the left span, as well as those at the left and right ends of the right span, are computed by the convenient conjugate-beam method. The equality of the rotation at the right end of the left span to the rotation at the left end of the right span ensures the compatibility check.

Example 21.5.1 Including shear-deformation effects, analyze the continuous beam resting on three supports shown in Fig. 21.5.1a by the force method. Solve by using the center reaction as the redundant. Make the complete statics and compatibility checks.

SOLUTION (a) *Basic simple beam under applied loads.* The shear and moment diagrams of the basic simple beam under applied loads are shown in Fig. 21.5.1b. By the conjugate-beam method, in reference to Fig. 21.5.1b, the bending deflection Δ_b at B is

$$\Delta_b \text{ at } B = \frac{1}{EI}[2349(9) - 729(5) - \tfrac{1}{2}(243)(3)(2) - \tfrac{1}{2}(256.5)(3)(1)]$$

$$= \frac{16{,}382.25}{EI}$$

or

$$\Delta_b \text{ at } B = \frac{1}{EI}[2511(15) - 891(11) - \tfrac{1}{2}(297)(9)(6) - \tfrac{1}{2}(256.5)(9)(3)]$$

$$= \frac{16{,}382.25}{EI} \quad \text{(Check)}$$

The shear-deflection diagram shown in Fig. 21.5.1c is analogous to the bending-moment diagram in Fig. 21.5.1b; the shear deflection Δ_s at B is

$$\Delta_s \text{ at } B = \frac{256.5\alpha}{GA}$$

752 INTERMEDIATE STRUCTURAL ANALYSIS

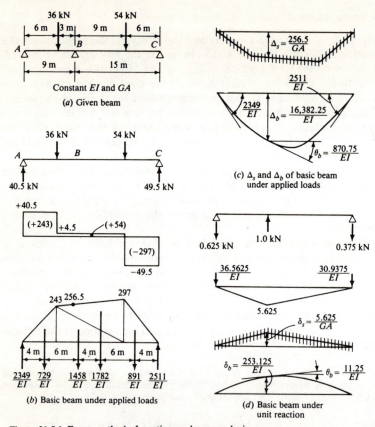

Figure 21.5.1 Force method of continuous-beam analysis.

The total bending and shear deflection at B is, then,

$$(\Delta_b + \Delta_s) \text{ at } B = \frac{16{,}382.25}{EI} + \frac{256.5\alpha}{GA} \quad \text{(downward)}$$

(b) *Basic simple beam under unit value of R_B.* The moment diagram of the basic simple beam under a unit value of R_B is shown in Fig. 21.5.1d. By the conjugate-beam method, in reference to Fig. 21.5.1d, the bending deflection δ_b at B is

$$\delta_b \text{ at } B = \frac{1}{EI}[36.5625(9) - \tfrac{1}{2}(5.625)(9)(3)] = \frac{253.125}{EI}$$

or

$$\delta_b \text{ at } B = \frac{1}{EI}[30.9375(15) - \tfrac{1}{2}(5.625)(15)(5)] = \frac{253.125}{EI} \quad \text{(Check)}$$

The shear-deflection diagram is also shown in Fig. 21.5.1d, from which

$$\delta_s \text{ at } B = \frac{5.625\alpha}{GA}$$

EFFECTS OF SHEAR DEFORMATIONS 753

The total bending and shear deflection at B is, then,

$$(\delta_b + \delta_s) \text{ at } B = \frac{253.125}{EI} + \frac{5.625\alpha}{GA} \quad \text{(upward)}$$

(c) *Obtaining the redundant from the compatibility equation.* The value of the redundant, or the reaction at B, can be obtained from the compatibility equation

$$(\delta_b + \delta_s)R_B = \Delta_b + \Delta_s$$

Using the values of $\Delta_b + \Delta_s$ and $\delta_b + \delta_s$ already found in parts (a) and (b), and calling $\beta = \alpha EI/GA$,

$$R_B = \frac{16{,}382.25 + 256.5\beta}{253.125 + 5.625\beta} = \frac{2912.4 + 45.6\beta}{45 + \beta}$$

The values of R_A and R_C may be obtained by superposing their values in Fig. 21.5.1b over R_B times their respective values in Fig. 21.5.1d; thus,

$$R_A = 40.5 - 0.625\frac{2912.4 + 45.6\beta}{45 + \beta} = \frac{2.25 + 12\beta}{45 + \beta}$$

$$R_C = 49.5 - 0.375\frac{2912.4 + 45.6\beta}{45 + \beta} = \frac{1135.35 + 32.4\beta}{45 + \beta}$$

The final shear and moment diagrams are shown in Fig. 21.5.2. The fact that the shear diagram closes satisfies the equation $\Sigma F_y = 0$ for the entire beam. The ordinates on the moment diagram are computed from summing up the shear areas; the fact that the moment diagram closes satisfies the equation $\Sigma M = 0$ for the entire beam. The shear-deflection diagram is analogous to the bending-moment diagram and is shown in Fig. 21.5.2.

(d) *Compatibility check.* Since the given beam is statically indeterminate to the first degree, there is to be only one compatibility check. This check may be made by seeing whether the rotation of the cross section due to bending only at the right end of the left span is equal to that at the left end of the right span.

Because there is an upward shear deflection at B, as shown in Fig. 21.5.2, there must be a like amount of bending deflection, in the downward direction, as shown in Fig. 21.5.3. The rotations of the cross sections at A, B, and C, due to bending only (there is no rotation due to shear, as shown in Fig. 21.5.2d), can be computed from the addition of three parts: one from an equivalent support settlement at B to neutralize the shear deflection, one due to the positive moment diagrams on spans AB and BC, and one due to the negative moment diagrams on spans AB and BC. Referring again to Fig. 21.5.3 and using the conjugate-beam method,

$$\theta_A = \theta_{A1} + \theta_{A2} - \theta_{A3}$$

$$= \frac{4839.75}{9GA(45 + \beta)} + \frac{4}{9}\left(\frac{324}{EI}\right) - \frac{1}{3}\left(\frac{1}{2EI}\right)\left(\frac{4839.75}{45 + \beta}\right)(9)$$

$$= \frac{1}{EI(45 + \beta)}(537.75\beta + 6480 + 144\beta - 7259.625)$$

$$= \frac{1}{EI(45 + \beta)}(-779.625 + 681.75\beta)$$

$$\theta_{BL} = \theta_{BL1} - \theta_{BL2} + \theta_{BL3}$$

$$= \frac{4839.75}{9GA(45 + \beta)} - \frac{5}{9}\left(\frac{324}{EI}\right) + \frac{2}{3}\left(\frac{1}{2EI}\right)\left(\frac{4839.75}{45 + \beta}\right)(9)$$

$$= \frac{1}{EI(45 + \beta)}(537.75\beta - 8100 - 180\beta + 14{,}519.25)$$

$$= \frac{1}{EI(45 + \beta)}(6419.25 + 357.75\beta)$$

$$\theta_{BR} = -\theta_{BR1} + \theta_{BR2} - \theta_{BR3}$$

$$= -\frac{4839.75}{15GA(45+\beta)} + \frac{7}{15}\left(\frac{1458}{EI}\right) - \frac{2}{3}\left(\frac{1}{2EI}\right)\left(\frac{4839.75}{45+\beta}\right) \quad (15)$$

$$= \frac{1}{EI(45+\beta)}(-322.65\beta + 30{,}618 + 680.4\beta - 24{,}198.75)$$

$$= \frac{1}{EI(45+\beta)}(6419.25 + 357.75\beta) \quad \text{(Check)}$$

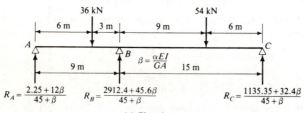

(a) Given beam

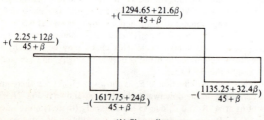

(b) Shear diagram

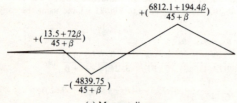

(c) Moment diagram

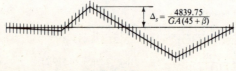

(d) Shear deflection

Figure 21.5.2 Results of analysis, Example 21.5.1.

EFFECTS OF SHEAR DEFORMATIONS 755

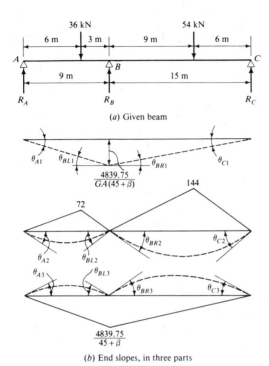

(a) Given beam

(b) End slopes, in three parts

Figure 21.5.3 Compatibility check, Example 21.5.1.

$$\theta_C = -\theta_{C1} - \theta_{C2} + \theta_{C3}$$

$$= -\frac{4839.75}{15GA(45+\beta)} - \frac{8}{15}\left(\frac{1458}{EI}\right) + \frac{1}{3}\left(\frac{1}{2EI}\right)\left(\frac{4839.75}{45+\beta}\right)(15)$$

$$= \frac{1}{EI(45+\beta)}(-322.65\beta - 34{,}992 - 777.6\beta + 12{,}099.375)$$

$$= \frac{1}{EI(45+\beta)}(-22{,}892.625 - 1100.25\beta)$$

Values of θ_A and θ_C are not needed for the compatibility check. They are conveniently computed here along with θ_{BL} and θ_{BR} so that all joint rotations are now available.

21.6 Displacement Method

When a member in a continuous beam or in a rigid frame has a small span-to-depth ratio, say 8, 6, 4, or less, the analyst may choose to consider the shear-deformation effects of such a member. This may be done by modifying the fixed-end moments and the member stiffness matrix for use in the displacement method.

So far as the fixed-end moments on a member subjected to uniform load are concerned, it will be shown that they are not affected by shear defor-

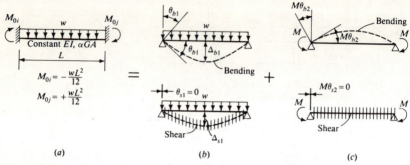

Figure 21.6.1 Bending and shear deformation of a fixed-end beam subjected to uniform load.

mation. As shown by Fig. 21.6.1, the redundant moment M is equal to

$$M = \frac{\theta_{b1} + \theta_{s1}}{\theta_{b2} + \theta_{s2}} = \frac{wL^3/(24EI) + 0}{L/(2EI) + 0} = \frac{wL^2}{12}$$

Thus, considering shear-deformation effects,

$$M_{0i} = -M = -\frac{wL^2}{12} \quad (21.6.1a)$$

$$M_{0j} = +M = +\frac{wL^2}{12} \quad (21.6.1b)$$

The fixed-end moments on a member subjected to a concentrated load, when shear-deformation effects are considered, may be obtained by multiplying the usual expression by the modifying factors shown in Fig. 21.6.2a. These modifying factors can be derived from the compatibility conditions as applied to a basic simple beam. It should be emphasized that it is the rotation of the cross section from the vertical direction, not the slope of the combined bending and shear-deflection curve, which should be zero at each of the two fixed ends.

Applying the conjugate-beam method to the bending-moment diagram of Fig. 21.6.2b,

$$\theta_{ib1} = \frac{Wab(L+b)}{6LEI} \quad \text{and} \quad \theta_{jb1} = \frac{Wab(L+a)}{6LEI} \quad (21.6.2)$$

The shear-deflection diagram of Fig. 21.6.2b is geometrically similar to the bending-moment diagram; thus

$$\theta_{is1} = 0 \quad \text{and} \quad \theta_{js1} = 0 \quad (21.6.3)$$

Applying the conjugate-beam method to the bending-moment diagrams of Fig.

EFFECTS OF SHEAR DEFORMATIONS

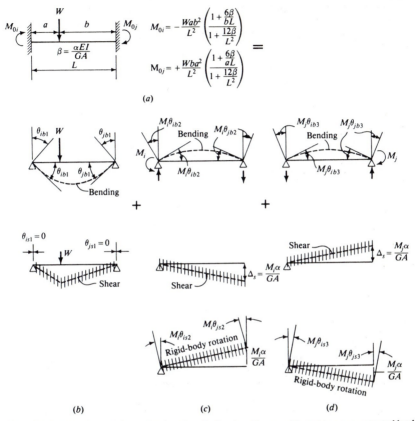

Figure 21.6.2 Bending and shear deformation of a fixed-end beam subjected to a concentrated load.

21.6.2c and d, respectively,

$$\theta_{ib2} = \frac{L}{3EI} \quad \text{and} \quad \theta_{jb2} = \frac{L}{6EI} \tag{21.6.4}$$

$$\theta_{ib3} = \frac{L}{6EI} \quad \text{and} \quad \theta_{jb3} = \frac{L}{3EI} \tag{21.6.5}$$

The shear-deflection diagrams shown in Fig. 21.6.2c and d are established on the basis of keeping not only the shear deflection at the left support at zero but also the rotation of the cross section there. Clearly the boundary condition at the right support, that of zero deflection, is not yet satisfied. A rigid-body rotation of the entire beam around the left support can neutralize the unwanted deflection at the right end, causing in the meantime the rotations of the cross sections at both the left and right ends of Fig. 21.6.2c,

$$\theta_{is2} = \theta_{js2} = \frac{\alpha}{LGA} \tag{21.6.6}$$

and, at both the left and right ends of Fig. 21.6.2d,

$$\theta_{is3} = \theta_{js3} = \frac{\alpha}{LGA} \tag{21.6.7}$$

Of course, the results of Eqs. (21.6.6) and (21.6.7) may also be obtained by the unit-load method; the physical explanation presented here is supplementary.

The absolute quantities M_i and M_j, as shown in Fig. 21.6.2b to d, may be obtained by solving simultaneously the two compatibility equations as shown below.

$$M_i(\theta_{ib2} + \theta_{is2}) + M_j(\theta_{ib3} - \theta_{is3}) = \theta_{ib1} \tag{21.6.8a}$$

$$M_i(\theta_{jb2} - \theta_{js2}) + M_j(\theta_{jb3} + \theta_{js3}) = \theta_{jb1} \tag{21.6.8b}$$

Substituting the expressions of Eqs. (21.6.2) to (21.6.7) into Eqs. (21.6.8a and b),

$$M_i\left(\frac{L}{3EI} + \frac{\alpha}{LGA}\right) + M_j\left(\frac{L}{6EI} - \frac{\alpha}{LGA}\right) = \frac{Wab(L+b)}{6LEI} \tag{21.6.9a}$$

$$M_i\left(\frac{L}{6EI} - \frac{\alpha}{LGA}\right) + M_j\left(\frac{L}{3EI} + \frac{\alpha}{LGA}\right) = \frac{Wab(L+a)}{6LEI} \tag{21.6.9b}$$

Solving Eqs. (21.6.9a and b) for the absolute quantities M_i and M_j, and using the sign conventions for M_{0i} and M_{0j} shown in Fig. 21.6.2a,

$$M_{0i} = -M_i = -\frac{Wab^2}{L^2}\left[\frac{1+6\beta/(bL)}{1+12\beta/L^2}\right] \qquad \beta = \frac{\alpha EI}{GA} \tag{21.6.10a}$$

$$M_{0j} = +M_j = +\frac{Wba^2}{L^2}\left[\frac{1+6\beta/(aL)}{1+12\beta/L^2}\right] \qquad \beta = \frac{\alpha EI}{GA} \tag{21.6.10b}$$

As shown by Fig. 21.6.3, the flexibility and stiffness matrices of a prismatic member express the relationships between the clockwise end moments M_i and M_j and the clockwise end rotations ϕ_i and ϕ_j. From the results assembled on the left sides of Eqs. (21.6.9a and b), the member

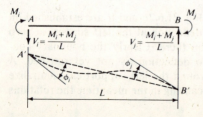

Figure 21.6.3 A prismatic member subjected to end moments only.

EFFECTS OF SHEAR DEFORMATIONS **759**

flexibility matrix, considering shear-deformation effects, takes the form

$$[D] = \begin{bmatrix} +\left(\dfrac{L}{3EI} + \dfrac{\alpha}{LGA}\right) & -\left(\dfrac{L}{6EI} - \dfrac{\alpha}{LGA}\right) \\ -\left(\dfrac{L}{6EI} - \dfrac{\alpha}{LGA}\right) & +\left(\dfrac{L}{3EI} + \dfrac{\alpha}{LGA}\right) \end{bmatrix} \quad (21.6.11)$$

The inversion of the $[D]$ matrix shown in Eq. (21.6.11) gives the member stiffness matrix, considering shear-deformation effects, as

$$[S] = \begin{bmatrix} +\dfrac{4EI}{L}\left(\dfrac{1+3k}{1+12k}\right) & +\dfrac{2EI}{L}\left(\dfrac{1-6k}{1+12k}\right) \\ +\dfrac{2EI}{L}\left(\dfrac{1-6k}{1+12k}\right) & +\dfrac{4EI}{L}\left(\dfrac{1+3k}{1+12k}\right) \end{bmatrix} \quad k = \dfrac{\alpha EI}{L^2 GA} = \dfrac{\beta}{L^2}$$
$$(21.6.12)$$

The expressions for the modified fixed-end moments, shown in Eqs. (21.6.1a and b) for the uniform load and in Eqs. (21.6.10a and b) for the concentrated load, and the expression for the modified stiffness matrix, shown in Eq. (21.6.12), are all that is needed for the inclusion of the shear-deformation effects of any particular member in the displacement method of analyzing continuous beams or rigid frames.

Example 21.6.1 Using the results of the analysis of the continuous beam by the force method shown in Fig. 21.5.2, verify numerically the equation

$$\{M^*\}_{2\times 1} = \{M_0\}_{2\times 1} + [S]_{2\times 2}\{\phi\}_{2\times 1}$$

first for member AB and then for member BC.

SOLUTION (a) *Member AB.* From Fig. 21.5.2, the final end moments at A and B of member AB are

$$M_A^* = 0$$

$$M_B^* = +\dfrac{4839.75}{45+\beta} \quad \text{(clockwise on } BA\text{)}$$

The modified fixed-end moments are

$$M_{0A} = -\dfrac{36(6)(3)^2}{9^2}\left(\dfrac{1+6\beta/27}{1+12\beta/81}\right) = -\dfrac{72(9+2\beta)}{27+4\beta}$$

$$M_{0B} = +\dfrac{36(3)(6)^2}{9^2}\left(\dfrac{1+6\beta/54}{1+12\beta/81}\right) = +\dfrac{144(9+\beta)}{27+4\beta}$$

The modified stiffness matrix is

$$[S] = \begin{bmatrix} \dfrac{4EI}{9}\left(\dfrac{1+3\beta/81}{1+12\beta/81}\right) & \dfrac{2EI}{9}\left(\dfrac{1-6\beta/81}{1+12\beta/81}\right) \\ \dfrac{2EI}{9}\left(\dfrac{1-6\beta/81}{1+12\beta/81}\right) & \dfrac{4EI}{9}\left(\dfrac{1+3\beta/81}{1+12\beta/81}\right) \end{bmatrix}$$

The member-end rotations are taken from part (d) in Example 21.5.1; they are

$$\phi_A = \dfrac{-779.625 + 681.75\beta}{EI(45+\beta)}$$

$$\phi_B = \dfrac{6419.25 + 357.75\beta}{EI(45+\beta)}$$

The equations $M_A^* = M_{0A} + S_{AA}\phi_A + S_{AB}\phi_B$ and $M_B^* = M_{0B} + S_{BA}\phi_A + S_{BB}\phi_B$ are numerically verified. There is a great deal of arithmetic involved in this verification, but it is considered worthwhile because one becomes more certain of the correctness of the formulas derived for the modified fixed-end moments and the modified stiffness matrix.

(b) *Member BC.* From Fig. 21.5.2, the final end moments at B and C of member BC are

$$M_B^* = -\frac{4839.75}{45 + \beta} \quad \text{(counterclockwise on } BC\text{)}$$

$$M_C^* = 0$$

The modified fixed-end moments are

$$M_{0B} = -\frac{54(9)(6)^2}{15^2}\left(\frac{1 + 6\beta/90}{1 + 12\beta/225}\right) = -\frac{388.8(15 + \beta)}{75 + 4\beta}$$

$$M_{0C} = +\frac{54(6)(9)^2}{15^2}\left(\frac{1 + 6\beta/135}{1 + 12\beta/225}\right) = +\frac{194.4(45 + 2\beta)}{75 + 4\beta}$$

The member-end rotations are taken from part (d) in Example 21.5.1; they are

$$\phi_B = \frac{6419.25 + 357.75\beta}{EI(45 + \beta)}$$

$$\phi_C = \frac{-22{,}892.625 - 1100.25\beta}{EI(45 + \beta)}$$

The equations $M_B^* = M_{0B} + S_{BB}\phi_B + S_{BC}\phi_C$ and $M_C^* = M_{0C} + S_{CB}\phi_B + S_{CC}\phi_C$ are then numerically verified.

21.7 Exercises

21.1 For the overhanging beam shown in Fig. 21.7.1, compute the deflection at the free end due to bending moment and shear. Check both parts of the deflection by the unit-load method and the geometric method. Use $G = 0.4E$ and $E = 1.05 \times 10^3 \text{ kN/cm}^2$.

21.2 The cross section of the beam shown in Fig. 21.7.2 is 30 cm wide and 120 cm deep, with a Poisson's ratio of 0.25. Considering shear-deformation effects, analyze by using the reaction at the right support as the redundant.

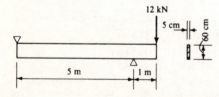

Figure 21.7.1 Exercise 21.1.

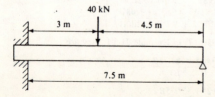

Figure 21.7.2 Exercises 21.2 and 21.3.

EFFECTS OF SHEAR DEFORMATIONS 761

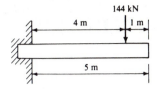

Figure 21.7.3 Exercise 21.4.

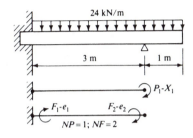

Figure 21.7.4 Exercise 21.5.

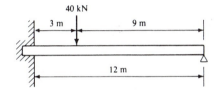

Figure 21.7.5 Exercise 21.6.

21.3 Solve Exercise 21.2 by the matrix displacement method using (*a*) one element of 7.5 m, and (*b*) two elements of 3 m and 4.5 m.

21.4 For the cantilever beam shown in Fig. 21.7.3, obtain the expressions for slope and deflection at the free end in terms of $\beta = \alpha EI/GA$, using the shear and moment diagrams from statics. Analyze the same cantilever beam by the matrix displacement method using one 5-m element. Show that the slope and deflection at the free end obtained earlier satisfy the equations $\{P\} = [ASB]\{X\}$ and $\{F^*\} = \{F_0\} + [SB]\{X\}$.

21.5 Considering shear deformation and using constant $\beta = \alpha EI/GA$ throughout, derive an expression for the reaction at the right support of the beam shown in Fig. 21.7.4 by the force method. Check the solution by the matrix displacement method using the *P*-*X* and *F*-*e* numbers as shown.

21.6 Considering shear deformation and using constant $\beta = \alpha EI/GA$ throughout, analyze the beam shown in Fig. 21.7.5 (*a*) using the reacting moment at the left support as the redundant, (*b*) using the reaction at the right support as the redundant, and (*c*) using one 12-m element in the matrix displacement method.

CHAPTER
TWENTY TWO

BEAMS ON ELASTIC FOUNDATION

22.1 General Description

Up to this point, treatment of structural members has been limited to those under the action of known transverse forces. In the displacement method of analysis, these transverse forces are transmitted to both ends of each member, in the fixed condition. One different kind of structural member is one which rests on a spongy material, hereafter called an *elastic foundation*, which offers a resistance proportional to the transverse deflection; a common example of this is a foundation beam resting on an elastic soil. Thus there are unknown transverse forces, equal to the product of the "stiffness modulus" of the supporting material and the yet unknown transverse deflection, acting on structural members on elastic foundation.

In the event that some member in a continuous beam or rigid frame is subjected to resistance offered by an elastic foundation, the displacement method of analysis can still be used, provided that expressions for the member stiffness matrix and for the fixed-end reactions and moments due to common types of transverse loads can be found as functions of the stiffness modulus. In this chapter, these needed expressions will be derived.

22.2 The Basic Differential Equation

Consider a structural member AB and its elastic curve $A'B'$ as shown in Fig. 22.2.1. It is subjected to a varying downward load of w per unit distance and to an upward force of ky per unit distance, wherein k is the stiffness modulus of the elastic foundation, measured in force per unit area. Sometimes the compression modulus of the medium is given in force per unit volume, in

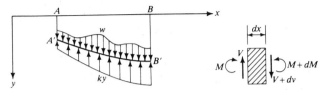

Figure 22.2.1 Structural member on elastic foundation.

which case the stiffness modulus is the product of the compression modulus and the width of the structural member.

From the equilibrium equations of resolution and rotation of an infinitesimal segment of the structural member,

$$\frac{dV}{dx} = ky - w \tag{22.2.1}$$

and

$$\frac{dM}{dx} = V \tag{22.2.2}$$

wherein the positive directions of shear V and bending moment M are as shown in Fig. 22.2.1. The change in slope between any two consecutive points at an infinitesimal distance dx apart is equal to

$$d\frac{dy}{dx} = -\frac{M}{EI} dx \tag{22.2.3}$$

the negative sign is due to the fact that the slope is decreasing in positive or concave bending. Combining Eqs. (22.2.1) to (22.2.3) gives the basic differential equation of the elastic curve as

$$\frac{d^4y}{dx^4} + \frac{k}{EI} y = +\frac{w}{EI} \tag{22.2.4}$$

The shear and bending moment become

$$V = -EI \frac{d^3y}{dx^3} \tag{22.2.5}$$

and

$$M = -EI \frac{d^2y}{dx^2} \tag{22.2.6}$$

22.3 General Solution of the Differential Equation

When there is no transverse load acting on the member, the basic differential equation (22.2.4) becomes

$$\frac{d^4y}{dx^4} + \frac{k}{EI} y = 0 \tag{22.3.1}$$

764 INTERMEDIATE STRUCTURAL ANALYSIS

of which the general solution is

$$y = A \cos \frac{\phi}{L} x \cosh \frac{\phi}{L} x + B \cos \frac{\phi}{L} x \sinh \frac{\phi}{L} x$$
$$+ C \sin \frac{\phi}{L} x \cosh \frac{\phi}{L} x + D \sin \frac{\phi}{L} x \sinh \frac{\phi}{L} x \quad (22.3.2a)$$

wherein
$$\phi = L \sqrt[4]{\frac{k}{4EI}} \quad (22.3.2b)$$

The correctness of the solution may be verified by substituting the derivatives of Eq. (22.3.2) into Eq. (22.3.1).

22.4 Boundary Conditions of an Unloaded Member

Since the general solution of the differential equation for an unloaded member resting on an elastic foundation includes four arbitrary constants, four boundary conditions are required for the evaluation of these constants. Two common approaches are (1) specifying the bending moments M_i, M_j and the shears V_i, V_j at the end points i and j; or (2) specifying the slopes θ_i, θ_j and the transverse deflections Δ_i, Δ_j. Positive directions for M_i, M_j, V_i, V_j and for θ_i, θ_j, Δ_i, Δ_j are as shown in Fig. 22.4.1. In this connection, it must be noted that the elastic foundation has been assumed to be capable of exerting either pull or push as if the structural member were securely attached to the medium. If pulling is not permitted and it does happen in the final solution of the problem, the segment which becomes free from the supporting material must be determined by trial and error.

It will be shown that the use of the first approach will yield a 4×4 flexibility matrix of a member on elastic foundation, and the second approach will produce the 4×4 stiffness matrix.

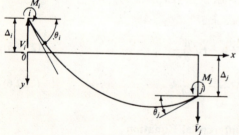

Figure 22.4.1 Boundary conditions of a structural member.

22.5 Flexibility Matrix of a Member on Elastic Foundation

Using the symbols s, s', c, and c' for $\sin \phi$, $\sinh \phi$, $\cos \phi$, and $\cosh \phi$, the four force quantities M_i, M_j, V_i, and V_j as defined in Fig. 22.4.1 may be expressed

BEAMS ON ELASTIC FOUNDATION

in terms of the four arbitrary constants A, B, C, and D as follows:

$$M_i = \left(-EI \frac{d^2y}{dx^2} \text{ at } x = 0\right) = -\frac{2EI\phi^2}{L^2} D \tag{22.5.1a}$$

$$M_j = \left(-EI \frac{d^2y}{dx^2} \text{ at } x = L\right)$$
$$= +\frac{2EI\phi^2}{L^2}(ss'A + sc'B - cs'C - cc'D) \tag{22.5.1b}$$

$$V_i = \left(-EI \frac{d^3y}{dx^3} \text{ at } x = 0\right) = +\frac{2EI\phi^3}{L^3}(B - C) \tag{22.5.1c}$$

$$V_j = \left(-EI \frac{d^3y}{dx^3} \text{ at } x = L\right)$$
$$= +\frac{2EI\phi^3}{L^3}[(sc' + cs')A + (ss' + cc')B + (ss' - cc')C + (sc' - cs')D] \tag{22.5.1d}$$

Conversely, the four arbitrary constants A, B, C, and D may be expressed in terms of the four force quantities M_i, M_j, V_i, and V_j by solving the four simultaneous equations (22.5.1); thus,

$$A = +\frac{L^2}{2EI\phi^2(s'^2 - s^2)}\left[-(s'^2 + s^2)M_i + 2ss'M_j\right.$$
$$\left. + \frac{L}{\phi}(sc - s'c')V_i - \frac{L}{\phi}(sc' - cs')V_j\right] \tag{22.5.2a}$$

$$B = +\frac{L^2}{2EI\phi^2(s'^2 - s^2)}\left[+(sc + s'c')M_i - (sc' + cs')M_j\right.$$
$$\left. + \frac{L}{\phi}s'^2 V_i + \frac{L}{\phi}(ss')V_j\right] \tag{22.5.2b}$$

$$C = +\frac{L^2}{2EI\phi^2(s'^2 - s^2)}\left[+(sc + s'c')M_i - (sc' + cs')M_j\right.$$
$$\left. + \frac{L}{\phi}s^2 V_i + \frac{L}{\phi}(ss')V_j\right] \tag{22.5.2c}$$

$$D = +\frac{L^2}{2EI\phi^2}(-M_i) \tag{22.5.2d}$$

But the four deformation quantities θ_i, θ_j, Δ_i, and Δ_j can be expressed in terms of the four arbitrary constants A, B, C, and D by

$$\theta_i = \left(+\frac{dy}{dx} \text{ at } x = 0\right) = +\frac{\phi}{L}(B + C) \tag{22.5.3a}$$

$$\theta_j = \left(-\frac{dy}{dx} \text{ at } x = L\right)$$

$$= +\frac{\phi}{L}[(sc' - cs')A + (ss' - cc')B - (ss' + cc')C - (sc' + cs')D]$$
(22.5.3b)

$$\Delta_i = (-y \text{ at } x = 0) = -A$$
(22.5.3c)

$$\Delta_j = (+y \text{ at } x = L) = +cc'A + cs'B + sc'C + ss'D$$
(22.5.3d)

Finally, the four deformation quantities θ_i, θ_j, Δ_i, and Δ_j may be expressed in terms of the four force quantities M_i, M_j, V_i, and V_j by substituting Eq. (22.5.2) into Eq. (22.5.3); the results can be arranged in matrix form as follows:

$$[D]_{4\times 4} = $$

	M_i	M_j	V_i	V_j
θ_i	$+\dfrac{L(sc+s'c')}{EI\phi(s'^2-s^2)}$	$-\dfrac{L(sc'+cs')}{EI\phi(s'^2-s^2)}$	$+\dfrac{L^2(s'^2+s^2)}{2EI\phi^2(s'^2-s^2)}$	$+\dfrac{L^2 ss'}{EI\phi^2(s'^2-s^2)}$
θ_j	$-\dfrac{L(sc'+cs')}{EI\phi(s'^2-s^2)}$	$+\dfrac{L(sc+s'c')}{EI\phi(s'^2-s^2)}$	$-\dfrac{L^2 ss'}{EI\phi^2(s'^2-s^2)}$	$-\dfrac{L^2(s'^2+s^2)}{2EI\phi^2(s'^2-s^2)}$
Δ_i	$+\dfrac{L^2(s'^2+s^2)}{2EI\phi^2(s'^2-s^2)}$	$-\dfrac{L^2 ss'}{EI\phi^2(s'^2-s^2)}$	$+\dfrac{L^3(s'c'-sc)}{2EI\phi^3(s'^2-s^2)}$	$+\dfrac{L^3(sc'-cs')}{2EI\phi^3(s'^2-s^2)}$
Δ_j	$+\dfrac{L^2 ss'}{EI\phi^2(s'^2-s^2)}$	$-\dfrac{L^2(s'^2+s^2)}{2EI\phi^2(s'^2-s^2)}$	$+\dfrac{L^3(sc'-cs')}{2EI\phi^3(s'^2-s^2)}$	$+\dfrac{L^3(s'c'-sc)}{2EI\phi^3(s'^2-s^2)}$

(22.5.4)

The $[D]$ matrix as expressed by Eq. (22.5.4) is the flexibility matrix of a member on elastic foundation. When the stiffness modulus k of the elastic medium is zero, then $\phi = 0$. In such a case, the flexibility matrix $[D]$ degenerates to become

$$[D]_{2\times 2} = $$

	M_i	M_j
θ_i	$+\dfrac{L}{3EI}$	$+\dfrac{L}{6EI}$
θ_j	$+\dfrac{L}{6EI}$	$+\dfrac{L}{3EI}$

(22.5.5)

because there can be only two conditions of statical equilibrium which are

(1) $M_i = +1$, $M_j = 0$, $V_i = -\dfrac{1}{L}$, $V_j = -\dfrac{1}{L}$, $\theta_i = +\dfrac{L}{3EI}$, $\theta_j = +\dfrac{L}{6EI}$, $\Delta_i = 0$, $\Delta_j = 0$;

and

(2) $M_i = 0$, $M_j = +1$, $V_i = +\dfrac{1}{L}$, $V_j = +\dfrac{1}{L}$, $\theta_i = +\dfrac{L}{6EI}$, $\theta_j = +\dfrac{L}{3EI}$, $\Delta_i = 0$, $\Delta_j = 0$.

22.6 Stiffness Matrix of a Member on Elastic Foundation

The four arbitrary constants A, B, C, and D may be expressed in terms of the four deformation quantities θ_i, θ_j, Δ_i, Δ_j, by solving the four simultaneous equations (22.5.3); thus,

$$A = -\Delta_i \tag{22.6.1a}$$

$$B = +\frac{1}{s'^2 - s^2}\left[\frac{-Ls^2}{\phi}\theta_i + \frac{Lss'}{\phi}\theta_j + (sc + s'c')\Delta_i + (sc' + cs')\Delta_j\right] \tag{22.6.1b}$$

$$C = +\frac{1}{s'^2 - s^2}\left[\frac{+Ls'^2}{\phi}\theta_i - \frac{Lss'}{\phi}\theta_j - (sc + s'c')\Delta_i - (sc' + cs')\Delta_j\right] \tag{22.6.1c}$$

$$D = +\frac{1}{s'^2 - s^2}\left[+\frac{L}{\phi}(sc - s'c')\theta_i + \frac{L}{\phi}(sc' - cs')\theta_j + (s'^2 + s^2)\Delta_i + 2ss'\Delta_j\right] \tag{22.6.1d}$$

The stiffness matrix $[S]$ of a member on elastic foundation can be obtained by substituting Eq. (22.6.1) into Eq. (22.5.1); thus,

$$[S]_{4\times 4} = \begin{array}{c|c|c|c|c} & \theta_i & \theta_j & \Delta_i & \Delta_j \\ \hline M_i & +\text{TEMP1} & -\text{TEMP2} & -\text{TEMP5} & -\text{TEMP6} \\ \hline M_j & -\text{TEMP2} & +\text{TEMP1} & +\text{TEMP6} & +\text{TEMP5} \\ \hline V_i & -\text{TEMP5} & +\text{TEMP6} & +\text{TEMP3} & +\text{TEMP4} \\ \hline V_j & -\text{TEMP6} & +\text{TEMP5} & +\text{TEMP4} & +\text{TEMP3} \end{array} \tag{22.6.2}$$

wherein

$$\text{TEMP1} = +\frac{2\phi(s'c' - sc)}{s'^2 - s^2}\frac{EI}{L} = +\frac{4EI}{L} \quad \text{at } \phi = 0$$

$$\text{TEMP2} = +\frac{2\phi(sc' - cs')}{s'^2 - s^2}\frac{EI}{L} = +\frac{2EI}{L} \quad \text{at } \phi = 0$$

$$\text{TEMP3} = +\frac{4\phi^3(sc + s'c')}{s'^2 - s^2}\frac{EI}{L^3} = +\frac{12EI}{L^3} \quad \text{at } \phi = 0$$

$$\text{TEMP4} = +\frac{4\phi^3(sc' + cs')}{s'^2 - s^2}\frac{EI}{L^3} = +\frac{12EI}{L^3} \quad \text{at } \phi = 0$$

$$\text{TEMP5} = +\frac{2\phi^2(s'^2 + s^2)}{s'^2 - s^2}\frac{EI}{L^2} = +\frac{6EI}{L^2} \quad \text{at } \phi = 0$$

$$\text{TEMP6} = +\frac{4\phi^2 ss'}{s'^2 - s^2}\frac{EI}{L^2} = +\frac{6EI}{L^2} \quad \text{at } \phi = 0$$

The degenerate values of TEMP1 to TEMP6 at $\phi = 0$ are the stiffness coefficients of an ordinary prismatic member as defined in Chaps. 11 and 12. These limiting values at $\phi = 0$ are here independently obtained by using L'Hospital's rule or by using the series expansions of $\sin \phi$, $\cos \phi$, $\sinh \phi$, and $\cosh \phi$, which are

$$\sin \phi = \phi - \frac{\phi^3}{3!} + \frac{\phi^5}{5!} - \frac{\phi^7}{7!} + \cdots$$

$$\cos \phi = 1 - \frac{\phi^2}{2!} + \frac{\phi^4}{4!} - \frac{\phi^6}{6!} + \cdots$$

$$\sinh \phi = \phi + \frac{\phi^3}{3!} + \frac{\phi^5}{5!} + \frac{\phi^7}{7!} + \cdots$$

$$\cosh \phi = 1 + \frac{\phi^2}{2!} + \frac{\phi^4}{4!} + \frac{\phi^6}{6!} + \cdots$$

Certainly the product of the flexibility matrix $[D]$ in Eq. (22.5.4) and the stiffness matrix $[S]$ in Eq. (22.6.2) should be a unit matrix. Although not shown, this multiplication has been carried out as a check.

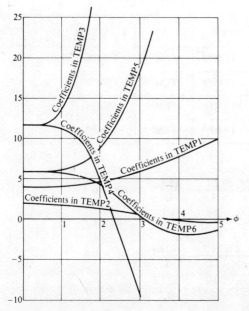

Figure 22.6.1 Variation of stiffness coefficients with $\phi = L\sqrt[4]{(k/4EI)}$.

Curves showing the values of TEMP1 to TEMP6 with ϕ from $\phi = 0$ to $\phi = 5$ are provided in Fig. 22.6.1.

22.7 Fixed-End Shears and Moments due to Uniform Load

In applying the displacement method of analysis, it is necessary to find the fixed-end shears and moments acting on the ends of a structural member due to known transverse loads, when the slope and deflection at both ends are held to zero. The boundary conditions are, therefore, $\theta_i = 0$, $\theta_j = 0$, $\Delta_i = 0$, and $\Delta_j = 0$. In this section, the fixed-end shears and moments for a member on elastic foundation will be derived for uniform load.

In reference to Eq. (22.2.4) the differential equation for the fixed-end beam shown in Fig. 22.7.1 is

$$\frac{d^4 y}{dx^4} + \frac{k}{EI} y = + \frac{q}{EI} \quad (22.7.1)$$

where q is a constant. The solution of this differential equation includes not only the general portion of Eq. (22.3.2a) but also a particular portion of $y_p = +q/k$; thus,

$$y = A \cos\frac{\phi}{L}x \cosh\frac{\phi}{L}x + B \cos\frac{\phi}{L}x \sinh\frac{\phi}{L}x$$
$$+ C \sin\frac{\phi}{L}x \cosh\frac{\phi}{L}x + D \sin\frac{\phi}{L}x \sinh\frac{\phi}{L}x + \frac{q}{k} \quad (22.7.2a)$$

where

$$\phi = L \sqrt[4]{\frac{k}{4EI}} \quad (22.7.2b)$$

Applying the boundary conditions

$$\theta_i = \left(+\frac{dy}{dx} \text{ at } x = 0\right) = +\frac{\phi}{L}(B + C) = 0 \quad (22.7.3a)$$

$$\theta_j = \left(-\frac{dy}{dx} \text{ at } x = L\right)$$
$$= +\frac{\phi}{L}[(sc' - cs')A + (ss' - cc')B - (ss' + cc')C - (sc' + cs')D] = 0$$
$$\quad (22.7.3b)$$

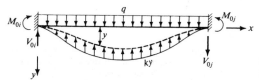

Figure 22.7.1 Fixed-end shears and moments due to uniform load.

$$\Delta_i = (-y \text{ at } x = 0) = -A - \frac{qL^4}{4\phi^4 EI} = 0 \tag{22.7.3c}$$

$$\Delta_j = (+y \text{ at } x = L)$$
$$= +cc'A + cs'B + sc'C + ss'D + \frac{qL^4}{4\phi^4 EI} = 0 \tag{22.7.3d}$$

and solving for the four integration constants,

$$A = -\frac{q}{k} \tag{22.7.4a}$$

$$B = +\frac{c'-c}{s'+s}\frac{q}{k} \tag{22.7.4b}$$

$$C = -\frac{c'-c}{s'+s}\frac{q}{k} \tag{22.7.4c}$$

$$D = +\frac{s'-s}{s'+s}\frac{q}{k} \tag{22.7.4d}$$

If the expressions for A, B, C, and D in Eq. (22.7.3) are substituted into Eq. (22.5.1), expressions for the fixed-end moments and shears M_{0i}, M_{0j}, V_{0i}, and V_{0j} (Fig. 22.7.1) may be obtained as follows:

$$M_{0i} = M_{0j} = -\frac{s'-s}{2\phi^2(s'+s)}qL^2 \tag{22.7.5a}$$

$$V_{0i} = -V_{0j} = +\frac{c'-c}{\phi(s'+s)}qL \tag{22.7.5b}$$

In the degenerate case of $\phi = 0$

$$M_{0i} = M_{0j} = -\tfrac{1}{12}qL^2 \tag{22.7.6a}$$

and
$$V_{0i} = -V_{0j} = +\tfrac{1}{2}qL \tag{22.7.6b}$$

22.8 Fixed-End Shears and Moments due to Straight Haunch Load

The differential equation for the fixed-end beam shown in Fig. 22.8.1 is

$$\frac{d^4y}{dx^4} + \frac{k}{EI}y = +\frac{px}{LEI} \tag{22.8.1}$$

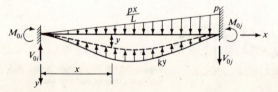

Figure 22.8.1 Fixed-end shears and moments due to straight haunch load.

where p is a constant. The particular solution of this differential equation, in addition to the general solution of Eq. (22.3.2a), is

$$y_p = +\frac{pL^3 x}{4\phi^4 EI} = +\frac{px}{kL} \tag{22.8.2a}$$

where

$$\phi = L \sqrt[4]{\frac{k}{4EI}} \tag{22.8.2b}$$

Applying the boundary conditions

$$\theta_i = \left(+\frac{dy}{dx} \text{ at } x = 0\right) = +\frac{\phi}{L}(B + C) + \frac{pL^3}{4\phi^4 EI} = 0 \tag{22.8.3a}$$

$$\theta_j = \left(-\frac{dy}{dx} \text{ at } x = L\right)$$

$$= +\frac{\phi}{L}[(sc' - cs')A + (ss' - cc')B - (ss' + cc')C$$

$$- (sc' + cs')D] - \frac{pL^3}{4\phi^4 EI} = 0 \tag{22.8.3b}$$

$$\Delta_i = (-y \text{ at } x = 0) = -A = 0 \tag{22.8.3c}$$

$$\Delta_j = (+y \text{ at } x = L) = +cc'A + cs'B + sc'C + ss'D + \frac{pL^4}{4\phi^4 EI} = 0 \tag{22.8.3d}$$

and solving for the four integration constants,

$$A = 0 \tag{22.8.4a}$$

$$B = +\frac{ss' + s^2 - \phi(cs' + sc')}{s'^2 - s^2} \frac{p}{k\phi} \tag{22.8.4b}$$

$$C = -\frac{ss' + s'^2 - \phi(cs' + sc')}{s'^2 - s^2} \frac{p}{k\phi} \tag{22.8.4c}$$

$$D = +\frac{(s' + s)(c' - c) - 2ss'\phi}{s'^2 - s^2} \frac{p}{k\phi} \tag{22.8.4d}$$

If the expressions for A, B, C, and D in Eq. (22.8.4) are substituted into Eq. (22.5.1), expressions for the fixed-end moments and shears M_{0i}, M_{0j}, V_{0i}, and V_{0j} (Fig. 22.8.1) may be obtained as follows:

$$M_{0i} = -\frac{1}{2\phi^3}\left(+\frac{c' - c}{s' - s} - \frac{2ss'\phi}{s'^2 - s^2}\right)pL^2 \tag{22.8.5a}$$

$$M_{0j} = -\frac{1}{2\phi^3}\left(-\frac{c' - c}{s' - s} + \frac{s'^2 + s^2}{s'^2 - s^2}\phi\right)pL^2 \tag{22.8.5b}$$

$$V_{0i} = +\frac{1}{2\phi^2}\left[+\frac{s' + s}{s' - s} - \frac{2(cs' + sc')\phi}{s'^2 - s^2}\right]pL \tag{22.8.5c}$$

$$V_{0j} = -\frac{1}{2\phi^2}\left[-\frac{s'+s}{s'-s} + \frac{2(s'c'+sc)\phi}{s'^2-s^2}\right]pL \qquad (22.8.5d)$$

In the degenerate case of $\phi = 0$,

$$M_{0i} = -\tfrac{1}{30}pL^2 \qquad (22.8.6a)$$

$$M_{0j} = -\tfrac{1}{20}pL^2 \qquad (22.8.6b)$$

$$V_{0i} = +\tfrac{3}{20}pL \qquad (22.8.6c)$$

$$V_{0j} = -\tfrac{7}{20}pL \qquad (22.8.6d)$$

22.9 The Local [SA^T] and [ASA^T] Matrices of a Segment in a Beam

The expressions for the 4×4 stiffness matrix of a member on elastic foundation, as shown in Eq. (22.6.2), are needed in establishing the local [SA^T] and [ASA^T] matrices in any direct stiffness application of the matrix displacement method. In this section specific application will be made to the case where this structural member is a segment in a beam. From the P-X and F-e relationships for such a segment as shown in Fig. 22.9.1, the statics matrix [A] and the deformation matrix [B] are as follows:

$$[A] = \begin{array}{|c|c|c|c|c|} \hline {}_{P}\diagdown {}^{F} & 1 & 2 & 3 & 4 \\ \hline 1 & +1 & 0 & 0 & 0 \\ \hline 2 & 0 & -1 & 0 & 0 \\ \hline 3 & 0 & 0 & -1 & 0 \\ \hline 4 & 0 & 0 & 0 & +1 \\ \hline \end{array} \qquad (22.9.1)$$

$$[B] = \begin{array}{|c|c|c|c|c|} \hline {}_{e}\diagdown {}^{X} & 1 & 2 & 3 & 4 \\ \hline 1 & +1 & 0 & 0 & 0 \\ \hline 2 & 0 & -1 & 0 & 0 \\ \hline 3 & 0 & 0 & -1 & 0 \\ \hline 4 & 0 & 0 & 0 & +1 \\ \hline \end{array} \qquad (22.9.2)$$

BEAMS ON ELASTIC FOUNDATION

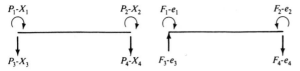

Figure 22.9.1 P-X and F-e relationships for a segment in a beam.

Using the $[S]$ matrix of Eq. (22.6.2),

$$[SA^T] = \begin{array}{|c|c|c|c|c|} \hline {}_F\backslash{}^X & 1 & 2 & 3 & 4 \\ \hline 1 & +\text{TEMP1} & +\text{TEMP2} & +\text{TEMP5} & -\text{TEMP6} \\ \hline 2 & -\text{TEMP2} & -\text{TEMP1} & -\text{TEMP6} & +\text{TEMP5} \\ \hline 3 & -\text{TEMP5} & -\text{TEMP6} & -\text{TEMP3} & +\text{TEMP4} \\ \hline 4 & -\text{TEMP6} & -\text{TEMP5} & -\text{TEMP4} & +\text{TEMP3} \\ \hline \end{array} \quad (22.9.3)$$

$$[ASA^T] = \begin{array}{|c|c|c|c|c|} \hline {}_P\backslash{}^X & 1 & 2 & 3 & 4 \\ \hline 1 & +\text{TEMP1} & +\text{TEMP2} & +\text{TEMP5} & -\text{TEMP6} \\ \hline 2 & +\text{TEMP2} & +\text{TEMP1} & +\text{TEMP6} & -\text{TEMP5} \\ \hline 3 & +\text{TEMP5} & +\text{TEMP6} & +\text{TEMP3} & -\text{TEMP4} \\ \hline 4 & -\text{TEMP6} & -\text{TEMP5} & -\text{TEMP4} & +\text{TEMP3} \\ \hline \end{array} \quad (22.9.4)$$

where the six values TEMP1 to TEMP6 are as defined in Sec. 22.6 and their coefficients are as plotted in Fig. 22.6.1.

It is well to recall, at this point, that in Eqs. (22.9.3) and (22.9.4), P_1 and P_2 are the clockwise acting moments at the left and right ends of the beam segment; P_3 and P_4, the downward acting forces; X_1 and X_2, the clockwise joint rotations; X_3 and X_4, the downward deflections; F_1 and F_2, the positive bending moments causing concave bending; and F_3 and F_4, the upward shear at the left end and the downward shear at the right end.

22.10 Analysis of Beams on Elastic Foundation by the Displacement Method

Having related the external P-X quantities to the internal F-e values of a segment in a beam on elastic foundation by the local $[SA^T]$ and $[ASA^T]$

774 INTERMEDIATE STRUCTURAL ANALYSIS

matrices in Sec. 22.9.1, the displacement method of analysis of such a beam with many segments follows the usual procedure of the direct stiffness method. This procedure includes (1) summing up the local $[ASA^T]$ matrices to form the external stiffness matrix of the total structure; (2) compiling the fixed-end moments and shears due to transverse loads acting on each segment, in addition to the concentrated forces acting at the junction points between segments, to form the $[P]$ matrix; (3) computing the external displacement matrix $[X]$ from

$$[X]_{NP \times NLC} = [ASA^T]^{-1}_{NP \times NP} [P]_{NP \times NLC} \qquad (22.10.1)$$

in which NP and NLC are the degree of freedom and the number of loading conditions, respectively; and (4) determining the final end moments and

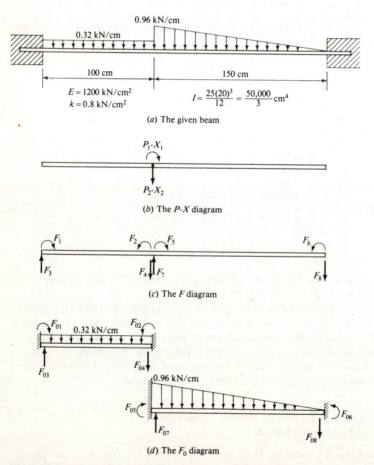

Figure 22.10.1 The beam on elastic foundation for Example 22.10.1.

BEAMS ON ELASTIC FOUNDATION 775

shears acting on each segment from

$$[F^*]_{4 \times NLC} = [F_0]_{4 \times NLC} + [SA^T]_{4 \times 4}[X]_{4 \times NLC} \qquad (22.10.2)$$

Although the details of the computations should best be handled by a general computer program, it may be desirable to illustrate the general concept by a simple numerical example.

Example 22.10.1 Analyze the beam on elastic foundation shown in Fig. 22.10.1a by the displacement method. The compression modulus of the elastic medium is $0.032 \, \text{kN/cm}^3$, and the timber beam is 25 cm wide and 20 cm deep with a modulus of elasticity of $1200 \, \text{kN/cm}^2$. Required are the external displacements X_1 and X_2 and the internal forces F_1^* through F_8^*, which include F_{01} through F_{08} in the fixed condition.

SOLUTION The stiffness modulus k is the product of the width of the beam and the compression modulus; thus

$$k = 25(0.032) = 0.8 \, \text{kN/cm}^2$$

For the first segment,

$$\phi_1 = L_1 \sqrt[4]{\frac{k}{4EI}} = 100 \sqrt[4]{\frac{0.8}{4(1200)(16,666.66)}} = 1.00$$

Using the notations $s = \sin \phi$, $c = \cos \phi$, $s' = \sinh \phi$, and $c' = \cosh \phi$,

$$s_1 = 0.8415 \qquad c_1 = 0.5403 \qquad s_1' = 1.1752 \qquad c_1' = 1.5430$$

Using Eq. (22.7.5),

$$F_{01} = -\frac{s_1' - s_1}{2\phi_1^2(s_1' + s_1)}(0.32)(100)^2 = -0.082735(0.32)(100)^2 = -264.75 \, \text{kN·cm}$$

$$F_{02} = -264.75 \, \text{kN·cm}$$

$$F_{03} = +\frac{c_1' - c_1}{\phi_1(s_1' + s_1)}(0.32)(100) = +0.4972(0.32)(100) = +15.910 \, \text{kN}$$

$$F_{04} = -15.910 \, \text{kN}$$

The contribution of the first segment to the global $[ASA^T]$ matrix would be the elements at (2, 2), (2, 4), (4, 2), and (4, 4) of Eq. (22.9.4); thus

Local $[ASA^T]$ at (2, 2) = global $[ASA^T]$ at (1, 1) = $+(\text{TEMP1})_1$
$$= +807,500 \, \text{kN·cm}$$

Local $[ASA^T]$ at (2, 4) = global $[ASA^T]$ at (1, 2) = $-(\text{TEMP5})_1$
$$= -12,417 \, \text{kN}$$

Local $[ASA^T]$ at (4, 2) = global $[ASA^T]$ at (2, 1) = $-(\text{TEMP5})_1$
$$= -12,417 \, \text{kN}$$

Local $[ASA^T]$ at (4, 4) = global $[ASA^T]$ at (2, 2) = $+(\text{TEMP3})_1$
$$= +269.60 \, \text{kN/cm}$$

For the second segment,

$$\phi_2 = L_2 \sqrt[4]{\frac{k}{4EI}} = 150 \sqrt[4]{\frac{0.8}{4(1200)(16,666.66)}} = 1.50$$

$$s_2 = 0.9975 \qquad c_2 = 0.0707 \qquad s_2' = 2.1293 \qquad c_2' = 2.3524$$

Although Eqs. (22.8.5) and (22.8.6) are derived for haunch at the right end, they may be used discretely for haunch at the left end; thus

$$F_{05} = -\frac{1}{2\phi_2^3}\left[-\frac{c_2'-c_2}{s_2'-s_2}+\frac{\phi_2(s_2'^2+s_2^2)}{s_2'^2-s_2^2}\right](0.96)(150)^2 = -1048.00 \text{ kN·cm}$$

$$F_{06} = -\frac{1}{2\phi_2^3}\left(+\frac{c_2'-c_2}{s_2'-s_2}-\frac{2s_2s_2'\phi_2}{s_2'^2-s_2^2}\right)(0.96)(150)^2 = -689.38 \text{ kN·cm}$$

$$F_{07} = +\frac{1}{2\phi_2^2}\left[-\frac{s_2'+s_2}{s_2'-s_2}+\frac{2\phi_2(s_2'c_2'+s_2c_2)}{s_2'^2-s_2^2}\right](0.96)(150) = +49.384 \text{ kN}$$

$$F_{08} = -\frac{1}{2\phi_2^2}\left[+\frac{s_2'+s_2}{s_2'-s_2}-\frac{2\phi_2(c_2s_2'+s_2c_2')}{s_2'^2-s_2^2}\right](0.96)(150) = -20.668 \text{ kN}$$

The contribution of the second segment to the global $[ASA^T]$ matrix would be the elements at (1, 1), (1, 3), (3, 1), and (3, 3) of Eq. (22.9.4); thus

Local $[ASA^T]$ at (1, 1) = global $[ASA^T]$ at (1, 1) = $+(\text{TEMP1})_2$
$= +558,200 \text{ kN·cm}$

Local $[ASA^T]$ at (1, 3) = global $[ASA^T]$ at (1, 2) = $+(\text{TEMP5})_2$
$= +6249.0 \text{ kN}$

Local $[ASA^T]$ at (3, 1) = global $[ASA^T]$ at (2, 1) = $+(\text{TEMP5})_2$
$= +6249.0 \text{ kN}$

Local $[ASA^T]$ at (3, 3) = global $[ASA^T]$ at (2, 2) = $+(\text{TEMP3})_2$
$= +114.82 \text{ kN/cm}$

The $[ASA^T]$ matrix of the entire structure is, by superposition,

$$[ASA^T] = \begin{bmatrix} (+807,500+558,200) & (-12,417+6249) \\ (-12,417+6249) & (+269.60+114.82) \end{bmatrix}$$

X \ P	1	2
1	+1,365,700 kN·cm	−6168 kN
2	− 6168 kN	+ 384.42 kN/cm

The $[ASA^T]^{-1}$ matrix is found by Gauss-Jordan elimination to be

	P \ X	1	2
$[ASA^T]^{-1} =$	1	$+ 0.78944 \times 10^{-6}$ rad/kN·cm	$+ 12.6666 \times 10^{-6}$ rad/kN
	2	$+12.6666 \times 10^{-6}$ cm/kN·cm	$+2804.6 \times 10^{-6}$ cm/kN

The elements in the $[P]$ matrix are:

$$P_1 = +F_{02} - F_{05} = +(-264.75) - (-1048.00) = +783.25 \text{ kN·cm}$$

$$P_2 = -F_{04} + F_{07} = -(-15.910) + (+49.384) = +65.294 \text{ kN}$$

From Eq. (22.10.1),

$$X_1 = +1445.4 \times 10^{-6} \text{ rad}$$
$$X_2 = +193,050 \times 10^{-6} \text{ cm}$$

From Eqs. (22.9.3) and (22.10.2),

$$\begin{Bmatrix} F_1^* \\ F_2^* \\ F_3^* \\ F_4^* \end{Bmatrix} = \begin{Bmatrix} F_{01} \\ F_{02} \\ F_{03} \\ F_{04} \end{Bmatrix} + \begin{bmatrix} +(TEMP2)_1 & -(TEMP6)_1 \\ -(TEMP1)_1 & +(TEMP5)_1 \\ -(TEMP6)_1 & +(TEMP4)_1 \\ -(TEMP5)_1 & +(TEMP3)_1 \end{bmatrix} \begin{Bmatrix} X_1 \\ X_2 \end{Bmatrix}$$

$$= \begin{Bmatrix} -264.75 \\ -264.75 \\ +15.910 \\ -15.910 \end{Bmatrix} + \begin{bmatrix} +394{,}300 & -11{,}755 \\ -807{,}500 & +12{,}417 \\ -11{,}755 & +229.82 \\ -12{,}417 & +269.60 \end{bmatrix} \begin{Bmatrix} +0.0014454 \\ +0.193050 \end{Bmatrix}$$

$$= \begin{Bmatrix} -264.75 \\ -264.75 \\ +15.910 \\ -15.910 \end{Bmatrix} + \begin{Bmatrix} -1699.38 \\ +1229.94 \\ +27.376 \\ +34.099 \end{Bmatrix}$$

$$= \begin{Bmatrix} -1964.13 \text{ kN·cm} \\ +965.19 \text{ kN·cm} \\ +43.286 \text{ kN} \\ +18.189 \text{ kN} \end{Bmatrix}$$

$$\begin{Bmatrix} F_5^* \\ F_6^* \\ F_7^* \\ F_8^* \end{Bmatrix} = \begin{Bmatrix} F_{05} \\ F_{06} \\ F_{07} \\ F_{08} \end{Bmatrix} + \begin{bmatrix} +(TEMP1)_2 & +(TEMP5)_2 \\ -(TEMP2)_2 & -(TEMP6)_2 \\ -(TEMP5)_2 & -(TEMP3)_2 \\ -(TEMP6)_2 & -(TEMP4)_2 \end{bmatrix} \begin{Bmatrix} X_1 \\ X_2 \end{Bmatrix}$$

$$= \begin{Bmatrix} -1048.00 \\ -689.38 \\ +49.384 \\ -20.668 \end{Bmatrix} + \begin{bmatrix} +558{,}200 & +6249.0 \\ -248{,}210 & -4801.5 \\ -6{,}249.0 & -114.82 \\ -4{,}801.5 & -56.448 \end{bmatrix} \begin{Bmatrix} +0.0014454 \\ +0.193050 \end{Bmatrix}$$

$$= \begin{bmatrix} -1048.00 & +2013.19 \\ -689.38 & -1285.69 \\ +49.384 & -31.198 \\ -20.668 & -12.837 \end{bmatrix}$$

$$= \begin{Bmatrix} +965.19 \\ -1975.07 \\ +18.186 \\ -38.505 \end{Bmatrix}$$

The two statics checks show up very well, as (1) the bending moment at the junction of the two segments is given by $F_2 = +965.19$ kN·cm and also by $F_5 = +965.19$ kN·cm, and (2) the shear at the same section is given by $F_4 = +18.189$ kN and also by $F_7 = +18.186$ kN.

22.11 Deflection, Shear, and Moment Ordinates within the Segment

The results of the displacement method of analysis, as described in the preceding section, give only the slope, deflection, moment, and shear at the

778 INTERMEDIATE STRUCTURAL ANALYSIS

junction points between segments, but not at, say, some equally divided points within each segment. However, the variation of deflection, moment, and shear within each segment is usually needed for purposes of design; in particular, if pulling the beam by the medium is not permitted, the beam must be reanalyzed by treating the distance in which deflection is upward as a separate segment with a zero stiffness modulus k.

Consider a typical beam segment on elastic foundation as shown in Fig. 22.11.1. It is required to derive the equations for deflection y, moment M, and shear V in terms of k, E, I, L, q P_L, P_R, F_1^*, F_2^*, F_3^*, and F_4^*.

The equation of the elastic curve (Fig. 22.11.1) is

$$y = +\frac{q}{k} + \frac{p_L}{k}\left(1 - \frac{x}{L}\right) + \frac{p_R}{k}\left(\frac{x}{L}\right) + A^* \cos\frac{\phi}{L}x \cosh\frac{\phi}{L}x + B^* \cos\frac{\phi}{L}\sinh\frac{\phi}{L}x$$
$$+ C^* \sin\frac{\phi}{L}x \cosh\frac{\phi}{L}x + D^* \sin\frac{\phi}{L}x \sinh\frac{\phi}{L}x \qquad (22.11.1a)$$

where

$$\phi = L\sqrt[4]{\frac{k}{4EI}} \qquad (22.11.1b)$$

The arbitrary constants A, B, C, and D may be obtained from the known values of F_1, F_2, F_3, and F_4 from Eqs. (22.5.2), where the notations M_i, M_j, V_i, and V_j have been used for F_1, F_2, F_3, and F_4; thus, letting $s = \sin\phi$, $c = \cos\phi$, $s' = \sinh\phi$, and $c' = \cosh\phi$,

$$A^* = +\frac{L^2}{2EI\phi^2(s'^2 - s^2)}\left[-(s'^2 + s^2)F_1^* + 2ss'F_2^* - \frac{L}{\phi}(s'c' - sc)F_3^* - \frac{L}{\phi}(sc' - cs')F_4^*\right]$$
$$(22.11.2a)$$

$$B^* = +\frac{L^2}{2EI\phi^2(s'^2 - s^2)}\left[+(sc + s'c')F_1^* - (sc' + cs')F_2^* + \frac{L}{\phi}s'^2F_3^* + \frac{L}{\phi}(ss')F_4^*\right]$$
$$(22.11.2b)$$

$$C^* = +\frac{L^2}{2EI\phi^2(s'^2 - s^2)}\left[+(sc + s'c')F_1^* - (sc' + cs')F_2^* + \frac{L}{\phi}s^2F_3^* + \frac{L}{\phi}(ss')F_4^*\right]$$
$$(22.11.2c)$$

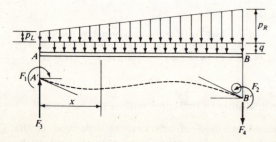

Figure 22.11.1 A typical beam segment on elastic foundation.

BEAMS ON ELASTIC FOUNDATION 779

$$D^* = +\frac{L^2}{2EI\phi^2}[-F_1^*] \tag{22.11.2d}$$

From Eqs. (22.2.5) and (22.11.1),

$$V = -EI\frac{d^3y}{dx^3}$$

$$= +\frac{2\phi^3 EI}{L^3}[(sc' + cs')A^* + (ss' + cc')B^* + (ss' - cc')C^* + (sc' - cs')D^*] \tag{22.11.3}$$

From Eqs. (22.2.6) and (22.11.1),

$$M = -EI\frac{d^2y}{dx^2}$$

$$= +\frac{2\phi^2 EI}{L^2}(ss'A^* + sc'B^* - cs'C^* - cs'D^*) \tag{22.11.4}$$

Summarizing, the deflection, shear, and moment ordinates within the segment can be obtained by first computing the arbitrary constants from Eq. (22.11.2) and then using Eqs. (22.11.1), (22.11.3), and (22.11.4), respectively.

Example 22.11.1 Compute the deflection, shear, and moment at every tenth point in each of the two segments in the beam of Example 22.10.1 (or Fig. 22.10.1) from the given loading and the moment and shear values at the junction points.

SOLUTION For the first segment,

$$\phi = 1.00 \quad L = 100 \text{ cm} \quad EI = 20 \times 10^6 \text{ kN·cm}^2$$
$$q = 0.32 \text{ kN/cm} \quad p_L = 0 \quad p_R = 0$$
$$s = 0.8415 \quad c = 0.5403 \quad s' = 1.1752 \quad c' = 1.5430$$
$$F_1^* = -1964.15 \text{ kN·cm} \quad F_2^* = +965.19 \text{ kN·cm}$$
$$F_3^* = +43.286 \text{ kN} \quad F_4^* = +18.189 \text{ kN}$$

Table 22.11.1 Deflection, shear, and moment in beam of Fig. 22.10.1

Segment	1			2		
Point	y, cm	V, kN	M, kN·cm	y, cm	V, kN	M, kN·cm
0	0	+43.292	−1964.10	+0.19304	+18.188	+ 965.10
0.1	+0.00456	+40.104	−1547.20	+0.20886	+ 6.932	+1151.50
0.2	+0.01686	+36.984	−1161.80	+0.21188	− 2.768	+1180.90
0.3	+0.03498	+33.988	− 807.10	+0.20174	−11.076	+1075.40
0.4	+0.05716	+31.152	− 481.50	+0.17960	−18.136	+ 854.80
0.5	+0.08176	+28.508	− 183.40	+0.14794	−24.084	+ 536.90
0.6	+0.10728	+26.064	+ 89.30	+0.11030	−29.008	+ 137.50
0.7	+0.13236	+23.824	+ 338.60	+0.07120	−32.960	− 328.50
0.8	+0.15576	+21.780	+ 566.40	+0.03584	−35.924	− 846.40
0.9	+0.17634	+19.908	+ 774.80	+0.01002	−37.824	−1401.00
1.0	+0.19304	+18.188	+ 965.10	0	−38.500	−1974.70

From Eq. (22.11.2),

$$A^* = -0.4000 \text{ cm}$$
$$B^* = +0.5412 \text{ cm}$$
$$C^* = -0.5412 \text{ cm}$$
$$D^* = +0.4910 \text{ cm}$$

For the second segment

$$\phi = 1.50 \quad L = 150 \text{ cm} \quad EI = 20 \times 10^6 \text{ kN} \cdot \text{cm}^2$$
$$q = 0 \quad p_L = 0.96 \text{ kN/cm} \quad p_R = 0$$
$$s = 0.9975 \quad c = 0.0707 \quad s' = 2.1293 \quad c' = 2.3524$$
$$F_1^* = +965.19 \text{ kN} \cdot \text{cm} \quad F_2^* = -1975.07 \text{ kN} \cdot \text{cm}$$
$$F_3^* = +18.186 \text{ kN} \quad F_4^* = -38.505 \text{ kN}$$

From Eq. (22.11.2)

$$A^* = -1.0068 \text{ cm}$$
$$B^* = +0.6994 \text{ cm}$$
$$C^* = +0.2448 \text{ cm}$$
$$D^* = -0.2412 \text{ cm}$$

The values of the deflection, shear, and moment at every tenth point in each of the two segments are computed from Eqs. (22.11.1), (22.11.3), and (22.11.4) and shown in Table 22.11.1. Some minor deviations in the numerical values presented are due to inaccuracies in hand computations.

22.12 Exercises

22.1 to 22.6 The timber beams shown in Figs. 22.12.1 to 22.12.6 are 25 cm wide and 20 cm deep, with $E = 1200 \text{ kN/cm}^2$. The beams rest on soil along the entire length; the compression modulus

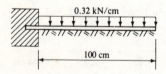

Figure 22.12.1 Exercise 22.1.

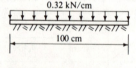

Figure 22.12.2 Exercise 22.2.

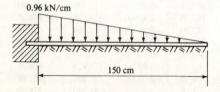

Figure 22.12.3 Exercise 22.3.

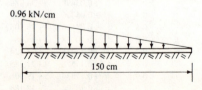

Figure 22.12.4 Exercise 22.4.

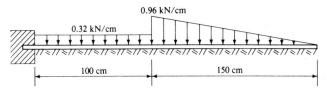

Figure 22.12.5 Exercise 22.5.

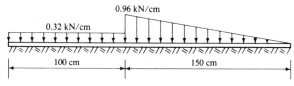

Figure 22.12.6 Exercise 22.6.

of the soil may be taken as $0.032\,\text{kN/cm}^3$. Analyze each beam by the matrix displacement method by showing the free-body diagram, shear diagram, moment diagram, and elastic curve. The degree of freedom is 2 for Exercises 22.1 and 22.3; 4 for Exercises 22.2 and 22.4; 4 for Exercise 22.5; and 6 for Exercise 22.6.

INDEX

INDEX

Absolute maximum bending moment, 476
Angle-weights method, 85
Arches, 649
Axial force in rigid frames, 20

Beam, definition of, 1
Beam deflections:
 by conjugate-beam method, 60
 by moment-area method, 51
 by partial-derivative method, 35
 by unit-load method, 25
Beam element, 585
Beam slopes:
 by conjugate-beam method, 60
 by moment-area method, 51
 by partial-derivative method, 35
 by unit-load method, 29
Beams, moments in, 15
 shears in, 15
Beams, statically indeterminate, 2
 analysis of: by force method, 101
  by matrix displacement method, 395
  by moment-distribution method, 267
  by slope-deflection method, 194
  by three-moment equation, 172
Beams on elastic foundation, 762
 analysis of, by displacement method, 773

Beams on elastic foundation (*Cont.*):
 fixed-condition forces for, 769, 770
 flexibility matrix of, 764
 stiffness matrix of, 767

Cantilever method, 514
Carry-over factor, 265, 534
 for members with variable moment of inertia, 326, 542
Castigliano's theorem, 33
Column-analogy method, 527
 analysis of: closed frames by, 555, 574
  fixed arches by, 654
  gable frames by, 560
  quadrangular frames by, 544, 565
 fixed-end moments by, 529, 536
 stiffness and carry-over factors by, 534, 542
Column matrix, 338
Combined truss and beam element, 585
 local stiffness matrix of, 628
Compatibility condition, 126
Composite structures, 585
 with truss and beam elements, 585, 590
 with truss and combined elements, 594, 607

786 INDEX

Conjugate-beam method:
  beam slopes and deflections by, 60
  slopes and deflections in rigid frames by, 64
Conjugate-beam theorems, 59
Consistent deformation:
  conditions of, 122
  method of, 8
Cross, Hardy, 262, 527
Curved members:
  fixed-condition forces for, 651
  flexibility matrix of, 690, 694
  rigid frames with, 649
  stiffness matrix of, 683

Deflections of beams:
  by conjugate-beam method, 60
  by moment-area method, 51
  by partial-derivative method, 35
  due to shear, 745
  by unit-load method, 25
Deflections of rigid frames:
  by moment-area/conjugate-beam method, 64
  by unit-load method, 39
Deflections of trusses:
  by angle-weights method, 85
  by graphical method, 93
  by joint-displacement-equation method, 90
  by unit-load method, 78
Deformation matrix:
  of beams, 399
  of trusses, 359
Deformation response:
  of beams and rigid frames, 13
  of trusses, 71, 357
Degree of freedom:
  of beams, 395
  of trusses, 357
Degree of indeterminacy:
  of beams, 101, 172, 397
  definition of, 2–4
  external, 5

Degree of indeterminacy (*Cont.*):
  internal, 5
  of multistory frames, 505
  of rigid frames, 126
  of trusses, 146, 357
Direct stiffness method, 371, 408
Displacement method:
  analysis of: beams on elastic foundation by, 773
    composite structures by, 590, 607
    horizontal grid frames by, 702, 710
    rigid frames with semirigid connections by, 728
    structures with shear connections by, 755
  definition of, 9, 11
Distribution factor, 266

Elastic-center method, analysis of fixed arches by, 654
Element flexibility matrix:
  for beam element, 402
  for beam element with variable moment of inertia, 457
  for members with semirigid end connections, 722
  for members with shear deformation, 759
Element stiffness matrix:
  for beam element, 402
  for beam element with variable moment of inertia, 456
  for members with semirigid end connections, 724
  for members with shear deformation, 759
  for truss members, 361

Fabrication errors, effect of, 387
First-order analysis, linear, 12

Fixed arches:
 analysis of: by column-analogy
  method, 654
  by elastic-center method, 654
 effects of: foundation yielding on,
  698
  rib shortening on, 698
  temperature and shrinkage on,
   698
 influence lines for, 656, 667
Fixed-condition forces:
 for curved members, 651
 for members on elastic foundation,
  769, 770
Fixed-end moments, 193, 529
 due to member-axis rotation,
  278
 for members with semirigid end
  connections, 725
 for members with shear deformations, 758
 for members with variable moment
  of inertia, 327, 536
 modified, 274
Flexibility coefficients, 243
Flexibility matrix:
 for curved members, 690, 694
 for members on elastic foundation,
  764
Force-displacement matrix:
 for beams, 403
 for trusses, 361
Force method:
 analysis of: composite structures
  by, 585, 594
  horizontal grid frames by, 705
  statically indeterminate beams
   by, 101
  statically indeterminate rigid
   frames by, 126
  statically indeterminate trusses
   by, 146
  structures with shear deformations by, 751
 definition of, 8, 11

Force response:
 of beams and rigid frames, 13
 of trusses, 71
Freedom, degree of:
 of beams, 395
 of trusses, 357

Gable frames, analysis of:
 by column-analogy method, 560
 by matrix displacement method,
  447
 by moment-distribution method,
  316
 by slope-deflection method, 228
Gauss-Jordan elimination method,
 349
Generalized force, definition of, 8
Global stiffness matrix:
 of beams, 404
 of trusses, 367

Horizontal grid frames:
 analysis of: with beam elements
  only, 705
  with combined beam and torsion
   elements, 710
 definition of, 702

Identity matrix, 344
Indeterminacy, degree of:
 of beams, 101, 172, 397
 definiton of, 2–4
 external, 5
 internal, 5
 of multistory frames, 505
 of rigid frames, 126
 of trusses, 146, 357
Influence lines:
 definition of, 458
 for statically determinate beams,
  459

Influence lines (*Cont.*):
  for statically determinate trusses, 483
  for statically indeterminate beams, 496
  for statically indeterminate trusses, 500
  for symmetrical fixed arches, 656
  for unsymmetrical fixed arches, 667
Inversion, matrix, 342
  pivot in, 346

Joint-displacement equation, 90
Joint-displacement-equation method, 90
Joint-force matrix, 375

Law of reciprocal deflections, 107
Law of reciprocal displacements, 374
Law of reciprocal forces, 374
Least work, theorem of, 116
Local stiffness matrix:
  of beam element, 405
  of beam element on elastic foundation, 772
  of combined beam and truss element, 628
  of truss element, 370

Matrix:
  column, 338
  definition of, 338
  identity, 344
  postmultiplier, 340
  premultiplier, 340
  singular, 345
  symmetric, 354
  unit, 344
Matrix displacement method:
  of beam analysis, 395
  of gable frames analysis, 447

Matrix displacement method (*Cont.*):
  of rigid-frame analysis, 421
  of truss analysis, 357
Matrix inversion, 342
  pivot in, 346
Matrix multiplication, 339
  inner product rule for, 341
Matrix transposition, 351
Maximum bending moment:
  absolute, 476
  in simple beams, 471
  in trusses, 484, 485
Maximum force in web members of trusses, 489
Maximum reaction and shear in simple beams, 467
Members with semirigid end connections:
  fixed-end moments for, 725
  flexibility matrix for, 722
  slip angles for, 721
  stiffness matrix for, 724
Members with shear deformation:
  fixed-end moments for, 755
  flexibility matrix for, 759
  stiffness matrix for, 759
Method of joints and sections, 72
Moment-area method:
  beam slopes and deflections by, 51
  slopes and deflections in rigid frames by, 64
Moment-area theorems, 49
Moment diagram, 17
Moment distribution, 264
  check on, 271, 290
  involving members with variable moment of inertia, 329
Moment-distribution method:
  analysis of: gable frames by, 316
  statically indeterminate beams by, 267
  statically indeterminate rigid frames by, 280, 289
  shear condition in, 290
  sidesway in, 289

Moment and shear distribution, 291, 505
Moments:
  in beams, 15
  in rigid frames, 20
Müller-Breslau influence theorem, 480, 493, 498
Multistory frames, degree of indeterminacy of, 505

Partial-derivative method, 33
  beam deflections by, 35
  beam slopes by, 35
Portal method, 510
Postmultiplier matrix, 340
Premultiplier matrix, 340
Principle of virtual work, 79, 364, 401

Reciprocal deflection, law of, 107
Reciprocal displacements, law of, 374
Reciprocal forces, law of, 374
Reciprocal virtual work theorem, general, 107, 311
Redundants, 100, 101, 146
Rigid frame, definition of, 1
Rigid frames:
  analysis of: by force method, 126
    by matrix displacement method, 421
    by moment-distribution method, 280, 289
    by slope-deflection method, 203, 210
  axial forces in, 20
  with curved members, 649
  moments in, 20
  with semirigid connections, 721
    analysis of, by displacement method, 728
  shears in, 20
  with sidesway, 429

Rigid frames (*Cont.*):
  without sidesway, 422
  slopes and deflections of: by moment-area/conjugate-beam method, 64
    by unit-load method, 39
  statically indeterminate, definition of, 3

Second-order analysis, 12
Secondary moments in trusses with rigid joints, 622, 637
Semirigid connections as rotational springs, 735
Shear conditions, 211
  in moment-distribution method, 290
Shear deflections of statically determinate beams, 745
Shear deformations:
  effects of: by displacement method, 755
    by force method, 751
  members with: fixed-end moments for, 758
    flexibility matrix for, 759
    stiffness matrix for, 759
  shape factors in, 743
Shear diagram, 17
Shears:
  in beams, 15
  in rigid frames, 20
Sidesway conditions, 211
Sidesway in moment-distribution method, 289
Singular matrix, 345
Slope-deflection equations for members:
  with member-axis rotation, 199
  without member-axis rotation, 192
  with variable moment of inertia, 241
Slope-deflection method, 190
  analysis of: gable frames by, 228

Slope-deflection method, analysis of (*Cont.*):
   statically indeterminate beams by, 194
   statically indeterminate rigid frames by, 203, 210
   structures with yielding of supports by, 200, 226, 254
Slopes of beams:
   by conjugate-beam method, 60
   by moment-area method, 51
   by partial-derivative method, 35
   by unit-load method, 29
Slopes of rigid frames:
   by moment-area/conjugate-beam method, 64
   by unit-load method, 39
Statics matrix:
   for beams, 398
   for trusses, 362
Stiffness coefficients, 243
Stiffness factor, 265
   for members with constant moment of inertia, 534
   for members with variable moment of inertia, 326, 542
   modified, 274
Stiffness matrix:
   for beams on elastic foundation, 767
   for curved members, 683
   element, 361, 402, 456
   global, 367, 404
   local, 370, 405
   for trusses, 362
Support settlements, 47
   in trusses, analysis for: by force method, 169
     by matrix displacement method, 389
   (*See also* Yielding of supports, analysis of)
Symmetric matrix, 354

Temperature changes, effect of, 387
Theorem of least work, 116

Three-moment equation, 172
Truss definition of, 1
Truss element, 585
Trusses:
   deflections of: by angle-weights method, 85
     by graphical method, 93
     by joint-displacement-equation method, 90
     by unit-load method, 78
   with rigid joints: primary axial forces in, 622, 623, 637
     secondary moments in, 622, 637
     third axial forces in, 624, 644
   statically indeterminate: analysis of: by force method, 146
     by matrix dispacement method, 357
   definition of, 4, 146

Unit-load method, 24
   beam deflections by, 25
   beam slopes by, 29
   rigid frame deflections by, 39
   rigid frame slopes by, 39
   shear deflections by, 745
   truss deflections by, 78
Unit matrix, 344

Virtual work, principle of, 79, 364, 401

Williot-Mohr diagram, 93

Yielding of supports, analysis of:
   by force method, 121, 138, 169
   by matrix displacement method, 389, 417, 444
   by moment-distribution method, 276, 312
   by slope deflection method, 200, 226, 254
   by three-moment equation, 183